Tasks for Vegetation Science

Volume VI

Series editor
Helmut Lieth, University of Osnabrück, Kürten Dürscheid, Germany

More information about this series at http://www.springer.com/series/6613

Bilquees Gul • Benno Böer
M. Ajmal Khan
Miguel Clüsener-Godt
Abdul Hameed
Editors

Sabkha Ecosystems

Volume VI: Asia/Pacific

Springer

Editors
Bilquees Gul
Institute of Sustainable Halophyte
Utilization (ISHU)
University of Karachi
Karachi, Pakistan

M. Ajmal Khan
Institute of Sustainable Halophyte
Utilization (ISHU)
University of Karachi
Karachi, Pakistan

Abdul Hameed
Institute of Sustainable Halophyte
Utilization (ISHU)
University of Karachi
Karachi, Pakistan

Benno Böer
UNESCO Bangkok Asia and Pacific
Regional Bureau for Education
Bangkok, Thailand

Miguel Clüsener-Godt
UNESCO Division of Ecological
and Earth Sciences, Man and the
Biosphere (MAB) Programme
Paris cedex 15, France

ISSN 0167-9406 ISSN 1875-130X (electronic)
Tasks for Vegetation Science
ISBN 978-3-030-04416-9 ISBN 978-3-030-04417-6 (eBook)
https://doi.org/10.1007/978-3-030-04417-6

Library of Congress Control Number: 2002024322

This Springer imprint is published by the registered company Springer Nature Switzerland AG.
The registered company address is: Gewerbestrasse 11, 6330 Cham, Switzerland

Foreword

The United Nations Sustainable Development Goals were launched in November 2015, with 17 goals, all interconnected, and each one of them as important as all the other goals.

With these developments in mind, it is clear that new ideas toward sustainable human living are of the utmost importance to meet the challenges the world's human population faces today. With 7.65 billion people inhabiting our common planet, and all of us in need of the basic necessities, such as access to food, clean water, clean air, carbon balance, education, medicinal care, shelter, and employment, we simply have to support all avenues that lead to improved situations. Scientific research is an absolutely essential ingredient to achieve the United Nations Sustainable Development Goals.

Some of the most productive ecosystems are coastal/marine, based on seawater, for example, high productive seagrass beds, marine macroalgal reefs, phytoplankton, mangroves, salt marshes, salt grasses, and other hypersaline ecosystems.

The Sabkha Ecosystems of the World book series provides the most comprehensive scientific documentation inspiring the way forward on how to utilize saline resources in the best interest of mankind. The next needed steps are to foster academia and involve land users, politicians, development banks, innovative farmers, investors, energy producers, and carbon off-setters to get involved.

Information provided in volumes I–VI clearly shows that there is a realistic expectation and scope to enhance biomass production and ecosystem conservation for carbon sequestration, food production, and the availability of renewable energy based on halophytes. This can potentially provide jobs, profitable business, and income for a large number of people.

Therefore, it is logical to foster scientific research, development, and monitoring into ecosystem functioning and services, with a view to make use of large reserves of salt water for primary productivity, in order to provide biomass for food, fodder, energy, and other cash crop halophytes.

The editors and authors of the Sabkha Ecosystems of the World series and their global colleagues are working since decades on the subject of halophyte research and development. They have already produced significant science-based knowledge and reasonable achievements, guiding the way forward to more scientific accomplishments based on the utilization of saline soils, salt water, and salt-tolerant plants and animals.

Knowing how best to utilize saline resources will be a blessing for mankind, in view of food security, as well as biofuel, and many other aspects of importance for human living.

This volume on the *Sabkha Ecosystems of Asia/Pacific* is the sixth and final volume of the Sabkha Ecosystems series, concluding the most comprehensive scientific documentation dealing with hypersaline coastal and inland ecosystems of the world. It contains numerous scientific and development aspects. The series started in 2002 and the following volumes have been published:

Volume I: *The Arabian Peninsula and Adjacent Countries*
Volume II: *West and Central Asia*
Volume III: *Africa and Southern Europe*
Volume IV: *Biodiversity Conservation and Cash Crop Halophyte Development*
Volume V: *The Americas*
Volume VI: *Asia/Pacific*

The UNESCO initiated this scientific documentation in 2000, and the well-known Kluwer and Springer science publishing houses took up the publication of this series in the Tasks for Vegetation Science. We are very thankful for this important scientific partnership.

I wish to express my thanks to the Springer Publishing House for its continuous support, as well as to the editors and numerous authors of this important publication. They worked relentlessly to make this highly important book series a reality.

I call on all interested stakeholders to continuously learn about halophytes and their importance and to link the academicians to governmental authorities, development banks, UN agencies, foundations, NGOs, philanthropists, and the private sector to get involved, support, and develop this highly important work. I congratulate the authors and editors of the Sabkha Ecosystems series for their collective accomplishment.

UNESCO Assistant Director-General
for Natural Sciences
Paris, France

Shamila Nair-Bedouelle

Preface

Globally, the total area of saline soil, as estimated by FAO/UNESCO soil map of the world, was 397 million ha, and that of sodic soils 434 million ha. This area may be divided into 230 million ha of irrigated land, of which 45 million ha (19.5%) were salt-affected soils; and of the almost 1500 million ha of dryland agriculture, 32 million ha (2.1%) were salt-affected soils. To ensure the future food security of 7.4 billion people when most of our water and soil resources are saline, the use of 2500 halophytes has become the absolute necessity and is in line with the UN Sustainable Development Goals (SDG).

Land degradation issues in the Asia-Pacific region are further complicated by the chronic poverty, political and social instability, and high rates of weathering. However, important insights into management and economic approaches to improve soil quality in the region await quality research in the area. Major soil degradation in the region is from salinity, erosion, and land pollution.

Sabkha is an Arabic term for flat salt-crusted desert. The local terminology of the Arabian Gulf region describes the extensive, barren, salt-encrusted, and periodically flooded coastal flats as well as inland salt flats. Nineteen years ago, Hans-Jörg Barth and Benno Böer, supported by UNESCO, initiated a global scientific book series "The Sabkha Ecosystems of the World," published initially at Kluwer, and later continued by Springer, in the Tasks for Vegetation Science Series. The first time the terminology of Sabkha was introduced for the international scientific community was through the Sabkha Ecosystems series.

According to Springers Annual Book Review Performance Report, the Sabkha Ecosystem series volumes I, II, III, IV, and V, and individual chapters therein, have been purchased and read by a large number of readers. It is undoubtedly the most comprehensive global scientific documentation on sabkha research, conservation, and development.

Five volumes have already been produced:

1. *The Arabian Peninsula and Adjacent Countries* (2002)
2. *West and Central Asia* (2006)
3. *Africa and Southern Europe* (2011)
4. *Cash Crop Halophyte and Biodiversity Conservation* (2014)
5. *The Americas* (2016)

The Asia and Pacific region presented here follows the specific UNESCO definition which does not strictly reflect the geographical region of the same name. The final volume deals with Asia/Pacific, where there is no strict definition of the region. Therefore, we include all of Asia and all the Pacific region. This includes Turkey, Russia, the Sinai Peninsula, the Arabian Peninsula, Central Asia, all of Asia, and the entire Pacific Ocean.

This final sixth volume concludes this most comprehensive global series dealing with the scientific documentation of salt-desert-ecosystems in Asia/Pacific. It is being published by Springer, spearheaded by the Institute of Sustainable Halophyte Utilization (ISHU) at the University of Karachi, and supported by UNESCO.

Articles in the ***Sabkha Book Volume VI: Asia/Pacific*** cover a broad range of topics from gene to ecosystem such as:

- Status of halophyte research and suggestions for improving science communication for halophyte conservation, research, and development
- Geological/sedimentological aspects of sabkhas
- Plant diversity and factors shaping diversity of Sabkhas
- Effects of environmental stresses and global climate change on plant performance in the context of ecology, ecophysiology, and physiology
- Mechanisms underlying salinity tolerance such as ion homeostasis and oxidative stress mitigation
- Ethnobotany and utilization of halophytes as non-conventional crops.
- Innovative approaches such as the use of drones and infrared thermography to study sabkha plants.

I could not have achieved my current level of success without the strong support of the UNESCO and ISHU. I am also very thankful to the editors and authors of this volume who worked relentlessly to make this highly important book series a reality. I wish to express my thanks to the colleagues from Springer Publishing House for their continuous support and interest in the series.

I would not like to conclude this preface without paying tribute to the late Prof. Helmut Lieth; it was under his wise leadership that made it all possible. It was his dream to ensure food security for the increasing population of the world particularly in arid regions using saline resources. It was in his spirit that the slogan "Cash Crop Halophytes" was minted during a halophyte meeting in Catania, in 1999. We have come a long way since then, and the Sabkha Ecosystems series has contributed immensely in popularization of the concept.

This book like several others before in the series is a good example for our efforts. I hope that this book receives the attention of the scientific community.

Karachi, Pakistan Bilquees Gul

Obituary for Prof. Ajmal Khan

Prof. Ajmal Khan passed away on Saturday, the 4th of May, 2019. Prof. Ajmal contributed to each of the six volumes of the "Sabkha Ecosystems of the World" series, published by Springer, and supported by UNESCO. His demise came to us as a great shock. We have unexpectedly lost a wonderful friend and an excellent professional partner. The world has lost one of the most important global drivers for halophyte research, conservation, and development. He worked with Springer and UNESCO for more than 20 years. It was based on his efforts that the Institute for Sustainable Halophyte Utilization (ISHU) was established at the University of Karachi, and ISHU's scientific contributions in halophyte research are truly remarkable. UNESCO and Springer are very thankful to him, for his friendship, which was paramount to foster this long-term professional partnership. He generously gave his time and expertise that benefitted numerous people in many different compartments of the world. He was an open-minded scientific visionary, thinking "out of the box"; he was clearly ahead of our time. He positively influenced many seniors and juniors in the scientific field by enhancing their capacity for constructive critical thinking. We will greatly miss him. We will keep him in our hearts, and keep many fond memories in our minds. May his soul rest in peace.

Acknowledgments

Technical and financial support of UNESCO and ISHU (Institute of Sustainable Halophyte Utilization, University of Karachi) is gratefully acknowledged for the publication of volume VI of the global Sabkha Ecosystems series.

Moreover, editors would like to thank the Springer Publishing House for the continuous cooperation and support through the publication process of the book series. Editors also thank all the authors who contributed their valued research work as chapters in all volumes of the Sabkha Ecosystems series.

Disclaimer

The designations employed and the presentation of the material throughout this publication do not imply the expression of any opinion whatsoever on the part of UNESCO and the Institute of Sustainable Halophyte Utilization (ISHU) of the University of Karachi concerning legal status of any country, territory, city, or area of its authorities, or concerning the delimitation of its frontiers or boundaries.

United Nations
Educational, Scientific and
Cultural Organization

Contents

Contributors

Chedly Abdelly Laboratory of Extremophile Plants, Centre of Biotechnology of Borj-Cedria, Hammam-Lif, Tunisia

Sourour Abidi Laboratoire des Productions Animales et Fourragères, Institut National de la Recherche Agronomique de Tunisie (INRAT), Université de Carthage, Ariana, Tunisia

M. Ajmal Khan Institute of Sustainable Halophyte Utilization (ISHU), University of Karachi, Karachi, Pakistan

Abdul Wali Al Khulaidi Al Baha University, Biljurashi, Saudi Arabia

Sara Alateeqi Group Seven (G7) Company, Kuwait City, Kuwait

Chingiz Alekperov Department of Molecular Biology and Ecology of Plants, Tel Aviv University, Tel Aviv, Israel

Md. Almujaddade Alfasane Ecology and Environment Laboratory, Department of Botany, University of Dhaka, Dhaka, Bangladesh

A. S. Alsharhan Middle East Geological & Environmental Establishment, Dubai, United Arab Emirates

Abdallah Atia Research Unit, Nutrition and Nitrogen Metabolism and Stress Protein, Department of Biology, Faculty of Sciences of Tunis, Campus Universitaire El Manar I, Tunis, Tunisia

Biology Department, King Khalid University, Abha, Saudi Arabia

Cherry Aung Department of Marine Science, Pathein University, Pathein, Myanmar

Yahya Awang Department of Crop Science, Faculty of Agriculture, Universiti Putra Malaysia, Serdang, Selangor, Malaysia

Irfan Aziz Institute of Sustainable Halophyte Utilization (ISHU), University of Karachi, Karachi, Pakistan

Zouhaier Barhoumi Laboratory of Extremophile Plants, Centre of Biotechnology of Borj-Cedria, Hammam-Lif, Tunisia

Karim Ben Hamed Laboratory of Extremophile Plants, Centre of Biotechnology of Borj Cedria, Hammam Lif, Tunisia

Murillo Amador Bernardo Center of Biological Researchers of Norwest, Instituto Politécnico Nacional, La Paz, Baja California Sur, México

Oscar Bianciotto Universidad Nacional de Tierra del Fuego A.I.A.S, Ushuaia, Tierra del Fuego, Argentina

Alan R. Biggs Davis College of Agriculture, Natural Resources and Design, West Virginia University, Morgantown, WV, USA

Benno Böer UNESCO Bangkok Asia and Pacific Regional Bureau for Education, Bangkok, Thailand

Elisabeth N. Bui CSIRO Land & Water, Black Mountain, Canberra, ACT, Australia

Isabel Caçador MARE – Marine and Environmental Sciences Centre, Faculty of Sciences, University of Lisbon, Lisbon, Portugal

Udo Censkowsky Bluesensus – Sustainability & Seafood Consulting, Weilheim, Germany

Bimal Kinkar Chand West Bengal University of Animal and Fishery Sciences, Kolkata, West Bengal, India

Miguel Clüsener-Godt UNESCO Division of Ecological and Earth Sciences, Man and the Biosphere (MAB) Programme, Paris cedex 15, France

Ahmed Debez Laboratory of Extremophile Plants, Centre of Biotechnology of Borj-Cedria, Hammam-Lif, Tunisia

Ajanta Dey Nature Environment & Wildlife Society, Kolkata, West Bengal, India

Bernardo Duarte MARE – Marine and Environmental Sciences Centre, Faculty of Sciences, University of Lisbon, Lisbon, Portugal

Sourabh Kumar Dubey Bluesensus – Sustainability & Seafood Consulting, Kolkata, West Bengal, India

Tamar Eilam Department of Molecular Biology and Ecology of Plants, Tel Aviv University, Tel Aviv, Israel

Ali El-Keblawy University of Sharjah, Sharjah, United Arab Emirates

Jesús López Elías Agriculture Department, Sonora University, Hermosillo, Sonora, Mexico

Amram Eshel Department of Molecular Biology and Ecology of Plants, Tel Aviv University, Tel Aviv, Israel

Graham Evans The School of Ocean and Earth Science, University of Southampton, St Brelades, Jersey

Saifolah Farmohammadi Academic Center for Education Culture and Research (ACECR) of Hormozgan, Iran and Province Department of Natural Resources, Plantation Office, Tehran, Hormozgan, Iran

Xiaohui Feng The Center for Agricultural Resources Research, Institute of Genetic and Developmental Biology, Chinese Academy of Sciences, Shijiazhuang, Hebei Province, China

Timothy J. Flowers School of Life Sciences, University of Sussex, Brighton, UK

School of Plant Biology, The University of Western Australia, Crawley, WA, Australia

Sanjay Gairola Sharjah Seedbank and Herbarium, Sharjah Research Academy, Sharjah, United Arab Emirates

David Gallacher Department of Interdisciplinary Studies, Zayed University, Dubai, United Arab Emirates

Ali Ghasemzadeh Department of Crop Science, Faculty of Agriculture, Universiti Putra Malaysia, Serdang, Selangor, Malaysia

Shahina A. Ghazanfar Royal Botanic Gardens Kew, Richmond, UK

Houda Gouia Research Unit, Nutrition and Nitrogen Metabolism and Stress Protein, Department of Biology, Faculty of Sciences of Tunis, Campus Universitaire El Manar I, Tunis, Tunisia

Bilquees Gul Institute of Sustainable Halophyte Utilization (ISHU), University of Karachi, Karachi, Pakistan

Salman Gulzar Institute of Sustainable Halophyte Utilization (ISHU), University of Karachi, Karachi, Pakistan

Abdul Hameed Institute of Sustainable Halophyte Utilization (ISHU), University of Karachi, Karachi, Pakistan

Chiraz Chaffei Haouari Research Unit, Nutrition and Nitrogen Metabolism and Stress Protein, Department of Biology, Faculty of Sciences of Tunis, Campus Universitaire El Manar I, Tunis, Tunisia

Ahmed Emdadul Haque Ecology and Environment Laboratory, Department of Botany, University of Dhaka, Dhaka, Bangladesh

Tahsina Haque Ecology and Environment Laboratory, Department of Botany, University of Dhaka, Dhaka, Bangladesh

Luis Guillermo Hernández-Montiel Center of Biological Researchers of Norwest, Instituto Politécnico Nacional, La Paz, Baja California Sur, México

Abdelaziz Hirich International Center for Biosaline Agriculture, Dubai, United Arab Emirates

Tabassum Hussain Institute of Sustainable Halophyte Utilization (ISHU), University of Karachi, Karachi, Pakistan

Sven Erick Jacobsen Department of Agriculture and Ecology, Faculty of Life Sciences, University of Copenhagen, Tastrup, Denmark

Nargis Jahan Ecology and Environment Laboratory, Department of Botany, University of Dhaka, Dhaka, Bangladesh

Tahmina Jamali Ecology and Environment Laboratory, Department of Botany, University of Dhaka, Dhaka, Bangladesh

Byoung Ryong Jeong Horticulture Major, Division of Applied Life Science (BK21 Plus), Graduate School, Gyeongsang National University, Jinju, South Korea

Institute of Agriculture and Life Science, Gyeongsang National University, Jinju, South Korea

Research Institute of Life Science, Gyeongsang National University, Jinju, South Korea

Mohammad Kafi Department of Agrotechnology, Ferdowsi University of Mashhad, Mashhad, Iran

C. G. St. C. Kendall School of the Earth, Ocean and Environment, University of South Carolina, Columbia, SC, USA

Anthony Kirkham Pen-yr-Allt, Nannerch, UK

Taek-Ryoun Kwon National Academy of Agricultural Sciences, Rural Development Administration, Suwon, Republic of Korea

Dafna Langgut The Laboratory of Archaeobotany and Ancient Environments at the Institute of Archaeology, Tel Aviv University, Tel Aviv, Israel

Guoqi Li Breeding Base of State Key Laboratory for Land Degradation and Ecological Restoration, Ningxia University, Yinchuan, China

Key Laboratory of Restoration and Recovery of Degraded Ecosystem in Northwest China of Ministry of Education, Ningxia University, Yinchuan, China

Jin Song Li The Center for Agricultural Resources Research, Institute of Genetic and Developmental Biology, Chinese Academy of Sciences, Shijiazhuang, Hebei Province, China

Xiaojing Liu The Center for Agricultural Resources Research, Institute of Genetic and Developmental Biology, Chinese Academy of Sciences, Shijiazhuang, Hebei Province, China

J. López-Portillo Instituto de Ecologia, A.C. (INECOL), Xalapa, Veracruz, Mexico

Sumedha Madawala Department of Botany, Faculty of Science, University of Peradeniya, Peradeniya, Sri Lanka

T. Mageswaran National Centre for Sustainable Coastal Management, MoEFCC, Chennai, Tamil Nadu, India

Tamer Mahmoud Sharjah Seedbank and Herbarium, Sharjah Research Academy, Sharjah, United Arab Emirates

Maniruzzaman Ecology and Environment Laboratory, Department of Botany, University of Dhaka, Dhaka, Bangladesh

Abinaya Manivannan Institute of Agriculture and Life Science, Gyeongsang National University, Jinju, South Korea

João Carlos Marques MARE – Marine and Environmental Sciences Centre, c/o Department of Zoology, Faculty of Sciences and Technology, University of Coimbra, Lisbon, Portugal

Pere Mutjé LEPAMAP Group, Department of Chemical Engineering, University of Girona, Girona, Spain

N. Nahar Ecology and Environment Laboratory, Department of Botany, University of Dhaka, Dhaka, Bangladesh

Kameswara Rao Nanduri International Center for Biosaline Agriculture, Dubai, United Arab Emirates

Nizar Nasri Faculty of Sciences of Tunis, University of Tunis El-Manar, Tunis, Tunisia

Saida Nasri Laboratoire des Productions Animales et Fourragères, Institut National de la Recherche Agronomique de Tunisie (INRAT), Université de Carthage, Ariana, Tunisia

A. K. M. Nazrul Islam Ecology and Environment Laboratory, Department of Botany, University of Dhaka, Dhaka, Bangladesh

Israel Oren The Robert H. Smith Faculty of Agriculture, Hebrew University of Jerusalem, Jerusalem, Israel

Soo Chul Park National Academy of Agricultural Sciences, Rural Development Administration, Suwon, Republic of Korea

Orathai Pongruktham Department of Marine and Coastal Resources, Mangrove Conservation Office, Bangkok, Thailand

Mokded Rabhi Laboratory of Extremophile Plants, Centre of Biotechnology of Borj-Cedria, Hammam-Lif, Tunisia

Megan Rafter ex-UNESCO Bangkok, Bangkok, Thailand

Plant Sciences, Rothamsted Research, Harpenden, UK

P. Ragavan SERB-National Post Doctoral Fellow, CSIR-National Botanical Research Fellow, Lucknow, Uttar Pradesh, India

Aysha Rasheed Institute of Sustainable Halophyte Utilization (ISHU), University of Karachi, Karachi, Pakistan

Sarwat Ghulam Rasool Institute of Sustainable Halophyte Utilization (ISHU), University of Karachi, Karachi, Pakistan

Malancha Roy Nature Environment & Wildlife Society, Kolkata, West Bengal, India

National Centre for Sustainable Coastal Management (NCSCM-Kolkata), Salt Lake Sector V, Kolkata, West Bengal, India

Edgar Omar Rueda-Puente Agriculture Department, Sonora University, Hermosillo, Sonora, Mexico

Saeed Saadat Soil and Water Research Institute, Education and Extension Organization (AREEO), Karaj, Iran

P. Saenger School of Environment, Science and Engineering, Southern Cross University, Lismore, NSW, Australia

Masoumeh Salehi National Salinity Research Center, Agricultural Research, Education and Extension Organization (AREEO), Yazd, Iran

Sahar Salem Laboratory of Extremophile Plants, Centre of Biotechnology of Borj Cedria, Hammam Lif, Tunisia

Faculty of Sciences of Tunis, University of Tunis El-Manar, Tunis, Tunisia

T. Senthil Kumar Department of Botany, Bharathidasan University, Tiruchirappalli, Tamil Nadu, India

Hatem A. Shabana Sharjah Seedbank and Herbarium, Sharjah Research Academy, Sharjah, United Arab Emirates

Wenshan Shao Breeding Base of State Key Laboratory for Land Degradation and Ecological Restoration, Ningxia University, Yinchuan, China

Key Laboratory of Restoration and Recovery of Degraded Ecosystem in Northwest China of Ministry of Education, Ningxia University, Yinchuan, China

C.-R. Sheue Department of Life Sciences and Center of Global Change Biology, National Chung Hsing University, Taichung, Taiwan

Hina Siddiqui Institute of Sustainable Halophyte Utilization (ISHU), University of Karachi, Karachi, Pakistan

Zamin Shaheed Siddiqui Stress Physiology Phenomic Lab., Department of Botany, University of Karachi, Karachi, Pakistan

Abderrazek Smaoui Laboratory of Extremophile Plants, Centre of Biotechnology of Borj Cedria, Hammam Lif, Tunisia

Prabhakaran Soundararajan Horticulture Major, Division of Applied Life Science (BK21 Plus), Graduate School, Gyeongsang National University, Jinju, South Korea

Sabiha Sultana Ecology and Environment Laboratory, Department of Botany, University of Dhaka, Dhaka, Bangladesh

Amin Tayebi-Meigooni Department of Crop Science, Faculty of Agriculture, Universiti Putra Malaysia, Serdang, Selangor, Malaysia

Gul Nilhan Tug Department of Biology, Faculty of Sciences, Ankara University, Ankara, Turkey

Muhammad Umar Stress Physiology Phenomic Lab., Department of Botany, University of Karachi, Karachi, Pakistan

Vadim Volkov Faculty of Life Sciences and Computing, London Metropolitan University, London, UK

Department of Plant Sciences, College of Agricultural and Environmental Sciences, University of California, Davis, CA, USA

Mecit Vural Department of Biology, Faculty of Sciences, Gazi University, Ankara, Turkey

Poonsri Wanthongchai Department of Marine and Coastal Resources, Mangrove Conservation Office, Bangkok, Thailand

Ahmet Emre Yaprak Department of Biology, Faculty of Sciences, Ankara University, Ankara, Turkey

J. W. H. Yong ARC Centre for Mine Site Restoration, School of Biological Sciences, University of Western Australia, Perth, Australia

M. Zaheer Ahmed Institute of Sustainable Halophyte Utilization (ISHU), University of Karachi, Karachi, Pakistan

Omid Zakeri Academic Center for Education Culture and Research (ACECR) of Hormozgan, Iran and Province Department of Natural Resources, Plantation Office, Tehran, Hormozgan, Iran

Xiumei Zhang The Center for Agricultural Resources Research, Institute of Genetic and Developmental Biology, Chinese Academy of Sciences, Shijiazhuang, Hebei Province, China

Panpan Zhao Breeding Base of State Key Laboratory for Land Degradation and Ecological Restoration, Ningxia University, Yinchuan, China

Key Laboratory of Restoration and Recovery of Degraded Ecosystem in Northwest China of Ministry of Education, Ningxia University, Yinchuan, China

Aviah Zilberstein Department of Molecular Biology and Ecology of Plants, Tel Aviv University, Tel Aviv, Israel

Christoph Zöckler Manfred Hermsen Foundation, Bremen, Germany

Introduction

This is the last of six volumes of this scientific book series. It started as an idea to provide and publish a definition of the Arab term *sabkha*, trying to allow for clarity in the scientific community. This was important because multidisciplinary researchers used the same term for significantly different ecosystems. Some researchers described systems that were completely devoid of vegetation, and others included salt marshes dominated by perennial *Arthrocnemum, Halocnemum, Halopeplis*, or *Aeluropus* as sabkha. A classification of the term *sabkha* was provided in volume I, with a description of different types and sub-types of *sabkha*.

Soon the idea was broadened, not only providing a definition and classification. *Sabkha* experts were contacted, and a new platform for global contributions on research results, nature conservation, and development of *sabkhat* was established.

The first and second volumes were quite large, in comparison to the relatively thin volume III, on Africa and Southern Europe, because it was difficult finding specialists working on this subject.

Volume IV covered biodiversity issues and developing cash crop halophytes in *sabkhat* under saline irrigation, and volumes V and VI conclude the series geographically, adding information about the Americas and Asia/Pacific.

We hope to have produced a useful global contribution toward nature conservation and development of *sabkha* ecosystems and that the series inspires young scientists to become active and conduct more research into the increasingly important subject of *sabkha* heritage conservation and making use of saline resources for food security and employment in halo-agriculture.

Many thanks to all who have assisted developing the series, especially the volume editors and authors and coauthors, and our most sincere thanks to the friends at the publishing house of Springer for their patience and support.

Halophyte Research, Conservation and Development: The Role of UNESCO Biosphere Reserves in Asia-Pacific

1

Benno Böer and Miguel Clüsener-Godt

Abstract

The short presentation discusses briefly the importance of halophytes in a world with freshwater shortages; it highlights the roles of coastal UNESCO Biosphere Reserves in Asia-Pacific and how they can contribute to achieve the UN Sustainable Development Goals via halophyte research, conservation and development.

Keywords

Conservation · Halophytes · Sustainable Development Goals · UNESCO Sphere Reserve

Background Information

Halophytes

Halophytes are salt-tolerant plants which can complete their life cycles in soils or waters with high salt concentrations. The term 'halophyte' is based on the ancient Greek language 'halas' meaning salt and 'phyton' meaning plant. Halophytes are salt plants. Some of the known halophytes are marginally salt-tolerant, and they can survive periods of an increase in soil and water salinity levels for a period of time, whereas others are fully adapted to full-strength seawater or above seawater salinity, and they complete their entire life cycle under seawater salinity. They do not need freshwater. In total, today, about 2600 halophytes are known to sciences. The highest salt tolerance is to be found in seagrasses, marine and coastal macro- and micro-algae, as well as among salt marshes, mangroves, and salt grasses.

UNESCO Biosphere Reserves

The concept of UNESCO Biosphere Reserves was developed and formulated between 1968 and 1971, in response to the ongoing unbridled exploitation of natural resources, already noticed by responsible visionaries almost 50 years ago. They are the oldest system of UN-accredited sites and perfectly suited to achieve the goals and targets of the 1972 Stockholm Declaration, the Millennium Development Goals and, since October 2015, the Sustainable Development Goals (SDGs).

UNESCO Biosphere Reserves are areas of terrestrial, marine or coastal ecosystems. Each reserve promotes solutions reconciling the conservation of biodiversity with its sustainable use. UNESCO Biosphere Reserves are 'Science for

B. Böer (✉)
UNESCO Bangkok Asia and Pacific Regional Bureau for Education, Bangkok, Thailand
e-mail: b.boer@unesco.org

M. Clüsener-Godt
UNESCO Division of Ecological and Earth Sciences, Man and the Biosphere (MAB) Programme, Paris cedex 15, France
e-mail: m.clusener-godt@unesco.org

B. Gul et al. (eds.), *Sabkha Ecosystems*, Tasks for Vegetation Science VI,
https://doi.org/10.1007/978-3-030-04417-6_1

Sustainability support sites' – special places for testing interdisciplinary approaches to understanding and managing changes and interactions between social and ecological systems, including conflict prevention and management of biodiversity. UNESCO Biosphere Reserves are nominated by national governments and remain under the sovereign jurisdiction of the states where they are located. Their status is internationally recognized. Today, there is a World Network of UNESCO Biosphere Reserves comprising a total of 669 sites, and 147 of them are located in Asia-Pacific.

UNESCO Biosphere Reserves have three interrelated zones that aim to fulfil three complementary and mutually reinforcing functions:

1. The core areas comprise strictly protected ecosystems that contribute to the conservation of landscapes, ecosystems, species and genetic variation.
2. The buffer zone surrounds or adjoins the core areas and is used for activities compatible with sound ecological practices that can reinforce scientific research, monitoring, training and education.
3. The transition area is the part of the reserve where the greatest activity is allowed, fostering economic and human development that is socioculturally and ecologically sustainable.

UNESCO Biosphere Reserves are often described as *sites for nature conservation and sustainable human living* or *sites for nature conservation in reconciliation with sustainable development*.

Considering the ongoing population dynamics and exploitation of natural resources, and climate change realities, UNESCO Biosphere Reserves are becoming globally increasingly more important and ideal places to focus and try to achieve the UN SDGs.

The UN Sustainable Development Goals

The UN SDGs were agreed by all Member States of the United Nations, and they were officially launched in October 2015. There are 17 goals in total, each considered of equal importance and interconnected to all of the other goals. Each goal in itself is broken down into more explicit targets, formulating more specifically on what precisely humanity tries to achieve until 2030, in order to foster sustainable human living on our common planet. Several goals and their targets directly relate to nature conservation, such as goals 6 (Clean Water and Sanitation), 12 (Responsible Consumption and Production), 13 (Climate Action), 14 (Life Below Water) and 15 (Life on Land); however, all of the goals relate to sustainable human living, meaning that all 17 SDGs related very well to the concept of UNESCO Biosphere Reserves and that makes them ideal places to try and achieve the SDGs. For UNESCO Biosphere Reserves goal 11 is of great relevance (Sustainable Cities and Communities) because UNESCO Biosphere Reserves have a very strong, essential human living dimension. There are no UNESCO Biosphere Reserves for stand-alone nature conservation; people are living in cities and communities inside UNESCO Biosphere Reserves. In order to achieve goal 11, all of the other goals have to be achieved simultaneously.

Halophyte Research, Conservation and Development in BRs Assist Achieving the UN SDGs

Advancing halophyte research, conservation and development assist achieving the SDGs in several different ways:

- Conservation, restoration and the sustainable use of ecosystems.
- Achieving the sustainable management and efficient use of natural resources.
- Enhancing the scientific and technological capacity to move towards more sustainable consumption and production.
- Upgrading technology for supplying modern and sustainable energy services for all.
- Achieve adequate sanitation for all.
- Substantially increase water-use efficiency.
- Substantially reduce the number of people suffering from water scarcity.
- Protect and restore water-related ecosystems.

Halophytes are very useful for nature-based solutions on water management, with a view to SDG 6 (Clean Water and Sanitation). Halophytes can and will be developed as elements in decentralized sewage systems (grey water and black water systems), contributing to achieve access to adequate and equitable sanitation and hygiene for all. Halophytes will be used to improve water quality by reducing pollution, for example, land-based marine pollution via floating mangroves that are believed to uptake land-based marine pollutants (N, P, K). Moreover, the more halophytes grow in anthropogenic systems as cash crop, the more the pressure on limited freshwater resources will be reduced. This will contribute to protect and restore water-related ecosystems, and it will also support and strengthen the participation of local communities in improving saltwater and sanitation management.

Halophytes support SDG 7 (Affordable and Clean Energy), because the conservation of blue carbon ecosystems, as well as halophyte research and development, contributes expanding the infrastructure and upgrade technology for supplying modern and sustainable energy services for all in many countries, including developing countries and coastal and small island developing states. Halophyte biomass can be developed and produced as clean and renewable biofuel.

Halophytes enhance the sustainable urbanization and capacity for participatory, integrated and sustainable human settlement planning and management in all countries, as formulated in SDG 11 (Sustainable Cities and Communities), because they will play an increasingly important role for providing important ecosystem services for human settlements, especially in saline ecosystems.

The local and indigenous populations living in UNESCO Biosphere Reserves will receive adequate training on consumption and production. This will contribute to the targets in SDG 12 (Responsible Consumption and Production): By 2030, achieve the sustainable management and efficient use of natural resources (target 12.2); By 2030, ensure that people everywhere have the relevant information and awareness for sustainable development and lifestyles in harmony with nature (target 12.8); Support developing countries to strengthen their scientific and technological capacity to move towards more sustainable patterns of consumption and production (target 12.9).

SDG 13 (Climate Action) will be supported by strengthening the resilience and adaptive capacity to climate-related hazards and natural disasters in all countries; education, awareness raising and human and institutional capacity on climate change mitigation, adaptation and impact reduction, and early warning will be improved by halophyte ecosystem restoration and development, as well as via training. Mechanisms for raising capacity for effective climate change-related planning and management in least developed countries and small island developing states will be promoted, including focusing on women, youth and local and marginalized communities.

SDG 15 (Life on Land) deals with the conservation, restoration and sustainable use of ecosystems and their services. That includes, of course, halophyte biodiversity and ecosystems.

These are only a few clear examples on how halophyte research, conservation and development contribute to the achievement of the SDGs.

What Are the Halophyte-Specific Roles of UNESCO Biosphere Reserves?

- Achieving halophyte conservation, development and logistic support and research.
- Establish appropriate zones with core areas and zones of sustainable development.
- Focusing on a multi-stakeholder and involvement of local communities in management.
- Fostering dialogue for conflict resolution of natural resource use.
- Demonstrating sound sustainable development practices and policies based on research and monitoring.
- Acting as sites of excellence for education and training.
- Participating in the World Network.

Apart from ongoing scientific halophyte research, and the normal halophyte ecosystem conservation, there are also some clear development opportunities. The application of *Sesuvium portulacastrum* and *Conocarpus* species has already demonstrated the capacity of halophytes for landscape greenification in hyper-arid landscapes.

It has also been suggested to establish at least one 'World Halophyte Garden', a living ex situ collection of all > 2600 halophyte species in the world, as a centre for halophyte biodiversity conservation, botanical education and a genetic pool for coastal and saline ecosystem restoration. When considering, for example, the loss of coastal and marine habitats in the Gulf, certainly, a genetic pool of halophyte germplasm would be very useful for the restoration of mangroves, salt marshes, seagrass meadows, etc. For the time being, such as garden does not exist, but it would surely be highly beneficial.

Another beneficial way of halophyte development is the establishment of a replicable demonstration farm that makes use of municipal food waste, waste water, and seawater for food security, especially in hot and dry deserts. Concepts and proposals have been developed and they are available. These suggestions remain to be applied, and they are expected to contribute to food security, water security, energy security and waste management and to generate jobs and income.

Floating mangroves for biofuel and carbon sequestration have been discussed in many different places in China, Djibouti, Ethiopia, France, Germany, Indonesia, Pakistan, Qatar, Saudi Arabia, Thailand, the United Arab Emirates and elsewhere. It has been demonstrated that it is possible to plant and produce mangroves afloat on top of the tropical oceans. An experimental plot has been established in Qatar (Fig. 1.1), and a feasibility study was launched in Pakistan. Floating mangroves will be produced with a positive carbon footprint, without the utilization of fossil fuel and without energy-intensive pumps. The employed construction materials are either naturally grown by the mangroves themselves or recycled/reused materials. The required energy to transport and maintain these floating mangrove

Fig. 1.1 *Avicennia marina* floating mangroves in Lusail City, Qatar (foto: Nabil Barazi)

devices will be supplied by wind and solar. The system will produce large amounts of environmentally friendly biofuel (charcoal, woodchips, wood pellets, other types of fuel), and, in addition, it has the capacity to sequester large amounts of atmospheric carbon, produce cash crop halophytes and generate jobs and income.

In conclusion, the utilization, scientific research, conservation and development of halophytes will be highly beneficial to achieve the SDGs. The best places to do so are UNESCO coastal Biosphere Reserves. Pilot studies should be developed and financed by the international donor community and climate fund managers.

Evidence for the Role of Salinity and Alkalinity in Plant Diversification in Australia

2

Elisabeth N. Bui

Abstract

Australia is the world's driest inhabited continent and has some of the world's most stable landscapes and some of the oldest flora, dating back to Gondwana. Two-third of the island continent experiences arid and semiarid climate. Under these climatic conditions where seasonal water deficits occur regularly, salts and carbonates accumulate in soils. Plant distributions have shifted and plants have evolved to adapt to these conditions. This paper summarizes the evidence for the role of soil salinity and alkalinity as drivers in plant diversification in Australia; there is good evidence that both have played an important role for grasses and acacias. Moreover adaptation to salinity may have facilitated the evolution of C_4 photosynthesis in *Neurachne*, an Australian endemic clade of grasses.

Keywords

Acacia · *Neurachne* · Plant diversification · Soil salinity · Alkalinity

E. N. Bui (✉)
CSIRO Land & Water, Black Mountain, Canberra, ACT, Australia
e-mail: Elisabeth.Bui@csiro.au

Introduction

Except for Lake MacLeod which is a coastal sabkha (Fig. 2.1), most of Australia's major salt lakes occur inland within desert dunefields and eolian sand plains and are thought to have coevolved with those landforms as the climate became drier during the Late Neogene and Quaternary: rivers and lakes dried up and evolved to salt lakes, commonly as chains of playas along riverine channels and extensive floodplains (Mernagh 2013; Mernagh et al. 2016; Morton et al. 2011; Warren 2016). Sabkhas are limited to playas and salt lakes, but saline and alkaline soils are common in southern Australia (Rengasamy 2006) and across the Nullarbor (Fig. 2.2), and they have played an important role in shaping plant distributions (e.g., Bui et al. 2014a, b). This paper presents evidence that supports soil salinity and alkalinity as drivers in plant biogeography, ecology, and diversification across Australia, not only in sabkhas.

Geographical Context of Soil Salinity and Alkalinity

Saline soils have an electrical conductivity (EC) > 4 dSm^{-1} in a saturated paste extract at 25C, a pH < 8.5, and Na^+ constitutes < 15% of total exchangeable cations (Richards 1954). In sodic soils, Na^+ is > 15% of total exchangeable

B. Gul et al. (eds.), *Sabkha Ecosystems*, Tasks for Vegetation Science VI,
https://doi.org/10.1007/978-3-030-04417-6_2

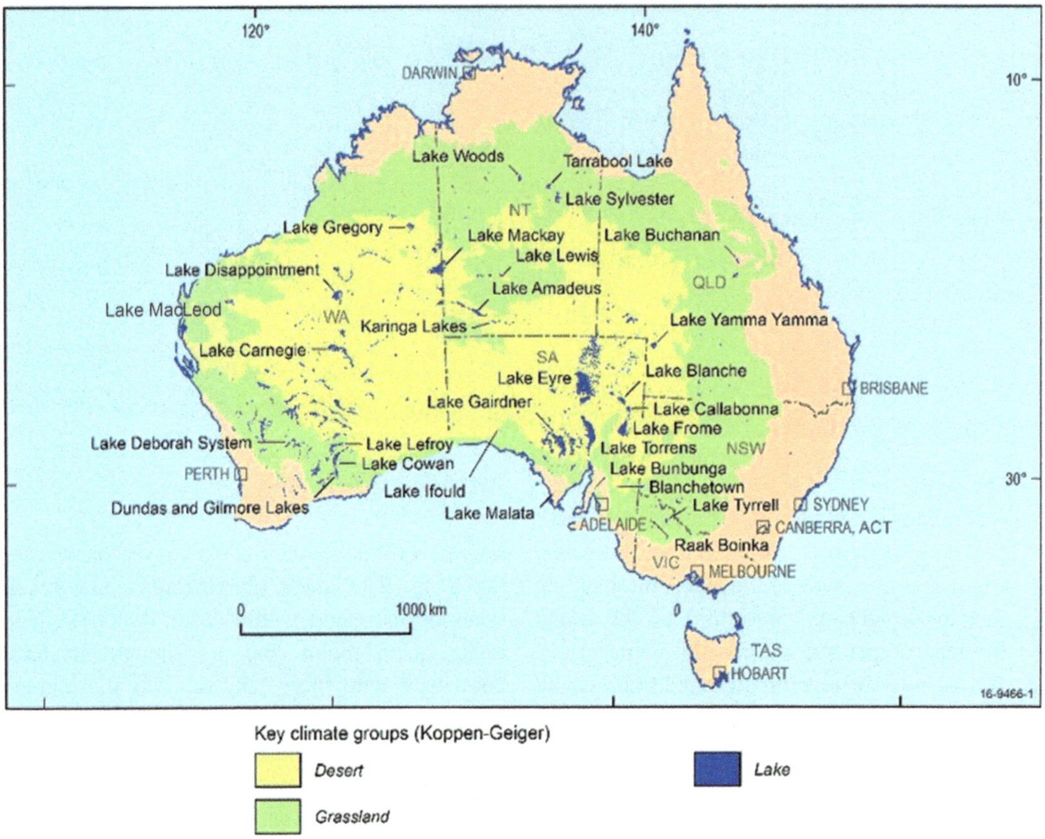

Fig. 2.1 Australia's major lakes, including salt lakes. (From Mernagh et al. 2016)

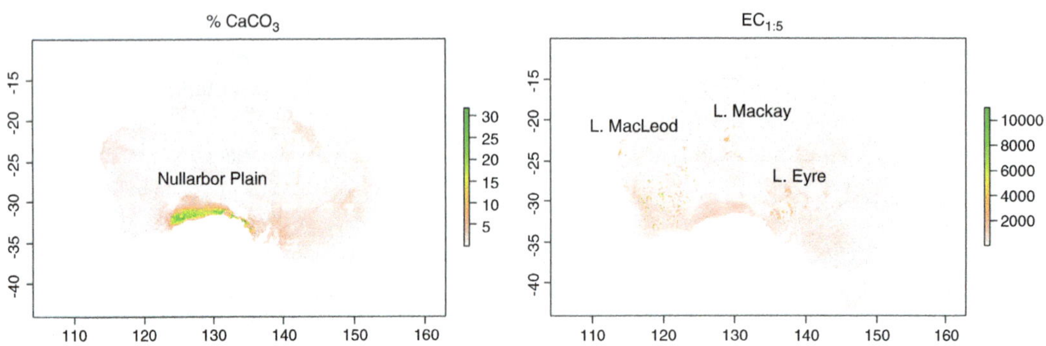

Fig. 2.2 Maps of average % calcium carbonate in soil at 0–80 cm depth (Wilford et al. 2015) and average electrical conductivity (μS cm^{-1}) in 1:5 soil/solution over 0–80 cm interval (Wilford et al. unpublished). The Nullarbor is an extensive limestone plain overlain by calcareous soils

cations, and generally, pH is > 8.5 (in older literature, these were called alkali soils). Salt-affected soils can be both saline and alkaline or saline-sodic. Soil salinity implies the presence of any salt, including chlorides (Cl^-), sulfates (SO_4^{2-}), nitrates (NO_3^-), borates (BO_3^{3-}), carbonates (CO_3^{2-}), and bicarbonates (HCO_3^-) of Na, calcium (Ca), magnesium (Mg), potassium (K), and iron (Fe) (Rengasamy 2006). The presence of sodium carbonate, Na_2CO_3, and sodium bicarbonate, $NaHCO_3$, implies a pH > 8.5. Calcareous soils are rich in carbonate minerals (calcite, $CaCO_3$, and dolomite, $Ca_{0.5}Mg_{0.5}CO_3$) and generally have a pH between 7 and 8.3; thus they are alkaline.

Salts and carbonates accumulate readily under climates where evaporation is high relative to precipitation (there is a seasonal water deficit) and leaching is insufficient to move salts out of the soil profile. Australia is blanketed by extensive areas of saline, calcareous, and sodic soils (Fig. 2.2). Gypsiferous soils are locally important at the southern end of Lake MacLeod, on dune systems along Lake Amadeus in central Australia (Fig. 2.1), and near coastal lakes in the Eyre Peninsula (Mernagh 2013). Gypsum ($CaSO_4 \cdot 2H_2O$) forms as massive crystalline gypsum at the capillary fringe of groundwater discharge zones and also occurs as fine-grained gypsum associated with lunettes and dunes derived from nearby playas (Mernagh 2013).

The arid Australian flora is largely similar to that in other arid zones worldwide with a mixture of grasses, chenopods, Asteraceae, Brassicaceae, and Malvaceae (Barlow 1981), but Myrtaceae, Mimosaceae, and Proteaceae woodlands and shrublands are also extensively represented (Morton et al. 2011).

Australia is the world's driest continent. Arid and semiarid conditions cover ~ 5 million km^2 or 2/3 of its area (Morton et al. 2011). These conditions correspond to "desert" and "grassland" classes in the Koppen climate classification (see http://www.bom.gov.au/jsp/ncc/climate_averages/climate-classifications/index.jsp?maptype=kpn#maps and Fig. 2.1). Nine major vegetation formations cover the desert and grassland climate classes of Australia (DEWR 2007): (i) hummock grasslands, dominated by spinifex (*Triodia* spp.); (ii) tussock grasslands, dominated by Mitchell grass (*Astrebla* spp.); (iii) *Acacia* woodlands and shrublands; (iv) mallee eucalypt woodlands; (v) casuarina forests and woodlands; (vi) heathlands and other plant communities dominated by Proteaceae; (vii) chenopod and samphire shrublands; (viii) other grasslands, herblands, sedgelands, and rushlands; and (ix) mangroves, with communities of stunted *Avicennia marina* around Lake MacLeod (Ellison and Simmonds 2003).

Plant communities dominated by Proteaceae, *Casuarina* forests, and woodlands are probably more adapted to very low P concentrations in sandy, quartz-rich soil and to drought (Lambers et al. 2014) than to salinity and alkalinity per se, but species assemblages and spatial distributions are responding to gradients in all these geochemical variables (e.g., Sander and Wardell-Johnson 2011). The role of salinity and alkalinity in structuring the rest of these vegetation formations will be discussed next in more detail.

Hummock and Tussock Grasslands

Hummock grasslands occupy almost 1/5 of Australia and are dominated by *Triodia* spp. often with acacia or eucalypt species in the sparse overstory, but *Zygochloa paradoxa* is sometimes important locally (DEWR 2007). Tussock grasslands occupy ~7% of the continent and in general are dominated by *Astrebla* spp. but sometimes *Dichanthium* spp. and *Chrysopon* spp. (DEWR 2007). A few Poaceae species found near the arcuate chain of saline lakes in central Australia are potentially associated with relatively high EC (and possibly gypsum given its prevalence in the area (Mernagh 2013)): *Neurachne tenuifolia*, *Triodia hubbardii*, *Triodia brizoides*, *Triodia spicata*, and *Austrostipa feresetacea*, but large uncertainties are associated with the EC estimates (Saslis-Lagoudakis et al. 2015). *Neurachne annularis* is predicted to be associated with high subsoil EC and has a narrow range restricted to southwest Western Australia. *Echinochloa lacunaria*, *Clausospicula extensa*, and *Austrostipa dongicola* are potentially associated with high pH (> 8) and *Neurachne tenuifolia*, *Triodia hubbardii*, and *Zygochloa paradoxa* with moderately alkaline pH (7.3–7.6) (Saslis-Lagoudakis et al. 2015). See http://ausgrass2.myspecies.info/content/fact-sheets for distributions and more botanical information on each species. Grass taxa found in areas of high predicted salinity also tend to be found in conditions of high predicted alkalinity which suggests a correlation between salt and alkalinity tolerance, possibly due to the presence of enabling traits that promote the evolution of stress tolerance in general (Saslis-Lagoudakis et al. 2015).

Neurachne tenuifolia and *N. annularis* are C3-photosynthesis species from a clade of Poaceae endemic to Australia that contains several other C3, two C4, and one C3–C4 intermediate species (Prendergast and Hattersley 1985; Christin et al. 2012). The C4 pathway is thought to have evolved in response to low atmospheric CO_2, heat, drought, and salinity (Sage et al. 2012). Recently, Christin et al. (2012) showed how the C4 pathway evolved more than once in this lineage, starting ~8–7 Ma (Fig. 2.3). The split that separates *N. annularis* from all the others is the oldest; the one that gives rise to *N. tenuifolia* and *N. minor*, a C3–C4 species, dates from ~7–6 Ma. The late Miocene (11–5 Ma) was a time of warm temperatures and low atmospheric CO_2 (Knorr et al. 2011). The current distribution of *Neurachne* species and the evolution of C4 species from an apparently salt-tolerant C3 ancestor suggest that the radiation is a response to low atmospheric CO_2, enabled by the presence of genes that confer a wide spectrum of stress tolerance – this interpretation differs somewhat with that of Christin et al. (2012).

The two C4 taxa and the C3–C4 intermediate species are all polyploids of apparently recent origin (Fig. 2.3), and the correlation of autopolyploidy and derived photosynthetic types suggests that photosynthetic diversification resulted from additional copies of genes encoding enzymes that fulfilled housekeeping functions in the C3 ancestors (Christin et al. 2012). *Neurachne* appears to provide empirical evidence that supports the hypothesis that a relaxation of selection due to

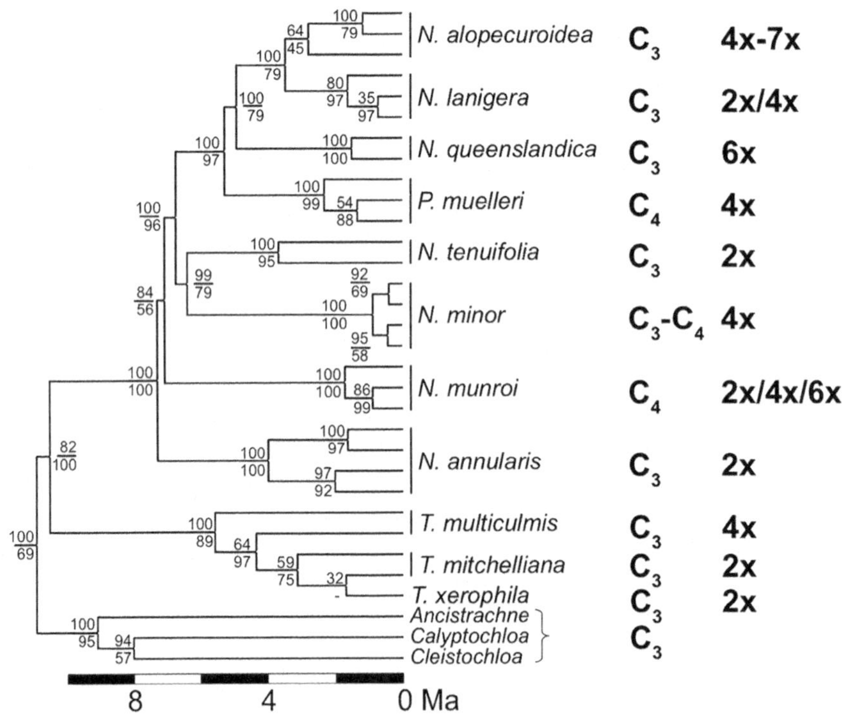

Fig. 2.3 Phylogeny of the Neurachninae, photosynthetic types, and ploidy levels. The tree was obtained through Bayesian inference using the plastid markers, ndhF, trnK/matK, trnLF, rpl16, and rpoC2, and the genomic region encoding the internal transcribed spacer 1, 5.8S rRNA, and internal transcribed spacer 2 (ITS). Bayesian support values and bootstrap values are indicated above and below branches, respectively. The tree was calibrated, and branch lengths are proportional to divergence times, in million years ago (Ma). Photosynthetic types are indicated on the right, as are ploidy levels. Multiple branches to the same species denote different genes or alleles. (From Christin et al. 2012)

genetic redundancy facilitates the recruitment of enzymes into novel photosynthetic functions (Monson 2003; Wang et al. 2009).

Acacia Woodlands and Shrublands

The best evidence for the role of soil salinity and alkalinity in plant diversification in Australia currently comes from the work on acacias (Bui 2013; Bui et al. 2014a, b; Joseph et al. 2013, 2015; Reid et al. 2016). Soil pH and salinity (EC) both play an important role in determining *Acacia* species assemblages across Australia, especially across southern Australia (ES and SW phytoregions in Fig. 2.4). Many *Acacia* have distributions that reflect high soil salinity and alkalinity and that preference is lineage specific. *Acacia harpophylla* and its sister species are associated with saline soils in Queensland (Bui 2013). *A. victoriae* and other species in its clade have been used to revegetate saline soils in Victoria (Joseph et al. 2013, 2015). Salinity and alkalinity tolerance have evolved repeatedly and often together over the 25 Ma of the *Acacia* radiation in Australia (Bui et al. 2014b). Similarly, sulfur accumulation and biomineralization of calcium sulfate have evolved within at least one clade of *Acacia*, *A. bivenosa*, and its sister species (Reid et al. 2016). *A. aneura* and other relatives precipitate large quantities of calcium oxalate crystals along the vascular bundles, possibly to regulate calcium levels and cope with aridity (Brown et al. 2013).

The *Acacia aneura* species complex, or mulga, is dominant in the *Acacia* woodland and shrubland plant communities that cover most of central Australia, occupying one 1/5 of the continent (Bui et al. 2014a). This species complex in the subgenus *Juliflorae* is estimated to have diverged from the rest of *Acacia* ~13 Ma (Miller et al. 2013), but within this clade, *A. aneura* itself is one of the more recently evolved species (see Supplementary Fig. S1 in Bui et al. 2014b), possibly in response to Quaternary aridity.

In the subgenus *Phyllodineae*, *Acacia robeorum*, native to the Great Sandy Desert in northwestern Australia, biomineralizes a large amount of calcium sulfate crystals in its phyllodes and branchlets (He et al. 2014). These crystals may regulate sulfur, calcium, and magnesium levels in the cytoplasm and serve as sinks for sulfur, magnesium, aluminum, and heavy metals (He et al. 2014). Biogenic formation of calcium sulfate crystals also explains the high foliar calcium and

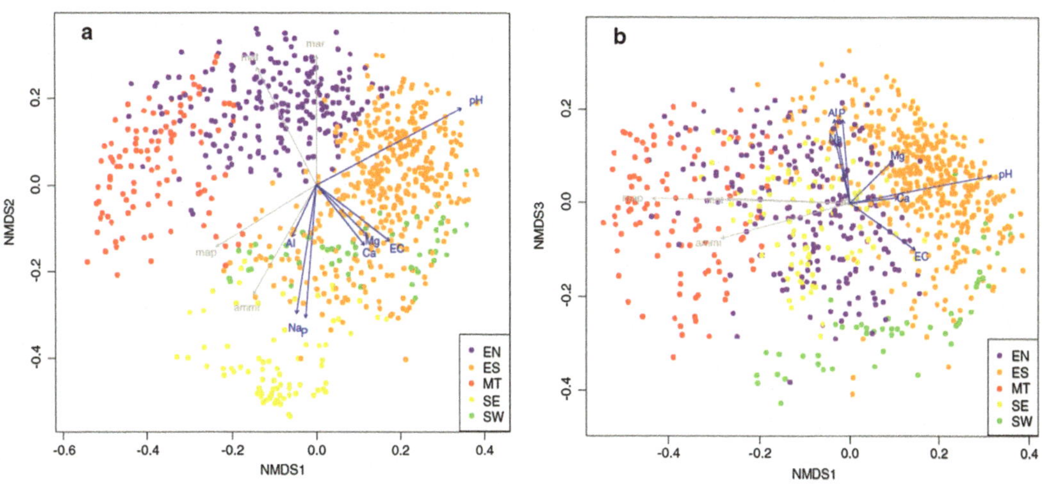

Fig. 2.4 Results of non-metric multidimensional scaling ordination of dissimilarity scores in *Acacia* compositional turnover in 868 grid cells, each 100-km on-edge, over Australia. Vectors for the topsoil (0–20 cm) geochemistry variables are scaled relative to each other, while vectors for climate variables show direction only. Points are grid cells; colors show five phytoregions: *MT* monsoonal tropics, *ES* Eremean south, *EN*, Eremean north, *SW* southwest, *SE* southeast. (From Bui et al. 2014a)

sulfur concentrations of A. *bivenosa*, *A. ligulata*, and other related thiophores in the *Phyllodineae* (Reid et al. 2016). This behavior is clearly independent of substrate because *A. robeorum* grows on sandy quartz-rich soil, not gypsiferous soil (Reid et al. 2016).

Mallee Eucalypt Woodlands

Mallee eucalypts are multi-stemmed trees that grow from a lignotuber and rarely exceed 6 m in height. The number of stems may be influenced by fire and/or high levels of salts, including borates, in soil. Many mallee and related species that occur in the Avon-Coolgardie regions in Western Australia are tolerant of salinity, e.g., *Eucalyptus sargentii* (Salt River Gum) (Nicolle 2005), *E. spathulata*, *E. mimica*, *E. alipes*, and *E. vegrandis* (Nicolle and Brooker 2005). Diversification in the subgenus *Symphyomyrtus* into species with mallee habit appears to have started around 1.25 Ma (Nicolle 2008) although *Symphyomyrtus* split from the rest of the eucalypts ~31.8 Ma (Andrew Thornhill, pers. comm.). Arid conditions have fluctuated over Australia throughout the Cenozoic (Martin 2006), and therefore there may have been several periods during which eucalypts responded to aridity and soil salinity.

Other Vegetation Formations

Australian sabkha ecosystems occur in the chenopod and samphire shrublands; other grasslands, herblands, sedgelands, and rushlands associated with saline lakes and the mangroves around Lake MacLeod. Salinity and soil chemistry in general reflect the zonation of mangroves around Lake MacLeod (Ellison and Simmonds 2003): a 10–50 m zone fringing the lake itself has tall, dense *Avicennia marina* shrubs with tall pneumatophores indicating low oxygen levels in the soil and interstitial water; behind the shoreline zone is a sharp demarcation from dense mangrove cover to domination by open salt flat with occasional very stunted mangrove shrubs. Samphires occur in both zones. The sharp demarcation between the two zones is controlled by micro-elevation and the relatively stable water levels of the perennial seepage ponds that in turn control the inflow of ocean water and hydraulic head through sinkholes in the lake floor (Ellison and Simmonds 2003; Warren 2016). Surrounding areas support low open samphire shrubland (Ellison and Simmonds 2003).

One samphire, *Tecticornia flabelliformis*, is listed as threatened under the Australian Government's Environment Protection and Biodiversity Conservation Act 1999 (Coleman and Cook 2009). The type location for the species is the Gulf St Vincent coast, near Adelaide in South Australia (Fig. 2.1), but there are additional records of the species from coastal sabkhas and playa lakes of the Eyre and Yorke Peninsulas in South Australia and from salt lakes in Victoria and Western Australia (Coleman and Cook 2009). A study of sabkhas in the Gulf St Vincent showed a significant difference in the means for soil pH ($p = 0.01$) and field moist Cl^- ($g\ kg^{-1}$) ($p = 0.05$) of two samphire associations, one dominated by *Tecticornia flabelliformis* and the other by *Tecticornia halocnemoides* (Coleman and Cook 2009): *T. flabelliformis* occurs on soils that have a higher average pH (8.02) and field moist Cl^- concentration (20.7 $g\ kg^{-1}$) than observed in the soils underlying *T. halocnemoides* (pH = 7.85, field moist $Cl^- = 16\ g\ kg^{-1}$). Moreover *T. flabelliformis* was restricted to locations where soils contain 31–37% clay on average, whereas *T. halocnemoides* occurred on soils with a much wider range of clay content. In Western Australia however, the pH of soils where *T. halocnemoides* occurred was even higher than 8.02 (Coleman and Cook 2009).

Diversification of *Tecticornia* and the evolution of photosynthetic pathways in the family Chenopodiaceae in general are currently the subject of active research (e.g., Kadereit et al. 2012; Voznesenskaya et al. 2008). The evolution of salt tolerance in the Amaranthaceae (into which

Chenopodiaceae is subsumed) is also a current research area (Steffen et al. 2015). Soon there could be more evidence of the role that soil salinity and alkalinity play in plant diversification.

References

Barlow BA (1981) The Australian flora: its origin and evolution. In Flora of Australia vol. 1. pp 25–75. Griffin Press, Netley.

Brown SL, Warwick NW, Prychid CJ (2013) Does aridity influence the morphology, distribution and accumulation of calcium oxalate crystals in *Acacia* (Leguminosae: Mimosoideae)? Plant Physiol Biochem 73:219–228

Bui EN (2013) Soil salinity: a neglected factor in plant ecology and biogeography. J Arid Environ 92:14–25

Bui EN, Gonzalez-Orozco CE, Miller JT (2014a) *Acacia*, climate, and geochemistry in Australia. Plant Soil 381:161–175

Bui EN, Thornhill A, Miller JT (2014b) Salt-and alkaline-tolerance are linked in *Acacia*. Biol Lett 10(7):20140278

Christin PA, Wallace MJ, Clayton H et al (2012) Multiple photosynthetic transitions, polyploidy, and lateral gene transfer in the grass subtribe Neurachninae. J Exp Bot 63(17):6297–6308

Coleman PSJ, Cook FS (2009) Habitat preferences of the Australian endangered samphire *Tecticornia flabelliformis*. Trans R Soc S Aust 133(2):300–306. https://doi.org/10.1080/03721426.2009.10887127

DEWR (2007) Australia's native vegetation: a summary of Australia's major vegetation groups. Australian Government, Canberra.

Ellison JC, Simmonds S (2003) Structure and productivity of inland mangrove stands at Lake MacLeod, Western Australia. J R Soc West Aust 86:21–26

He H, Veneklaas EJ, Kuo J, Lambers H (2014) Physiological and ecological significance of biomineralization in plants. Trends in Plant Science 19(3):166–174

Joseph S, Bhave M, Miller JT, Murphy DJ (2013) Rapid identification of *Acacia* species with potential salt tolerance by using nuclear ribosomal DNA markers. Sustain Agric Res 2(4):77

Joseph S, Murphy DJ, Bhave M (2015) Identification of salt tolerant *Acacia* species for saline land utilisation. Biologia 70(2):174–182

Kadereit G, Ackerly D, Pirie MD (2012) A broader model for C4 photosynthesis evolution in plants inferred from the goosefoot family (Chenopodiaceae ss). Proc R Soc B 279(1741):3304–3311. https://doi.org/10.1098/rspb.2012.04401471e2954.

Knorr G, Butzin M, Micheels A, Lohmann G (2011) A warm Miocene climate at low atmospheric CO_2 levels. Geophys Res Lett 38:L20701. https://doi.org/10.1029/2011GL048873

Lambers H, Shane MW, Laliberté E et al (2014) Plant mineral nutrition. In: Plant life on the sandplains in Southwest Australia, a global biodiversity hotspot. UWA Publishing, Crawley, Crawley, pp 101–127

Martin HA (2006) Cenozoic climatic change and the development of the arid vegetation in Australia. J Arid Environ 66:533–563

Mernagh TP (ed) (2013) A review of Australian salt lakes and assessment of their potential for strategic resources. Record 2013/39. Geoscience Australia, Canberra

Mernagh TP, Bastrakov EN, Jaireth S et al (2016) A review of Australian salt lakes and associated mineral systems. Aust J Earth Sci 63(2):1–27

Miller JT, Murphy DJ, Ho SY, Cantrill DJ, Seigler D (2013) Comparative dating of *Acacia*: combining fossils and multiple phylogenies to infer ages of clades with poor fossil records. Aust J Bot 61(6):436–445

Monson RK (2003) Gene duplication, neofunctionalization, and the evolution of C4 photosynthesis. Int J Plant Sci 164:S43–S54

Morton SR, Smith DS, Dickman CR et al (2011) A fresh framework for the ecology of arid Australia. J Arid Environ 75(4):313–329

Nicolle D (2005) A rare and endangered new subspecies of *Eucalyptus sargentii* (Myrtaceae) with high potential for revegetation of saline sites from South-Western Australia and notes on *E. diminuta* and *E. sargentii* subsp. *fallens*. Nuytsia 15:395–402

Nicolle D (2008) Systematic studies of the mallees, *Eucalyptus* series Subulatae (Myrtaceae). Dissertation, Flinders University of South Australia.

Nicolle D, Brooker MIH (2005) Reassessment of the saline-dwelling *Eucalyptus spathulata* complex (Myrtaceae) from southern Western Australia. Nuytsia 15:403–429

Prendergast HDV, Hattersley PW (1985) Distribution and cytology of Australian *Neurachne* and its allies (Poaceae), a group containing C3, C4 and C3-C4 intermediate species. Aust J Bot 33(3):317–336

Reid N, Robson TC, Radcliffe B, Verrall M (2016) Excessive sulphur accumulation and ionic storage behaviour identified in species of *Acacia* (Leguminosae: Mimosoideae). Ann Bot 117(4):653–666

Rengasamy P (2006) World salinization with emphasis on Australia. J Exp Bot 57:1017–1023

Richards LA (ed) (1954) Diagnosis and improvement of saline and alkali soils, USDA handbook 60. USDA, Washington, DC

Sage RF, Sage TL, Kocacinar F (2012) Photorespiration and the evolution of C4 photosynthesis. Annu Rev Plant Biol 63:19–47

Sander J, Wardell-Johnson G (2011) Impacts of soil fertility on species and phylogenetic turnover in the high-rainfall zone of the Southwest Australian global biodiversity hotspot. Plant Soil 345:103–124

Saslis-Lagoudakis CH, Hua X, Bui E et al (2015) Predicting species' tolerance to salinity and alkalinity using distribution data and geochemical modelling: a case study using Australian grasses. Ann Bot 115(3):343–351

Steffen S, Ball P, Mucina L, Kadereit G (2015) Phylogeny, biogeography and ecological diversification of *Sarcocornia* (Salicornioideae, Amaranthaceae). Ann Bot 115(3):353–368

Voznesenskaya EV, Akhani H, Koteyeva NK et al (2008) Structural, biochemical, and physiological characterization of photosynthesis in two C4 subspecies of *Tecticornia indica* and the C3 species *Tecticornia pergranulata* (Chenopodiaceae). J Exp Bot 59(7):1715–1734

Wang X, Gowik U, Tang H et al (2009) Comparative genomic analysis of C4 photosynthetic pathway evolution in grasses. Genome Biol 10:R68

Warren JK (2016) Evaporites: a geological compendium. Springer, Cham

Wilford J, de Caritat P, Bui E (2015) Modelling the abundance of soil calcium carbonate across Australia using geochemical survey data and environmental predictors. Geoderma 259–260:81–92. https://doi.org/10.1016/j.geoderma.2015.05.003

3 Aspects of the Abu Dhabi Sabkha

Anthony Kirkham and Graham Evans

Abstract

The Flandrian transgression inundated an embayed Abu Dhabi coastline defined by headlands dominated by Miocene and Quaternary continental strata. Subsequent coastal development has produced the sabkha where Holocene marine carbonate dominates the outer reaches of the coastal plain whilst the inner parts are dominated by Pleistocene aeolian sediment which also underlie the Holocene sequence. The distribution of the Holocene carbonates, mainly formed under a forced regression, has been controlled by the formation of two barrier lagoon systems, the first of which is represented by a severely deflated storm beach system and the second by the present-day barrier island complex. The early storm beaches are still active along the seaward margin of Sabkha Matti, beyond the western end of the protective barrier island complex. The small shallow lagoons behind the deflated storm beaches have been infilled with marine and aeolian carbonates and early Holocene evaporites including transgressive and regressive anhydrite. The currently active lagoons behind the frontal barrier island complex have been progressively infilled by both mainland coastal accretion and barrier island enlargement through leeward accretion. Such leeward accretion has sometimes dominated the two processes. The modern intertidal microbial mat is separated from, and partly overlaps, the buried regressional microbial mat and seems to represent a renewed eustatic transgression in response to global warming. Whilst the mainland coastal sabkhas are expanding seawards, they are also extending landwards as a result of aeolian deflation. Marine and aeolian processes are therefore important in the bidirectional enlargement of the coastal sabkhas. Late Holocene peritidal and sabkha evaporites are forming along the mainland coastline seawards of the deflated storm beaches, but they are very rare on the barrier islands. Some banded anhydrites with planar bases and containing bouquets of anhydrite which replaced vertical gypsum crystals, plus interbedded microbial mat, indicate former salinas on the sabkhas. Holocene dolomite is difficult to detect in the field. Much of the dolomite is likely to have formed inorganically, perhaps encouraged by the presence of Mg-Si clay such as palygorskite, but there is increasing evidence of microbially induced dolomitization within the microbial mats. There is also evidence, from sulphide concentrations, of dolomite resulting from the activity

A. Kirkham (✉)
Pen-yr-Allt, Nannerch, UK

G. Evans
The School of Ocean and Earth Science,
University of Southampton, St Brelades, Jersey

B. Gul et al. (eds.), *Sabkha Ecosystems*, Tasks for Vegetation Science VI,
https://doi.org/10.1007/978-3-030-04417-6_3

of sulphate-reducing bacteria, possibly within former salinas. Whilst several Holocene hardgrounds are known to exist, they tend to display remarkably flat tops and irregular bases and may extend over large areas. They may act as chronological markers although it is possible that they may be diachronous. The once pristine sabkhas are now so spoiled by petroleum and civil engineering activities that Abu Dhabi and the global community are in grave danger of completely losing the best natural example of coastal sabkhas in the world although as yet there are regrettably only very limited efforts to try and preserve a small area as a site of special scientific interest.

Keywords
Aeolian deflation · Holocene · Evaporite · Dolomite · Hardground · Sabkha

Introduction

Sabkhas are a geological phenomenon which, although described by famous early desert explorers of parts of North Africa, Red Sea, Indian Ocean and Arabian/Persian Gulf such as Bertram Thomas, Wilfred Thesiger, etc., attracted little attention from the contemporary geological community. One of the earliest geological descriptions of sabkhas, from North Africa, is that of Tricart (1954). The first geological description of the Abu Dhabi sabkha was by Evans et al. (1964). Purser (1985) later presented a review of modern sabkhas and their geological significance.

The essential conditions necessary for the development of coastal sabkhas are an arid environment, a porous body of sediment adjacent to the coast and a strong wind regime. They are defined by Walker (2012) as "a supratidal environment of sedimentation formed under arid to semi-arid conditions on restricted coastal plains just above normal high tide level. It is gradational between the land surface and the intertidal environment. Sabkhas are characterized by evaporite salts, tidal flow and also aeolian deposits". All the necessary conditions exist in Abu Dhabi.

In the general geological literature, 'sabkha deposits' or 'sabkha sequence' refer to the progradational environment and subsequent deposits such as of the outer part of the Abu Dhabi sabkha, whereas the term 'sabkha' is used more generally by physical geographers and others for the flat salty coastal plain which is only partly produced by coastal progradation but is largely a deflational plain underlain by aeolian sediments.

Sabkhas extend for 300 km along the Abu Dhabi Emirate Arabian/Persian Gulf coast with maximum widths of approximately 15 km. The combination of cemented Pleistocene seif dunes and Miocene headlands created an embayed coastline though many of the embayments have been backfilled by sedimentary processes during the Holocene (Kirkham 1998a; Evans and Kirkham 2002). Irrespective of human interference through building and other civil engineering projects, the coastal sabkhas between Abu Dhabi Island and Bu Labyad (Fig. 3.1) provide a virtually unbroken expanse of extremely flat topography. Considerable Pleistocene sediments and Miocene inliers originally located between Abu Dhabi Island and Bu Labyad have been removed by Holocene deflation as indicated by 'winged spit' developments within a parallel storm beach system. Further west as far as Jebel Barakah, 25 km west of Jebel Dhanna, the sabkhas are compartmentalised to a large extent by surviving Miocene inliers. Beyond Jebel Barakah the mainland coastline is dominated by the mainly continental Sabkha Matti which is 60 kms wide and extends inland for over 200 kms.

The Abu Dhabi coastal sabkha consist of two parts: (1) an inner, salt-encrusted plain underlain by Pleistocene aeolian dune sand and (2) an outer part dominated by Holocene marine sediment produced mainly by coastal progradation over the last 6000–5000 years. It is the latter part which has mostly attracted the attention of the geological community since its description in the 1960s as it provided an alternative anhydrite formation model to the subaqueous model for understanding carbonate-evaporite sequences – a discovery

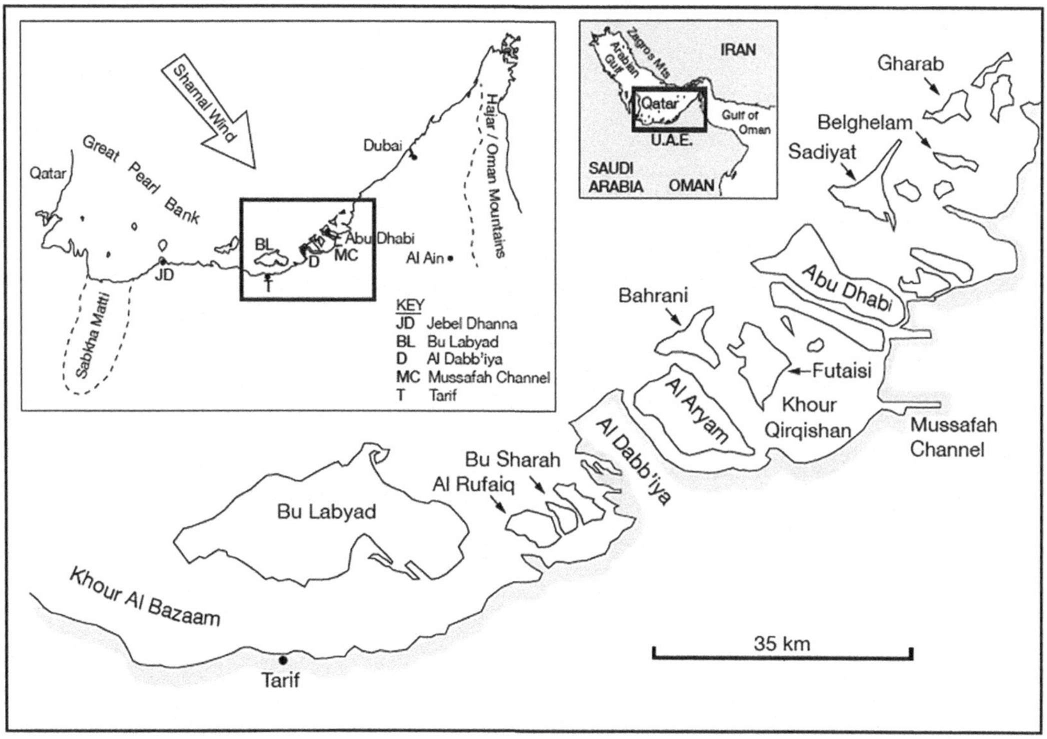

Fig. 3.1 General location map

which led to reinterpretation of many ancient geological sequences.

Much has been written (Evans 2011) about the sedimentology of the arid coastal strip of western Abu Dhabi Emirate which, since the description of the 'Evans line', has been generally accepted as the classical stereotype of a coastal sabkha (Evans et al. 1969; Figs. 3.2, 3.3, and 3.4). This transect is no longer identifiable due to intense civil engineering activity but was used by McKenzie et al. (1980, 1981) during their study of Holocene dolomitization, and the various sampling points along the line were still identifiable in 1995 when the present authors walked with Dr. Peter Bush (a co-author of Evans et al. 1969) along most of the 'line' length. Evans et al. (1969) demonstrated how the tripartite upward succession of subtidal, intertidal and supratidal sediments capped by an aeolian deflation surface on these sabkhas provides the classic indicators of the idealised coastal sabkha sequence. Some of the most interesting features of this sedimentological regime are reviewed such as: the arid climate; the northwesterly Shamal wind; and a physiography dominated by barrier islands where an interesting suite of sediments is accumulating.

Barrier Island Complex

Pleistocene carbonates are exposed at the seaward extremities of most of the barrier islands and may extend southwards towards the mainland coastline as island 'spines' (e.g. Abu Dhabi Island before development; Belghelam, (Evans and Kirkham 2009a). Elsewhere they form peninsulas extending seawards from the mainland, and these have been identified as palaeo-seifs (Kirkham 1998a). These Pleistocene carbonates comprise aeolianite of the Ghayathi Formation, often capped by Pleistocene shallow marine carbonates of the Fuwayrit Formation (Hadley et al. 1998). Both of these Pleistocene formations

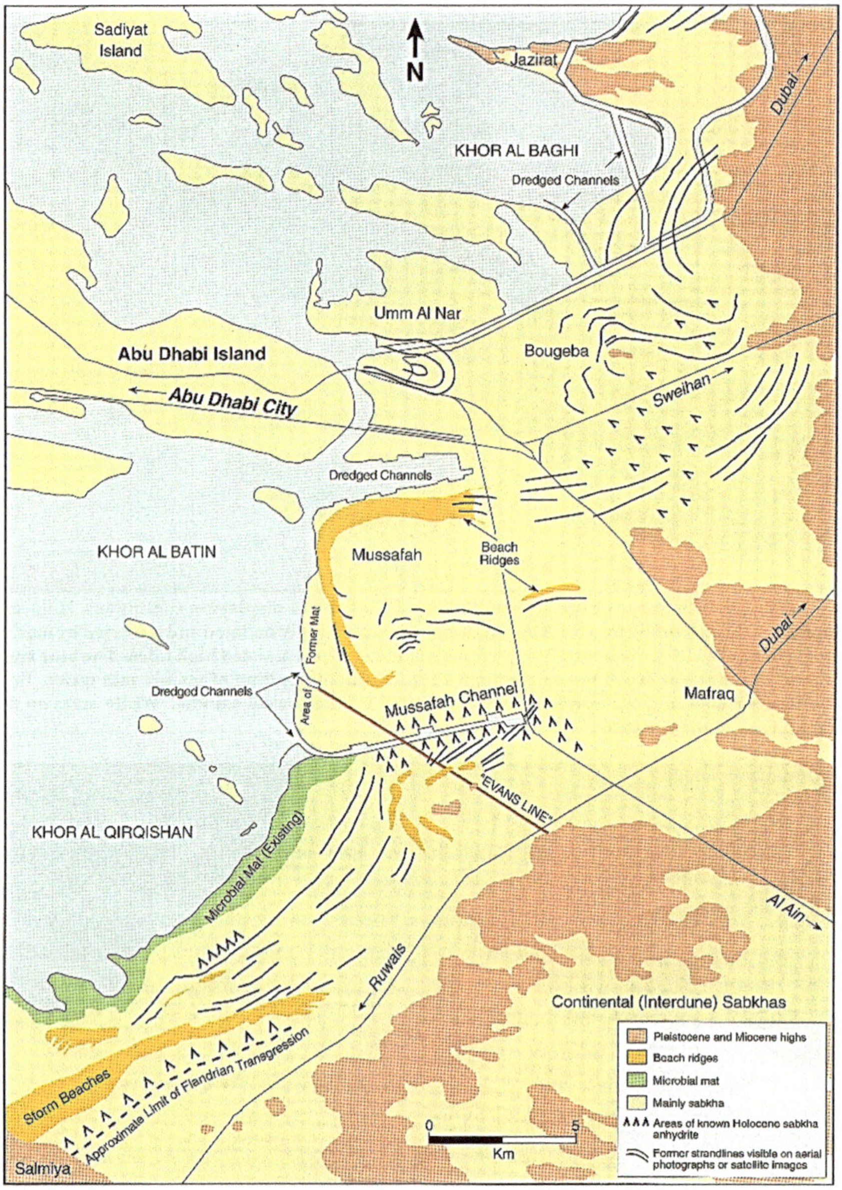

Fig. 3.2 A summary of the Holocene shoreline positions on the mainland adjacent to Abu Dhabi Island. Pleistocene and Miocene outcrops, to the north and south, respectively, created an embayed early Holocene coastline that prograded seawards as a series of spit-lagoon complexes. The areas around Mussafah and northeast of Umm Al Nar are reconstructed largely from aerial photographs of approximately 1960 vintage. Virgin conditions no longer exist there due to interference from civil engineering projects, but anhydrite distributions are mainly after Butler (1970). Details south of Mussafah are based on satellite images and field work. (After Kirkham 1997). Published with permission of GeoArabia

were once continuous along the Abu Dhabi coastline.

Although severely deflated on the sabkha plain by aeolian processes, the Pleistocene carbonates exist today as low isolated outcrops or zeugen (Kirkham 1998a). Purser and Evans (1973) demonstrated how individual barrier islands, which were embryonic during the early-

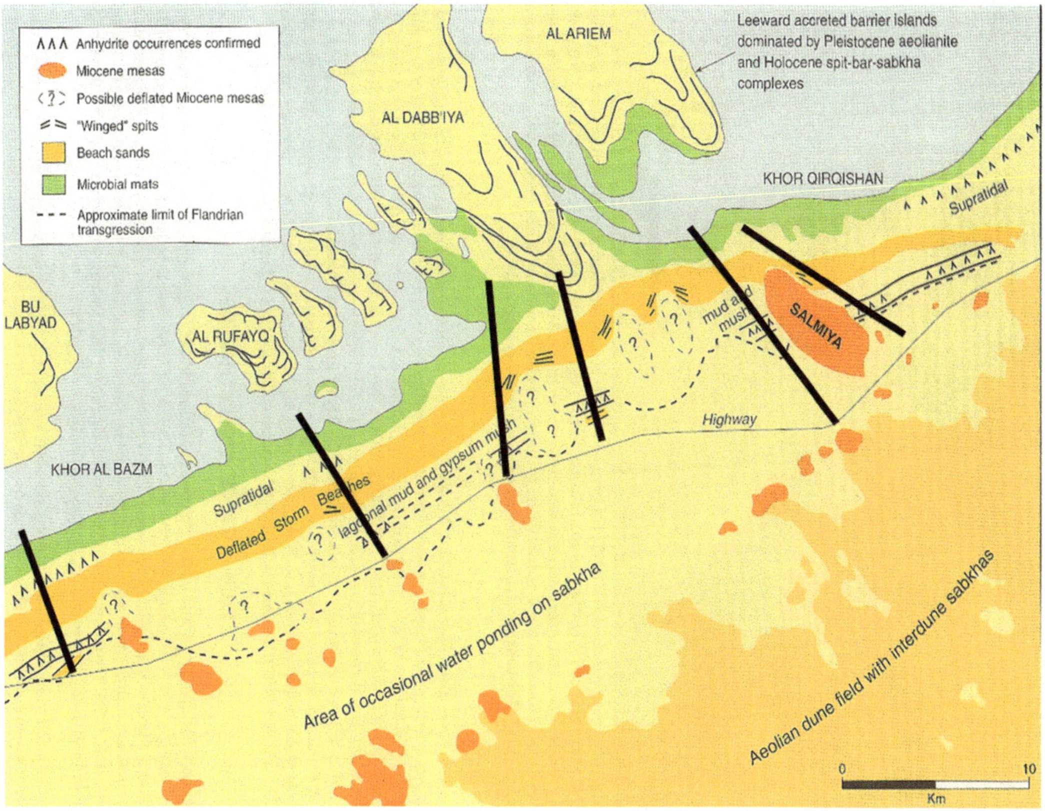

Fig. 3.3 A sketch of the general sabkha between Bu Labyad and Al Ariem (Aryam). Thick black lines indicate field transects. Early and late Holocene anhydrite zones are separated by the deflated storm beaches. The early Holocene anhydrite zone is detectable landwards of the storm beaches with the aid of suitably processed satellite images and is about 0.5 km wide. Note the beach sands in two localities immediately landwards of the early Holocene anhydrite. After Kirkham (2011). Published with permission of GeoArabia

mid Holocene, increased in size partly by longshore drift but mainly by leeward accretion towards the mainland coast. Two of these islands, Abu Dhabi and Al Zabb'iya, have now effectively merged with the mainland coast. Leeward accretion was usually dominated by pairs of carbonate spits extending laterally alongshore and landwards from the seaward ends of the islands, whilst the central parts of the accreting islands were infilled with aeolian sediment or storm washover sands (Kirkham 1997).

During the last glaciation, eustatic sea level was ~110–120 m lower than current level. The Arabian/Persian Gulf was exposed to fluvial, aeolian and probable lacustrine depositional processes as the sea retreated to the Straits of Hormuz. At approximately 18 years Ka, the start of the current interglacial period ushered in the gradual eustatic sea level rise assigned to the Flandrian transgression. However, this transgression was discontinuous as Kassler (1973) demonstrated several still-stands which gave rise to erosional notches separating 'platform levels' on the sea bed of the southern Arabian/Persian Gulf. This eustatic sea level rise has been suggested to explain the biblical legend of Noah's Ark and the Great Flood (Teller et al. 2000). It is further postulated that the penultimate still-stand gave rise to an erosional notch or palaeo-cliff line that defines the alignment of the seaward extremities of the main barrier islands (Kirkham 1998b). The line of this notch was degraded by tidal currents

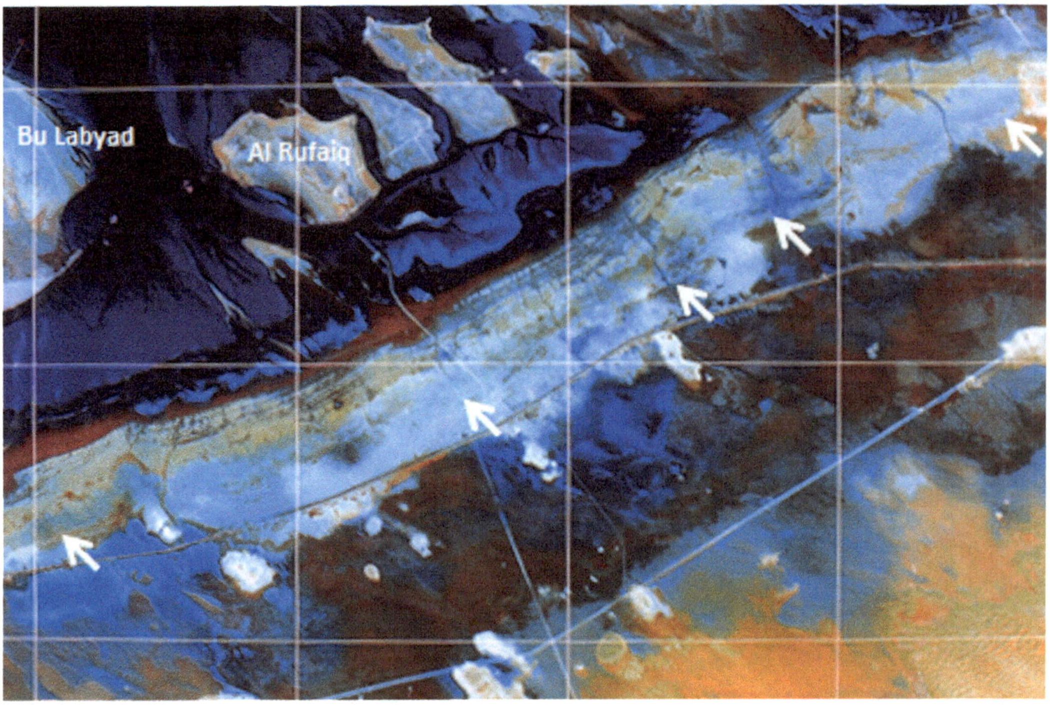

Fig. 3.4 Satellite image with strips of early Holocene anhydrite indicated by arrows. Dark areas immediately south of the anhydrite are where sea- or rainwater gets impounded. Note the parallel storm beaches, crossed by narrow tidal channels, immediately north of the anhydrite. Graticule scale is ~10 km

moving through the channels that currently separate the barrier islands. The Flandrian transgression culminated ~6 years Ka with sea level ~1–2 m higher than at present day. This culmination is reflected by a notch in the form of a wave-cut platform on the Pleistocene zeugen (Kirkham 1998b) at the northern extremities of the barrier islands and elsewhere surrounds small islets (zeugen) in the lagoons of the barrier island complex. The subsequent 1–2 m fall to present-day sea level led to yet another (lower) wave-cut platform being cut into the Pleistocene carbonates (Evans and Kirkham 2005).

Lagoon Developments and Rates of Infill

The present-day tidal ranges are *ca* 2 m in the open waters of the Gulf and *ca* 1 m at the inner extremities of the protected lagoons behind the barrier island complex. During the early-mid Holocene embryonic stages of the protective barrier, the mainland coastline would have experienced higher tides and higher energy wave activity similar to those experienced today beyond the western limit of the barrier island complex. The result was the formation of numerous parallel storm beach ridges (now severely deflated) which extend along the mainland coastal sabkhas inland of the present-day lagoons. One such set of parallel storm beaches end as a compound spit at their eastern end south of Mussafah. 'Winged' spits formed within this storm beach system where former Miocene mesas once protruded as peninsulas but have since been removed by deflation (Kirkham 1997). Since the development of the barrier island complex, these storm beaches and 'winged' spits are no longer 'active' and have also been severely deflated except further west along the shoreline of Sabkha Matti, where they persist because present-day wave activity is still strong beyond the protective barrier islands and where headlands of Miocene strata still protrude seawards (Fig. 3.5). An excellent vertical exposure (now buried) of a few of these 'winged spit' storm beaches with washover sedimentary structure was described from the eastern end of the Mussafah Channel by Kirkham (1998b) and Kenig (2011) (Fig. 3.6).

Very shallow early Holocene lagoons existed behind the storm beach ridges and were supplied by marine waters crossing the ridges via numerous narrow tidal channels that are clearly visible on satellite images (Fig. 3.4; Kirkham 1998b). These shallow early Holocene lagoons are now completely infilled with carbonate muds, aeolian sediment, buried microbial mats and early Holocene evaporites, including anhydrite. Farrant et al. (2012) mapped a strip of overwash deposits behind the storm beach ridges which had not been recognised by earlier workers. The approximate inland limit of the Flandrian transgression has been detected in two places near the landward edge of the infilled lagoon (Kirkham 1997).

Evans et al. (1969) showed that progradation along the Evans line from the inner storm beaches to the modern lagoon had occurred at ~0.73 m/year starting at ~3750 Ka and continued possibly to significantly less than 1000 Ka. Patterson and Kinsman (1977, 1981) estimated a progradation of 1–2 m/year to the east of Abu Dhabi Island. Lokier and Steuber (2008) estimated a progradation of 0.75 m/year between Bu Labyad and Al Rufaiq starting 2258–1897 Ka. Progradation therefore seems to have started at slightly different times along the mainland coastline. The actual rate probably depended on the orientation of the coastline relative to the Shamal wind direction and the degree of protection by the barrier islands. It has been mainly assumed that such progradation has dominated the lagoon infill process – irrespective of barrier island leeward accretion. The 'winged' storm beaches provide a good indication of where the main embayments once existed and

Fig. 3.5 A 'winged spit' with small lagoon to its left containing thin microbial mat. The spit is essentially a mollusc shell bank with a variety of fauna represented. East side of Jebel Dhanna. Open sea to the right

Fig. 3.6 Easterly prograding shelly washover sediment (bracketed) within a 'winged spit', south bank of eastern Mussafah Channel. See also Fig. 3.2

wherein the greatest progradation rates occurred (i.e. in bays between headlands).

None of the above estimates include the infill of the early Holocene lagoon behind the deflated parallel storm beaches. Furthermore, although specific data is lacking, it is evident that rates of leeward accretion of some barrier islands may have been greater than mainland progradation. Leeward accretion of the easternmost barrier islands has been greatest on Abu Dhabi and Al Aryam Islands and the Al Zabb'iya tombolo. To a certain extent, the rates of leeward accretion may reflect the original depths of the modern lagoons.

Sea Level Changes and Current Sea Level Rise

Evans et al. (1969) and Patterson & Kinsman (1982) showed that the level of the regressive microbial mats fell seawards implying a falling sea level of at least 1 m since the establishment of the storm beach ridges on the sabkha. Later workers have supported this, but, other than this early work, only a few of the subsequent traverses described have been based on actual surveyed profiles. Farrant et al. (2012), for example, used spot elevation data from airborne LIDAR for topographic control on a sampled transect in the vicinity of Bu Labyad and Al Rufaiq, and Kenig (2011) carefully determined elevations by reference of features to low tide slack water level.

The interpretation of progradational late Holocene sediments in the area is complicated by what appears to be the current eustatic rise of sea level due to global warming (Cazenave and Nerem 2003; Leuliette et al. 2004), but the effects are subtle and only detectable in a limited way. On the Evans line, the contemporary intertidal microbial mat is separated from a slightly older buried mat by intervening carbonate sediment, and it was suggested later that it was a product of a modern transgression (Kirkham 1998b; Evans and Kirkham 2002, 2005). The gypsum mush layer is frequently overlain by modern, flat, transgressive microbial mat which may be attributed to the current eustatic sea level rise in response to global warming.

Lokier and Steuber (2010) stated that an extensive subsurface microbial mat on their transect south of Al Rufaiq was part of a progradational sequence and yet curiously claimed that it was not regressive. It is overlain by a gypsum mush, and its upper parts are degraded by the precipitation of gypsum within it, thus showing typical signs of regressive progradation. However, they also stated that the subsurface microbial mat had been traced to 7 cm below the landward limit of the recent microbial mat and therefore appears to validate the transgressive nature of the latter. They therefore apparently encountered the same initial regressive and current transgressive microbial mats discussed above. Lokier and Steuber (2008) identified a possible recent relative sea level rise affecting Abu Dhabi as suggested by Evans and Kirkham (2005), and Lokier et al. (2015) later indicated this likelihood.

Wood et al. (2012) have questioned this pattern of sea level behaviour. They argued that the Abu Dhabi area has suffered an uplift of 125 m during the last 18,000 years based on the optically stimulated luminescence (OSL) dating technique. However, Stevens et al. (2011, 2014) later disputed this interpretation with support of subsequent OSL results. Lokier et al. (2015) similarly disputed the claims of Wood et al. and concluded that their own OSL-based datings were also unreliable. Reasons for such misleading OSL data were discussed in detail by Stevens et al. (2014) and Lokier et al. (2015).

Whilst it has generally been assumed that the lagoons between the barrier island complex and the mainland coast would eventually be infilled with progradational carbonate-evaporite complex, a rising sea level may instead lead to a slow enlargement and retreat of the mainland shoreline. The current eustatic sea level rise should be a major concern of local government with low lying areas of the UAE prone to eventual flooding and coastal settlements having a limited lifespan (Evans and Kirkham 2002).

Coastal Sabkha Enlargement

The current coastal sabkhas were originally subjected to planation of earlier desert dune sands by marine erosion (discussed later) during the Pleistocene interglacial sea level highstands before being further modified by aeolian deflation and coastal progradation during the Holocene. In general, they are =/<15kms wide. However, only the seaward reaches of these coastal sabkhas contain marine sediments including Pleistocene limestones (Fuwayrit Formation), plus Holocene carbonates and evaporites displaying lateral accretion/progradation of the mainland coast into the lagoons. The inner reaches of the coastal sabkhas in contrast are dominated by mainly Pleistocene aeolian sediment which also underlies the Holocene marine carbonates and evaporites on the outer sabkha plain. The landward limits of the coastal sabkhas are defined by a scarp of Miocene sediments and/or Pleistocene aeolianites with a veneer of modern aeolian dune sands. The scarp is currently retreating very slowly landwards due to aeolian deflation (Fig. 3.7). As a result, the sabkhas have expanded in two opposing directions: seawards by progradation and landwards by deflation.

Since the Flandrian transgression climaxed at ~2 m higher than present-day sea level ~6000–5000 Ka, there has been a fall in sea level corresponding with the coastal progradation, i.e. there has been a forced regression over most of the last ~6000 years. This fall of sea level was accompanied by the development of red banding in the sediments which is possibly related to successively lower still-stands during the fall of the palaeo-water tables (Kirkham 1998a). The sediments represent offlap geometry as the sabkha trinity of successive subtidal, intertidal and supratidal facies advanced seawards.

The elevation of the sediments exposed in the Mussafah Channel banks gradually falls seawards due to the forced regression. Along the Evans line, which intersects the Mussafah Channel, the surface of the sabkha slopes at ~0.4 m/km (Evans et al. 1969). At the eastern extremity of the Channel, the sabkha surface is 2–3 m above present-day beach level, and a full Holocene transgressive-regressive parasequence has been observed above the Pleistocene deposits (Fig. 3.8). Near the seaward extremity, the sabkha surface merges with the present-day intertidal zone; the regressive microbial mat occurs at a much lower elevation (Fig. 3.9); and the overlying zones of

Fig. 3.7 Bedding features of a Pleistocene barchan dune (Ghayathi Formation) are exposed by deflation at the landward edge of a coastal sabkha. The distant dark area (top left) is the sabkha, beyond which, and to the right, is the edge of the Miocene hinterland overlain by less deflated Pleistocene aeolian dunes

Fig. 3.8 Pleistocene-Holocene sequence near the landward end of the 8-km-long Mussafah Channel. This exposure is no longer accessible due to the Channel having been infilled at this point. The modern beach level provides an indication of the height of the sabkha surface at 2–3 m above low tide level. Tidal range is about 1 m

Fig. 3.9 Late Holocene sediments including flat-topped regressive hardground (H), microbial mat (M) and gypsum mush (G) near the seaward extremity of the Mussafah Channel, west of the 'Evans line'. At this point, the sabkha surface (SS) has reduced in elevation almost to intertidal flat level (IF) due partly to a forced regression over the last 5–3 thousand years. This locality no longer exists due to dredging and widening of the seaward end of the Channel (southern side)

gypsum mush and anhydrite have not yet prograded that far. The various Holocene transgressive and regressive facies are diachronous. However, two hardgrounds (discussed below) which dip seawards were regarded essentially as chronostratigraphic markers (Kirkham 1998b).

The sabkha undergoes the competing effects of jacking up by displacive evaporites versus deflation. Aeolian deflation is a major denudation agent affecting dry sabkha sediment above the capillary zone. This is well demonstrated, for instance, when large formerly phreatic gypsum crystals litter the sabkha surface (Evans et al. 1964; Evans and Kirkham 2005) and when contorted anhydrite bands are exposed and truncated at the sabkha surface (Kirkham 2011). Such denudation should be considered when estimating rates of sediment accumulation although the effects may be masked especially in the peritidal zone by: the current sea level rise elevating the capillary zone; moistening the sediment surface; stabilising the surface grains; and resisting aeolian deflation. Such an alert was highlighted by Kirkham and Evans (2010) with regard to sediment accumulation rates discussed by Lokier and Steuber (2008, 2010).

Abu Dhabi Formation

Wood and Sanford (2002), Wood et al. (2002) and Wood (2011) introduced the term 'Abu Dhabi Formation' to describe what they interpreted as Holocene sediments including marine reworked Pleistocene. It overlies Pleistocene and Miocene sediments. Their Abu Dhabi Formation below the water table was described as extremely uniform fine sand composed of detrital carbonate (60%) and quartz (35%) grains, plus other minor constituents including anhydrite. Their illustrations indicate that it onlaps Miocene scarps although it is unclear whether they are Miocene inliers or the scarp defining part of the landward margin of the coastal sabkhas. They regarded the entire Abu Dhabi Formation as marine, and the implication is that the entire width of the coastal sabkhas could have been inundated (planed) by the Holocene transgression.

It is possible that parts of the inner coastal sabkha were planed by a Pleistocene transgression but, as there is no evidence of Pleistocene marine deposits there, it is questionable whether the transgression of the last interglacial (depositing the Fuwayrit Formation) extended as far as the present-day Miocene scarp at the inner margin of the sabkha.

It is more likely that most of the Abu Dhabi Formation, as defined by Wood and his co-authors, is aeolian Pleistocene rather than marine reworked and its total Holocene age is therefore highly debatable. The only marine sediments within their Abu Dhabi Formation are those containing 'algal remnants' (stromatolites) in a seaward wedge of sediments which Wood and his co-authors designated the progradational Evans Member of the Abu Dhabi Formation. In contrast, Farrant et al. (2012) confined the term 'Abu Dhabi Formation' to that same marine Evans Member.

Kirkham (1997) defined the approximate Holocene limit of the Flandrian transgression as occurring about midway across the coastal sabkha. The transgression reworked the uppermost Pleistocene aeolian carbonates (Ghayathi Formation) as seen at the eastern end of the Mussafah Channel (Kirkham 1998b). Transgressive and regressive anhydrite nodules formed in the aeolianite sediments near the transgressive limit (discussed below).

Aeolian Sediments

There are various units of aeolian sediments in the area as well as the unconsolidated Recent aeolian sands that migrate across the Arabian peninsula. Hadley et al. (1998) assigned the Pleistocene siliciclastic plus the carbonate ('miliolite') aeolian sediments to the Ghayathi Formation. These sediments are more calcareous near the coast and become increasingly siliciclastic inland. This change is accompanied by a landward colour change from white to red as the haematite coatings of increasing quartz grains impose their imprint. Farrant et al. (2012) later retricted the Ghayathi Formation to the carbonate aeolianites.

Using data from numerous boreholes located on the coastal sabkha, Wood and Sanford (2002), Wood et al. (2002) and Wood (2011) illustrated carbonate-cemented Pleistocene sands between the Abu Dhabi Formation, as they defined it, and underlying Miocene strata. They showed cemented Pleistocene sands exposed as inliers which presumably include zeugen (Kirkham 1998a) containing the Ghayathi Formation of Farrant et al. (2012). The present authors dispute this stratigraphic model on the basis that the Ghayathi Formation carbonates rest above red aeolian Pleistocene quartzose sands which occur across the sabkha.

Farrant et al. (2012) distinguished three separate stratigraphic units of aeolian sediments: (1) carbonate aeolianites restricted to the coastal reaches and assigned to the Ghayathi Formation, (2) the siliciclastic aeolian sands in the newly established Pleistocene Madinat Sayed Formation which comprises unconsolidated to friable aeolian calcareous quartzose sandstone and (3) a unit of 'unconsolidated, undivided aeolian sands', the Quaternary age of which is uncertain. The Madinat Sayed Formation is confined to the inland areas of the UAE, whilst the 'undivided' sand is several metres thick and occurs in two regions: (a) a 15-km-wide strip between Abu Dhabi and Ruwais occupying the area particularly between the storm beaches and the Miocene escarpment, where they are not calcareous enough to be classified as Ghayathi Formation, and (b) beneath interdune sabkhas and seem to equate to the Aradah Formation of Hadley et al. (1998).

Evans et al. (1969) and Evans and Kirkham (2009b) described a ubiquitous brown aeolian sand beneath the coastal sabkhas which is generally unconsolidated, but occasionally it is cemented by evaporites, especially by gypsum. Within the Holocene supratidal sediments, horizontally bedded, light-coloured aeolian carbonate sand containing anhydrite nodules unconformably overlie brown aeolian sand with high-angled cross-bedding (Fig. 3.10). The brown sand is continuous seawards beneath the prograding wedge of carbonates bordering the lagoons – when it becomes grey in colour (Farrant et al. 2012). Park (2011) recorded extensive grey aeolian sand immediately beneath the Holocene marine sediments northeast of Abu Dhabi Island. This may equate to the 'undivided' aeolian sands of Farrant et al. (2012). It appears to be more

Fig. 3.10 Light-coloured, supratidal aeolian carbonate sand containing early Holocene anhydrite nodules and unconformably overlying steeply crossbedded brown Pleistocene quartzose aeolian sand

siliciclastic, but there are few detailed analyses available. Evans et al. (1973) and Evans and Kirkham (2002) indicated very little change in composition of the aeolian sands which extended to over 9 m depth at locations along the Evans line. Peebles et al. (1994) encountered Quaternary (Holocene and (?)Pleistocene) sediments extending to Miocene strata at ~6 m depth in boreholes along the Mussafah Channel. However, within the exposed Pleistocene aeolian sands at the extreme eastern of the Mussafah Channel, there is visually a sharp increase in siliciclastic content beneath Pleistocene carbonate dune sands (Fig. 3.8). The grey colour of these siliciclastic sands is probably due to chemical reduction under marine influence.

Evans and Kirkham (2005) and Evans (2011) recognised a sharp contact between unconsolidated brown silica-dominated aeolian sand and overlying Ghayathi aeolianites. It was observed from trenching beneath zeugen west of Tarif and elsewhere at the inner reaches of the sabkha landwards of the Holocene marine sediments. Farrant et al. (2012) concluded with the aid of OSL dating that the Ghayathi Formation at zeugen near Tarif predated unconsolidated quartzose aeolian sand. They also found Ghayathi Formation overlying unconsolidated quartzose aeolian sand. However, dating of the Ghayathi Formation at the Tarif zeugen is contentious (Stevens et al. 2011). It is unclear whether all the unconsolidated aeolian sands mentioned above are the same. Farrant et al. (2012) illustrated the Ghayathi Formation underlying the Pleistocene Fuwayrit Formation and, with the aid of borehole and seismic data and excavations, underlying the marine Holocene Abu Dhabi Formation and/or aeolian sand which, according to its definition by Wood (2011), comprised marine facies of their Holocene Abu Dhabi Formation even across the inner coastal sabkha. Of course, much of the Holocene supratidal sediment is aeolian despite lacking cross-bedding. OSL dating led Farrant et al. (2012) to postulate at least four periods of Ghayathi Formation dune deposition.

Oolite Formation

Classical ooliths are currently forming in the area especially on oolith deltas at the seaward ends of tidal channels that separate the barrier islands (Purser and Evans 1973; Evans et al. 1973). However, they are also forming today on rela-

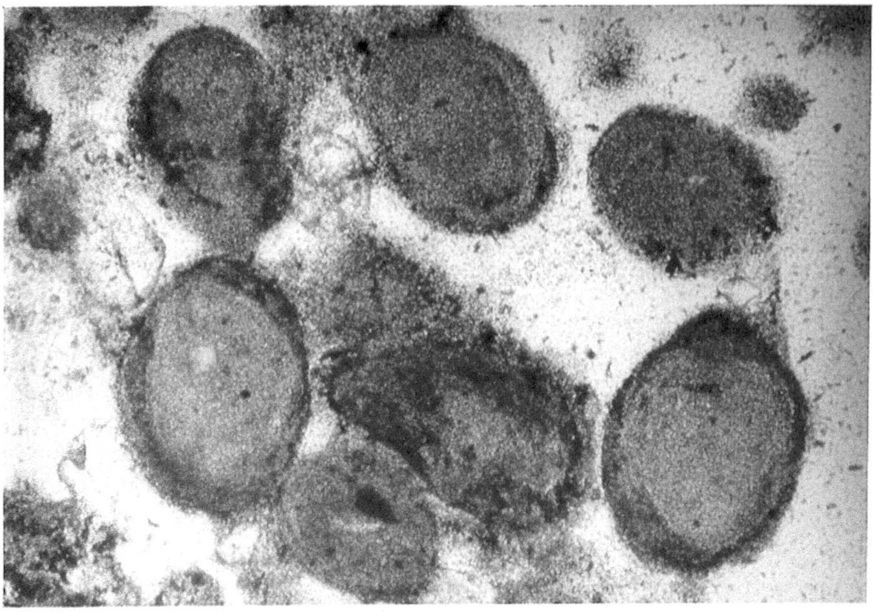

Fig. 3.11 Superficial ooliths in sediment from the current mainland intertidal flats, inner Khor Qirqishan

tively exposed beaches such as those near Jebel Dhanna and further west (e.g. Jebel Barakah and Sila; Loreau and Purser 1973) and may be blown onshore to form local aeolian dunes. They are forming in the marine settings in response to relatively high sea water salinity, wave agitation and tidal currents. During the earlier Holocene, prior to the establishment of the present barrier island complex and when the mainland coast was more exposed to wave activity, ooliths also formed extensively in localities such as along the Mussafah Embayment (Kirkham 1998b). Even today, superficial ooliths may be forming on the barrier-protected mainland beach of Khor Qirqishan between Mussafah Channel and Al Aryam (Fig. 3.11). These may be equivalent to the 'coated grains' or 'unpolished lagoonal ooids' described by Loreau and Purser (1973) and Purser and Loreau (1973) from the lagoon, southwest of Jebel Dhanna.

Evaporite Distribution

Curtis et al. (1963) and Evans et al. (1969) identified four main Holocene evaporites within the coastal sabkhas: surface halite, upper intertidal gypsum mush, coarsely crystalline gypsum and supratidal anhydrite.

A halite-rich crust covers the sabkha, but the main halite deposits form in impersistent pans of impounded standing water. The halite is ephemeral and dissolves and reprecipitates with each marine or meteoric flood.

The gypsum mush of the progradational sequence precipitates rapidly as crystals up to a centimetre in size as a synsedimentary layer within the upper intertidal zone but may continue to precipitate and dehydrate to anhydrite during early burial beneath and after supratidal banded anhydrite has formed. Such late $CaSO_4$ precipitation creates vuggy diapirs that jack up the anhydrite layers and expose them to deflation at the sabkha surface (Kirkham 2011).

Coarsely crystalline gypsum forms slowly at slightly greater depth and lower in the intertidal zone or phreatically below the water table or beneath aquicludes as large isolated crystals several centimetres in size or as intergrown crystals reminiscent of desert roses and commonly poikilitically incorporate carbonate sediment or microbial mat. The intergrown crystals may be extremely abundant, and deflation may expose

Fig. 3.12 A lag of 'phreatic' gypsum crystals on the sabkha surface exposed by deflation. They form a protective cover against further deflation and could dehydrate to an anhydrite layer during burial

them as concentrates (lags) on the sabkha surface to form a continuous cover (Fig. 3.12).

Anhydrite, which has attracted the most interest as it is the only locality where it is known to be forming in substantial amounts today, is the most famous evaporite. It forms either by dehydration of gypsum or precipitates directly within the capillary zone. Primary precipitation of anhydrite has traditionally been regarded as essentially a supratidal feature (Evans et al. 1969; Butler et al. 1982; Kendall et al. 2002; Park 2011; Kendall and Alsharhan 2011; Strohmenger et al. 2011), but it may also occur in the present-day upper intertidal zone (Fig. 3.13; Kendall and Alsharhan 2011; Lokier and Steuber (2008, 2010). Marine reworking or erosion of the supratidal zone may lead to exposure of anhydrite in the intertidal zone under transgressive conditions. The exact demarcation of upper intertidal and supratidal zones needs much better definition.

Kirkham (1997) described how the anhydrite distribution can be defined as early and late Holocene between Abu Dhabi and Bu Labyad Islands. The early Holocene anhydrite occurs landwards of the parallel storm beaches and can be mapped with the aid of suitably processed satellite images (Figs. 3.2, 3.3 and 3.4). It formed during the first phase of shallow lagoon development behind the now deflated storm beaches along the mainland shore. Some of this early Holocene anhydrite was interpreted as transgressive in origin (Kirkham 1997) as predicted by Evans (1970). The late Holocene anhydrite which occurs seawards of the parallel storm beaches is a consequence of the current lagoonal system. The primary anhydrites are displacive and tend to jack up the sabkha surface to the extent that the early Holocene anhydrite zone causes extensive ephemeral linear lakes to accumulate on the sabkha landwards of it after heavy rainfall. These lakes quickly become saline as the water dissolves pre-existing halite on and within the surficial sabkha sediments. Such lakes may last for many weeks as they slowly evaporate and reprecipitate the halite largely as snow-white crusts on the sabkha surface.

From reconnaissance studies made in the restricted coastal sabkhas between the headlands of Miocene rocks between Bu Labyad and Jebel Dhanna, the anhydrite occurrences are restricted to the most landward parts and are assumed to equate to the early Holocene anhydrites discussed above. No anhydrite has been discovered yet along the coastal regions of Sabkha Matti.

Fig. 3.13 A central core and circular bands of anhydrite in a truncated diapir exposed in the upper intertidal zone on the mainland coast south of Bu Labyad. The truncation may be due to aeolian deflation but most likely to marine erosion during the current marine transgression. Scale shown by a centimetre ruler and surrounding footprints

The anhydrite normally has a nodular habit, and the nodules frequently merge and give rise to contorted or ptygmatically folded enterolithic bands. Elsewhere it forms continuous sheets with relatively flat bases; is interbedded with microbial mat; and includes bouquets of anhydritised vertical gypsum crystals which Kirkham (2011) interpreted as subaqueous in origin and formed within salinas on a Holocene sabkha surface. No gypsum salinas exist today along the Abu Dhabi sabkhas although primary subaqueous gypsum has been reported as currently forming in Khor Odaid, southeast Qatar (Loreau and Purser 1973). Farrant et al. (2012) recognised gypsum that had probably precipitated in marginal lagoons (salinas) which were thought to have formed during the initial stages of the Holocene forced regression in the vicinity of Bu Labyad-Al Rufaiq.

Although gypsum and halite can be found within sabkhas on many of the barrier islands, Al Aryam is the only barrier island where anhydrite is known to occur (Evans and Kirkham 2009b).

Holocene Dolomite

Dolomite has often been quoted as occurring in the sabkha sediments of the Abu Dhabi coastline. Indeed there have been fundamental research projects on the area attempting to understand the origin of penecontemporaneous dolomitization in the Holocene strata (e.g. McKenzie et al. 1981; Patterson and Kinsman 1981, 1982; Wood 2011). This dolomitization was originally regarded as a purely inorganic chemical process, but Bontognali et al. (2010) showed that microbially induced dolomitisation is an important factor in these arid coastline environments. Bahniuk et al. (2015) recorded a similar association between microbial matter and dolomite in a hypersaline coastal lagoon near Rio de Janeiro. Strohmenger et al. (2010) discovered significant dolomite volumes throughout the Pleistocene and Holocene sediments of Mussafah Channel which they attributed to this type of origin although they stressed that it is microcrystalline and below resolution of conventional petrographic thin section analysis. Patterned dolomites resulting from subaqueous activity of sulphate-reducing bacteria under highly saline conditions (salinas) can be inferred from black sulphide blemishes in the carbonate sediment (Kirkham 2004; Wright and Kirkham 2011). Such blemishes have been observed in the banks of the Mussafah Channel in very close proximity to probable subaqueous gypsum and therefore provide a field criterion for recognition of the presence of dolomite in the sedimentary section (Kirkham 1998b). Although Kenig (2011) recorded significant amounts of dolomite in the subtidal lagoonal sediments of the Mussafah Holocene sequence, he did not speculate on their origin.

Magnesium-Silicate Clays

Palygorskite, a Mg-Si clay, forms in both marine and continental saline settings including soils and evaporitic lakes under arid climates. Peebles et al. (1997) reported the occurrence of palygorskite in the shallow subsurface of continental Sabkha Matti. Wood and Sanford (2002), Wood et al. (2002) and Wood (2011) recorded palygorskite in the Abu Dhabi Formation of coastal sabkhas but did not elaborate further on its distribution. The contentious nature of that formation has already been discussed. Evans and Kirkham (2005) recorded Mg-Si clay, sepiolite, from an interdune continental sabkha 50 km south of Abu Dhabi Island, and Huggett et al. (in press) have since reconfirmed its presence plus palygorskite at the same locality. Kirkham and Huggett (in prep) describe probable pedogenic palygorskite from Sabkha Matti and possible salina palygorskite from Mussafah Channel. In both facies, there is an intimate association with authigenic dolomite. Bontognali et al. (2010) discovered an amorphous Mg-Si phase accumulated within extracellular polymeric substances (EPS) of microbial mats of Abu Dhabi although they were unable to recognise a specific clay type, but an association with dolomite has already been mentioned. In modern microbial mats from Qatar, Perri et al. (2018) described the initial stages of mineral precipitation taking place within EPS and upon viruses, whereby an amorphous Ca-Mg-Si material which evolved into palygorskite was precipitated within the top millimetre of the mat and was again associated with dolomite. Various workers (e.g. Wright and Barnett (2015)) have recorded the association of Mg-Si clays with dolomitization in Aptian lacustrine sediments of the southern Atlantic. Mg-Si clays are perhaps widely distributed in a range of facies across Abu Dhabi, and its distribution is of particular interest because Mg-Si clays appear to initiate dolomite authigenesis.

Holocene Hardgrounds

Considerable interest was aroused by the discovery in the 1950s and 1960s of cemented carbonate sediments (lithified horizons/bored surfaces) named hardgrounds on the sea floor (see review in Bricker 1971). Shinn (1969, 1971) showed that subaqueous cemented layers covered large areas of the southern Arabian/Persian Gulf. Others including Purser and Evans (1973) and Evans et al. (1973) discussed cemented sediments in the intertidal and subtidal zones along the Abu Dhabi coastline. Such cemented layers are a common and important component of the sabkha sequences and have a contentious origin.

Thin (rarely >15 cms thick) extensive hardgrounds are common from Jebel Dhanna to Abu Dhabi. They tend to be remarkably flat-topped, often displaying fossilised ripples, but with irregular bases. At the north end of Al Rufaiq Island, Kirkham has observed many tens of square metres of Holocene sediment having been manually stripped to expose a continuous hardground. Tidal currents in the channels between the barrier islands have apparently stripped the soft sediment from above a flat-topped hardground, the presence of which limited the depth of channel erosion (Fig. 22 in Evans and Kirkham 2005). In addition, shingled hardgrounds extend several kilometres within the carbonate sediments exposed along the Mussafah Channel (Kirkham 1998b). Evans et al. (1969) described a buried hardground extending seawards to the lagoon in front of the beach ridge along the Evans line. Kirkham (1998b) recognised a landward limit of a transgressive hardground just seawards of winged storm beach deposits at the eastern end of Mussafah Channel, and Farrant et al. (2012) implied that the hardgrounds do not extend inland beyond the storm beaches into the early Holocene lagoons behind.

Shinn (1969) described flat-topped hardgrounds around Qatar that were bored, encrusted and eroded near the shore where a protective overlying sediment was absent and that submarine cementation is most pronounced at the sediment-water interface in sand-sized sediment and decreases downwards but ceases 10–15 cms below the interface and that cementation depth is much reduced in fine-grained sediments due to lower permeability. He discovered that cementation graded upwards into uncemented sediment further offshore where the hardgrounds had irregular tops and lacked signs of boring. He was of the opinion that the hardgrounds he described all originated subtidally. Shinn (2011) observed that the subtidal hardgrounds may merge imperceptibly into beachrocks forming at low tide level, the cementation process of which was significantly more rapid due to evaporation during exposure.

Bathurst (1971) described how hardgrounds form subtidally from marine cements precipitating out of circulating marine waters beneath a mobile sediment veneer under conditions of exceptionally slow sediment accumulation. Evans et al. (1973) and Evans (1995) regarded hardgrounds as diastems. According to Bathurst, such hardgrounds would have evolved from coalescing nodules. The remarkably planar top surface of subtidal hardgrounds must have resulted from later erosion by wave or current action or by abrasion from mobile sediment.

In contrast, Strohmenger et al. (2011) suggested that the hardgrounds of Abu Dhabi formed from rapid cementation at the vadose/phreatic interface of the groundwater table, implying that the carbonate precipitation had occurred in the subsurface above mean sea level. However, as the water table fluctuates by ~1 m twice daily in the supratidal zone, it is difficult to understand how such thin and distinctive hardgrounds could have resulted. A study of vadose and phreatic cements would assist in their interpretation.

Alsharhan et al. (1995) described Holocene beachrocks formed during the (final) stages of the Flandrian transgression on the east side of Jebel Dhanna. They extended from 2 m deeper than low water mark to above high water mark where they were overlain by regressive intertidal deposits and microbial mat. Kendall and Alsharhan (2011) interpreted present-day subtidal limestones as beachrocks formed during the Holocene (Flandrian) transgression in the Khor Al Bazaam. Farrant et al. (2012) recognised a unit containing three hardgrounds resting on a transgressive surface of 'undivided' aeolian sands or the Ghayathi Formation in the vicinity of Bu Labyad-Al Rufaiq. Shinn (2011) also claimed at least some hardgrounds in Doha Bay, Qatar, as part of an early Holocene transgressive sequence tract.

The best and nearest approach to a three-dimensional continuous exposure of hardgrounds beneath the sabkha has been within the banks of the Mussafah Channel where two seaward-dipping hardgrounds have been recorded. Both Kirkham (1998b) and Lokier et al. (2015) regarded the extensive lower hardground as transgressive as it directly overlies a Flandrian transgressive microbial mat at the eastern end of Mussafah Channel. As microbial mats develop in the mid-upper intertidal zones, according to Walther's law, it is likely that its overlying hardground formed at least partly in the lower intertidal zone as a beachrock. This hardground dips seawards and falls 1–1.5 m over a distance of ~3 kms, which could have been within lower intertidal range during the Flandrian transgression, when overall tidal range is likely to have been ~2 m in the absence of the yet to be developed barrier island system. A beachrock origin is therefore entirely possible for this transgressive hardground. Kirkham (1998b) interpreted flat-topped hardgrounds as beachrocks reminiscent of the extensive areas of cemented intertidal flats east of Jebel Dhanna (Fig. 3.14). At the eastern extremity of Mussafah Channel, they had been secondarily wind-blasted to reveal internal cavities of cerithid gastropod shells. Such windblasting of gastropod shells was also recorded on Marawah Island, though unrelated to a hardground (Evans

Fig. 3.14 Erosional remnant of flat-topped modern beachrock, intertidal flats, east side Jebel Dhanna. The top of the hardground shows faint ripples but is less rippled than the surrounding soft beach sediment around it

and Kirkham (2002). Excavations during fish farm preparations south of Bu Labyad have revealed Holocene intertidal(?) barnacles encrusting the flat-topped hardgrounds. Lokier et al. (2015) suggested that the transgressive hardground in eastern Mussafah Channel may have formed within a lower intertidal pond, but it is difficult to explain all such extremely flat-topped hardgrounds originating in ponds when the best correlated ones seem to be so extensive and clearly dip seawards.

One flat-topped hardground along the mainland coast of the lagoon south of Al Rufaiq/Al Qanatir can be traced seawards at progressively shallower burial depths below regressive and progradational Holocene subtidal muds and their successively overlying intertidal microbial mats, gypsum mush and supratidal anhydrite. The seawards thinning of this regressive and diachronous sedimentary sequence above could represent either downlapping onto an earlier Holocene transgressive hardground or downlapping onto a regressive hardground which itself is diachronous and extending seawards. Whittle et al. (1998) and Farrant et al. (2012) regarded this hardground as transgressive. The hardgrounds described by Strohmenger et al. (2011) in the same vicinity were interpreted as regressive, although with evidence of 'minor sea level rises'.

Evans et al. (1969) found no conclusive proof of either a subaerial or subaqueous origin for the hardground along the Evans line but conjectured that its sheet-like extent over a vast part of the lagoons rendered it more likely to be subtidal and lower intertidal in origin and regarded it as part of a prograding regressive unit (i.e. a final lagoonal stage of relative nondeposition, a diastem) before it was covered by the prograding coastal wedge of intertidal and supratidal sediments. It is unknown whether this hardground was flat-topped, but it possibly correlates with the upper, regressive hardground in the Mussafah Channel which intersects the Evans line. Most of the hardgrounds along the eastern Qatar coast are parts of a net offlap accretion process and can be regarded as regressional (Shinn 2011).

In the mainland area south of Bu Labyad and Al Rufaiq, Lokier and Steuber (2008, 2010) recorded a single hardground which apparently undulated according to their correlation, but Kirkham and Evans (2010) considered the possibility of there being at least two hardgrounds or that the undulation reflected a lack of survey control. Kirkham and Evans (2010) further considered the possibility of hardground shingling as occurs along the Mussafah Channel. South of Al Rufaiq, Strohmenger et al. (2011) identified up to five hardgrounds, and their illustrations suggested overlapping geometry. Also in the same vicinity, Farrant et al. (2012) recorded five flat-topped hardgrounds of which three were described as having been encrusted by unspecified organisms. Their illustration showed the five hardgrounds as areally separated features (i.e. without overlap or shingling), but with a perplexing attitude within the sedimentary sequence.

Strohmenger et al. (2010) referred to two types of hardground in the banks at the eastern extremity of the Mussafah Channel. One type apparently formed as beachrock on the edges of tidal channels. The other (main) type of hardground is certainly the same as the transgressive hardground described by Kirkham (1998b) and Lokier et al. (2015). A distinctive feature of this hardground is its flat top and highly irregular base. Lokier et al. 2015) described its 'remarkably planar (top) surface' although it does display fossilised ripples (Kirkham 1998b) and may be extensively burrowed and bored by *Lithophaga* (Strohmenger et al. 2010). The flat upper and irregular basal surfaces are typical of all other hardground occurrences examined by the present authors.

Again, Evans et al. (1973) and Whittle et al. (1998) regarded the hardgrounds of the area as diachronous, but without substantiative evidence. From radiocarbon dating, Strohmenger et al. (2011) concluded that the several hardgrounds encountered near Al Qanatir are diachronous, but their summary illustrations are contradictory. Close examination of their radiocarbon dates leaves scope for a different correlation and so their evidence for diachroneity is suspect. They stressed the importance of the radiocarbon dating to correlation, but miscorrelation can still occur when dealing with pits rather than the continuous outcrop available along the Mussafah Channel. Certainly, the late Holocene regressive deposits of the coastal sabkhas are diachronous as shown by radiocarbon dating of both the microbial mat along the Evans line and those along the transect described by Lokier and Steuber (2008, 2010). However, Kirkham (1998b) regarded the two shingled hardgrounds exposed in the banks of Mussafah Channel as essentially isochronous clinoforms.

Tepee Structures

The tidal channel between Al Rufaiq and Al Qanatir was floored by an extensive hardground displaying tepee-defined polygons *ca* 50 m across. Hardgrounds with tepee-defined polygons <=400 m (Kendall and Skipwith 1969a, b) and <=166 m across (Lokier and Steuber 2009) are known from the intertidal-lagoonal area southeast of Bu Labyad and Al Rufaiq (Kendall et al. 1994; Alsharhan et al. 1995; Kendall and Alsharhan 2011).

Lokier and Steuber (2009) described a polygonally cracked hardground apparently formed in a very shallow subtidal setting. Only the uplifted polygon margins are exposed briefly at low tide. The fact that their flat tops may be bored by *Lithophaga* supports a submarine origin or at least a return to subtidal setting. They concluded that bacterially induced, displacive carbonate cementation beneath several centimetres of unconsolidated sediment was operating in polygonally cracked hardgrounds. The base of the hardground was sharp and irregular, but the top was characterised by a transition from the poorly consolidated grainstone to the underlying hardground, and so a direct comparison with the flat-topped hardgrounds cannot be made. It is possible that this hardground had a common origin with flat-topped hardgrounds, and indeed it is probably correlatable with flat-topped hardgrounds encountered at localities a few kilometres to the east, but ponding of carbonate sediment within individual late-stage polygons has modified the depositional and diagenetic setting. Although the largest teepees occur in the inner lagoon southeast of Bu Labyad, the presence of teepee structures is not diagnostic of subtidal carbonate hardground formation as teepees can also occur intertidally (Alsharhan et al. 1995; Evans and Kirkham 2005). Evamy (1973) recorded intertidal tepee structures in the Sabkha Matti area.

Conservation

As already mentioned, the Abu Dhabi part of the UAE has been a classic field area for arid carbonate-evaporite studies for many years, but it is regrettable that most of the natural coastal reaches and barrier islands have since been destroyed or severely modified by civil engineering works and oilfield operations or are inaccessible due to privatisation and for security reasons (Evans and Kirkham 2002; Lokier 2013; Lokier and Fiorini 2016). For instance, the area east of Abu Dhabi Island where Kinsman (1969) and Butler (1973) made earlier detailed studies of the coastal sabkha is now completely urbanised. Satellite imagery clearly demonstrates that both western and eastern Jebel Dhanna areas have either been severely interfered with or lost as a geological record due to industrial expansion. It also shows that the Mussafah industrial estate has obliterated evidence of a spit system that existed immediately north of the Mussafah Channel (Butler 1973). Whilst the Mussafah Channel temporarily produced a magnificent and unique continuous geological exposure in its banks, it is being gradually lost to further industrial development and completely refilled progressively at its landward eastern end, thus masking excellent tidal channel and spit deposits. Strohmenger et al. (2010) recommended that the Mussafah Channel be protected for future visits and studies, but have been ignored to no avail as has all appeals to the UAE authorities by the authors over many years, and by Lokier and others.

During the mid-1990s, the Society of Explorationists in the Emirates (SEE) requested their Secretary (A. Kirkham) to recommend conservation areas under the realisation that the classic coastal sabkhas were being irreparably damaged and lost from the natural heritage of Abu Dhabi. Many areas were consequently identified and ranked in terms of their areal coverage. They were presented to various influential personnel in the hope that a limited area could be conserved. The main areas identified were in the vicinity of the Al Dabb'iya Oilfield and were therefore highly susceptible to interference from oilfield operations, but at least the smallest and highest ranked area was located on the mainland along the southern periphery of the field where it was assumed to be least likely interfered with by the oilfield activities (Fig. 3.15). Unfortunately, extensive dredging operations were later initiated even within this perceived relatively 'safe' area – thus destroying the best of the then surviving microbial mat developments.

It is lamentable that local environmental agencies and operating oil companies, in spite of appeals by the writers and others, do not appreciate the importance and uniqueness of the Abu Dhabi sabkha as a model in the understanding of ancient carbonate sequences throughout the world. It is frequently referenced in petroleum-related professional journals, and is a significant part of the natural heritage of the UAE. The eradication of classic, virgin sabkha has reached a critical juncture, and there will be almost complete destruction without urgent intervention by the UAE government. At least in recent years, a sign advertising an area of special geological interest has been erected due to valiant efforts of Lokier, Strohmenger and others at the southern end of the rough track that leads from the main highway to the jetty from which ferries may depart for Al Rufaiq Island. The area immediately east of the jetty was the focus of detailed analysis by Strohmenger et al. (2011) and has long been a favoured geological excursion locality for 'Friends of the Sabkhas' such as Alsharhan et al. (1995). Lokier (2013) made a valiant attempt to identify areas of geoconservation including Marawah Island (Evans et al. 2002) and, coincidentally, the highest ranked area defined by Kirkham in the late 1990s. He proposed the construction of a visitor centre,

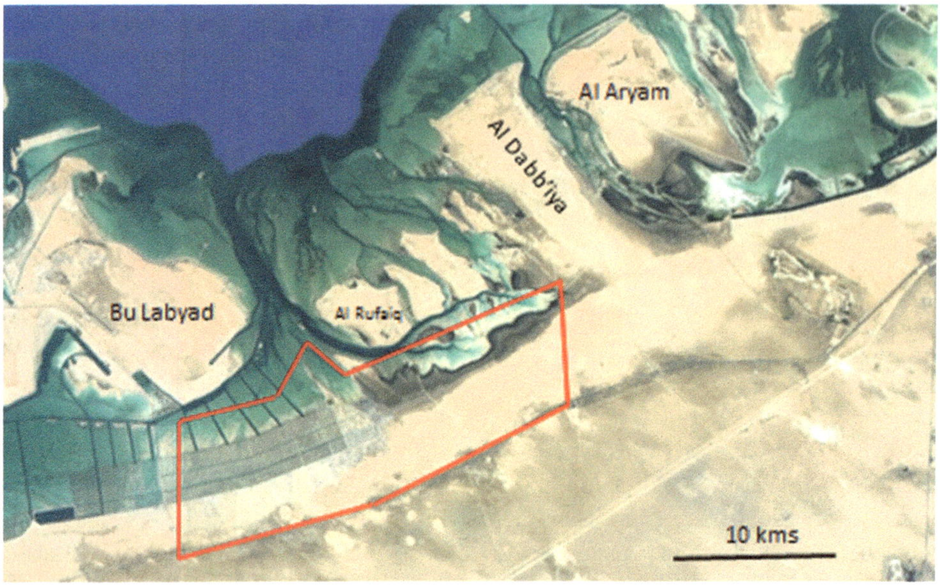

Fig. 3.15 Highest ranked area for sabkha preservation, 1997, as presented to the Society for Exploration in the Emirates. It was defined to include parallel storm beaches (along its southern half), microbial mats and hardground polygons. It is evident that a regular grid of extensive dredging has since ruined the eastern half of the area

raised boardwalks and observation platform. The present authors certainly support these proposals in principle although extreme care and careful planning would be needed to avoid irreparable damage to the sensitive sediments when constructing these facilities. The present authors have long been advocating a programme of shallow coring, core resination, slabbing and detailed sedimentological description of the cores with photographs such as those of Strohmenger et al. (2011) as a basis for museum display which could be housed in a place such as Lokier's proposed visitor centre or in a planned museum on Sadiyat Island as part of an exhibit illustrating the evolution of the Abu Dhabi coastline. However, the pace of development in this classic area is so fast that the geological community has precious little time to conserve such valuable localities.

Conclusions

Two lagoon systems, separated by a storm beach system, developed during and since the climax of the Flandrian transgression. The initial, smaller, shallower lagoon system was the site of early Holocene evaporite precipitation and has been completed infilled, whereas the later, current, major lagoon protected by the barrier island complex has been only partially infilled by late Holocene evaporitic sabkha progradation from the mainland under a forced regression and by leeward accretion of the barrier islands. Mainland progradation has progressed at different rates. In places, leeward accretion has been more efficient at lagoon infill than mainland coastal progradation. Both lagoon systems were characterised by anhydrite precipitation.

The coastal sabkhas have enlarged in opposing directions: (1) mainly by seaward progradation under the influence of the late Holocene forced regression and (2) by aeolian deflation at their landward edges.

It appears that the area is currently experiencing a renewed transgression, and so the modern lagoon infill is presumably being slowed or may ultimately be reversed and thus endangering the extensive urban developments on the sabkha with flooding.

Further detailed study is needed to test the exact relationship of the zone of anhydrite development and the tidal regime.

The Quaternary aeolian sediments which underlie the coastal sabkhas are still not fully understood in terms of their lithological variation and precise age. Confusion remains regarding their stratigraphic relationships and nomenclature. In fact, there has been confusion over the relatively new usage of the terms 'unconsolidated undivided aeolian sands'. Any confusion regarding the occurrence of perceived aeolian sands (not marine reworked) within the original definition of 'Abu Dhabi Formation' is partly resolved as its most recent usage by the British Geological Survey has limited the 'Abu Dhabi Formation' to the Holocene marine sediments of the outer sabkha.

It is becoming increasingly possible that much of the Holocene dolomite is microbial in origin rather than inorganic. Patterned dolomites resulting from sulphate-reducing bacteria within salinas are also suspected. Palygorskite appears to have a strong relationship with dolomite in both coastal and continental sabkhas although the distribution of this Mg-Si clay is poorly understood.

The origins of remarkably flat-topped Holocene hardgrounds are probably the most contentious subject. Clearly there are both transgressive and regressive hardgrounds, but there is still no consensus about their depositional setting (intertidal/subtidal/subsurface), their cementation mechanism(s), how they may correlate, or the degree to which they are diachronous. Further studies are required to clarify these issues.

Most of the virgin sabkhas have been irreparably lost or interfered with, so the UAE and the global geological community have almost completely lost a valuable natural heritage. It is imperative that the small areas of the last vestiges of relatively untouched sabkhas can be conserved.

Acknowledgements The authors wish to thank the following for their support in the field at various times in recent years: Bryan Twombley, Philippe Lapointe, Hugh Nicholson, Richard (Dick) Hornby and Veryan Pappin, Peter Hellyer and The Natural History Society of Abu Dhabi. Mrs. Kate Davis kindly assisted with the cartography.

References

Alsharhan AS, Kendall CGStC, Whittle GL (1995) Field trip guide to examine the 'Holocene carbonate/Evaporites of Abu Dhabi, United Arab Emirates. International Conference on Quaternary Deserts and Climatic Change, Al Ain, UAE, 46p

Bahniuk A, McKenzie JA, Perri E, Bontagnali TRR, Vögeli N, Rezende CE, Rangel TP, Vasconcelos C (2015) Characterization of environmental conditions during microbial Mg-carbonate precipitation and early diagenetic dolomite crust formation: Brejo do Espinho, Rio de Janeiro, Brazil. In: Bosence DWJ, Gibbons KA, Le Heron DP, Morgan WA, Prichard T, Vining BA (eds) Microbial carbonates in space and time, Geological Society Special Publication, vol 418. Geological Society, London, pp 243–259

Bathurst RGC (1971) Carbonate sediments and their diagenesis. Developments in Sedimentology 12. Elsevier, Amsterdam

Bontognali TRR, Vasconcelos CG, Warthmann RJ, Bernasconi SM, Dupraz C, Strohmenger CJ, Mckenzie JA (2010) Dolomite formation within microbial mats in the coastal sabkha of Abu Dhabi (United Arab Emirates). Sedimentology 57:824–844

Bricker OP (ed) (1971) Carbonate cements. John Hopkins Press, Baltimore/London, p 376

Butler GP (1970) Recent gypsum and anhydrite of the Abu Dhabi sabkha, Trucial Coast: an alternative explanation of origin. In: Third Symposium on Salt. Geological Society of North Ohio, vol 1, pp 120–152

Butler GP (1973) Strontium geochemistry of modern and ancient calcium sulphate minerals. In: Purser BH (ed) The Persian Gulf: Holocene carbonate sedimentation and diagenesis in a shallow epicontinental sea. Springer, Berlin, pp 423–452

Butler GP, Harris PM, Kendall CGSC (1982) Recent evaporites from Abu Dhabi coastal flats. In: Hanford CR, Louks RG, Davies GR (eds) Deposition and diagenetic spectra of evaporites, vol 3. Society of Economic Palaeontology and Mineralogy Core Workshop, Tulsa/Okla, pp 33–64

Cazenave A, Nerem RS (2003) Present-day sea level change: observations and causes. Rev Geophys 42:RG3001

Curtis R, Evans G, Kinsman DJJ, Shearman DJ (1963) Association of dolomite and anhydrite in the recent sediments of the Persian Gulf. Nature 197:679–680

Evamy BD (1973) The precipitation of aragonite and its alteration to calcite on the Trucial Coast of the Persian Gulf. In: Purser BH (ed) The Persian Gulf. Springer, Berlin, pp 329–341

Evans G (1970) Coastal and nearshore sedimentation: a comparison of clastic and carbonate deposition. Proc Geol Assoc 81:493–508

Evans G (1995) The Arabian Gulf: a modern carbonate-evaporite factory: a review. Cuadernos Geologicos Ibérica 19:61–96

Evans G (2011) An historical review of the Quaternary sedimentology of the Gulf (Arabian/Persian and its geological impact). In: Kendall CGSC, Alsharhan AS (eds) Quaternary carbonate and evaporite sedimentary facies and their ancient analogues, International Association of Sedimentology Special Publication, vol 43. Wiley-Blackwell, Oxford, pp 11–44

Evans G, Kirkham A (2002) The Abu Dhabi sabkha. In: Barth HJ, Böer BB (eds) Ecosystems. Kluwer Academic Publishers, Dordrecht, pp 7–20

Evans G, Kirkham A (2005) The Quaternary deposits. In: Hellyer P, Aspinall S (eds) The Emirates: a natural history. Trident Press, London

Evans G, Kirkham A (2009a) A geological description of Belghelam Island, North-East Abu Dhabi, U.A.E. Tribulus 18:4–9

Evans G, Kirkham A (2009b) A report on a geological reconnaissance of Al Aryam Island, March 2004, Abu Dhabi. Tribulus 18:10–17

Evans G, Kendall CGSC, Skipwith PAd'E (1964) Origin of the coastal flats, the sabkha, of the Trucial coast, Persian Gulf. Nature 202:759–761

Evans G, Schmidt V, Bush P, Nelson H (1969) Stratigraphy and geologic history of the sabkha Abu Dhabi, Persian Gulf. Sedimentology 12:145–159

Evans G, Murray JW, Biggs HEJ, Bate R, Bush PR (1973) The oceanography, ecology, sedimentology and geomorphology of parts of the Trucial Coast Barrier Island complex, Persian Gulf. In: Purser BH (ed) The Persian Gulf. Springer, Berlin, pp 233–278

Evans G, Kirkham A, Carter RA (2002) Quaternary development of the United Arab Emirates coast: new evidence from Marawah Island, Abu Dhabi. GeoArabia 7(3):441–458

Farrant AR et al (2012) The geology and geophysics of the United Arab Emirates. Vol. 6. Geology of the Western and Central United Emirates. Ministry of Energy, United Arab Emirates

Hadley DG, Brouwers EM, Bown TM (1998) Quaternary palaeodunes, Arabian Gulf Coast, Abu Dhabi Emirate: age and palaeoenvironmental evolution. In: Alsharhan AS, Glennie KW, Whittle GL, Kendall CGSC (eds) Quaternary deserts and climatic change. Balkema, Rotterdam, pp 123–140

Huggett J, Evans G, Kirkham A (in press) An occurrence of Palygorskite in the U.A.E. Tribulus

Kassler P (1973) The structural and geomorphic evolution of the Persian Gulf. In: Purser BH (ed) The Persian Gulf. Springer, Berlin, pp 11–32

Kendall CGSC, Alsharhan AS (2011) Holocene geomorphology and recent carbonate-evaporite sedimentation of the coastal region of Abu Dhabi, United Arab Emirates. In: Kendal CGSC, Alsharhan AS (eds) Quaternary carbonate and evaporite sedimentary facies and their ancient analogues, International Association of Sedimentology Special Publication, vol 43. Wiley-Blackwell, Oxford, pp 45–88

Kendall CGStC, Skipwith SPAd'E (1969a) Geomorphology of a recent shallow water carbonate province: Khor Al Bazam, Trucial Coast, southwest Persian Gulf. Geol Soc Am Bull 80:865–892

Kendall CGStC, Skipwith SPAd'E (1969b) Holocene shallow-water carbonate and evaporite sediments of Khor al Bazam, Abu Dhabi, southwest Persian Gulf. Am Assoc Pet Geol Bull 53:841–869

Kendall CGSC, Sadd JL, Alsharhan A (1994) Holocene marine cement coatings on beach-rocks of the Abu Dhabi coastline (UAE): analogs for cement fabrics in ancient limestones. Carbonates Evaporites 9:119–131

Kendall CGStC, Alsharhan ARS, Cohen A (2002) The Holocene tidal flat complex of Arabian Gulf coast of Abu Dhabi. In Barth HJ, Böer BB (eds) Ecosystems, pp 21–36

Kenig F (2011) Distribution of organic matter in the transgressive and regressive Holocene sabkha sediments of Abu Dhabi, United Arab Emirates. In: Kendall CGSC, Alsharhan AS (eds) Quaternary carbonate and evaporite sedimentary facies and their ancient analogues, International Association of Sedimentology Special Publication, vol 43. Wiley-Blackwell, Oxford, pp 277–298

Kinsman DJJ (1969) Modes of formation, sedimentary associations and diagnostic features of shallow-water and supratidal evaporites. Am Assoc Pet Geol Bull 53:830–840

Kirkham A (1997) Shoreline evolution, aeolian deflation and anhydrite distribution of the Holocene, Abu Dhabi. GeoArabia 2(4):403–416

Kirkham A (1998a) Pleistocene carbonate seif dunes and their role in the development of complex past and present coastlines of the U.A.E. GeoArabia 3(1):19–32

Kirkham A (1998b) A quaternary proximal foreland ramp and its continental fringe, Arabian Gulf, U.A.E. In: Wright VP, Burchette TP (eds) Carbonate ramps, Geological Society Special Publication, vol 149. Geological Society, London, pp 15–41

Kirkham A (2004) Patterned Dolomites: microbial origins and clues to vanished evaporites in the Arab Formation, Upper Jurassic, Arabian Gulf. In: Braithwaite CJR, Rizzi G, Darke G (eds) The geometry and petrogenesis of dolomite hydrocarbon reservoirs, Geological Society Special Publication, vol 235. Geological Society, London, pp 301–308

Kirkham A (2011) Halite, sulphates, sabkhat and Salinas of the coastal regions and Sabkha Matti of Abu Dhabi, United Arab Emirates. In: Kendall CGSC, Alsharhan AS (eds) Quaternary carbonate and evaporite sedimentary facies and their ancient analogues, International Association of Sedimentology Special Publication, vol 43. Wiley-Blackwell, Oxford, pp 265–277

Kirkham A, Evans G (2010) Quantification of carbonate ramp sedimentation and progradation rates for the late Holocene Abu Dhabi shoreline – discussion. J Sediment Res 80:300–301

Kirkham A, Huggett J (in prep) Dolomite and palygorskite occurrences in Abu Dhabi Emirate

Leuliette E, Nerem R, Mitchum G (2004) Calibration of TOPEX/Poseiden and Jason altimeter data to construct a continuous record of mean sea level change. Mar Geod 27:79–94

Lokier SW (2013) Coastal sabkha preservation in the Arabian Gulf. Geoheritage 5:11–22

Lokier SW, Fiorini F (2016) Temporal evolution of a carbonate coastal system, Abu Dhabi, United Arab Emirates. Mar Geol 381:102–113

Lokier SW, Steuber T (2008) Quantification of carbonate ramp sedimentation and progradation rates for the late Holocene Abu Dhabi shoreline. J Sediment Res 78:423–431

Lokier SW, Steuber T (2009) Large-scale intertidal polygonal featuresof the Abu Dhabi coastline. Sedimentology 56:609–622

Lokier SW, Steuber T (2010) Quantification of carbonate ramp sedimentation and progradation rates for the late Holocene Abu Dhabi shoreline – a reply. J Sediment Res 80:302

Lokier SW, Bateman MD, Larkin NR, Rye P, Stewart JR (2015) Late quaternary sea-level changes of the Persian Gulf. Quat Res 84:69–81

Loreau J-P, Purser BH (1973) Distribution and ultrastructure of Holocene ooids in the Persian Gulf. In: Purser BH (ed) The Persian Gulf: Holocene carbonate sedimentation and diagenesis in a shallow Epicontinental Sea. Springer, Berlin, pp 279–328

McKenzie JA, Hsü K, Schneider JF (1980) Movement of subsurface waters under the sabkha, Abu Dhabi, U.A.E. and its relation to evaporite dolomite genesis. SEPM Special Publications, 28, Concepts and Models of Dolomitization, pp 11–30

McKenzie JA, Hsü K, Schneider JF (1981) Holocene dolomitization of calcium carbonate sediments from the coastal sabkhas of Abu Dhabi, U.A.E.: a stable isotope study. J Geol 89:185–197

Park RK (2011) The impact of sea-level change on ramp margin deposition: lessons from the Holocene sabkhas of Abu Dhabi, United Arab Emirates. In: Kendall CGSC, Alsharhan AS (eds) Quaternary carbonate and evaporite sedimentary facies and their ancient analogues, International Association of Sedimentology Special Publication, vol 43. Wiley-Blackwell, Oxford, pp 89–112

Patterson RJ, Kinsman DJJ (1977) Marine and continental ground-water source in a Persian Gulf coastal sabkha. In: Reefs and related carbonates-ecology and sedimentology. Am Assoc Pet Geol Bull 4:381–397

Patterson RJ, Kinsman DJJ (1981) Hydrologic framework of a sabkha along the Arabian Gulf. Am Assoc Pet Geol Bull 65:1457–1475

Patterson RJ, Kinsman DJJ (1982) Formation of diagenetic dolomite in coastal sabkha along the Arabian (Persian) Gulf. Am Assoc Pet Geol Bull 66(1):28–43

Peebles RG, Suzuki M, Shaner M (1994) The effects of long-term shallow-burial diagenesis on carbonate-evaporite successions. In: Al-Husseini MI (ed) Middle East petroleum geosciences Geo'94 conference, vol 2. Gulf Petrolink, Manama, pp 761–769

Peebles R, Kirkham A, Shaner M (1997) A field guide to the Abu Dhabi Coast, United Arab Emirates. Arid Coastline Depositional Environments. An American Association of Petroleum Geologists International Field Seminar, 2–7 November

Perri E, Tucker ME, Slowakiewicz M, Whitaker F, Bowen L, Perrotta ID (2018) Carbonate and silicate biomineralization in a hypersaline microbial mat (Mesaieed sabkha, Qatar): roles of bacteria, extracellular polymeric substances and viruses. Sedimentology 65:1213–1245

Purser BH (1985) Coastal evaporite systems. In: Friedman GM, Krumbein WE (eds) Hypersaline ecosystems. Springer, Berlin, pp 72–102

Purser BH, Evans G (1973) Regional sedimentation along the Trucial coast, SE Persian Gulf. In: Purser BH (ed) The Persian Gulf. Springer, Berlin, pp 211–231

Purser BH, Loreau J-P (1973) Aragonitic supratidal encrustations on the Trucial coast, Persian Gulf. In: Purser BH (ed) The Persian Gulf. Springer, Berlin, pp 343–376

Shinn EA (1969) Submarine lithification holocene carbonate sediments in the Persian Gulf. Sedimentology 12(1/2):109–144

Shinn EA (1971) Holocene submarine cementation in the Persian Gulf. In: Bricker OP (ed) Carbonate cements. John Hopkins Press, Baltimore/London, pp 63–65

Shinn EA (2011) Interplay between Holocene sedimentation and diagenesis, and implications for hydrocarbon exploitation: return to the sabkha of Ras Said, Qatar. In: Kendall CGSC, Alsharhan AS (eds) Quaternary carbonate and evaporite sedimentary facies and their ancient analogues, International Association of Sedimentology Special Publication, vol 43. Wiley-Blackwell, Oxford, pp 133–148

Stevens T, Kirkham A, Evans G (2011) Quaternary sea levels: recent evidence from Abu Dhabi. Tribulus 19:158–159

Stevens T, Jestico MJ, Evans G, Kirkham A (2014) Eustatic control of late Quaternary sea level changes in the Arabian/Persian Gulf. Quat Res 82:175–184

Strohmenger CJ, Al-Mansoori A, Al-Jeelani O, Al-Shamry A, Al-Hosani I, Al-Mehsin K, Shebl H (2010) The sabkha sequence at Mussafah Channel (Abu Dhabi, United Arab Emirates): facies stacking patterns, microbial-mediated dolomite and evaporite overprint. GeoArabia 15:49–90

Strohmenger CJ, Shebl H, Al-Mansoori A, Al-Mehsin K, Al-Jeelani O, Al-Hosani I, Al-Shamry A, Al-baker S (2011) Facies stacking patterns in a modern arid environment: a case study of the Abu Dhabi sabkha in the vicinity of Al-Qanatir Island, United Arab Emirates. In: Kendal CGSC, Alsharhan AS (eds) Quaternary carbonate and evaporite sedimentary facies and their ancient analogues, International Association of Sedimentology Special Publication, vol 43. Wiley-Blackwell, Oxford, pp 149–182

Teller JT, Glennie KW, Lancaster N, Sinhvi AK (2000) Calcareous dunes of the United Arab Emirates and Noah's flood: the postglacial reflooding of the Persian (Arabian) Gulf. Quat Int 68–71:297–308

Tricart J (1954) Une forme de relief climatique: les sabkhas. Rev Géomorphol Dynam 5:97–101

Walker MJ (2012) Hot deserts: engineering geology and geomorphology. Engineering group working party report. Geological Society Engineering Geology Special Publication No. 25

Whittle GL, Alsharhan AS, Kendall CGSC (1998) Petrography of Holocene beachrock and hardgrounds, Abu Dhabi, United Arab Emirates. In: Alsharhan AS, Glennie KW, Whittle GL, Kendall CGSC (eds) Quaternary deserts and climatic change. Balkema, Rotterdam, pp 57–70

Wood WW (2011) An historical odyssey: the origins of solutes in the coastal sabkha of Abu Dhabi, United Arab Emirates. In: Kendall CGStC, Alsharhan AS (eds) Quaternary carbonate and evaporite sedimentary facies and their ancient analogues. A tribute to Douglas James Shearman. Special Publication, 43 of International Association of Sedimentologists, pp 243–254

Wood WW, Sanford WE (2002) Hydrogeology and solute chemistry of the coastal-sabkha aquifer in the Emirate of Abu Dhabi. In: Boer B, Barth H-J (eds) Sabkha ecosystems. Kluwer Academic Publishers, pp 173–185

Wood WW, Sanford WE, Al Habschi ARS (2002) The source of solutes in the coastal sabkha of Abu Dhabi. Geol Soc Am Bull 114:259–268

Wood WW, Bailey RM, Hampton BA, Kraemer TF, Lu Z, Clark DW, James RHR, Al Ramadan K (2012) Rapid late Pleistocene/Holocene uplift and coastal evolution of the southern Arabian (Persian) Gulf. Quat Res 77:215–220

Wright P, Barnett AJ (2015) An abiotic model for the development of textures in some South Atlantic early Cretaceous lacustrine carbonates. In: Bosence DW, Gibbons KA, Le Heron DP, Morgan WA, Prichard T, Vining PA (eds) Microbial carbonates in space and time, Geological Society Special Publication, vol 418. Geological Society, London, pp 209–220

Wright D, Kirkham A (2011) The role of bacterial sulphate reduction in carbonate replacement of vanished evaporites: examples from the Holocene, Jurassic and Neoarchean. In: Kendall CGStC, Alsharhan AS (eds) Quaternary carbonate and evaporite sedimentary facies and their ancient analogues. A Tribute to Douglas James Shearman. Special Publication 43 of International Association of Sedimentologists, pp 299–314

Comparative Seed Germination Ecology of Sabkha and Playa Halophytes of Pakistan

4

Aysha Rasheed, M. Zaheer Ahmed, Bilquees Gul, M. Ajmal Khan, and Abdul Hameed

Abstract

Seed germination is a complex transition from quiescence stage of dry seeds to metabolically active seedling stage, hence is considered a critical time in the life history of plants. A number of factors such as salinity, temperature, photoperiod, and their interactions reportedly influence germination at temporal and spatial scale. However, seed germination responses may vary among species. In this study, we compared seed germination and recovery responses of two sabkha (*Suaeda fruticosa* and *Salsola drummondii*) and two playa (*Suaeda heterophylla* and *Halogeton glomeratus*) halophytes to variations in salinity, temperature, and photoperiod. There were many commonalities and differences in responses of seeds. Seeds of all test species were nondormant and positively photoblastic during germination. Salinity increments decreased seed germination of all test species. However, two Salsoloideae species (*S. drummondii* and *H. glomeratus*) showed higher seed germination than the two Suaedoideae species (*S. fruticosa* and *S. heterophylla*) under nonsaline conditions, irrespective of the habitat type. Under nonsaline conditions, seed germination of Salsoloideae species was temperature-independent, while seeds of Suaedoideae species germinated optimally at moderate temperature of 20/30 °C. Generally, the salinity tolerance limit of sabkha perennials was greater as compared to the playa annuals. Light improved seed germination of Suaedoideae and Salsoloideae members in nonsaline and saline condition, respectively. Recovery responses were generally species-specific. Ecophysiological significance of the findings has also been discussed.

Keywords

Halophyte · Photoperiod · Salinity · Seed dormancy · Seed germination · Temperature

Introduction

Seed germination is an important event in the life cycle of plants that ensures establishment of seedlings in suitable community gaps when conditions are conducive for plant growth (Gul et al. 2013; Estrelles et al. 2015). Therefore, spatial and temporal distribution of plant species is dependent on the seed germination responses to environmental conditions (Baskin et al. 2001; Huang et al. 2003; El-Keblawy and Al-Rawai 2006). In saline habitats, salinity, temperature,

A. Rasheed · M. Z. Ahmed · B. Gul · M. A. Khan · A. Hameed (✉)
Institute of Sustainable Halophyte Utilization (ISHU), University of Karachi, Karachi, Pakistan
e-mail: ahameed@uok.edu.pk

B. Gul et al. (eds.), *Sabkha Ecosystems*, Tasks for Vegetation Science VI,
https://doi.org/10.1007/978-3-030-04417-6_4

and photoperiod are reportedly the key factors influencing the seed germination responses (Gul et al. 2013). Despite several commonalities in seed germination responses, stress tolerance limits and specific germination requirements are often species-specific (Khan and Gul 2006). This knowledge is therefore essential for developing effective *ex situ* conservation and/or mass-scale cultivation strategies for the threatened and economically important species (Andreou et al. 2011; Gul et al. 2013). However, information on seed germination ecology of halophytes is very limited. For instance, a literature search on Thomson Reuters' Web of Science™ with keywords "halophyte and seed germination" returned just 389 articles (accessed on October 25, 2017), whereas there are 1469 known species of halophytes worldwide (http://www.sussex.ac.uk/affiliates/halophytes/).

Despite high salinity tolerance during growth stage, most halophytes are sensitive to salinity during seed germination (Gul et al. 2013; Kranner and Seal 2013). Hence, seeds of most halophytes germinate maximally under nonsaline conditions, and increases in salinity decrease their germination (Ungar 1978; Hameed et al. 2006; Ahmed and Khan 2010). However, there is substantial variation in upper limit (~10% germination) of salinity tolerance of halophyte seeds, which ranges from ~200 to 1700 mM NaCl (Gul et al. 2013). Another salient feature of halophyte seeds is their ability to maintain viability under high salinity by entering in a state of enforced dormancy and subsequent recovery of germination when salinity is removed/diluted following rains (Ungar 1978; Pujol et al. 2000; Ahmed and Khan 2010; Orlovsky et al. 2011). Both salinity tolerance and germination recovery of halophyte seeds are often influenced significantly by temperature and photoperiod (El-Keblawy 2004; Gul et al. 2013). Generally, sub-/supraoptimal temperatures and dark cause reductions in both seed germination and recovery responses to salinity (Khan and Gul 2006). However, interactive effects of temperature and photoperiod with salinity on seed germination and recovery may vary among species.

Generally, there are few studies on comparisons of seed germination and recovery responses of halophytes from contracting habitats (Gul et al. 2013; Barton et al. 2016; Ruiz et al. 2016). To the best of our knowledge, no such comparison has been made yet for halophytes from coastal sabkha versus mountainous playa habitats. Sabkhas are saline ecosystems along coastal areas of arid region and have a high diversity of perennial succulent halophytes (Khan et al. 2002; Akhani 2006). Whereas, playas are saline pans of the mountainous areas, which harbor unique diversity of mostly annual/herbaceous halophytes (Gul and Weber 2001; Khan and Qaiser 2006; Ahmed and Khan 2010). These habitats possess contrasting environmental conditions and the vegetation with contrasting selection pressures. In this study, we compared the seed germination and recovery responses of sabkha (*Suaeda fruticosa* and *Salsola drummondii*) and playa (*Suaeda heterophylla* and *Halogeton glomeratus*) halophytes to variations in salinity, temperature, and photoperiod. More specifically, we seeked answer to the following questions: (1) Is salinity tolerance limit of sabkha halophytes greater than that of playa halophytes? (2) Are seeds of playa halophytes more sensitive to variations in temperature than those of sabkha halophytes? (3) Is there any difference between sabkha and playa halophytes for light sensitivity during seed germination? (4) Are there any commonalities in seed germination and recovery responses of the halophytes from two contrasting habitats?

Material and Methods

Seed Collection and Study Site

Mature seeds of *Halogeton glomeratus* and *Suaeda heterophylla* were collected from a salt playa located in the upper Hunza, Pakistan (elevation: 2569 m, 36° 25.983′ N and 74°51.775′ E). Whereas, the seed of *Salsola drummondii* and *Suaeda fruticosa* were harvested from populations located in sabkhas of Winder, Balochistan (elevation: 12 m; 24° 25′ 07.16″ N and 66° 37′ 32.38″ E), and Karachi, Sindh (Elevation: 51 m; 24° 52′ 21.87″ N and 66° 51′ 24.58″ E), respectively. Seeds were separated from inflorescence, surface sterilized by using 1% commercial bleach (sodium hypochlorite), rinsed with distilled water, air dried, and stored in clear plastic Petri plates. Freshly collected seeds were used in germination experiments.

Seed Germination Experiments

Germination was carried out in tight-fitting plastic Petri plates (5 cm diameter) with 7 ml of test solution (0, 100, 200, 300, 400, 500 mM NaCl for *H. glomeratus* and *S. heterophylla*; 0, 200, 400, 600 mM NaCl for *S. fruticosa*; 0, 200, 400, 600, 800, 1000 mM NaCl for *S. drummondii*). These NaCl concentrations were selected after a preliminary test for salt tolerance. Four replicates of 25 seeds each were used for all treatments. Seeds were considered to be germinated at the emergence of the radicle (Bewley and Black 1994). Seeds were germinated in a programmed incubator (Percival Scientific, Boone, Iowa, USA) for environmental parameters, light (12 h photoperiod; 25 μmol m^{-2} s^{-1}, 400–700 nm; Philips cool-white fluorescent lamps) and temperature (10/20, 15/25, 20/30, and 25/35 °C, dark/light). To determine the effect of photoperiod on germination response, one set was placed in a 12 h photoperiod, while the second one in complete darkness (24 h) using photographic envelops as light shields. Seed germination percentage of 12 h photoperiod set was recorded on every alternate day for up to 20 days and for 24 h-dark only on the 20th day. Rate of germination was calculated using a modified Timson index of germination velocity, i.e., $\sum G/t$, where G is germination percentage after every 48 h and t is the total germination period (Khan and Ungar 1984). Maximum value possible for this index is 50 for our data (i.e., 1000/20). The higher the value, the more rapid is the germination.

Seed Recovery Experiments

Recovery from Salinity After completion of germination experiment, all ungerminated seeds from various NaCl concentrations were transferred to distilled water for another 20 days to study recovery of germination. Germination recovery percentage was calculated by using the following formula: [(a–b)/(c–b) × 100], where a is the number of seeds germinated after transfer to distilled water, b is the number of seeds germinated in saline solution, and c is the total number of seeds.

Recovery from 24 h Darkness Recovery from dark was also studied by transferring the Petri plates with seeds to 12 h photoperiod keeping test solution conditions unchanged.

Statistical Analyses The data was transformed using arcsine transformation before the statistical analysis. Analyses of variance (ANOVA) were performed to determine the significance of the effects of salinity, temperature, light on seed germination, and recovery data of selected species. Bonferroni post hoc tests ($p < 0.05$) were carried out to determine significant difference between individual means. Statistical analyses were done by using SPSS for Windows version 11 (Anon 2001).

Results

Seed Germination Responses

ANOVA showed species, photoperiod, thermoperiod, salinity, habitat, and their interactions had significant effects ($P < 0.0001$) on mean final germination (MFG) and rate of germination (GR) (Tables 4.1 and 4.2). Mean final germination (MFG) percentage and rate of seed germination were maximum in nonsaline condition at 12 h photoperiod for all test plant species (Fig. 4.1). Moreover, seed germination of *H. glomeratus* and *S. drummondii* was >80% in

Table 4.1 Four-way ANOVA indicating effects of species (Sp), photoperiod (P), thermoperiod (T), salinity (S), and their interactions on mean final germination (MFG), rate of germination (GR), and germination recovery (R) of test species

Parameters	MFG	GR	R
Sp	645.97***	179.87***	3.13E+03***
P	588.40***	–	605.06***
T	27.86***	22.39***	430.93***
S	538.39***	191.84***	36.91***
Sp × P	26.97***	–	82.33***
Sp × T	21.80***	13.01***	34.86***
Sp × S	12.05***	11.47***	211.99***
Sp × P × T	14.13***	–	1.88*
Sp × P × S	12.43***	–	15.82***
Sp × T × S	8.16***	4.68***	17.22***
Sp × P × T × S	3.66***	–	2.50***

Numbers represent F- values. *$P < 0.01$, **$P < 0.001$, and ***$P < 0.000$

Table 4.2 Four-way ANOVA indicating effects of habitat (H), photoperiod (P), thermoperiod (T), salinity (S), and their interactions on mean final germination (MFG), rate of germination (GR), and germination recovery (R) of test species

Parameters	MFG	GR	R
H	710.29***	75.33***	74.03***
P	243.45***		15.70***
T	16.06***	6.18***	4.19**
S	208.72***	68.83***	1.84*
H × P	0.09^{ns}		2.39^{ns}
H × T	28.63***	5.95**	0.54^{ns}
H × S	2.53*	3.76**	3.29**
H × P × T	3.96**		0.03^{ns}
H × P × S	5.50***		0.55^{ns}
H × T × S	2.64**	2.56*	0.50^{ns}
H × P × T × S	1.42^{ns}		0.079^{ns}

Numbers represent F- values. *$P < 0.01$, **$P < 0.001$, and ***$P < 0.000$

the absence of salinity at all temperature regimes (Fig. 4.1). In nonsaline condition, maximum seed germination of *S. heterophylla* (60%) and *S. fruticosa* (80%)was found at 20/30 °C. Seed germination percentage and rate of germination decreased with the increment of salinity for all test species (Fig. 4.1). The salt tolerance decreased in the following order of species: *S. drummondii* (1000 mM NaCl) > *S. fruticosa* (600 mM NaCl) > *H. glomeratus* (500 mM NaCl) = *S. heterophylla* (500 mM NaCl) (Fig. 4.1). In general, rate of seed germination was higher in *S. drummondii* and lower in *S. heterophylla* (Fig. 4.1). Optimal temperature regime was 20/30 °C for seed germination of all species under saline condition (Fig. 4.1). Complete dark inhibited seed germination of all test species in the absence of salinity, and this response was temperature dependent. Germination inhibition due to complete dark was not found in all species at 25/35 °C.

Seed Recovery Responses

ANOVA showed species, photoperiod, thermoperiod, salinity, habitat, and their interactions had significant effects ($P < 0.0001$) on germination recovery (R) (Tables 4.1 and 4.2). Germination recovery from salinity was higher in *H. glomeratus* (100%) and *S. fruticosa* (60%), but only ≤20% in case of *S. drummondii* and *S. heterophylla* (Fig. 4.2). Recovery of germination from complete darkness to 12 h photoperiod was found in *H. glomeratus*, *S. fruticosa*, and *S. heterophylla* when seeds were present in nonsaline and lower salinity treatments. However, no seed recovery was observed in *S. drummondii* during transfer from 24 dark to 12 h photoperiod (Fig. 4.2).

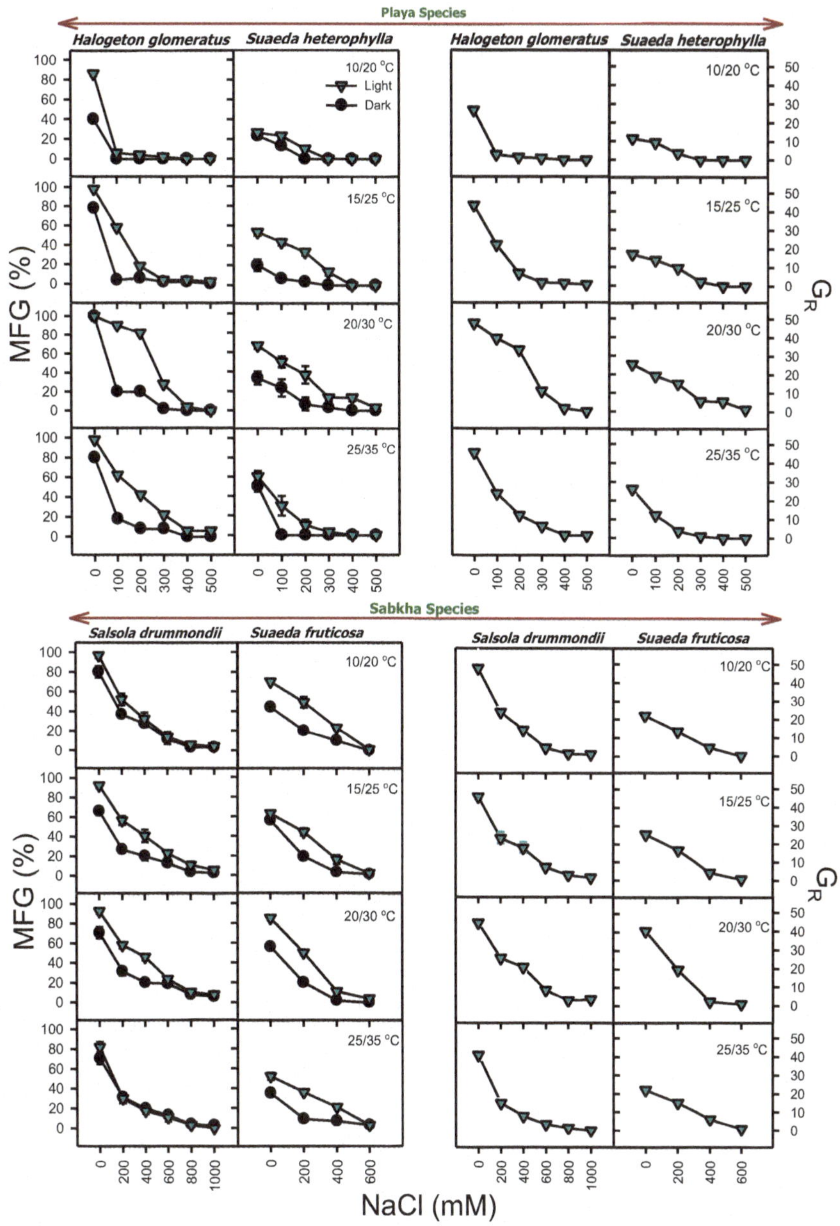

Fig. 4.1 Effect of salt (NaCl), temperature regimes (10/20, 15/25, 20/30, 25/35 °C), and photoperiods (triangle and circle represent 12 h light-12 h dark and 24 h dark, respectively) on the seed germination of different halophytes from contrasting habitats (mountainous playa and coastal sabkha). Symbols represent mean ± standard errors

Discussion

Seed Germination Responses Under Nonsaline Conditions

Seeds of all test species were nondormant and germinated maximally in distilled water. Generally, seeds of two Salsoloideae species (*S. drummondii* and *H. glomeratus*) showed higher (>95%) germination than the seeds of two Suaedoideae species (*S. fruticosa* and *S. heterophylla*) under nonsaline conditions irrespective of the habitat. Likewise, seeds of many Salsoloideae species such as *Haloxylon*

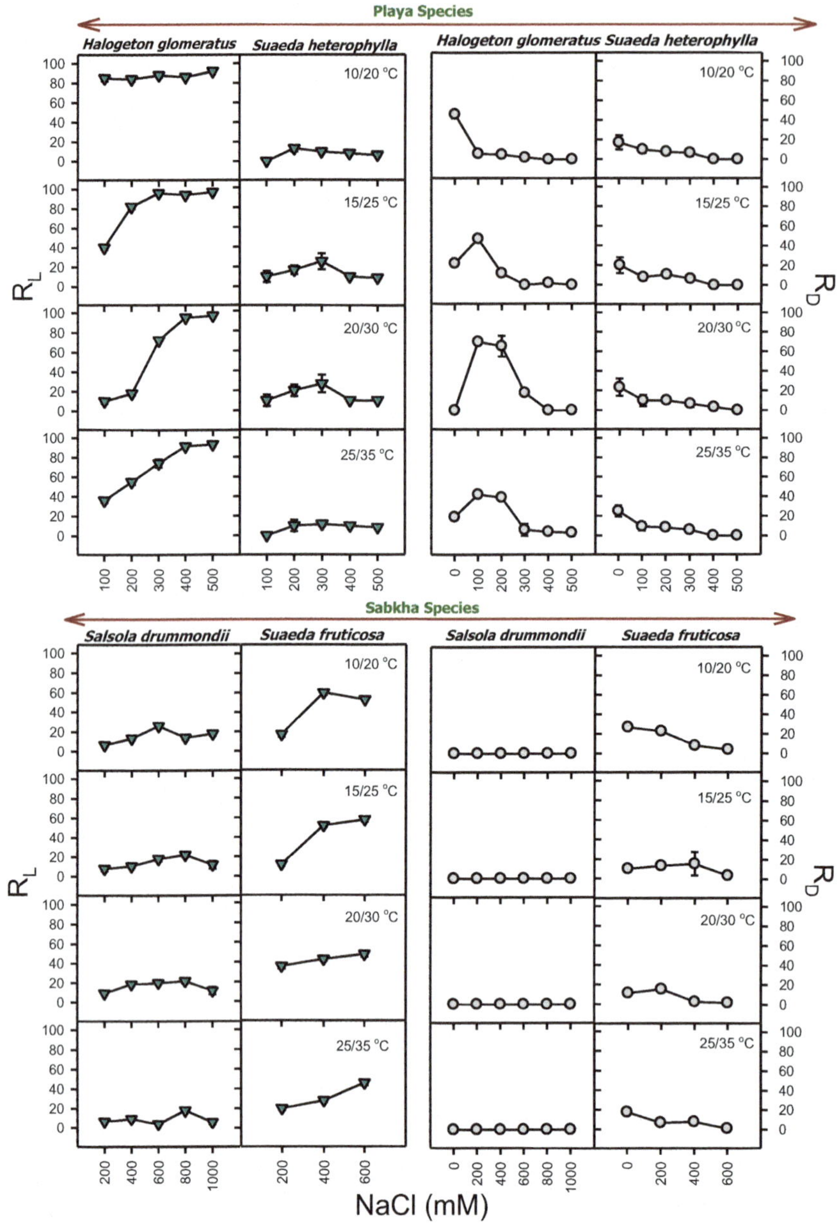

Fig. 4.2 Effect of salt (NaCl) and temperature regimes (10/20, 15/25, 20/30, 25/35 °C) on the germination recovery from salinity (triangle and circle represent 12 h light-12 h dark and 24 h dark, respectively) of different halophytes from contrasting habitats (mountainous playa and coastal sabkha). Symbols represent mean ± standard errors

salicornicum (yellow dewinged seeds; Bhatt et al. 2017), *Salsola affinis* (without perianth; Wei et al. 2008), and *Salsola imbricata* (Mehrun-Nisa et al. 2007) showed >90% germination in distilled water under optimal conditions. Whereas, seeds of many Suaedoideae species such as *Suaeda aegyptiaca* (~50%; El-Keblawy et al. 2017), *S. salsa* (~75%; Song et al. 2012), and *S. physophora* (~80%; Song et al. 2005) showed germination in range of 50–80% under nonsaline conditions. Parsons (2012) reviewed that a number of Salsoloideae species including

Halogeton glomeratus show very fast germination, which he ascribed as a strategy to rapidly exploit temporarily favorable conditions under stressful habitats.

Light is an important factor influencing seed germination of halophytes by imposing conditional dormancy of various degrees (Baskin and Baskin 1998; Qu et al. 2008; Khan and Gul 2006). In this study, seeds of all test species generally germinated better in presence of light than in complete dark. Gul et al. (2013) reviewed that the seeds of most halophytes are positively photoblastic during germination and this attribute plays an important role in prevention of seed germination under burial in deep soil layers and also in determining the time of germination.

Under nonsaline condition, seeds of two Salsoloideae species (*S. drummondii* and *H. glomeratus*) showed >80% germination in temperature-independent manner. Wei et al. (2008) reported that the seeds of *S. affinis* can germinate in a wide range (5–30 °C) of temperatures and exhibit "opportunistic" strategy to produce seedlings whenever conditions are conducive for seedling survival. In contrast, germination of two Suaedoideae species (*S. fruticosa* and *S. heterophylla*) was generally higher at moderate temperature regime of 20/30 °C. Similarly, seeds of *Suaeda moquinii* (Khan et al. 2001b) also germinated optimally under moderate (20/30 °C) temperature.

Seed Germination Responses Under Saline Conditions

Both rate and mean final germination percentage of the seeds of all test species decreased with increases in salinity. Salinity tolerance was in the following order: *S. drummondii* (1000 mM NaCl) > *S. fruticosa*(600 mM NaCl) > *H. glomeratus* (500 mM NaCl) > *S. heterophylla* (500 mM NaCl). Hence, salinity tolerance of sabkha perennials was greater than that of playa annuals. Similarly, sabkha halophytes *Salsola imbricata* (Mehrun-Nisa et al. 2007) and *S. vermiculata* (Guma et al. 2010) also showed high tolerance to the level of seawater salinity (≥600 mM NaCl). Whereas, seeds of playa halophytes *Lepidium latifolium* and *Peganum harmala* could germinate only in <600 mM NaCl (Ahmed and Khan 2010). However, many sabkha species such as *Atriplex stocksii* (Khan and Rizvi 1994) and *Zygophyllum simplex* (Khan and Ungar 1997) are low (<350 mM NaCl) salinity tolerant, while seeds of some playa halophytes like *Kochia scoparia* (Khan et al. 2001a) and *Allenrolfea occidentalis* (Gul and Weber 1999) are highly tolerant during germination. Based on literature review, we found that the number of species capable of seed germination in up to 0.6 M NaCl is higher from mountainous playa than the coastal sabkha and seeds of more Salsoloideae than the Suaedoideae halophytes reportedly germinate in up to 0.6 M NaCl (Fig. 4.3). Hence despite aforementioned general trends observed in this study, salinity tolerance of halophyte seeds is a variable trait.

Often interaction of salinity with different co-occurring factors such as sub-/supraoptimal temperature(s) and dark further aggravates its inhibitory effects on germination of halophyte seeds (Gul et al. 2013). In the current study, low temperature (10/20 °C) caused substantial reduction in germination of playa species under saline conditions. Similarly, adverse effects of high salinity were further intensified at lower than the optimal temperatures in many halophytes such as *Zostera marina* (Xu et al. 2016), *Zostera japonica* (Kaldy et al. 2015), and *Medicago arborea* (Nedjimi et al. 2014). A literature search also revealed that the seed germination of playa halophytes is more sensitive to low temperatures (Fig. 4.4). Whereas, seed germination of sabkha species under saline conditions was lowest at high (25/35 °C) temperature. Similarly, germination of many other sabkha species such as *Limonium stocksii* (Zia and Khan 2004) and *Suaeda moquinii* (Black seeds; Khan et al. 2001b) was also inhibited more at high temperature. A combination of moderate salinity and dark also caused greater inhibition of germination than the effects of respective salinity dose in the presence of light in most test species. Similar results were obtained for *Limonium stocksii* (Zia and Khan 2004) and *L. tabernense* (Fernández et al. 2016).

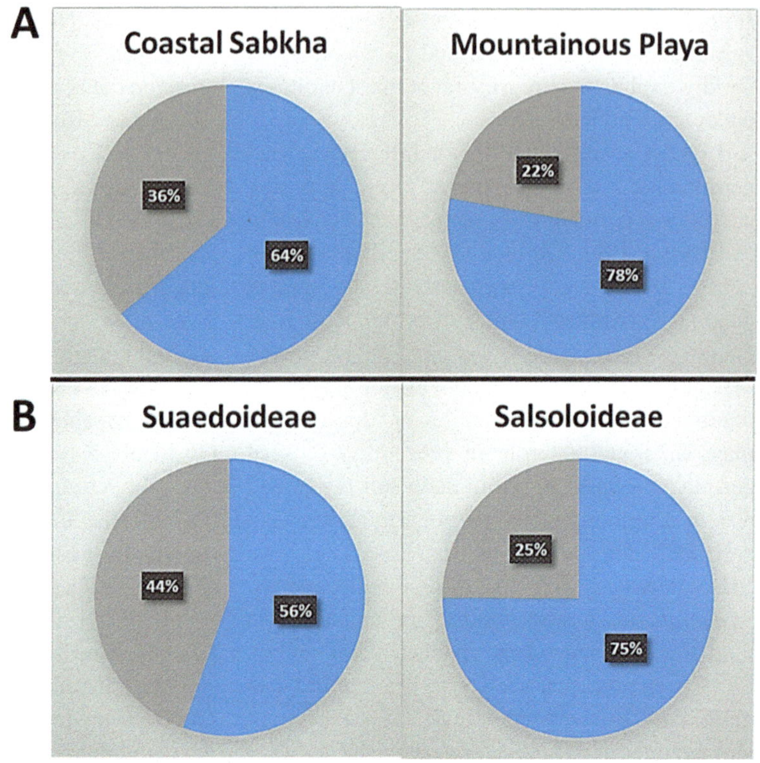

Fig. 4.3 Trends regarding salt tolerance during seed germination of Amaranthaceae species (as listed in Table 4.3). Comparison between (**a**) costal sabkha and mountainous playa habitat and (**b**) subfamilies Suaedoideae and Salsoloideae. Blue color represents the percentage of high salinity-resistant species (> 0.6 M NaCl), while gray color represents low salinity-resistant species (< 0.6 M NaCl)

Based on literature search, we found that the species of playa habitat are more sensitive for light during germination than those from sabkha under both absence and presence of salinity (Fig. 4.5). Literature search also indicated that the light is necessary for seed germination of Suaedoideae members in nonsaline condition but not in saline medium, while reverse trend was true for Salsoloideae members (Fig. 4.5).

Recovery of Germination from Salinity

In this study, seed recovery from salinity was completely independent of habitat type, plant nature, and their taxonomical relationship (subfamily of Amaranthaceae). A literature search indicates that the species of Amaranthaceae from mountainous playa have better ability to recover from high salinity in comparison of coastal sabkha species, while recovery responses of Salsoloideae and Suaedoideae halophytes are generally comparable (Fig. 4.6). Among test species, seeds of *H. glomeratus* (playa annual – Salsoloideae) and *S. fruticosa* (sabkha perennial – Suaedoideae) had higher ability to recover from salinity. It seems that the ability of seed recovery is linked with the seed germination response of species particular under salinity. Both *H. glomeratus* and *S. fruticosa* showed better seed germination percentage under salinity in comparison of *S. heterophylla* and *S. drummondii*. Similar results were reported previously for high salt-resistant species like *Arthrocnemum macrostachyum* and *Salicornia ramosissima* (Khan et al. 1998; Rubio-Casal et al. 2003). Seed recovery from salinity in *S. fruticosa* was higher in cooler temperature regimes than the high temperatures, which could be a strategy of this plant to prevent seedling formation in early summer when high temperature alongside high salinity would minimize the chances of seedling survival. Whereas, low recovery from salinity during seed germination of *S. heterophylla* and *S. drummondii* is probably due to ion toxicity, which needs to be tested through detailed studies in future.

Table 4.3 Seed germination responses (salinity tolerance, light sensitivity, temperature sensitivity and recovery) of different Amaranthaceae species (from coastal sabkha and mountain playa habitats)

Species name	NaCl tolerance (molar)	Light sensitivity		Temperature sensitivity		Recovery from NaCl		References
		Nonsaline	saline	Low T (<20/30)	High T (>20/30)	< 0.6 M	> 0.6 M	
Coastal sabkha								
Sarcocornia perennis	1.0	–	–	–	–	–	–	Redano et al. (2004)
Sarcocornia fruticosa	1.0	–	–	–	–	–	–	Redando et al. (2004)
Salicornia bigelovii	1.0	–	–	–	–	–	–	Rivers and Weber (1971)
Suaeda japonica (SU)	0.9	–	–	–	–	–	–	Yokoishi and Tanimoto (1994)
Kochia prostrata	0.8	–	–	–	–	–	–	Orlovsky et al. (2011)
Salicornia europaea	0.8	–	–	–	–	–	–	Ungar (1962, 1967)
Salsola imbricate (SA)	0.8	LS	LS	HS	HS	LR	LR	Mehrun-Nisa et al. (2007)
Haloxylon salicornicum (SA)	0.8	LS	–	LS	HS	HR	MR	El-Keblawy and Al-Shamsi (2008)
Sarcocornia quinqueflora	0.7	–	–	–	–	–	–	Patridge and Wilson (1987)
Salicornia pacifica	0.7	–	–	–	–	–	–	Khan and Weber (1986)
Salicornia brachiata	0.6	–	–	–	–	–	–	Joshi and Iyengar (1982)
Salsola drummondii (SA)	0.8	LS	LS	LS	HS	LR	LR	Rasheed et al. (2015)
Salsola vermiculata (SA)	0.6	–	–	–	–	HR	HR	Guma et al. (2010)
Suaeda maritima (SU)	0.6	–	–	–	–	–	–	Boucaud and Ungar (1976)
Bienertia cycloptera (SU) – *red*	0.6	HS	LS	HS	LS	LR	LR	Bhatt et al. (2016)
Bienertia cycloptera - yellow	0.6	LS	LS	HS	HS	HR	LR	Bhatt et al. (2016)
Anabasis setifera (SA)	0.6	LS	LS	–	–	HR	LR	El-Keblawy et al. (2016)
Suaeda fruticosa (SU)	0.5	HS	LS	HS	HS	HR	LR	Khan and Ungar (1998)
Haloxylon stocksii (SA)	0.5	–	–	LS	HS	LR	LR	Khan and Ungar (1996)
Atriplex lentiformis	0.4	–	–	–	–	–	–	Mikheil et al. (1992)
Arthrocnemum halocnemoides	0.4	–	–	–	–	–	–	Malcolm (1964)
Atriplex stocksii	0.3	–	–	LS	HS	HR	LR	Khan and Rizvi (1994)
Atriplex patula	0.3	–	–	–	–	–	–	Ungar (1996)

(continued)

Table 4.3 (continued)

Species name	NaCl tolerance (molar)	Light sensitivity		Temperature sensitivity		Recovery from NaCl		References
		Nonsaline	saline	Low T (<20/30)	High T (>20/30)	< 0.6 M	> 0.6 M	
Salicornia brachystachya	0.2	–	–	–	–	–	–	Huiskes et al. (1985)
Salicornia dolistachya	0.2	–	–	–	–	–	–	Huiskes et al. (1985)
Suaeda nudiflora (SU)	0.2	–	–	–	–	–	–	Joshi and Iyengar (1982)
Mountain playa								
Salsola iberica (SA)	1.0	–	–	HS	LS	HR	HR	Khan et al. (2002a)
Sarcobatus vermiculatus	1.0	–	–	HS	HS	LR	LR	Khan et al. (2002b)
Suaeda moquinii (SU)	1.0	–	–	HS	HS	LR	LR	Khan et al. (2001b)
Atriplex rosea	1.0	–	–	–	–	–	–	Khan et al. (2004)
Salicornia rubra	1.0	–	–	HS	HS	HR	HR	Khan et al. (2000)
Suaeda depressa (SU)	0.8	–	–	–	–	HR	HR	Ungar (1962)
Allenrolfea occidentalis	0.8	HS	HS	–	–	HR	HR	Gul and Weber (1999)
Suaeda heterophylla (SU)	0.4	HS	LS	HS	HS	LR	LR	Hameed et al. (2013)
Halogeton glomeratus (SA)	0.4	LS	HS	HS	HS	HR	LR	Ahmed and Khan (2010)

LS low sensitive, *HS* high sensitive, *LR* low recovery, *HR* high recovery. While SA and SU represent two subfamilies of Amaranthaceae, Salsoloideae and Suaedoideae, respectively

Recovery of Germination from Dark

Based on the literature, species of mountainous playa habitat appear to have greater light sensitivity during germination in comparison with species of coastal habitat, irrespective of the presence and absence of salinity. Moreover, light seems to be more important for seed germination of Suaedoideae members in nonsaline condition but not in the presence of salinity, while reverse trend was true for Salsoloideae members. Among test species, recovery from dark was observed only in *H. glomeratus* (Salsoloideae member of mountainous playa habitat) at warmer temperature regimes. The ability of *H. glomeratus* seeds to recover from dark could help this plant to germinate at the time of warmer temperature when sufficient moisture is available that dilutes the substrate salinity. Seed recovery from dark was not observed in sabkha species probably due to their perennial nature, which usually do not depend entirely on seed germination to maintain their population (Khan et al. 2017). The response of other perennial subtropical halophytes like *Cyperus conglomeratus*, *Haloxylon stocksii*, *Salsola imbricate*, and *Sporobolus ioclados* (references mentioned in Gul et al. 2013) is in agreement with our findings. In addition, the

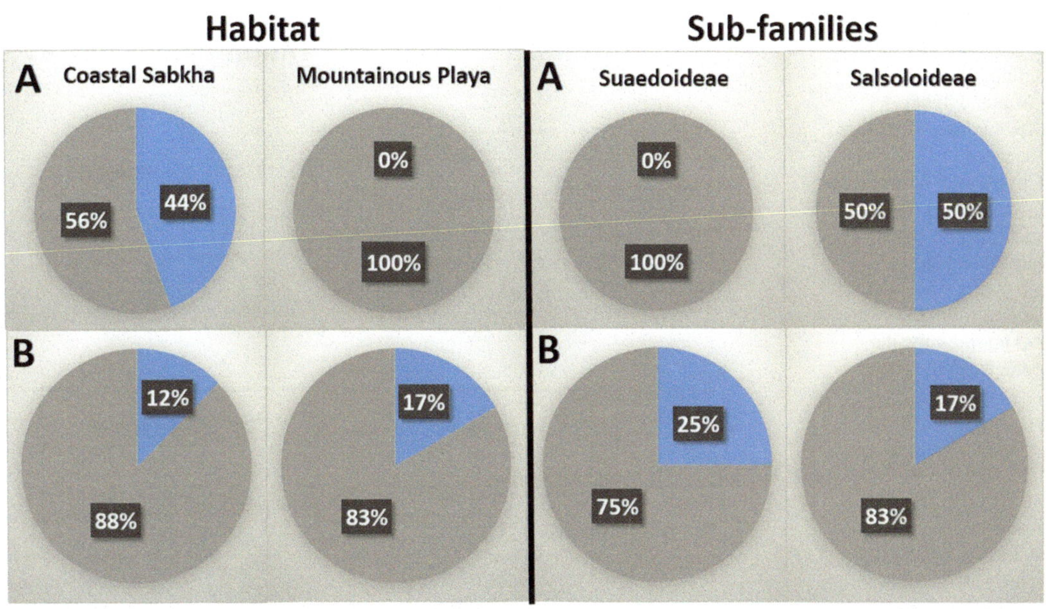

Fig. 4.4 Trends regarding temperature influence on seed germination of Amaranthaceae species (as listed in Table 4.3). Comparison between two temperature regimes, low temperature regimes (< 20/30 °C; **a**-upper panel) and warmer temperature regime (>20/30 °C; **b**-lower panel); two habitats, costal sabkha and mountainous playa (**a** and **b** of left panel), two subfamilies, Suaedoideae and Salsoloideae (**a** and **b** of right panel). Blue color represents the percentage of low light-sensitive species, while gray color represents the data of high light-sensitive species

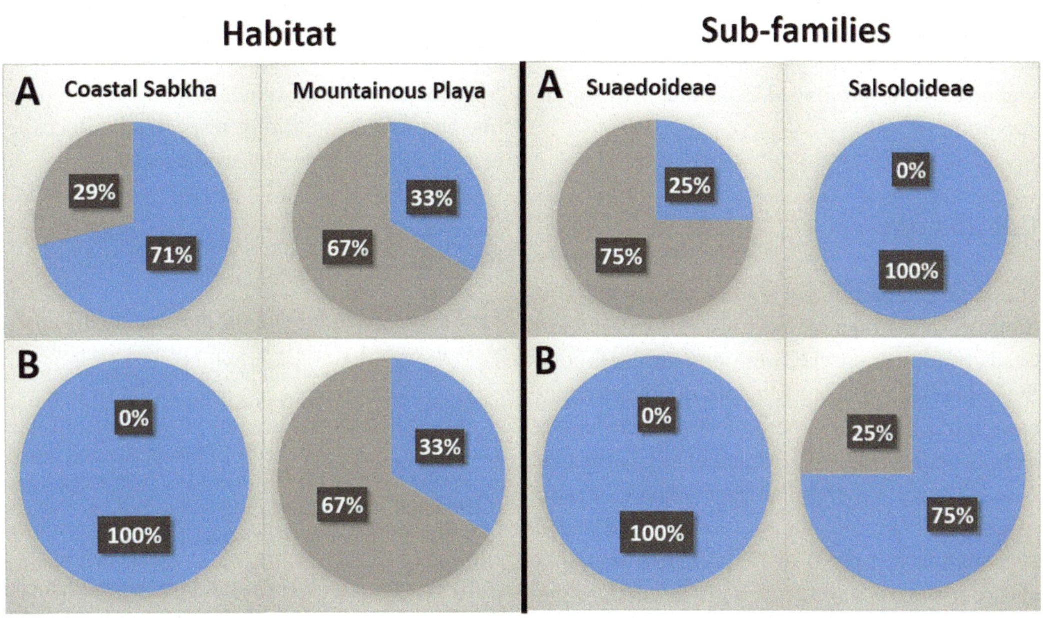

Fig. 4.5 Trends regarding light requirement during seed germination of Amaranthaceae species (as listed in Table 4.3). Comparison between two salinity levels, nonsaline (**a**-upper panel) and saline (**b**-lower panel); two habitats, costal sabkha and mountainous playa (**a** and **b** of left panel), two subfamilies, Suaedoideae and Salsoloideae (**a** and **b** of right panel). Blue color represents the percentage of low light-sensitive species, while gray color represents the data of high light-sensitive species

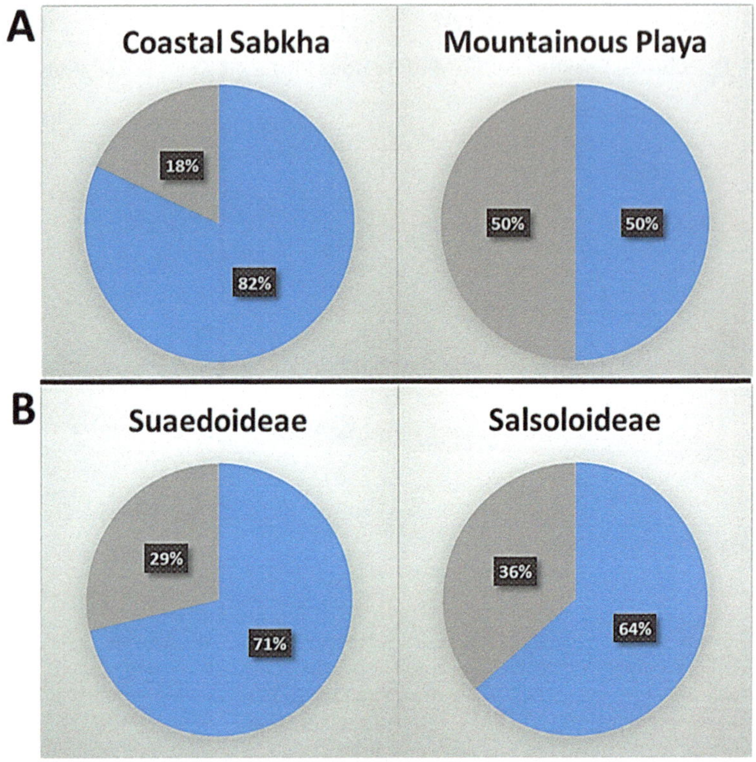

Fig. 4.6 Trends regarding recovery from salinity during seed germination of Amaranthaceae species (as listed in Table 4.3). Comparison between (**a**) costal sabkha and mountainous playa habitat and (**b**) subfamilies Suaedoideae and Salsoloideae. Blue color represents the percentage of species recovered when salinity is less than seawater (<0.6 M), while gray color represents the percent data of recovered species when salinity is greater than seawater (>0.6 M)

failure of seed recovery from dark in *Suaeda heterophylla* is possibly the result of marshy habitat, in which deeply buried seeds lose their viability due to prolonged exposure with anoxic condition.

Conclusions

Our data indicates many commonalities and differences in germination responses of halophytes. Seeds of all test species were nondormant, positively photoblastic, and generally preferred moderate temperatures for germination. Increases in salinity decreased seed germination of all test species. However, two Salsoloideae species (*S. drummondii* and *H. glomeratus*) showed higher seed germination compared to two Suaedoideae species (*S. fruticosa* and *S. heterophylla*) under nonsaline conditions irrespective of the habitat. Under nonsaline conditions seed germination of two Salsoloideac species was temperature-independent, while seeds of Suaedoideae species germinated optimally at moderate temperature of 20/30 °C. Generally, the salinity tolerance limit of sabkha perennials was greater than that of playa annuals. Light improved seed germination of Suaedoideae and Salsoloideae members in nonsaline and saline conditions, respectively. Recovery responses were generally species-specific.

References

Ahmed MZ, Khan MA (2010) Tolerance and recovery responses of playa halophytes to light, salinity and temperature stresses during seed germination. Flora-Morphology, Distribution, Functional Ecology of Plants 205:764–771

Akhani H (2006) Biodiversity of halophytic and sabkha ecosystems in Iran. In: Khan MA, Böer B, Kust GS, Barth HJ (eds) Sabkha ecosystems. Tasks for vegetation science. Springer, Dordrecht, pp 71–88

Andreou M, Delipetrou P, Kadis C, Tsiamis G, Bourtzis K, Georghiou K (2011) An integrated approach for the conservation of threatened plants: the case of *Arabis kennedyae* (Brassicaceae). Acta Oecol 37:239–248

Anon (2001) SPSS for windows. SPSS, Chicago

Barton ML, Medel ID, Johnston KK, Whitcraft CR (2016) Seed collection and germination strategies for common wetland and coastal sage scrub species in Southern California. Bull South Calif Acad Sci 115:41–71

Baskin CC, Baskin JM (1998) Seeds: ecology, biogeography, and, evolution of dormancy and germination. Elsevier, Amsterdam

Baskin CC, Milberg P, Andersson L, Baskin JM (2001) Seed dormancy-breaking and germination requirements of *Drosera anglica*, an insectivorous species of the Northern Hemisphere. Acta Oecol 22:1–8

Bewley JD, Black M (1994) Seeds. In: Seeds. Springer, Boston, pp 1–33

Bhatt A, Pérez-García F, Phondani PC (2016) Foliage colour influence on seed germination of *Bienertia cycloptera* in Arabian deserts. Nordic J Bot 34:428–434

Bhatt A, Phartyal SS, Nicholas A (2017) Ecological role of distinct fruit-wing perianth color in synchronization of seed germination in *Haloxylon salicornicum*. Plant Species Biology 32:121–133

Boucaud J, Ungar IA (1976) Hormonal control of germination under saline conditions of three halophytic taxa in the genus *Suaeda*. Physiol Plant 37:143–148

El-Keblawy A (2004) Salinity effects on seed germination of the common desert range grass, *Panicum turgidum*. Seed Sci Technol 32:873–878

El-Keblawy A, Al-Rawai A (2006) Effects of seed maturation time and dry storage on light and temperature requirements during germination in invasive *Prosopis juliflora*. Flora-Morphology, Distribution, Functional Ecology of Plants 201:135–143

El-Keblawy A, Al-Shamsi N (2008) Salinity, temperature and light affect seed germination of *Haloxylon salicornicum*, a common perennial shrub of the Arabian deserts. Seed Sci Technol 36:679–688

El-Keblawy A, Gairola S, Bhatt A (2016) Maternal salinity environment affects salt tolerance during germination in *Anabasis setifera*: a facultative desert halophyte. J Arid Land 8:254–263

El-Keblawy A, Gairola S, Bhatt A, Mahmoud T (2017) Effects of maternal salinity on salt tolerance during germination of *Suaeda aegyptiaca*, a facultative halophyte in the Arab gulf desert. Plant Species Biology 32:45–53

Estrelles E, Biondi E, Galiè M, Mainardi F, Hurtado A, Soriano P (2015) Aridity level, rainfall pattern and soil features as key factors in germination strategies in salt-affected plant communities. J Arid Environ 117(1–9):1–9

Fernández ICD, Luque EG, Mercado FG, Pedrosa W (2016) Influence of temperature and salinity on the germination of *Limonium tabernense* Erben from Tabernas Desert (Almería, SE Spain). Flora-Morphology, Distribution, Functional Ecology of Plants 218:68–74

Gul B, Weber DJ (1999) Effect of salinity, light, and temperature on germination in *Allenrolfea occidentalis*. Can J Bot 77:240–246

Gul B, Weber DJ (2001) Seed bank dynamics in a Great Basin salt playa. J Arid Environ 49:785–794

Gul B, Ansari R, Flowers TJ, Khan MA (2013) Germination strategies of halophyte seeds under salinity. Environ Exp Bot 92:4–18

Guma IR, Padrón-Mederos MA, Santos-Guerra A, Reyes-Betancort JA (2010) Effect of temperature and salinity on germination of *Salsola vermiculata* L. (Chenopodiaceae) from Canary Islands. J Arid Environ 74:708–711

Hameed A, Ahmed MZ, Khan MA (2006) Comparative effects of NaCl and seasalt on seed germination of coastal halophytes. Pak J Bot 38:1605–1612

Huang Z, Zhang X, Zheng G, Gutterman Y (2003) Influence of light, temperature, salinity and storage on seed germination of *Haloxylon ammodendron*. J Arid Environ 55:453–464

Hameed A, Ahmed MZ, Gulzar S, Gul B, Alam J, Hegazy AK, Alatar ARA, Khan MA (2013) Seed germination and recovery responses of *Suaeda heterophylla* to abiotic stresses. Pak J Bot 45:1649–1656

Huiskes AHL, Stienstra AW, Koutstaal BP, Markusse MM, Van Soelen J (1985) Germination ecology of *Salicornia dolichostachya* and *Salicornia brachystachya*. Acta Bot Neerl 34:369–380

Joshi AJ, Iyengar ERR (1982) Effect of salinity on the germination of *Salicornia brachiata* Roxb [Angiosperm tree]. Indian J Plant Physiol 25:65–70

Kaldy JE, Shafer DJ, Ailstock MS, Magoun AD (2015) Effects of temperature, salinity and seed age on induction of *Zostera japonica* germination in North America, USA. Aquat Bot 126:73–79

Khan MA, Gul B (2006) Halophyte seed germination. In: Khan MA, Weber DJ (eds) Ecophysiology of high salinity tolerant plants. Springer, Dordrecht, pp 11–30

Khan MA, Qaiser M (2006) Halophytes of Pakistan: characteristics, distribution and potential economic usages. In: Khan MA, Kust GS, Barth HJ, Boer B (eds) Sabkha ecosystems. Springer, Dordrecht, pp 129–153

Khan MA, Rizvi Y (1994) Effect of salinity, temperature, and growth regulators on the germination and early seedling growth of *Atriplex griffithii* var. *stocksii*. Can J Bot 72:475–479

Khan MA, Ungar IA (1984) Seed polymorphism and germination responses to salinity stress in *Atriplex triangularis* Willd. Bot Gaz 145:487–494

Khan MA, Weber DJ (1986) Factors influencing seed germination in *Salicornia pacifica var. utahensis*. Am J Bot 73:1163–1167

Khan MA, Ungar IA (1996) Influence of salinity and temperature on the germination of *Haloxylon recurvum* Bunge ex. Boiss. Ann Bot 78:547–551

Khan MA, Ungar IA (1997) Alleviation of seed dormancy in the desert forb *Zygophyllum simplex* L. from Pakistan. Ann Bot 80:395–400

Khan MA, Ungar IA (1998) Germination of salt tolerant shrub *Suaeda fruticosa* from Pakistan: salinity and temperature responses. Seed Sci Technol 26:657–667

Khan MA, Gul B, Weber DJ (2000) Germination responses of *Salicornia rubra* to temperature and salinity. J Arid Environ 45:207–214

Khan MA, Gul B, Weber DJ (2001a) Influence of salinity and temperature on the germination of *Kochia scoparia*. Wetl Ecol Manag 9:483–489

Khan MA, Gul B, Weber DJ (2001b) Germination of dimorphic seeds of *Suaeda moquinii* under high salinity stress. Aust J Bot 49:185–192

Khan MA, Gul B, Weber DJ (2002a) Seed germination in the Great Basin halophyte *Salsola iberica*. Can J Bot 80:650–655

Khan MA, Gul B, Weber DJ (2002b) Seed germination in relation to salinity and temperature in *Sarcobatus vermiculatus*. Biol Plant 45:133–135

Khan MA, Gul B, Weber DJ (2004) Temperature and high salinity effects in germinating dimorphic seeds of *Atriplex rosea*. West North Am Nat 64:193–201

Khan MA, Shaikh F, Zehra A, Ahmed MZ, Gul B, Ansari R (2017) Role of chemicals in alleviating salinity and light related seed dormancy in sub-tropical grasses. Flora 233:150–155

Khan MA, Ungar IA, Gul B (1998) Action of compatible osmotica and growth regulators in alleviating the effect of salinity on the germination of dimorphic seeds of *Arthrocnemum indicum* L. Int J Plant Sci 159:313–317

Kranner I, Seal CE (2013) Salt stress, signaling and redox control in seeds. Funct Plant Biol 40:848–859

Malcolm CV (1964) Effect of salt, temperature and seed scarification on germination of two varieties of *Arthrocnemum halocnemoides*. J R Soc West Aust 47:72–74

Mehrun-Nisa, Khan MA, Weber DJ (2007) Dormancy, germination and viability of *Salsola imbricata* seeds in relation to light, temperature and salinity. Seed Sci Technol 35:595–606

Mikheil GS, Meyer SE, Pendleton RL (1992) Variation in germination response to temperature and salinity in shrubby *Atriplex species*. J Arid Environ 22:39–49

Nedjimi B, Mohammedi N, Belkheiri S (2014) Germination responses of medic tree (*Medicago arborea*) seeds to salinity and temperature. Agric Res 3:308–312

Orlovsky NS, Japakova UN, Shulgina I, Volis S (2011) Comparative study of seed germination and growth of *Kochia prostrata* and *Kochia scoparia* (Chenopodiaceae) under salinity. J Arid Environ 75:532–537

Parsons RF (2012) Incidence and ecology of very fast germination. Seed Sci Res 22:161–167

Partridge TR, Wilson JB (1987) Salt tolerance of salt marsh plants of Otago, New Zealand. New Zealand J Bot 25:559–566

Pujol JA, Calvo JF, Ramirez-Diaz L (2000) Recovery of germination from different osmotic conditions by four halophytes from southeastern Spain. Ann Bot 85:279–286

Qu X, Baskin JM, Wang L, Huang Z (2008) Effects of cold stratification, temperature, light and salinity on seed germination and radicle growth of the desert halophyte shrub, *Kalidium caspicum* (Chenopodiaceae). Plant Growth Regul 54:241–248

Rasheed A, Hameed A, Khan MA, Gul B (2015) Effects of salinity, temperature, light and dormancy regulating chemicals on seed germination of *Salsola drummondii* ulbr. Pak J Bot 47:11–19

Redano S, Rubio-Casal AE, Castillo JM, Luque CJ, Alvarez AA, Luque T, Figueroa ME (2004) Influence of salinity and light on germination of three taxa with contrasted habitats. Aquat Bot 78:255–264

Rivers WG, Weber DJ (1971) The influence of salinity and temperature on seed germination in *Salicornia bigelovii*. Physiol Plant 24:73–75

Rubio-Casal AE, Castillo JM, Luque CJ, Figueroa ME (2003) Influence of salinity on germination and seeds viability of two primary colonizers of Mediterranean salt pans. J Arid Environ 53:145–154

Ruiz KB, Biondi S, Martínez EA, Orsini F, Antognoni F, Jacobsen SE (2016) *Quinoa*–a model crop for understanding salt-tolerance mechanisms in halophytes. Plant Biosyst 150:357–371

Song HL, Sun ZG, Mou XJ, Zhao JY (2012) Salinity effect and seed priming treatments on the germination of *Suaeda salsa* in the tidal marsh of the Yellow River estuary. Afr J Biotechnol 11:7923–7933

Song JIE, Feng GU, Tian C, Zhang F (2005) Strategies for adaptation of *Suaeda physophora*, *Haloxylon ammodendron* and *Haloxylon persicum* to a saline environment during seed-germination stage. Ann Bot 96:399–405

Ungar IA (1962) Influence of salinity on seed germination in succulent halophytes. Ecology 43:763–764

Ungar IA (1967) Influence of salinity and temperature on seed germination. Ohio J Sci 67:120–123

Ungar IA (1978) Halophyte seed germination. Bot Rev 44:233–264

Ungar IA (1996) Effect of salinity on seed germination, growth, and ion accumulation of *Atriplex patula* (Chenopodiaceae). Am J Bot 83:604–607

Wei Y, Dong M, Huang ZY, Tan DY (2008) Factors influencing seed germination of *Salsola affinis* (Chenopodiaceae), a dominant annual halophyte inhabiting the deserts of Xinjiang, China. Flora 203:134–140

Xu S, Zhou Y, Wang P, Wang F, Zhang X, Gu R (2016) Salinity and temperature significantly influence seed germination, seedling establishment, and seedling growth of eelgrass *Zostera marina* L. Peer J 4:e2697

Yokoishi T, Tanimoto S (1994) Seed germination of the halophyte *Suaeda japonica* under salt stress. J Plant Res 107:385–388

Zia S, Khan MA (2004) Effect of light, salinity, and temperature on seed germination of *Limonium stocksii*. Can J Bot 82:151–157

Plants of Sabkha Ecosystems of the Arabian Peninsula

5

Shahina A. Ghazanfar, Benno Böer, Abdul Wali Al Khulaidi, Ali El-Keblawy, and Sara Alateeqi

Abstract

Sabkhas are unique ecosystems that are highly saline and where specially adapted plants are able to grow, flower, and fruit. In general, saline environments are poor in species – for the Arabian Peninsula about 120 taxa are recorded as halophytes which constitute about 4% of the total flora of the Arabian Peninsula. Key halophytes of Arabia are nearly always perennial; predominant life-forms are somewhat succulent, semiwoody dwarf shrubs belonging to the families Amaranthaceae, Zygophyllaceae, and Plumbaginaceae and hemicryptophytes belonging to the Poaceae, Cyperaceae, and Juncaceae; annuals are exceptions. Coastal species are either obligate halophytes or salt-tolerant genera from unspecialized families, such as *Sporobolus* and *Aeluropus* (Poaceae), or salt-secreting species such as *Avicennia* (Acanthaceae) and *Limonium* (Plumbaginaceae). The submerged coastal vegetation, e.g., seagrasses, is one of the most important vegetation types of the Gulf coast and is of great importance to marine fauna. The north-south distribution of coastal species is more distinct on the Red Sea coast, with the border lying near Jeddah, than on the Persian Gulf coast where there is a broad transitional zone between Qatar and northern Oman. The east-west distribution of coastal species is not as distinct. The eastern elements are either restricted to the coasts around the Arabian Gulf or are Irano-Turanian species extending into the Gulf region. Several vicariant species groups of halophytes are represented in the Arabian Peninsula. Halophytes have developed strategies for seed germination such as high germination levels and fast germination speed. These traits are found in the sabkha plants of the Arabian Peninsula. Some halophytes have been investigated for their potential for phytoremediation in their ability to survive weathered oil-contaminated soils. They have been found to have a set of micoorganisms around their root system that are related to the degradation of oil in contaminated soils. Sabkha ecosystems are being degraded and altered throughout the Gulf countries as they appear to be nonproductive. Over the last two decades, there has been a growing concern in protecting and restoring mangroves, and programs do to so have seen

S. A. Ghazanfar (✉)
Royal Botanic Gardens Kew, Richmond, UK
e-mail: S.Ghazanfar@kew.org

B. Böer
UNESCO Bangkok Asia and Pacific Regional Bureau for Education, Bangkok, Thailand

A. W. Al Khulaidi
Al Baha University, Biljurashi, Saudi Arabia

A. El-Keblawy
University of Sharjah, Sharjah, United Arab Emirates

S. Alateeqi
Group Seven (G7) Company, Kuwait City, Kuwait

B. Gul et al. (eds.), *Sabkha Ecosystems*, Tasks for Vegetation Science VI,
https://doi.org/10.1007/978-3-030-04417-6_5

promising results. But, on the whole, coastal and inland sabkhas are neglected, and these unique ecosystems require urgent protection.

Keywords
Annual · Arabia · Halophytes · Life-form · Perrenial · Sabkha

Introduction

Sabkhas are unique ecosystems that are highly saline and where specially adapted plants are able to grow. These plants, the true halophytes such as the mangroves, seagrasses, or some species of Amaranthaceae (in the former Chenopodiaceae), are able to complete their life cycle under saline conditions, where salt concentration is at least 200 mM NaCl (Flowers et al. 1986). Other halophytic plants growing on sabkhas depend on rain for their seed to germinate, but they can flower and fruit in saline habitats, and their seeds can survive saline conditions for considerable long periods of time. All halophytes need to regulate their cellular Na^+, Cl^-, and K^+ concentrations as they adjust to the external water potential. However, species differ in the succulence (water content per unit area of leaf; Flowers et al. 1986) and in the solutes accumulated. Detailed account of salinity tolerance in flowering plants is given in Flowers and Colmer (2008).

Nearly all terrestrial salt-tolerant plants belong to angiosperms (flowering plants), although a few are ferns (in families Pteridaceae and Ophioglossaceae) and several are marine algae. Worldwide, salt-tolerant flowering plants are found in about a third of the total plant families (Heywood et al. 2007), in about 500 genera of which about half belong to only 20 families (Table 5.1). Among monocotyledons, the Poaceae contain more halophytic genera than any other family (7% of the family); in Cyperaceae 14% of the genera are salt tolerant. Among the eudicots, Amaranthaceae (Chenopodioideae) have the highest proportion of halophytic genera followed by Asteraceae, Aizoaceae, Leguminosae, Apiaceae, Euphorbiaceae, Brassicaceae, Plantaginaceae, and Caryophyllaceae (Flowers et al. 1986) (Table 5.1).

Table 5.1 Families of flowering plants in which halophytic genera worldwide occur most frequently

Family	Percentage of halophytic genera
Chenopodiaceae	44
Aizoaceae	15
Cyperaceae	14
Caryophyllaceae	11
Poaceae	7
Arecaceae	6
Asteraceae	3
Fabaceae	3

Adapted from Flowers et al. (1986)

Sabkhas of the Arabian Peninsula

The Arabian Peninsula, including Bahrain, Kuwait, Oman, Qatar, Saudi Arabia, Yemen, United Arab Emirates, and Yemen, lies between the Red Sea and the Persian Gulf (Fig. 5.1). It is covered mostly by sandy and gravelly plains with escarpment mountains in the southwest in Saudi Arabia, Yemen, and southern Oman and in the north in Oman and the UAE. Large areas of the Arabian Peninsula are covered by sand deserts which are (mostly) uninhabited. Over all the Arabian Peninsula is arid and lacks overground water. Few springs and oases exist which are used for subsistence agriculture. Climatically most areas in the plains and mountains receive, on average, beween 50 and 400 mm of precipitation a year (Fig. 5.2).

The Arabian Gulf coastal plain is a narrow strip bordering the northern part of the Arabian Peninsula. It is continuous with the depression in western and southern Iraq which constitutes the flood plain and deltas of the two rivers, Tigris R. and Euphrates R. (Chapman 1978). The southern part of the depression includes the western half of the Gulf and the Arabian Gulf coastal region. The low coastal flats extend to several kilometers inland and are periodically inundated by the sea. Coastal sabkhas are present along the coastline from Kuwait to the end of the Persian Gulf in Oman. Sabkha Maṭṭī, southwest of Qatar, is the largest of these with an area of about 6000 km²

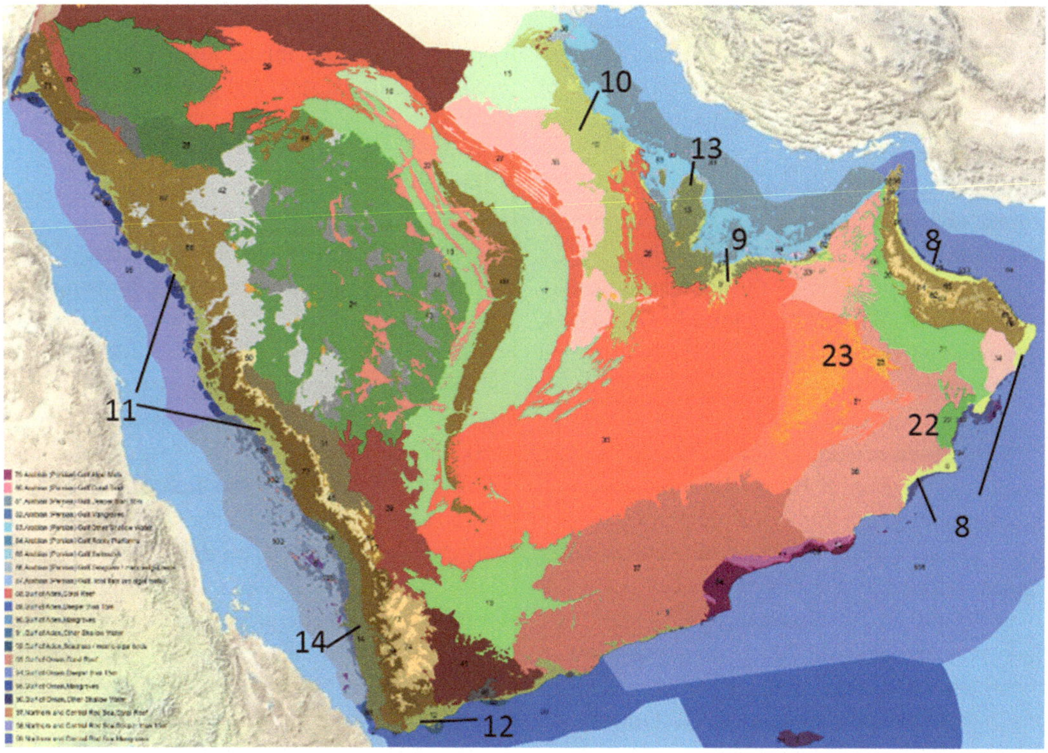

Fig. 5.1 Integrated terrestrial and marine habitat map of the Arabian Peninsula
Major coastal and inland sabkhas of the Arabian Peninsula: 8, Oman Coastal Plain; 9, Gulf Coastal and Sabkha Matti; 10, Northern Gulf Coastal Plain and Sabkha; 11, Red Sea Coastal Plain and Sabkha; 12, Southern Coastal Plain; 13, Southern Gulf Coastal Plain; 14, Tihama Coastal Plain; 22, Inland Sabkha Huqf; 23, Inland Sabkha Umm as Samim.
Source: Abu Dhabi Global Environment Data Information (AGEDI 2013). Reproduced with permission

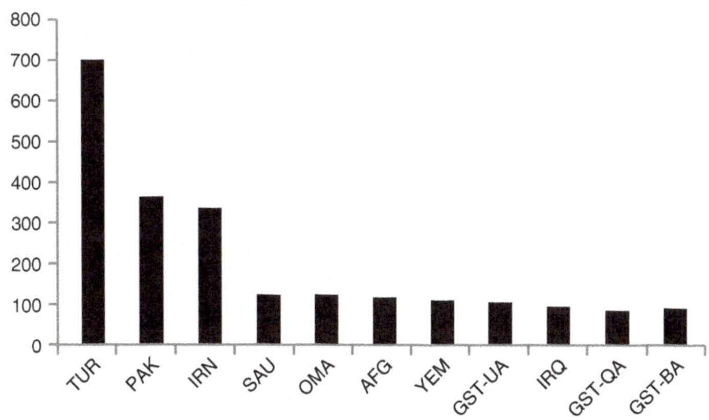

Fig. 5.2 Number of halophytic species in countries of SW Asia. TUR Turkey; PAK Pakistan; IRN Iran; SAU Saudi Arabia; OMA Oman; AFG Afghanistan; YEM Yemen; GST-UA United Arab Emirates; IRQ Iraq; GST-QA Qatar; GST-BA Bahrain. (Data from published sources – see references)

(Chapman 1978) (Fig. 5.1). It is characterized by a thin crust of salt (halite) and a mat of algae underlain by sand, silt, or clay, with a layer of gypsum about 50 cm below the surface. Coastal sabkhas have probably formed because of postglacial flooding of the Arabian Gulf which cut the supply of sand to dunes further south, and deflation removed the sand to the water table, which evaporated due to the increasing aridity over the past 5000 years (Glennie 1987). Coastal sabkhas are frequently flooded during storms and spring tide.

Coastal sabkhas are also present in the coastal region of central Oman. On the eastern coast lies the Barr al Hikman Peninsula which is a flat featureless highly saline plain from sea level to about 20 m asl and low sandy coastal dunes. Formed as a result of a fall in sea level, it is unique in its quaternary sediments (the largest in the Arabian Peninsula) with alluvial sandy gravel overlain with a thin layer of shifting aeolian sand (Glennie 1987; Gubba and Glennie 1998). This and other low, flat coastal landscapes are composed of layers of sand, silt, mud, and salt to a depth of several meters. Evaporation brings up salts which form a crust on the surface. The sabkhas become firm during dry weather and are often covered with mud polygons.

Inland sabkhas in the Arabian Peninsula are present where wadis flow and terminate and whose drainage is frequently blocked by constantly shifting dunes and where former lakes existed. Two large inland sabkhas in the northern part of the Peninsula are Sabkha Maṭṭī in the UAE which also extends southward into Saudi Arabia and Umm as Samim, a large inland salt plain in western Oman (Fig. 5.1). Umm as Samim is fed by a few wadis originating in the western Hajar mountain range of northern Oman. It consists of a main zone of salt crust, including heaved crust. Fresh salt is continually precipitated as a result of evaporation, and the expansion breaks the surface into polygonal plates bounded by rims of fresh salt. Vegetation present around the fringes and runnels leading to the sabkha is highly salt tolerant.

The Red Sea coastal plain consists of a narrow coralline plain and inland eroded bedrock covered with alluvial sand and gravel.

Climate

The climate of the Arabian Peninsula ranges from hyperarid to semiarid and is markedly influenced by topography. The hyperarid areas receive <100 mm rainfall, the arid areas 100–250 mm rainfall, the semiarid plains and foothills 250–500 mm rainfall, and mountains and summits >500 mm average rainfall. The western mountains influence rainfall along the Red Sea coast and coasts, and the Zagros Mountains of western Iran play an important part in rainfall over the extreme east of the Peninsula. Winter is the rainfall period for the north, the eastern coast, and coastal areas of northern Oman of the Peninsula which receive about 50% of rainfall and then the remainder coming in the spring months. Spring rain is also received by Central Oman and landward areas of the southern mountains. The southern regions receive summer rainfall, this being entirely due to the influence of the southwest monsoon. (Fisher and Membery (1998) give a detailed account of the climate of the Arabian Peninsula; Fig. 5.3).

Distribution and Biogeography of the Halophytic Flora of the Arabian Peninsula

A fair amount of literature exists on halophytes of the countries of SW Asia including the Arabian Peninsula, and a number of papers are present on the physiology and germination studies of halophytes (Böer 2004; Flowers 1986 and references therein; Flowers and Comer 2008 and references therein, Ghazanfar 2011 and references therein). More recently, several studies have concentrated on the phylogeny of halophytes mainly in the Family Chenopodiaceae leading to revised classifications and changes in the nomenclature of species (Akhani et al. 2007; Kadereit et al. 2006, 2006a, 2007; Kadereit and Freitag 2011;

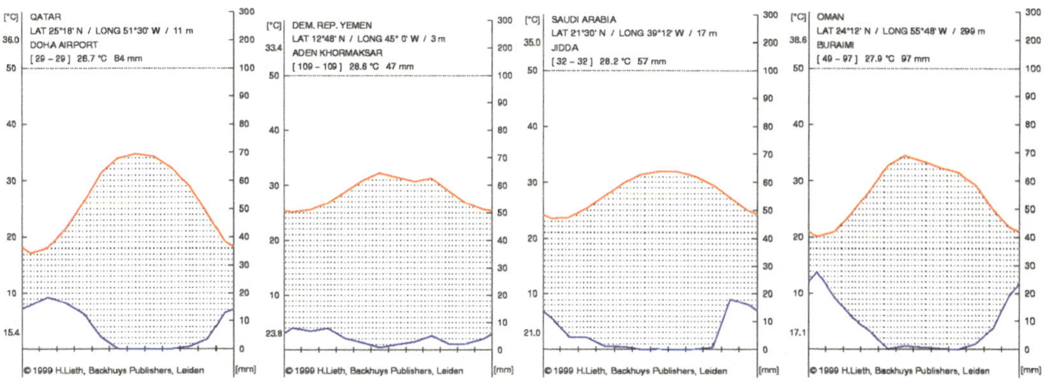

Fig. 5.3 Climate diagrams of countries of the Arabian Peninsula. Doha (Qatar), Aden (Yemen), Jeddah (Saudi Arabia), Buraimi (Oman)

Table 5.2 Number of halophytic taxa in plant families in the Arabian Peninsula

Family	No. halophytic species	Family	No. halophytic species
Acanthaceae	1	Juncaceae	2
Aizoaceae	4	Liliaceae	1
Apocynaceae	2	Mimosaceae	2
Arecaceae	1	Najadaceae	1
Asteraceae	2	Orobanchaceae	1
Boraginaceae	2	Plumbaginaceae	6
Chenopodiaceae	41	Poaceae	17
Caryophyllacae	8	Portulacaceae	1
Ceratophyllaceae	1	Potamogetonaceae	1
Convolvulaceae	2	Rhizophoraceae	2
Cymodoceaceae	3	Ruppiaceae	1
Cynomoriaceae	1	Salvadoraceae	1
Cyperaceae	4	Tamaricaceae	3
Fabaceae	5	Typhaceae	1
Frankeniaceae	1	Zygophyllaceae	11
Hydrocharitaceae	2		

From Ghazanfar (2011)

Sukhorukov et al. 2016). The following section is adapted from halophytes of SW Asia by Ghazanfar et al. (2014).

In general, saline and arid environments are poor in species. Of the total 415 plant families (APG III 2009 and update APG IV 2016), halophytes of SW Asia are recorded in 68 families (117 plant families worldwide as recorded by Aronson 1989). The majority of halophytes belong to the families Chenopodiaceae, Poaceae, Fabaceae, Asteraceae, and Cyperaceae. Chenopodiaceae has the largest number of species and genera of all families only exceeded by Poaceae which has more genera (but fewer species) than Chenopodiaceae (Table 5.1). These data are in accordance with that found for halophytes of the world (Flowers et al. 1986; Table 5.2).

Floristically, the Arabian Peninsula mainly falls within the Saharo-Sindian and Irano-Turanian floristic regions (Leonard 1981–89; Zohary 1973), to which the majority of the halophytic communities belong.

Halophytes can be obligate or facultative. Whereas obligate halophytes survive only in saline habitats, facultative halophytes grow equally well in saline and nonsaline habitats.

Important and frequent halophytes in the sabkha ecosystems of the Arabian Peninsula are mostly perennial hemicryptophytes, succulents, subshrubs, and stoloniferous perennial herbs. The most salt-tolerant obligate halophytes in the Arabian Peninsula include *Arthrocnemum macrostachyum*, *Caroxylon* spp., *Cyperus aucheri*, *Halocnemum strobilaceum*, *Halopeplis perfoliata*, *Limonium* spp., *Salicornia perennans* (=*Salicornia europaea* sensu auctt.), *Seidlitzia rosmarinus*, *Suaeda* spp., *Tamarix* spp., and *Tetraena* spp.; grasses and sedges include *Aeluropus lagopoides*, *Juncus rigidus*, *Odyssea mucronata*, *Sporobolus spicatus*, *S. consimilis*, *Urochondra setulosa*, and mangroves *Avicennia marina*. The most important facultative halophytes include *Salsola drummondii*, *Suaeda vermiculata*, *Suaeda aegyptiaca*, *Anabasis setifera*, and *Tetraena qatarense*.

About 120 taxa are recorded as halophytes in the Arabian Peninsula (see Appendix). This constitutes about 4% of the total flora of the Arabian Peninsula (±3500 taxa) (Ghazanfar et al. 2014), third in the world after Turkey, Iran, and Pakistan (Tables 5.1 and 5.2; Fig. 5.2). Halophytes in SW Asia constitute about half the number of halophyte taxa (and families) recorded for the world by Aronson (1989).

Not surprisingly the majority of halophytes belong to the families Amaranthaceae, Poaceae, Zygophyllaceae, Fabaceae, and Plumbaginaceae. Table 5.1 shows the distribution of halophytic taxa and their families in the Arabian Peninsula (see Abbas 2002; Abed 2002; Al-Gifri and Gabali 2002; Al-Turki et al. 2000; Barth 2002; Böer and Al Hajiri 2002; Böer and Gliddon 1998; Brown et al. 2008; Ghazanfar 2002, 2003, 2006, 2007, 2011, 2015; Omar et al. 2002).

Key species in saline habitats of Arabia are nearly always perennial. The predominant life-forms are succulent, semiwoody dwarf shrubs belonging to the families Amaranthaceae, Zygophyllaceae, and Plumbaginaceae and hemicryptophytes with runners and spiny leaves belonging to the families Poaceae and Juncaceae. Annual succulents such as *Bienertia cycloptera* and *Tetraena simplex* are exceptions. Coastal species are either obligate halophytes like the representatives of the families Amaranthaceae, Frankeniaceae, and Plumbaginaceae or salt-tolerant genera from unspecialized families, such as *Sporobolus* and *Aeluropus* (Poaceae) or salt-secreting species such as *Avicennia* (Acanthaceae) and *Limonium* (Plumbaginaceae). The most common coastal and salt-tolerant species are *Arthrocnemum macrostachyum*, *Halocnemum strobilaceum*, *Halopeplis perfoliata*, *Caroxylon* spp., and *Suaeda* spp., (Amaranthaceae); *Aeluropus lagopoides*, *Odyssea mucronata*, *Sporobolus spicatus*, and *S. consimilis* (Poaceae); *Juncus rigidus* (Juncaceae); *Tetraena* spp. (Zygophyllaceae); *Limonium* spp. (Plumbaginaceae); and *Avicennia marina* (Acanthaceae) (Deil 1998; Ghazanfar et al. 2014 and references therein).

The submerged coastal vegetation of the Arabian Peninsula, especially that of the Gulf, has been well studied owing to the rapid coastal development. The submerged seagrass beds are one of the most important vegetation types and highly productive ecosystems of great importance to the marine fauna especially the marine turtles, shrimps, and numerous species of fish and are highly important carbon sinks. Sheppard et al. (1992) report five species of seagrasses from the Gulf, *Halodule uninervis*, *H. wrightii*, *Halophila stipulacea*, *H. ovalis*, and *Syringodium isoetifolium*. *Ruppia maritima* is also reported in several coastal lagoons (Mandaville 1990). *Halodule uninervis*, *Halophila stipulacea*, and *H. ovalis* are most widespread and the most common.

Biogeographical limits of the coastal and saline vegetation of the Arabian Peninsula have been adapted from Deil (1998). Biogeographically Vesey-Fitzgerald (1957) was the first to recognize the difference between the salt marsh flora on either side of the Tropic of Cancer, and Freitag (1991) showed that the tropical and extra-tropical distribution of the halophytic coastal species of the Chenopodiaceae is similar to that of the non-halophytic species.

Halopeplis perfoliata is an example of a species with a typical circum-Arabian distribution in the Nubo-Sindian zone of the Sahara-Sindian phytochorion. *Arthrocnemum macrostachyum* is a bi-regional species with a Sahara-Sindian/

Mediterranean distribution. *Halocnemum strobilaceum* is a pluri-regional species occurring in the Mediterranean/Saharo-Sindian/Irano-Turanian phytochoria with a distinct southern distributional boundary. *Salsola schweinfurthii* is a Saharo-Arabian species, and *Seidlitzia rosmarinus* has a Saharo-Sindian and Irano-Turanian distribution, not occurring south of Jeddah or Musandam. *Suaeda monoica* is a tropical Saharo-Sindian species commonly distributed in Sudan and Eritrea and with its northernmost limit on the Diimaniyat Islands off the coast of Muscat (Ghazanfar 1992); it is replaced by *Nitraria retusa* further north (Freitag 1991; Kassas and Zahran 1967). The distributional limits of *Seidlitzia rosmarinus* delimit to a large extent the extra-tropical from tropical coastal vegetation complexes. It occurs in seasonally wet inland saline habitats, often replacing the *Halocnemum* community on drier habitats, and usually forming a community of its own, which in the Irano-Turanian region includes several halophytic annuals. *Arthrocnemum macrostachyum* is replaced by the truly tropical *Halosarcia indica* in southeast Pakistan and western India and by *Halopeplis perfoliata* in the southern coasts of the Arabian Peninsula. *Odyssea mucronata* is not distributed north of Jeddah, and similarly *Limonium axillare* is replaced by *L. pruinosum* north of the Tropic of Cancer. Other extra-tropical taxa include *Cornulaca ehrenbergii*, *Gymnocarpos decander*, *Anabasis setifera*, and *Halopyrum mucronatum*.

The north-south distribution of coastal species is more distinct on the Red Sea coast, with the border lying near Jeddah, than on the Arabian Gulf Coast where there is a broad transitional zone lying between Qatar and northern Oman. The east-west distribution of coastal species is not as distinct as that of the north-south distribution. The eastern elements are either restricted to the coasts around the Arabian Gulf (e.g., *Salsola drummondii*) or are Irano-Turanian species extending into the Gulf region (e.g., *Bienertia cycloptera* and *Seidlitzia rosmarinus*). Some east-west species are closely related vicariants, such as *Salsola drummondii* restricted to eastern Arabia and extending eastward to India and *S. schweinfurthii* distributed mainly from eastern Saudi Arabia to Jordan, with an outlier recorded from Oman (Miller and Cope 1996).

There are several vicariant groups of halophytic species in the Arabian Peninsula. These include species in the genera *Cornulaca*, with *C. monacantha* distributed from southwest Asia eastward to Pakistan (Boulos 1992) and *C. aucheri* distributed in the eastern regions of the Peninsula, Iraq, Iran, and southwest Pakistan; *Salsola*, with *Caroxylon vermiculatum* (=*Salsola chaudharyi*, treated as *Salsola villosa* in Miller and Cope 1996) in central Saudi Arabia (Botschantzev 1984) and *S. omanensis* in the coastal plains of Dhofar (Boulos 1991); and *Suaeda* (Freitag 1991) with *Suaeda moschata* restricted to the Barr al Hikman Peninsula and Hallaniyat Islands in Oman (Scott 1981). Other examples include the *Cyperus conglomeratus* complex (*C. aucheri*, *C. conglomeratus*; Kukkonen 1991) and the *Limonium axillare* group (*L. axillare*, *L. stocksii*, *L. carnosum*, and *L.* cf. *stocksii*). East-west and littoral-inland vicariance is well illustrated in the genus *Tetraena* section Mediterranean, with *T. coccineum* mainly distributed in the northern coasts of the Red Sea, *T. qatarense* in the Arabian Gulf and Gulf of Oman (Boulos 1987), *T. hamiense* in the southwestern corner of the Arabian Peninsula, *T. mandavillei* in the southern Rub' al Khali and Hadhramaut, and *T. migahidii* in the Nafud (El-Hadidi 1977, 1980). *Sevada schimperi*, a monotypic genus within Chenopodiaceae, is endemic to the coastal habitats around Bab al Mandab, Yemen (Freitag 1989).

Terrestral Sabkha Vegetation and Plant Communities

Vegetation, plant communities, and the zonation of plants of coastal and inland sabkhas have been well described for all countries of the Arabian Peninsula in varied detail. A summary of the vegetation of the coastal and inland sabkhas of the countries is given here with references to the studies made for each country (adapted from Deil 1998). Detailed accounts of vegetation types are also given in Barth (2002).

Saudi Arabia

Refs: Coastal vegetation of the Red Sea coast (El-Shourbagy et al. 1987; Chaudhary 1998); around springs (El-Sheikh and Youssef 1981; El-Sheikh et al. 1985); eastern coast (Mandaville 1990); Shaltout et al. (1997); Arabian Gulf coast in the vicinity of Jubail (Böer 1994, 1996; Böer and Warnken 1992).

The coastal regions have extensive stands of *Suaeda monoica*, *S. fruticosa*, and *S. vermiculata*. Where some freshwater is available, *Tamarix nilotica* and thickets of *Salvadora persica* can be found. The littoral salt marsh communities on the Red Sea coast north of Jeddah consist of *Avicennia marina* in the first zone followed by a *Halopeplis perfoliata* zone in the moist but not waterlogged soil fringing the shoreline. On soft aeolian deposits overlaying mudflats, *Aeluropus massauensis* occurs in the third zone, and on coarse soils where the water table is below 1.5 m, a *Tetraena coccineum* or a *Limonium axillare-Suaeda pruinosa* zone is present. Mangroves are also found on the west coast and the Farasan Islands consisting of *Avicennia marina* and *Rhizophora mucronata*, while on the east coast, only *A. marina* is found.

Sabkhas are also present around springs. The vegetation of sabkhas around Al Khari springs southeast of Riyadh and that of Al Qassim in the Nefud consist of *Seidlitzia rosmarinus* where water is present at depths of 35–75 cm and salinity of 50,000 μS; a *Tetraena decumbens-Caroxylon imbricatum* community is found where the water is at a depth of 60–120 cm and salinity 500 μS.

The inland sandy saline plains have *Caroxylon* spp. and *Hammada salicornica*, associated with *Acacia tortilis*.

Yemen

Refs: Yemen coastal vegetation (Al Khulaidi et al. 2010; Al Khulaidi 2013); Tihama Coast (El-Demerdash et al. 1995); Gulf of Aden (Al-Gifri and Gabali 2002; Kürschner et al. 1998), Hadhramaut coast at Felek, east of Mukalla (Kürschner et al. 1998).

The coastal plains of Yemen show a number of vegetation types: mangrove (*Avicennia marina*) swamp occurs along the Red Sea coastal fringe, mainly north of the wadi Siham outlet; isolated swamps are also seen south Al Mukha, north Yakhtol (southern Tihama), and around Bir Ali (west of Al Mukalla). Occasionally other plants such as *Aeluropus lagopoides*, *Suaeda* spp., and others may occur with the mangroves. These form a transition to other vegetation types found further inland. A *Suaeda vermiculata* shrubland is found along the coast on flat, often bare muddy ground and covers from shoreline to about 5 km inland. *Suaeda vermiculata* and *Aeluropus lagopoides* are the most common species in this sabkha.

Northward from Wadi Siham the *Avicennia marina* zone is followed by a *Limonium cylindrifolium-Suaeda fruticosa-Limonium axillare* community which forms hummocks. A sterile sabkha is present, after which raised beaches above the high tide level are covered by *Atriplex farinosa*, *Tetraena hamiense*, *Aeluropus lagopoides*, and *Halopyrum mucronatum*. Sand dunes toward the seaward side are colonized by *Suaeda monoica* and *Caroxylon spinescens* and the inland dunes by *Odyssea mucronata*, *Jatropha pelargoniifolia*, and *Leptadenia pyrotechnica*.

The southwestern corner of the Arabian Peninsula is characterized by *Odyssea mucronata*, endemic to this part of Arabia. *O. mucronata* is a clump-forming, spiny, rhizomatous perennial which colonizes semimobile dunes and flat sandy areas. Depending on the depth of sand, an *Odyssea mucronata-Suaeda monoica* community can be distinguished on flat sandy layers overlying saline silts and an *Odyssea mucronata-Panicum turgidum* community on deeper sand.

The Hadhramaut coast is situated in the transition zone from the southeastern to the southwestern vegetation type. This is seen from the *Cyperus conglomeratus/C. aucheri* associations, where the Omano-Makranian element (Kürschner 1986), *Coelachyrum piercei*, and the Eritreo-Arabian element *Odyssea mucronata* are common members. The coastal vegetation shows a strong phytogeographical relationship with the coasts of northeast Africa. The species zones are (1) coastal dunes colonized by sedges and grasses (*Cyperus aucheri*, *Halopyrum mucronatum*,

Odyssea mucronata, *Coelachyrum piercei*, and *Panicum turgidum*); (2) sandy-salty depressions colonized by the endemic *Urochondra setulosa* association, with the codominant *Arthrophytum macrostachyum*, *Limonium cylindrifolium*, and *Crotalaria saltiana*; (3) clayey-salty, relatively wet areas colonized by monospecific stands of *Arthrophytum macrostachyum*; (4) sandy coastal plains colonized by the endemic *Anabasis ehrenbergii-Pulicaria hadramautica-Tetraena hamiense* association; and (5) the karstic limestone plateau colonized by *Stipagrostis paradisea*, *Commiphora gileadensis*, and *Euphorbia rubriseminalis*.

Oman

Refs: Coastal, inland sabkha, and saline and brackish water vegetation (Ghazanfar 1992, 1993, 1995, 1998, 1999, 2002, 2006); coastal vegetation of the islands of Masirah and Shagaf (Ghazanfar and Rappenhöner 1994); vegetation of the Qurm Nature Reserve near Muscat (Kürschner 1986); inland sabkha Umm as Samim (Heathcote and King 1998).

The coastal vegetation in the northern Oman can be classified into four plant communities: (1) a *Limonium stocksii-Tetraena qatarense* community in northern Oman where the coasts are mainly sandy and interspersed with rocky limestone headlands; (2) a *Limonium sarcophyllum-Suaeda aegyptiaca* community characteristic of rocky shores with narrow beach areas and a wide spray zone; (3) an *Atriplex-Suaeda* community characteristic of the vegetation of offshore islands, flat sandy beaches, and coastal sabkhas (dominant and associated species are *Atriplex coriacea*, *A. farinosum*, *A. leucoclada*, *Arthrocnemum macrostachyum*, *Suaeda aegyptiaca*, *S. vermiculata*, *S. monoica*, *S. moschata*, and *Halocnemum strobilaceum*) and a *Limonium axillare-Sporobolus-Urochondra* community characteristic of the vegetation of the southern coasts, with *Limonium axillare*, *Urochondra setulosa*, and *Sporobolus* spp. associated with several other species depending on coastal geomorphology; and (4) coastal lagoons with *Sporobolus virginicus*, *S. iocladus*, and *Paspalum vaginatum* as the main bordering species and *Phragmites australis* and *Typha* spp. forming the bordering reeds. In addition, *Avicennia marina* occurs throughout coastal Oman in discontinuous patches and over a wide range of water salinities.

On the Barr al Hikman Peninsula and the offshore island of Masirah, *Avicennia marina* is present in sheltered lagoons, a halophytic shrub community dominated by *Atriplex farinosa* and *Suaeda moschata* occurs on low coastal dunes which receive salt spray, and a *Halopyrum mucronatum-Urochondra setulosa* community occurs on more or less stabilized dunes. An *Arthrocnemum macrostachyum-Suaeda vermiculata* community occurs on the saline, silt plains and a *Limonium stocksii-Cyperus auheri-Sphaerocoma aucheri* community on shallow sands.

The inland sabkha, Umm as Samīm, is the largest in the Arabian Peninsula. It lies in northwest Oman bordering the sand desert of the Rub' al Khali and covers an area of *c.* 5000 km^2. Much of the sabkha is too saline to support any vegetation, but plants exist on the edges. The fringing vegetation of Umm as Samim is very sparse since rainfall is scanty (<50 mm per year) and temperatures high. The few species present are *Aeluropus lagopoides*, *Cornulaca monacantha*, *Hammada salicornica*, *Salsola* cf. *drummondii*, *Suaeda aegyptiaca*, and *Tetraena qatarense.* Therophytes consist of *Tetraena simplex* and *Tribulus longipetalus*.

United Arab Emirates

Refs: Coastal and sabkha flora (Böer 2002); coastal vegetation near Dubai (Deil and Müller-Hohenstein 1996); coastal vegetation and conservation (Brown et al. 2008); biogeography (Böer and Saenger 2006).

A transect through the coastal dunes and sabkha in the UAE shows the typical dry haloseries within the Omano-Makranian region of the Arabian Gulf; four plant communities are present associated in the *Limonium stocksii-Tetraena qatarense* vegetation complex: (1) the seaward dunes colonized by the *Cornulaca monacantha-Sphaerocoma aucheri* community (the Salsolo-Suaedetalia of Knapp 968); (2) the landward dunes colonized by *Halopyrum mucronatum* (stabilizing the sand), *Atriplex leucoclada*, and *Suaeda aegyptiaca*; (3) salty depressions which may be temporarily inundated with seawater colonized by *Halopeplis perfoliata*; and (4) an ephemeral, salt-tolerant *Frankenia pulverulenta-Tetraena simplex* plant community growing in depressions with sandy overlays. The landward dunes, away from the influence of salt spray, are also dominated by *Cornulaca monacantha* and *Sphaerocoma aucheri*. They are associated here with glycophytic (i.e., non-halophytic) dune species such as *Panicum turgidum*, *Crotalaria persica*, *Lotus garcinii*, *Taverniera spartea*, and *Indigofera intricata*. Similar is as well the halophytic vegetation of Qatar which along a transect from the mangrove zone to the sabkha plain shows a distinct floristic and edaphic gradient with the following zonation: (1) *Avicennia marina*, (2) *Arthrocnemum macrostachyum*, (3) *Halocnemum strobilaceum*, and (4) *Juncus rigidus-Aeluropus lagopoides*. Associated species are *Tetraena qatarense*, *Halopeplis perfoliata*, and *Anabasis setifera*.

The vegetation of the Arabian Gulf coast in the vicinity of Jubail is similar to that of Bahrain. The zonation of species within the intertidal zone from the sea landward is given as follows: (1) *Avicennia marina*, (2) *Salicornia perennans*, (3) *Arthrocnemum macrostachyum*, (4) *Halocnemum strobilaceum*, (5) *Halopeplis perfoliata* (>2 m and above the intertidal zone), (6) *Limonium axillare* (>2 m and above the intertidal zone), and (7) *Tetraena qatarense* (>2 m and above the intertidal zone). The outer fringe consists of the *Seidlitzia rosmarinus* community on small dunes followed by *Rhanterium epapposum*, *Hammada salicornica*, *Panicum turgidum*, and *Calligonum comosum* on nonsaline sands.

Bahrain

Refs: Abbas 2002; Abbas and El-Oqlah 1992.

Sabkhas and coastal lowlands represent about 40% of the area of Bahrain. These are the western and northeastern coastal plains, and the southwest and south sabkhas, all of which support halophytic vegetation. In the western coastal plain, four plant communities can be distinguished: *Aeluropus lagopoides* community, *Tetraena qatarense* community, *Halopeplis perfoliata* community, and *Sporobolus ioclados* community. These are to large extent interconnected with each other. The northeastern coast supports the mangrove, *Avicennia marina*, in the intertidal zone followed landward by *Arthrocnemum macrostachyum* and inland by *Suaeda vermiculata*. *Cressa cretica* is found in small scattered depressions, and more inland from the mangroves, *Halocnemum strobilaceum* dominates on a reclaimed area that used to be part of the mangrove swamp.

The southwest sabkha is a large salt pan and subject to tidal inundation. Algal mats dominate this coastal sabkha, and where the salinity decreases, *T. qatarense* and *Cyperus aucheri* are present on small sand dunes. The south sabkha is more diverse with *H. perfoliata* in very saline soils associated with few *H. strobilaceum*, *A. macrostachyum*, and *Seidlitzia rosmarinus*.

Qatar

Refs: Halophytic vegetation (Abdel-Razik 1991; Abdel-Razik and Ismail 1990; Babikir 1984; Batanouny 1981; Batanouny and Turki 1983; Babikir and Kürschner 1992; Böer and Hajiri 2002.

The major sabkha vegetation zones can be best described along a transect from the mangroves to the sabkha plain which show distinct floristic gradient: (1) *Avicennia marina*, (2) *Arthrocnemum macrostachyum*, (3) *Halocnemum strobilaceum*, and (4) *Juncus rigidus-Aeluropus lagopoides*. Associated species are *Tetraena qatarense*, *Halopeplis perfoliata*, and *Anabasis setifera*. In a coastal littoral plain in southwestern Qatar around the Gulf of Salwa, seven interconnected halophytic plant communities form a mosaic. These are the following:

1. The *Halopeplis perfoliata* community on sandy beaches along the Gulf shore and surrounding depressions, not inundated by the sea.

2. The *Halocnemum strobilaceum* community, which colonizes the depressions.
3. The *Halopyrum mucronatum-Sporobolus consimilis* community on calcareous sands.
4. *Limonium axillare*, *Suaeda vermiculata*, and *Cistanche tubulosa* forming sandy mounds.
5. The *Tetraena qatarense* community growing in shallow depressions and runnels on coarse-textured soils, associated with *Cornulaca monacantha*, *Robbairea delileana*, and *Stipagrostis*.
6. Inland, *Panicum turgidum* and *Pennisetum divisum* tussocks are present on fine sand and A*nabasis setifera* on coarse sand.
7. The *Suaeda vermiculata* community on fine-textured soils but restricted to the southwestern area.

In northwestern Qatar (ad Dakhira), an *Avicennia marina* association is present in the supralittoral border followed by an ephemeral halophytic community, the *Salicornia perennans-Suaeda maritima* association, in the intertidal zone. An *Arthrocnemum macrostachyum* association is present in the supra-tidal area with *Halopeplis perfoliata* sometimes associated with it. The *Aeluropus lagopoides-Tamarix passerinoides* association is present on dunes and the *Caroxylon cyclophyllum-Panicum turgidum-Anabasis setifera* association on windblown, sandy accumulations at the foot of limestone cliffs. The limestone plateau itself is colonized by a xeromorphic, very open dwarf shrubland of *Tetraena qatarense*, *Helianthemum lippii*, and *Lycium shawii*.

Kuwait

Refs: Coastal vegetation (Omar et al. 2002; Omar 2007); coastal vegetation and zonation (Halwagy and Halwagy 1977; Halwagy et al. 1982; Halwagy 1986).

In Kuwait, *Salicornia perennans* (=*Salicornia europaea*) grows on low, frequently inundated mud banks or along creeks, sometimes associated with *Aeluropus lagopoides* and *Bienertia sinuspersici* (=*Bienertia cycloptera*) or with *Juncus rigidus* on the fringes of creeks. A *Halocnemum strobilaceum* community occupies the lower marshes along the shoreline with the seaward edge inundated very frequently by tides. A *Seidlitzia rosmarinus* community occurs further inland, followed by *Nitraria retusa* above the high tide mark dominating the middle marshes, and finally the *Tetraena qatarense* community on elevated, coarse sandy sites on the landward edge of the marsh. The salt marshes are fringed by non-halophytic communities such as the *Cyperus aucheri* community, the *Rhanterium epapposum-Convolvulus oxyphyllus-Stipagrostis plumosa* community, and the *Hammada salicornica* community, the latter covering most of the territory of Kuwait.

Intertidal Vegetation

Mangroves

Mangroves occur throughout the coasts of the Arabian Peninsula, bordering bays and creeks, some offshore islands, and on several sea lagoons. Of the three recorded species, *Avicennia marina* is by far the commonest and most abundant (Frey and Kürschner 1989; Sheppard et al. 1992), being tolerant of low air temperatures (12 °C) and high water salinities (40–50%) (Böer 1996; Sheppard et al. 1992). The distribution of mangroves indicates that cold winter temperatures rather than salinity limit their northernmost extent, and mangroves may formerly have been more common in the Gulf and Red Sea than they are at present (Sheppard et al. 1992, and references therein).

Avicennia marina, originally described from Al Luhayyah on the Red Sea coast of Yemen, occurs southward from latitude 26°N along the Red Sea coast and in the Gulfs of Aden and Oman. The northernmost populations of *A. marina* are recorded from *c.* 27°N in the Jubail Marine Wildlife Sanctuary on the Arabian Gulf coast of Saudi Arabia (Böer and Warnken 1992) and the Gulf of Suez and Sinai coast in the Gulf of Eilat (Danin 1983). Dense stands of this species occur on Mahout Island in Gubbat Al Hashish in central Oman, where the trees are up to 4 m in height and where the mangroves sustain shrimp, crab, and other fisheries of commercial importance (Fouda and Al-Muharrami 1996). *Rhizophora mucronata* is known from Gizan (south of Jeddah) and the Farasan Islands (El-Demerdash 1996) and from the Gulf of Aqaba and Bahrain and an isolated stand of about 200 trees on Jazeerat Al Mubarraz in Abu Dhabi. *Bruguiera gymnorhiza* has been recorded from the offshore islands near Hodeida (Zahran 1975), though its presence there is unconfirmed (Sheppard et al. 1992). In the last decade, attempts have been made for the restoration of *Avicennia marina* especially in the UAE (El Amry 1998; de Soyza et al. 2002), and attempts on restoration coupled with conservation, sustainable use, and as carbon sinks and the UAE have resulted in positive results (Bhat et al. 2004; Böer 1996; Böer et al. 2014).

Salt Marshes

Only a handful of perennial species are found in the intertidal zone of the coasts of the Arabian Peninsula.

On coastal flats such as those found in Bahrain, Kuwait, Qatar, and Oman, the dominant (and forming monospecific stands) are *Salicornia perennans* and *Arthorcnemum macrostachyum*. *Salicornia perennans* reproduces mainly by vegetative growth in Kuwait (Brown in obs.). *Atriplex farinosa* and *A. leucoclada* are also found, associated with *Suaeda* spp. (*S. monoica*, *S. moschata*) in Oman. South of UAE, *Salicornia perennans* is not found on the southern coasts of Arabian Peninsula. On rocky and pebbly shores, such as those found in northern and southern Oman, a *Limonium sarcophyllum-Suaeda aegyptiaca* (northern Oman) and a *Limonium axillare-Sporobolus-Urochondra* (southern Oman) community is found. *Limonium cylindrifolium* with the endemic *Urochondra setulosa* and codominant *Arthrophytum macrostachyum* community is found in saline depressions on Yemen coasts.

Sub-tidal Vegetation

Seagrasses

About 11 species of seagrasses have been recorded from the Arabian Peninsula. Their distribution is controlled by a complex of environmental factors which include substrate, depth, temperature, salinity, and light penetration (Sheppard et al. 1992). Shallow coastal bays (<10 m deep) often have well-developed seagrass beds, such as along the shallow southeast coasts of Bahrain, where the species are restricted to shallow waters with good light penetration. Relatively dense seagrass beds occur in central and southern Oman (Jupp et al. 1996) and the Gulf of Aden. Four species are recorded from southeast Arabia and the Gulf (Jupp et al. 1996, Sheppard et al. 1992), with most communities dominated by the smaller-bodied species *Halodule uninervis*, *Halophila ovalis*, and *H. stipulacea*. The larger *Syringodium isoetifolium* occurs in the Gulf but is relatively rare. In contrast, several larger-bodied and wide-leaved seagrasses such as *Thalassadedron ciliatum*, *Thalassia hemprichii*, *Cymodocea rotunda*, and *C. serrulata* occur in the Red Sea (Aleem 1979, Jupp et al. 1996). It has been suggested that the effects of seasonal upwelling along the southeastern coasts of the Arabian Peninsula, which causes large fluctuations in sea temperature, are responsible for the impoverished seagrass beds (Basson et al. 1977; De Clerck and Coppejans 1994) and the occurrence of only small-bodied hardy species (Jupp et al. 1996).

Seed Dispersal and Germination Strategies in Halophytes

Halophytes of arid and hyperarid deserts of the Gulf regions are facing several natural stresses such as high temperatures, salinity, and drought. The scarcity of rainfall received in many years in Arab Gulf region (Böer 1997) coupled with high evaporation due to high temperatures, especially during summer, resulted in the formation of what are called sabkha ecosystems or hypersaline salt marshes (Khan and Gul 2006). In order to enhance survival and fitness in such stressful environments, halophytes developed complementary sets of adaptation and survival strategies during different stages of their life cycle (Gutterman 1994; El-Keblawy 2004). The success of halophytes in highly saline soils is greatly dependent on their success in germination and seedling establishment, which are the most sensitive stages in a plant life (El-Keblawy 2013; El-Keblawy and Bhatt 2015; El-Keblawy et al. 2015). In addition, other factors such as seed morphology, mass, wing size, and persistence can all greatly affect the seed dispersal, dormancy, and germination behavior and consequently affect fitness of many desert halophytes (El-Keblawy et al. 2014, 2016a; El-Keblawy and Bhatt 2015; Xing et al. 2013).

Dispersal and Seed Bank

Seeds are either stored in the soil (i.e., soil seed bank) or retained above ground on maternal plants until they are released (i.e., aerial seed bank) (Gunster 1992). Seeds of halophytes are usually stored in saline soils and consequently exposed to salinity stress (Aziz and Khan 1996; El-Keblawy 2014). Persistent seed banks of halophytes carry seeds over a predictable dry or hypersaline period after which germination occurs. As seeds of many of the halophytes could not germinate in salinity level more than seawater salinity, germination usually happens when saline habitats receive effective rainfalls that dilute soil salinity (El-Keblawy 2014). Still, seeds of some halophytes are very sensitive to salinity and cannot germinate above 300 mM NaCl (Khan and Gul 2006). Such plants retain their seeds on the plant canopy as aerial seed bank (El-Keblawy and Bhatt 2015). Retention of seeds in the aerial seed bank may protect them from the lethal effects of salt in the soil. El-Keblawy and Bhatt (2015) compared salinity tolerance in two species with aerial seed bank (*Halocnemum strobilaceum* and *Halopeplis perfoliata*). They found that *H. strobilaceum*, which has a short-term aerial seed bank (less than 9 months), is more tolerant to salinity, but *H. perfoliata*, which has a long-term aerial seed bank (more than 17 months), is less salt tolerant. This result suggests that aerial seed bank protects salt-sensitive seeds from effects of high soil salinity (El-Keblawy and Bhatt 2015). The maintenance of aerial seed bank as a strategy to avoid detrimental soil salinity effects in less tolerant species is especially important during summer, when soil salinity increases.

The distribution of different plant species is the result of their strategies of seed dispersal, dormancy, and germination behavior (Kos et al. 2012). Under the unpredictable heterogeneous environments, such as saline habitats of arid deserts, plants develop multiple strategies through producing offspring that differ in time and place of germination and tolerance to environmental stresses (Baskin and Baskin 1998; El-Keblawy 2003). Fruits of many halophytic plants have winged perianths that help their dispersal and determine the proper place of seed storage and time of germination (Wei et al. 2008; Xing et al. 2013). In the Arabian Peninsula, fruits of many desert halophytes, such as *Hammada salicornica*, *Haloxylon persicum*, *Salsola drummondii*, and *Caroxylon imbricatum*, have winged perianths that help them to disperse and also regulate their dormancy and seed bank dynamics (El-Keblawy 2013). However, seeds of other halophytes, such as *Halopeplis perfoliata* and *Halocnemum strobilaceum*, do not have any dispersal structures and consequently have the chance to bury in the soils (El-Keblawy and Bhatt 2015; El-Keblawy et al. 2015). Still, some other halophytes, such as *Anabasis setifera*, produce nonpersistent wings that could help fruits in dispersal but usually disintegrate within few months after seed landing. The dispersal structures of the last group should help fruit dispersal, but their degradation could help seeds to bury in the salty soil (El-Keblawy et al. 2016a, b). The presence of winged perianths has been considered as an important trait that helps seed to disperse and regulate dormancy and seed bank dynamics (El-Keblawy 2014).

Fruits with winged perianths are able to explore habitats away from their maternal sites. In addition, as winged fruits usually land over soil surface, they face diurnal fluctuations in temperatures and are exposed to intense light during storage (Zalamea et al. 2015). The diurnal soil surface temperature during the summer in the UAE fluctuates between 20 and 60 °C for more than 4 h between noon and midnight (El-Keblawy and Al-Hamadi 2009). Several studies have reported that diurnal fluctuations resulted in breaking seed dormancy of some halophytes, such as *Sporobolus ioclados*, *Diplachne fusca*, *Limonium axillare*, *Halocnemum strobilaceum*, and *Halopeplis perfoliata* (El-Keblawy 2013; Morgan and Myers 1989). However, exposure of seeds to diurnal fluctuation in natural conditions under the very high temperatures of the Arabian Peninsula causes seed death in other halophytes, such as *Caroxylon imbricatum* (El-Keblawy et al. 2007) and *Hammada salicornica* (El-Keblawy and Al-Shamsi 2008). Seeds of plants that have winged perianths usually have a transient seed bank, but those without wings form persistent transient seed bank (El-Keblawy 2013).

Salinity and Tolerance During Germination

Survival of halophyte seeds in the belowground seed banks depends on their capacity for salt tolerance at the germination stage, their ability to tolerate hypersaline conditions during storage in the soil, and/or their ability to avoid salinity (Kozlowski and Pallardy 2002; Ungar 2001). Several studies have concluded that seeds of a few halophytes, such as *Salicornia rubra* (Khan et al. 2000), *Salicornia pacifica* (Khan and Weber 1986), *Salicornia herbacea* (Chapman 1960), *Halocnemum strobilaceum* (Qu et al. 2008; El-Keblawy and Bhatt 2015), and *Salsola drummondii* and *Suaeda vermiculata* (El-Keblawy, unpublished data), can germinate at salinities above that of seawater (*c.* 500 mM NaCl). Conversely, seeds of other halophytes, including *Halopeplis perfoliata* (Mahmoud et al. 1983; El-Keblawy et al. 2015), *Salicornia brachystachya*, and *Salicornia dolistachya* (Huiskes et al. 1985), cannot tolerate seawater at germination stage.

The absence of perianth structures provide halophytes a greater chance to bury in very saline soils. These seeds are exposed to very high salt concentrations, especially during summer, when water evaporates leaving a salt crust near and on the soil surface. The small buried seeds of halophytes have to survive these hypersaline conditions and be able to germinate once the salinity level is reduced, which usually happens after heavy rainfall (El-Keblawy 2004). The ability of halophyte seeds to maintain their viability after an extended period of exposure to salinity has been recorded in several species (see Khan and Gul 2006).

Khan and Gul (2006) reviewed the germination recovery of salt-treated seeds of many halophytes of the Great Basin desert and found substantial recovery in distilled water of seeds treated with up to 600 mM NaCl in *Halogeton glomeratus*, *Sarcobatus vermiculatus*, *Suaeda moquinii*, and *Triglochin maritima*. Similarly, high salinity did not permanently injure seeds, and germination is fully recovered when seeds were transferred to distilled water in many halophytes of subtropical regions, such as *Atriplex patula* (Ungar 2001); *Suaeda fruticosa* (Khan and Ungar 1997); *Arthrocnemum macrostachyum*, *Sarcocornia fruticosa*, and *Salicornia ramoissim* (Pujol et al. 2000); *Salicornia rubra* (Khan et al. 2000); and *Limonium stocksii* (Zia and Khan 2004). Khan and Gul (2006) indicated that species from temperate area (e.g., Great Basin desert) tolerated higher salinities and were able to recover their germination than those from subtropical region, such as the Arabian Peninsula and Pakistan. Such data indicates that seeds of the Great Basin halophytes can tolerate higher salinity when present in the seed bank. The ability of seeds of many halophytes to maintain their viability during exposure to high salinity levels and to recover their germination after transfer to distilled water indicates that the effect of NaCl is more likely to be a reversible osmotic inhibition of germination, rather than ion specific toxicity (El-Keblawy and Al-Shamsi 2008).

The germination recovery of salt-treated seeds of several halophytes is dependent on the temperature regime of incubation. In several halophytes including *Salsola imbricata* (El-Keblawy et al. 2007), *Salsola vermiculata* (Guma et al. 2010), *Hammada salicornica* (El-Keblawy and Al-Shamsi 2008), and *Limonium stocksii* (Zia and Khan 2004), recovery was seen to be higher at lower temperatures and consistent with greater rainfall in Arab Gulf regions. The recovery was greater at moderate temperatures, compared to lower and higher temperatures, in other halophytes such as *Urochondra setulosa* (Gulzar et al. 2001) and *Puccinellia nuttalliana* (Macke and Ungar 1971). Seeds of *Aeluropus lagopoides* exposed to higher salinity recovered quickly at warmer compared to moderate and lower temperatures (Gulzar and Khan 2001).

Fast Germination of Halophytes

Several species of halophytes in subtropical regions produce seeds that germinate very fast and to high levels immediately after maturation. Typically, the time of maturation of these seeds coincides with the onset of rainfall and cooler temperatures, which are favorable for seed germination and seedling recruitment. However, seeds of these species die within few month of dispersal (i.e., form a transient seed bank). For example, fresh seeds of both *Caroxylon imbricatum* and *Hammada salicornica* have high germination levels and germination speed. However, room temperature and warm storage for 9 months resulted in complete death of the seeds (El-Keblawy 2014). Similarly, Khan (1990) and Zaman and Khan (1992) studied temporal dynamics of seed bank of four perennial halophytes (*Cressa cretica*, *Haloxylon stocksii*, *Caroxylon imbricatum*, and *Sporobolus ioclados*) and found that the high germination observed for fresh seeds was gradually reduced with time until they finally died in few months after dispersal. In addition, seeds of *Aeluropus lagopoides* were not dormant and showed 100% germination at the optimal temperature at the time of seed maturation and maintained a transient seed bank (Gulzar and Khan 2001).

Parsons (2012) reviewed the speed of germination and concluded that there are a group of plants, especially those from arid or saline habitats that germinate in less than 24 h from imbibition. A total of 20 species were recorded from the Amaranthaceae (15 of them are from the subfamily Salsoloideae), which most of them are known to survive saline habitats. Seeds of the Salsoloideae contain fully differentiated spiral embryos that immediately uncoil and rupture the thin seed coat once water imbibition takes place (Parsons 2012). The fast germination has been reported for many halophytes of subtropical climate of the Arab Gulf region and East Asia. These include *Hammada salicornica*, *Haloxylon recurvum* (Sharma and Sen 1989), *Limonium axillare* (Mahmoud et al. 1983) and *L. stocksii* (Zia and Khan 2004), *Caroxylon imbricatum* (El-Keblawy et al. 2007), and *H. salicornica* (El-Keblawy and Al-Shamsi 2008). Fast germination has been considered as a strategy to utilize the brief period of water availability and ensure rapid seedling growth early in the growing season. Earlier emergence usually produces more vigorous seedlings that are characterized by greater competitive advantages, compared with late-emerged seedlings. Similarly, halophyte seeds stored in saline soils recover their germination shortly after rainfall. The fast germination after rainfall confers seedling longer growing period for establishment, before salinity increases with evaporation (El-Keblawy et al. 2016b).

Halophytes in Bioremediation

Phytoremediation and phytovolatization are very useful tools to clean up polluted environments. These techniques require suitable plants that can extract metals from soil that they either accumulate them or volatize them through their foliage (Padmavathiamma et al. 2014). In the Arabian Peninsula, among the halophytes, *Phragmites australis* has been used extensively to clean contaminated wastewater as it absorbs large amounts of water, preventing the spread of contaminated wastewater into adjacent uncontaminated areas. *Phragmites australis* has also been used to treat oil production water (OPWs) in Oman, with significant reduction of the concentration of toxic heavy metals (80%) and total hydrocarbons (96%). The quality of treated water conformed to Omani wastewater standards for agricultural reuse (Mahruki et al. 2006).

Prosopis cineraria, *Acacia senegal*, and *Acacia nilotica* were used in a study to stimulate microbial degradation of soil pollutants in desert soil that was contaminated with 2.5–2.6% crude petroleum oil (Mathur et al. 2010). The rhizosphere of these plants was tested for their abilities to degrade the pollutants. The results showed that a highest reduction (26%) of total petroleum hydrocarbons (TPHs) was observed in the rhizosphere soil of *P. cineraria*, a facultative halophytic tree in the Arabian Peninsula, as compared to 15.6% and 12.8% reduction in the rhizosphere soil of *A. senegal* and *A. nilotica*, respectively. The results clearly revealed the efficiency of *P. cineraria* for phytoremediation of TPHs in a contaminated desert soil when compared to the other two legume trees (Mathur et al. 2010).

Tamarix aphylla has also been used as a vegetation filter to "clean" soils polluted with heavy metals around petrochemical and detergent factories (Al-Taisan 2009).

Among the Amaranthaceae, Chenopodioideae, *Hammada salicornica* was studied by Al-Ateeqi (2014) to test its tolerance for weathered oil-contaminated soils as a potential phytoremediator on polluted Kuwait soils. In chenopods, the rhizosphere mainly supports the existence of bacteria but not so much fungi, yet some species are known to do so (Gawronski and Gawronska 2007), and apparently *Hammada* is one of them. In the rhizosphere of *H. salicornica*, few species of bacteria and fungi were found (Al-Ateeqi 2014). Both species of bacteria, *Inquilinus* sp. and *Streptomyces*, were present in the rhizosphere of *H. salicornia* growing on oil-contaminated soil in Kuwait. *Inquilinus* sp. is related to petroleum degradation (Tuan et al. 2011), and *Streptomyces* is known to consume n-octadecane, kerosene, n-hexadecane, and crude oil as a sole carbon source (Tuan et al. 2011). Another species of bacteria, *Rhodococcus*, is also related with oil degradation (Auffret et al. 2009). Several other species such as *Agrobacterium tumefaciens*, *Nocardia cyriacigeorgica*, *Sphingopyxis* sp., and *Gordonia lacunae*/*Gordonia terrae* (Nolvak et al. 2012; Steliga 2012) are all related with oil degradation.

For the presence of fungi in the rhizosphere of *Hammada*, Steliga (2012) talks in general about the usefulness of the *Penicillium* as a species that would be good for preparation of bioremediation strategies which would enhance the result of cleaning up contaminants. *Penicillium simplicissimum* has been found in the *Hammada* rhizosphere as well.

Hammada salicornica has been investigated by Brown and Porembsky (2000) as one of the plants that had survived in an oil-contaminated area on the northern side of Kuwait Bay. In their study, they found that where as tar-like oil tracks remained largely unvegetated 7 years after oil release, a number of *Hammada* shrubs survived oil contamination mainly due to the presence of phytogenic hillocks (*nebkhas*) around their bases. These phytogenic hillocks provided "safe sites" for a number of plant species. This also applied to blowouts, former phytogenic hillocks on the oil tracks that had been subjected to severe sand deflation in recent years. Laboratory studies showed that the seed bank under the oil tracks had been completely damaged but a number of seedlings emerged from soil samples on the phytogenic hillocks and blowouts, even though their numbers were lower.

Phragmites australis, *Tamarix aphylla*, *Prosopis cineraria*, and *Hammada salicornica* are seen as useful halophytes that have great potential in phytoremediation; the latter two can tolerate weathered oil contamination and have a set of micoorganisms around their root system that are related to the degradation of oil in contaminated soils (Al-Ateeqi 2014; Mathur et al. 2010).

Conservation of Sabkha Ecosystems

Sabkha ecosystems are unique ecosystems which support plants that are not only specialized in their physiology and morphology but have also developed strategies in their life cycles and seed dispersal and have potential as bioremediators for water and contaminated soils. These sandy and saline ecosystems with their specialized flora and fauna are living laboratories that offer unique opportunities for research into salinity tolerance and best survival of plants in arid and hyperarid environments.

The coastal areas on the Arabian Peninsula are being transformed rapidly for amenity and resort building. Sabkha ecosystems are being degraded and altered throughout the Gulf countries as they appear to be nonproductive. Only a few coastal areas in the Arabian Peninsula are designated as nature areas and are protected. These protected areas are designated mainly for the protection of birds (e.g., Bar al Hikman, Oman), turtles (e.g., Ras al Had, Oman), and marine fauna (Aspinall 1995, 1996a, b; Baldwin 1996; Baldwin and Kiyumi 1999), which provide a degree of protection to the plants as well; a few are designated solely for the protection of mangroves (e.g., Qurm Nature Reserve, Oman; Khor Kalba, Sharjah; Bul Syayeef, Abu Dhabi; Ras Ghanada, Abu Dhabi). However, in the last two decades, there has been a growing concern in protecting and restoring mangroves, and programs do to so have seen promising results.

It is our wish and hope that sabkha ecosystems get the same protection as other unique ecosystems in the Arabian Peninsula.

Appendix: Halophytes of the Arabian Peninsula. Accepted Names in Bold; Synonyms in *Italics*

AIZOACEAE

Mesembryanthemum nodiflorum L., Sp. Pl. 480 (1753).

Syn. *Chlorophytum nodiflorum* L. (1753).

Sesuvium sesuvioides (Fenzl) Verdc., Kew Bull. 1957, 349 (1957).

Basionym. *Diplochonium sesuvioides* Fenzl (1839).

Sesuvium portulacastrum Syst. Nat., ed. 10. 2: 1058 (1759).

Trianthema triquetra Willd., Ges. Naturf. Fr. Berlin Neue Schriften 4: 181 (1803).

Zaleya pentandra (L.) Jeffery, Kew Bull. 14 (2): 238 (1960).

Syn. *Trianthema pentandra* L. (1767).

APOCYNACEAE: ASCLEPIADOIDEAE

Pentatropis nivalis (J.F.Gmel). D.V.Field & J.R.I.Wood, Kew Bull. 38(2): 215 (1983).

Basionym. *Asclepias nivalis* J.F.Gmel. (1791).

ASTERACEAE

Pluchea dioscorides DC., Prodr. 5: 450 (1836).

Pulicaria hadramautica Edinb. J. Bot. 50(1): 79 (1993).

ACANTHACEAE

Avicennia marina Vierh., Denkschr. Kaiserl. Akad. Wiss. Wien. Math.-Naturwiss. Kl. lxxi. 435 (1907).

BORAGINACEAE

Heliotropium bacciferum Forssk., Fl. Aegypt.-Arab. 38 (1775) sensu lato.

Syn. *Heliotropium undulatum* Vahl var. *ramosissimum* Lehm. (1831); *H. ramosissimum* (Lehm.) DC. (1845); *H. kotschyi* Bunge (1869) nom. Nud.; *H. tuberculosum* (Boiss.) Boiss. (1879); *H. persicum* auct.: Boiss. (1879), non Lam. (1789); *H. lignosum* Bornm. (1937) nomen nudum; *H. fartakense* O.Schwartz (1939); *H. bacciferum* Forssk. subsp. *lignosum* (Vatke) Kazmi var. *fartakense* (O.Schwartz) Kazmi (1970).

AMARANTHACEAE: CHENOPODIOIDEAE

Agathophora iraqensis Botsch., in Bot. Zhurn. 62(10): 1451 (1977).

Syn. *Halogeton alopecuroides* Moq., Chenop. Monogr. Enum. 161 (1840).

Anabasis setifera Moq., Chenop. Monogr. Enum. 164 (1840).

Syn. *Salsola setifera* (Moq.) Akhani (2007) nom. Illegit.

Arthrocnemum macrostachyum (Moric.) K.Koch, Hort. Dendrol. 96, no. 3 (1853).

Basionym. *Salicornia glauca* Delile (1813) non Stocks (1812); *Salicornia macrostachya* Moric (1820);

Arthrocnemem glaucum (Delile) Ung.-Sternb. (1876).

Atriplex farinosa Forssk., Fl. Aegypt.-Arab. 123 (1775).

Syn. *A. hastata* Forssk. (1775) non Linn.

Atriplex stocksii Boiss., Diagn. Pl. Or. Nov. ser. 2(4): 73 (1859).

Atriplex griffithii Moq. var. **stocksii** (Boiss.) Boiss., Fl. Or. (1879).

Syn. *A. sokotranum* Vierh. (1903); *A griffithii* Moq. subsp. *stocksii* (Boiss.) Boulos (1991).

Atriplex leucoclada Boiss., Diagn. Pl. Or. Nov. ser. 2 (12): 95 (1853) var. **inamoena** (Allen) Zohary, Fl. Palest. 1: 147 (1966).

Syn. *A. inamoena* Allen (1939).

Bassia muricata (L.) Asch., Beitr. Fl. Aethiop. 1: 289 (1867).

Syn. *Salsola muricata* L. (1767); *Kochia muricata* (L.) Schrad. (1809).

Bassia eriophora (Schrad.) Asch., Beitr.. Fl. Aethiop. 187 (1867).

Syn. *Kochia eriophora* Scghrad. (1909).

Bienertia cycloptera Bunge, Trudy Imp. S.-Petersb. Bot. Sada vi, ii, 425 (1879) & Boiss., Fl. Or. 4: 945 (1879).

Caroxylon cyclophyllum (Baker) Akhani & E.H.Roalson, Int. J. Pl. Sci. 168(6): 947 (2007).

Basionym. *Salsola cyclophylla* Baker (1894).

Caroxylon imbricatum (Forssk.) Moq., Prodr. (DC.) 13(2): 177 (1849).

Basionym. *Salsola imbricata* Forssk. (1775); *Chenopodium baryosmon* Schult. ex Roem. & Schult. (1820); *Salsola baryosma* (Roem. & Schult.) Dandy (1950); *Caroxylon imbricatum* (Forssk.) Akhani & E.H.Roalson (2007) nom. superfl. Later homonym.

Caroxylon spinescens (Moq.) Akhani & E.H.Roalson, Int. J. Pl. Sci. 168(6): 948 (2007).

Basionym. *Salsola spinescens* Moq. (1849).

Caroxylon villosum (Schult.) Akhani & E.H.Roalson, Int. J. Pl. Sci. 168(6): 948 (2007).

Basionym. *Salsola villosa* Schult. (1820).

Cornulaca aucheri Moq., Chenopodium Monogr. Enum. 163 (1840).

Syn. *Cornulaca leucacantha* Charif & Aellen (1950).

Cornulaca monacantha Delile, Fl. Aegypt., Ill. 206, t 22, f.3 (1814).

Halocnemum strobilaceum (Pallas) M.Bieb., Fl. Taur.-Caucas. 3: 3 (1819).

Basionym. *Salicornia strobilacea* Pallas (1771).

Halopeplis perfoliata (Forssk.) Bunge ex Schweinf. & Aschers, Fl. Aethiop. 289, nomen; et ex Ung.-Sternb. in Atti. Congr. Bot. Firenze, 874,329 (1876).

Basionym. *Salicornia perfoliata* Forssk. (1775).

Halothamnus bottae Jaub. & Spach, Ill. Pl. Orient. 2: 50, t. 136 (1845).

Syn. *Caroxylon bottae* (Jaub. & Spach) Moq. (1849); *Salsola bottae* (Jaub. & Spach) Boiss. (1879).

Haloxylon persicum Bunge ex Boiss. & Bushe, Nouv. Mém. Soc. Imp. Naturalistes Moscou 12: 189 (1860).

Hammada salicornica (Moq.) Iljin (1948).

Syn. *Haloxylon salicornicum* (Moq.) Bunge ex Boiss., (1879); *Caroxylon salicornicum* Moq. (1849); *Hammada elegans* (Bunge) Botsch (1964).

Kaviria rubescens (Franch.) Akhani, Int. J. Pl. Sci. 168(6): 948 (2007).

Basionym. *Salsola rubescens* Franch., Sert. Somal. 60 (1882); *Salola hadramautica* Baker (1894); *Salsola leucophyla* Baker (1894).

Salicornia perennans Willd., Sp. Pl. 1: 24 (1797).

Syn. *Salicornia europaea* auctt. Non L., Sp. Pl. 3 (1753).

Salsola drummondii Ulbr., Nat. Pflanzenfam. 2, 16C: 256 (1934).
Syn. *Salsola obpyrifolia* Botsch & Akhani (1989).
Salsola schweinfurthii Solms-Laub., Bot. Zeit. 59: 173 (1901).
Suaeda aegyptiaca (Hasselq.) Zohary, J. Linn. Soc. Bot. 55: 635 (1957).
Basionym. *Chenopodium aegyptiacum* Hasselq. (1757).
Syn. *Suaeda hortensis* Forssk. ex J.F.Gmel. (1791); *Suaeda baccata* Forss. Ex J.F. Gmelin (1791); *Schanginia hortensis* (Forssk. ex Gmelin) Moq. (1840); *S. aegyptiaca* (Hasselq.) Aellen (1964).
Seidlitzia rosmarinus Ehrenb. ex Boiss., Fl. Or. 4: 951 (1879).
Syn. *Salsola rosmarinus* (Ehrenb. ex Boiss.) Akhani (2007).
Suaeda moschata A.J.Scott, Kew Bull. 36(3) 558 (1981).
Suaeda monoica Forssk. ex J.F.Gmel., Syst. Nat. ed. 1791: 2, 503 (1791).
Sevada schimperi Moq. in DC., Prodr. 13(2): 154 (1849).

CERATOPHYLLACEAE

Ceratophyllum demersum L., Sp. Pl. 992 (1753).

CARYOPHYLLACEAE

Herniaria maskatensis Bornm., Mitth. Thuring. Bot. Vereins 6: 51 (1894).
Polycarpaea spicata Wight & Arn. in Ann. Nat. Hist. ser. 1 (3): 91 (1831).
Polycarpaea jazirensis R.A. Clement, Edinb. J. Bot. 51(1): 53–54 (1994).
Sphaerocoma aucheri Boiss., Fl. Or. 1: 739 (1867).
Xerotia arabica Oliver in Hk., Icon. Pl. 24, t. 2359 (1895).
Polycarpon succulentum J. Gay in Rev. Bot. Bull. Mens. 2: 372 (1846).
Spergularia diandra (Guss.) Heldr. et Sart. in Heldr., Herb. Graec. Norm. no. 492 (1855).
Syn. *Arenaria diandra* Guss. (1827).
Spergularia marina (L.) Gris., Spic. 1: 213 (1843).
Syn. *Arenaria rubra* L. var. *marina* L. (1753).

CONVOLVULACEAE

Cressa cretica L. Sp. Pl. 223 (1753).
Ipomoea pes-caprae (L.) R.Br., Narr. Exped. R. Zaire 477 (1818).
Basionym. *Convolvulus pes-caprae* L. (1753).

CYMODOCEACEAE

Halodule uninervis Boiss., Fl. Or. 5: 24 (1882).
Syringodium isoetifolium (Asch.) Dandy, J. Bot. 77: 116 (1939).
Thalassodendron ciliatum (Forssk.) Hartog, Verh. Kon. Ned. Akad. Wet., Afd. Nat. Sect. 2, 59(1): 88 (1970).

CYNOMORIACEAE

Cynomorium coccineum L., Sp. Pl. 2: 970 (1753).

CYPERACEAE

Cyperus arenarius Salzm. Ex Steud., Syn. Pl. Glumac. 2(7): 46 (1854) publ. (1855).
Cyperus conglomeratus Vahl, Enum. Pl. 2, 334 (1805).
Cyperus laevigatus L., Mant. 179 (1771).
Schoenoplectus littoralis Palla, Sitz. Zool.-Bot. Ges.Wien. 38: 49 (1888).

FABACEAE

Alhagi graecorum Boiss., Diagn. Pl. Or. Ser. 1, 9: 114 (1848).
Syn. *A. maurorum* DC. (1825) non Medik.
Lotus garcinii DC., Prodr. 2: 212 (1825).
Taverniera lappacea (Forssk.) DC., Prodr. 2: 339 (1852).
Basionym. *Hedysarum lappaceum* Forssk. (1775).
Taverniera spartea (Burm.f.) DC., Prodr. 2: 339 (1852).
Basionym. *Hedysarum spartium* Burm.f. (1768).

Crotalaria saltiaina T.Anders., Bot. Rep. T. 648 (1812).

FRANKENIACEAE

Frankenia pulverulenta L., Sp. Pl. 332 (1753).

HYDROCHARITACEAE

Halophila ovalis (R.Br.) Hook.f., Bot. Antart. Voy. III, 2: 45 (1858).

Halophila stipulacea Asch., Sitz. Ges. Naturf. Freunde Berlin 3(1867).

Najas flexilis (Willd.) Rostk. & W.L.E.Schmidt, Fl. Sedin. 382 (1824).

Najas graminea Delile Descript. Egypte, Hist. Nat. 2: 282 (1813).

Najas marina L., Sp. Pl. 2: 1015 (1753).

JUNCACEAE.

Juncus rigidus Desf., Fl. Atlant. 1: 312 (1798).

Juncus acutus L., Sp. Pl. 1: 325 (1753).

LILIACEAE

Dipcadi biflorum Ghaz., Kew Bull. 51(4): 805 (1996).

MIMOSACEAE

Acacia tortilis (Forssk.) Hayne, Arzneigew. 10: I, t. 31 (1827).

Basionym. *Mimosa tortilis* Forssk. (1775).

Prosopis cineraria (L.) Druce, Rep. Bot. Soc. Exch. Cl. Brit. Isles 1913, 3: 422 (1914).

Basionym. *Mimosa cineraria* L. (1753).

OROBANCHACEAE

Cistanche phelypaea (L.) Cout., Fl. Portugal: 571 (1913).

Basionym. *Lathraea phelypaea* L. (1753).

Syn. *Orobanche tinctoria* Forssk. (1775); *Phelypaea tubulosa* Schrenk (1840); *Cistanche tubulosa* (Schrenk) Hook.f. (1884); *Cistanche tinctoria* (Forssk.) Beck (1904).

PLUMBAGINACEAE

Limonium axillare (Forssk.) Kuntze, Rev. Gen. Pl. 2: 395 (1891).

Syn. *Statice axillaris* Forssk. (1775).

Limonium carnosum (Boiss.) O. Kuntze, Rev. Gen. Pl. 2: 395 (1891).

Syn. *Statice carnosum* Boiss. (1879).

Limonium cylindrifolium Verdc. ex Cufod., Bull. Jard. Bot. Natl. Belg. 30 (Suppl.) 661 (1960).

Limonium milleri Ghaz. & J.R.Edm., Edinb.J. Bot. 60(1): 15 (2003).

Limonium sarcophyllum Ghaz. & J.R.Edm., Edinb.J. Bot. 60(1): 13 (2003).

Limonium stocksii (Boiss.) Kuntze, Rev. Gen. Pl. 2: 396 (1891).

Syn. *Statice arabicum* Jaub. & Spach (1844); *S. stocksii* Boiss. in DC. (1848); Boiss. (1879).

POACEAE

Aeluropus lagopoides (L.) Trin. Ex Thwaites, Enum. Pl. Zeyl.: 374 (1864).

Syn. *Aeluropus littoralis* auct. non (Gouan) Parl.

Aristida abnormis Chiov., Pirotta, Fl. Eritrea 48 (1903).

Arundo donax L., Sp. Pl., 1: 81 (1753).

Echinochloa crusgalli (L.) P.Beauv., Ess. Agrostogr.: 53: 161 (1812).

Basionym. *Panicum crusgalli* L. (1753).

Halopyrum mucronatum (L.) Stapf, Hook.f., Icon. Pl. 25: t. 2448 (1896).

Basionym. *Uniola mucronatum* L. (1762).

Panicum antidotale Retz., Observ. Bot. (Retzius) iv. 17 (1786).

Panicum turgidum Forssk., Fl. Aegypt.-Arab. 18 (1775).

Paspalidum desertorum (A.Rich.) Stapf, Fl. Trop. Afr. 9(4): 585 (1920).

Basionym. *Panicum desertorum* A. Rich. (1850).

Paspalum distichum L., Syst. Nat. Ed. 10, 2: 855 (1759).

Paspalum vaginatum Sw., Prodr.: 21 (1788).

Phragmites australis (Cav.) Trin. ex Steud., Nomencl. Bot. Ed. 2, 2: 324 (1841).

Syn. *P. communis* Trin. (1820); *Arundo donax* Forrsk., (1775) non L.

Sporobolus consimilis Fresen., Mus. Senckenberg. 2: 140 (1837).

Sporobolus helvolus (Trin.) T.Durand & Schinz., Consp. Fl. Afric. 5: 820 (1895).
Basionym. *Vilfa helvola* Trin. (1837).
Sporobolus ioclades (Nees ex Trin.) Nees, Fl. Afric. Austr. Ill. 1: 161 (1841).
Syn. *S. arabicus* Boiss., (1853); *S. jemenicus* Pilg. Ex Schwartz (1939); *S. kentrophyllus* (K.Schum.) Calyton (1997).
Sporobolus spicatus (Vahl.) Kunth, Revis. Gramin. 1: 67 (1829).
Syn. *Agrostis virginica* Forssk. (1775).
Sporobolus virginicus (L.) Kunth, Revis. Gramin. 1: 67 (1829).
Urochondra setulosa (Trin.) C.E.Hubb., Hook., Icon. Pl. 35: t. 3457 (1947).
Basionym. *Vilfa setulosa* Trin. (1840).

PORTULACCACEAE
Portulaca oleracea L., Sp. Pl. 445 (1753).

POTAMOGETONACEAE
Potamogeton pectinatus L., Sp. Pl. 127 (1753).

RHIZOPHORACEAE
Rhizophora mucronata Lam., Encycl. 6(1): 189 (1804).
Bruguiera gymnorrhiza (L.) Sav., Ecycl. 4: 696 (1798).

RUPPIACEAE
Ruppia maritima L., Sp. Pl. 1: 127 (1753).

SALVADORACEAE
Salvadora persica L., Sp. Pl. 1: 122 (1753).

TAMARICACEAE
Tamarix mascatensis Bunge, Tentamen 60 (1852).
Tamarix aphylla (L.) G. Karsten, Deutsch. Fl.: 641 (1882).
Syn. *Thuja aphylla* L. (1755) p.p.; *Tamarix orientalis* Forssk. (1775); *T. articulata* Vahl (1791), nom. illegit.
Tamarix aucheriana (Decne.) Baum, Monogr. Rev. Tamarix: 148 (1978).
Syn. *Trichaurus aucherianus* Decne. ex Walpers (1843); *T. passerinoides*, Boiss. (1867), non Del. ex Desv.

TYPHACEAE
Typha domingensis Pers., Syn. Pl. 2(2): 532 (1807).

ZANNICHELLIACEAE
Zannichellia palustris L., Sp. Pl. 969 (1753).

ZYGOPHYLLACEAE
Fagonia indica Burm.f., Fl. Indica 102, t. 34, f.1 (1768).
Fagonia luntii Bak., Kew Bull. 1894: 330 (1894).
Fagonia ovalifolia Hadidi, Fl. Iran. 98: 2, t.1, (1972).
Fagonia schweinfurthii (Hadidi) Hadidi, Oester. Bot. Z. 121: 272 (1973).
Syn. *Fagonia arabica* Edgeworth & Hook. F. (1874) non L.; *F. indica* Burm.f. var. *schweinfurthii* Hadidi (1972).
Nitraria retusa Asch., Verh. Biot. Prov. Barndenberg 18: 94 (1876).
Tetraena alba (L.f.) Beier & Thulin, Pl. Syst. Evol. 240: 35 (2003).
Basionym. *Zygophyllum album* L.f. (1762).
Tetraena hamiensis (Schweinf.) Beier & Thulin, Pl. Syst. Evol. 240: 35 (2003).
Basionym. *Zygophyllum hamiensis* Schweinf. (1899).
Tetraeana qatarensis (Hadidi) Beier & Thulin, Pl. Syst. Evol. 240(1–4): 36 (2003).
Basionym. *Zygophyllum qatarense* Hadidi (1978); *Z. coccineum* auct., non Linn.; *Z. smithii* Hadidi, nom. nud.; *Z. hamiensis* var. *qatarense* (Hadidi) Jac. Thomas & Chaudhary (2001).
Tetraena simplex (L.) Beier & Thulin, Pl. Syst. Evol. 240: 36 (2003).
Basionym. *Zygophyllum simplex* L. (1767).
Tribulus arabicus Hosni, Bot. Not. 130: 261. (1977).
Syn. *T. omanensis* Hosni (1978).

References

Abbas J (2002) Plant communities bordering the sabkhat of Bahrain Island. In: Barth H-J, Böer B (eds) Sabkha ecosystems vol. 1: the Arabian Peninsula and adjacent countries. Kluwer Academic, Dordrecht, pp 51–62

Abbas JA, El-Oqlah AA (1992) Distribution and communities of halophytic plants in Bahrain. J Arid Environ 22:205–218

Abdel-Razik MS (1991) Population structure and ecological performance of the mangrove *Avicennia marina* (Forssk.) Vierh. On the Arbian Gulf coast of Qatar. J Arid Environ 20:331–338

Abdel-Razik MS, Ismail AM (1990) Vegetation composition of a maritime salt marsh in Qatar in relation to edaphic factors. J Veg Sci 1:85–88

Abed AM (2002) An overview of an inland sabkha in Jordan: the Taba Sabkha in southern Wadi Araba. In: Barth H-J, Böer B (eds) Sabkha ecosystems vol. 1: the Arabian Peninsula and adjacent countries. Kluwer Academic, Dordrecht, pp 83–98

AGEDI (2013) Systematic conservation planning assessments and spatial privatizations for the Emirate of Abu Dhabi, the United Arab Emirates and the Arabian Peninsula. Abu Dhabi

Akani H, Edwards G, Roalson EH (2007) Diversification of the Old World Salsoleae s.l. (Chenopodiaceae): molecular phylogenetic analysis of nuclear and chloroplast data sets and a revised classification. Int J Plant Sci 168(6):931–956

Al-Ateeqi S (2014) Phytoremediation of oil-polluted desert soil in Kuwait using native plant species. PhD thesis, unpublished. University of Glasgow, UK

Al-Gifri AN, Gabali SA (2002) The coastal sabkhat of Yemen. In: Barth H-J, Böer B (eds) Sabkha ecosystems vol. 1: the Arabian Peninsula and adjacent countries. Kluwer Academic, Dordrecht, pp 141–146

Al Khulaidi AA (2013) Flora of Yemen. The Sustainable Natural Resource Management Project (SNRMP II) EPA and UNDP, Republic of Yemen

Al Khulaidi AA, Miller AG, Furley P (2010) Environmental and human determinates of vegetation distribution: in the Hadhramaut region. LAP Lambert Academic Publishing, Saarbrücken, p 420

Al-Taisan WA (2009) Suitability of using *Phragmites australis* and *Tamarix aphylla* as vegetation filters in industrial areas. Am J Environ Sci 5(6):740–747

Al-Turki TA, Omer S, Ghafoor A (2000) A synopsis of the genus Atriplex L. (Chenopodiaceae) in Saudi Arabia. Feddes Repert 111(5–6):261–293

Aleem AA (1979) A contribution to the study of seagrasses along the Red Sea coast of Saudi Arabia. Proc Saudi Biol Soc 3:113–136

APG IV (2016) An update of the Angiosperm Phylogeny Group classification for the orders and families of flowering plants: APG IV. Bot J Linn Soc 181:1–20

Aronson J (1989) HALOPH; salt tolerant plants for the world – a computerized global data base of halophytes with emphasis on their economic uses. University of Arizona Press, Tucson

Aspinall S (1995) Why the Socotra Cormorant, Phalacrocorax nigrogularis should be protected. Tribulus 5(2):10–12

Aspinall S (1996a) Status and conservation of the breeding birds of the United Arab Emirates. Hobby, Liverpool/Dubai

Aspinall S (1996b) Time for a protected area network in the UAE. Tibulus 6(1):5–9

Auffret M, Labbé D, Thouand G et al (2009) Degradation of a mixture of hydrocarbons, gasoline, and diesel oil additives by Rhodococcus aetherivorans and Rhodococcus wratislaviensis. Appl Environ Microbiol 75(24):7774–7782

Aziz S, Khan MA (1996) Seed bank dynamics of a semiarid coastal shrub community in Pakistan. J Arid Environ 34:81–87

Babikir AA (1984) Vegetation and envrironment on the coastal sand, dunes and playas of Khor El-Odaid Area, Qatar. Geo J 9:377–385

Babikir AA, Kürschner H (1992) Vegetational patterns within a coastal saline of NE-Qatar. Arab Gulf J Sci Res 10:61–75

Baldwin R (1996) Marine reptiles. In: Vine PJ (ed) Natural Emirates – wildlife and environment of the United Arab Emirates. Trident Press Ltd., London, pp 136–149

Baldwin R, Al Kiyumi A (1999) The ecology and conservation status of sea turtles of Oman. In: Fisher M, Ghazanfar SA, Spalton JA (eds) The natural history of Oman: a festschrift for Michael Gallagher. Backhuys Publishers, Leiden, pp 89–98

Barth HJ (2002) The sabkhat of Saudi Arabia – an introduction. In: Barth H-J, Böer B (eds) Sabkha ecosystems vol. 1: the Arabian Peninsula and adjacent countries. Kluwer Academic, Dordrecht, pp 37–50

Baskin CC, Baskin JM (1998) Seeds: ecology, biogeography, and evolution of dormancy and germination. Elsevier, San Diego

Basson PW, Burchard JE, Hardy JT, Price ARG (1977) Biotopes of the Western Arabian Gulf. Aramco, Dhahran

Batanouny KH (1981) Ecology and flora of Qatar. Qatar University Press, Doha

Batanouny KH, Turki AA (1983) Vegetation of South-Western Qatar. Arab Gulf J Sci Res 1:5–19

Bhat NR, Suleiman MK, Shahid SA (2004) Mangrove, *Avicennia marina*: establishment and growth under the arid climate of Kuwait. Arid Land Res Manag 18(2):127–139

Böer B (1994) Status, environmental factors and recovery of the intertidal and terrestrial vegetation between Ras as-Zaur and Abu Ali Island after the Gulf war oil spill. In: Establishment of a marine habitat and wildlife sanctuary for the Gulf region. Final report for phase II. CEC/NCWCD, Frankfurt/Jubail, pp 229–253

Böer B (1996) Trial planting of mangroves (*Avicennia marina)* and salt marsh plants (*Salicornia europaea*) in oil-impacted soil in Jubail area, Saudi Arabia. In: Krupp F, Abuzinada AA, Nader JA (eds) A marine wildlife sanctuary for the Arabian Gulf.

National Commission for Wildlife Conservation and Development, Riyadh, pp 186–192
Böer B (1997) An introduction to the climate of the United Arab Emirates. J Arid Environ 35:3–16
Böer B (2002) The coastal and sabkha flora of the United Arab Emirates. Short communication In: Barth HJ, Böer B (eds) 2002: Sabkah ecosystems vol. I: the Arabian peninsula and adjacent countries. Afghanistan, Pakistan, Iran, Jordan, Kuwait, Saudi Arabia, Bahrain, Qatar, United Arab Emirates, Oman, Yemen, Egypt, Sudan, Eritrea, Ethiopia, Djibouti, Somalia. Tasks for Vegetation Science 36, Kluwer Academic Publishers, pp 303–309
Böer B (2004) Halophyte development in the Gulf Arab countries – UNESCO Doha's activities 2001–2003. Trop Ecol 45(1):187–189
Boer B, Al Hajiri S (2002) The coastal and sabkha flora of Qatar: an introduction. In: Barth H-J, Böer B (eds) Sabkha Ecosystems vol. 1: the Arabian Peninsula and adjacent countries. Kluwer Academic, Dordrecht, pp 63–70
Böer B, Gliddon D (1998) Mapping of coastal ecosystems and halophytes (case study of Abu Dhabi, United Arab Emirates). Mar Freshw Res 49(4):297–301
Böer B, Saenger P (2006) The biogeography of the coastal vegetation of the Abu Dhabi gulf coast. In: Khan A, Böer B, Kust GS, Barth H-J (eds) 2006: Sabkah ecosystems vol. II: West and Central Asia. Tasks for vegetation science 42. Springer, Heidelberg, pp 31–36. 259p
Böer B, Warnken J (1992) Qualitative analysis of the coastal and inland vegetation of the Dawkat ad-Dafi and Dawkat al-Mussalamiya region. In: Establishment of a marine habitat and wildlife sanctuary for the Gulf region. Final report for phase I. CEC/NCWCD, Frankfurt/Jubail, pp 81–101
Böer B, Huot C, Sutcliffe M (2014) Floating mangroves: the solution to reduce atmospheric carbon levels and land-based marine pollution? In: Khan MA, Böer B, Öztürk M, Al Abdessalaam TZ, Clüsener-Godt M, Gul B (eds) 2014: Sabkha ecosystems vol IV: cash crop halophyte and biodiversity conservation. Tasks for vegetation science 47. Springer, Heidelberg, pp 327–333. 339p
Botschantzev VP (1984) Two new species of the genus *Salsola* from Saudi Arabia. Bot J 69:686–688
Boulos L (1987) A contribution to the flora of Kuwait. Candollea 42:263–275
Boulos L (1991) A new species of *Salsola* from Oman. Studies in the Chenopodiaceae of Arabia: 3. Kew Bull 46:297–299
Boulos L (1992) Notes on *Agathophora* (Fenzl) Bunge and *Cornulaca* Del. Studies in the Chenopodiaceae of Arabia V. Kew Bull 47:283–287
Brown G, Porembsky S (2000) Phytogenic hillocks and blow-outs as 'safe sites' for plants in an oil contaminated area of northern Kuwait. Environ Conserv 27(3):242–249
Brown G, Böer B, Sakkir S (2008) The coastal vegetation of the western and southern Gulf – characterisation and conservation aspects. In: Abuzinada AH, Barth H-J, Krupp F, Boer B, Al Abdessalam TZ (eds) Protecting the Gulf's Marine Ecosystems from Pollution. Birkhauser Verlag, Basel, pp 23–44
Chapman VJ (1960) Salt marshes and salt deserts of the world. Inter-science Publishers, New York
Chapman RW (1978) In: Al-Sayari SS, Zötl JG (eds) Quaternary period in Saudi Arabia. Spinger, New York
Chaudhary SA (1998) Flora of the Kingdom of Saudi Arabia 1: 14–15. Ministry of Agriculture and Water. National Agriculture and Water Research Centre, Riyadh
Danin A (1983) Desert vegetation of Israel and Sinai. Cana Publishing House, Jerusalem
De Clerck O, Coppejans EV (1994) Status of the macroalgae and seagrass vegetation after the 1991 Gulf war oil spill. Cour Forschungsinst Senckenberg 166:18–21
de Soyza AG, Böer B, Vistro N (2002). Sustainable development of mangroves for coastal sabkhat environments in Abu Dhabi, UAE. In: Barth HJ, Böer B (eds) Sabkah ecosystems vol. I: the Arabian peninsula and adjacent countries. Afghanistan, Pakistan, Iran, Jordan, Kuwait, Saudi Arabia, Bahrain, Qatar, United Arab Emirates, Oman, Yemen, Egypt, Sudan, Eritrea, Ethiopia, Djibouti, Somalia. Tasks for Vegetation Science 36. Kluwer Academic Publishers, Dordrecht
Deil U (1998) Coastal and sabkha vegetation. In: Ghazanfar SA, Fisher M (eds) Vegetation of the Arabian Peninsula. Kluwer Academic, Dordercht, pp 209–228
Deil U, Müller-Hohenstein K (1996) An outline of the vegetation of Dubai (UAE). Verhandlungen der Gesellschaft für Ökologie 25:77–95
El Amry M (1998) Population structure, demography and life tables of Avicennia marina (Forssk.) Vierh. at sites on the eastern and western coasts of the United Arab Emirates. Mar Freshw Res 49(4):303–308
El-Demerdash MA (1996) The vegetation of the Farasan Islands, Red Sea, Saudi Arabia. J Veg Sci 7:81–88
El-Demerdash MA, Hegazy AK, Zilay MA (1995) Vegetation-soil relationship in Tihamah coastal plains of Jazan region, Saudi Arabia. J Arid Environ 30:161–174
El-Hadidi MN (1977) Two new *Zygophyllum* species from Arabia. Publ Cairo Univ Herbarium 7/8:327–331
El-Hadidi MN (1980) On the taxonomy of *Zygophyllum* section *Bipartita*. Kew Bull 35:335–340
El-Keblawy A (2003) Effects of achene dimorphism on dormancy and progeny traits the two ephemerals *Hedypnois cretica* (L.) Dum.-Cours. and *Crepis aspera* L. (Asteracea). Can J Bot 81:550–559
El-Keblawy A (2004) Salinity effects on seed germination of the common desert range grass, Panicum turgidum. Seed Sci Technol 32:873–878
El-Keblawy A (2013) Effects of seed storage on germination of two succulent desert halophytes with little dormancy and transient seed bank. Acta Ecol Sin 33:338–343
El-Keblawy A (2014) Effects of seed storage on germination of desert halophytes with transient seed bank. In:

Khan MA, Böer B, Kust GS, Barth H-J (eds) Sabkha ecosystem IV. Springer, Dordrecht, pp 193–203

El-Keblawy A, Al-Hamadi F (2009) Assessment of the differential response of weeds to soil solarization by two methods. Weed Biol Manage 9:72–78

El-Keblawy A, Al-Shamsi N (2008) Salinity, temperature and light affect seed germination of *Haloxylon salicornicum*, a common perennial shrub of the Arabian deserts. Seed Sci Technol 36:679–688

El-Keblawy A, Bhatt A (2015) Aerial seed bank affects germination behaviour of two small seeded halophytes in the Arabian deserts. J Arid Environ 115:10–17

El-Keblawy A, Al-Ansari F, Hassan N, Al-Shamsi N (2007) Salinity, temperature and light affect germination of *Salsola imbricata*. Seed Sci Technol 35:272–281

El-Keblawy A, Bhatt A, Gairola S (2014) Perianth colour affect germination behavior in the wind pollinated *Salsola rubescens* in the Arabian deserts. Botany 92:69–75

El-Keblawy A, Bhatt A, Gairola S (2015) Storage on maternal plants affects light and temperature requirements during germination in two small seeded halophytes in the Arabian deserts. Pak J Bot 47:1701–1708

El-Keblawy A, Gairola S, Bhatt A (2016a) Maternal habitat affects germination requirements of *Anabasis setifera*, a succulent shrub of the Arabian deserts. Acta Bot Bras 30:35–40

El-Keblawy A, Gairola S, Bhatt A (2016b) Maternal salinity environment affects salt tolerance during germination in *Anabasis setifera*: a facultative desert halophyte. J Arid Land 8:254–263

El-Sheikh AM, Youssef MM (1981) Halophytic and xerophytic vegetation near al Kharj springs. Bull Fac Sci King Saud Univ 12:5–21

El-Sheikh MA, Mahmoud A, El-Tom M (1985) Ecology of the inland salt marsh vegetation at Al-Shiggat in Al-Qassim district, Saudi Arabia. Arab Gulf J Sci Res 3:165–182

El-Shourbagy MN, Al-Eidaros OH, Al-Zahrani HS (1987) Distribution of *Halopeplis perfoliata* (Forssk.) Bge. ex Schweinf. In the Red Sea coastal salt marshes: phytosociological relations and respones to soil. J Coast Res 3:179–187

Fisher M, Membery D (1998) Climate. In: Ghazanfar SA, Fisher M (eds) Vegetation of the Arabian Peninsula. Kluwer Academic, Dordrecht, pp 5–38

Flowers TJ, Colmer TD (2008) Salinity tolerance in halophytes. Tansley Rev – New Phytologist Trust 179:945–963

Flowers TJ, Hajibagheri MA, Clipson NJW (1986) Halophytes. Q Rev Biol 61(3):313–337

Fouda MM, Al-Muharrami MA (1996) Significance of mangroves in the arid environment of the Sultanate of Oman. Oman J Agric Sci 1:41–49

Freitag H (1989) Contributions to the chenopod flora of Egypt. Flora 183:149–173

Freitag H (1991) The distribution of some prominent Chenopidiacae in SW Asia and their phytogeographical significance. Flora et Vegetatio Mundi 9:281–292

Frey W, Kürschner H (1989) Die Vegetation im Vorderen Orient. Erläuterungen zur Karte A VI 1 Vorderer Orient. Vegetation des Tübinger Atlas des Vorderen Orients. Beihefte zum Tübinger Atlas des Vorderen Orients, Reihe A (Naturwissenschaften) Nr. 30. Dr Ludwig Reichert Verlag, Wiesbaden

Gawronski S, Gawronska H (2007) Plant taxonomy for phytoremediation, vol 75. Springer, Dordrecht

Ghazanfar SA (1992) Quantitative and biogeographic analysis of the flora of the Sultanate of Oman. Glob Ecol Biogeogr Lett 2:189–195

Ghazanfar SA (1993) Vegetation of the khawrs and fresh water Springs of Dhofar. Part E. In: Khawrs and Springs of the Dhofar Governorate. Survey and monitoring studies. Unpublished report, Planning Committee for Development and Environment in the Governorate of Dhofar, Oman

Ghazanfar SA (1995) Coastal sabkhas: an analysis of the vegetation of Barr al Hikman. In: Khan MA, Ungar IA (eds) The biology of salt tolerant plants. Department of Botany, University of Karachi, Karachi, pp 277–283

Ghazanfar SA (1998) Water Vegetation. In: Ghazanfar SA, Fisher M (eds) Vegetation of the Arabian Peninsula. Kluwer Academic Press, Dordrecht, pp 229–240

Ghazanfar SA (1999) Coastal vegetation of Oman. Estuar Coast Shelf Sci 49:21–27

Ghazanfar SA (2002) The sabkha vegetation of Oman. In: Barth H-J, Böer B (eds) Sabkha ecosystems vol. 1: the Arabian Peninsula and adjacent countries. Kluwer Academic, Dordrecht, pp 99–108

Ghazanfar SA (2003) Flora of the Sultanate of Oman, vol. 1, Piperaceae-Primulaceae (Text + photo CD-ROM). Scripta Botanica Belgica Series 25. National Botanic Garden of Belgium, Meise, Belgium, 262pp

Ghazanfar SA (2006) Saline and alkaline vegetation of NE Africa and the Arabian Peninsula: an overview. In: Orzturk M, Waisel Y, Khan MA, Gork G (eds) Biosaline agriculture and salinity tolerance in plants. Birkhaeuser Publishing Ltd, Basel, pp 101–108

Ghazanfar SA, Rappenhöner D (1994) Vegetation and flora of the islands of Masirah and Shaghaf, Sultanate of Oman. Arab Gulf J Scientific Res 12(3):509–524

Ghazanfar SA (2007) Flora of the Sultanate of Oman, vol. 2, Crassulaceae-Apiaceae (Text + photo CD-ROM). Scripta Botanica Belgica. National Botanic Garden of Belgium, Meise, Belgium, 220pp

Ghazanfar SA (2011) Restoring saline habitats: indentification and name changes in the halophytes of the Arabian peninsula. In: Özturk M, Mermut AR, Celik A (eds) Urbanisation, land use, land degradation and environment. Springer, Dordrecht, pp 315–329

Ghazanfar SA (2015). Flora of the Sultanate of Oman, vol. 3, Loganiaceae-Asteraceae (Text + photo CD-ROM). Scripta Botanica Belgica. National Botanic Garden of Belgium, Meise, Belgium, 371pp

Ghazanfar SA, Altundag E, Yaprak AE, Osborne J, Tug GN, Vural M (2014) Halophytes of SW Asia. In: Khan MA et al (eds) Sabkha ecosystems: vol IV: cash crop halophyte and biodiversity conservation, tasks for vegetation science 47. ©Springer Science+Business Media, Dordrecht, pp 105–133

Glennie KW (1987) Desert sedimentary environments, present and past: a summary. Sediment Geol 50:135–165

Gubba I, Glennie KN (1998) Geology. In: Ghazanfar SA, Fisher M (eds) Vegetation of the Arabian Peninsula. Kluwer Academic Press, Dordrecht, pp 1–4

Gulzar S, Khan MA (2001) Seed germination of a halophytic grass Aeluropus lagopoides. Ann Bot 87:319–324

Gulzar S, Khan MA, Ungar IA (2001) Effect of salinity and temperature on the germination of Urochondra setulosa (Trin.) C. E. Hubbard. Seed Sci Technol 29:21–29

Guma IR, Padrón-Mederos MA, Santos-Guerra A, Reyes-Betancort JA (2010) Effect of temperature and salinity on germination of Salsola vermiculata L.(Chenopodiaceae) from Canary Islands. J Arid Environ 74:708–711

Gunster A (1992) Aerial seedbanks in the central Namib: distribution of serotinous plants in relation to climate and habit. J Biogeography 563–572

Gutterman Y (1994) Strategies of seed dispersal and germination in plants inhabiting deserts. Bot Rev 60:373–425

Halwagy R (1986) On the ecology and vegetation of Kuwait. In: Kürschner H (ed) Contributions to the vegetation of Southwest Asia. Beihefte zum Tübinger Atlas des Vorderen Orients, Reihe A (Naturwissenschaften) Nr. 24. Dr Ludwig Reichert Verlag, Wiesbaden, pp 81–109

Halwagy R, Halwagy M (1977) Ecological studies on the desert of Kuwait. III. The vegetation of the coastal salt marshes. J Univ Kuwait (Sci) 4:33–73

Halwagy R, Moustafa AF, Kamal S (1982) On the ecology of the desert vegetation in Kuwait. J Arid Environ 5:95–107

Heathcote JA, King S (1998) Umm as Samim, Oman: a sabkha with evidence for climatic change. In: Alsharhan AS, Glennie KW, Whittle GL, Kendall CGSC (eds) Quaternary deserts and climatic change. Balkema, Rotterdam

Heywood VH, Brummitt RK, Culham A, Seberg O (2007) Flowering plant families of the world. Royal Botanic Gardens, Kew. 424 pp

Huiskes AHL, Schat H, Elenbaas PFM (1985) Cytotaxonomic status and morphological characterisation of *Salicornia dolichostachya* and *Salicornia brachystachya*. Acta Bot Neerl 34:271–282

Jupp BP, Durako MJ, Kenworthy WJ, Thayer GW, Schillak L (1996) Distribution, abundance and species composition of seagrasses at several sites in Oman. Aquat Bot 53:199–213

Kadereit G, Freitag H (2011) Phylogeny of Camphorosmoideae. Taxon 1:51–78

Kadereit G, Hohmann S, Kadereit JW (2006) A synopsis of Chenopodiaceae subfam. Betoideae and notes on the taxonomy of Beta. Willdenowia 36:9–20

Kadereit G, Mucina L, Freitag H (2006a) Phylogeny of Salicornioideae (Chenopodiaceae): diversification, biogeography, and evolutionary trends in leaf and flower morphology. Taxon 55(3):617–642

Kadereit G, Ball P, Beer S, Mucina L, Sokoloff D, Teege P, Yaprak AE, Freitag H (2007) A taxonomic nightmare come true: phylogeny and biogeography of glassworts (Salicornia L., Chenopodiaceae). Taxon 56(4):1134–1170

Kassas M, Zahran MA (1967) On the ecology of the Red Sea littoral saltmarsh, Egypt. Ecol Monogr 37:297–316

Khan MA (1990) The relationship of seed bank to vegetation in a saline desert community. In: Sen DN, Mohammed S (eds), Marvel of seeds. Proceeding of international seed symposium, Jodhpur, pp 87–92

Khan MA, Gul B (2006) Halophyte seed germination. In: Ecophysiology of high salinity tolerant plants. Springer, Dordrecht, pp 11–30

Khan M, Ungar I (1997) Effects of thermoperiod on recovery of seed germination of halophytes from saline conditions. Am J Bot 84:279–279

Khan MA, Weber DJ (1986) Factors influencing seed germination in *Salicornia pacifia* var. utahensis. Am J Bot 73:1163–1167

Khan MA, Gul B, Weber DJ (2000) Germination responses of *Salicornia rubra* to temperature and salinity. J Arid Environ 45:207–214

Kos M, Baskin CC, Baskin JM (2012) Relationship of kinds of seed dormancy with habitat and life history in the Southern Kalahari flora. J Veg Sci 23:869–879

Kozlowski TT, Pallardy SG (2002) Acclimation and adaptive responses of woody plants to environmental stresses. Bot Rev 68:270–334

Kukkonen I (1991) Problems in *Carex* section Physodae and *Cyperus conglomeratus* within the Flora Iranica area. Flora et Vegetatio Mundi 9:63–73

Kürschner H (1986) A study of the vegetation of the Qurm Nature Reserve, Muscat area, Oman. Arab Gulf J Sci Res 4:23–52

Kürschner H, Al-Gifri AN, Al-Subai MY, Rowaished AK (1998) Vegetation pattern within coastal salines of southern Yemen. Feddes Repert

Leonard J (1981–89) Contribution a l'étude de la flore et de la végétation des deserts d'Iran. Fasc. Jardin botanique national de Belgique. Meise, pp 1–9

Macke AJ, Ungar IA (1971) The effects of salinity on germination and early growth of Puccinellia nuttalliana. Can J Bot 49:515–520

Mahmoud A, El Sheikh AM, Abdul Baset S (1983) Germination of two halophytes: Halopeplis perfoliata and Limonium axillare from Saudi Arabia. J Arid Environ 6:87–98

Mahruki AA, Alloway B, Patzelt H (2006) The use of reed-bed technology for treating oil-production waters in the Sultanate of Oman. SPE 98548, Society of Petroleum Engineers International Conference on Health, Safety and the Environment in Oil and Gas Exploration and Production, Abu Dhabi, UAE, 2–4 April 2006

Mandaville JP (1990) Flora of Eastern Saudi Arabia. Kegan Paul International, London, p 482

Mathur N, Joginder S, Sachendra B, Avinash B, Mohnish V, Anil V (2010) Phytoremediation potential of some multipurpose tree species of Indian Thar desert in oil contaminated soil. Adv Environ Biol 4(2):131–137

Miller AG, Cope TA (1996) Flora of the Arabian Peninsula and Socotra, vol 1. Edinburgh Univeristy Press, Edinburgh

Morgan WC, Myers BA (1989) Germination of the salt-tolerant grass *Diplachne fusca*. I. Dormancy and temperature responses. Aust J Bot 37:225–237

Nolvak H, Sildvee T, Knipsalu M, Truu J (2012) Application of microbial community profiling and functional gene detection for assessment of natural attenuation of petroleum hydrocarbons in boreal subsurface. Boreal Environ Res 17(2):113–127

Omar AS (2007) Vegetation of Kuwait: a comprehensive illustrated guide to the flora and ecology of the desert of Kuwait. Kuwait Institute of Scientific Research (KISR), Kuwait

Omar AS, Misak RF, Shahid S (2002) Sabkhat and halophytes of Kuwait. In: Barth HJ, Böer B (eds) Sabkha ecosystems vol. 1: the Arabian Peninsula and adjacent countries. Kluwer Academic, Dordrecht, pp 70–82

Padmavathiamma PK, Ahmed M, Rahman AR (2014) Phytoremediation – a sustainable approach for contaminant remediation in arid and semi-arid regions – a review. Emir J Food Agric 26(9):757–772. https://doi.org/10.9755/ejfa.v26i9.18202

Parsons RF (2012) Incidence and ecology of very fast germination. Seed Sci Res 22:161–167

Pujol JA, Calvo JF, Ramirez-Diaz L (2000) Recovery of germination from different osmotic conditions by four halophytes from southeastern Spain. Ann Bot 85:279–286

Qu XX, Huang ZY, Baskin JM, Baskin CC (2008) Effect of temperature, light and salinity on seed germination and radical growth of the geographically widespread halophyte shrub Halocnemum strobilaceum. Ann Bot 101:293–299

Scott AJ (1981) A new *Suaeda* (Chenopodiaceae) from Dhofar. Kew Bull 36:558

Shaltout KH, El-Halawagy EF, El-Garawany MM (1997) Coastal lowland vegetation of eastern Saudi Arabia. Biodivers Conserv 6:1027–1040

Sharma TP, Sen DN (1989) A new report on abnormally fast germinating seeds of *Haloxylon* spp.: an ecological adaptation to saline habitat. Curr Sci 58:382–385

Sheppard CRC, Price ARG, Roberts CM (1992) Marine ecology of the Arabian region: patterns and processes in extreme tropical environments. Academic, London, p 359

Steliga T (2012) Role of Fungi in Biodegredation of petroleum hydrocarbon in drill waste. Pol J Environ Stud 21(2):471–479

Sukhorukov A, Aellen A, Edmondson J, Townsend C (2016). Chenopodiaceae. In: Ghazanfar SA, Edmondson JR (eds), Flora of Iraq. 5(1): 164–256. Kew Publishing Surrey

Tuan N, Hsieh H et al (2011) Analysis of bacterial degradation pathways for long-chain alkylphenols involving phenol hydroxylase, alkylphenol monooxygenase and catechol dioxygenase genes. Bioresour Technol 102:4232–4240

Ungar IA (2001) Seed banks and seed population dynamics of halophytes. Wetl Ecol Manag 9:499–510

Vesey-Fitzgerald DF (1957) The vegetation of the Red Sea coast north of Jedda, Saudi Arabia. J Ecol 45:547–562

Wei Y, Dong M, Huang ZY, Tan DY (2008) Factors influencing seed germination of *Salsola affinis* (Chenopodiaceae), a dominant annual halophyte inhabiting the deserts of Xinjiang, China. Flora 203:134–140

White F, Léonard J (1991) Phytogeographical links between Africa and Southwest Asia. Flora et Vegetatio Mundi 9:229–246

Xing J, Cai M, Chen S, Chen L, Lan H (2013) Seed germination, plant growth and physiological responses of *Salsola ikonnikovii* to short-term NaCl stress. Plant Biosys 147:285–297

Zahran MA (1975). Biogeography of mangrove vegetation along the Red Sea coast. In: Proceedings of the international symposium on the biology and management of mangroves, Gainsville. pp 45–51

Zalamea PC, Sarmiento C, Arnold AE, Davis AS, Dalling JW (2015) Do soil microbes and abrasion by soil particles influence persistence and loss of physical dormancy in seeds of tropical pioneers? Front Plant Sci 5:799

Zaman AU, Khan MA (1992) The role of buried viable seeds in saline desert community. Bangladesh J Bot 21:1–10

Zia S, Khan MA (2004) Effect of light, salinity and temperature on seed germination of *Limonium stocksii*. Can J Bot 82:151–157

Zohary M (1973) Geobotanical foundations of the Middle East. 2 vols. Gustav Fischer Verlag, Stuttgart

RETRACTED CHAPTER: Desert Halophytes: Alleviating the Pressure on the Scarce Resources of Arable Soil and Freshwater

6

Amram Eshel, Chingiz Alekperov, Israel Oren, Tamar Eilam, and Aviah Zilberstein

Abstract

The utilization of plants for mitigating carbon dioxide accumulation in the atmosphere in Clean Development Mechanism (CDM) projects and biofuel production causes a severe burden on the limited sources of arable land and freshwater. This research is aimed at finding alternative plant types for biomass and biofuel production among desert halophytes. Such plants have the advantages of being naturally adapted to grow under the harsh desert conditions and on nonarable soils irrigated with reclaimed sewage or other types of brackish water. Exceptionally fast growing salt-resistant genotypes were identified among native populations of *Tamarix* in Israel. These may serve for future CDM projects and short rotation forestry for biomass production. Another plant that originated from East Africa, *Euphorbia tirucalli*, was also shown to be able to grow under desert conditions and saline water irrigation. This plant has been named in the literature as a potential source of biofuel.

Keywords

Biofuel · Biomass · Clean Development Mechanism · Desert · Halophytes · Salt tolerance

Introduction

Plants were recognized as an important instrument in mitigating the global climate change by sequestering carbon dioxide in the context of CDM projects and biofuel production as outlined in the Marrakech Accords (2001), which was a result of the Kyoto Protocol (IPCC 2000; Ellis et al. 2007; Fargione et al. 2008). As a result, a worldwide movement of diverting common agricultural assets toward forestation and biofuel production took place. However, after the first surge of activities, the economic realities of the effects of such activities on the scarce resources of arable soil and freshwater and its inevitable effect on the agricultural product markets were realized (Dagoumas et al. 2006; Dalla Marta et al. 2010; Harto et al. 2010; Kullander 2010). This has become especially evident in developing countries where the "low carbon society" concept, promoted following the Kyoto Protocol, has become a difficult target to achieve. Utilization of edible

This chapter has been retracted.
The retraction note to this chapter is available at https://doi.org/10.1007/978-3-030-04417-6_33

A. Eshel · C. Alekperov (✉) · T. Eilam
A. Zilberstein
Department of Molecular Biology and Ecology of Plants, Tel Aviv University, Tel Aviv, Israel
e-mail: chingiza@post.tau.ac.il

I. Oren
The Robert H. Smith Faculty of Agriculture, Hebrew University of Jerusalem, Jerusalem, Israel

B. Gul et al. (eds.), *Sabkha Ecosystems*, Tasks for Vegetation Science VI,
https://doi.org/10.1007/978-3-030-04417-6_6

feedstock such as palm oil and sugarcane for biofuel has disrupted the fragile industry due to the fluctuations of feedstock prices (Goh and Lee 2010). The alternative proposed was the use of the, so-called, "second-generation" crops which produce inedible products and were supposed to grow on marginal soils and saline water. Jatropha (*Jatropha curcas*) had a central position among these plants, but a thorough study of its economic and environmental feasibility as an energy crop had not been performed beforehand. Later it was found that the salinity of soil and limited freshwater sources restricted the production of jatropha oil, which required large amounts of water during the cultivation of young plants (Maes et al. 2009; Goh and Lee 2010). Instead of the sugar and starch crops that were competing with food and feed crops for high-quality land, the cultivation of lignocellulosic crops on marginal and set-aside lands was considered a more environmentally sound and sustainable option for renewable energy production (Fargione et al. 2008; Carroll and Somerville 2009; Frigon and Guiot 2010). In a previous article (Eshel et al. 2010), we have shown that desert halophytes adapted to grow on nonarable soils under extreme desert conditions can be used as new sources for energy crops. Such uses of halophytes will not compete with conventional agriculture for valuable resources of fertile soil and freshwater (Ruan et al. 2010). In order to obtain the high yields necessary for an economically viable operation, the plants received plentiful amounts of reclaimed sewage and brackish water. Under such conditions the old-world desert trees *Tamarix* spp. yielded 52 to 26 ton/ha/y of organic biomass. Another desert plant, *Euphorbia tirucalli* was mentioned in the literature as a potential biofuel plant about 30 years ago and was estimated then to be a highly economical source of biofuel (Nielsen et al. 1977; Calvin 1980; Duke 1983). The origin of *E. tirucalli* is the arid regions of East Africa, yet it was grown as an ornamental plant in other parts of the world. Here we report on a continuation of this research aimed at identifying salt-resistant *Tamarix* types that will be highly efficient as a source of biomass when grown under desert conditions and irrigated with saline water. We also include the first results of growing *E. tirucalli* under such nonagricultural environment.

Materials and Methods

Tamarix Collection and Field Experiment

In order to find salt-adapted trees of *Tamarix* spp., we collected cuttings from natural populations in salt-affected habitats along the Mediterranean coast of Israel and in inland habitats along the Jordan valley, by the Dead Sea, and the Aravah desert (Fig. 6.1). They were grown in a common garden experimental field located in Yotvata at the southern part of the Aravah Valley in the Negev desert in Israel (29°53′43.90″ N 35°04′[illegible]4.[illegible]5″ E). The local air temperatures vary between a monthly average of 40 °C in the hot summer month (August) and 20 °C in the winter (January). Average relative humidity is 20% and annual rainfall 25 mm. Average potential evapotranspiration is 11 mm/day. The soil is coarse alluvial aridisol. Irrigation with water from a local desalination plant (EC [illegible] ~ 12 dS/m) was applied daily through drip irrigation system at a rate that compensated for evapotranspiration (average 10 mm/day). Ten cuttings of each accession were planted directly to the field plot in the spring (February–March) of 2008. Tree height was measured every 6 months in order to identify the fastest-growing trees. In August 2010, a single tree of nine selected accessions was felled, and its biomass was measured.

Tamarix Sand Culture Experiment

A sand culture experiment was carried out at Tel Aviv University campus. The sand culture experiment was designed to determine the comparative response of the tested lines to the salt treatments. Obviously other growth conditions were not equal to those in the Negev because of the differences of climatic conditions between the two locations. Cuttings from both fast- and slow-growing accessions grown in the field experiment at Yotvata were planted in 2-L plastic pots filled with coarse sand. For a 6-week period, all pots were irrigated with control solution (tap water +0.75 g/L of combined fertilizer [N20%: P20%:

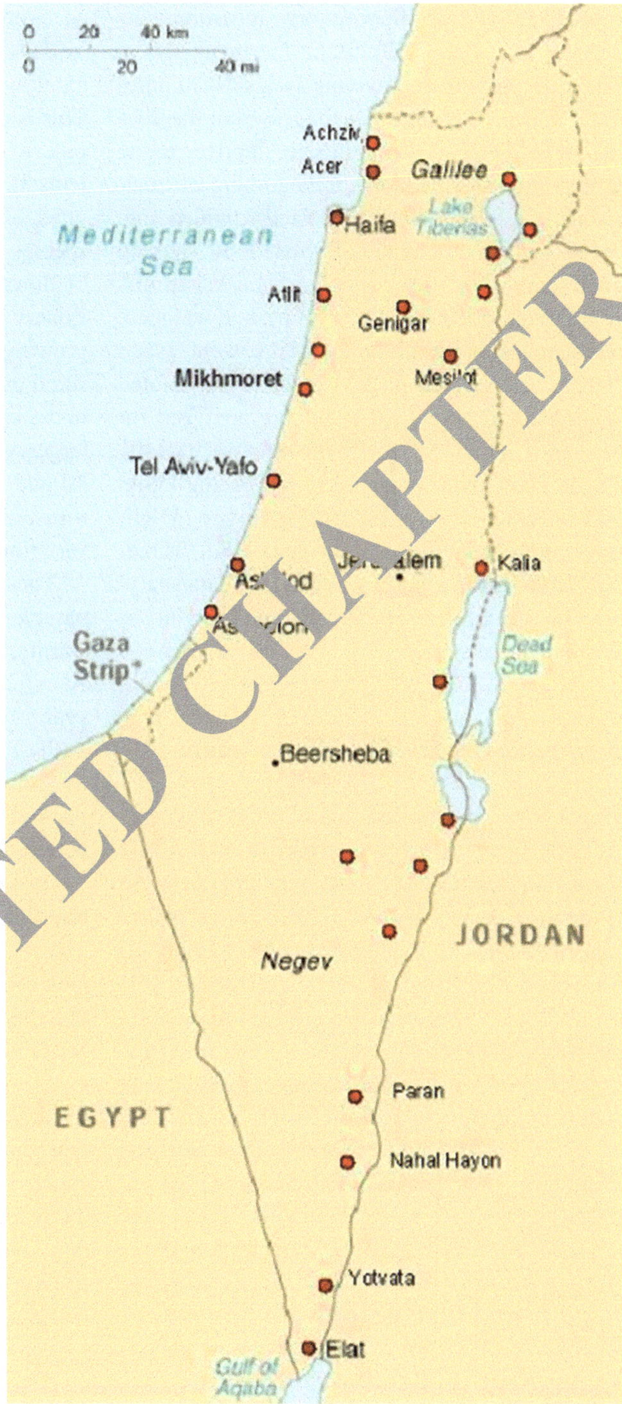

Fig. 6.1 A map of Israel showing the locations of *Tamarix* collection sites along the Mediterranean coast and inland saline and desert habitats. The field experiment is at Yotvata

K20% Haifa fertilizers, Haifa, Israel]). After this initial period, three treatments were applied: fertilizers only (control), 100 mM NaCl plus the fertilizers, and 200 mM NaCl plus the fertilizers (N20: P20: K20). The pots were irrigated every other day by drip irrigation at a rate that prevented salt concentration buildup in the sand. The experiment was set up in five blocks and with a fully randomized design. Two months after beginning of the treatments, pulse amplitude modulated (PAM) fluorometry measurements (PAM2000, Walz, Effeltrich, Germany) were taken on dark adapted young twigs 10 minutes before the measuring, since the leaves of the trees are scale like about 1 mm in size, apprised to the twigs. Four months after the beginning of treatments, the plants were harvested, divided into green and woody material, rinsed by shaking each sample for 30 sec in a 0.5 L beaker of distilled water, dried (80 °C for 48 h), and weighed. Subsamples of dried green twigs were ashed (500 °C), dissolved in 2 M HNO3, and diluted with distilled water. Extracts were analyzed for Na and K ion content by flame photometry (401 Flame Photometer, Corning Inc., Corning USA) and for chloride by a chloride analyzer (MKII, Sherwood Scientific, Cambridge, UK). Three-way analysis of variance was used for data analysis with accession no., treatment, and block no. as the main effects.

Euphorbia tirucalli Field Experiment

Plants were propagated by cuttings collected from bushes growing in the garden at Tel Aviv University campus. The trees were first grown in a nursery located in the Mediterranean coastal plane in Israel (29°22′58.00″ N 34°52′24.00″ E) on xeralf soil irrigated with freshwater (EC 1 ~ 2 dS/m). Propagated material from the nursery was used for establishment of a 0.5 ha experimental plot in the southern part of the Aravah Valley in the Negev desert in Israel, 10 km south of the one used for the *Tamarix*. This plot was drip-irrigated daily with saline sewage (EC 8–12 dS/m).

Results and Discussion

Tamarix spp. Collection and Field Experiment

Our wide-ranging sampling covered the six common *Tamarix* species growing in Israel: *T. aphylla*, *T. aravensis*, *T. jordanis*, *T. nilotica*, *T. palaestina*, and *T. tetragyna*. Cuttings of 65 accessions were planted in a common garden in Yotvata under extreme desert conditions and irrigated with brine from a local desalination plant. This is the only use found for these waters which otherwise would cause severe salinization problems if poured at a nearby site or mixed with other effluents. Systematic identification of the species was based on microscopic examination of flowers collected in the garden, according to Zohary (1976). In order to identify the fastest-growing types within our collection, we measured tree heights every 6 months. As can be seen in the variability within the tested accessions was large. After 6 months, tree height varied between 30 and 180 cm. Heights of the 15 best trees as measured 1 year after planting (Fig. 6.2) showed variation between 180 and 400 cm.

These are extremely rapid growth rates comparable to fast-growing *Eucalyptus* species in warm climates (cf. Callister et al. 2007; Calvo-Alvarado et al. 2007; Watanabe et al. 2009). A year and a half after planting, an average tree of each of nine selected accessions was felled, and its biomass was measured. The results (Fig. 6.3) show that tree weights at that stage varied between 30 and 80 kg.

At a density of 2500 trees per ha, this would translate to 75 to 200 ton/ha (FW) or 25 to 67 ton/ha/y (DM). Further evaluation of these trees under the conditions used in commercial plantations is underway. The response of selected accessions to salinity was tested in a sand culture experiment. The accessions were chosen so they would include both fast- and slow-growing genotypes in order to see whether salt response is associated with growth rate. As can be seen from the data presented in Fig. 6.4, the slow growers (#46, #47, #48) were most salt sensitive and exhibited reduced growth under the salt treatments as compared with the control.

The other accessions showed a small decrease or even a slight increase under the 100 mM treatment and a relatively small decrease under 200 mM salt. The decrease in growth under the salt treatment was not associated with reduction in the efficiency of the photosynthetic system as indicated by the PAM measurements of Fv/Fm (Fig. 6.5).

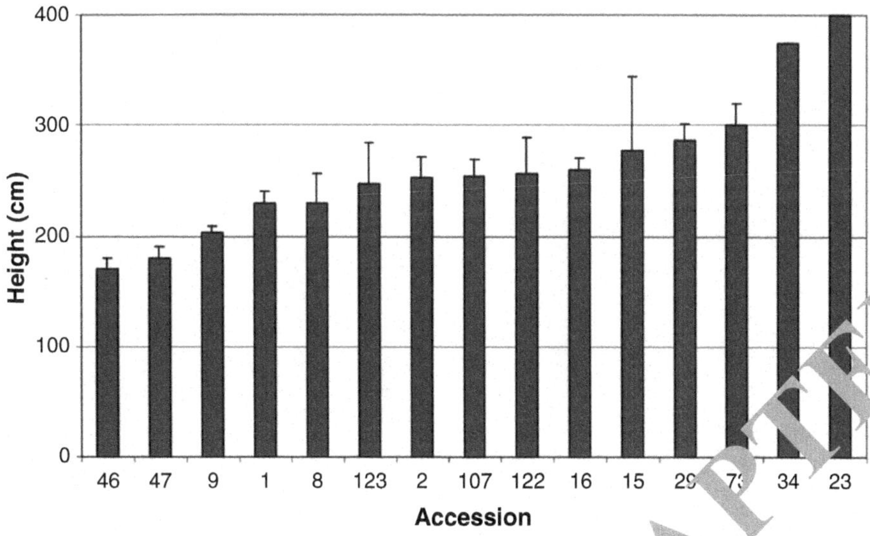

Fig. 6.2 Height (average ± SD of 3 replicates) of *Tamarix* spp. trees. Trees of selected accessions propagated from cuttings in the spring of 2008 and were irrigated with saline water under extreme desert conditions (10–25 mm precipitation per year) at Yotvata, Israel. Measured in Feb. 2009

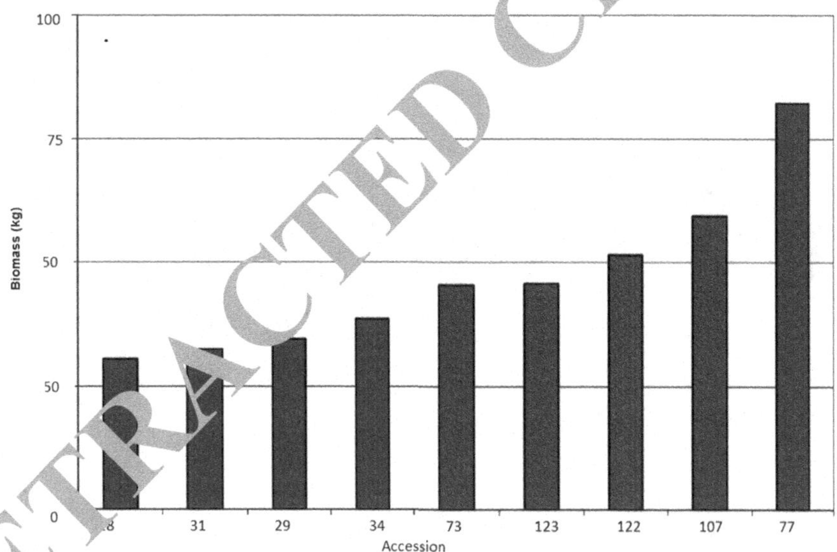

Fig. 6.3 Biomass of single *Tamarix* spp. trees. Trees of selected accessions propagated from cuttings in the spring of 2008 and were irrigated with saline water under extreme desert conditions at Yotvata, Israel. Measured in August 2010

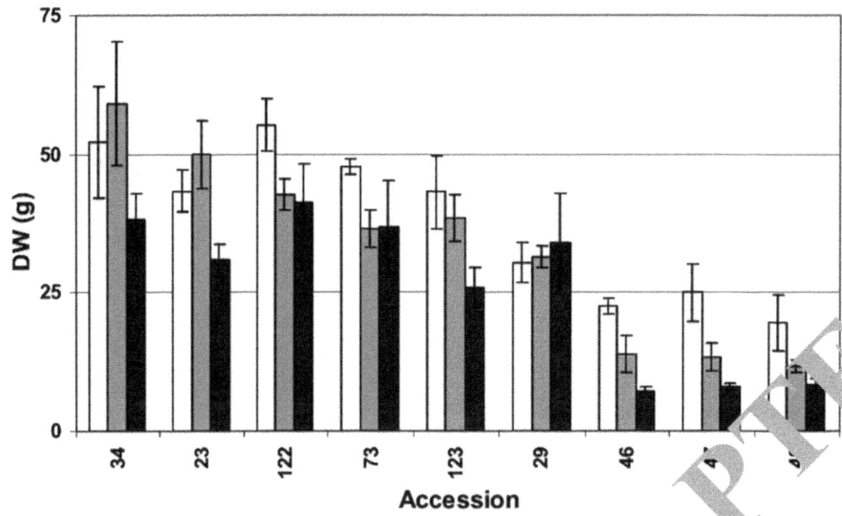

Fig. 6.4 Dry weight per plant (average ± SEM of 5 replicates) of selected *Tamarix* accessions. Plants were grown in sand culture for 4 months under tap water with no added salt [open bars], tap water + 100 mM NaCl [gray bars], and tap water + 200 mM NaCl [black bars]

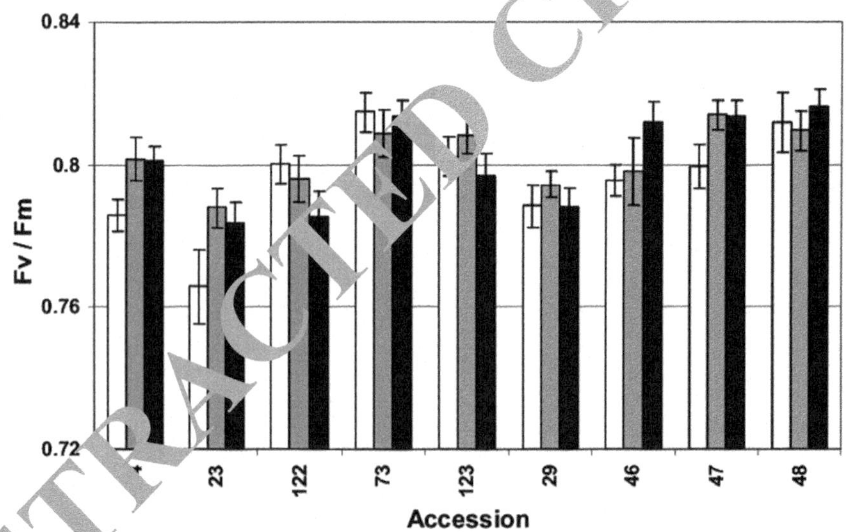

Fig. 6.5 PAM measurements (average ± SEM of 5 replicates) of selected *Tamarix* accessions. Plants were grown in sand culture for 4 months under tap water with no added salt [open bars], tap water + 100 mM NaCl [gray bars], and tap water + 200 mM NaCl [black bars]

On the contrary, even the species that showed a marked decrease in growth rate under salt did not exhibit reduction in this photosynthetic efficiency parameter. As these plants rely on salt secretion (Waisel 1972) as a major adaptation mechanism, Na content did not increase very much under the 100 and 200 mM treatments (Fig. 6.6).

Even the slow-growing lines that were affected severely by the salt did not show excessive Na accumulation. The main difference between the salt-sensitive and the more tolerant lines was in the Na/K ratio. Apparently inhibition of K uptake by the presence of NaCl was stronger in the slow-growing, more sensitive accessions.

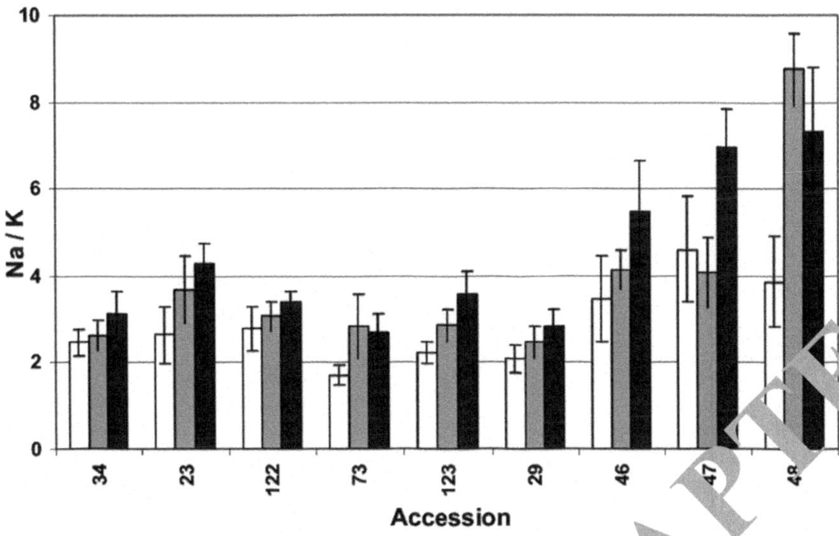

Fig. 6.6 Na ion content (average ± SEM of 5 replicates) of selected *Tamarix* accessions. Plants were grown in sand culture for 4 months under tap water with no added salt [open bars], tap water + 100 mM NaCl [gray bars], and tap water + 200 mM NaCl [black bars]

Euphorbia tirucalli Growth

Cuttings taken from our nursery plantation were planted (500 plants/ha; average cutting weight 0.08 ± 0.01 kg) in the extreme desert condition at the Southern Aravah Valley in May 2009. These plants are amenable to in vitro mass vegetative propagation (Ma et al. 2011). The plot was irrigated with saline sewage from a local cowshed after a primary treatment in a constructed wetland. The salinity of the water was too high for any common agricultural use and was dispersed in the area to evaporate, causing salinization and pollution. The *Euphorbia* bushes grew well under these conditions, and at a sampling in June 2010, a year after planting, average plant weight was 2.3 ± 1.6 kg, showing an average increase in FW of almost 30-fold. This would correspond to a yield of ca 11 ton/ha/y FW or 2.2 ton/ha/y DW. The content of latex and potential biofuel precursors in these desert-grown plants will be determined in the future.

Conclusions

The utilization of desert halophytes as energy crops can alleviate the pressure on the scarce resources of arable soil and freshwater. Such plants have the advantages of being naturally adapted to grow under the harsh desert climatic conditions of high temperature, low humidity, and high radiation intensity, on nonarable soils. Their growth is enhanced when irrigated with reclaimed sewage or other types of brackish water. Intensive studies can reveal, as we have demonstrated here, highly resistant and fast-growing genotypes (Ruan and Teixeira da Silva 2011). The desert environment is characterized by high radiation intensities and high temperatures throughout most of the year; such conditions are conducive for high biomass production. Another benefit of utilizing nonarable desert soils for biomass production is the low amounts of carbon dioxide and nitrogen compounds that will be released factors that threaten to offset the benefits of such operation in tropical rainforests, peatlands, savannas, or grasslands (Fargione et al. 2008).

Acknowledgments This work was supported by funds from the Italian Ministry of Environment, Land and Sea and Boeing Corp., USA.

Appendix

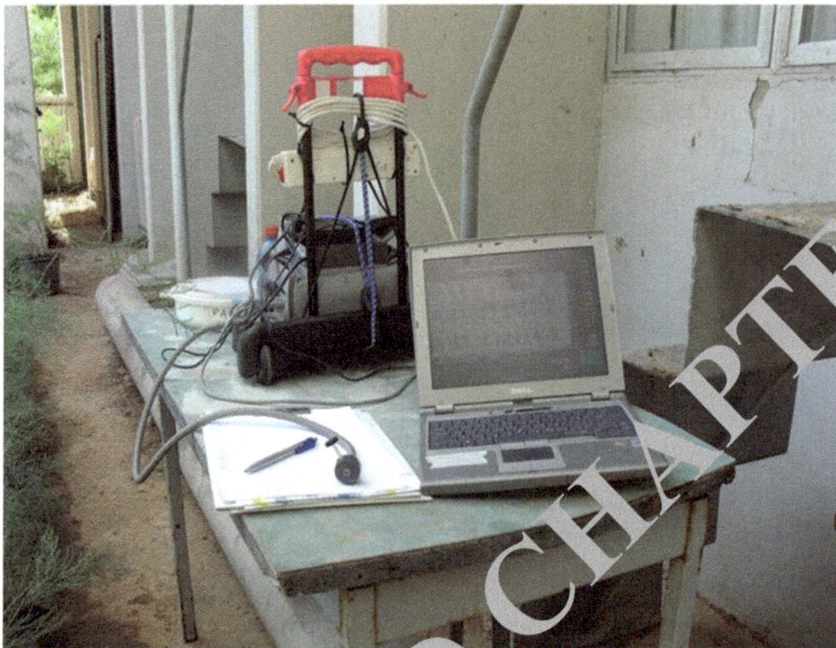

Supplementary Fig. 6.7 Figure of PAM

References

Calvin M (1980) Hydrocarbons from plants: analytical methods and observations. Naturwissenschaften 67:525–533

Callister AN, Ades PK, Arndt SK, Adams MA (2007) Clonal variation in shoot respiration and tree growth of *Eucalyptus* hybrids. Can J For Res 37:1404–1413

Calvo-Alvarado JC, Arias D, Richter DD (2007) Early growth performance of native and introduced fast growing tree species in wet to sub-humid climates of the southern region of Costa Rica. For Ecol Manag 242:227–235

Carroll A, Somerville C (2009) Cellulosic biofuels. Annu Rev Plant Biol 60:165–182

Dagoumas AS, Papagiannis GK, Dokopoulos PS (2006) An economic assessment of the Kyoto protocol application. Energy Policy 34:26–39

Dalla Marta A, Mancini M, Ferrise R, Bindi M, Orlandini S (2010) Energy crops for biofuel production: analysis of the potential in Tuscany. Biomass Bioenergy 34:1041–1052

Duke JK (1983) Handbook of energy crops. Available online: http://www.hort.purdue.edu/newcrop/duke_energy/Euphorbia_tirucalli.html

Ellis J, Winkler H, Corfee-Morlot J, Gagnon-Lebrun F (2007) CDM: taking stock and looking forward. Energy Policy 35:15–28

Eshel A, Zilberstein A, Alekparov C, Eilam T, Oren I, Waisel Y (2010) Biomass production by desert halophytes: alleviating the pressure on food production. World Scientific and Engineering Academy and Society (WSEAS)Recent Advances in Energy & Environment (EE′10) Cambridge, UK, February 23–25, 2010, pp 362–367 Available online: http://www.wseas.us/e-library/conferences/2010/Cambridge/EE/EE-60.pdf

Fargione J, Hill J, Tilman D, Polasky S, Hawthorne P (2008) Land clearing and the biofuel carbon debt. Science 319:1235–1238

Frigon J-C, Guiot SR (2010) Biomethane production from starch and lignocellulosic crops. Biofuels Bioprod Biorefin 4:447–458

Goh CS, Lee KT (2010) Will biofuel projects in Southeast Asia become white elephants? Energy Policy 38:3847–3848

Harto C, Meyers R, Williams E (2010) Life cycle water use of low-carbon transport fuels. Energy Policy 38:4933–4944

IPCC (Intergovernmental Panel on Climate Change) (2000) Special report on land use, land-use change and forestry. Available online: http://www.ipcc.ch/pub/srlulucf-e.pdf

Kullander S (2010) Food security: crops for people not for cars. AMBIO 39:249–256

Ma G-H, Teixeira da Silva JA, Wu G-J (2011) Direct adventitious shoot formation from apical shoot explants of *Euphorbia tirucalli*. J Plant Growth Regul 30(1):114–116

Maes WH, Achten WMJ, Ruebens B, Raes D, Samson R, Muys B (2009) Plant–water relationships and growth strategies of *Jatropha curcas* L. seedlings under different levels of drought stress. J Arid Environ 73:877–884

The Marrakech Accords (2001) Seventh session of the Conference of Parties, 29 October–9 November, Marrakech, Morocco. Available online: http://unfccc.int/cop7/index.html

Nielsen PE, Nishimura H, Otvos JW, Calvin M (1977) Plant crops as a source of fuel and hydrocarbon-like materials. Science 198:942–944

Ruan C-J, Teixeira da Silva JA, Qin P, Lutts S, Gallagher JL (2010) Halophyte improvement for a salinized world. Crit Rev Plant Sci 29(6):329–359

Ruan C-J, Teixeira da Silva JA (2011) Metabolomics: creating new potentials for unraveling mechanisms in response to salt and drought stress and for biotechnological improvement of xero-halophytes. Crit Rev Biotechnol 31(2):153–169

Waisel Y (1972) Biology of halophytes. Academic, New York, 395 pp

Watanabe Y, Masunaga T, Fashola OO, Agboola A, PK O, Wakatsuki T (2009) *Eucalyptus camaldulensis* and *Pinus caribaea* growth in relation to soil physicochemical properties in plantation forests in Northern Nigeria. Soil Sci Plant Nutr 55:132–141

Zohary M (1976) Tamarix. Flora Palaestina, vol II. Israel Academy of Science and Humanities, Jerusalem, pp 350–362

Drone-Based Vegetation Assessment in Arid Ecosystems

7

David Gallacher

Abstract

Proof of long-term vegetation change in arid rangelands is often insufficient to influence policy, even when the change is clear to ecologists. Drones provide a way to collect unbiased evidence of plant spatiotemporal distribution at a dramatically reduced cost for the scales needed in these habitats. Assessment of phytomass spatial distribution by drone has become a routine, but further analysis requires advanced skills in data collection and post-flight processing. Accurate assessment of phytomass temporal change will require protocols to be developed for data collection and analysis. Biodiversity assessment by drone is unreliable, but there is potential for assessing phytomass change within and among taxonomic groups in arid rangelands, by repeatedly sampling areas in which perennial plants have been classified manually.

Keywords

Arid environments · Unmanned aerial vehicle · Grazing

D. Gallacher (✉)
Department of Interdisciplinary Studies, Zayed University, Dubai, United Arab Emirates

Introduction

Proving Vegetation Change

Arid rangelands throughout Asia and the Middle East are modified ecosystems, resembling the ecosystems of past centuries but with sometimes dramatic differences in ecological structure. Timeless wildernesses have become fragmented habitats that face pressures from changing demographics and livestock management. Routine monitoring of these habitats is an essential part of conservation and rehabilitation (Mitchell 2010) but is often too infrequent or insufficient to convincingly prove long-term change. This is particularly true in the Middle East. Overgrazing is often cited as a leading cause of rangeland decline (Batanouny 1990; Assaeed 1997; Abed and Hellyer 2001; Ferguson and McCann 1998), but data to prove to stakeholders the extent that palatable species have declined is generally lacking. There are many livestock exclosure studies that show a predictable increase in vegetation when protected from hoofed herbivores (Gallacher 2006), but exclosures are artificial habitats. Localized extinction of a plant species is seldom reported, if ever, and the impact of reduced vegetation on native fauna is insufficiently quantified or discussed. Rangeland decline may seem obvious to ecologists who measure health in terms of biodiversity, but it isn't obvious to pastoralists who are interested in livestock numbers and

B. Gul et al. (eds.), *Sabkha Ecosystems*, Tasks for Vegetation Science VI,
https://doi.org/10.1007/978-3-030-04417-6_7

livelihoods. Decline occurs over a timeframe of decades and is therefore detectable in the living memory of only the oldest members of a society.

Initial surveying and routine monitoring of biodiversity is one of the key tasks of rangeland management (Holechek et al. 2010). Any protected area with a management plan and a core budget should therefore have biodiversity data, but many protected areas in the Middle East are lacking data in sufficient detail to detect long-term ecological change. Reasons for this lack of data are varied:

- Lack of a detailed management plan that is routinely followed
- Cost of maintaining a trained botanist or of attracting a suitable botanist to a remote location
- A land management agenda that is influenced by both anthropocentric and ecocentric goals

Local communities, and sometimes even protected area managers, often favor grazing and hunting rights (anthropocentric) over the preservation of natural ecosystems (ecocentric). This reduces the political motivation to fund basic ecological research, as well as the ability to act on its findings. Ground-based (traditional) monitoring is more expensive in arid rangelands than other habitats. Time cost is high because vegetation is both sparse and spatially heterogeneous. Sampling plots must therefore be large to include a sufficiently representative number of plants, numerous enough across sometimes enormous reserves to sample all habitats, and frequent enough within habitats to include spatially heterogeneous vegetation. These factors combined means that an effective ground-based monitoring requires an enormous amount of walking.

Drones Add Value

Ecological monitoring involves studying the distribution of organisms over space and time. Different ecological processes occur at different spatiotemporal scales; tree clustering occurs at a larger spatial scale than lizard territory definition, and tree populations change at a larger temporal scale than biennial grasses. Spatial monitoring has previously been conducted at the scale of ground-based observation, or satellite (sometimes light aircraft) imagery, but the former is expensive and the latter inaccurate (Walton et al. 2013). Temporal monitoring over longer periods is fraught with inaccuracies due to changes in observer bias, method, and the purpose for the observation. Comparing current observations to written or oral records can be reliable for individual species when such records exist but is less reliable for providing proof of changes to an ecological system. Low-altitude aerial data capture is more objective than ground-based methods, which are influenced by the botanical knowledge, preferences, and integrity of the observer (Donahue 1999).

Drones supply information in a previously difficult-to-reach spatiotemporal range and therefore have great potential for the study of spatial ecology (Anderson and Gaston 2013). Inexpensive, backpack-sized drones can survey a 1 km^2 sample area to 2 cm ground resolution. Larger areas and finer resolutions are possible, but the volume of data produced can be daunting. In this spatial scale, data acquisition by drone is less invasive and more time efficient than ground-based methods and provides greater detail with less atmospheric interference than other aerial methods. Temporally, data acquisition by drone is more flexible than other aerial methods, though is still limited by adverse weather. Data can be collected more often since the cost of another drone flight is marginal. Hence, annual surveys may become quarterly, and intraday monitoring becomes feasible.

Photographic data is more convincing to non-scientists. It is not influenced by the ability and bias of the botanist, though it is also less taxonomically detailed. For determining phytomass distribution at the hectare scale, drone data is both more reliable and more convincing to outsiders. Land surface imagery can become a permanent, indisputable record of rangeland status at a single point in time. Image quality will likely improve over time, but this change can be accounted for in time series analyses, whereas changes in human bias and competence cannot.

Imaging Desert Vegetation Is New

The spatial characteristics of desert vegetation are quite different to habitats that are more commonly monitored from the air. Arid rangeland contains a low density of phytomass that is heterogeneously distributed in response to soil type, water availability, and sand movement. Vegetation is dominated by woody perennial species, though annuals and non-woody perennials may contribute significantly to herbivore diets. Most plant canopies are spatially separated from other plant canopies by bare ground. Physical and chemical herbivory defenses are pronounced, such that much of the phytomass present in an area may not be available to a specific herbivore.

There has been minimal use of data from satellites or conventional aircraft to assess arid rangeland vegetation. Data collection cost is high and benefits obscure, when compared to other habitats. Vegetation indices such as NDVI (normalized difference vegetation index) have been impractical for several reasons:

- Accuracy is poor. Measurement error is substantial when overall phytomass readings are low, thus obscuring ecologically significant change.
- Financial drivers are absent. Phytomass evaluation is often supported by rural industries, such as seasonal grazing rights, beekeeping access, or recreational hunting.
- Anthropocentric value is obscured. The majority of phytomass in some habitats is contained in tree canopies or chemically defended plant tissue. Since livestock consume palatable phytomass within reach, a change in phytomass within these habitats may not correlate with changed grazing potential.
- Ecocentric value is poorly understood. A small change in phytomass that is unavailable to livestock may have a large impact on one or more native fauna, but this effect is difficult to quantify.

The application of drones for agriculture and horticulture has received much attention worldwide since 2012, but methods are not directly applicable to ecological study. Imagery is mainly used to identify growth inconsistencies within a monoculture, with the aim of delivering a faster and more targeted response to patches of poorer growth. Reduced growth can be detected by NDVI regardless of whether caused by predation (disease, vermin), water (excess or deficit), inconsistent sowing, or any other factor. Yield estimates, necessary for postharvest management, can be improved with the same data and also by plant height measures.

Software for analyzing drone-acquired data is yet to be refined for ecological purposes (Crutsinger et al. 2016). Current software can evaluate plant cover (percent of bare ground) and provide NDVI estimates. However, NDVI is a relative measure of plant growth, not an absolute measure of phytomass, that is most suited to the spatial analysis of a single species at a single phenological stage. Changes in the spatiotemporal distribution of multiple species are possible, but not yet straightforward with existing commercial software. In arid systems, the dominance of perennial plants makes it feasible to track changes in individual plants over time.

Requirements

Skills

Using drone-acquired imagery to answer ecological questions requires a knowledge of ecology, drones, and post-flight processing. The task of selecting, operating, and maintaining a drone was until recently a major barrier. In 2012 the hardware and software were commercially available to produce orthorectified imagery from drones, but the learning curve was prohibitively daunting to anyone without a background in electronics or geographic information systems (GIS). Today the commercial options have vastly improved, and operators need only a basic knowledge of drone hardware and operational software to achieve the same result.

It is now relatively easy for an ecologist to purchase and operate a system (drone, software, computer) to observe spatial distribution of habitats, vegetation, and some animals. This simplifies some tasks, such as obtaining population estimates of clustered animals and quantifying the effect of a disturbance (e.g., road, tourist campsite) on phytomass. However, many potential ecological applications still require a deeper knowledge of post-flight processing. An example is assessing vegetation change over time, which is confounded by numerous sources of data collection error, and requires the curation of large data sets.

Drone-based monitoring requires a team that has each of these skills and good communication among members. The ability to assemble a successful team is probably the greatest current limitation to applying this technology to ecological work.

Aerial Hardware

Research of drones for ecology prior to 2012 used larger fixed-wing aircraft deployed with a catapult launcher. These drones still dominate research in inhospitable environments such as for Arctic wildlife surveys. Since 2012 a booming market demand for small multirotor drones drove rapid improvements in the <5 kg size category. The market for larger size categories was muted by legal restrictions. Multirotor drones that fit in a backpack are highly portable, easy to use, and quick to deploy, and their range is sufficient for most purposes. Nevertheless, the future may see the rise of other drone classes. This author has previously predicted the emergence of regionally based midsized drones that produce data for sale, just as satellites do today (Gallacher 2016).

Drone hardware can be thought of as a combination of sensor and platform. Sensors are most often cameras within and beyond the visible light range, but other types exist, such as the lasers that are present in LIDAR and atmospheric sensors. Platforms have mostly been either multirotor (helicopter style) or fixed-wing (airplane style), but again there are exceptions, such as lighter-than-air (balloon style) and hybrids. In general, multirotor drones are more stable while fixed wing drones have a longer range. Improvements in platform/sensor combinations continue to be announced regularly, and it would be unwise to predict combinations that will dominate a few years from now.

Analysis

Producing orthorectified composite images of land surfaces from drone-acquired images has become a routine and can be done with NDVI data as easily as with visible light. Cloud-based services are available and a good option if analyses are infrequent. Otherwise, analysis requires sufficient processing speed and memory capacity which is also available retail, at cost. Good results are possible without prior GIS knowledge. The end product is a higher-resolution equivalent of satellite data.

There is currently no software package for systematically addressing ecological questions beyond the orthorectified composite image. Further analysis requires the use of multiple software packages and therefore an understanding of the assumed inputs and outputs. This is complex. Converting files between raster (dot matrix) and vector (points, lines, and polygons) formats incurs losses of accuracy, and comparing different raster files also incurs assumptions of the data. Clustering algorithms collate points in a raster file that share common features, based on underlying assumptions. The optimal way to handle data for the sake of generic ecological questions is yet to be established.

Here I will discuss three broad areas of vegetation/ecological analysis that are not yet standardized and thus require expertise to be performed well: temporal vegetation change, volume/height assessment, and automated recognition.

Temporal Vegetation Change

Assessing the change in vegetation over time is straightforward when the changes are large, as may occur in a revegetation program or in assessing the size change of an easily detectable habitat, such as a patch of bare ground. These changes can be visualized by importing composite images into a GIS program as separate layers. Assessing change becomes more complex when monitoring for subtle changes, particularly in arid habitats with low phytomass.

NDVI readings can be significantly affected by light and shadow conditions. Readings are likely to be maximized on cloudless days when the sun is directly overhead, though the author has not seen any research to quantify the change. Standardizing the time of day is not always possible due to the constraints of weather (cloud, wind, rain), and standardizing across seasons is even more difficult. White balance of images can, and should, be adjusted in post-flight processing, by calibrating to a whitish object that is present in images from each flight.

The spatial distribution of vegetation in arid rangelands enables plants to be measured individually. Size distribution and survival could then be monitored over a square kilometer containing perhaps several thousand plants. Change in the relative dominance of species could be traced by classifying each perennial shrub. While all this is technically possible, it would currently require moving among four software packages to (1) produce the composite images, (2) align different composite images together, (3) identify plant positions and sizes via a clustering algorithm, and (4) collate the information into a database.

Volume/Height Assessment

Adding height information to canopy size data improves the estimation of canopy surface area. Height is significant for some ecological processes, such as forage access of ground-based herbivores and nesting sites for birds. However, height estimation by GPS is less accurate than horizontal, due to satellites being aboveground only. Consequently, height measures should be calibrated against a known height, such as a tree, a pole, or a custom-placed object.

Volume measurements of vegetation are tricky to obtain and equally difficult to interpret. Photogrammetry software creates three-dimensional shapes by creating a point cloud from points seen in more than one image. This works well for buildings with clearly defined edges, but not for plants that contain thousands of similar-looking points, each swaying in the breeze. Nevertheless, with some perseverance, it is possible to obtain 3D images of plant communities that are reasonably accurate. The next step is to interpret the ecological significance of a change in canopy volume.

Automated Recognition

Algorithms to extract and classify objects from a mountain of data have the potential to revolutionize some forms of ecological work. Automated sifting through millions of images for the two or three containing evidence of a rare species (plant, animal, or artifact) would make viable the use of drones to monitor populations at risk of extinction. Automated plant recognition has been attempted, with some success for classifying larger plants to family and in recognizing micro-environments. The task is easier in arid systems due to the relatively low biodiversity.

Object-based image analysis was developed for handling satellite data but is computationally demanding. Pixels are clustered according to similarity, and clusters are then classified by color consistency, size, and shape. Both visible and non-visible light values of a plant component change with lighting conditions, phenology and water status. Hence classification rule sets must be broad if they are to be applied at different times and in different habitats. The application of artificial intelligence (AI) algorithms to drone imagery is set to remove these limitations. Machine- and deep-learning AI algorithms are still computationally demanding to train, but once developed can be run with standard computers on endless datasets. AI chipsets can be mounted on drones for in-flight object recognition. Postflight AI promises to achieve the discriminatory power of human observation when sufficiently trained. Ecologists can expect to see new hardware/software combinations become available that will enable aerial counting of organisms and artifacts, but at time of writing the application of AI to drone imagery is still in its infancy.

Detecting Vegetation Change

Vegetation monitoring primarily involves the within-season estimation of phytomass and the multi-season assessment of plant community structure (Holechek et al. 2010). Phytomass responds to season, weather, and herbivory; therefore an increase does not necessarily mean a restored ecological system. Persistent ecological stresses cause a change in community structure. For example, intense grazing reduces palatable species (decreasers) but favors species shunned by herbivores (increasers), which face reduced competition for water and nutrients. A recent comprehensive study identified 26 biophysical indicators of sustainability (Mitchell 2010), a subset of which are sufficient for substantial, if not comprehensive, monitoring within a specific habitat (Booth and Tueller 2003).

Phytomass

Phytomass cannot be measured directly by UAV, but it is rarely measured directly in ground-based monitoring either, since to do so requires time-consuming destructive sampling of vegetation. Proxy measures include:

- Cover, or its inverse, bare ground. This measure is particularly useful for assessing soil stability from wind or water erosion.
- NDVI, which is an approximation of chlorophyll based on its absorption spectrum, where red light is absorbed but near infrared (NIR) light is not (the "red edge"). Cameras vary enormously on the specificity and accuracy of the red and NIR bands. Other vegetation indexes also exist.
- Height. Vegetation height can be measured by LIDAR or assessed from point clouds. Techniques using LIDAR have become well established to service insurance of commercial forestry.

NDVI is widely used but is yet to be verified for ecological purposes at the spatiotemporal scales that drones make available. Historically, the index was developed as a way to create habitat maps from satellite data and then adapted for crop monitoring from drone data. For habitat maps, chlorophyll distribution is measured across plant communities (multiple individuals of multiple species and phenological stages), which gives a good indication of spatial plant productivity at this scale. For crops, chlorophyll distribution is measured across monocultures (multiple individuals of a single species and phenological stage), which identifies areas of slower growth. In contrast, ecological systems contain multiple species at multiple phenological scales, which drones can measure to the level of individual plant. Spatial variation in NDVI at this scale may represent factors other than phytomass and growth rates.

Basic research is needed to understand how well NDVI values can be used to predict phytomass (or growth rates) among different species, under different measuring conditions, and at different plant phenologies. Shaded vegetation and soil moisture also affect readings.

Plant Diversity

It is unlikely that a reliable system for evaluating plant diversity by drone will emerge in the foreseeable future. Plant identification from composite drone images is significantly less accurate than ground-level identification (Gallacher et al. 2016). Nevertheless, desert ecosystems are more suitable than other habitats since plant classification accuracy is inversely proportional to biodiversity within a size class (Féret and Asner 2012). Positive reports of classification have focused on large monospecific vegetation patches and fixed plant phenological stages and are inconsistent across scales (Ustin and Santos 2010). Three-dimensional shape data could improve classification but is computationally expensive (Hung et al. 2012).

Drones can survey large areas with greater frequency and lower cost when compared to ground-based methods. Imagery is suitable for assessment of ground cover and broad vegetation categories (Laliberte 2012). Therefore, the reduced accuracy of plant diversity measurement could be compensated by larger samples observed more frequently. Drone data can therefore be used to evaluate diversity of broad plant classes or to track the spread of easily identifiable species.

Phytomass Within Plant Taxa

Arid rangelands are dominated by perennial woody shrubs and trees which can be measured individually in drone-acquired aerial data. It is unnecessary to reclassify perennial plants across multiple surveys; thus ground-truthing is only needed for plants that have appeared since the previous survey. Therefore, it is feasible to monitor vegetation change within and among perennial species over a wide area (e.g., 1 km^2), with non-woody species treated as a single group. The resulting measurements would have much richer ecological meaning. However, while technically possible, it is currently difficult and untested (see previous section "Temporal Vegetation Change").

The output of this proposed method would be a database of individual perennial plants, each classified taxonomically. Repeated size measurements, such as canopy size, plant height, and average canopy NDVI, could be added to the database periodically (e.g., every 1/4, 1, or 5 years). Interpretation of the data would include the long-term assessment of absolute and relative abundance of perennial plants, including assessments of survival and life span.

Conclusion

Drones have great potential for advancing our understanding of spatiotemporal vegetation ecology in desert ecosystems. The visual imagery collected by drones is less affected by observer bias and is more convincing evidence for nonscientists than tables and figures of data. At the spatial scales that are relevant to arid rangeland vegetation, drone-acquired data is far cheaper than satellite, manned aircraft, and ground-based methods.

Spatial distribution of vegetation can be visualized with relative ease, using off-the-shelf hardware and post-flight software. Showing a change in distribution over time is equally straightforward if the change is dramatic, such as with a revegetation program.

Automated biodiversity assessment using drones is currently unfeasible. Object recognition is computationally expensive and requires specialized training of the software. Ground-based observation is necessary for accurate assessment of plant diversity.

Assessment of phytomass change within and among taxonomic groups is feasible in arid rangelands due to the dominance of perennial woody species, though research is needed to develop protocols in data collection and analysis. The data produced would have much greater ecological significance than is currently available by any method. Relative and absolute abundance of perennial plant species provides information on the species directly and also on animal species that require specific plants within their habitat. Size distributions also inform survival and life span.

Acknowledgments I dearly thank my colleagues Tamer Khafaga and Greg Simkins of the Dubai Desert Conservation Reserve for their invaluable support and advice. I also thank Zayed University for financial support.

References

Abed I, Hellyer P (2001) United Arab Emirates: a new perspective. Trident Press Ltd, London

Anderson K, Gaston KJ (2013) Lightweight unmanned aerial vehicles will revolutionize spatial ecology. Front Ecol Environ 11(3):138–146. https://doi.org/10.1890/120150

Assaeed AM (1997) Estimation of biomass and utilization of three perennial range grasses in Saudi Arabia. J Arid Environ 36(1):103–111. https://doi.org/10.1006/jare.1996.0200

Batanouny KH (1990) Rangeland ecology of the Arab gulf countries. In: Halwagy R, Taha FK, Omar SA (eds) Proceedings of the 1st international conference on range management in the Arabian Gulf. Kegan Paul International Ltd, London, pp 33–55

Booth DT, Tueller PT (2003) Rangeland monitoring using remote sensing. Arid Land Res Manag 17(4):455–467. https://doi.org/10.1080/713936105

Crutsinger GM, Short J, Sollenberger R (2016) The future of UAVs in ecology: an insider perspective from the Silicon Valley drone industry. J Unmanned Veh Syst 4(3):161–168. https://doi.org/10.1139/juvs-2016-0008

Donahue, DL (1999). The Western Range Revisited: Removing Livestock from Public Lands to Conserve Native Biodiversity. Oklahoma: University of Oklahoma Press

Féret J-B, Asner GP (2012) Semi-supervised methods to identify individual crowns of lowland tropical canopy species using imaging spectroscopy and LiDAR. Remote Sens 4(8):2457–2476. https://doi.org/10.3390/rs4082457

Ferguson M, McCann I, Manners G. (1998) Less water, more grazing. ICARDA Caravan (ICARDA). http://agris.fao.org/agris-search/search.do?recordID=QV2002000053

Gallacher DJ (2016) Drones to manage the urban environment. Risks, rewards, alternatives. J Unmanned Veh Sys. https://doi.org/10.1139/juvs-2015-0040

Gallacher DJ, Hill JP (2006) Effects of camel grazing on the ecology of small perennial plants in the Dubai (UAE) Inland Desert. J Arid Environ 66(4):738–750. https://doi.org/10.1016/j.jaridenv.2005.12.007

Gallacher DJ, Khafaga T, Mahmoud Ahmed T, Shabana HA (2016) Plant species identification via drone images in an arid shrubland. In Iwaasa A, Schellenberg M, Willms W, Larson K. (Ed.) Proceedings of the 10th International Rangeland Congress, pp 981–982. Saskatoon, Canada: Society for Range Management. https://doi.org/10.13140/RG.2.1.4755.5047

Holechek JL, Pieper RD, Herbel CH (2010) Range management: principles and practices, 6th edn. Prentice Hall, Boston

Hung C, Bryson M, Sukkarieh S (2012) Multi-class predictive template for tree crown detection. ISPRS J Photogramm Remote Sens 68:170–183. https://doi.org/10.1016/j.isprsjprs.2012.01.009

Laliberte A (2012, Spring) Mapping rangelands with unmanned aircraft. Professional Surveyor. Retrieved from http://www.profsurv.com/magazine/article.aspx?i=71086

Mitchell JE (2010) Criteria and indicators of sustainable rangeland management. University of Wyoming, Cooperative Extension Service. Retrieved from http://www.sustainablerangeland.org/pdf/SM56.pdf

Ustin SL, Santos MJ (2010) Spectral identification of native and non-native plant species. In Proceedings of ASD and IEEE GRS; art, science and applications of reflectance spectroscopy symposium, Vol. II (p. 17). Boulder, CO.

Walton KM, Spalinger DE, Harris NR, Collins WB, Willacker JJ (2013) High spatial resolution vegetation mapping for assessment of wildlife habitat. Wildl Soc Bull 37(4):906–915. https://doi.org/10.1002/wsb.344

Salt-Induced Changes in Growth and Damage Avoidance Mechanisms of Hydroponically Grown Chinese Kale (*Brassica alboglabra* L.)

8

Amin Tayebi-Meigooni, Yahya Awang, Alan R. Biggs, and Ali Ghasemzadeh

Abstract

Accretion of nonessential ions such as sodium and chloride due to the selective ion-uptake process into the root and/or the use of poor-quality irrigation water may adversely affect the production potential of crops such as Chinese kale in closed-loop hydroponic systems. This research identifies the effects of salt on the plant's growth and the damage avoidance mechanisms in foliage of hydroponically grown Chinese kale. Twenty-one-day-old plants (at the 4-leaf stage) were exposed to salinity (0, 25, 50, 75, and 100 mM NaCl) for 16 days. Results indicated that growth parameters, leaf water status, photosynthetic pigments, and photosystem II efficiency declined as salinity increased in the rhizosphere, whereas proline biosynthesis was stimulated progressively. Salinity induced higher H_2O_2 endogenesis and lipid peroxidation while triggering antioxidant enzymes activity in the leaf tissues. Activity of ascorbate peroxidase and peroxidase gradually increased with salinity, although peroxidase activity was decreased at high salinity (>75 mM NaCl). Catalase, a crucial component of the free radical neutralization process, was also observed at high salinity (>75 mM NaCl). Chinese kale plants were tolerant to salinity at 25 mM and employed a complex mechanism composed of an excess energy dissipation system along with the enzymatic antioxidant system, thus making it relatively tolerant to salinity beyond 75 mM NaCl.

Keywords

Salinity · Chlorophyll fluorescence · Oxidative stress · Lipid peroxidation · Antioxidant enzymes

Introduction

Salinity is one of the major environmental factors limiting plant productivity worldwide. It affects plant growth and development through osmotic and ionic disruption (Munns and Tester 2008). Accumulation of toxic ions such as sodium and chloride in plant tissues can adversely affect plant

A. Tayebi-Meigooni (✉) · Y. Awang
A. Ghasemzadeh
Department of Crop Science, Faculty of Agriculture, Universiti Putra Malaysia,
Serdang, Selangor, Malaysia
e-mail: yahya_awg@putra.upm.edu.my; alighasemzadeh@upm.edu.my

A. R. Biggs
Davis College of Agriculture, Natural Resources and Design, West Virginia University,
Morgantown, WV, USA
e-mail: alan.biggs@mail.wvu.edu

B. Gul et al. (eds.), *Sabkha Ecosystems*, Tasks for Vegetation Science VI,
https://doi.org/10.1007/978-3-030-04417-6_8

biomass production partially due to restriction in the uptake of essential elements (Ghoulam et al. 2002). At higher salinity, ionic cytotoxicity may be the predominant cause of crop susceptibility due to replacement of Na^+ instead of K^+ and binding Na^+ and Cl^- ions with amino acids (D'Souza and Devaraj 2010) that can negatively affect major processes including photosynthesis, protein and lipid metabolisms, as well as energy conservation (Parida and Das 2005). Imbalanced metabolism under such condition may result in oxidative stress (D'Souza and Devaraj 2010).

Limited CO_2 availability under stress-induced stomatal closure restricts carbon reduction, which implies lower NADP accessibility to act as the specific electron acceptor from photosystem I. This could trigger O_2 reduction to superoxide by the Mehler reaction and initiate chain reactions that generate more noxious reactive oxygen species (ROS). Singlet oxygen as an unusual product might also be generated at photo system II owing to ineffective energy dissipation processes (Sekmen et al. 2007; Gill and Tuteja 2010).

Accelerated ROS formation causes oxidative damage to biomolecules such as membrane lipids, proteins, and nucleic acids and may lead to plant death (D'Souza and Devaraj 2010; Merati et al. 2014). Plants undergo some changes in their normal metabolism to cope with the unbalanced conditions caused by salinity in their environment (Daneshmand et al. 2010). They have evolved specific cellular and subcellular preventative-protective complexes consisting of an energy dissipation mechanism and an antioxidative system to mitigate oxidative damage. For example, superoxide dismutase (SOD) converts superoxide (O^{2-}) to H_2O_2 in various subcellular compartments. H_2O_2 is eliminated by ascorbate peroxidase (APX) and catalase (CAT) with a lower affinity than APX (Asada 2006; Sekmen et al. 2007). Different classes of peroxidase (POX) also are associated with scavenging H_2O_2 generated in chloroplasts and physiological processes such as growth and cell formation (Sekmen et al. 2007). The capability to neutralize ROS and reduce their harmful effects on macromolecules is an important trait of stress tolerance in higher plants (Xiong and Zhu 2002) (Fig. 8.1).

Plants exposed to osmotic stress may accumulate low molecular weight compatible solutes such as proline, which is considered to have multiple roles in osmoregulation, protein and membrane protection, energy conservation, and free radical scavenging, as well as being a component of stress-related signaling processes (Daneshmand et al. 2010; Tayebi-Meigooni et al. 2014).

In closed-loop hydroponic systems, increasing salinity can be caused by differential ion uptake by the crop itself and/or use of low-quality irrigation water containing nonessential ions such as sodium and chloride and may negatively affect salt-sensitive hydroponically grown plants. Chinese kale (*Brassica alboglabra* L.) can be successfully planted in hydroponic systems and be used as fresh or cooked vegetables in the human diet. Recent research shows a high level of potential anticancer capability of Chinese kale due to high levels of glucosinolates and ascorbic acid (Hecht et al. 2004). Despite the vast literature on salinity, no attention has been particularly devoted to investigate the physio-biochemical mechanisms contributing to salinity tolerance in Chinese kale. Therefore, the aim of this study was to determine the effect of NaCl salinity on subsequent changes on growth, leaf water status, energy harvesting/dissipating complex, and elucidation of antioxidative defense mechanisms. Results obtained in this study may contribute to a better understanding of the responses of Chinese kale plants to salinity and breeding program development.

Materials and Methods

Planting Materials and Growth Conditions

The seeds of Chinese kale (*B. alboglabra* L.) cv. 'Standard Kailan' were sown in trays maintained with frequent water applications until seedling emergence. Young seedlings were then placed in a plastic film pool and fed with basic nutrient

Fig. 8.1 Effect of salinity on growth of Chinese kale plants. Plants were subjected to different concentrations of NaCl as indicated for 12 days

solution containing 250, 67, 239, 160, 30, 80, 3, 0.62, 0.44, 0.02, 0.11, and 0.048 mg L^{-1} of N, P, K, Ca, Mg, S, Fe, Mn, B, Cu, Zn, and Mo, respectively.

Salt Stress Treatment

Fifteen-day-old seedlings of uniform size were installed in a static aerated hydroponic system containing the basic nutrient solution. Six days later, plants at the 4-leaf stage were exposed to salinity by adding 0, 25, 50, 75, and 100 mM NaCl to the basic solution. Osmotic shock was avoided by gradually increasing the NaCl salt (25 mM each day) until reaching the desired concentrations. Salt treatments were continued for 16 days. Plants grown on nutrient solution without NaCl served as the control.

Collecting Samples

Five plants from each experimental unit were used to estimate growth parameters, including leaf surface area as well as the dry mass of leaf, shoot and root tissues. Fresh samples for biochemical assays were taken from new fully expanded leaves and frozen immediately in liquid N on-site and kept at −80 °C. Leaf relative water content (RWC) was also determined using fresh leaf samples.

Determination of Relative Water Content

The relative water content (RWC) was measured according to the following formula described by Smart and Bingham (1974).

$$\text{RWC} = [(\text{fresh weight} - \text{dry weight}) / (\text{turgid weight} - \text{dry weight})] \times 100 \quad (8.1)$$

Determination of Chlorophyll Fluorescence

Two weeks after initiation of salinity increase, maximum efficiency of photosystem II (Fv/Fm) in dark-adapted leaves (new fully expanded) of five plants for each experimental unit was randomly measured using a portable Plant Efficiency Analyzer (Hansatech, UK). Measurements were done at 10.00am-12.00md with the approximate solar irradiance 1100 to 1300 μmolm^{-2} s^{-1}. Fv/Fm was calculated according to Schreiber et al. (1995).

Determination of Photosynthetic Pigments

Chlorophylls *a* and *b* and carotenoids were extracted in 95% (v/v) acetone and calculated using the following relationships described by Lichtenthaler and Buschmann (2001).

$$\text{Chlorophyll}a\left(\mu g/\text{ml}\right)=11.24A_{661.6}-2.04A_{644.8} \quad (8.2)$$

$$\text{Chlorophyll}b\left(\mu g/\text{ml}\right)=20.13A_{644.8}-4.19A_{661.6} \quad (8.3)$$

$$\text{Carotenoids}\left(\mu g/\text{ml}\right)=\left(1000A_{470}-1.90Ca-63.14Cb\right)/214 \quad (8.4)$$

Determination of Proline

Proline content was measured according to the procedure of Bates et al. (1973). Leaf samples (0.2 g) were homogenized in 3 ml sulfosalicylic acid (3% w/v) and then centrifuged. The supernatant was gently mixed with glacial acetic acid and ninhydrin solution, by volume in a ratio of 1:1:1, respectively, and then incubated at 95 °C for 1 h. The cooled mixture was then treated with toluene, and the absorbance of the upper phase (chromophore-containing) was read at 520 nm. Proline content was estimated using a standard curve and was expressed as μmol g^{-1} fresh weight.

Determination of Hydrogen Peroxide and Lipid Peroxidation

Content of hydrogen peroxide was measured based on Velikova et al. (2000). Fresh leaf samples (0.5 g) were homogenized in 5 ml 0.1% (w:v) trichloroacetic acid (TCA) and centrifuged at 12,000 g for 15 min. 0.5 ml (10 mM) phosphate buffer and 1 ml (1 M) KI were added to 0.5 ml of the supernatants. The absorbance of the final solution was read at 390 nm. The content of H_2O_2 was estimated using a standard curve.

Lipid peroxidation was estimated by measuring the level of malondialdehyde (MDA) according to the method of Du and Bramlage (1992). Leaf samples (0.25 g) were homogenized in 5 ml of 0.1% (w/v) (TCA) solution and centrifuged at 15000 × g for 15 min. The supernatant (1.0 ml) was added to 4 ml 0.5% (w/v) thiobarbituric acid (TBA) in 20% TCA (w/v). The mixture was then heated at 95 °C for 30 minutes and immediately cooled to stop the reaction. The absorbance of the colored solution was recorded at 532 nm and corrected for nonspecific turbidity by subtracting the absorbance at 600 nm. MDA concentrations were calculated from the extinction coefficient of 155 mM^{-1} cm^{-1} using the following formula.

$$\text{MDA}\left(\mu\text{mol/g fresh weight}\right)=\left[\left(A_{532}-A_{600}\right)/155\right]\times10^{3}\times\text{Dilution factor} \quad (8.5)$$

Extraction for Protein and Enzyme Assays

The frozen samples (0.2 g) were homogenized with 3 ml ice-cold extraction buffer (50 mM potassium phosphate buffer, pH 7.8, 0.5% (w/v) PVP, 0.1 mM EDTA, and 0.2 (v/v) triton X-100) and centrifuged at 18,000 × g for 20 min at 4 °C. Extraction media for ascorbate peroxidase (APX) was prepared with inclusion of ascorbate (2 mM) in the main buffer. The supernatants were used as crude extract in the protein and enzymes assays.

Assay of Protein and Antioxidant Enzymes

Total protein content was determined using bovine serum albumin (BSA) as a standard following the method described by Bradford (1976).

Catalase (CAT) activity was determined following the method of Cakmak et al. (1993). The assay mixture contained 25 mM phosphate buffer (pH 7.0), 10 mM H_2O_2, and enzyme extract (100 μl) in a total volume of 1.0 ml. The reduction of absorbance at 240 nm as a result of H_2O_2 degradation was recorded every 30 s with spectrophotometer (Model Varian Cary 50 UV-Vis Spectrophotometer, USA). CAT activity was calculated using the extinction coefficient for H_2O_2 (39.4 mM^{-1} cm^{-1}).

Ascorbate peroxidase (APX) activity was estimated according to the method of Webb and Allen (1995) with a slight modification. The reaction mixture (1 ml) was made up of enzyme extract (100 μl) along with 50 mM phosphate buffer (pH 7.0), 0.2 mM EDTA, 0.25 mM H_2O_2, and 0.5 mM sodium ascorbate. The decrease in absorbance at 290 nm was recorded for 1 min. The APX activity was expressed as units mg^{-1} protein (extinction coefficient 2.8 mM^{-1} cm^{-1}).

Peroxidase (POX) activity was assayed by the guaiacol oxidation method as described by Ghanati et al. (2002). The reaction mixture contained 100 μl crude extract as well as 50 mM potassium phosphate buffer (pH 7.0), 28 mM guaiacol, and 10 mM H_2O_2 in the final volume of 3.0 ml. Changes in optical density of the reacting samples were monitored every 30 s by an increase in absorbance at 470 nm. The APX activity was expressed as units per mg soluble protein (extinction coefficient 26.6 mM^{-1} cm^{-1}).

Experimental Design and Statistical Analysis

A randomized complete block design (RCBD) with five replications was used. Each experimental unit was composed of a hydroponic tray with 12 plants. Data were analyzed using SAS 9.1 software (SAS Institute Inc. 2004). Means comparisons were conducted using least significant differences (LSD) at 5% level of probability.

Results

Increased salt levels in the root zone for 2 weeks resulted in significant changes for all morphological and physio-biochemical traits (Table 8.1).

Table 8.1 Mean square of NaCl salinity level (NaCl) for leaf area; leaf, stem, and root dry weights; plant total dry weights; root/shoot ratio; concentrations of Chl*a*, Chl*b*, total chlorophyll, carotenoids and Chl*a*:*b* ratio, and RWC; chlorophyll fluorescence and content of hydrogen peroxide; MDA and proline; contents of total soluble protein; and activities of CAT, APX, and POX

	Independent variable
Dependent variable	NaCl
Leaf area	109027.545***
Leaf dry mass	1.698***
Root dry mass	0.0066**
Stem dry mass	0.613***
Total dry mass	4.820***
Root/shoot ratio	0.0380***
Relative water content	228.0123***
Chlorophyll *a*	0.1773**
Chlorophyll *b*	0.0467***
Total chlorophylls	0.4040**
Chlorophyll *a*:*b* ratio	0.229*
Carotenoids	0.00812**
Chlorophyll fluorescence	0.00115*
Proline	12342.017***
Hydrogen peroxide	38.973**
MDA	19.242***
Protein	5.252**
APX	5.910**
POX	8.271***
CAT	0.418***

* $P < 0.05$. ** $P \leq 0.01$. *** $P \leq 0.001$

Plant Growth

Leaf area was significantly reduced by 65% in 100 mM salinity compared to the control. The corresponding reductions were approximately 5.4% for 25, 37.7% for 50, and 52.5% for 75 mM, respectively (Table 8.2). Leaf area IC50 (half the maximum inhibition concentration) occurred at about 75.31 mM NaCl (calculated using linear regression (y = 533.83–3.6566x)).

Reductions were 14.4%, 34.9%, 47.2%, and 58.3% for leaf dry mass (LDM) and 11.7%, 37.2%, 51.7%, and 57.7% for stem dry mass (SDM) as NaCl concentration increased from 0 to 25, 50, 75, and 100 mM, compared to the non-salinized plants, respectively. Root dry mass (RDM) was also significantly reduced by 30.6% for 100 mM salt compared to the control. RDM

Table 8.2 Effect of different NaCl concentrations on leaf area, leaf dry weight, stem dry weight, root dry weight, total plant weight, and root/shoot ratio

NaCl (mM)	Leaf area (cm^2 plant^{-1})	Leaf dry weight (g plant^{-1})	Shoot dry weight (g plant^{-1})	Root dry weight (g plant^{-1})	Total dry weight (g plant^{-1})	Root/shoot ratio
0	516.90 ±43.34a	2.45 ± 0.06a	1.40 ± 0.05a	0.25 ± 0.02a	4.09 ± 0.10a	0.18 ± 0.01c
25	488.90 ± 50.19a	2.10 ± 0.12b	1.23 ± 0.10b	0.21 ± 0.01b	3.72 ± 0.23b	0.17 ± 0.02c
50	322.19 ± 21.87b	1.59 ± 0.04c	0.88 ± 0.06c	0.18 ± 0.01bc	2.68 ± 0.08c	0.21 ± 0.02c
75	245.52 ± 23.73bc	1.29 ± 0.05d	0.67 ± 0.06d	0.25 ± 0.02a	2.22 ± 0.06d	0.38 ± 0.05a
100	181.51 ± 10.85c	1.02 ± 0.04e	0.59 ± 0.01d	0.17 ± 0.02c	1.78 ± 0.02e	0.29 ± 0.03b

Values are means ± SE. based on five replicates ($n = 5$). Means followed by the same letters are not significantly different at $P < 0.05$

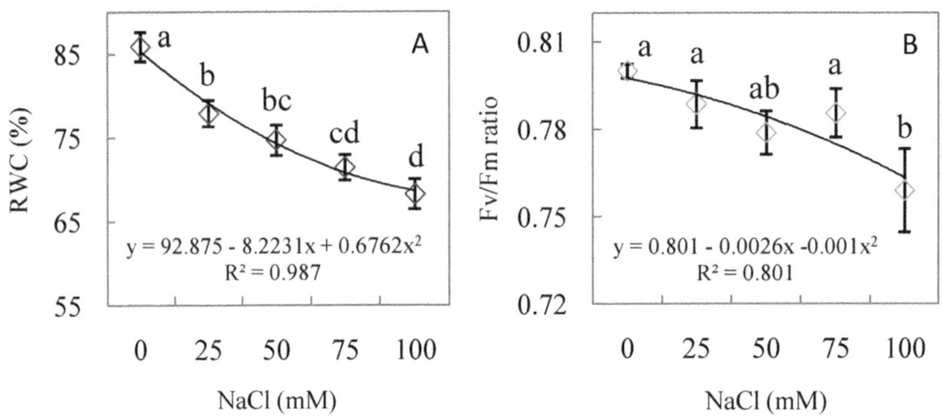

Fig. 8.2 Effect of different NaCl concentrations on relative water content (**A**) and Fv/Fm ratio (**B**) (means ± SE, $n = 5$). Different letters indicate significant differences according to least significant difference (LSD) multiple range test ($P < 0.05$)

increased about 2.7% by salt stress at 75 mM (Table 8.2).

Total dry mass (TDM) was significantly reduced by 52.7% at the highest salinity compared to the control. The corresponding reductions were approximately 9.8%, 24.1%, and 38.4% for 25, 50, and 75 mM, respectively (Table 8.2). TDM (IC50) calculated using linear regression ($y = 4.1227–0.0245x$) was at 84.73 mM NaCl. The negative impact of salinity was greater on SDM than RDM; hence a significant increase in root/shoot ratio was revealed in comparison to the control (Table 8.2).

Relative Water Content

RWC markedly declined in response to increasing salinity. RWC was reduced by 7.92, 11.14, 14.42, and 17.58% as NaCl concentration increased from 0 to 25, 50, 75, and 100 mM, respectively (Fig. 8.2A).

Chlorophyll Fluorescence

Maximum quantum yield (Fv/Fm) was affected by salinity ($P < 0.05$) and a slight reduction in Fv/Fm (5.2%) observed at 100 mM NaCl compared to the control (Fig. 8.2B).

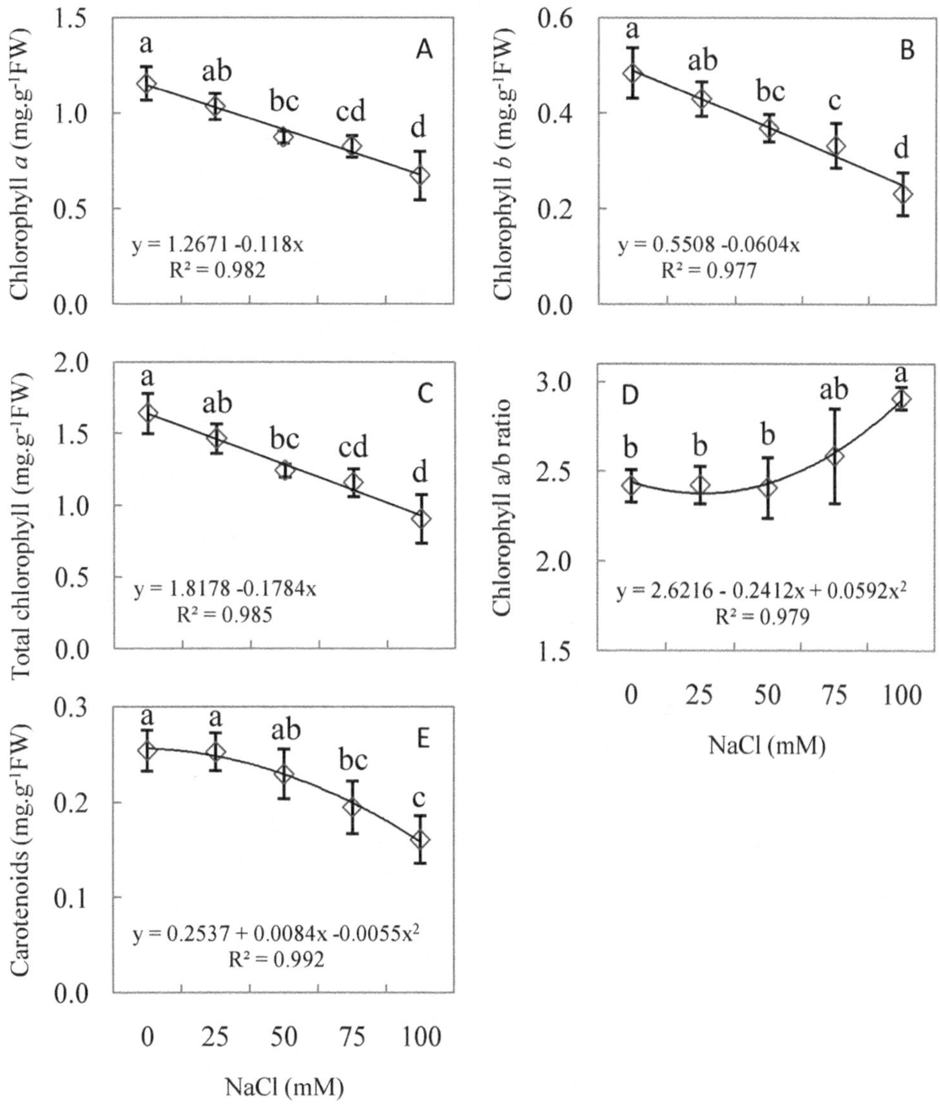

Fig. 8.3 Effect of different NaCl concentrations on the contents of chlorophyll a (**A**), chlorophyll b (**B**), total chlorophyll (a + b) (**C**), Chla:b ratio (**D**), and carotenoid (**E**) in leaves of Chinese kale (means ± SE, *n* = 5). Different letters indicate significant differences according to least significant difference (LSD) multiple range test ($P < 0.05$)

Photosynthetic Pigments

Chlorophylls *a* and *b* and their total were reduced by 41.9, 52.2, and 44.9%, respectively, at 100 mM NaCl (Fig. 8.3A–C). The Chlorophyll *a*:*b* ratio increased at higher salinity. At the highest salinity level, Chlorophyll *a*:*b* ratio was 1.2-fold greater than the control (Fig. 8.3D). Content of carotenoids decreased about 36.7% at 100 mM NaCl compared to the control (Fig. 8.3E).

Proline, Hydrogen Peroxide, and Lipid Peroxidation

Changes in proline content were about 2.17-, 4.2-, 5.3-, and 7.6-fold compared to the control as

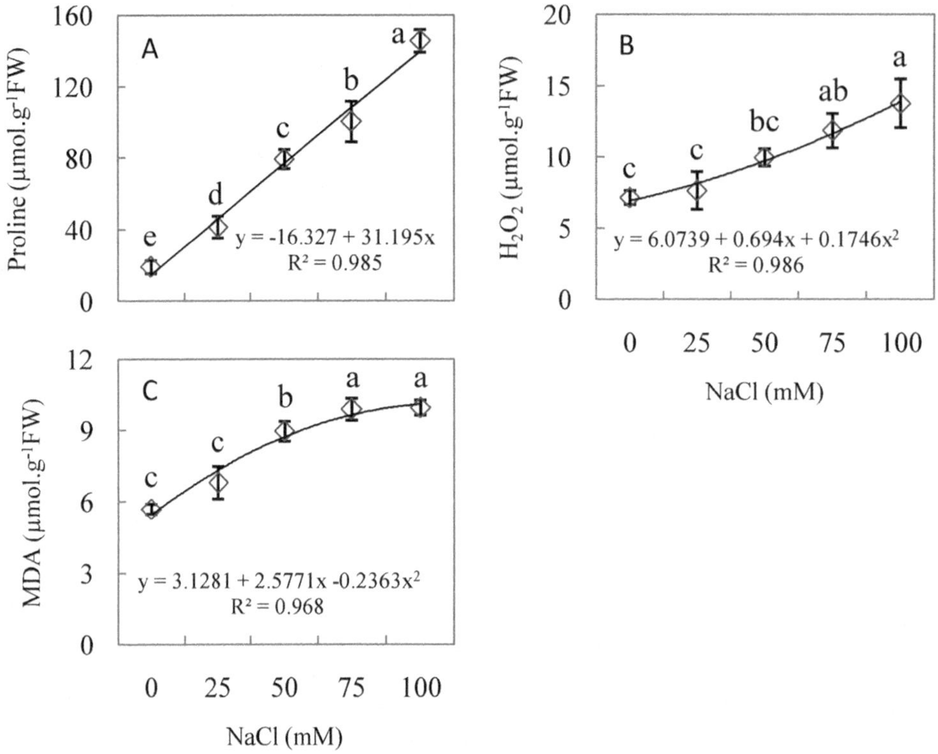

Fig. 8.4 Effect of different NaCl concentrations on the contents of proline (**A**), hydrogen peroxide (**B**), and MDA (**C**) in leaves of Chinese kale (means ± SE, *n* = 5). Different letters indicate significant differences according to least significant difference (LSD) multiple range test ($P < 0.05$)

salinity increased to 25, 50, 75, and 100 mM NaCl, respectively (Fig. 8.4A).

Content of H_2O_2 was boosted from 7.16 (μmol. g^{-1}FW) in control plants to 7.65, 9.96, 11.85, and 13.77 (μmol.g^{-1}FW) in plants subjected to 25, 50, 75, and 100 mM of NaCl, respectively (Fig. 8.4B). MDA level enhancement was about 1.19-, 1.57-, 1.73-, and 1.74-fold as salinity increased from 0 to 25, 50, 75, and 100 mM NaCl, respectively (Fig. 8.4C).

Protein and Antioxidant Enzymes

Protein content decreased with increasing salinity ($P \leq 0.01$). These reductions were about 4.1, 6.4, 25.8, and 28.5% for the 25, 50, 75, and 100 mM NaCl, respectively (Fig. 8.5A).

CAT activity remained constant in response to salinity up to 50 mM NaCl, although higher salinity (75 and 100 mM NaCl) caused a notable increase ($P \leq 0.01$) in CAT activity with as much as 1.65- and 2.35-fold, respectively, compared to the control (Fig. 8.5B).

APX activity showed a significant increase in plants subjected to salinity compared to the control ($P \leq 0.01$). APX activity progressively increased from 1.3- to 2.2-fold for 25 to 100 mM NaCl salinity. No significant changes in APX activity between 75 and 100 mM were observed (Fig. 8.5C). A similar increase in POX activity was found as well. POX activity increased ($P \leq 0.01$) by 6.6-, 12.1-, 16.3-, and 14.8-fold when salinity increased from 0 to 25, 50, 75, and 100 mM NaCl (Fig. 8.5D).

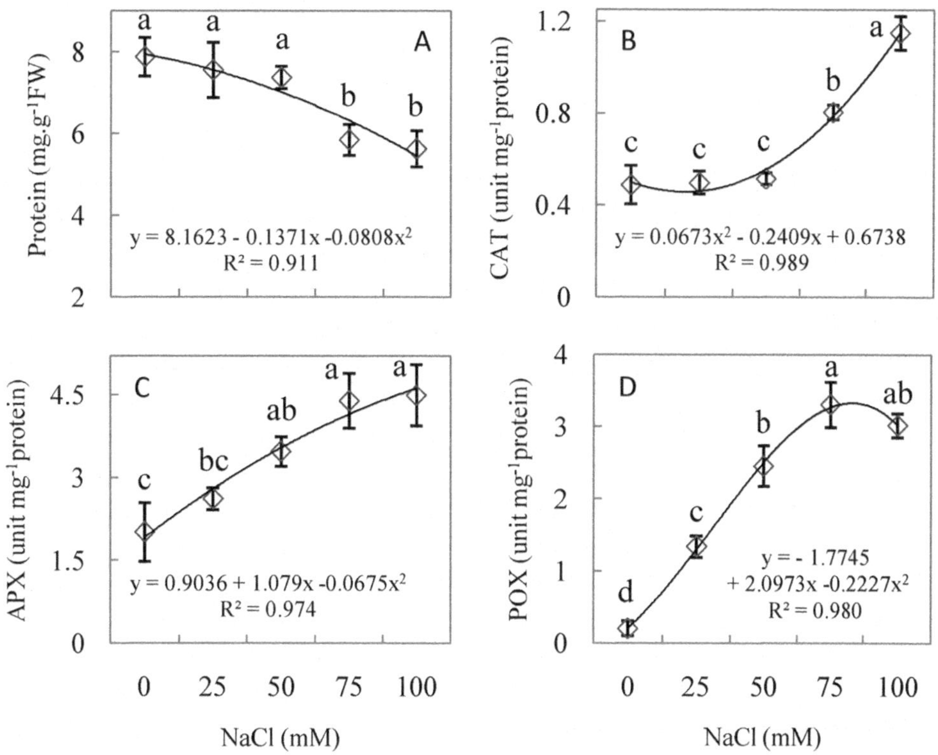

Fig. 8.5 Effect of different NaCl concentrations on the contents of protein (**A**) and antioxidant enzymes activity of CAT (**B**), APX (**C**), and POX (**D**) in leaves of Chinese kale (means ± SE, $n = 5$). Different letters indicate significant differences according to least significant difference (LSD) multiple range test ($P < 0.05$)

Discussion

In this study, we investigated the effects of NaCl salinity on growth, water status, content of pigments, proline, hydrogen peroxide, and magnitude of lipid peroxidation, as well as manifestations of adaptive mechanisms in foliage of Chinese kale plant. Increasing level of NaCl in the rhizosphere results in the osmotic stress, ion cytotoxicity, and imbalanced nutrition which can decline plant growth and development (Kholova et al. 2009).

RWC reflects plant water status generally, and it is positively correlated with rates of cell enlargement and leaf elongation (Singh et al. 2007). RWC decreased under salinity and this reduction increased with increasing NaCl concentration in the root zone as reported in previous studies (Singh et al. 2007; Kholova et al. 2009). Lower RWC implies smaller photosynthetic surface under salinity, which can be manifested as lower growth and developments on the whole plant.

Salinity resulted in a significant reduction in growth parameters such as number of leaves, leaf surface area, and plant organ dry masses. Our results (Table 8.2) showed that salinity reduced dry mass of leaf, shoot, and root, which is consistent with earlier reports for various plant species (Ashraf and Orooj 2006; Psarras et al. 2008). Unequal reduction in dry weights of shoot and root under salinity resulted in root/shoot ratio magnification, which is in agreement with the findings of Essa (2002).

Maximum quantum efficiency of PSII photochemistry (Fv/Fm) is usually used as a photoinhibition index (Krause and Weis 1991; Netondo et al. 2004). Several reports have noted that PSII photochemistry is tolerant of moderate salt conditions and few changes in light utilization have been reported (Ma et al. 1997; Lima et al. 2002; Santos 2004). In sorghum, the efficiency of PSII photochemistry was considerably impacted only at high salinity beyond 200 mM (Notondo et al. 2004). In this study, the Fv/Fm ratio was measured to investigate the possible photoinhibition occurrence at different salt levels. Based on the results, downregulation of PSII photochemistry in plants subjected to high salinity implies the existence of a mechanism for keeping excessive energy away from the PSII photo apparatus (Weis and Berry 1987; Krause and Weis 1991). The results show that efficiency of PSII photochemistry changed under moderate and high salt treatments (Fig. 8.2B).

Pigments content was decreased as salinity increased in root media. A positive correlation between chlorophyll degradation and concentration of NaCl has been reported previously (Hernandez et al. 1995, 2000). The rate of chlorophyll degradation under salinity is linked with the size of ROS generation, chlorophyllase activity, and chlorophyll synthesis (Renault 2005). Increase in Chlorophyll *a*:*b* ratio (Fig. 8.3A–C) under salt stress implies more vulnerability of Chlorophyll *b* to salinity than Chlorophyll *a* (Fig. 8.3D), which is in agreement with result reported by Kholova et al. (2009) and Daneshmand et al. (2010). Decrease in the content of carotenoids under salt stress (Fig. 8.3E) may be attributed to degradation of β-carotene and formation of zeaxanthins (Sultana et al. 1999). The process is thought to provide protection of photosynthetic apparatus by quenching excessive energy of chlorophyll excitation (Daneshmand et al. 2010).

Content of proline was progressively amplified with the increase in salinity (Fig. 8.4A). This phenomenon is considered to facilitate the retention of water and balancing the osmotic potential of the cytoplasm (Hasegawa et al. 2000; Kholova et al. 2009). Osmolytes like proline also are believed to protect the structure and function of various macromolecules related to higher membrane stability and enzymatic antioxidant activity (Kholova et al. 2009).

Salt stress may cause molecular damage to plant cells either directly or indirectly through the formation of ROS. They can react with a target substance, such as lipids, proteins, and/or nucleic acids. H_2O_2, as well as other ROS, disturbs metabolic functions and causes oxidative damage at sites where it accumulates (Foyer et al. 1997). H_2O_2 is also responsible for lipid peroxidation (Parida and Das 2005). Estimating malondialdehyde (MDA) is used as an indicator of lipid peroxidation to evaluate the extent of oxidative damage to biomembranes (Lin and Kao 2000). Our results show that content of H_2O_2 progressively increased with salinity (Fig. 8.4B) while statistically remaining constant at 75–100 mM NaCl salinity. A similar trend for MDA content was also observed (Fig. 8.4C). Changes in H_2O_2 and MDA contents under salt stress have been reported in several studies (Kholova et al. 2009; Daneshmand et al. 2010; Tayebimeigooni et al. 2012). However, small changes in MDA level and membrane stability index in tolerant plants facing salinity or drought (Egert and Tevini 2002; Sairam et al. 2005; Ashraf and Ali 2008) support the existence of a high potential scavenging system to overcome oxidative stress.

Decrease in the protein content of plants subjected to salinity was observed in the present study (Fig. 8.5A). Reduction of protein content was probably linked with the stimulating effect of ROS on protein degradation (Davies 1987). Faced with salinity, plants employ a number of nonenzymatic and enzymatic antioxidants to prevent and overcome oxidative stress (Sekmen et al. 2007; Manchanda and Garg 2008). The existence of a positive correlation between the activity of the antioxidant enzymes and salinity tolerance has been reported (Sekmen et al. 2007; Garg and Manchanda, 2009; Chawla et al. 2013). SOD dismute superoxide to hydrogen peroxide. H_2O_2 is later eliminated through conversion to H_2O in subsequent reactions by CAT and APX. APX has a higher affinity than CAT for scavenging H_2O_2 and uses ascorbate as its specific electron donor (Asada 2006; Sekmen et al. 2007). APX association with monodehydroascorbate reductase, dehydroascorbate reductase, and glutathione reductase (GR) appeared to be the most important peroxidase to remove H_2O_2, through the Foyer-Halliwell-Asada pathway (Hernandez et al. 2010).

Based on our results, CAT and APX activities in leaves of plants subjected to salinity increased (Fig. 8.5B, C). Our results suggest that CAT and APX most likely were significant in the detoxification of hydrogen peroxide in this cultivar. Similar to our results, salt-induced activity of CAT and APX in salt-tolerant plantago and rice have been observed (Turkan et al. 2013; Chawla et al. 2013).

POX involvement in growth and development, the lignification and suberization processes, as well as H_2O_2 scavenging process has been reported (Dionisio-Sese and Tobita 1998). Under salinity stress, POX activity has been found to be stimulated in salt-tolerant plants (Ashraf and Ali 2008; Turkan et al. 2013; Chawla et al. 2013). In this study, POX activity significantly increased in leaf tissues of plants exposed to NaCl salinity. Induction of POX activity continued to 75 mM NaCl and showed a slight reduction as salinity reached 100 mM NaCl (Fig. 8.5D). Increasing POX activity is in agreement with previous results reported on Indian mustard (Yusuf et al. 2008), plantago (Turkan et al. 2013), and jatropha (Hishida et al. 2014).

Various enzymatic antioxidants participate in ROS detoxification process. SOD counteracts superoxide radical and converts it to H_2O_2 (Asada 2006). H_2O_2 is detoxified by scavengers such as APX, CAT, and POX, and in the absence or lack of scavengers, H_2O_2 accumulates in tissues. Several researches proved that NaCl salinity tolerance is closely correlated to the antioxidant capacity of plants as observed in purslane, canola, Indian mustard, plantago, and rice (Yazici et al. 2007; Ashraf and Ali 2008; Yusuf et al. 2008; Turkan et al. 2013; Chawla et al. 2013)

Conclusions

In this study, changes in pigment content, as well as a downregulation in PSII photochemistry, are manifestations to maintain a balance between harvesting and utilizing energy in order to reduce the risk of salt-induced oxidative stress. Accumulation of osmoregulators such as proline eased water stream continuance throughout the salinized plants, thus facilitating gas exchange and higher PSI performance and less ROS formation. The induction of antioxidant enzyme activity, including APX, CAT, and POX, by NaCl treatment suggests their pivotal role in the detoxification process in leaves of stressed plants. Our result shows that cv. 'Standard Kailan' is moderately tolerant to NaCl salinity, whereas higher salinity (>75 mM) suppressed the oxidative defense system. The energy regulation system and enzymatic antioxidant system are prominently contributing to salt tolerance in this study. Chinese kale was less affected by salinity up to 25 mM and employed a complex defense system making it relatively tolerant to high NaCl salinity. Our research potentially provides salinity guidelines for hydroponic production of standard Chinese kale.

Acknowledgment The authors gratefully acknowledge the Universiti Putra Malaysia for financial support.

References

Asada K (2006) Production and scavenging of reactive oxygen species in chloroplasts and their functions. Plant Physiol 141:391–396. https://doi.org/10.1104/pp.106.082040

Ashraf M, Ali Q (2008) Relative membrane permeability and activities of some antioxidant enzymes as the key determinants of salt tolerance in canola (*Brassica napus* L.). Environ Exp Bot 63:266–273. https://doi.org/10.1016/j.envexpbot.2007.11.008

Ashraf M, Orooj A (2006) Salt stress effects on growth, ion accumulation and seed oil concentration in an arid zone traditional medicinal plant ajwain (*Trachyspermum ammi* [L.] Sprague). J Arid Environ 64:209–220. https://doi.org/10.1016/j.jaridenv.2005.04.015

Bates LR, Waldrenand RP, Teare ID (1973) A rapid determination of free proline for water stress studies. Plant Soil 39:205–207. https://doi.org/10.1007/BF00018060

Bradford M (1976) A rapid and sensitive method for the quantitation of protein utilizing the principle of protein-dye binding. Annu Rev Biochem 72:248–254. https://doi.org/10.1016/0003-2697(76)90527-3

Cakmak I, Strbac D, Marschner H (1993) Activities of hydrogen peroxide scavenging enzymes in germinated wheat seeds. J Exp Bot 44:127–132. https://doi.org/10.1093/jxb/44.1.127

Chawla S, Jain S, Jain V (2013) Salinity induced oxidative stress and antioxidant system in salt-tolerant and salt-sensitive cultivars of rice (*Oryza sativa* L.). J Plant Biochem Biotechnol 22:27–34. https://doi.org/10.1007/s13562-012-0107-4

D'Souza MR, Devaraj VR (2010) Biochemical responses of Hyacinth bean (*Lablab purpureus*) to salinity stress. Acta Physiol Plant 32:341–353. https://doi.org/10.1007/s11738-009-0412-2

Daneshmand F, Arvin MJ, Kalantari KM (2010) Physiological responses to NaCl stress in three wild species of potato in vitro. Acta Physiol Plant 32:91–101. https://doi.org/10.1007/s11738-009-0384-2

Davies KJA (1987) Protein damage and degradation by oxygen radicals. I. General aspects. J Biol Chem 262:9895–9901. http://www.jbc.org/content/262/20/9895.full.pdf+html

Dionisio-Sese ML, Tobita S (1998) Antioxidant responses of rice seedlings to salinity stress. Plant Sci 135:1–9. https://doi.org/10.1016/S0168-9452(98)00025-9

Du Z, Bramlage WJ (1992) Modified thiobarbituric acid assay for measuring lipid oxidation in sugar-rich plant tissue extracts. J Agric Food Chem 40:1566–1570. https://doi.org/10.1021/jf00021a018

Egert M, Tevini M (2002) Influence of drought on some physiological parameters symptomatic for oxidative stress in leaves of chives (*Allium schoenoprasum*). Environ Exp Bot 48:43–49. https://doi.org/10.1016/S0098-8472(02)00008-4

Essa TA (2002) Effect of salinity stress on growth and nutrient composition of three soybean (*Glycine max* L. Merrill) cultivars. J Agron Crop Sci 188:86–93. https://doi.org/10.1046/j.1439-037X.2002.00537.x

Foyer CH, Lopez-Delgado H, Dat JF, Scott IM (1997) Hydrogen peroxide- and glutathione-associated mechanisms of acclimatory stress tolerance and signalling. Physiol Plant 100:241–254. https://doi.org/10.1034/j.1399-3054.1997.1000205.x

Garg N, Manchanda G (2009) ROS generation in plants: boon or bane? Plant Biosyst 143:81–96. https://doi.org/10.1080/11263500802633626

Ghanati F, Morita A, Yokota H (2002) Induction of suberin and increase of lignin content by excess Boron in Tobacco cell. Soil Sci Plant Nutr 48:357–364. https://doi.org/10.1080/00380768.2002.10409212

Ghoulam C, Foursy A, Fares K (2002) Effects of salt stress on growth, inorganic ions and proline accumulation in relation to osmotic adjustment in five sugar beet cultivars. Environ Exp Bot 47:39–50. https://doi.org/10.1016/S0098-8472(01)00109-5

Gill SS, Tuteja N (2010) Reactive oxygen species and antioxidant machinery in abiotic stress tolerance in crop plants. Plant Physiol Biochem 48:909–930. https://doi.org/10.1016/j.plaphy.2010.08.016

Hasegawa PM, Bressan R, Zhu JK, Bohnert H (2000) Plant cellular and molecular responses to high salinity. Annu Rev Plant Physiol Plant Mol Biol 51:463–499. https://doi.org/10.1146/annurev.arplant.51.1.463

Hecht SS, Carmella SG, Kenney PMJ, Low SH, Arakawa K, Yu MC (2004) Effects of cruciferous vegetable consumption on urinary metabolites of the tobacco-specific lung carcinogen 4-(Methylnitrosamino)-1-(3-Pyridyl)-1-Butanone in Singapore Chinese. Cancer Epidemiol Biomark Prev 13:997–1004

Hernandez JA, Jimenez A, Mullineaux P, Sevilla F (2000) Tolerance of pea (*Pisum sativum* L.) to long-term salt stress is associated with induction of antioxidant defences. Plant Cell Environ 23:853–862. https://doi.org/10.1046/j.1365-3040.2000.00602.x

Hernandez JA, Olmos E, Corpas FJ, Sevilla F, del Río LA (1995) Salt-induced oxidative stress in chloroplasts of pea plants. Plant Sci 105:151–167. https://doi.org/10.1016/0168-9452(94)04047-8

Hernandez M, Fernandez-Garcia N, Diaz-Vivancos P, Olmos E (2010) A different role for hydrogen peroxide and the antioxidative system under short and long salt stress in *Brassica oleracea* roots. J Exp Bot 61:521–535. https://doi.org/10.1093/jxb/erp321

Hishida M, Ascencio-Valle F, Fujiyama H, Orduño-Cruz A, Endo T, Larrinaga-Mayoral JÁ (2014) Antioxidant enzyme responses to salinity stress of *Jatropha curcas* and *J. cinerea* at seedling stage. Russ J Plant Physiol 61:53–62. https://doi.org/10.1134/S1021443714010063

Kholova J, Sairam RK, Meena RC, Srivastava GC (2009) Response of maize genotypes to salinity stress in relation to osmolytes and metal-ions contents, oxidative stress and antioxidant enzymes activity. Biol Plantarum 53:249–256. https://doi.org/10.1007/s10535-009-0047-6

Krause GH, Weis E (1991) Chlorophyll fluorescence and photosynthesis: the basics. Annu Rev Plant Physiol Plant Mol Biol 42:313–349. https://doi.org/10.1146/annurev.pp.42.060191.001525

Lichtenthaler HK, Buschmann C (2001) Chlorophylls and carotenoids: measurement and characterization by UV–vis spectroscopy. In: Wrolstad RE, Acree TE, An H, Decker EA, Penner MH, Reid DS, Schwartz SJ, Shoemaker CF, Sporns P (eds) Current protocols in food analytical chemistry. Wiley, New York, pp F4.3.1–F4.3.8. https://doi.org/10.1002/0471142913.faf0403s01

Lima ALS, DaMatta FM, Pinheiro HA, Totola MR, Loureiro ME (2002) Photochemical responses and oxidative stress in two clones of *Coffea canephora* under water deficit conditions. Environ Exp Bot 47:239–247. https://doi.org/10.1016/S0098-8472(01)00130-7

Lin CC, Kao CH (2000) Effect of NaCl stress on H_2O_2 metabolism in rice leaves. Plant Growth Regul 30:151–155. https://doi.org/10.1023/A:1006345126589

Ma H, Lindsay F, Wang S, Altman A, Hüttermann A (1997) Photosynthetic response of *Populus euphratica* to salt stress. Forest Ecol Manag 93:55–61. https://doi.org/10.1016/S0378-1127(96)03943-6

Manchanda G, Garg N (2008) Salinity and its effects on the functional biology of legumes. ActaPhysiol Plant 30:595–618. https://doi.org/10.1007/s11738-008-0173-3

Merati MJ, Hassanpour H, Niknam V, Mirmasoumi M (2014) Exogenous application of penconazole regulates plant growth and antioxidative responses in salt-stressed *Mentha pulegium* L. J Plant Interact 9:791–801. https://doi.org/10.1080/17429145.2014.948084

Munns R, Tester M (2008) Mechanism of salinity tolerance. Annu Rev Plant Biol 59:651–681. https://doi.org/10.1146/annurev.arplant.59.032607.092911

Netondo GW, Onyango JC, Beck E (2004) Sorghum and salinity II: gas exchange and chlorophyll fluorescence of sorghum (*Sorghum bicolor* L.) under salt stress. Crop Sci 44:806–811. https://doi.org/10.2135/cropsci2004.7970

Parida AK, Das AB (2005) Salt tolerance and salinity effects on plants: a review. Ecotoxicol Environ Saf 60:324–349. https://doi.org/10.1016/j.ecoenv.2004.06.010

Psarras G, Bertak M, Chartzoulakis K (2008) Response of greenhouse tomato to salt stress and K^+ supplement. Plant Biosyst 142:149–153. https://doi.org/10.1080/11263500701872903

Renault S (2005) Tamarack response to salinity: effects of sodium chloride on growth and ion, pigment, and soluble carbohydrate levels. Can J For Res 35:2806–2812. https://doi.org/10.1139/x05-194

Sairam RK, Srivastava GC, Agarwal S, Meena RC (2005) Differences in antioxidant activity in response to salinity stress in tolerant and susceptible wheat genotypes. Biol Plant 49:85–91. https://doi.org/10.1007/s10535-005-5091-2

Santos CV (2004) Regulation of chlorophyll biosynthesis and degradation by salt stress in sunflower leaves. Sci Hort 103:93–99. https://doi.org/10.1016/j.scienta.2004.04.009

SAS Institute Inc (2004) SAS/STAT user's guide, version 9.1. Cary, North Carolina

Schreiber U, Bilger W, Neubauer C (1995) Chlorophyll fluorescence as a nonintrusive indicator for rapid assessment of in vivo photosynthesis. In: Schulze ED, Caldwell MM (eds) Ecophysiology of photosynthesis. Springer, Berlin, pp 49–70. https://doi.org/10.1007/978-3-642-79354-7_3

Sekmen AH, Türkan I, Takio S (2007) Differential responses of antioxidative enzymes and lipid peroxidation to salt stress in salt-tolerant *Plantago maritime* and salt-sensitive *Plantago media*. Physiol Plantarum 131:399–411. https://doi.org/10.1111/j.1399-3054.2007.00970.x

Singh MP, Singh DK, Rai M (2007) Assessment of growth, physiological and biochemical parameters and activities of antioxidative enzymes in salinity tolerant and sensitive basmati rice varieties. J Agron Crop Sci 193:398–412. https://doi.org/10.1111/j.1439-037X.2007.00267.x

Smart RE, Bingham GE (1974) Rapid estimates of relative water content. Plant Physiol 53:258–260. https://doi.org/10.1104/pp.53.2.258

Sultana N, Ikeda T, Itoh R (1999) Effect of NaCl salinity on photosynthesis and dry matter accumulation in developing rice grains. Environ Exp Bot 42:211–220. https://doi.org/10.1016/S0098-8472(99)00035-0

Tayebimeigooni A, Awang Y, Mahmood M, Selamat A, Wahab Z (2012) Leaf water status, proline content, lipid peroxidation and accumulation of hydrogen peroxide in salinized Chinese kale (*Brassica alboglabra*). J Food Agric Environ 10:371–374

Tayebi-Meigooni A, Awang Y, Biggs AR, Rosli M, Madani B, Ghasemzadeh A (2014) Mitigation of salt-induced oxidative damage in Chinese kale (*Brassica alboglabra* L.) using ascorbic acid. Acta Agric Scand Sect B 64:13–23. https://doi.org/10.1080/09064710.2013.869347

Turkan I, Demiral T, Sekmen AH (2013) The regulation of antioxidant enzymes in two Plantago species differing in salinity tolerance under combination of waterlogging and salinity. Funct Plant Biol 40:484–493. https://doi.org/10.1071/FP12147

Velikova V, Yordanov I, Edreva A (2000) Oxidative stress and some antioxidant systems in acid rain-treated bean plants Protective role of exogenous polyamines. Plant Sci 151:59–66. https://doi.org/10.1016/S0168-9452(99)00197-1

Webb RP, Allen RD (1995) Isolation and characterization of a cDNA for spinach cytosolic ascorbate peroxidase. Plant Physiol 108:13–25. https://doi.org/10.1104/pp.108.3.1325

Weis E, Berry JA (1987) Quantum efficiency of photosystem II in relation to energy-dependent quenching of chlorophyll fluorescence. Biochim Biophys Acta Bioenerg 894:198–208. https://doi.org/10.1016/0005-2728(87)90190-3

Xiong L, Zhu JK (2002) Molecular and genetic aspects of plant responses to osmotic stress. Plant Cell Environ 25:131–139. https://doi.org/10.1046/j.1365-3040.2002.00782.x

Yazici I, Turkan F, Sekmenah DT (2007) Salinity tolerance of purslane (*Portulaca oleracea* L.) is achieved by enhanced antioxidative system, lower level of lipid peroxidation and proline accumulation. Environ Exp Bot 61:49–57. https://doi.org/10.1016/j.envexpbot.2007.02.010

Yusuf M, Hasan SA, Ali B, Hayat S, Fariduddin Q, Ahmad A (2008) Effect of salicylic acid on salinity-induced changes in *Brassica juncea*. J Integr Plant Biol 50:1096–1102. https://doi.org/10.1111/j.1744-7909.2008.00697.x

Growth Dynamic of *Tamarix chinensis* Plantations in High Salinity Coastal Land and Its Ecological Effect

9

Xiaohui Feng, Xiaojing Liu, Xiumei Zhang, and Jin Song Li

Abstract

Tamarisk (*Tamarix chinensis*), a pioneer native halophyte from the coastal region of North China, has a high potential to establish in high salinity soil. The ecological benefits from planting tamarisk include changes to the structure and function of the saline land ecosystem. However, field experiments on the ecological benefit of using tamarisk in the restoration of saline lands along the coastal areas of China are lacking. In the current study, tamarisk was planted in 2005, 2007, and 2009 to assess the effectiveness of the plant in vegetation and saline soil restoration. The individual biomass, carbon sequestration, species diversity, and soil characteristics of tamarisk plantations in the different planting years were investigated. The results showed that tamarisk grew well with increasing individual biomass during the experimentation period. Carbon density increased from 0.98 t(C) hm^{-2} (land without tamarisk) to 4.78 t(C) hm^{-2} in the 3-year plots, 5.56 t(C) hm^{-2} in the 5-year plots, and 6.89 t(C) hm^{-2} in the 7-year plots. Significant soil quality improvement was indicated by lower soil salinity and higher organic matter. Salt content in the 0- to 10-cm soil layer of barren land was 10.53 g kg^{-1}, while it was only 3.35 g kg^{-1}, 1.86 g kg^{-1}, and 5.54 g kg^{-1} in the 3-, 5-, and 7-year plots with tamarisk, respectively. Potassium (K) availability in the soil was increased. In addition, tamarisk played a significant "eco-engine" role in terms of increasing the species of surrounding grasses. Where only 8 grass species existed in this area before tamarisk planting, a total of 15 grass species existed in the 7-year plot. However, the importance value of salt-tolerant species decreased in tamarisk plots. Clonal plants such as *Aeluropus sinensis* and *Sonchus arvensis* became more dominant than seed propagation plants such as *Suaeda salsa*. This suggested that tamarisk plants progressively alleviated salt stress in the coastal environment. The study demonstrated that tamarisk was effective in vegetation regeneration in saline lands, improving soil quality and increasing carbon sequestration.

Keywords

Coastal saline land · Biomass · Carbon storage · Productivity · Biodiversity · *Tamarix chinensis*

X. Feng · X. Liu (✉) · X. Zhang · J. S. Li
The Center for Agricultural Resources Research, Institute of Genetic and Developmental Biology, Chinese Academy of Sciences, Shijiazhuang, Hebei Province, China
e-mail: xjliu@sjziam.ac.cn

B. Gul et al. (eds.), *Sabkha Ecosystems*, Tasks for Vegetation Science VI,
https://doi.org/10.1007/978-3-030-04417-6_9

Introduction

Saline land in the west bank of Bo Sea Bay of North China is characterized by a shallow groundwater table and high soil salt content, which is usually more than 5 g kg^{-1}. Therefore, a large area of the high salinity land in the area was abandoned (Liu et al. 2010) due to low land productivity. How to improve the land productivity of coastal saline land has become an urgent problem that needs to be addressed.

The planting of native flora in such areas was reported to combat salinity- related issues in these types of soil (Shan et al. 2011, 2012). Tamarisk (*Tamarix chinensis*) is one of the native halophytes from the coastal area of China. It has a strong potential to utilize phytoremediation to improve soil fertility, which not only helps in reducing soil salinity but also improves biodiversity of plants and microbes. Soil organic matter, nitrogen (N), and phosphorus (P) increased and soil pH decreased (Guan et al. 2009; Hou et al. 2008; Lei et al. 2011; Wang et al. 2006). Hence, tamarisk is widely used in the vegetation restoration of saline land (Zhao et al. 2011).

At present, most studies focus on the application of tamarisk in coastal saline land greening projects. It has been proved that tamarisk is a good greening species because it is a fast-growing shrub and has a high carbon sequestration ability. However, it is difficult to form a tamarisk-dominant community only using natural diffusion, in which case the effect of carbon sequestration and saline soil recovery is not obvious. It is therefore necessary to manually plant tamarisk in saline land to study the changes in structure and function of the saline land ecosystem.

The interaction between plants and the environment is the driving force for plant community succession. After tamarisk planting, the species composition and structure of the plant community changes. Therefore, it is important to study the diversity of herbaceous plant species after planting tamarisk to understand the process of vegetation restoration in high salinity land. There has been little to no research on coastal saline tamarisk planting and its biomass, carbon reserve dynamics, and influence on plant community diversity. The alterations in community biodiversity directly affect the productivity of the saline land and the soil fertility.

In this study, we investigated the properties of the 3 different tamarisk plantations (3-year, 5-year, and 7-year plantations). We studied the biomass dynamic of individual tamarisk plants and tamarisk communities and their distribution pattern. These data were used to evaluate the carbon sequestration potential of planted tamarisks and their influence on the vegetation diversity of the entire community and on the salt soil content.

Study Site

The study site was in the high salinity coastal land in Haixing County of Hebei Province, North China (E117.85°, N38.17°), as shown in Fig. 9.1. The region is a coastal low-lying plain in the west bank of Bo Sea Bay. The soil salt content is composed mainly of chloride (Cl^-,45–50% of the total ion content), with sodium (Na^+) as the major cation. The groundwater level is 0.3–1.0 m below the surface with groundwater salinity 7–27 g L^{-1}. The climate is a warm, temperate, semi-humid, continental, monsoon one, with an annual average temperature of 12.1 °C. January is the coldest month with an average temperature of −4.5 °C. Annual average precipitation is 582.3 mm, with the precipitation mainly concentrated in July and August. There is little precipitation in winter (November, December, and January).

Due to the distinctive wet and dry seasons, the soil salt content also has a distinct seasonal pattern. High evaporation in spring and autumn cause high salinization while high precipitation washes/dilutes the soil salt in the summer. The winter is cold and dry and salt movement through the soil stops at this time. For this study, the tamarisks were planted in 2005, 2007, and 2009.

Planting method: In March, cuttings of annual tamarisk branches of 0.8~1.2 cm diameter and 12 cm long were planted in plastic bags in a nursery. When the seedlings were approximately 40 cm tall, they were transplanted into the research field and with a spacing of 0.5 m between each seedling. The seedlings were planted at the end of June or before the rainy season.

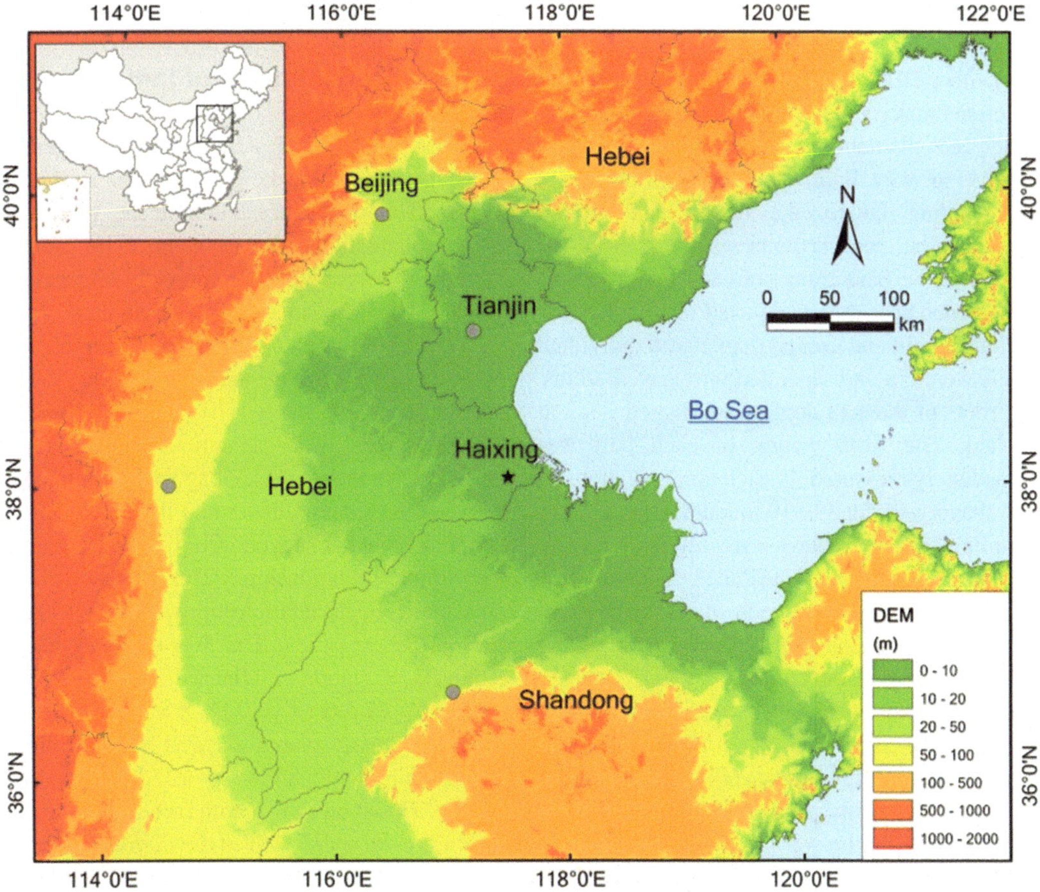

Fig. 9.1 Study site of the tamarisk plantation in Haixing County, Hebei Province, China, in the west bank of Bo Sea Bay

Fig. 9.2 The abandoned land and the tamarisk plantation in high salinity soil

Methods

In October 2011, a survey was conducted on the different tamarisk plantations (i.e., 3-year, 5-year, and 7-year plantations) and the abandoned land (CK), (Fig. 9.2). Following a random pattern in the planting lattice, the height and ground diameter (the measure point was 5 cm from the ground at the base trunk) of each tamarisk was measured. These data were used to calculate the average height and ground diameter. 6 individual plants were cut and separated into leaves, stems, and roots of the 1 m depth were digged out and washed. The fresh weight of each part was immediately measured. In each tamarisk plot and the abandoned land , 10 quadrats (1 × 1 m sampling square frames) were set randomly to investigate community attributes. The number, height, and coverage of each plant species was determined in the quadrats. The fresh weight of these herbs was also recorded.

Soil samples were taken from 4 soil layers (0–10 cm, 10–20 cm, 20–30 cm, 30–50 cm). The salt content, total nitrogen, available phosphorus, available potassium, and organic matter were determined from the soil samples. The tamarisk and herb samples were dried at 85 °C in an oven until a constant weight was obtained. According to Xu et al.'s (2003) study, the carbon content of different tamarisk organs was used. The carbon content of the above-ground parts was 44.3% and the carbon content of the root was 42.6%. The carbon content of the reference herb was 35.0% (Zheng et al. 2007). The species importance value was used to compare the superiority of different species in the community: importance value = (*relative coverage* + *relative frequency* + *relative density*)/3 × 100 (Song 2001). The redundancy analysis (RDA) of the importance value, the soil salt content, and the planting time of different species was used to explain the relationship between the species and the environment.

Results and Analysis

Growth Characteristics of Tamarisk

Tamarisk height and stem diameter increased with age. The mean height was 150 ± 29 cm, 176 ± 25 cm, and 201 ± 33 cm in the 3-, 5-, and 7-year plantations, respectively. The annual increase in height was approximately 12.5 cm per year. The mean stem diameter was 18.09 ± 4.70 mm, 21.95 ± 4.29 mm, and 26.96 ± 7.75 mm in the 3-, 5-, and 7-year plantations, respectively. The plant diameters in the 5-year to 7-year plantations showed rapid growth. With the increase of age, the relationship between the stem diameter and the height of the tamarisk plant changed from a linear to a logarithmic relationship. The 3-year plants presented a linear relationship; $y = 5.62x + 48.74$ ($R^2 = 0.79$, $n = 53$). The 5-year plants had a relatively poor linear relationship; $y = 4.51 + 77.59x$ ($R^2 = 0.58$, $n = 43$), while the 7-year tamarisk plants showed a logarithmic relationship; $y = 99.83 \ln(x) - 122.75$ ($R^2 = 0.71$, $n = 48$). This showed that with increasing age, the height growth speed slowed down gradually, and the radial growth of the stem diameter accelerated. This illustrated that the growth of tamarisk shifted activity from the apical meristem to the radial meristem (Figs. 9.3 and 9.4).

Changes in Biomass and Carbon Storage of Individual Tamarisk Plants

Biomass

With the increase in age, the biomass of different organs increased gradually. The allocation of biomass to different organs was observed to occur at different rates. Tamarisk plants aged 3 years, 5 years, and 7 years had biomass measurements of 640.24 g, 758.97 g, and 970.63 g, respectively. The shoot to root biomass

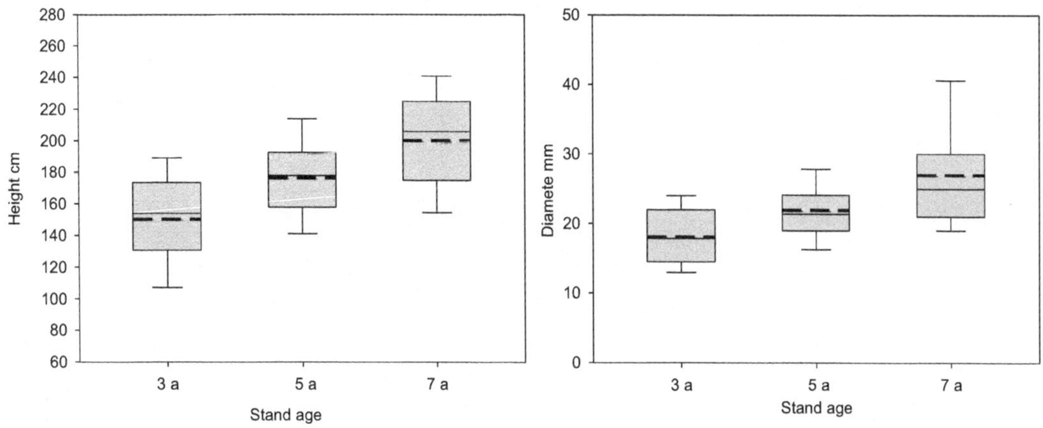

Fig. 9.3 Height and diameter of tamarisk plants in the different age plantations

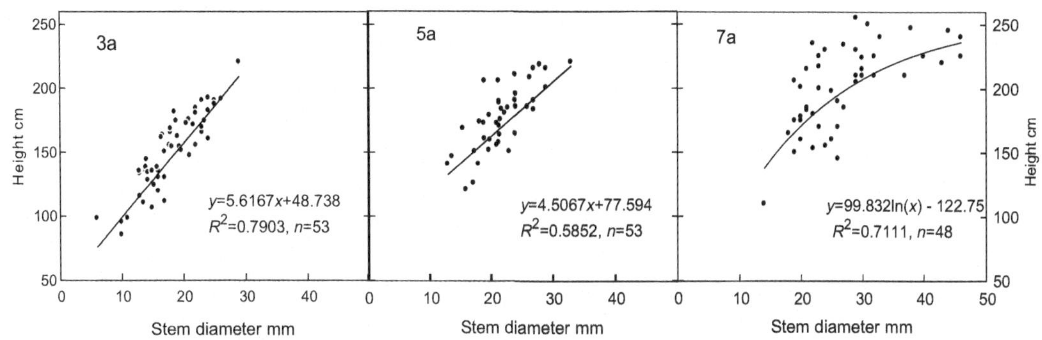

Fig. 9.4 The relationship between the diameter and the height in different age plantations

ratio generally remained unchanged. When comparing the data between the 3-year and 7-year plantations, the biomass of the branches gradually increased (from 33.64% to 48.23%) out of the total plant's biomass, while the proportion of the root biomass reduced from 52.80% to 42.50%. The ratio between the above-ground and root biomass increased from 0.87 to 1.44. Tamarisk growth in the first 3 years was mainly seen in root elongation and plant height increase, but the radial growth (i.e., stem diameter) accelerated significantly after 3 years and biomass accumulation was mainly focused on the branches. The biomass differences between individual plants gradually increased with age. The biomass of different organs and their proportion in the total biomass is shown in Table 9.1.

Community Carbon Storage

The plants coverage was higher in areas where tamarisk was planted in comparison to land where other herbs were growing. The herb biomass gradually declined over time (Table 9.2). In the bare land of the control plot, the herb biomass had a larger standard deviation but the coverage in the contral plot was higher than the biomass of the herb quadrat taken from the tamarisk stands. This was because tamarisk coverage caused a decrease in the number of herbs.

The mean carbon storage of the abandoned land (CK) was 0.98 t(C) hm^{-2}, but the carbon storage of the 3-year, 5-year, and 7-year tamarisk plantations was 4.78 t(C) hm^{-2}, 5.56 t(C) hm^{-2}, and 6.89 t(C) hm^{-2}, respectively. The trend changed with the change in biomass seen within

Table 9.1 Biomass of different plant organs in the different aged *Tamarix chinensis* individual plants

	3 years[a]		5 years[a]		7 years[a]	
	Biomass (g plant^{-1})	Rate %	Biomass (g plant^{-1})	Rate %	Biomass (g plant^{-1})	Rate %
Leaves	77.67 ± 32.22[a]	12.13	82.17 ± 46.89[a]	10.83	126.42 ± 63.32[a]	13.02
Branches	215.40 ± 61.72[a]	33.64	329.63 ± 124.47[ab]	43.43	468.16 ± 182.68[b]	48.23
Roots	338.03 ± 12.05[a]	52.80	356.32 ± 26.73[a]	46.95	412.51 ± 268.02[a]	42.50
Above-ground	293.07 ± 92.67[a]	45.78	411.80 ± 167.04[ab]	54.26	594.58 ± 241.63[b]	61.26
Total biomass	640.24 ± 105.65[a]	–	758.97 ± 173.71[ab]	–	970.63 ± 398.29[b]	–
Shoot to root ratio	–	0.87	–	1.16	–	1.44

[a,b,c] indicate significant differences among different aged tamarisk, $p<0.05$

Table 9.2 Carbon storage of different age *Tamarix chinensis* plantations (t(C) hm^{-2})

Treatment	CK	3[a]	5[a]	7[a]
Herb	0.98 ± 1.20[a]	1.12 ± 0.36[a]	1.09 ± 0.38[a]	1.08 ± 0.46[a]
Leaf	–	0.46 ± 0.19[a]	0.49 ± 0.28[a]	0.73 ± 0.33[a]
Branch	–	1.27 ± 0.36[a]	1.95 ± 0.73[ab]	2.73 ± 1.20[b]
Root	–	1.97 ± 0.11[a]	2.02 ± 0.16[ab]	2.34 ± 0.89[b]
T. chinensis	–	3.65 ± 0.53[a]	4.46 ± 0.97[ab]	5.81 ± 2.24[b]
Community	0.98	4.78	5.56	6.89

[a,b] indicate significant differences among different aged tamarisk plantation, $p<0.05$

the individual tamarisk plant. The carbon storage of leaves and branches increased significantly with plant age. The carbon storage increased more slowly in the later period than in the early stage, but its rate of increase was lower than that of the total tamarisk plant. This was because as the canopy density increased gradually, competition between individual plants grew fiercer and some individual plants in a dominant position grew rapidly. Some individual plants grew under the shade of superior plants, and thus their growth was restrained, and they might have even died, so that the density of the tamarisk community declined slightly. Thus, the individual biomass increased at a higher rate than the increase rate of the community's carbon storage. When viewed from the dynamic process of the tamarisk's growth, the fastest growth was in the first 3 years, when the carbon accumulation rate was also high, but 3 years later and as a result of increased community canopy density and competition between individual plants, the growth rate and carbon accumulation rate both slowed down.

Soil Salt Content

Tamarisk planting brought about a reduction in the surface evaporation, which slowed the upward movement of soil salt and significantly reduced the soil salt content (Fig. 9.5). In the contrast plot, the soil salt content in the 0- to 10-cm depth of soil was 10.53 g kg^{-1}, which was much higher than that of the land covered by other vegetation. The soil of the 5-year plantation had the lowest soil salinity of 1.86 g kg^{-1}, but it did not decrease further with an increase in time. The 7-year plantation had higher soil salinity than the soil of the 3-year and 5-year plantations. The soil salt content was slightly higher in the 0- to 10-cm layer than in the 10- to 20-cm layer of soil.

With increasing plant age, soil organic matter, total nitrogen, and available phosphorus content increased slightly, but no significant difference was observed between the communities. Available potassium content in the soil was significantly increased, as shown in Fig. 9.6.

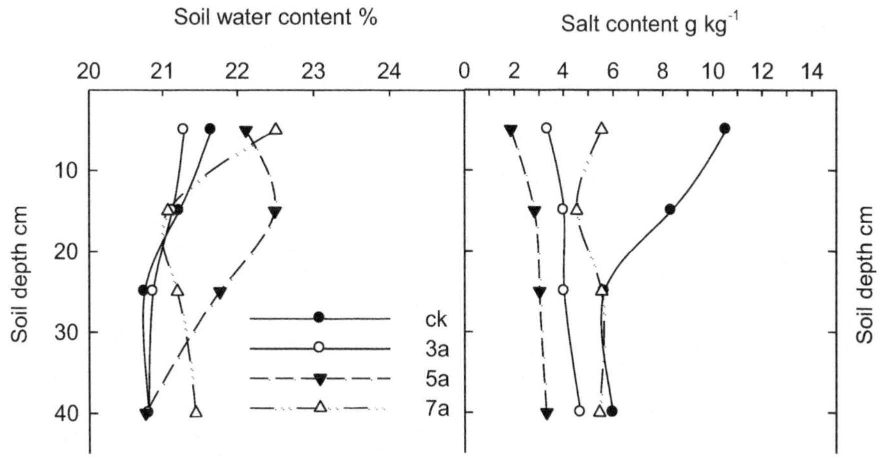

Fig. 9.5 Soil water and salt content of the different layers

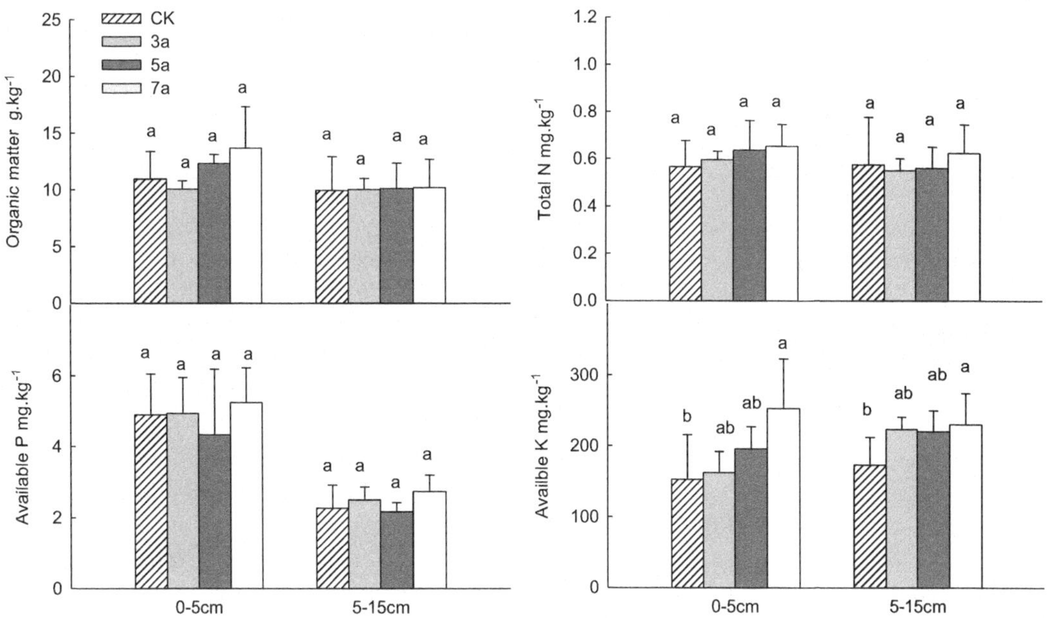

Fig. 9.6 Organic matter, total N, available P, and available K soil in the 0- to 5-cm and 5- to 10-cm layers in different age plantations. a,b indicate signiifcant differences among different treatments, $p<0.05$

Herb Species

In the community, there were 17 herbaceous species, belonging to 6 families and 17 genera. There were 7 species of *compositae*, belonging to 7 different genera, and 6 species of *gramineae*, belonging to 7 genera. The proportion of the 2 families makes up two-thirds of all species; among them 9 are perennial, 4 are biennial, and 4 are annual. The importance value of each species is shown in Table 9.3. There were only 8 species in the herb community of the contrast plot. The dominant species were halophyte *Suaeda salsa* and *Imperata cylindrica*, whose importance values were 34.8 and 36.1 respectively. In the 3-year tamarisk plantation, there were 10 species, with

Table 9.3 Lifestyle, regeneration manner, and importance value of the herb plants in the different tamarisk communities

Latin name	Life style	Regeneration style	Treatment			
			CK	3[a]	5[a]	7[a]
Heteropappus altaicus	Perennial	Seed	0	5.3	0	1.4
Imperata cylindrical	Perennial	Rhizomes, seed	36.1	21.1	17.7	8.1
Echinochloa crusgali	Annual	Seed	0	0	1.2	0
Plantago asiatica	Biennial	Rhizomes, seed	0	0	0	1
Cirsium setosum	Perennial	Rhizomes, seed	0	0	0	1.2
Cynanchum chinense	Perennial	Seed	0	0	1.8	4.7
Limonium sinense	Biennial	Seed	1.3	2.7	3.6	0
Setaria viridis	Annual	Seed	1.8	4.2	11.4	9.3
Tripolium vulgare	Biennial	Seed	0	0	2.6	0.8
Sonchus arvensis	Perennial	Rhizomes, seed	4	19	20	21.3
Phragmites australis	Perennial	Rhizomes, seed	6.7	1.9	2.1	4.8
Scorzonera mongolica	Perennial	Rhizomes, seed	10.7	10	10.2	12.2
Taraxacum mongolicum	Perennial	Seed	0	0	0	0.9
Artemisia capillaris	Perennial	Seed	0	0	0	0.9
Puccinellia tenuiflora	Perennial	Seed	0	10.3	1.7	1.4
Suaeda salsa	Annual	Seed	34.8	6.6	3.2	0.7
Aeluropus sinensis	Perennial	Rhizomes, seed	13.1	18.8	24.6	31.3

the dominant species being *I. cylindrical, Sonchus arvensis,* and *Aeluropus sinensis.* Their importance values were 21.1, 19.0, and 18.8 respectively. In the 5-year tamarisk plantation, there were 12 species, with the dominant species being *A. sinensis, S. arvensis,* and *I. cylindrical.* Their importance values were 24.6, 20.0, and 17.7 respectively. In the 7-year tamarisk plantation, there were 15 species. The dominant species were *A. sinensis* and *S. arvensis*, whose importance values were 31.3 and 21.3, respectively. It can be seen that tamarisk planting changed the undergrowth environment, promoted the settlement of plants that were not tolerant to saline environments, and increased the species abundance of the entire community. With the increase of tamarisk age, the importance value of *A. sinensis, S. arvensis,* and *Setaria viridis* increased. The importance value of saline *S. salsa* and *I. cylindrical* gradually reduced, and the importance value of *Scorzonera mongolica* generally remained unchanged. Plants with a poor salt tolerance, such as *Plantago asiatica, Cirsium setosum, Taraxacum mongolicum,* and *Artemisia capillaris*, appeared in the 7-year tamarisk plantation. With regard to the reproduction method of herbaceous plants, with an increase in tamarisk age, the perennial herbaceous plants that reproduced with both seeds and rhizomes were the best suited. This was because the different stages of the plant growth could occupy the undergrowth and use rhizome propagation to become dominant in the community.

RDA analysis on the importance values of species, the soil salinity of each soil layer, and the planting time showed that the importance values of species were mainly affected by the planting time and the soil salinity. As shown in Fig. 9.7, with the horizontal axis as the first axis, the variance was 0.883. With the vertical axis as the second axis, the variance was 0.090 and accumulation was 0.973, and this explains the salt content. According to the species position and direction in the ordination axes, the species can be divided into three types: those with a positive correlation with salt content, those with a positive correlation with time, and those that were not sensitive, which corresponds to salt-tolerant species, species without salt tolerant ability, and neutral species.

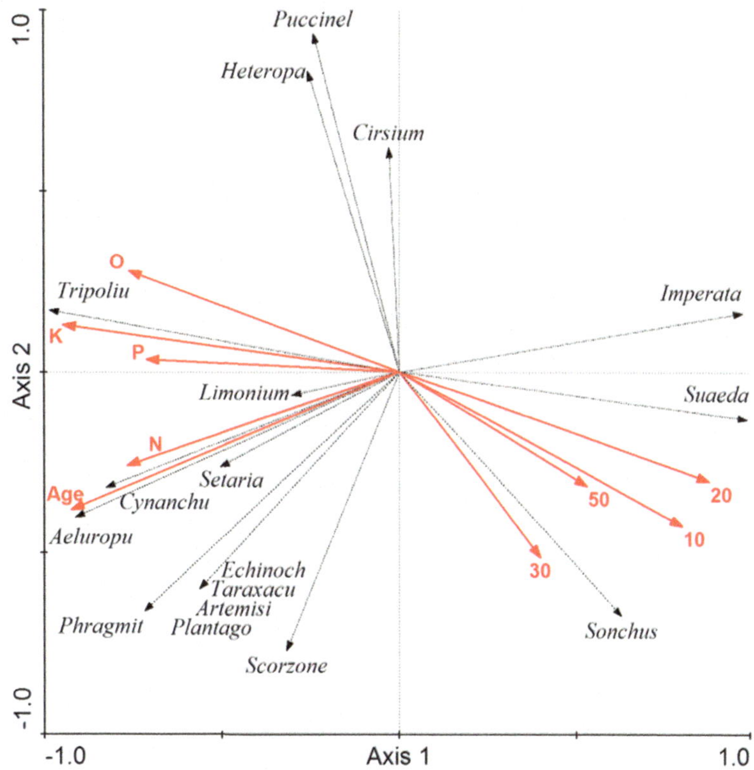

Fig. 9.7 The RDA analysis of the importance values and the soil salinity

Discussion

Tamarisk Growth, Biomass, and Community Carbon Storage

The 3-year tamarisk plantation exhibited a rapid growth phase that reached up to 150 cm in height with an average increment of 50 cm per year. Tamarisk cuttings with a height of 40 cm can grow to 80 cm 100 days after cutting during spring (Guo and Wang 2010), while after stumping the new branch, it can grow more than 150 cm with a stem diameter of >8 mm (Fu et al. 2010). This suggests that in their first year, tamarisks grow quickly after cutting. This is similar to the growing process of *Caragana microphylla* seedlings in arid zones (Zhu et al. 2011).

The 5-year tamarisk plantation presents a relatively poor relationship between stem diameter and plant height. This indicates that after 3 years, the height growth rate slowed down. For the 7-year tamarisk plantation, the relationship between its height and diameter was no longer linear. This indicates that with an increase in age, the growth of tamarisk slows down. Because tamarisk is drought-tolerant, the growth of its seedling root takes precedence over its above-ground part, which is a type of adaptive growth strategy in drought environments (Wei et al. 2007). Tamarisk biomass changes with the growth of the individual plant in the first 3 years, and biomass is mainly allocated in the root. With time, the proportion of stem biomass increases.

Tamarisk planting decreased the biomass of surrounding herbs. This was mainly due to the gradually increased canopy cover from the tamarisk. According to our results, if one harvests the above-ground part of the tamarisk, the highest productivity age should be 3 years. Compared with other vegetation, tamarisk has a biomass equivalent to the planted *Hippophae rhamnoides* in the Loess Plateau (Liu and Liu 1994), and it is higher than the biomass (780 g m^{-2}) of the natural tamarisk community at the lower reaches of the Heihe River (Peng et al. 2010). This is due to the scattered distribution in nature versus the evenly distributed planting pattern in this study. The biomass of the tamarisk in this study was higher than the theoretical production rate (310 g m^{-2}) of global saline land (Wicke et al. 2011). This illustrates that tamarisk planting in coastal saline land is an effective method of carbon sequestration.

Changes in Tamarisk Community Species Diversity

The change of the salt content in the saline soil was the limiting factor that determined the community succession (Karlin et al. 2011). The increased abundance of herbaceous plants in the community is an important sign of progress succession. This has been previously observed in the succession from saline to non-saline areas in the muddy beach near our study site (Zhao et al. 2008). Two years after the tamarisk planting, herbaceous species had increased in the community and there was an interaction between saline soil vegetation and the saline soil. In northeast China, the reverse succession of saline land was a result of the destruction of vegetation (Wang et al. 2003). With the passing of time, the importance values of salt-tolerant plants such as *S. salsa* declined. Similarly, after planting *Caragana Fabr* in Datong, Shanxi Province, the amount of salt-tolerant plants fell sharply in the community (Zhang et al. 2009). *S. salsa* is a strongly salt-tolerant plant, but it is not as competitive as *A. sinensis* in soil with low salinity, thus the coverage and number of *A. sinensis* increased. This result also indicates the sharp decline of competitiveness of saline plants in environments with lower salinity (Ungar 1998). An increase in vegetation coverage inhibits soil salinization in the winter. Thus, low salt-resistant species enhance the diversity of the community, which further decreases the soil salinity. In the 5-year tamarisk plantation, the importance value of *A. sinensis* was higher than in the 3-year plantation, and species such as *Setaria viridis*, a low salt-resistant plant, appeared in the community. In the 7-year tamarisk plantation, the number of species was even greater. RDA analysis shows that of the 3 types of plants, salt-tolerant plants mainly included *S. salsa* and *Phragmites australis*; non-salt-tolerant plants were *S. viridis* and *Asiatic plantain*; plants that were not sensitive to salt included *Scorzonera mongolica*. As the age of tamarisk increased in the community, the advantages of root-breeding perennial plants increased gradually. This was because of the niche genetic effect of the root, which gave those plants an advantage in the competition. In the progressive succession of the saline land in the Songnen Plain of northeast China, vegetative propagation occupies a dominant status (Yang and Zheng 1998). Tamarisk planting promotes the progress of coastal saline land succession, gradually reduces the soil salinity, and increases the vegetation community's richness.

The Effect of Tamarisk on Soil Salinity

Soil salt content changes with the season in coastal saline land. Due to the shallow underground saline water level, surface evaporation forces the salt to move up and accumulate in the soil surface in spring and early summer. Rainfall in the summer causes salt leaching. Increased vegetation coverage and salt leaching during the rainy season are two important ways to reduce the soil salt. Tamarisk planting increases the surface coverage and decreases the surface soil salt in spring by reducing evaporation, which allows more herbaceous plants to grow under the canopy, thereby further increasing the vegetation coverage and reducing the surface water losses and the salinity. This is the cycle of decreasing soil salinity. Therefore, after tamarisk planting, soil salt content was significantly lower than in the contrasting land, which was the result of the interaction between plants and the environment. In different communities, soil salinity increased with the depth of soil, and this was because the sampling time was in early October, which was after the rain and consequent leaching, and the salt content in the soil had leached from the surface to the deeper layers, increasing soil salinity with soil depth. According to a study in Ningxia, tamarisk reduced the soil pH (Lei et al. 2010), but soil pH in this study did not change. The soil in Ningxia was saline-alkaline and plantation tamarisk reduced the pH, while the coastal saline soil already had a low pH, therefore there were no significant changes in the pH in the current study. The study in the desert region of Xinjiang showed an increase in soil salinity below the tamarisk canopy due to the salt-secreting function of the soil between the tamarisk plants (Yin et al. 2008). In addition, multiple factors increase soil salinity in deserts, such as a naturally low salt soil with low rainfall but high evaporation, which accumulates higher salt on the surface. This study demonstrated a significant decrease of soil salt in the 5-year plantation but a slight increase in the 7-year plantation. This may be associated with the background salt content before planting as well as the salt-secreting properties of leaves, although this needs to be investigated further. Without a doubt, tamarisk planting significantly reduced the soil salinity compared with the saline land without tamarisk planting.

The returning of plant branches and leaves to the soil increased the soil nutrient content. In this study, the potassium content of the surface soil significantly increased with plantation age. Although planting tamarisk could promote nutrient accumulation in the surface soil, this saline soil still lacks organic matter, nitrogen, and phosphorus.

Conclusions

Planting tamarisk in the saline soil created a shrub layer and promoted the progressive succession of the plant community. The biomass of tamarisk increased with time. Carbon storage of the 7-year plantation was 6.89 t(C) hm^{-2}, while the carbon storage of bare saline land was 0.98 t(C) hm^{-2}. Salt content of the surface soil significantly decreased after planting tamarisk and the available potassium significantly increased. Herbaceous plant abundance increased and the dominance of salt-tolerant species in the community gradually decreased with time, while perennial herbaceous plants that reproduced with both seeds and rhizomes gradually dominated. *T. chinensis* is the "ecosystem engine species" in high salinity coastal land, and planting tamarisk is an effective way to improve the utilization of high salinity coastal land.

References

Fu RS, Guan CZ, Yin LR et al (2010) Primary research on growth and development of tow species of Tamarix and application. Tianjin Agric Sci 16(2):112–114

Guan HB, Wang XL, Ju D et al (2009) Soiled modification and application of *Tamarix Chinensis* on the saline soil. Resour Dev Market 25(10):918–921

Guo L, Wang QL (2010) Cutting experiment of *Tarnarix chinensis* Lour. on alkaline soil. East China For Manag 24(1):25–27

Hou BD, Ma FY, Wu HY et al (2008) Characteristics of soil nematodes communities at different succession stages of wetland in the Yellow River Delta. Chin J Appl Environ Biol 14(2):202–206

Karlin MS, Bachmeier OA, Dalmasso A et al (2011) Environmental dynamics in salinas grandes, Catamarca, Argentina. Arid Land Res Manag 25(4):328–350

Lei JY, Ban NR, Zhang YH (2011) Effects and partition characteristics of *Tamarix ramosissima* on nutrients and salt of saline-alkali soils. Bull Soil Water Conserv 31(4):73–78

Liu ZD, Liu ZW (1994) Study on biomass of *Hippophae rhamoides* and *Caragana microphyllal.* on loess plateau. Acta Agriculturae Boreali-Occidentalis Sinica 3(2):92–96

Liu XJ, Li XJ, Chen LN et al (2010) Study on the adaptive farming system in saline soils—a case study in saline area of strand plain in Hebei Province. Chin J Eco-Agric 18(4):911–913

Peng SZ, Zhao CY, Peng HH et al (2010) Spatial distribution of *Tamarix ramosissima* aboveground biomass and water consumption in the lower reaches of Heihe River, Northwest China. Chin J Appl Ecol 20(9):1940–1946

Shan QH, Zhang JF, Ruan WJ et al (2011) Response of soil quality indicators to comprehensive amelioration measures in coastal salt-affected land. Acta Ecol Sin 31(3):6072–6079

Shan QH, Zhang JF, Shen LM et al (2012) Effects of forestry ecological engineering on herb community in coastal saline-alkali land. Chin J Ecol 31(6):1411–1418

Song YC (2001) Vegetation ecology. East China Normal University Press, Shanghai, pp 45–46

Ungar IA (1998) Are biotic factors significant in influencing the distribution of halophytes in saline habitats. Bot Rev 64(2):176–199

Wang HJ, Sheng LX, Chen P et al (2003) Study on influencing factors and control system of reversal evolvement in Western Songnen plain. J Northeast Norm Univ (Natural Science Edition) 35(3):60–65

Wang YZ, Liu YX, Wei CL et al (2006) Improvement of salt-affected soils with six halophytes. J Anhui Agric Sci 34(2):951–957

Wei J, Zhang XM, Maw D et al (2007) Seedling growth dynamics of *Tamarix austromongolica* and its acclimation strategy in hinterland of desert. Arid Land Geogr 30(2):666–673

Wicke B, Smeets E, Dornburg V et al (2011) The global technical and economic potential of bioenergy from salt-affected soils. Energy Environ Sci 4(8):2669–2681

Xu YR, Zhang WJ, Feng ZW et al (2003) Caloric values, elemental contents and correlations between them of some plants on sea-beach salinity soil in Tianjin. Acta Ecol Sin 23(3):450–455

Yang YF, Zheng HY (1998) Comparison analysis on the experimental communities during progressive succession on alkaline patches in the Songnen plain of China. J Plant Ecol 22(1):214–221

Yin CH, Feng G, Tian CY et al (2008) Enrichment effects of soil organic matter and salinity under the tamarisk shrubs in arid area. Chin J Eco-Agric 16(1):263–265

Zhang LZ, Niu W, Niu Y et al (2009) Impact of *Caragana* Fabr. plantation on plant community and soil properities of saline-alkali wasteland. Acta Ecol Sin 29(9):4693–4699

Zhao LY, Huang MY, Yuan JJ et al (2008) Research on floral succession in plants protecting engineer of muddy sea wall. Tianjin Agric Sci 14(3):56–60

Zhao KF, Zhang WJ, Fan H et al (2011) Biological measures for utilization and development of salinized soil. Chin J Soil Sci 32(S1):115–119

Zheng WJ, Bao WK, Gu B et al (2007) Carbon concentration and its characteristics in terrestrial higher plants. Chin J Ecol 26(1):307–313

Zhu YL, Wang S, Lin YG et al (2011) Development of *Caragana microphylla* seedling root system in hilly regions of loess plateau. Bull Soil Water Conserv 31(2):232–237

Mechanisms of Ion Transport in Halophytes: From Roots to Leaves

10

Vadim Volkov and Timothy J. Flowers

Abstract

The chapter describes peculiarities of ion transport in halophytic plants, aiming to help understand the mechanisms important for their tolerance of salt. An initial introduction to methods for studying ion transport is followed by analysis of ion transport from a broad thermodynamic point of view. Further detailed survey of ion channels and ion transporters in plants adds to the picture of ion transport pathways through cell membranes. A typical 'generalised' plant cell is depicted to illustrate the variety of ion transport systems known so far for all plants. This serves as a basis for a comparison of ion transport in salt-sensitive glycophytes and salt-tolerant halophytes. Next, there is a description of what we know of transport systems in halophytes, beginning from the thermodynamics of ion transport under salinity. In halophytes, low negative stable plasma membrane potentials and cytoplasmic Na^+ concentrations that are often higher than in glycophytes are important for their life under salinity. Comparison of similar pairs of plants with contrasting halophytic and glycophytic habits allows us to find specific features of ion transport essential for high salinity tolerance. Mechanisms of high- and low-affinity sodium transport in halophytes are briefly characterised to explain and stress the increased accumulation of Na^+ by halophytes compared to glycophytes. Description of ion channels and transporters in halophytes and pathways of ion transport from nutrient solution to their roots, then to the xylem and finally to leaves completes the chapter. Problems and unsolved questions are proposed for the future study of ion transport in halophytes.

Keywords

Glycophyte · Halophyte · Ion transport · Salt tolerance

V. Volkov (✉)
Faculty of Life Sciences and Computing, London Metropolitan University, London, UK

Department of Plant Sciences, College of Agricultural and Environmental Sciences, University of California, Davis, CA, USA

T. J. Flowers
School of Life Sciences, University of Sussex, Brighton, UK

School of Plant Biology, The University of Western Australia, Crawley, WA, Australia

Halophytes: A General Introduction

The surprising and striking capability of halophytic plants is to survive under salinity and even flourish on salinised soils or in salt solutions. This feature of halophytes poses numerous questions about their physiology and biochemistry,

B. Gul et al. (eds.), *Sabkha Ecosystems*, Tasks for Vegetation Science VI,
https://doi.org/10.1007/978-3-030-04417-6_10

Fig. 10.1 Halophytes and glycophytes in natural habitats and under experimental conditions. (**a**) Halophytes *Salicornia* spp. (larger and bright green plants in the picture) and *Suaeda maritima* (smaller and greyish-green plants) growing in a saline estuarine mud near the river Medway in the UK, where the area is flooded with seawater under high tides. The photograph was taken in early June 2014. The size of the largest specimen is about 15 cm. From (Volkov 2015b). (**b**) The growth of wheat, a glycophyte, is prevented by a salt-affected soil in northern Spain, in sharp contrast to the ability of *Salicornia* and *Suaeda* seen in Fig. 10.1a. (**c**) Growth of (*a*) the glycophyte *Arabidopsis thaliana* and (*b*) the genetically and morphologically similar halophyte *Eutrema halophilum* (synonym, *Thellungiella halophila*)[1] in saline medium. Five- to 7-week-old *Arabidopsis thaliana* and *Eutrema halophilum* plants were grown hydroponically in 'control' medium, before plants were exposed to an additional 100 mm NaCl for 1 week. Plants were grown in plastic cylinders with a diameter of about 1 cm. (From Volkov et al. 2004, with permission of John Wiley and Sons)

about their water and ion relations and about properties of their ion and water transport systems. Many halophytes are able to grow at high concentrations of salt (Fig. 10.1), under irrigation by seawater (euhalophytes: Flowers and Colmer 2008) and, exceptionally, under several times seawater salt concentrations (e.g. see English and Colmer 2013). The majority of halophytes are sodium-tolerant plants. Indeed, Na^+ is the main cation in salinised soils with Cl^- being the most commonly found anion, although sulphates, carbonates and bicarbonates of calcium, magnesium and potassium sometimes contribute to soil and water salinisation (Flowers et al. 1977). Concentrations of sodium chloride over 100 mM severely inhibit growth of most plants (Fig. 10.1b) yet produce little or no inhibition in halophytes: in fact, concentrations of 100–400 mM stimulate growth in many euhalophytes (Fig. 10.2).

The aim of this chapter is to characterise molecular ion transport systems, mostly cation transport systems, of halophytes and pose questions for further study. The cation transport systems known so far in halophytes are presumed to be the same as those in glycophytes due to common ancestry and evolution (Flowers et al. 2010). Indeed, the trait of salt tolerance emerged independently over 70 times in different groups of grasses (Bennett et al. 2013). Therefore, a short introduction starts with a summary of methods available to study ion transport in general and

[1]We have used names as in The Plant List (http://www.theplantlist.org) and added in parentheses the name used in the publications we cite. However, we have NOT changed the names of genes or proteins.

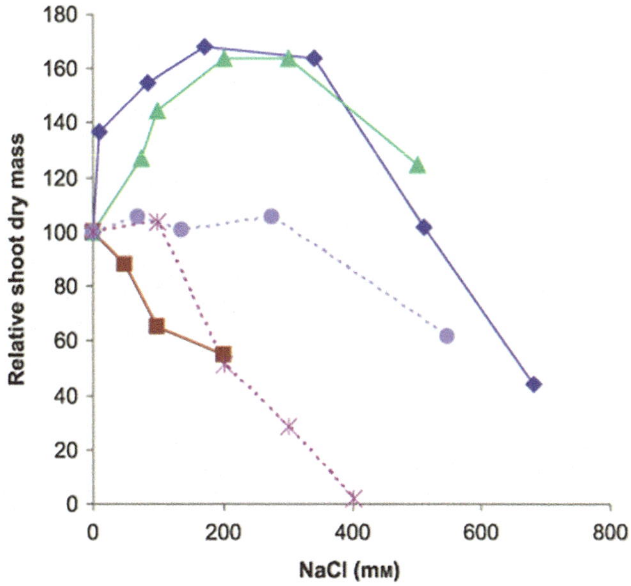

Fig. 10.2 Effect of increasing salinity on the growth (relative to that in the absence of, or in very low, NaCl) on shoot dry mass of a range of halophytes. The length of exposure to salt is given in parentheses. Solid lines, dicotyledonous species; broken lines, monocotyledonous species. Diamonds, *Suaeda maritima* (35 d; Yeo and Flowers 1980); squares, *Eutrema halophilum* (14 d; M'Rah et al. 2007); triangles, *Disphyma australe* (60 d; Neales and Sharkey 1981); stars, *Puccinellia peisonis* (42 d; Stelzer and Läuchli 1977); circles, *Distichlis spicata* (21 d; Parrondo et al. 1978). (From Flowers and Colmer 2008, with permission of John Wiley and Sons)

describes a 'generalised plant cell' with its cation transport systems. Since most of our existing knowledge in ion transport comes from glycophytes (plants that are not halophytes), the 'generalised plant cell' is of a glycophytic plant; the known cation transport systems for halophytes will be introduced in turn.

Basics of Ion Transport in Plants

Methods to Measure Ion Fluxes

Progress in methodologies is indispensable for progress in research. A brief summary of current methods for measuring basic properties of ion fluxes is given in Table 10.1. The fluxes are determined as changes in ion concentrations or amount of ions per unit of time per a normalised unit of surface area or of weight or volume. These fluxes, obtained by different methods, can be expressed per surface area of plant organ or tissue, per surface area of plasma membrane of cells or protoplasts, per unit of dry weight or per unit of fresh weight; recalculations and comparisons require translation from one set of units to the other ones (e.g. Yeo and Flowers 1986; Flowers and Yeo 1986; translations for electrophysiology and uptake by root cells are considered in White and Lemtiri-Chlieh 1995; Volkov 2015b).

Mechanisms of Ion Transport

Transport of ions is driven by physicochemical forces, primarily differences of ion concentrations (to be more precise, activities of ions) and differences in electric potential across plasma or vacuolar membranes.

Membrane potentials of plant cells (electric potential across the plasma membrane between the cytoplasm and outside of the cell) can be routinely measured after impalement with microelectrodes with tiny sharp tips around 0.1 μm in diameter (e.g. described in Blatt 1991). Voltage-sensitive fluorescent proteins and dyes (reviewed in Mutoh et al. 2012) can also be used for, at least, indications of membrane potentials in tis-

Table 10.1 Comparative summary for methods to measure characteristics of ion fluxes in plants, their tissues and cells

Method or group of methods	Principle of the method	Spatial and temporal resolution; advantages	Disadvantages
Kinetic measurements of ion concentrations	Ion concentrations in plant tissues or in nutrient solution are measured over time; the concentration differences are plotted against time	Resolution is at the level of whole plant or plant tissues; usually tens of minutes and hours are needed to register changes. Simple methods without a need of special equipment, convenient for most ions	Measure net fluxes, not separate influx and efflux. Low level of resolution
MIFE: microelectrode ion flux estimate/ measurements	Tiny ion-selective microelectrode with a tip diameter of around 1 μm vibrates with typical frequencies 0.1–1 Hz in the vicinity of a cell or plant tissue and measures ion concentrations. Changes in concentrations over time are recalculated as ion fluxes	Resolution is at the level of individual cells within seconds; measurements may last for hours	Require special equipment and measure net fluxes. Organic compounds often interfere with ion-selective electrodes, so extra calibrations and checks are needed. Reliable ion-selective resins are available so far for a few major ions only
Measurements of unidirectional ion fluxes and short-term influx	Plant tissues or organs are loaded with radioactive ions or rare ions to imitate ions of interest. Unidirectional outward (or inward for short-term influx) ion fluxes of the isotope or rare ion are measured as changes in concentrations against time	Spatial resolution is at the level of whole plant or plant tissues; recordings from minutes lasting to hours are needed to register changes	Require radioactive isotopes and often complicated calculations with several proposed pools of ions
Electrophysiological methods	Isolated cell membrane, a piece of membrane or single cell within a plant preparation is subjected to different voltages (wide range of −300 mV to +200 mV is usually sufficient and used), and ion current is registered in the form of electric current	Spatial resolution of single molecules of ion channels or single cells. Temporal resolution from μs to minutes. High accuracy and the possibility to find specific molecules for transport of specific ions	Require special equipment, indirect measurements, measure total electric current carried out by several ion species
Fluorescent indicators	The concentration of an ion is estimated by a fluorescent ion-selective indicator within a cell or cell compartment	Resolution at the level of individual cells within tens of seconds to minutes, recordings over tens of minutes to hours	Require specific protocols for loading fluorescent probe. Intrinsic autofluorescence and nonspecific adsorption of fluorescent indicators could be drawbacks

The distinction between the methods is partially arbitrary. From (Volkov 2015b) with slight modifications

sues and populations of cells (Matzke and Matzke 2013). Membrane potentials between −220 mV and −70 mV are frequently recorded by microelectrodes, although values around −300 mV have been reported; more negative values are often measured in root cells, while more positive potentials are found in leaf cells (apart from leaf guard cells) (Higinbotham 1973; L'Roy and Hendrix 1980; Blatt 1987; Walker et al. 1995,1998; Carden et al. 2001, 2003; Shabala and Lew 2002; Fricke et al. 2006; Murthy and Tester 2006; Shabala et al. 2006; Volkov and Amtmann 2006; Armengaud et al. 2009; Hammou et al. 2014). Vacuolar membrane potentials (determined here as the electric potential between the vacuole and the cytoplasm across the vacuolar membrane) are around 0 mV or −10 to +40 mV. So, the vacuolar interior is usually 0 to +40 mV more positive than the cytoplasm with pH values about 2 units lower than in the cytoplasm: pH values are about 5.0–6.1 or less in vacuoles compared to 7.0–7.7 in the cytoplasm (e.g. Walker et al. 1995; Carden et al. 2003; Cuin et al. 2003; Martinoia et al. 2012).

The thermodynamics of ion transport is described by several equations that are based on the fact that the movement of substances depends on gradients in their free energy. Free energy is described in thermodynamics by the term 'chemical potential', which for an ion is a function if its activity (effective concentration) and of its electric charge under electric fields that are present. So, the diffusion of an ion depends not just on its concentration gradient but also on any gradient of electric charge (more detail of these concepts can be found in Flowers and Yeo 2007 or Nobel 2005). It is possible to describe the balance of ions and charge across a membrane when an equilibrium has been reached – when there is no difference in chemical potential (free energy) of the ion across that membrane. This balance is quantified by the Nernst equation, applied to selectively permeable membranes, and links ion concentrations on either side of the membrane to the electric potential across that membrane (under equilibrium conditions, when net flux of ions via the membrane is absent):

$$E_s = E_1 - E_2 = \frac{RT}{Z_s F} \ln\left(\frac{S_2}{S_1}\right)$$

(Hille 2001). Here E_s is the electric potential difference between the two sides (E_1 and E_2) of the membrane (in volts then converted to mV), R is the universal gas constant equal to 8.31 J K^{-1} $mole^{-1}$, T is temperature in K, Z_S is the charge of ion S, F is the Faraday constant equal to 96,500 s ampere*$mole^{-1}$ (charge of 1 mole of monovalent cations), and $[S_1]$ and $[S_2]$ are concentrations of ion S on the two sides of the membrane. To be correct, the activities of the ion S should be taken rather than its concentration: this is especially important for halophytes where the ion concentrations can be high enough for the activity to be just 70% of the concentration.

Basically, the diffusion of a monovalent ion such as K^+, Na^+ or Cl^- is such that a tenfold difference in concentration (S) across the ion-selective membrane equilibrates with an electric potential, which is about ±60 mV (slightly depending on temperature) with sign determined by the ion charge. This means by generating a membrane potential of −60 mV, (inside negative) a cation will equilibrate across the plasma membrane with an internal concentration that is ten times the external concentration. So, for K^+, which might typically have a concentration of 100 mM in the cytoplasm of an epidermal root cell, the membrane potential to ensure uptake of K^+ from a soil concentration in the soil solution of 100 μM should be below −180 mV to satisfy the electrochemically downhill transport of K^+ ions. Lower concentrations of K^+ outside the cells at the membrane potential of −180 mV may require cotransport of K^+ with other ions (e.g. with H^+).

Where several ion species with different permeabilities through the membrane cross that membrane, a more complicated Goldman–Hodgkin–Katz voltage equation is applicable; it takes into account the individual permeabilities for the ions. For Na^+, K^+, H^+ and Cl^- (obviously more ions could be considered and more components be added, but these are the major monovalent ions in saline soils), the equation will be:

$$E_{\text{membrane}} = \frac{RT}{F} \ln\left(\frac{P_{Na^+}\left[Na^+\right]_{out} + P_{K^+}\left[K^+\right]_{out} + P_{H^+}\left[H^+\right]_{out} + P_{Cl^-}\left[Cl^-\right]_{in}}{P_{Na^+}\left[Na^+\right]_{in} + P_{K^+}\left[K^+\right]_{in} + P_{H^+}\left[H^+\right]_{in} + P_{Cl^-}\left[Cl^-\right]_{out}}\right),$$

where $E_{membrane}$ is the potential across the membrane (or $E_{reversal}$ where there is zero net ion current via the membrane), P are permeabilities for the corresponding ions and the terms in square brackets stand for the concentrations of the ions (e.g. $[Na^+]_{out}$ is the external concentration of Na^+) (Hille 2001). Usually K^+ permeability dominates, and the membrane potential is close to $E_{reversal}$ of K^+ (when P_{K^+} dominates in the Goldman–Hodgkin–Katz equation and permeabilities for Na^+, H^+ and Cl^- are negligibly low compared to the permeability for K^+, we can use the Nernst equation for K^+), though this may change and depend on cell type and membrane voltages. An active plasma membrane proton pump (H^+-ATPase, reviewed, e.g. in Palmgren 2001) shifts the potential across the plasma membrane to more negative values compared to the calculated $E_{reversal}$ for K^+; K^+ and Na^+ pumps are not known for plants.

The existence of a cell wall with ion-selective properties is a peculiarity of plant cells; it generates a charge difference between the cell wall and the cytoplasm inside the plasma membrane (usually adds −40 to −10 mV to the cytoplasm relative to the outside of the cell; it is due to the presence of fixed negative charges in the cell wall and depends on the external ion concentrations). This so-called Donnan potential (see Briggs and Robertson 1957; Briggs and Hope 1958; Shomer et al. 2003) influences the membrane potential difference between the cytoplasm and the exterior of the cell and slightly changes the composition of the solution that is adjacent to the plasma membrane (briefly reviewed in Flowers and Yeo 1986; Volkov 2015a).

Pathways for Ion Transport Through Membranes

Transport of most ions including Na^+ and K^+ in plants occurs passively (following the electrochemical forces) via ion-selective proteinaceous pores of ion channels. Most ion channels can change their conformation from open to closed states and vice versa (so-called gating) under applied voltages or after binding ligands and regulators. Another pathway is via proteinaceous transporters with slower transport rates (see below for more details about ion channels and ion transporters). Ion transport via ion channels is electrogenic since ions carry electric charge, while transporters realise electrogenic or non-electrogenic transport, transporting one ion (or molecule) or cotransporting/antiporting several charged ions in one or opposite directions, correspondingly. Cotransport of several ions or even small molecules may provide the energy required for the transport of other ions (Fig. 10.3). While transport of one ion species via a transporter could be against its electrochemical potential gradient, the total ion flux of several ions or molecules via a transporter is down the overall gradient of electrochemical potentials. The energy for transferring one ion against its electrochemical potential is derived from co- or antiport of the other ion species. For example, HAK transporters cotransport K^+ together with H^+ (Bañuelos et al. 1995; Rodriguez-Navarro 2000; Grabov 2007), which generates several orders of concentration difference for the K^+ as there is a large inward gradient for H^+ generated by a plasma membrane ATPase that transports protons out of the cell (see below). A membrane potential of −180 mV potentially allows K^+ uptake from a medium with a similar external pH to that of the cytoplasm, but when cotransported with H^+ with a stoichiometry of 1:1, against 10^6 differences in K^+ concentrations (e.g. as illustrated above from 0.1 μM outside to 100 mM inside; Rodriguez-Navarro 2000) (Fig. 10.3). A higher concentrative capacity could be achieved using different pH differences across the membrane or greater number of protons per K^+; a cytoplasmic pH of about 7.5 and external low pH of 4 would add over three more orders of concentration. Real transporters, however, are not ideal molecular machines with 100% efficiency; extra noise and heat production are unavoidable during their functioning. However, the theoretical considerations are the basis for further analysis of transport processes.

The only active transport systems of plant plasma membranes are membrane ATPases, which hydrolyse ATP and use the released energy

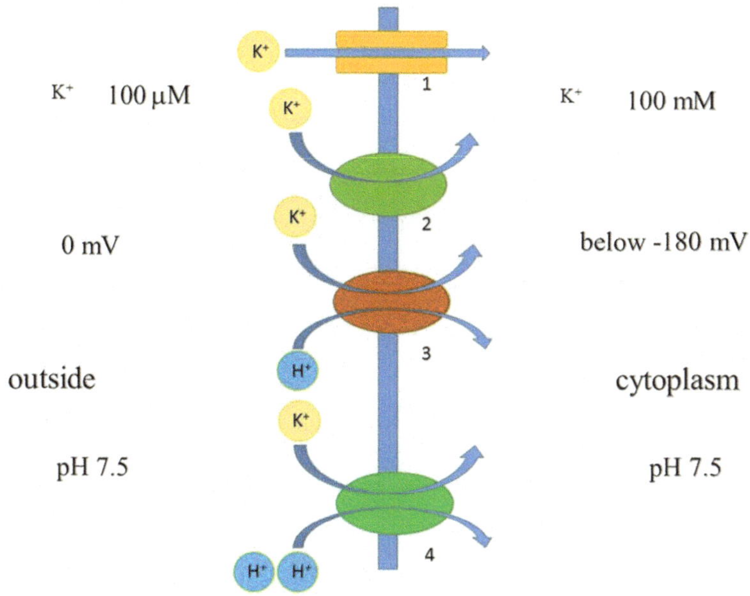

Fig. 10.3 A simplified scheme demonstrating the principles of ion transport across a membrane, using the potassium ion as an example and an ion channel and three hypothetical transporters together with often encountered pH values and concentrations. Voltage difference below −180 mV allows potassium transport against 1000-fold concentration difference via potassium-selective pore of ion channel *1*. Similar thermodynamically favourable potassium transport with lower rates and specific mechanism is facilitated by potassium transporter *2*. Transporters numbered *3* and *4* are H^+/K^+-cotransporters; they cotransport one or two H^+ per K^+; H^+ is transported according to voltage difference, hence adding energy for K^+ transport. Transporters *3* and *4* can potentially lead to inward transport of K^+ against over 1000,000 concentration difference; their functioning depends also on pH difference across membrane. More details are in the text. (From Volkov 2015b, with modifications)

for directed transport of ions. Plant plasma membranes contain H^+-ATPase (reviewed, e.g. in Palmgren 2001): this ATPase pumps H^+ out of the cells, thus acidifying the external medium of the cells, keeping pH of cytoplasm slightly alkaline at around 7.5 and shifting the membrane potential more negative than the reversal potential for K^+ (that membrane potential where there is zero net ion current via the membrane). H^+-ATPase is the only plasma membrane ATPase of higher plants, while Na^+-ATPase was discovered in the moss *Physcomitrella patens* (Benito and Rodríguez-Navarro 2003; Lunde et al. 2007). Plant vacuolar membranes are also energised by an ATPase (V-type H^+-ATPase) as well as by a vacuolar H^+-pyrophosphatase. Membrane ATPases, via several indirect steps, restore slowly dissipating electrochemical gradients for most ions and also shift the membrane potential to the desired range of voltages. The plasma membrane H^+-ATPase is able to hyperpolarise cells below −200 mV; it is possible and important to determine the components of membrane potential related to (1) direct activity of H^+-ATPase and (2) asymmetric distribution of the ions permeable through the membrane. It is interesting to note that animals have an electrogenic Na^+/K^+-ATPase in their plasma membranes (Skou 1998), and so there is a tighter link between Na^+ and K^+ fluxes, than in plants, where the link is an indirect one via the activity of the plasma membrane H^+-ATPase.

Sets of transporters and ion channels are specific for different cell types and organisms, with tens of distinct proteins characterised so far; the proteins often form heteromers with variable properties and regulation (Fig. 10.4). Detailed analysis of genome sequences of the salt-sensitive model glycophytic plant *Arabidopsis thaliana* revealed that about 5% of about 25,000 genes of the plant potentially encode membrane transport proteins; the genes of about 880 proteins are classified in 46 unique groups, while genes of cation

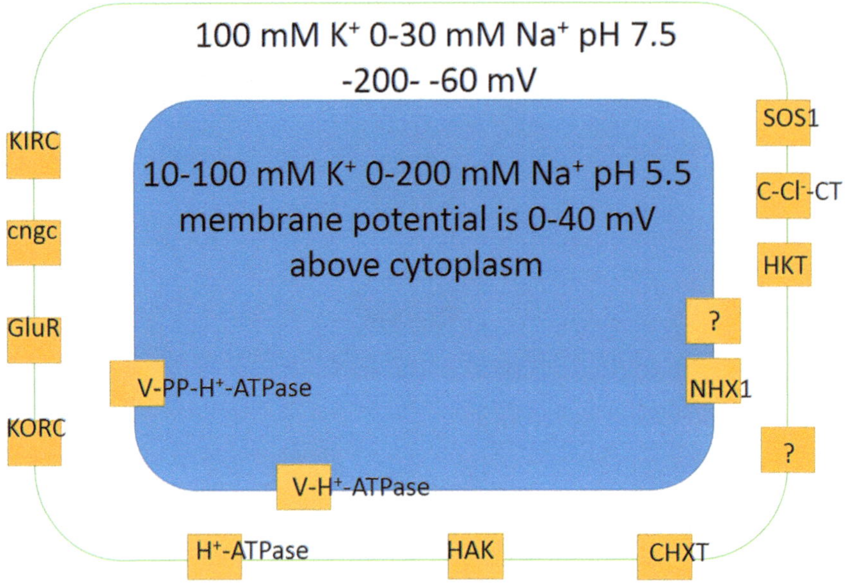

Fig. 10.4 Basic scheme of membrane potentials, K^+ and Na^+ concentrations and pH values in a generalised plant cell together with the main ion transport systems ensuring K^+ and Na^+ transport according to electrochemical forces. Since different cell types are usually specialised for more specific ion transport functions, all the depicted ion channels and transporters may not be active in a given cell type. Rather the concentrations and membrane potentials are indicative and change depending on conditions of mineral nutrition: they are not the same for all cell types (see text and references for more details). KIRC are inward-rectifying K^+ channels (e.g. Hirsch et al. 1998); KORC are outward-rectifying K^+ channels; GluR are glutamate receptors; cngc are cyclic nucleotide-gated ion channels; HAK are high-affinity potassium transporters; CHXT are cation H^+ exchange transporters (e.g. Evans et al. 2012); HKT are high K^+ affinity transporters; C-Cl^--CT are cation chloride cotransporters; SOS1 is a well-studied sodium-proton antiporter; H^+-ATPase is the proton pump of the plasma membrane; V-H^+-ATPase is the vacuolar proton pump; V-PP-H^+-ATPase is the vacuolar pyrophosphatase, another vacuolar proton pump; and NHX1 is vacuolar sodium (cation)/proton antiporter. For more details and description, see text. (From Volkov 2015b, with modifications)

channels/transporters predict, from their coding, over 150 proteins (Mäser et al. 2001). Special databases include information about transport proteins, e.g. plant membrane transport database http://aramemnon.uni-koeln.de/; http://www.yeastgenome.org/ is a useful source of information for yeast proteins including yeast membrane transport proteins.

Ion Channels and Ion Transporters in More Detail

Numerous methods of biophysics, biochemistry, genetics and molecular biology are used to study the properties of individual ion channels and transporters, their structure and molecular interactions: the methods are not described in detail here, but can be found in classic books on ion transport (e.g. Hille 2001) and further numerous reviews (see Volkov 2015b for these). Here we summarise what is known of some of the channels and transporters found in plants.

One controversial question in ion transport is the relative role of ion channels and transporters. It is commonly accepted that ion channels in an open state/conformation allow passage of over 10^6–10^8 ions per second via a selective pore formed within a protein molecule. The diameter of the pore is determined by the molecular structure of the ion channel, from 1.2 nm for the potassium channel KcsA with a narrow part of 0.4 nm in diameter (e.g. Doyle et al. 1998; Jiang et al. 2002; MacKinnon 2004) to 1.5 nm and even

2.8 nm diameters of the general bacterial porins with low selectivity that allow the passage of small hydrophilic molecules (about 0.6 nm pores for the highly selective porins; e.g. Galdiero et al. 2012). The diameter of the pore and the nature of the amino acids lining it essentially determine the ion selectivity of the ion channel and the potential number of ions passing per unit of time. To put these dimensions in context, the diameter of a non-hydrated Na^+ ion is about 0.2 nm and of K^+ is 0.3 nm, while effective diameters of the hydrated ions are estimated as 0.72 nm for Na^+ and 0.62 nm for K^+ (Nightingale 1959; Collins 1997; Mähler and Persson 2012). In spite of the similarities of size, the selectivity could, for example, be over 1000 for K^+ over Na^+ in potassium-selective ion channels or over 10 for Na^+ over K^+ in sodium-selective channels – all due to selectivity filters defined by conserved amino acids for specific channel types. Often the amino acid sequence glycine-tyrosine-glycine (GYG) indicates selectivity for K^+. Introducing mutations into the pore to change the amino acids can convert a potassium-selective ion channels to a nonselective one (Heginbotham et al. 1992). The interactions of ions with the protein molecule of ion channels are not yet well understood, but probably involve non-electrostatic ion–ion interactions, van der Waals forces, interaction with water molecules and numerous other interactions. Several methods of modelling and of simulations of molecular dynamics have been applied within at least the last 30 years; a sharp increase in computing power has allowed the lipid environment of membranes, pH and known biochemical factors and regulators to be included in the models (e.g. reviewed in Maffeo et al. 2012).

Direct measurements (such as patch clamp and several less sensitive electrophysiological and other methods; Table 10.1) are the basis for investigating ion fluxes via ion channels; these methods provide information about permeating ions, number of ions crossing the membrane per second, selectivity and gating during the transport processes. For example, a small current of 1 pA can be measured by patch clamp. This corresponds to an ion (monovalent) current of 10^{-12} A/(1.6×10^{-19} coulombs) $\approx 6 \times 10^6$ ions/second ($+1.6 \times 10^{-19}$ coulombs is the elementary charge – a charge of a monovalent cation or opposite to the negative charge of an electron). Most ion channels demonstrate much larger electric currents with complex voltage-dependent patterns of open–closed states (Fig. 10.5).

Transporters can be considered as enzymes where conformational changes of a protein molecule are required for a complete transport cycle of ions (e.g. Gadsby 2009; Ashcroft et al. 2009). The turnover rate of the transporter is the number of complete transport cycles performed per second (e.g. Longpré and Lapointe 2011). The plant plasma membrane H^+-ATPase pumps about 100 ions per a second (Sze et al. 1999), which is at the lower end of estimates for turnover rate of transporters. The value is, however, comparable to 20 to 100 H^+ per second by yeast plasma membrane ATPase Pma1 (Serrano 1988) and a turnover rate of 160 s^{-1} of animal Na^+/K^+-ATPase (Skou 1998). Similar or even lower turnover rates, from 3 to 60 s^{-1}, were shown for a human sodium/glucose cotransporter (Longpré and Lapointe 2011 and references therein), while a turnover rate of about 500 s^{-1} was estimated for sucrose/H^+ cotransporter from maize ZmSUT1 (Carpaneto et al. 2010). The highest possible turnover rate for activity of ion transporters has been assessed from studies of protein structure and frequency of conformational changes with an estimated upper limit of 10^6 s^{-1} (Chakrapani and Auerbach 2005), although this seems, from actual measurements, to be an overestimated value.

Mechanisms of ion transport by transporters are less well understood than transport via pores of ion channels, although the genes of many transporters have been sequenced and well-studied. Briefly, several mechanisms are predicted for different transporters. HKT transporters with at least eight transmembrane domains could be similar to ion channels: the transmembrane domains form a specific ion-selective pore with properties distinct from the pore of ion channels according to basic crystal structure analysis (Cao et al. 2011; reviewed in Yamaguchi et al. 2013). HKT transporters can resemble ion channels electrically with similar current-voltage (IV)

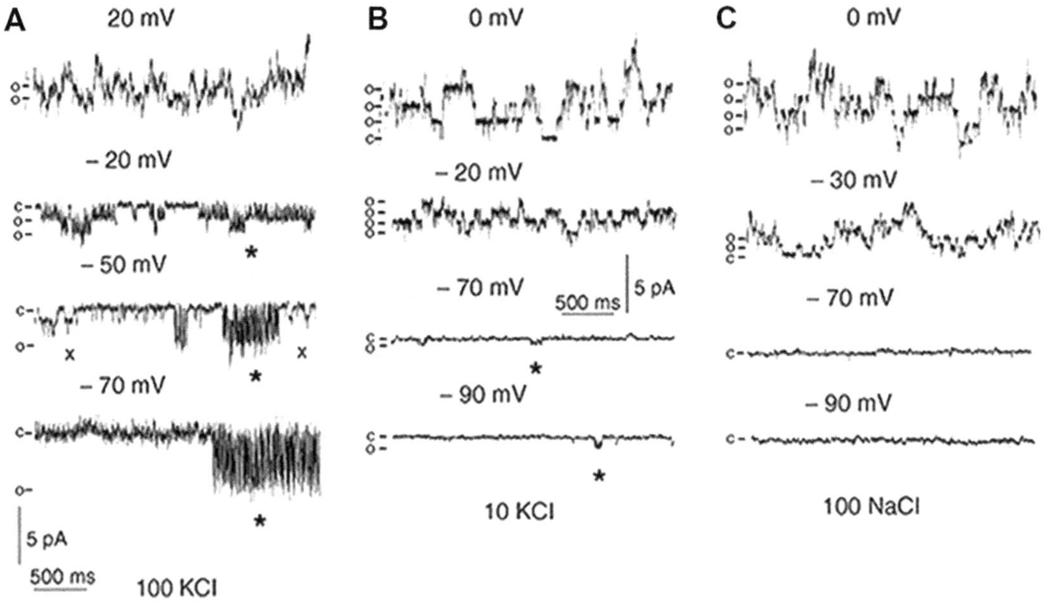

Fig. 10.5 Single-channel electrophysiological recordings in outside-out membrane patches from root protoplasts of the halophyte *Eutrema halophilum*. The pipette solution was 100 mM KCl. The bath solution contained 100 mM KCl (**a**), 10 mM KCl (**b**) or 100 mM NaCl (**c**). *c*, current level with no open channels; *o*, current levels of single-channel openings. Spiky openings of outward-rectifying channels allowing inward current are indicated with asterisks. Openings of a second type of channel are indicated with crosses. (Reproduced from Volkov and Amtmann 2006, with the permission from the publisher John Wiley and Sons)

curves. The reversal potential of ion current mediated by HKT can shift following ion concentrations inside and outside the cell. Furthermore, the pores of HKT transporters have selectivity filters in the first transmembrane domain with conserved glycine for K^+-selectivity and serine for Na^+-selectivity (reviewed in Yamaguchi et al. 2013; de Almeida 2014). An extra amino acid constriction with an arginine residue in the last transmembrane domain makes an additional energy barrier for ion transport (Kato et al. 2007; Cao et al. 2011; reviewed in Yamaguchi et al. 2013; Benito et al. 2014; de Almeida 2014). Some HKT transporters symport K^+ with Na^+ at low K^+ concentrations (reviewed in Waters et al. 2013); presumably Na^+ adds energy for cotransport 'pushing' K^+ via the amino acid constrictions (Benito et al. 2014), although the exact mechanisms of transport are still to be elucidated. Some other monovalent ion transporters found in plants (e.g. cation H^+ exchange transporters and cation chloride cotransporters: see Fig. 10.4 for more details) are less well-studied than HKT.

HAK transporters, which also transport K^+, are not found in animals and Protista (Grabov 2007). Presumably they cotransport K^+ together with H^+ (Bañuelos et al. 1995; Rodriguez-Navarro 2000; Grabov 2007), but have not been crystallised so far, while attempts to express them in *Xenopus* oocytes failed; therefore the mechanisms of ion transport by HAK transporters are not yet well understood. Gene sequences and comparison of HAK transporters predict 10–14 transmembrane domains (Greiner et al. 2011). Amino acid substitutions within the region between the second and third putative transmembrane domains of *Arabidopsis thaliana* HAK5 transporter essentially changed ion transport selectivity indicating a selectivity filter within the region (Alemán et al. 2014). Lack of a putative-specific pore similar to HKT transporters suggests that HAK transporters realise a specific mechanism for K^+/H^+ symport (Alemán et al. 2014).

NHX1 and SOS1 are cation/H^+ antiporters. SOS1 has 12 predicted transmembrane domains at the N-terminal part and a long C-terminal tail composed of 700 amino acids (Shi et al. 2000): the protein forms homodimers (Núñez-Ramírez et al. 2012). The molecular mechanisms of Na^+/H^+ antiport by SOS1 are under investigation, but the known crystal structure of bacterial Na^+/H^+ antiporter NhaA (Hunte et al. 2005) could be a basis for understanding transport by both NHX1 and SOS1. NhaA has 12 transmembrane domains, existing as a dimer; amino acid helices of the protein form negatively charged funnel-like structure, which leads to the cytoplasm from the centre of the membrane and selectively attracts cations. Nha1 in conformation with bound cation follows transformation and releases cation to the outer side being protonated at the same time at aspartate moieties. Deprotonation releases H^+ to the cytoplasmic side and return of the protein to its initial conformation completes the transport cycle (Hunte et al. 2005). The mechanism, with conformation changes limited to a part of the protein, makes NhaA one of the fastest transporters; NhaA has a turnover rate (activity of catalytic centre) of about 89,000 s^{-1} (reviewed in Hunte et al. 2005). Tonoplast NHX1 could share the same transport mechanism.

Ion Concentrations in Cells

Ion concentrations in cells are a reflection of net ion fluxes across the plasma membrane and tonoplast, with concentrations differing between subcellular compartments. Within these compartments, certain ranges of ion concentrations, especially of K^+, Na^+ and Ca^{2+}, are vital for cell and protein function. Typical K^+ concentrations in the cytoplasm of plant cells have been measured independently by several methods (including ion-selective electrodes, fluorescent dyes and X-ray microanalysis) and range around 60–140 mM (Pitman et al. 1981; Hajibagheri et al. 1988; Hajibagheri and Flowers 1989; Walker et al. 1995, 1998; Korolev et al. 2000; Cuin et al. 2003; Halperin and Lynch 2003; Shabala et al. 2006; Hammou et al. 2014), though concentrations above 200 mM have been estimated by efflux analysis (reviewed in Britto and Kronzucker 2008) and 12 mM or lower in root cells of potassium-deprived plants of *Arabidopsis thaliana* (Armengaud et al. 2009). Higher K^+ concentrations of 200–350 mM measured in cell sap by X-ray microanalysis or capillary electrophoresis are likely to reflect vacuolar concentrations accumulated under a sufficient potassium supply (Malone et al. 1991; Fricke et al. 1994; Bazzanella et al. 1998; Volkov et al. 2004). It is worth mentioning that in some animal cells, K^+ seems to be among regulators of apoptotic enzymes, activating them at concentrations below 50 mM (Hughes and Cidlowski 1999). This concentration is under half the 140–160 mM of typical intracellular K^+ in most animals (e.g. Hinke 1959; Campion 1974; Hughes and Cidlowski 1999; Sejersted and Sjøgaard 2000).

Cytoplasmic concentrations of Na^+ in the cells of glycophytes growing in nonsaline substrates (<5 mM concentrations of Na^+) have rarely been measured in plants but are likely to be low (the activity of Na^+ in the cytoplasm of barley roots only reached 1.7 mM over 5 d of salinisation when the external Na^+ concentration had been increased to 200 mM: Carden et al. 2003). The Na^+ concentrations in freshwater algae (growing in the absence of salt or at low Na^+) are also generally low at 1 to 50 mM (see Flowers et al. 2015 for references). Where glycophytes have been exposed to salinity (more than 50 mM Na^+), then data are still rare. After several days of NaCl treatment, the cytoplasmic Na^+ in barley root cortical cells only reached 20–50 mM (Carden et al. 2003) and 30–60 mM in the root hairs of *Arabidopsis thaliana* (Halperin and Lynch 2003). Cytoplasmic Na^+ concentrations in algae growing in salt water are a little, but not much, higher at 20–60 mM (see Flowers et al. 2015). In halophytes, estimates of cytoplasmic Na^+ concentrations are also few, but values of 100–200 mM were deduced by Flowers et al. (2015) in their analysis of the variety of data available. However it is not known if the cytosolic Na^+ itself at a uniform concentration (100–200 mM with activities of 76–140 mM) or perhaps compartmentalised in microvesicles derived from a pinocytotic transport system (Field et al. 1980; Balnokin et al. 2007; Flowers et al. 2018).

High Na^+ concentrations over 100 mM often have inhibiting effect on protein synthesis at least in salt-sensitive glycophytes (Hall and Flowers 1973; Wyn Jones and Pollard 1983; Gibson et al. 1984; Flowers and Dalmond 1992). Na^+ also (1) competes with K^+ for allosteric sites of enzymes and (2) interacts with ion channels (e.g. Na^+ changes the gating of K^+ outward-rectifying currents in root protoplasts of the halophytic plant *Eutrema halophilum*, currents which are most likely carried by Shaker-type K^+ channels: Volkov and Amtmann 2006). Moreover, at the cellular level, salt stress induces apoptosis (Katsuhara and Kawasaki 1996; Huh et al. 2002; briefly reviewed in Shabala 2009; Demidchik et al. 2010). Under salt stress, at least for glycophytes, it is important to maintain a high (a minimum of 1) cytoplasmic K^+/Na^+ ratio (Maathuis and Amtmann 1999; Chen et al. 2007). It is essential, however, to mention that some proteins (due to specific amino acid composition or structural peculiarities) and processes in halophytes are able to withstand higher Na^+ concentrations without loss of activity (e.g. Flowers and Dalmond 1992; Premkumar et al. 2005); this seems to be a secondary evolutionary adaptation. It is also interesting that cell-wall proteins of halophytes and glycophytes do not change their activity within a wide range of Na^+ concentrations, often from 0 to over 0.5–1 M (Thiyagarajah et al. 1996).

Much higher Na^+ concentrations (around 500 mM or more) are tolerated in vacuoles than in the cytoplasm: one of the functions of vacuoles is to sequester and isolate Na^+. Concentrations of Na^+ in vacuoles may exceed 0.5–2 M being up to ten times the cytoplasmic Na^+ concentrations (e.g. Flowers et al. 1977; Zhao et al. 2005; Flowers and Colmer 2008) achieved by the activities of specific ion transport systems (reviewed in Martinoia et al. 2012). Under salt treatment of 2 M NaCl for 85 days, shoot tissue concentrations of Na^+ in species of the halophyte *Tecticornia* were about 2 M, so presumably vacuolar Na^+ concentrations were over 2 M in the leaves (English and Colmer 2013).

Ion Transport in Halophytes

Halophytes need all the minerals required by other plants but manage to acquire these essential elements in the face of very high external concentrations of (normally) Na^+ and Cl^-. Halophytes also use Na^+ and Cl^- for osmotic adjustment, but most do not appear to have an obligate requirement for Na^+; Cl^- is required in low concentrations for photosynthesis as in other plants. The high demands of Na^+ and Cl^- for osmotic adjustment in halophytes lead to considerable fluxes of Na^+ and Cl^- – from 1 to 10 nmol Na^+ g^{-1} fresh mass roots s^{-1}: the range of values for influx is similar, between 2 and 13 nmol g^{-1} fresh mass roots s^{-1} with external NaCl concentrations of 25–500 mM (Flowers and Colmer 2008; Flowers et al. 2018). These influx values may, however, be lower than those reported for glycophytes, suggesting halophytes are able to regulate Na^+ influx (Essah et al. 2003; Wang et al. 2006). The difference between glycophytes and halophytes is that the high fluxes are lethal in glycophytes.

Unfortunately, far fewer molecular data are currently available for halophytes compared to well-studied glycophytes. A few halophytic species have attracted more attention than others and, hence, allow us to come to some conclusions; these species are *Suaeda maritima*, *Mesembryanthemum crystallinum*, *Thellungiella halophila* (recently reclassified to *Eutrema halophilum*), *Spartina* spp., *Zostera marina*, *Rhizophora mangle*, *Avicennia marina* and several more species from the Poaceae and Amaranthaceae (a table with numbers of publications per specific halophyte between 1997 and 2007 is given in Flowers and Colmer 2008). In the following section, we address distinct features of ion transport systems of halophytes and their glycophytic relatives.

Ionic Environment Encountered by Halophytes with a Special Attention to Seawater

In order to understand ion transport in halophytes, we need a brief characterisation of their physicochemical environment in order to (1) predict which ion transport systems could be used for transport and (2) to reveal potential strategies of the plants for ion homeostasis. We consider a typical example of seawater with Na^+ concentration about 500 mM, Cl^- concentration about 600 mM, K^+ concentration about 10 mM and pH buffered to 8.3. Under these conditions, the reversal potential for Na^+ (assuming even high range of 200 mM cytoplasmic Na^+, see above) is over +20 mV, for K^+ (assuming 100 mM cytoplasmic K^+) is about −60 mV, for Cl^- (assuming 60 mM cytoplasmic Cl^-; e.g. Flowers et al. 2015) is about −60 mV and for H^+ (assuming cytoplasmic pH as 7.4–7.5) is about −50 mV. This means that with a membrane potential of the cell below +20 mV, passive transport of Na^+ would occur into the cell and the transport be limited by the conductance of the membrane for Na^+. Passive transport of K^+ and H^+ would be inward at membrane potentials below −60 mV and −50 mV, respectively. In contrast, passive fluxes of negatively charged Cl^- would be outward at membrane potentials below −60 mV. Therefore, a reasonable strategy for a halophyte root cell in seawater would be to keep its membrane potential below −60 mV and to minimise Na^+ uptake by reducing the Na^+ permeability of its plasma membrane. Moreover, in order to avoid further increase in the Na^+ concentration in the cytoplasm, fast Na^+ transport to the central vacuole and/or Na^+ efflux out of the cell are needed.

The role of each process (Na^+ efflux or transport to vacuole) differs between halophytes. *Suaeda maritima*, a salt marsh species, showed little or no Na^+ efflux after pre-loading with $^{22}Na^+$ (Yeo 1981), while roots of *Eutrema halophilum* (*Thellungiella halophila*, a plant of dry saline soils) demonstrated high $^{22}Na^+$ efflux equivalent to about 75–80% of Na^+ taken up (Wang et al. 2006). *Eutrema halophilum* also exhibited decreased salt tolerance after reduced expression of plasma membrane Na^+/H^+ antiporter ThSOS1 (Oh et al. 2009). The essential role of Na^+ efflux was demonstrated also for the marine halophyte *Zostera marina* (the efflux helped to keep low cytoplasmic Na^+ concentrations about 10 mM for this species (Rubio et al. 2011)), which grows submerged in seawater. H^+-ATPase could be used to lower the membrane potential and keep it below −60 mV. Since in seawater pH is higher outside of the cell, Na^+ efflux by Na^+/H^+ antiporters would require several H^+ per single Na^+ for Na^+ efflux, for thermodynamical reasons. An initial estimate indicates that at a membrane potential around −150 mV and the abovementioned reversal potentials of +20 mV for Na^+ and −50 mV for H^+, the minimal energy for efflux of one Na^+ would be equal to the total voltage difference multiplied by the transferred electric charge. The assumption is based on energy conservation law and definition of voltage as energy in electric field per a unit of charge. Hence, the energy for Na^+ efflux would be (150 + 20 = 170) mV * electric charge of Na^+, while antiport of one H^+ can provide just (150–50 = 100) mV * electric charge of H^+. Therefore, antiport of at least 2 H^+ is required per efflux of one Na^+; more complex analysis would include the mechanism of transport with electroneutral or electrogenic components. Potentially cells may use layers of mucilage to maintain the acidity of their surroundings, hence shift the reversal potential for H^+ (it is +60 mV at external pH 6.5 and cytoplasmic pH 7.5) and still use Na^+/H^+ antiporters with a stoichiometry 1:1. Indeed, pericellular acid zones were found in the marine angiosperm *Zostera marina* (Hellblom et al. 2001; Mercado et al. 2003) and in the marine brown macroalga *Laminaria saccharina* (Axelsson et al. 2000; Mercado et al. 2006); the generation of the acid zones in *Zostera marina* depended on respiration (Carr and Axelsson 2008). The pericellular acid zones are important for photosynthesis and potentially may serve for facilitated Na^+ efflux. Not surprisingly, the important role of Na^+ efflux (which likely depends on active Na^+/H^+ exchange transport systems in the plasma membrane) was shown for *Zostera marina* (Rubio et al. 2011). Fast transport of excessive Na^+ to leaves and stems could also be implied for halophytes. The strategies of Na^+ efflux and localisation of Na^+ in vacuoles or transport to leaves would require metabolic energy, but essentially do not differ from the strategies of glycophytes in coping with high Na^+. However, marine halophytes are not limited by K^+ (the concentration of which in seawater is at least 10- and maybe 100-fold that in most soils) and whose uptake could use other sets of ion transport systems. Based on theory and plasma membrane location of HAK transporters, they are not needed for K^+ uptake from seawater. Indeed, HAK transporters were suggested to be important under low K^+ concentrations, and their genes are mostly found in land glycophytes, though also in halophyte *Mesembryanthemum crystallinum* (Grabov 2007).

Ion Concentrations and Membrane Potential

Ion Concentrations in Halophytes in Comparison to Glycophytes

Some halophytes can tolerate or may even require for optimal growth much higher cytoplasmic Na^+ concentrations compared to glycophytes. Cytoplasmic concentrations around 150–200 mM were estimated for *Suaeda maritima*, *Atriplex nummularia* and several other halophytic species (summarised in Flowers et al. 2015). Direct measurements using sodium-selective electrodes or fluorescent dyes are still required to confirm the estimates since much lower cytoplasmic concentrations around 10 mM were measured for *Zostera marina* (Rubio et al. 2011). However, a concentration of 200 mM is presumably 3–10 times over the measured Na^+ cytoplasmic concentrations of 20–60 mM after several days of mild NaCl treatment reported for glycophytes (for barley, Carden et al. 2001 2003; for *Arabidopsis*, Halperin and Lynch 2003). This makes ion transport properties different, at least quantitatively, in halophytes compared to glycophytes. This also suggests that halophytes have specific Na^+ uptake systems and specific regulation of ion transport compared to glycophytes. Indeed, cytoplasmic Na^+ is a regulator of ion transport systems. For example, in *Arabidopsis thaliana* the potassium-selective ion channel AKT1 is important for K^+ uptake from dilute nutrient solutions at the expenses of negative cell membrane potential: the activity of the channel is associated with inward-rectifying electric currents in electrophysiological patch clamp experiments. Interestingly, addition of 10 mM Na^+ to the cytoplasmic side of the plasma membrane for root protoplasts from *Arabidopsis thaliana* inhibited the current (Qi and Spalding 2004). It was proposed that the K^+-deficient phenotype of SOS1 mutants of *Arabidopsis thaliana* is explained by this inhibition, since mutants lack SOS1 transporters to remove excessive Na^+ from cytoplasm (Wu et al. 1996). Presumably, halophytes have different regulation of channels and transporters by Na^+.

One more example is regulation by intracellular Na^+ of ion currents in guard cells of the halophyte *Tripolium pannonicum* subsp. *tripolium* (synonym *Aster tripolium*) and its glycophytic relative *Aster amellus* (Robinson et al. 1997; Véry et al. 1998). An inward-rectifying potassium-selective electric current in guard cells was strongly inhibited by internal 100 mM Na^+ in the halophyte, while unaffected in the glycophyte. The inhibition was evident after 10 min of treatment, developed in time and depended on free Ca^{2+} concentration: buffering free Ca^{2+} abolished the effect (Véry et al. 1998). Presumably, the inhibition of inward current was the reason why the halophyte decreased stomatal aperture under increased Na^+ in preparations of epidermal strips, while the glycophyte, in contrast, increased stomatal aperture. Here the similar ion currents with resembling ion selectivities, conductances and current-voltage dependencies demonstrated different regulation by intracellular Na^+ in the halophyte and related glycophyte.

High vacuolar Na^+ concentrations are encountered both in halophytes and in glycophytes, though the values are still several times higher in halophytes. However, the ratios of vacuolar to cytoplasmic Na^+ concentrations are not strikingly different. Obviously, more information and direct measurements are required to get the exact numbers for further considerations and conclusions.

Membrane Potentials in Halophytes in Comparison to Glycophytes

The membrane potential of a cell is an integral parameter characterising membrane ion conductances, the activity of plasma membrane H^+-ATPase and electrogenic transporters. Moreover, a stable negative membrane potential is essential for taking up charged cations via ion channels. It seems that, in general, halophytes do not alter their membrane potential under different salt concentrations when cells of glycophytes strongly depolarise under salt treatment. Comparison of the membrane potential in epidermal root cells of the glycophyte *Arabidopsis thaliana* and the halophyte *Eutrema halophilum* (*Thellungiella halophila*) did not reveal significant differences (Volkov and Amtmann 2006). The values were −105 mV and −119 mV in 1 mM KCl for *Arabidopsis thaliana* and *Eutrema halophilum*, respectively. However, addition of 100 mM NaCl depolarised cells of the glycophyte by 63 mV but the halophyte by only 37 mV. After several successive cycles of changes from 100 mM NaCl +1 mM KCl to 1 mM KCl, cells of *Arabidopsis thaliana* depolarised to −25 mV, while cells of *Eutrema halophilum* kept the same initial membrane potential of −138 mV (Volkov and Amtmann 2006).

Addition of 100 mM NaCl to roots of sunflower (a glycophyte) plants depolarised their outer cortical cells from about −160 mV to about −50 mV after 24 h of treatment (Cakirlar and Bowling 1981): the effect of 67 mM Na_2SO_4 was even more pronounced. A low concentration of 1 mM NaCl, on the other hand, induced hyperpolarisation to −180 mV within 4 h of treatment. Interestingly, negative values of the membrane potential linearly correlated with the fresh weight of plants over the period of growth for 2 weeks with added NaCl (Cakirlar and Bowling 1981). High concentration of Cl^- also induced depolarisation for root epidermal cells of the aquatic glycophyte *Trianea bogotensis* (Akhundova et al. 1990).

A different situation has been observed for halophytes. The membrane potential for cortical root cells of the halophyte *Atriplex hastata* was about −130 mV in nutrient solution with added 100 mM NaCl (Anderson et al. 1977). The value was relatively stable around −140 to −130 mV when NaCl concentration changed across a surprisingly wide range (100–600 mM) of external NaCl concentrations. Addition of 1 mM KCN, however, depolarised the membrane by 60 mV for all the NaCl concentrations indicating an essential role of energy-dependent component, presumably H^+-ATPase fuelled by ATP provided by cell metabolism (Anderson et al. 1977). Similar and even more pronounced results have been obtained with the coastal halophyte, *Salicornia bigelovii* (L'Roy and Hendrix 1980). The plants were adapted to several concentrations of NaCl for 2 weeks, and within the period cells changed ion concentrations and membrane potential in a complex manner. For short-term treatments, the membrane potential of root cortical cells in *Salicornia bigelovii* hyperpolarised with increasing external salt concentration, which was applied in the form of artificial seawater (ASW). 10% ASW corresponded to −115 mV; for 50% ASW and 100% ASW, the membrane potential of root cells was around −150 mV. The behaviour of the plasma membrane potential of root cortical cells was unusual at low (10% ASW) and high (100% ASW) salt: the membrane was depolarised by the uncoupler CCCP at low salt and hyperpolarised at high salt. CO gas depolarised the membrane by about 50 mV under both conditions suggesting an essential electrogenic respiration-dependent component of the membrane potential (L'Roy and Hendrix 1980). The measured membrane potential was below $E_{reversal}$ for K^+ by 30 to 60 mV also supporting the presence of electrogenic component (L'Roy and Hendrix 1980) (see above in "Mechanisms of Ion Transport" and in "Ionic Environment Encountered by Halophytes with a Special Attention to

Seawater" where the reversal potentials of K^+ which constitute the potassium-dependent component of membrane potential are discussed: the lower values of membrane potential are usually due to activity of plasma membrane H^+-ATPase). In line with these findings is the observed sharply increased gene expression of plasma membrane H^+-ATPase in the halophyte *Atriplex nummularia*, especially in the roots of the plant, after salt treatment while a much lower increase was found in the glycophyte *Nicotiana tabacum* under the same conditions (Niu et al. 1993). Similarly, membrane potentials of leaf cells in the marine halophyte *Zostera marina* were about −160 to −150 mV in natural seawater. The membrane potential was composed of two components (Fernández et al. 1999; Rubio et al. 2011): the first component of about −90 mV was associated with electrogenic activity of plasma membrane H^+-ATPase and the second was due to asymmetric ion distribution. The membrane potential of *Zostera marina* was influenced by cyanide (membrane depolarised), fusicoccin (membrane hyperpolarised), CO_2 (membrane hyperpolarised) and responded to changes in illumination (Fernández et al. 1999). Rapid upregulation of plasma membrane H^+-ATPase activity in cells from mature and elongation zones of roots was suggested to be the reason for lower depolarisation of the cells in halophytes *Atriplex lentiformis* and *Chenopodium quinoa* after salt treatment (Bose et al. 2015).

Comparable Pairs of Halophytes and Glycophytes

Comparison of ion fluxes via membranes between halophytes and glycophytes often demonstrates lower sodium uptake for halophytes (reviewed in Flowers and Colmer 2008; compare Table 2 from Flowers and Colmer 2008 with Table 1 from Kronzucker and Britto 2011). However, an evident problem in comparison is the high variability in ion transport between plant species because of growth rate, tissue-specific variability and other physiological factors. Consequently, it is more useful to consider similar plants in order to achieve comparable values. Recent results on ion fluxes between the glycophyte *Arabidopsis thaliana* and the halophyte *Eutrema halophilum*, two species that are similar from the point of genome and morphology, demonstrated lower Na^+ fluxes (per surface membrane area in root protoplasts and for uptake per dry weight) and higher K^+/Na^+ selectivity of ion currents in the roots and root protoplasts of the halophyte under salt treatment (Figs. 10.3 and 10.4 and in Volkov et al. 2004; Wang et al. 2006; Wang 2006; Volkov and Amtmann 2006; Amtmann 2009). Similar results for roots of the two plants were confirmed later in Alemán et al. (2009). Lower Na^+ accumulation and higher K^+/Na^+ ratio under salt treatment were also found in roots of leguminous halophyte *Melilotus indicus* compared to similar herbaceous glycophytic species, *Medicago intertexta* (Zahran et al. 2007).

Other strategies may be used by halophytes depending on the level of salinity tolerance, plant morphology, habitat and the other environmental factors and evolutionary history. For example, salt-tolerant *Plantago maritima* had similar Na^+ uptake rates by roots compared to salt-sensitive *Plantago media* (Erdei and Kuiper 1979; de Boer 1985); rather, salt tolerance in the pair was associated with xylem transport and sodium accumulation in vacuoles of leaf cells.

Vacuolar membranes of several halophytes have also been a subject of special investigation. Electrophysiological patch clamp study of vacuoles from leaves of *Suaeda maritima* did not find any unusual features to support the high salt tolerance of the halophyte (Maathuis et al. 1992). Patch clamp experiments to compare vacuoles from roots of *Plantago maritima* and *Plantago media* also did not reveal striking differences apart from an extra smaller ion channel conductance in the tonoplast of the halophyte; salt stress essentially reduced the open probability of larger nonselective between K^+ and Na^+ ion channel conductance in both species but did not change the properties of the conductance (Maathuis and Prins 1990).

Sodium Uptake in Halophytes

External sodium concentrations between 50 and 400 mM often have a positive effect on the growth of halophytes (reviewed in Flowers and Colmer 2008; Shabala and Mackay 2011; also Yue et al. 2012; Pan et al. 2016), which is in sharp contrast to its effect on glycophytes where such salt concentrations are inhibitory and, generally, in the long run, lethal (though Na^+ can have beneficial effect for several glycophytes at concentrations below 1 mM; reviewed in Jennings 1976; Subbarao et al. 2003; Kronzucker et al. 2013). The question then arises as to what aspects of the membrane and cellular pathways for sodium uptake by roots of halophytes enable them to flourish in saline substrates.

For glycophytes, HKT-type and probably HAK transporters are suggested to be the high-affinity pathway for Na^+ uptake (reviewed in Kronzucker et al. 2013; Maathuis et al. 2014; Nieves-Cordones et al. 2016). Cyclic nucleotide-gated channels (Maathuis and Sanders 2001), other nonselective cation channels (Demidchik and Tester 2002; Essah et al. 2003; Demidchik and Maathuis 2007) and glutamate receptors (Davenport 2002) are for low-affinity pathways of Na^+ transport. The distinction between high affinity (or functioning at lower Na^+ concentrations, usually with $K_{1/2}$ below 100 μM for Na^+ uptake, e.g. Haro et al. 2010) and low affinity (or functioning at higher Na^+ concentrations, within the millimolar range) is somewhat arbitrary and depends on plant species, range of membrane potentials and the composition of nutrient solution.

Again, surprisingly little is known for halophytes. Nonselective cation currents with higher selectivity for K^+ over Na^+ (about 6.7 from $E_{reversal}$ and 1.7 from conductance) seem to be the main entry pathway for Na^+ to root protoplasts and presumably roots of *Eutrema halophilum*; the K^+/Na^+ selectivity of the pathway was much lower (about 1.9 from $E_{reversal}$ and 1.4 from conductance) in *Arabidopsis* (Volkov and Amtmann 2006).

For *Suaeda maritima*, the low-affinity pathway was analysed using several blockers of ion channels, physiologically active compounds and several other treatments under two levels of external NaCl, 25 mM or 150 mM (Wang et al. 2007). Neither external 0.5 mM cAMP nor 10 mM Ca^{2+} had an effect on Na^+ concentrations in plants after 24 h of treatment and NaCl exposure; 5 mM Ba^{2+} decreased Na^+ concentrations in plants and inhibited root $^{22}Na^+$ influx at both NaCl concentrations. Each of 10 mM TEA^+ and 3 mM Cs^+, which block outward and inward K^+ channels, respectively, decreased Na^+ concentrations under 150 mM NaCl only. Moreover, TEA^+ and Cs^+ inhibited the root $^{22}Na^+$ influx at 150 mM NaCl, but not at 25 mM NaCl. The experiments led to conclusions that at lower (25 mM) NaCl concentration, the main pathway of Na^+ uptake is via HKT-like transporters; that at higher (150 mM) NaCl concentrations, Na^+ is taken up via AKT1-like (Shaker type) ion channels; and that cyclic nucleotide-gated ion channels were not involved in Na^+ uptake in *Suaeda maritima* (Wang et al. 2007). Unfortunately, so far this is the only relatively detailed research on mechanisms of Na^+ uptake in halophytes, and, evidently, electrophysiological analysis together with membrane potential measurements would be very helpful for deciphering the exact mechanisms of Na^+ transport.

Comparable Ion Transport Systems

Again little is known about halophytes and the molecular identities of their ion transport systems. The genome of *Eutrema parvulum* (*Thellungiella parvula*) has been sequenced and analysed (Dassanayake et al. 2011). As with *Arabidopsis thaliana*, genes of ion channels and transporters were found, though *Eutrema parvulum* has more genes encoding transporters than *Arabidopsis thaliana*. Attempts to find a determinant of salinity tolerance in *Eutrema parvulum* compared to *Arabidopsis thaliana* led to the discovery of SOS1, a plasma membrane Na^+/H^+ antiporter (Oh et al. 2009) and to HKT1, an electrogenic Na^+ or K^+ transporter (Ali et al. 2013). RNA interference was used to reduce expression of ThSOS1, which resulted in decreased salt tolerance of the halophyte and faster Na^+ accumulation in shoots (Oh et al. 2009). It should be mentioned, however, that results from experiments with seedlings should not be generalised to older plants and the whole transcriptome is often influenced by the nature of treatments. A similar approach of RNA interference allowed reduction of expression of TsHKT1;2 in *Eutrema salsugineum* (*Thellungiella salsuginea*): the *TsHKT1;2-RNAi* plants also had lower salinity tolerance and exhibited severe symptoms of potassium deficiency (Ali et al. 2013). TsHKT1;2 is involved in K^+ uptake and is important for maintaining the K^+/Na^+ ratio, thus alleviating the effects of high Na^+. Interestingly, at the transcriptional level, *E. salsugineum* differs from *Arabidopsis thaliana*: *TsHKT1;2* is dramatically induced upon salt stress, whereas *AtHKT1* is downregulated. Moreover, the amino acid sequences of TsHKT1;2 and AtHKT1 differed slightly rendering higher K^+ affinity to TsHKT1;2 than AtHKT1 (Ali et al. 2013).

Several ion transporters from another halophyte, *Mesembryanthemum crystallinum*, have also been successfully studied (Su et al. 2002, 2003). Expression of K^+ transporters McHAK1 and McHAK3 was induced by treatment with 400 mM NaCl in a tissue-specific manner, mainly in roots and leaves, with a distinct temporal pattern for each transporter. Presumably, the increased expression helps to improve K^+ nutrition of plants under NaCl treatment (Su et al. 2002). A HKT-like transporter, McHKT1, was also revealed in *Mesembryanthemum crystallinum* (Su et al. 2003). The gene *McHKT1* was cloned and heterologously expressed in oocytes of *Xenopus laevis*, where corresponding ion currents demonstrated low selectivity between K^+ and Na^+. The expression pattern of *McHKT1* indicated plasma membrane location of the transporter and transient induction by 400 mM NaCl treatment: expression peaked after 6 h of treatment and later decreased (Su et al. 2003). The exact role of McHKT1 remains uncertain although the transporter could be important for initial Na^+ uptake providing a signal for NaCl adaptation and for unloading Na^+ from xylem to leaves (Su et al. 2003). At the same time, transcripts of Shaker-type K^+ root-specific ion channel MKT1 significantly dropped under NaCl treatment (Su et al. 2001). It might be that the halophyte under salt treatment switches ion transport from ion channels to ion transporters to aid regulation of Na^+ transport and with lower fluxes that could be easier to direct and locate. Generally speaking, sets of several different transporters are expressed in a coordinated manner after NaCl treatment in *Mesembryanthemum crystallinum* (Su et al. 2003).

It would be interesting to express heterologously the cation transporters and ion channels from halophytes and study their selectivities and regulation, a difficult and labour-intensive aim. It is worth noting, however, that often activity of ion channels and transporters requires extra proteins under heterologous expression. For example, activity of an inward K^+ channel from *Arabidopsis thaliana* AKT1 was recorded when extra Ca^{2+}-sensing calcineurin B-like protein AtCBL1 (or AtCBL9) and interacting with it protein kinase AtCIPK23 were co-expressed together with AKT1 (Li et al. 2006; Xu et al. 2006). Interestingly, a regulatory Ca^{2+} sensor AtCBL1 and kinase AtCIPK23 from *Arabidopsis thaliana* were sufficient to record activity of inward potassium ion channel HvAKT1 from barley under heterologous expression (Boscari et al. 2009). However, no method is free of limitations and potential erroneous conclusions (e.g. discussed in Rodríguez-Navarro and Rubio 2006 and Wang et al. 2007). Interactions between ion transport systems and numerous feedbacks can alter the expression of the other transport system in knockout and overexpressing mutants in an unpredicted manner. Any biological system is very complex with interwoven and interacting elements; our present knowledge led to systems biology where an organism is studied as a whole with specific topology of networks for thousands of interacting genes, proteins and metabolites. Moreover, a heterologous expression system may not reproduce kinetic characteristics of transporters *in planta* (Garciadeblás et al. 2003; Haro et al. 2005). There is no perfect system currently available to investigate K^+ and Na^+ uptake in halophytes.

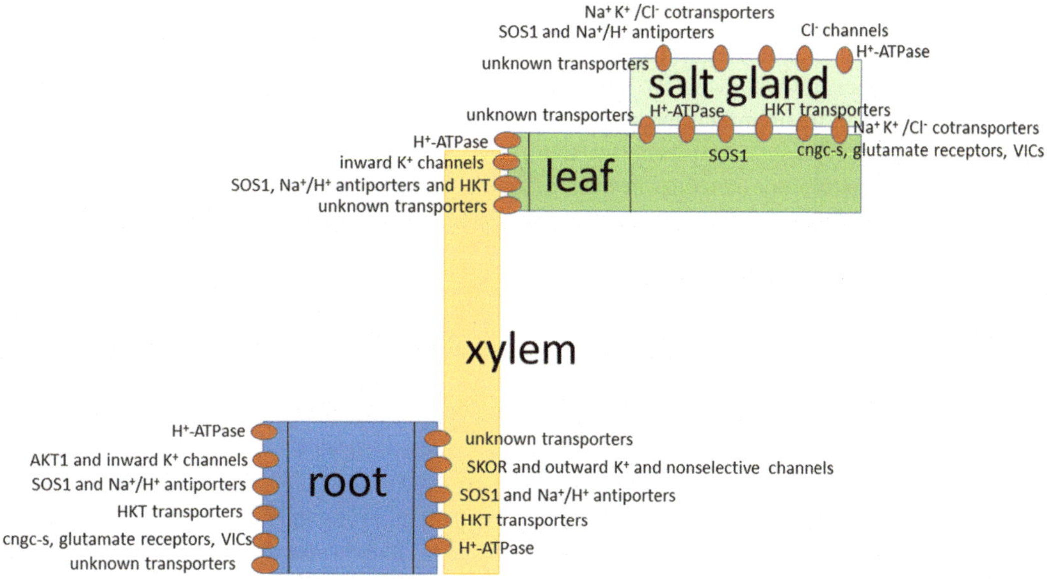

Fig. 10.6 Main determinants of ion transport within whole plants. Uptake of ions by root epidermal cells depends on ion concentrations in the external solution, on relatively stable ion concentrations in the cytoplasm and on plasma membrane potential. H^+-ATPase can shift the membrane potential to more negative values. AKT1 and inward K^+ channels are the pathway for selective K^+ uptake, but some of them could non-selectively transport Na^+ into cells. SOS1 and Na^+/H^+ transporters are mostly for Na^+ efflux; HKT transporters are able to transport K^+ or Na^+. Glutamate receptors, cngc-s (cyclic nucleotide-gated ion channels) and VICs (voltage-independent currents) are mostly for slow Na^+ uptake. Xylem loading also depends on physicochemical parameters in xylem parenchyma cells and the xylem, including the pH of xylem sap. SKOR (stellar K^+ outward-rectifying ion channel) and nonselective outward-rectifying ion channels activate on cell depolarisation and release K^+ and Na^+ to the xylem. SOS1 releases Na^+ depending on the pH of the xylem sap controlled by H^+-ATPases; HKT is also involved in Na^+ and K^+ loading and reuptake. Complex regulation of xylem loading by phytohormones, Ca^{2+} signalling and 14-3-3 proteins among others is reviewed in De Boer and Volkov (2003) and Köhler (2007). Ion transport to salt glands, which are present in a small number of halophytes, requires activity of ion transporters including Na^+ and K^+/Cl^- cotransporters. (Reviewed in Yuan et al. 2016)

Crucial Points for Sodium Transport in Halophytes

Uptake and redistribution of Na^+ and K^+ are regulated at the levels of (1) uptake of ions via the plasma membrane of root cells, (2) xylem loading and unloading to transport the ions to leaves and redistribute them within the plant and (3) ion secretion via salt glands or transport to salt bladders if any. Each of the processes is essential for determining the final result of halophytism (Fig. 10.6).

Xylem loading and unloading also depend on electrochemical differences for specific ions at the surfaces of xylem parenchyma cells (e.g. reviewed in De Boer and Volkov 2003). The physicochemical forces for ion transport depend on membrane potential of cells lining the xylem vessel, on ion concentrations in the cytoplasm of the cells and on concentrations of ions in the xylem sap. Assuming xylem Na^+ concentrations around 1 mM without NaCl treatment and around 50 mM under NaCl treatment (e.g. estimated by Clipson and Flowers 1987 for *Suaeda maritima*; Su et al. 2001 for *Mesembryanthemum crystallinum*; reviewed in Shabala and Mackay 2011), it is possible to understand that presumably different transporters operate within the range of concentrations. SOS1 transporters and the other Na^+/H^+ antiporters are important for loading Na^+ against the concentration differences but in a pH-dependent manner: the pH of xylem sap influences the Na^+ transport, and the plasma membrane H^+-ATPase of xylem parenchyma cells is indirectly involved in the process, providing energy for loading (other pH-regulating systems do also participate; De Boer and Volkov 2003). HKT-like transporters seem to be important for loading or retrieval of Na^+ according to any concentration gradient. Unfortunately, not much has been studied in detail for halophytes. Xylem K^+ concentrations are relatively stable (e.g. were reported as 50 mM independently of NaCl treatment for *Mesembryanthemum crystallinum*; Su et al. 2001). Compared to Na^+, K^+ loading depends more on outward ion channels (Wegner and Raschke 1994; Wegner and De Boer 1997a 1997b; De Boer and Volkov 2003). Recently an outward-rectifying ZmSKOR ion channel was shown to be important for K^+ homeostasis in the xerohalophyte *Zygophyllum xanthoxylum* under salinity and drought (Hu et al. 2016), while ZxSOS1 (Ma et al. 2014) and ZxNHX (Yuan et al. 2015) were important for controlling long-distance transport and K^+ and Na^+ homeostasis at the whole plant level for this xerohalophyte.

Salt bladders and glands are present in a relatively small number of halophytes; probably less than 25% of halophytes are equipped with salt glands to extrude salts (reviewed in Yuan et al. 2016; Dassanayake and Larkin 2017). Interestingly, some halophytes (e.g. *Tamarix ramosissima*) are able to increase sharply Na^+ secretion via salt glands and even increase Na^+/K^+ ratio in secreted liquid under high compared to lower salinity (Shuyskaya et al. 2017). The basic ion channels and transporters discovered so far in salt glands are the same as described for glycophytes including nonselective cation channels, SOS1, HKT-like and the other transporters. High activity of the plasma membrane ATPase is required to drive the unidirectional ion transport and exclude the salts. Proteomics of salt bladders of *Mesembryanthemum crystallinum* revealed many proteins responsible for H^+-transport, especially a large number of vacuolar V-ATPase subunits (Barkla et al. 2012). However, the isolation of individual genes of channels and transporters and understanding the properties and potentially peculiarities of ion transport systems is still required to understand the mechanisms of ion transport in salt glands and to salt bladders.

Problems and Knowns–Unknowns

So far, we have many more unknowns than knowns and a rather patchy collection of data for different halophytes. To complete a comprehensive view of ion transport in individual halophytes, it is reasonable to start from several 'model' species. The choice of the plants is suggested in Flowers and Colmer (2008; Table 6) based on taxonomical group, presence or absence of salt glands, type of photosynthetic carbon assimilation, level of tolerance and a few more features. Some of the list of model halophytes, *Mesembryanthemum crystallinum*, *Avicennia marina*, *Eutrema halophilum*, *Limonium* spp., *Hordeum marina* and *Spartina* spp., *Rhizophora mangle, Suaeda maritima* and *Zostera marina* (Table 6 in Flowers and Colmer 2008), have been mentioned above.

We now need direct measurements of ion concentrations, ion fluxes and electric membrane potentials, genome sequencing, sequencing of noncoding regulatory RNAs, analysis of individual genes in heterologous expression systems, analysis of protein regulation and protein properties as well as cytological studies of vesicular transport. This whole story needs to be composed for specific halophytes starting from membrane potentials, electrophysiological understanding of ion conductances and ion fluxes of Na^+ and K^+ via membranes, with further searches for individual ion channels and transporters responsible for the conductances and ion fluxes. The next step is to understand peculiarities of the ion channels and transporters at the molecular level, ion selectivities and regulation. Basically, the approach of system biology is to understand the network of involved genes, small regulatory RNAs and transcriptional factors and compose an integral description of ion transport for specific species of halophytes. There is still a long story to discover, though the way of completing complex integral schemes for the range of selected 'model' halophytes would form cornerstones for our understanding of ion transport in most and probably all halophytes.

One more complex question that could be posed is whether or not any other mechanisms of ion transport could be important for halophytes. There is, for example, evidence that vesicular transport of ions could be important under some circumstances (e.g. in salt glands of halophytes, e.g. reviewed in Yuan et al. 2016): further investigation is needed to clarify the role (if any) of vesicular transport of ions in halophytes (the topic of potential vesicular transport is reviewed in depth in Flowers et al., 2018).

References

Akhundova TS, Mardanov AA, Ali Zade VM (1990) Effect of kinetin on the membrane potential of root epidermal cells of *Trianea bogotensis* in a nutrient medium with an excessive amount of chloride. Izvestiya Akademii Nauk Azerbaidzhana Seriya Biologicheskikh Nauk 0(1):12–17

Alemán F, Nieves-Cordones M, Martínez V, Rubio F (2009) Potassium/sodium steady-state homeostasis in *Thellungiella halophila* and *Arabidopsis thaliana* under long-term salinity conditions. Plant Sci 176:768–774

Alemán F, Caballero F, Ródenas R, Rivero RM, Martínez V, Rubio F (2014) The F130S point mutation in the *Arabidopsis* high-affinity K^+ transporter AtHAK5 increases K^+ over Na^+ and Cs^+ selectivity and confers Na^+ and Cs^+ tolerance to yeast under heterologous expression. Front Plant Sci 5:430

Ali A, Park HC, Aman R, Ali Z, Yun D-J (2013) Role of HKT1 in *Thellungiella salsuginea*, a model extremophile plant. Plant Signal Behav 8:e25196

Amtmann A (2009) Learning from evolution: *Thellungiella* generates new knowledge on essential and critical components of abiotic stress tolerance in plants. Mol Plant 2:3–12

Anderson WP, Willcocks DA, Wright BJ (1977) Electrophysiological measurements on root of *Atriplex hastata*. J Exp Bot 28:894–901

Armengaud P, Sulpice R, Miller AJ, Stitt M, Amtmann A, Gibon Y (2009) Multilevel analysis of primary metabolism provides new insights into the role of potassium nutrition for glycolysis and nitrogen assimilation in *Arabidopsis* roots. Plant Physiol 150:772–785

Ashcroft F, Gadsby D, Miller C (2009) Introduction. The blurred boundary between channels and transporters: we dedicate this volume to the memory of Peter Läuger, a pioneer of the link between channels and pumps. Phil Trans R S B: Biol Sci 364:145–147

Axelsson L, Mercado J, Figueroa F (2000) Utilization of HCO_3^- at high pH by the brown macroalga *Laminaria saccharina*. Eur J Phycol 35:53–59

Balnokin YV, Kurkova EB, Khalilova LA, Myasoedov NA, Yusufov AG (2007) Pinocytosis in the root cells of a salt-accumulating halophyte and its possible involvement in chloride transport. Russ J Plant Physiol 54:797–805

Bañuelos MA, Klein RD, Alexander-Bowman SJ, Rodriguez-Navarro A (1995) A potassium transporter of the yeast *Schwanniomyces occidentalis* homologous to the Kup system of *Escherichia coli* has a high concentrative capacity. EMBO J 14:3021–3027

Barkla BJ, Vera-Estrella R, Pantoja O (2012) Protein profiling of epidermal bladder cells from the halophyte *Mesembryanthemum crystallinum*. Proteomics 12:2862–2865

Bazzanella A, Lochmann H, Tomos AD, Bächmann K (1998) Determination of inorganic cations and anions in single plant cells by capillary zone electrophoresis. J Chromatogr 809:231–239

Benito B, Rodríguez-Navarro A (2003) Molecular cloning and characterization of a sodium-pump ATPase of the moss *Physcomitrella patens*. Plant J 36:382–389

Benito B, Haro R, Amtmann A, Cuin TA, Dreyer I (2014) The twins K^+ and Na^+ in plants. J Plant Physiol 171:723–731

Bennett TH, Flowers TJ, Bromham L (2013) Repeated evolution of salt-tolerance in grasses. Biol Lett 9:20130029. https://doi.org/10.1098/rsbl.2013.0029

Blatt MR (1987) Electrical characteristics of stomatal guard cells: the ionic basis of the membrane potential and the consequence of potassium chlorides leakage from microelectrodes. Planta 170:272–287

Blatt MR (1991) A primer in plant electrophysiological methods. In: Hostettmann K (ed) Methods in plant biochemistry. Academic, London, pp 281–321

Boscari A, Clément M, Volkov V, Golldack D, Hybiak J, Miller AJ, Amtmann A, Fricke W (2009) Potassium channels in barley: cloning, functional characterization and expression analyses in relation to leaf growth and development. Plant Cell Environ 32:1761–1777

Bose J, Rodrigo-Moreno A, Lai D, Xie Y, Shen W, Shabala S (2015) Rapid regulation of the plasma membrane H^+-ATPase activity is essential to salinity tolerance in two halophyte species, *Atriplex lentiformis* and *Chenopodium quinoa*. Ann Bot 115:481–494

Briggs GE, Hope AB (1958) Electric potential differences and the Donnan equilibrium in plant tissues. J Exp Bot 9:365–371

Briggs GE, Robertson RN (1957) Apparent free space. Ann Rev Plant Physiol 8:11–30

Britto DT, Kronzucker HJ (2008) Cellular mechanisms of potassium transport in plants. Physiol Plant 133:637–650

Cakirlar H, Bowling DJF (1981) The effect of salinity on the membrane potential of sunflower roots. J Exp Bot 32:479–485

Campion DS (1974) Resting membrane potential and ionic distribution in fast- and slow-twitch mammalian muscle. J Clin Invest 54:514–518

Cao Y, Jin X, Huang H, Derebe MG, Levin EJ, Kabaleeswaran V et al (2011) Crystal structure of a potassium ion transporter, TrkH. Nature 471:336–340

Carden DE, Diamond D, Miller AJ (2001) An improved Na^+-selective microelectrode for intracellular measurements in plant cells. J Exp Bot 52:1353–1359

Carden DE, Walker DJ, Flowers TJ, Miller AJ (2003) Single-cell measurements of the contributions of cytosolic Na^+ and K^+ to salt tolerance. Plant Physiol 131:676–683

Carpaneto A, Koepsell H, Bamberg E, Hedrich R, Geiger D (2010) Sucrose- and H^+-dependent charge movements associated with the gating of sucrose transporter ZmSUT1. PLoS One 5:e12605

Carr H, Axelsson L (2008) Photosynthetic utilization of bicarbonate in *Zostera marina* is reduced by inhibitors of mitochondrial ATPase and electron transport. Plant Physiol 147:879–885

Chakrapani S, Auerbach A (2005) A speed limit for conformational change of an allosteric membrane protein. Proc Natl Acad Sci U S A 102:87–92

Chen ZH, Pottosin II, Cuin TA, Fuglsang AT, Tester M, Jha D, Zepeda-Jazo I, Zhou MX, Palmgren MG, Newman IA, Shabala S (2007) Root plasma membrane transporters controlling K^+/Na^+ homeostasis in salt-stressed barley. Plant Physiol 145:1714–1725

Clipson NJW, Flowers TJ (1987) Salt tolerance in the halophyte *Suaeda maritima* (L) Dum – the effect of salinity on the concentration of sodium in the xylem. New Phytol 105:359–366

Collins KD (1997) Charge density-dependent strength of hydration and biological structure. Biophys J 72:65–76

Cuin TA, Miller AJ, Laurie SA, Leigh RA (2003) Potassium activities in cell compartments of salt-grown barley leaves. J Exp Bot 54:657–661

Dassanayake M, Larkin J (2017) Making plants break a sweat: the structure, function, and evolution of plant salt glands. Front Plant Sci 8:406. https://doi.org/10.3389/fpls.2017.00406

Dassanayake M, Oh DH, Haas JS, Hernandez A, Hong H, Ali S, Yun DJ, Bressan RA, Zhu JK, Bohnert HJ, Cheeseman JM (2011) The genome of the extremophile crucifer *Thellungiella parvula*. Nat Genet 43:913–918

Davenport R (2002) Glutamate receptors in plants. Ann Bot 90:549–557

de Almeida PMF (2014) The role of HKT transporters in salinity tolerance of tomato. Ph.D. thesis, Vrije Universiteit, Amsterdam, p 219

De Boer AH (1985) Xylem/symplast ion exchange: mechanism and function in salt tolerance and growth. Ph.D. thesis, Rijksuniversiteit Groningen, Groningen

De Boer AH, Volkov V (2003) Logistics of water and salt transport through the plant: structure and functioning of the xylem. Plant Cell Environ 26:87–101

Demidchik V, Maathuis FJ (2007) Physiological roles of nonselective cation channels in plants: from salt stress to signalling and development. New Phytol 175:387–404

Demidchik V, Tester M (2002) Sodium fluxes through nonselective cation channels in the plasma membrane of protoplasts from Arabidopsis roots. Plant Physiol 128:379–387

Demidchik V, Cuin TA, Svistunenko D, Smith SJ, Miller AJ, Shabala S et al (2010) *Arabidopsis* root K^+-efflux conductance activated by hydroxyl radicals: single-channel properties, genetic basis and involvement in stress-induced cell death. J Cell Sci 123:1468–1479

Doyle DA, Cabral MJ, Pfuetzner RA, Kuo A, Gulbis JM, Cohen SL et al (1998) The structure of the potassium channel: molecular basis of K^+ conduction and selectivity. Science 280:69–77

English JP, Colmer TD (2013) Tolerance of extreme salinity in two stem-succulent halophytes (*Tecticornia* species). Funct Plant Biol 40:897–912

Erdei L, Kuiper PJC (1979) The effect of salinity on growth. Cation content, Na^+-uptake and translocation in salt-sensitive and salt-tolerant *Plantago* species. Physiol Plant 47:95–99

Essah PA, Davenport R, Tester M (2003) Sodium influx and accumulation in Arabidopsis. Plant Physiol 133:307–318

Evans A, Hall D, Pritchard J, Newbury HJ (2012) The roles of the cation transporters CHX21 and CHX23 in the development of *Arabidopsis thaliana*. J Exp Bot 63:59–67

Fernández JA, García-Sánchez MJ, Felle HH (1999) Physiological evidence for a proton pump and sodium exclusion mechanisms at the plasma membrane of the marine angiosperm *Zostera marina* L. J Exp Bot 50:1763–1768

Field CD, Flowers TJ, Hall JL (1980) Rubidium transport in membrane vesicles from the halophyte *Suaeda maritima*. Ann Bot 46:401–407

Flowers TJ, Colmer TD (2008) Salinity tolerance in halophytes. New Phytol 179:945–963

Flowers TJ, Dalmond D (1992) Protein synthesis in halophytes: the influence of potassium, sodium and magnesium in vitro. Plant Soil 146:153–161

Flowers TJ, Yeo AR (1986) Ion relations of plants under drought and salinity. Aust J Plant Physiol 13:75–91

Flowers T, Yeo AR (2007) The driving forces for water and solute movement. In: Yeo AR, Flowers TJ (eds) Plant solute transport. Blackwell, Oxford, pp 29–46

Flowers TJ, Troke PF, Yeo AR (1977) The mechanism of salt tolerance in halophytes. Annu Rev Plant Physiol 28:89–121

Flowers TJ, Galal HK, Bromham L (2010) Evolution of halophytes: multiple origins of salt tolerance in land plants. Funct Plant Biol 37:604–612

Flowers TJ, Munns R, Colmer TD (2015) Sodium chloride toxicity and the cellular basis of salt tolerance in halophytes. Ann Bot 115:419–431

Flowers TJ, Glenn EP, Volkov V (2018) Could vesicular transport of Na^+ and Cl^- be a feature of salt tolerance in halophytes? Ann Bot https://doi.org/10.1093/aob/mcy164

Fricke W, Leigh RA, Tomos AD (1994) Epidermal solute concentrations and osmolality in barley leaves studied at the single-cell level. Planta 192:317–323

Fricke W, Akhiyarova G, Wei W, Alexandersson E, Miller A, Kjellbom PO et al (2006) The short-term growth response to salt of the developing barley leaf. J Exp Bot 57:1079–1095

Gadsby DC (2009) Ion channels versus ion pumps: the principal difference, in principle. Nat Rev Mol Cell Biol 10:344–352

Galdiero S, Falanga A, Cantisani M, Tarallo R, Della Pepa ME, D'Oriano V et al (2012) Microbe-host interactions: structure and role of Gram-negative bacterial porins. Curr Protein Pept Sci 13:843–854

Garciadeblás B, Senn ME, Bañuelos MA, Rodríguez-Navarro A (2003) Sodium transport and HKT transporters: the rice model. Plant J 34:788–801

Gibson TS, Speirs J, Brady CJ (1984) Salt-tolerance in plants. II. In vivo translation of m-RNAs from salt-tolerant and salt-sensitive plants on wheat germ ribosomes. Responses to ions and compatible organic solutes. Plant Cell Environ 7:579–587

Grabov A (2007) Plant KT/KUP/HAK potassium transporters: single family – multiple functions. Ann Bot 99:1035–1041

Greiner T, Ramos J, Alvarez MC, Gurnon JR, Kang M, Van Etten JL et al (2011) Functional HAK/KUP/KT-like potassium transporter encoded by chlorella viruses. Plant J 68:977–986

Hajibagheri MA, Flowers TJ (1989) X-ray microanalysis of ion distribution within root cortical cells of the halophyte *Suaeda maritima* (L.) Dum. Planta 177:131–134

Hajibagheri MA, Flowers TJ, Collins JC, Yeo AR (1988) A comparison of the methods of X-ray microanalysis. J Exp Bot 39:279–290

Hall JL, Flowers TJ (1973) The effect of salt on protein synthesis in the halophyte *Suaeda maritima*. Planta 110:361–368

Halperin SJ, Lynch JP (2003) Effects of salinity on cytosolic Na^+ and K^+ in root hairs of *Arabidopsis thaliana*: in vivo measurements using the fluorescent dyes SBFI and PBFI. J Exp Bot 54:2035–2043

Hammou KA, Rubio L, Fernández JA, García-Sánchez MJ (2014) Potassium uptake in the halophyte *Halimione portulacoides* L. Aellen. Environ Exp Bot 107:15–24

Haro R, Bañuelos MA, Senn ME, Barrero-Gil J, Rodríguez-Navarro A (2005) HKT1 mediates sodium uniport in roots. Pitfalls in the expression of HKT1 in yeast. Plant Physiol 139:1495–1506

Haro R, Bañuelos MA, Rodríguez-Navarro A (2010) High-affinity sodium uptake in land plants. Plant Cell Physiol 51:68–79

Heginbotham L, Abramson T, MacKinnon R (1992) A functional connection between the pores of distantly related ion channels as revealed by mutant K^+ channels. Science 258:1152–1155

Hellblom F, Beer S, Björk M, Axelsson L (2001) A buffer sensitive inorganic carbon utilisation system in *Zostera marina*. Aquat Bot 69:55–62

Higinbotham N (1973) Electropotentials of plant cells. Annu Rev Plant Physiol 24:25–46

Hille B (2001) Ion channels and excitable membranes, 3rd edn. Sinauer Associates, Sunderland. 814 p

Hinke JAM (1959) Glass micro-electrodes for measuring intracellular activities of sodium and potassium. Nature 184:1257–1258

Hirsch RE, Lewis BD, Spalding EP, Sussman MR (1998) A role for the AKT1 potassium channel in plant nutrition. Science 280:918–921

Hu J, Ma Q, Kumar T, Duan H-R, Zhang J-L, Yuan H-J, Wang Q, Ali Khan S, Wang P, Suo-Min Wang S-M (2016) *ZxSKOR* is important for salinity and drought tolerance of *Zygophyllum xanthoxylum* by maintaining K^+ homeostasis. Plant Growth Regul 80:195–205. https://doi.org/10.1007/s10725-016-0157-z

Hughes FM Jr, Cidlowski JA (1999) Potassium is a critical regulator of apoptotic enzymes in vitro and in vivo. Adv Enzym Regul 39:157–171

Huh GH, Damsz B, Matsumoto TK, Reddy MP, Rus AM, Ibeas JI et al (2002) Salt causes ion disequilibrium-induced programmed cell death in yeast and plants. Plant J 29:649–659

Hunte C, Screpanti E, Venturi M, Rimon A, Padan E, Michel H (2005) Structure of a Na^+/H^+ antiporter and insights into mechanism of action and regulation by pH. Nature 435:1197–1202

Jennings DH (1976) The effects of sodium chloride on higher plants. Biol Rev 51:453–486

Jiang Y, Lee A, Chen J, Cadene M, Chait BT, MacKinnon R (2002) The open pore conformation of potassium channels. Nature 417:523–526

Kato N, Akai M, Zulkifli L, Matsuda N, Kato Y, Goshima S et al (2007) Role of positively charged amino acids in the M2D transmembrane helix of Ktr/Trk/HKT type cation transporters. Channels (Austin) 1:161–171

Katsuhara M, Kawasaki T (1996) Salt stress induced nuclear and DNA degradation in meristematic cells of barley roots. Plant Cell Physiol 37:169–173

Köhler B (2007) Step by step: deciphering ion transport in the root xylem parenchyma. Plant Signal Behav 2:303–305

Korolev AV, Tomos AD, Bowtell R, Farrar JF (2000) Spatial and temporal distribution of solutes in the developing carrot taproot measured at single-cell resolution. J Exp Bot 51:567–577

Kronzucker HJ, Britto DT (2011) Sodium transport in plants: a critical review. New Phytol 189:54–81

Kronzucker HJ, Coskun D, Schulze LM, Wong JR, Britto DT (2013) Sodium as nutrient and toxicant. Plant Soil 369:1–23

L'Roy A, Hendrix DL (1980) Effect of salinity upon cell membrane potential in the marine halophyte, *Salicornia bigelovii* Torr. Plant Physiol 65:544–549

Li L, Kim B-G, Cheong YH, Pandey GK, Luan S (2006) A Ca^{2+} signaling pathway regulates a K^+ channel for low-K response in *Arabidopsis*. PNAS 103:12625–12630

Longpré JP, Lapointe JY (2011) Determination of the Na^+/glucose cotransporter (SGLT1) turnover rate using the ion-trap technique. Biophys J 100:52–59

Lunde C, Drew DP, Jacobs AK, Tester M (2007) Exclusion of Na^+ via sodium ATPase (PpENA1) ensures normal growth of *Physcomitrella patens* under moderate salt stress. Plant Physiol 144:1786–1796

M'Rah S, Ouerghi Z, Eymery F, Rey P, Hajji M, Grignon C, Lachaal M (2007) Efficiency of biochemical protection against toxic effects of accumulated salt differentiates *Thellungiella halophila* from *Arabidopsis thaliana*. J Plant Physiol 164:375–384

Ma Q, Li YX, Yuan HJ, Hu J, Wei L, Bao AK, Zhang JL, Wang S-M (2014) ZxSOS1 is essential for long-distance transport and spatial distribution of Na^+ and K^+ in the xerophyte *Zygophyllum xanthoxylum*. Plant Soil 374:661–676

Maathuis FJM, Amtmann A (1999) K^+ nutrition and Na^+ toxicity: the basis of cellular K^+/Na^+ ratios. Ann Bot 84:123–133

Maathuis FJ, Prins HB (1990) Patch clamp studies on root cell vacuoles of a salt-tolerant and a salt-sensitive *Plantago* species. Plant Physiol 92:23–28

Maathuis FJM, Sanders D (2001) Sodium uptake in Arabidopsis roots is regulated by cyclic nucleotides. Plant Physiol 127:1617–1625

Maathuis FJM, Flowers TJ, Yeo AR (1992) Sodium-chloride compartmentation in leaf vacuoles of the halophyte *Suaeda maritima* (L) Dum. and its relation to tonoplast permeability. J Exp Bot 43:1219–1223

Maathuis FJM, Ahmad I, Patishtan J (2014) Regulation of Na^+ fluxes in plants. Front Plant Sci 5:467. https://doi.org/10.3389/fpls.2014.00467

MacKinnon R (2004) Potassium channels and the atomic basis of selective ion conduction (Nobel Lecture). Angew Chem Int Ed Engl 43:4265–4277

Maffeo C, Bhattacharya S, Yoo J, Wells D, Aksimentiev A (2012) Modeling and simulation of ion channels. Chem Rev 112:6250–6284

Mähler J, Persson I (2012) A study of the hydration of the alkali metal ions in aqueous solution inorganic chemistry. Am Chem Soc 51:425–438

Malone M, Leigh RA, Tomos AD (1991) Concentrations of vacuolar inorganic ions in individual cells of intact wheat leaf epidermis. J Exp Bot 42:305–309

Martinoia E, Meyer S, De Angeli A, Nagy R (2012) Vacuolar transporters in their physiological context. Annu Rev Plant Biol 63:183–213

Mäser P, Thomine S, Schroeder JI, Ward JM, Hirschi K, Sze H et al (2001) Phylogenetic relationships within cation transporter families of *Arabidopsis*. Plant Physiol 126:1646–1667

Matzke AJ, Matzke M (2013) Membrane "potential-omics": toward voltage imaging at the cell population level in roots of living plants. Front Plant Sci 4:311. https://doi.org/10.3389/fpls.2013.00311

Mercado JM, Niell FX, Silva J, Santos R (2003) Use of light and inorganic carbon acquisition by two morphotypes of *Zostera noltii* Hornem. J Exp Mar Biol Ecol 297:71–84

Mercado JM, Andría JR, Pérez-Llorens JL, Vergara JJ, Axelsson L (2006) Evidence for a plasmalemma-based CO_2 concentrating mechanism in *Laminaria saccharina*. Photosynth Res 88:259–268

Murthy M, Tester M (2006) Cation currents in protoplasts from the roots of a Na^+ hyperaccumulating mutant of *Capsicum annuum*. J Exp Bot 57:1171–1180

Mutoh H, Akemann W, Knöpfel T (2012) Genetically engineered fluorescent voltage reporters. ACS Chem Neurosci 3:585–592

Neales TF, Sharkey PJ (1981) Effect of salinity on growth and on mineral and organic constituents of the halophyte *Disphyma australe* (Soland.) J.M. Black. Aust J Plant Physiol 8:165–179

Nieves-Cordones M, Martinez V, Benito B, Rubio F (2016) Comparison between Arabidopsis and rice for main pathways of K^+ and Na^+ uptake by roots. Front Plant Sci 7:992. https://doi.org/10.3389/fpls.2016.00992

Nightingale ER Jr (1959) Phenomenological theory of ion solvation. Effective radii of hydrated ions. Ions J Phys Chem 63:1381–1387

Niu X, Narasimhan ML, Salzman RA, Bressan RA, Hasegawa PH (1993) NaCl regulation of plasma membrane H^+-ATPase gene expression in a glycophyte and a halophyte. Plant Physiol 103:713–718

Nobel PS (2005) Physicochemical and environmental plant physiology, 3rd edn. Academic, Burlington. 540 p

Núñez-Ramírez R, Sánchez-Barrena MJ, Villalta I, Vega JF, Pardo JM, Quintero FJ et al (2012) Structural insights on the plant salt-overly-sensitive 1 (SOS1) Na^+/H^+ antiporter. J Mol Biol 424:283–294

Oh DH, Leidi E, Zhang Q, Hwang SM, Li Y, Quintero FJ, Jiang X, D'Urzo MP, Lee SY, Zhao Y, Bahk JD, Bressan RA, Yun D-J, Pardo JM, Bohnert HJ (2009) Loss of halophytism by interference with SOS1 expression. Plant Physiol 151:210–222

Palmgren MG (2001) PLANT PLASMA MEMBRANE H^+-ATPases: powerhouses for nutrient uptake. Annu Rev Plant Physiol Plant Mol Biol 52:817–845

Pan Y-Q, Guo H, Wang S-M, Zhao B, Zhang J-L, Ma Q, Yin H-J, Bao A-K (2016) The photosynthesis, Na^+/K^+ homeostasis and osmotic adjustment of *Atriplex canescens* in response to salinity. Front Plant Sci 7:848. https://doi.org/10.3389/fpls.2016.00848

Parrondo RT, Gosselink JG, Hopkinson CS (1978) Effects of salinity and drainage on the growth of three salt marsh grasses. Bot Gaz 139:102–107

Pitman MG, Läuchli A, Stelzer R (1981) Ion distribution in roots of barley seedlings measured by electron probe x-ray microanalysis. Plant Physiol 68:673–679

Premkumar L, Greenblatt HM, Bagashwar UK, Savchenko T, Gokhman I, Sussman JL et al (2005) 3D structure of a halotolerant algal carbonic anhydrase predicts halotolerance of a mammalian homolog. Proc Natl Acad Sci U S A 102:7493–7498

Qi Z, Spalding EP (2004) Protection of plasma membrane K^+ transport by the salt overly sensitive Na^+–H^+ antiporter during salinity stress. Plant Physiol 136:2548–2555

Robinson MF, Véry A-A, Sanders D, Mansfield TA (1997) How can stomata contribute to salt tolerance? Ann Bot 80:387–393

Rodriguez-Navarro A (2000) Potassium transport in fungi and plants. BBA Rev Biomembr 1469:1–30

Rodríguez-Navarro A, Rubio F (2006) High-affinity potassium and sodium transport systems in plants. J Exp Bot 57:1149–1160

Rubio L, Belver A, Venema K, García-Sánchez MJ, Fernández JA (2011) Evidence for a sodium efflux mechanism in the leaf cells of the seagrass *Zostera marina* L. J Exp Mar Biol Ecol 402:56–64

Sejersted OM, Sjøgaard G (2000) Dynamics and consequences of potassium shifts in skeletal muscle and heart during exercise. Physiol Rev 80:1411–1481

Serrano R (1988) Structure and function of proton translocating ATPase in plasma membranes of plants and fungi. Biochim Biophys Acta 947:1–28

Shabala S (2009) Salinity and programmed cell death: unravelling mechanisms for ion specific signalling. J Exp Bot 60:709–712

Shabala SN, Lew RR (2002) Turgor regulation in osmotically stressed *Arabidopsis* epidermal root cells. Direct support for the role of inorganic ion uptake as revealed by concurrent flux and cell turgor measurements. Plant Physiol 129:290–299

Shabala S, Mackay A (2011) Ion transport in halophytes. Adv Bot Res 57:151–199

Shabala S, Demidchik V, Shabala L, Cuin TA, Smith SJ, Miller AJ et al (2006) Extracellular Ca^{2+} ameliorates NaCl-induced K^+ loss from *Arabidopsis* root and leaf cells by controlling plasma membrane K^+-permeable channels. Plant Physiol 141:1653–1665

Shi H, Ishitani M, Kim C, Zhu JK (2000) The *Arabidopsis thaliana* salt tolerance gene SOS1 encodes a putative Na^+/H^+ antiporter. Proc Natl Acad Sci U S A 97:6896–6901

Shomer I, Novacky AJ, Pike SM, Yermiyahu U, Kinraide TB (2003) Electrical potentials of plant cell walls in response to the ionic environment. Plant Physiol 133:411–422

Shuyskaya EV, Rakhamkulova ZF, Lebedeva MP, Kolesnikov AV, Safarova A, Borisochkina TI, Toderich KN (2017) Different mechanisms of ion homeostasis are dominant in the recretohalophyte *Tamarix ramosissima* under different soil salinity. Acta Physiol Plant 39:1–12

Skou JC (1998) Nobel Lecture. The identification of the sodium pump. Biosci Rep 18:155–169

Stelzer R, Läuchli A (1977) Salt- and flooding tolerance of *Puccinellia peisonis*. I The effect of NaCl- and KCl-salinity on growth at varied oxygen supply to the root. Zeitschrift Fur Pflanzenphysiologie 83:35–42

Su H, Golldack D, Katsuhara M, Zhao C, Bohnert HJ (2001) Expression and stress-dependent induction of potassium channel transcripts in the common ice plant. Plant Physiol 125:604–614

Su H, Golldack D, Zhao C, Bohnert HJ (2002) The expression of HAK-type K^+ transporters is regulated in response to salinity stress in common ice plant. Plant Physiol 129:1482–1493

Su H, Balderas E, Vera-Estrella R, Golldack D, Quigley F, Zhao C, Pantoja O, Bohnert HJ (2003) Expression of the cation transporter McHKT1 in a halophyte. Plant Mol Biol 52:967–980

Subbarao GV, Ito O, Berry WL, Wheeler RM (2003) Sodium – a functional plant nutrient. Crit Rev Plant Sci 22:391–416

Sze H, Li X, Palmgren MG (1999) Energization of plant cell membranes by H^+-pumping ATPases. Regulation and biosynthesis. Plant Cell 11:677–690

Thiyagarajah M, Fry SC, Yeo AR (1996) In vitro salt tolerance of cell wall enzymes from halophytes and glycophytes. J Exp Bot 47:1717–1724

Véry A-A, Robinson MF, Mansfield TA, Sanders D (1998) Guard cell cation channels are involved in Na^+ induced stomatal closure in a halophyte. Plant J 14:509–521

Volkov V (2015a) Quantitative description of ion transport via plasma membrane of yeast and small cells. Front Plant Sci 6:425. https://doi.org/10.3389/fpls.2015.00425

Volkov V (2015b) Salinity tolerance in plants. Quantitative approach to ion transport starting from halophytes and stepping to genetic and protein engineering for manipulating ion fluxes. Front Plant Sci 6:873. https://doi.org/10.3389/fpls.2015.00873

Volkov V, Amtmann A (2006) *Thellungiella halophila*, a salt-tolerant relative of *Arabidopsis thaliana*, has specific root ion-channel features supporting K^+/Na^+ homeostasis under salinity stress. Plant J 48:342–353

Volkov V, Wang B, Dominy PJ, Fricke W, Amtmann A (2004) *Thellungiella halophila*, a salt-tolerant relative of *Arabidopsis thaliana*, possesses effective mechanisms to discriminate between potassium and sodium. Plant Cell Environ 27:1–14

Walker DJ, Smith SJ, Miller AJ (1995) Simultaneous measurement of intracellular pH and K^+ or NO_3^- in barley root cells using triple-barreled, ion-selective microelectrodes. Plant Physiol 108:743–751

Walker DJ, Black CR, Miller AJ (1998) The role of cytosolic potassium and pH in the growth of barley roots. Plant Physiol 118:957–964

Wang B (2006) Physiological and molecular strategies for salt tolerance in Thellungiella halophila, a close relative of *Arabidopsis thaliana*. Ph.D. thesis, Division of Biochemistry and Molecular Biology, Institute of Biomedical and Life Sciences, University of Glasgow, Glasgow, p 235

Wang B, Davenport RJ, Volkov V, Amtmann A (2006) Low unidirectional sodium influx into root cells restricts net sodium accumulation in *Thellungiella halophila*, a salt-tolerant relative of *Arabidopsis thaliana*. J Exp Bot 57:1161–1170

Wang S-M, Zhang J-L, Flowers TJ (2007) Low-affinity Na^+ uptake in the halophyte *Suaeda maritima*. Plant Physiol 145:559–571

Waters S, Gilliham M, Hrmova M (2013) Plant high-affinity potassium (HKT) transporters involved in salinity tolerance: structural insights to probe differences in ion selectivity. Int J Mol Sci 14:7660–7680

Wegner LH, De Boer AH (1997a) Properties of two outward-rectifying channels in root xylem parenchyma cells suggest a role in K^+ homeostasis and long-distance signaling. Plant Physiol 115:1707–1719

Wegner LH, De Boer AH (1997b) Two inward K^+ channels in the xylem parenchyma cells of barley roots are regulated by G-protein modulators through a membrane-delimited pathway. Planta 203:506–516

Wegner LH, Raschke K (1994) Ion channels in the xylem parenchyma of barley roots. Plant Physiol 105:799–813

White PJ, Lemtiri-Chlieh F (1995) Potassium currents across the plasma membrane of protoplasts derived from rye roots: a patch clamp study. J Exp Bot 46:497–511

Wu SJ, Ding L, Zhu JK (1996) SOS1, a genetic locus essential for salt tolerance and potassium acquisition. Plant Cell 8:617–627

Wyn Jones RGA, Pollard A (1983) Proteins, enzymes and inorganic ions. In: Lauchli A, Person A (eds) Encyclopedia of plant physiology, New series, vol 15B. Springer, New York, pp 528–562

Xu J, Li H-D, Chen L-Q, Wang Y, Liu L-L, He L, Wu W-H (2006) A protein kinase, interacting with two calcineurin B-like proteins, regulates K^+ transporter AKT1 in *Arabidopsis*. Cell 125:1347–1360

Yamaguchi T, Hamamoto S, Uozumi N (2013) Sodium transport system in plant cells. Front Plant Sci 4:410

Yeo AR (1981) Salt tolerance in the halophyte *Suaeda maritima* (L.) Dum.: intracellular compartmentation of ions. J Exp Bot 32:487–497

Yeo AR, Flowers TJ (1980) Salt tolerance in the halophyte *Suaeda maritima* (L.) Dum.: evaluation of the effect of salinity upon growth. J Exp Bot 31:1171–1183

Yeo AR, Flowers TJ (1986) Ion transport in *Suaeda maritima*: its relation to growth and implications for the pathway of radial transport of ions across the root. J Exp Bot 37:143–159

Yuan H-J, Ma Q, Wu G-Q, Wang P, Hu J, Wang S-M (2015) ZxNHX controls Na^+ and K^+ homeostasis at the whole-plant level in *Zygophyllum xanthoxylum* through feedback regulation of the expression of genes involved in their transport. Ann Bot 115:495–507

Yuan F, Leng B, Wang B (2016) Progress in studying salt secretion from the salt glands in Recretohalophytes: how do plants secrete salt? Front Plant Sci 7:977. https://doi.org/10.3389/fpls.2016.00977

Yue LJ, Li SX, Ma Q, Zhou XR, Wu GQ, Bao AK, Zhang JL, Wang SM (2012) NaCl stimulates growth and alleviates water stress in the xerophyte *Zygophyllum xanthoxylum*. J Arid Environ 87:153–160

Zahran HH, Marín-Manzano MC, Sánchez-Raya AJ, Bedmar EJ, Venema K, Rodríguez-Rosales MP (2007) Effect of salt stress on the expression of NHX-type ion transporters in *Medicago intertexta* and *Melilotus indicus* plants. Physiol Plant 131:122–130

Zhao KF, Fan H, Song J, Sun MX, Wang BZ, Zhang SQ et al (2005) Two Na^+ and Cl^- hyperaccumulators of the Chenopodiaceae. J Integr Plant Biol 47:311–318

Interpretations of Holocene Carbonate-Evaporites of Coastal and Inland Sabkhas of Abu Dhabi (United Arab Emirates) from Landsat Satellite Images and Field Survey

11

A. S. Alsharhan and C. G. St. C. Kendall

Abstract

Sedimentary features that characterize the coastline of Abu Dhabi include beach ridges, sabkha flats, algal mats, and mangroves associated with carbonate-evaporites that are tied field descriptions, photography, and satellite imagery. Knowledge of the facies content and geometries of these bodies is intended for as analogs of similar features recognized in subsurface regional sequence stratigraphy. This paper is directed at the local and international geological and hydrocarbon exploration community, especially those starting their career.

Through much of the Quaternary, the local climate and depositional settings favored the accumulation of bedded carbonates associated with minor clastics and evaporates. Accommodation for these sediments was controlled by cycles of transgression and regression through minor epeirogenic movements and the glacial eustatic sea-level fluctuations common to the Pleistocene and Holocene.

Holocene sedimentary facies of the coastal area include offshore pelecypod shell gravels, onshore coral reefs, coralgal sands, oolite shoals, inshore grapestones and pelleted lime muds, and cyanobacterial mats and mangrove swamps flanked by sabkha salt flats filled by a mix of calcareous silts and muddy calcareous sands with some aeolian quartz, displacive sulfate nodules, and crystals capped by halite crusts. Storm and spring tides and infrequent wadi runoff cause periodic flooding of these salt flats. The Holocene coastal facies complex onlaps Pleistocene aeolian cross-bedded carbonate sands rich in miliolids and shell fragments. Landward these in turn onlap exposures of Miocene mixed siliciclastics, carbonates. From southeast to west, the inland area of desert is dominated by a mix of extensive gravel plains onlapped by the product of prevailing winds, a variety of aeolian sand dunes of different sizes and forms, enclosing inland sabkhas, and scattered mesas of Neogene. Sabkhat Matti, a large inland lowland sabkha, extends inland from the coast some 120 km, reaching a height of over 40 m above sea level at its southern tip.

A. S. Alsharhan
Middle East Geological & Environmental Establishment, Dubai, United Arab Emirates
e-mail: sharhana@emirates.net.ae

C. G. St. C. Kendall (✉)
School of the Earth, Ocean and Environment, University of South Carolina, Columbia, SC, USA
e-mail: kendall@geol.sc.edu

B. Gul et al. (eds.), *Sabkha Ecosystems*, Tasks for Vegetation Science VI,
https://doi.org/10.1007/978-3-030-04417-6_11

Keywords
Recent carbonate-evaporites · Landsat images · Coastline of Abu Dhabi

Introduction

This paper provides a review of interpretations of the coastal and inland areas of Abu Dhabi Emirate determined from the iterative use of field studies and Landsat satellite images. The scantly vegetated arid coastal setting of Abu Dhabi is ideal for studies which mix remote sensing and ties to sediment samples to establish field verification. The Abu Dhabi Emirate is on the southwest side of the Arabian peninsula between latitudes 22° 40′ and 25° 00′ and longitudes 51° 00′ and 56° 00′ and with an area of some 67,350 km^2 and an east-west coastline of 400 km and is the largest emirate of the United Arab Emirates (Fig. 11.1).

Adjacent to the Arabian Gulf, the generalized geology of coastal Abu Dhabi is that of a Holocene coastal carbonate-evaporite complex bounded to the south by desert areas characterized by siliciclastics, minor carbonate sands, and local sabkhas (Fig. 11.2). Near the mountainous areas to the east, large areas of fluviatile gravel extend beneath the aeolian sand, transported here by the prevailing wind. Much of the area is covered by sand dunes of varying morphology. The aeolian sand dunes include transverse, barchan, and seif dunes complexes and cover large areas and are highly varied in type and orientation.

Geologic Setting

The United Arab Emirates forms the southwestern margin of the semi-enclosed Arabian Gulf, overlying the depressed margin of the unstable shelf of the Arabian foreland. The coastlines rimming the Gulf vary in character:

- The southwestern margin is a Holocene complex of coastal barrier islands and intertidal to supratidal evaporite-dominated coastal plains or sabkhas, on-lapping Pleistocene cross-bedded carbonate sands, which is the focus of this paper. Landward are lines of outcropping hills of mixed siliciclastic and carbonate Miocene sediment flanking desert areas to the south.
- At the extreme northwest of the Gulf is the confluence of the Tigris, Euphrates, and Karun rivers form the siliciclastic Shatt al Arab delta.
- On the western Arabian side, carbonates accumulate with some aeolian siliciclastics along a linear shoreline broken by embayments and the Qatar Peninsula that modify the tidal and current patterns on the southwest side of the Gulf.
- The Iranian shoreline is linear and rocky, with estuaries and coastal plains, associated with rivers draining in the Zagros Mountains (Purser and Seibold 1973), and represents the site of terrigenous sedimentation with some carbonate deposition.

The asymmetric Arabian Gulf has a gently sloping bathymetry with a long deeper axis separating the Arabian stable foreland stratigraphy characterized by late Pliocene to Pleistocene folding, faulting, and salt diapirism and the unstable Iranian Tertiary fold belt formed by the Plio-Pleistocene Zagros folding. This axis is part of a syncline tied to the folding of the Zagros (Kassler 1973). The Arabian Gulf can be subdivided sub-basins and swells that include the "Western" and "Central" sub-basins of the Iranian side of the Gulf divided by the shallow ridge of the "Central Swell" and the third sub-basin known as "Arabian Homocline" on the Arabian side (Purser and Seibold 1973). In the northern Arabian Gulf, the Shatt Al Arab Delta forms the Mesopotamian shallow shelf. South of the Strait of Hormuz, where the Arabian Gulf connects with the Gulf of Oman, the east swell of the Arabian shelf separates the Arabian homocline from the Central sub-basin on the Iranian side.

The morphology of the Arabian Gulf is dominated by Plio-Pleistocene tectonics that mixed with deposition and erosion modified the local relief during the Quaternary (Kassler 1973). A

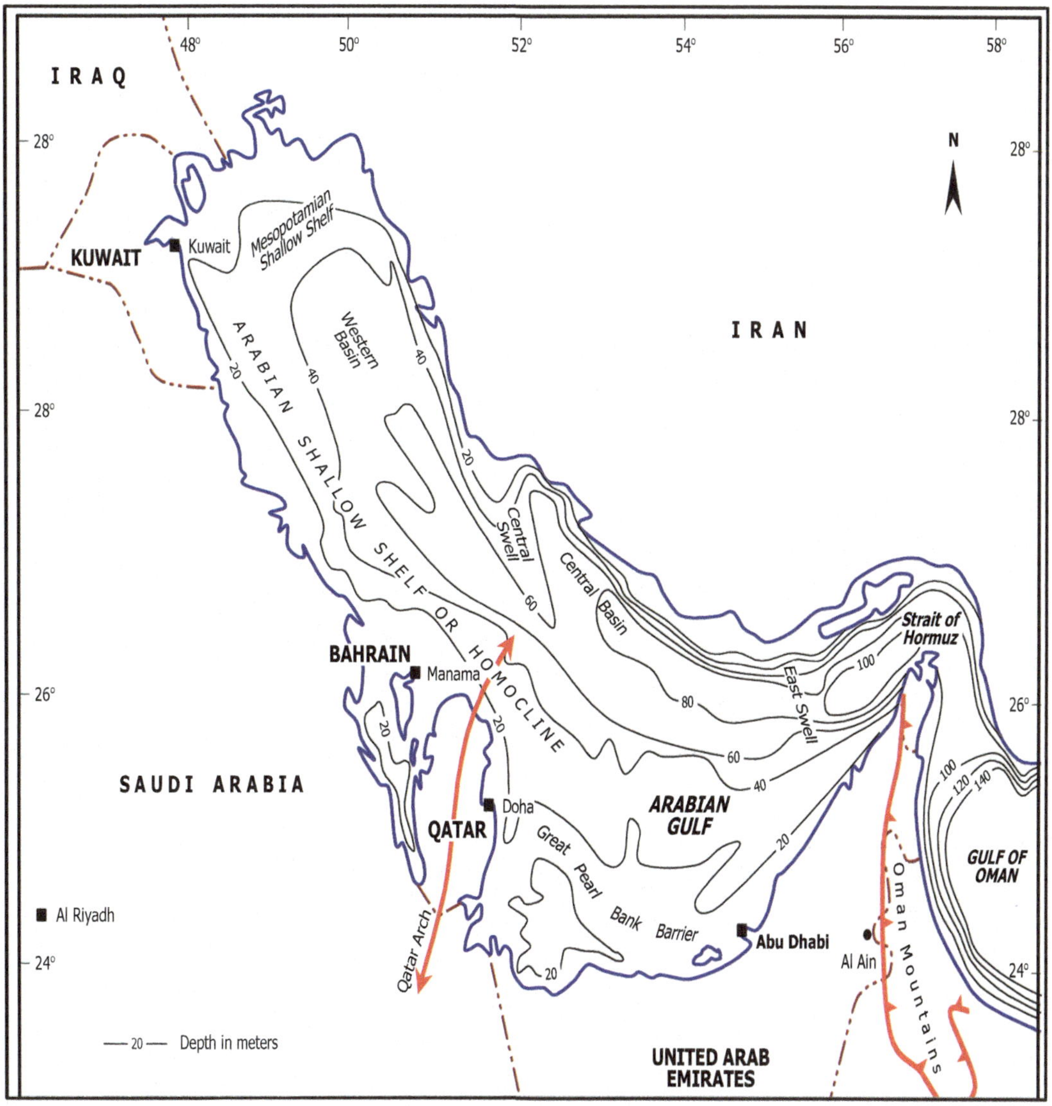

Fig. 11.1 Location map of the United Arab Emirates and adjacent area showing principal bathymetric provinces and depth of water (in meters) of the Arabian Gulf. (Modified from Seibold and Vollbrecht (1969), Alsharhan and Kendall (2003))

sea-level fall of around 120 m during the Pleistocene exposes the Arabian Gulf to river channel into its flanks. During this maximum fall and the subsequent rise, a series of platforms were cut into the pre-existing surface (Kassler 1973; Weijermars 1999). Dramatic climatic changes drive Late Pleistocene and recent sea-level changes, so there is a broad coincidence between the deduced sea-level and temperature curves (Fig. 11.3).

Through much of the Quaternary, the local climate and depositional setting favored the accumulation of bedded carbonates associated with minor clastics and evaporates. Cycles of transgression and regression accommodated these sediments, controlled by epeirogenic movement and glacial eustatic sea-level fluctuations common to the Pleistocene and Holocene.

The Holocene carbonate/evaporite sedimentary complex of Abu Dhabi coast (Fig. 11.2)

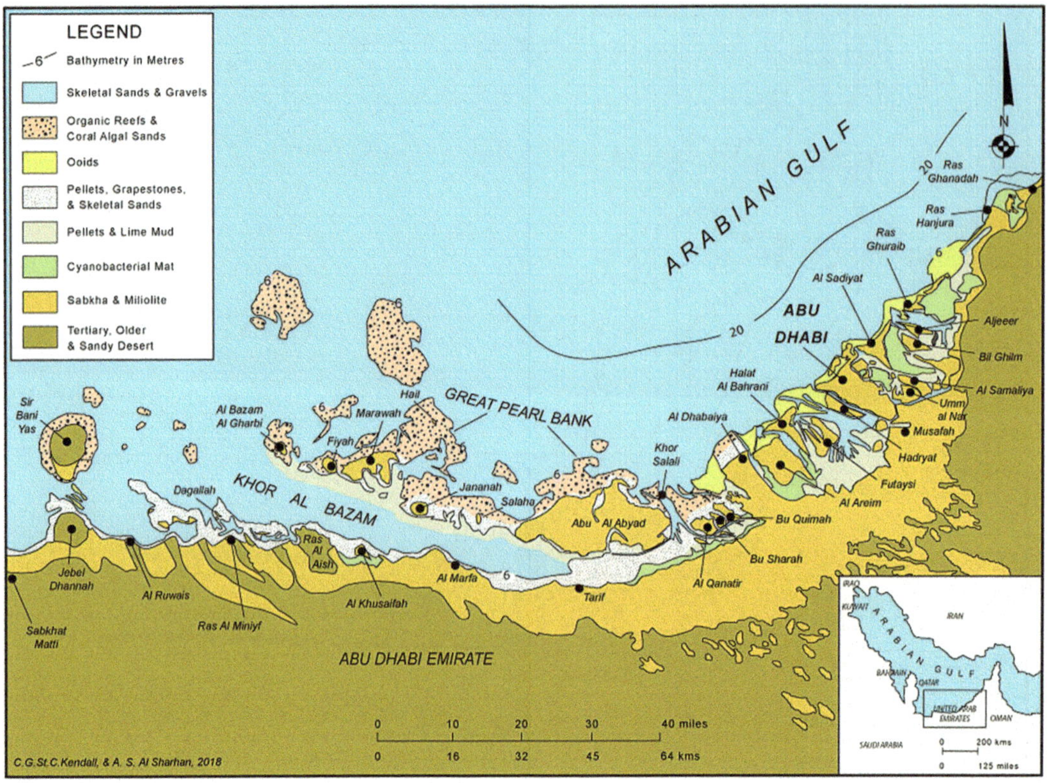

Fig. 11.2 Simplified map of Holocene facies and subaerial surface geology of coastal Abu Dhabi, United Arab Emirates. (After Alsharhan and Kendall 2002).

accumulated over Quaternary and Neogene sedimentary rocks. Outcrops of the Quaternary are known locally as miliolite (Kirkham 1997; Kendall and Alsharhan 2011). Miliolite, a field term used throughout the Arabian Gulf region, describes a lightly cemented aeolian calcareous cross-bedded sand of post-Miocene age with a locally high miliolid and shell fragment content. Miliolite accumulated during the last major glacial eustatic lowstand in the Arabian Gulf (Kirkham 1998a). Exposures are common to the numerous and widespread barrier islands, including Abu Dhabi, Al Saadiyat, Al Qanatir, Abu Al Abyad, and Marawwah. These islands and the peninsula of Al Dhabaiya are nucleated around miliolite outcrops. At the northern coastal tip of the Al Dhabaiya peninsula, Pleistocene miliolite often forms isolated mesa-shaped hills. These protect the coast from erosion, while its flanks are dissected by tidal creeks lined by stands of the black mangrove. These locally elevated Quaternary outcrops are often as much as 2–4 m above sea level (Kirkham 1998b). Beneath the coastal plain, they cover Neogene rocks, while at the southern margin of the coastal plain, they are banked against these rocks. Where the miliolite projects peninsulas into the sea, seawater, boring algae, and animal activities etch the rock surface, much as the cay rock of the Bahamas is affected (Newell et al. 1960). Commonly in inland areas, wind and marine flooding have truncated the miliolite flush with the stoke surface that coincides with the water table of the surrounding sabkha (Kirkham 1997).

Where the Neogene and Quaternary rocks crop out close to the coast, they form low mesas and buttes surrounded by alluvial fans (Figs. 11.4 and 11.5). These hills are remnants of a much more continuous cover of almost horizontal rock, eroded by the combined action of flash floods and wind (Kirkham 1997). The Neogene section is largely Miocene and composed of marls,

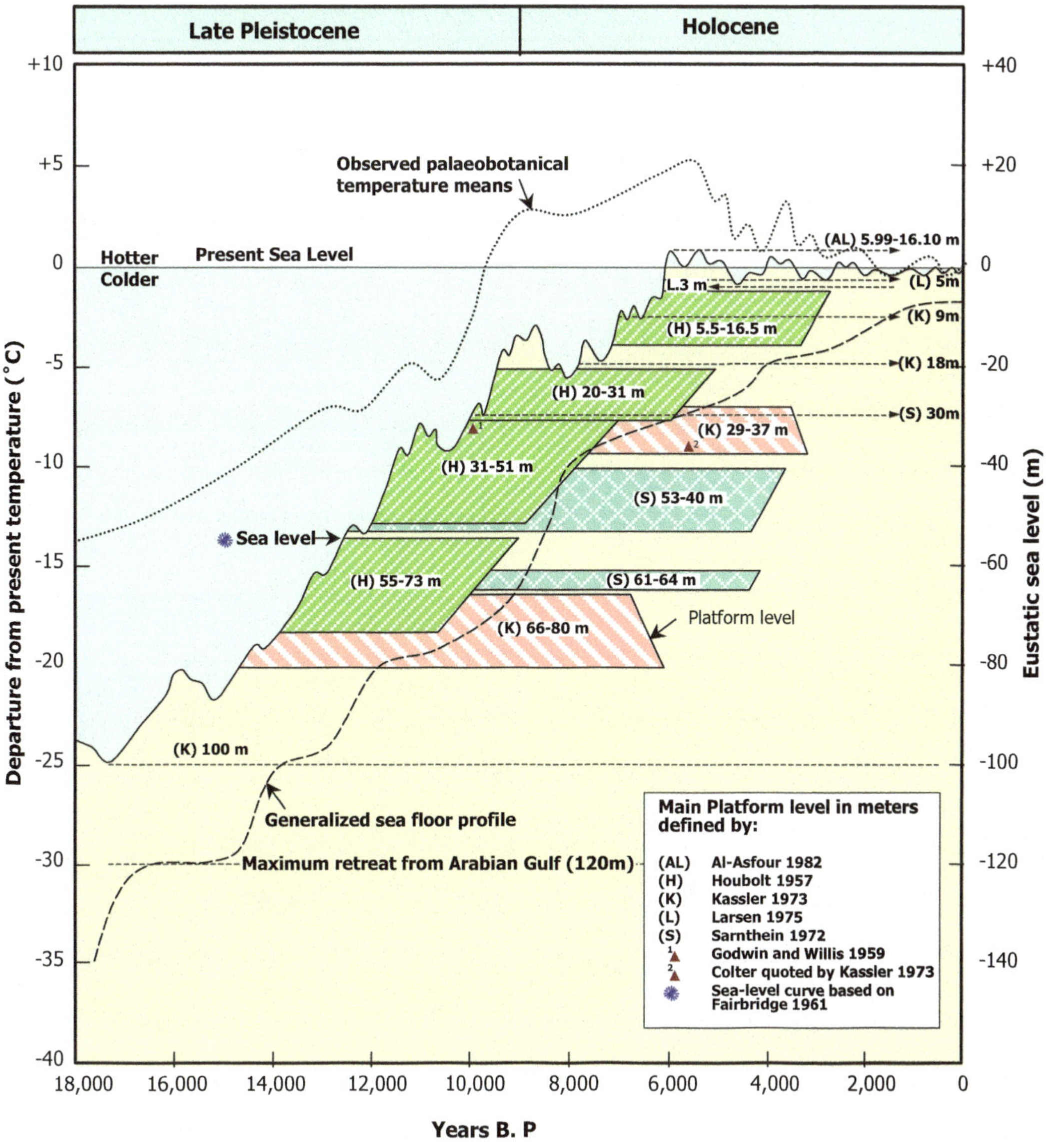

Fig. 11.3 Pleistocene-Holocene temperatures, sea levels, and terrace heights of the Arabian Gulf area. (Compiled from many sources by Alsharhan and Kendall (2003))

sandstones, limestones, and evaporites dipping gently southward cropping out as a low northeast-southwest escarpment of low mesas seldom higher than 35 m. This escarpment parallels the Abu Dhabi coast and is cut by valleys that trend northwest-southeast. The orientation of the ridges and valleys is similar to that of many of the local islands and lagoons, suggesting a combination of structural control and the dominant wind blowing from the northwest. It is difficult to distinguish between these structural and wind controls.

Climatic and Hydrographic Controls

As the Arabian Gulf is surrounded by land, it exhibits extreme seasonal fluctuations (Purser and Seibold 1973); Abu Dhabi has an arid, sub-

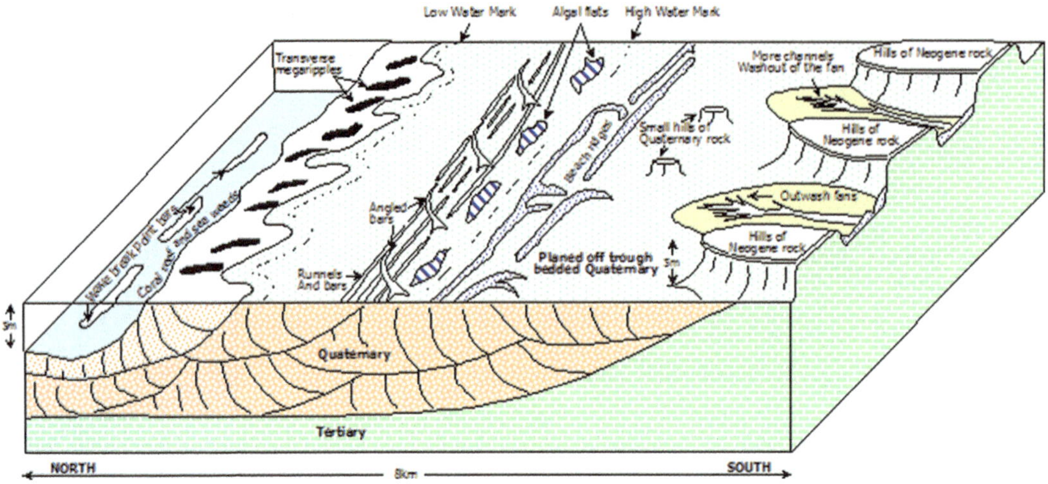

Fig. 11.4 General block diagram of the gross surface stratigraphy and morphology of coastal Abu Dhabi

Fig. 11.5 (**a**) Ras Al Miniyf Neogene outcrop rimmed by coastal terraces, tidal channels, and offshore bars where grapestones accumulate. (**b**) Late Pleistocene miliolite outcrop at northern end of the Dhabaiya Peninsula with cross-bedded aeolianite and miliolid carbonates capped by a marine horizon. (**c**) Outcrop of late Pleistocene miliolite just south and west of the Abu Dhabi sabkha salt flats composed of cross-bedded aeolianite carbonate blown from the Gulf at the last lowstand of sea level. (**d**) Strokes surface exposed by wind-induced deflation of cross-bedded late Pleistocene miliolite aeolianite. This surface is at or just above the water table. Outcrop adjacent to that of (**c**)

tropical continental climate. The narrow Strait of Hormuz enables the exchange of marine water into the Gulf, and the water temperature and salinity vary widely. Since the Arabian Gulf is close to the Tropic of Cancer, during the summer its temperatures are high, and air temperatures often reach 45–50 °C in the summer but are as low as 0 °C in the winter. The combined effect of strong winds, high temperature, and low rainfall results in significant evaporation and high salinities (Kendall and Skipwith 1969). The high temperature, coupled with the aridity of the Abu Dhabi coast, explains the widespread occurrence of carbonates and evaporites in its coastal plain. Sabkha surface temperatures reach over 40 °C in the winter months, but do not reach much above 55 °C in the summer. Water temperatures of the Arabian Gulf tend to increase away from its entrance, particularly in the shallow coastal areas and lagoons (Purser and Seibold 1973).

Evaporation rates in the southern Arabian Gulf were estimated by Privett (1959) to be as much as 124 cm per annum. High summer salinities recorded by Sugden (1963) for parts of the southern Gulf suggest that evaporation is greatest in summer, especially in restricted lagoons. Correspondingly humidity along the coast is also higher in the summer. Salinities of the Arabian Gulf suggest the hydrographic barrier created by oceanic upwelling at the Strait of Hormuz slows the entering of the Indian Ocean waters. The water movement is counterclockwise driven by density currents (Fig. 11.6) (Kendall and Alsharhan 2011; Sheppard et al. 1992). Salinities range from 37% near the Strait of Hormuz to greater than 65% in the lagoons of the Arabian Coast (Bathurst 1975). Water lost to high rates of evaporation is not compensated by fluvial input and rainfall. The result is that nearshore waters of the Gulf have average salinities ranging from 40‰ to 45‰; lagoonal waters have salinity's ranging from 54‰ to 67‰; and the central part of the Gulf has salinity's that average 35–40‰. More specifically the shallow waters of the Abu Dhabi Emirate coast have salinities that average 40–50‰, while the lagoonal areas and embayments of Abu Dhabi can average 60–70‰ (Evans et al. 1969).

In the Abu Dhabi coastal waters, salinity changes with the state of the tide; the more restricted the circulation, the greater the salinity variation (Kinsman and Park 1976). Brines collected from the groundwaters of the sabkha both before and after marine storm- and rain- induced flooding produce small difference in the concentrations of the various salts (Bush 1973; Butler et al. 1982), and the surface and near evaporites were affected. Halite was washed from the surface sediments, while anhydrite on the surface above the intertidal zone was eroded. In Khor al Bazam lagoon of western Abu Dhabi, pH values from a wide number of stations and from different times of day were all found to be approximately 7.8 at a temperature of 25 °C.

The composition of the groundwater entering the coastal sediments of Abu Dhabi is derived from continental Omani brines (Wood et al. 2002) and from the coastal lagoons (Butler 1965). Most of the common elements present in the marine waters maintain a constant ratio. The exception is calcium, which is depleted in the inner lagoons, particularly in summer (Evans et al. 1969). The coastal and lagoonal waters are also found to contain lower concentrations of phosphate and nitrate but a higher concentration of silicates than the waters of the open sea. Seasonal changes in nutrient concentration occur in the coastal and lagoonal waters. The phosphate concentration is higher in winter, while the silicate concentration is higher in summer. The nitrate concentration shows no significant seasonal variation (Evans et al. 1969).

Winds blow dominantly from the northwest throughout the year. The strongest winds are the northwest gale-force "shamals" which occur during the winter. When these winds are prolonged and coincide with spring tides, they cause flooding of extensive portions of the coastal plain. Shamals are probably as important a geological agent as the hurricanes described by Pray (1966). These shamals carry sediment onto the supratidal flat, breach, and flatten coastal dunes; initiate intertidal spits and beach ridges at the top of the intertidal flat; transport sediment landward from the seaward edge of the offshore bank and the coastal terraces; construct wave-break-point bars in front of the offshore bank and coastal terrace; and initiate transverse megaripples on the shoal areas of tidal deltas and coastal terraces. Kirkham (1997) suggests the dominant wind direction has been constant through the Holocene and Quaternary, so that the wind-blown Pleistocene miliolite sediment of the area has cross-bedding which invariably dips southeast.

The morphology of the Abu Dhabi coastline reflects the shamals. East of the Al Dhabaiya pen-

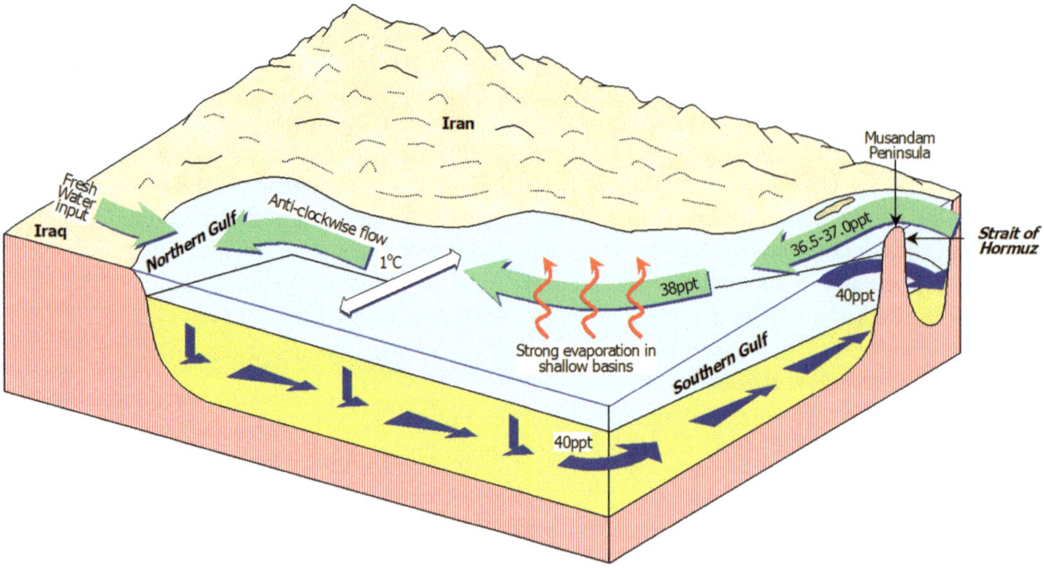

Fig. 11.6 Counterclockwise circulation patterns of the Arabian Gulf that are driven by density currents. Salinities in the Gulf are slightly higher than the Indian Ocean. Note the narrow connection to the Indian Ocean at Strait of Hormuz. (Modified from Sheppard et al. (1992). Green arrows are incoming surface water from the Gulf of Oman, and blue arrows are a denser deeper water flow)

insula, the coast is formed by a complex series barrier islands, including Abu Dhabi and Al Saadiyat, that protect lagoons from the waves of the open Arabian Gulf but provide access to the open sea via tidal channels and their deltas. Westward of the peninsula is a continuous open body of water, the elongate lagoon of the Khor al Bazam, whose western end is connected to the Arabian Gulf (Fig. 11.2). Offshore is a shallow bank protecting this lagoon, flanked by coral banks, capped by extensive sandy shoals, and incised by tidal channels. Here waves from the open Arabian Gulf, generated by the northwestern shamal winds, exceed 2.5 m. These waves are dissipated by northern flank of the offshore bank before they can reach the Khor al Bazam lagoon. A series of widely spaced barrier islands occur on the shoal and are more widely spaced than the barrier islands of the eastern Abu Dhabi. The eastern areas of the coast trend northeast-southwest, have a narrower barrier/lagoon complex, and have more restricted circulation than that of the Khor al Bazam.

To the south of the Khor al Bazam, waves seldom exceed 1 m, and tides and tidal currents in the lee of the northern shoals are more effective in this lagoon generating sand bars, sand waves, and scour marks. On the intertidal and terrace areas of the southern shore of the Khor al Bazam, landward of the break point of waves, a steady longshore drift is driven by southeastern directed waves, created by the northwest winds (Kirkham 1997) moving material to the east except when interrupted by headlands. These waves break obliquely on the east-west trending coastline. Within the shoal and channel area east of Abu al Abyad Island, wave and tidal action cause sediment movement.

Waves and currents are the most important means of sediment transport in the shallow areas of the Arabian parts of the basin (Purser and Seibold 1973). The northeasterly trending "shamal" winds are responsible for wave activity along the United Arab Emirates coast, with the tidal deltas receiving most of the wave energy (Evans et al. 1969). They wave energy causes turbulence in the lagoonal settings, winnow offshore the sediments not protected by islands or banks. As indicated earlier, the general circulation pattern of the Arabian Gulf consists of a slow circulatory surface current, moving anticlockwise along the Iranian coast (Emery 1956). While this current has little effect on sediment textures, it does control the temperature, salinity, and nutrient concentration of the Arabian Gulf.

Tidal currents are aligned approximately parallel to the axis of the Arabian Gulf. The tides are mixed diurnal, ranging from 2.5 m seaward of the islands to approximately 1 m in the protected lagoons. At the coastal barriers of eastern Abu Dhabi, the tidal range is approximately 1.5 m but only 1 m in the adjacent lagoons (Evans 1970), though, as indicated earlier, prolonged winds may cause the tide to rise several meters, flooding coastal areas. Evans et al. (1969) report tidal current speeds measured at three locations: (a) seaward of the tidal deltas of eastern Abu Dhabi at 0.25 m/sec at the sea surface and 0.15 m/s on the seafloor, flowing approximately parallel to the shore; (b) within the tidal deltas at 0.65 m/s at the sea surface and 0.40 m/s at the seafloor; and (c) in the inner southern lagoons at 0.25 m/s at the sea surface and 0.20 m/s on the seafloor.

The tidal currents of western Abu Dhabi reach high velocities where they are constricted by channels, forming deltas that are often up to 8 km^2 in diameter. Seaward of the offshore bank, tidal movement is in an east-west direction, but within the Khor al Bazam lagoon, the regime is more complicated. Here the offshore barrier/lagoon complex provides mixed protection to the extensive Khor al Bazam lagoon; there is less restricted circulation than the lagoons to the east. This is because its western end is up to 20 m deep, connects with the Arabian Gulf, and is exposed to greater wave action. Thus, the western Khor al Bazam is subjected to some wave energy and gently sloping beach faces result, so on the south shore, small dunes are forming to the landward margins of the beaches (Kendall and Skipwith 1969). Meanwhile the eastern Khor al Bazam is at its most restricted, and it becomes intertidal just south of the island of Abu Al Abyad. The dominant tidal movement in the Khor al Bazam is east-west.

As with the lagoons south of Abu Dhabi Island, restricted access to open waters in the Khor al Bazam (Figs. 11.7a, 11.7b, and 11.8a) cause both a time lag and a distinct decrease in tidal range, southward across the lagoon from the seaward, northern entrance. At the restricted eastern head of the Khor al Bazam, particularly along the south coast of Abu al Abyad Island, westerly flowing ebb currents are dominant, and sand from here is moved into the deeper parts of the lagoon to the west (Kendall and Alsharhan 2011).

North and seaward of the offshore bank, the action of waves and the resultant movement of material are complicated by the presence of offshore shoals and coral banks. Here, tidal currents flow in two directions but tend not to be channeled as they are in the Khor al Bazam. Material is moved both to the east and to the west. There is movement to the east when the ebbing current carries suspended sediment discharged from tidal channels, as at the mouth of Khor Salali. Alternatively, where waves stir up bottom material, westward sediment movement may occur at low tide. This is then transported by the flooding tide. In many cases wind-generated currents are probably the controlling factor, with respect to the overall longshore movement of material. These have an easterly trend, where the offshore bank and the mainland coast are aligned east-west and a southwest trend, where the coast is aligned northeast-southwest. The Khor al Bazam deepens westward (Figs. 11.7a, 11.7b, and 11.8a) and shallows eastward where it locally becomes intertidal, with a tidal range of 1–2 m. This body of water, or khor, probably represents a wadi valley system, formed during the Pleistocene sea lows that is currently being filled by bioclastic sediments.

Holocene S.W. Arabian Gulf Offshore, Coastal Complex, and Barrier Islands

Introduction

Located in the low-latitude tropics, the sediment distribution of the southwest Arabian Gulf reflects a mix of arid climate, low- or high-wave energy, northwesterly shamal winds, and the presence or absence of offshore barriers (Wagner

Fig. 11.7a Satellite image of the coastal area of Abu Dhabi Emirate. (*A*) Bu Tini bathymetric high; (*B*) Mubarras bathymetric high; (*C*) Offshore barrier bar complex that extends roughly parallel to the coastline and defined northern flank of the Khor Al Bazam; (*D*) Khor al Salali and Al Qanatir/Bu Shara shelf channel system; and (*E*) Abu Dhabi Barrier Islands, ebb deltas, and their protected macrotidal lagoons

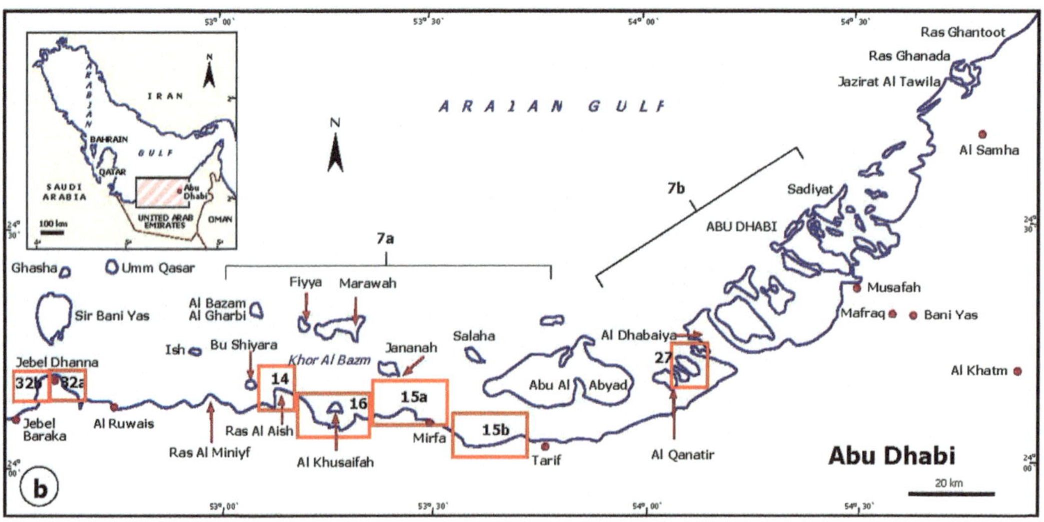

Fig. 11.7b Location of satellite images from the Abu Dhabi Emirate that have different geomorphological features and are listed by figure number as follows

7a. Offshore bank with dispersed islands from Al Bazam Al Gharbi to Abu Al Abyad

7b. Al Qanatir and Bu Shara channeled shelf, and Abu Dhabi and Al Saadiyat barrier island complex

14. Crescentic bars at Ras Al Aish

15a. Intertidal terrace in Al Mirfa area

15b. Intertidal terrace west of Tarif area

16. Al Khusaifah cyanobacteria flats

27. Cyanobacteria flats south of Al Qanatir

32a. Coastal terraces east of Jebel Dhanna

32b. Bay southwest of Jebel Dhanna

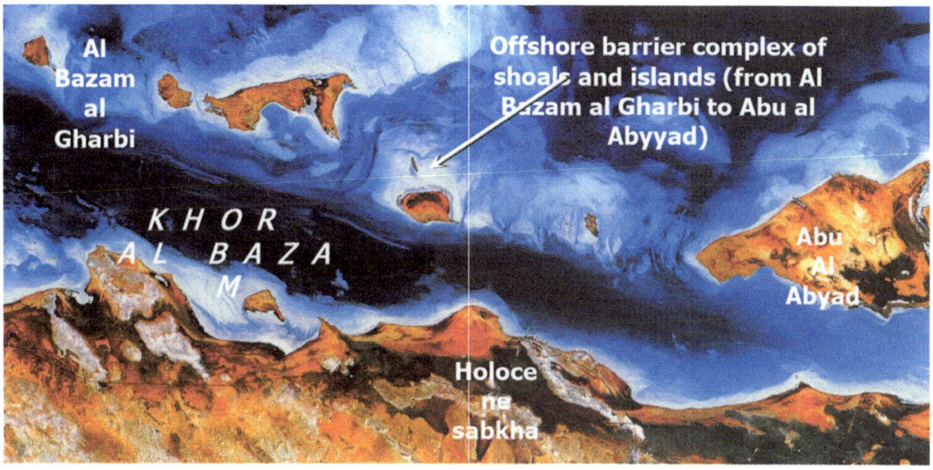

Fig. 11.8a Landsat image of Khor al Bazam showing offshore barrier island complex and oolitic-bioclastic carbonate sands extend from Al Bazam al Gharbi to Abu Al Abyad island while along the coast are supratidal flat and algal mats

and Van der Togt 1973). The southwest coastal sediments of the Arabian Gulf from Jebel Dhanna to Abu Dhabi and eastward accumulate on a classic carbonate ramp (Evans 1994; Williams and Walkden 2001, 2002). The waters of the outer shelf, deeper than 20 m, with low to moderate energy conditions exist to the north and east and are the site of the accumulation of micritic and some argillaceous sediment. The depths to which algae grow suggest that the euphotic zone extends to a depth of 20 m in the southern part of the Gulf on the Arabian side and is as deep as 30 m in the clear water portions of the axis (Purser and Seibold 1973). Southward and landward, there are accumulations of pelecypod sand and gravel on the gently sloping seafloor of the inner shelf (Houblot 1957). Here in waters as deep as 20 m, bioclastic sediments collect suggesting that the storm wave base extends to at least this depth (Purser and Seibold 1973). Updip locally thin bands of grainstone occur, with compositions reflecting the adjacent shallow shelf shoals, representing storm washover fans from these shoals. In the vicinity of Abu Dhabi Island, at the crest of the ramp where wave and tidal energy is concentrated, the offshore shelf skeletal sediments pass landward into oolite tidal delta shoals, whereas westward skeletal banks and reefs form the crest of the ramp (Fig. 11.2).

The close to 20-km-wide broad coastal strip of islands and peninsulas from Ras Ghanada to the northeast to western of Jebel Baraka in the west encloses shallow tidal embayments. Part of this strip is formed by the southwest shore of the Arabian Gulf and is captured in a Landsat image as a coastal complex of barrier island lagoons and sabkhas (Figs. 11.7a and 11.7b). The main sedimentary facies and geomorphic setting of coastal Abu Dhabi is listed in Table 11.1. Here this coastline is divided by the At Al Dhabaiya peninsula into two distinct geomorphic provinces with different facies distributions.

These Holocene carbonate-evaporite sediments of the coastal complex and inland sabkhas of Abu Dhabi were mapped using field descriptions, photography, and satellite imagery. As illustrated in Figs. 11.7a, 11.7b through 11.8a, 11.8b, the coastline of Abu Dhabi was analyzed using Landsat Thematic Mapper (TM) scenes of the region. These were tied to geographic information system (GIS) developed to map and assess the sediment distribution of the area (Althausen et al. 2003; Kendall et al. 2003a, b, and Lakshmi et al. 2003).

The area of the relatively protected waters of the Khor Al Bazam was selected by Althausen et al. (2003) as the type study area which they studied using field descriptions, photography,

Table 11.1 Sediment facies and their equivalent geomorphic setting, Abu Dhabi coastal complex

Sediment facies	Geomorphic unit
Molluscan sand	Lagoon and open sea
Coral and coralline algal sand	Seaward shoals and outer edge of offshore bank
Ooid sand	Tidal deltas and coastal terrace
Pellet aggregate and pellet sand	Coastal terrace and south edge of offshore bank
Fecal pellet sand	Protected lagoon and mangrove swamp
Cyanobacterial mat	Upper intertidal
Evaporites (anhydrite, gypsum, and halite)	Supratidal
Aggregate grains (friable aggregates, lumps, grapestone, botryoidal lump, encrusted lump, worm tubes, and shell infillings)	Lagoon, coastal terrace, and offshore bank

and satellite imagery. This lagoon and a complex coast are prone to environmental and natural changes. The eastern end of this lagoon was classified and mapped for its bathymetric geologic features using digital image processing techniques. Past beach ridges, sabkha flats, algal mats, and black mangroves, associated with the Khor Al Bazam, were identified in Fig 11.7a and 11.7b through 11.8a and 11.8b.

The images were evaluated using a hybrid combination of band ratioing, iterative-statistical clustering, and principal components analysis to produce classification maps (Althausen et al. (2003)). The findings identified land-cover features, in the uplands, and bathymetric features in the Landsat data sets. Digital image processing algorithms and techniques enhanced the imagery, so features nearly undetectable in visual examination were discernible after classification, i.e., oolitic ridges and black mangroves. The func-

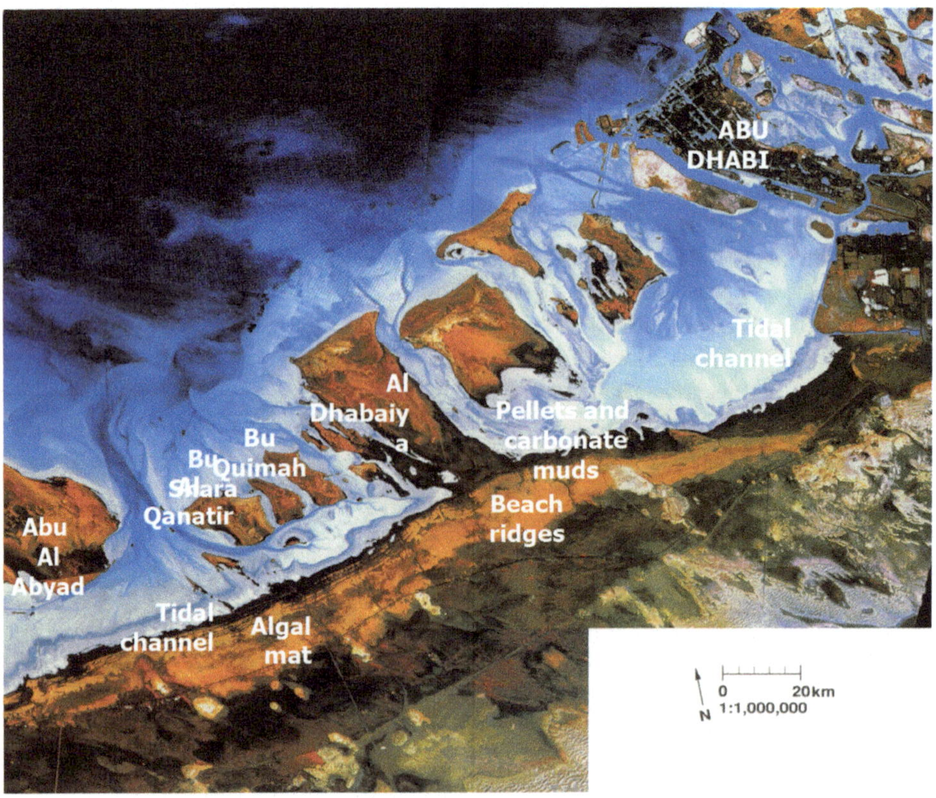

Fig. 11.8b Landsat image showing tidal channels, pellets, and carbonate muds between the islands. Beach ridges, algal mat, and coastal sabkha forming along the coast of western Abu Dhabi Emirate

tionality of combining raw TM bands with principal components and band ratios in a hybrid spectral analysis was demonstrated by this research. Some digital image processing techniques were tested and should help to identify areas that are sensitive to environmental stress, i.e., algal mats, corals, and black mangroves.

The lagoon offers a series of dynamic ecological habitats, including algal mats, coral reefs, and black mangroves that are home to numerous species of fish and shrimp (Sabock and Gurr 1969). Sediment samples, along transects, from each of the facies were collected by Dr. Gregory Whittle of the UAE Desert Marine Environment Research Center in 1993 and 1994. The samples were analyzed to establish field verification for the Khor Al Bazam. Ancillary data, consisting of sediment and facies maps, developed by Kendall and others (1963–present), were also used as information for the region. Such data allowed the remotely sensed data to be calibrated and better utilized to assess what the surface environment is like.

As the result of this research, it can be seen that west of the Dhabaiya peninsula are the "Qanatir/Bu Shara shelf channel system" (Figs. 11.7a, 11.7b, 11.8b), Al Abyad Island, and westward beyond this are extensive bioclastic sandy shoals and coral banks cut by tidal channels. The "Qanatir/Bu Shara shelf channel system" consists of the Khor al Salali (a wide deep tidal channel immediately east of Abu Al Abyad) and a series of smaller north-south tidal channels flanked by mangrove swamps that dissect a system of north-south linear islands. Coral reefs grow along the seaward margin of the offshore bank, and oolites are restricted to a few coastal strips. This complex of shoals and islands bound the Khor al Bazam lagoon and protect it from heavy wave action. The barrier islands on these shoals are more widely separated from each other than those to the east of Al Dhabaiya (Fig. 11.7a). Landward and westward the southern shore of the Khor Al Bazam though protected from heavy wave action is subject to some wave reworking and has sandy shores, stranded high-energy beach ridges, and reworked late Holocene aeolian and outwash fans enveloping headlands of Tertiary rocks. This contrasts with the region south of Island of Abu Al Abyad and the adjacent Qanatir/Bu Shara shelf channel system where a protected linear belt of intertidal cyanobacterial flats flanking supratidal sabkha evaporites occurs.

Macrotidal Coast of Barrier Islands and Tidal Deltas

East of the Al Dhabaiya peninsula is the Abu Dhabi barrier islands, divided from each other by tidal channels and deltas. These barrier islands include Al Bahrani, Abu Dhabi, Al Saadiyat, and Ghurab (Fig. 11.9). Ebb channels between the islands initially have a dendritic pattern, coalescing to form one or more major channels, before reaching the open sea (Kendall and Skipwith 1969). Coral reefs occur in small patches north of the islands at their center and along the seaward margins of tidal channels, while oolites collect on the inter-island tidal deltas (Evans et al. 1964a, b) (Fig. 11.10). The "drum stick" barrier islands, Al Bahrani, Abu Dhabi, and Al Saadiyat, have long leeward tails broken up by a series of ebb tidal deltas that lie between and just north of these islands and are the site of the accumulation of ooids. These barrier islands protect macrotidal carbonate-mud prone lagoons, mangrove swamps, flanked to the south by protected intertidal cyanobacterial flats, and supratidal sabkha evaporites. Landward are stranded high-energy beach ridges and reworked late Holocene aeolian and outwash fans, matching those of the eastern Khor al Bazam.

The barriers islands, including Abu Dhabi, lie on a narrow shelf open to heavy seas when the shamals blow. They have steep beach faces, and landward are flanked large aeolian dunes that prior the building boom of the emirate often had a height greater than 5 m (Fig. 11.10). At right angles to the dominant wind direction, these barrier islands have "T" shapes, whose tops lie seaward and parallel to the trend of the coast. The stems of the "T" extend south and lagoonward (Fig. 11.9). The sediment tails are products of aeolian transport of carbonate sand, and the lateral accretion of sand spits driven by longshore

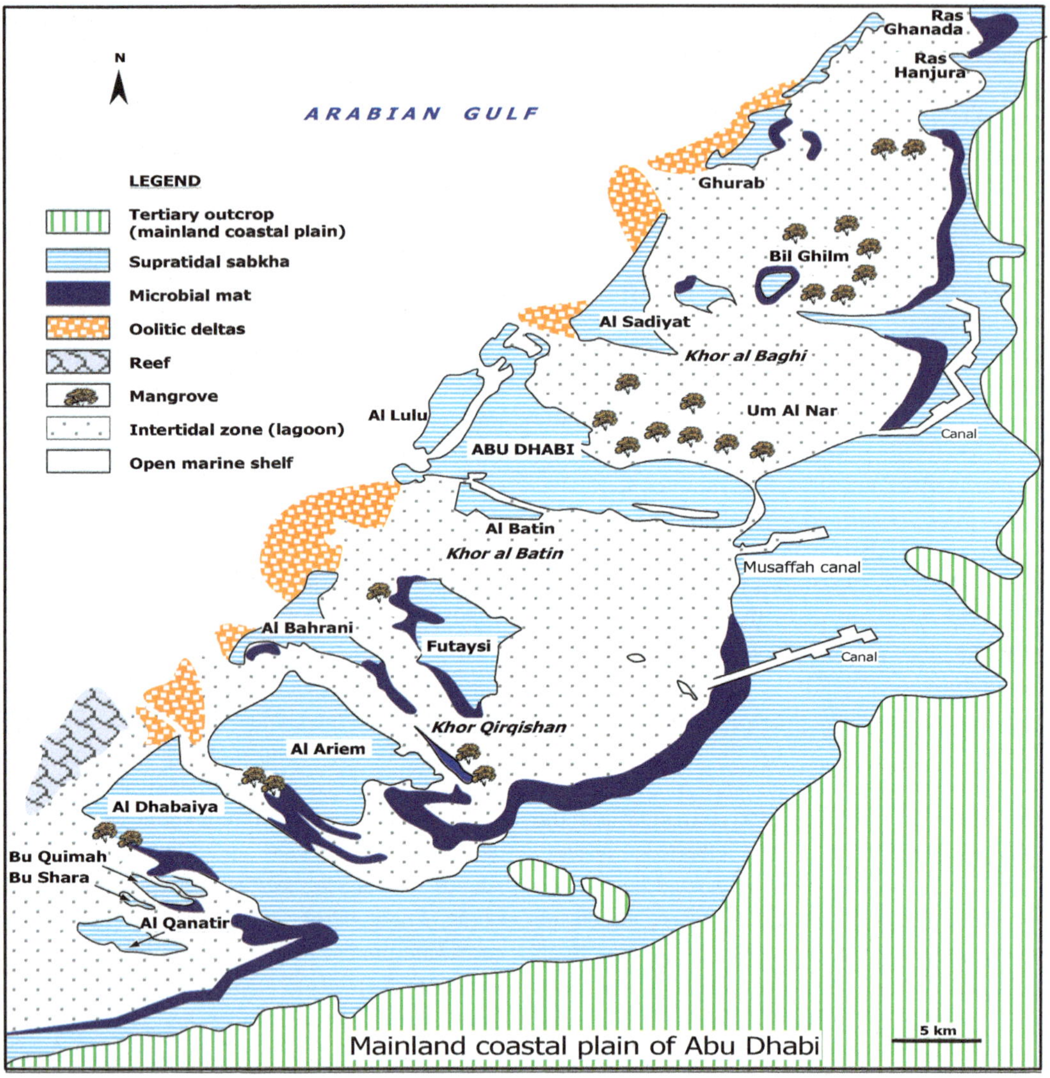

Fig. 11.9 Sedimentary facies distribution around Abu Dhabi Island and adjacent islands. (Modified from Alsharhan and Kendall (2003))

currents. These tails restrict the circulation, in the series of coastal lagoons formed in their protection. At the seaward margins of most of the barrier islands' oolitic sands and locally, skeletal grainstones collect on relatively high-energy shoals, beach barriers, tidal deltas, and channels. These latter channels are the site of the accumulation of well-abraded mollusks, echinoids, foraminifera, and corals. Laterally discontinuous reefs occur just north of the central portions of the barrier islands (Evans et al. 1964a, b). Interestingly coral reefs also develop along the seaward flanks of the inter-island tidal channels and in the seaward portions of the lagoons, where tidal channel waters provide enough circulation to enable coral growth (Kinsman 1964).

Interestingly at Ras Ghanada, a sabkha and lagoon complex just to the northeast of Abu Dhabi Island (Fig. 11.11), there is a great example of one of the more mature sediment-filled systems along the Abu Dhabi coast. It includes a nearly closed lagoon connected to the Arabian

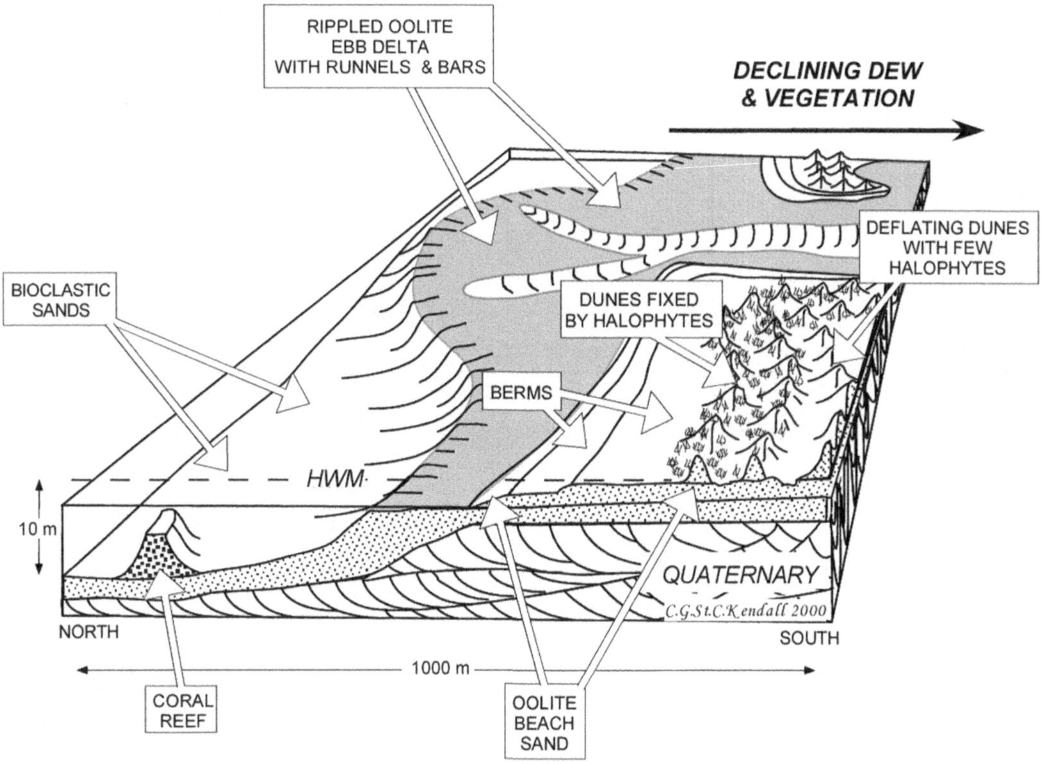

Fig. 11.10 Block diagram of ooid delta morphology with runnels and bars and aeolian dunes eastern Abu Dhabi

Gulf by narrow channels and is being enveloped by a large sabkha constituting the supratidal part of the lagoon (Baltzer et al. 1994).

Macrotidal Restricted Lagoons

In the restricted lagoons to the south and landward of the barrier islands, the tidal movement tends to be at right angles to the coast. The sediments of the lagoons are commonly dominated by lime mudstones with scattered skeletal fragments which are the remains of in situ organisms. Where sand-size carbonates occur in the seaward portions of the lagoons, these are often composed of fecal pellets. Close to the edges of the oolite deltas, where the deltas are subject to flood tides, grapestones accumulate.

At low tide extensive parts of the lagoon floor are exposed, and evaporation of the intertidal waters drives up salinities. This is matched by a decrease in echinoids, corals, and algae and an increase in gastropods and imperforate foraminifera. Here, the accumulation of lagoonal sediments is being encroached on and terminated by sediments, from tidal flat settings (often lined by mangroves and cyanobacterial mats), that in turn are flanked and being encroached on by supratidal salt flat (sabkha) sediments.

Coastal Terraces

A subtidal to intertidal series of fairly continuous coastal terraces flank the southern margin of the western Khor al Bazam, from Tarif to Jebel Dhanna. The edge of these coastal terraces has an abrupt drop off which is colonized by green, brown, and red calcareous algae and corals.

At the Jebel Dhanna, the coastal terrace sediments accumulate on the eroded surface of what appears to be a Late Pleistocene or Early Holocene reef flat and wave-cut benches of pre-Holocene rocks (Figs. 11.12 and 11.13). The

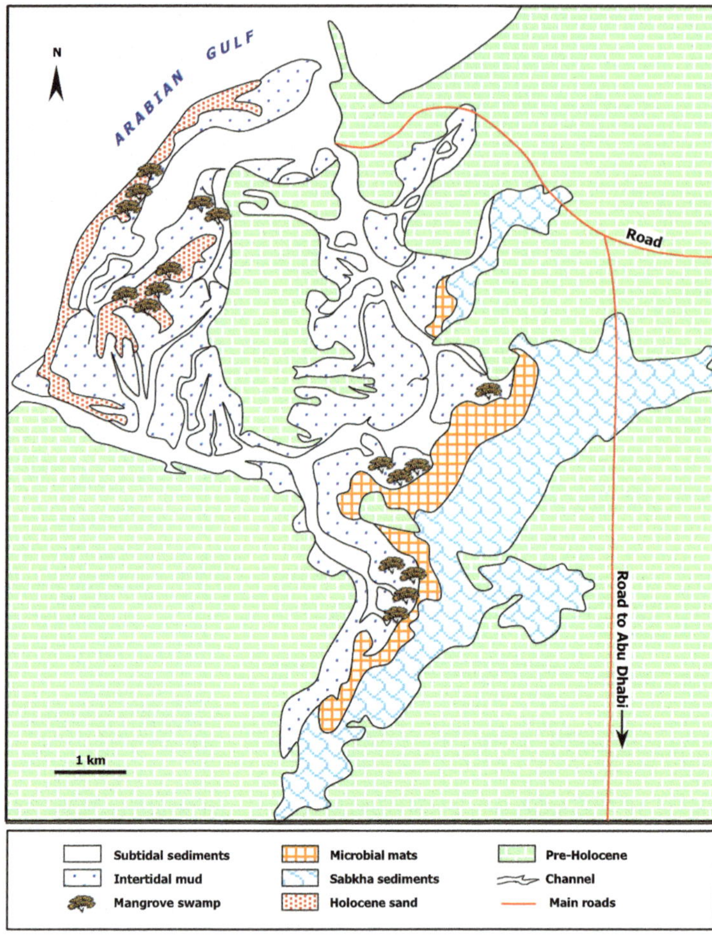

Fig. 11.11 Sedimentary facies of Ras Ghanada, eastern Abu Dhabi Island. (Modified from Baltzer et al. (1994), Alsharhan and Kendall (2003))

sediments have varying thickness of unconsolidated sands that may be bioclastic, oolitic, or grapestones. Locally, as Kendall and Skipwith (1969) note, alternating layers of poorly cemented indurated cemented crust (1–3 cm thick) may underlie these sands. The cemented layers have flat upper surfaces and irregular lower surfaces. Some layers are contemporaneous beachrock and are forming from the unconsolidated sediment of the sand flat cemented by calcium carbonate precipitated from evaporating capillary water during low tide. This latter process is probably active during the heat of summer. Before additional sediment accumulates, the upper surface of the beachrock is often truncated by storm action. Similar crusts were found at the west end of Abu Dhabi Island.

Not all the layers are contemporaneous, and some are of different composition than the loose sand above them. These horizons often extend from 2 m below the low water mark to a few meters above the high-water mark (in the latter case they are often overlain by beach and algal sediments) (Kendall and Skipwith 1969; Kirkham 1998a). These limestones probably represent sediments deposited as sea-level transgressed, across the shelf during the last rise in sea level. They were cemented to form beachrock that was contemporaneous with the change in sea level.

At the seaward edge of the coastal terrace and offshore bank, sand ridges lie parallel to the shore and are covered by the highest tides in the region. These match the break point bars of King (1959).

Fig. 11.12 Landsat images of Jebel Dhanna Peninsula (**a**) reefs and coastal terraces east of Jebel Dhanna and Dagallah. (**b**) Enclosed bay facies southwest of Jebel Dhanna. Note the small embayment rimmed by spits. The western margin of this embayment is the site of precipitation of marine travertines

Field inspection of the bars of the Khor al Bazam visible on aerial photographs suggests that they are produced during storms by plunging breakers, as they crossed the edge of the offshore shoal areas and coastal terrace. The spill of the breakers transported the sand across the top of the reef fronts and terrace. These sands are composed of bioclastic debris. Parallel ridges of bioclastic carbonate sand separated by gullies occur within the intertidal zone of the coastal terrace. Three types of runnels and bars are distinguished in this area: those with frequent crosscutting channels, those with few crosscutting channels, and those which are angled (Fig. 11.14). Similar bars are believed by Williams (1960) to be formed by breaking waves.

Intertidal spits and beaches on which locally oolites collect delimit the landward edge of the coastal terrace. In the vicinity of Jebel Dhanna peninsula (Fig. 11.12a, b), oolitic grains are accumulating on the coastal terraces to its west and offshore to the north in the vicinity of Sir Bani Yas Island. Closer to the shoreline, the surface of the terraces may be divided on the basis of fauna, into *Cerithium* sp. flats on the lower area and crab flats on the upper. *Scopimera* sp., the crab characteristic of the latter zone, produces radial patterns of feeding balls. Crabs, worms, and mollusks extensively burrow the sediments of the flats so that primary sedimentary structures are destroyed. Ginsburg (1957) reported the same phenomenon in carbonate sediments of Florida.

Intertidal spits are one of the most common features associated with the shoreline of the western Khor al Bazam and can be seen just west of Jebel Dhanna to the Tarif-Al Mirfa areas of western Abu Dhabi (Figs. 11.12b, 11.15, 11.16, and 11.17). They delineate the upper limit of the intertidal flats (Kendall and Skipwith 1969; Kirkham 1997) and trend from west to east, forming by longshore drift. The seaward side of each spit has a beach-face slope, with a gradient of 1:6, while the slope of the adjacent seaward intertidal flat has a gradient of approximately 1:5000. A line of brown algae or bladder wrack

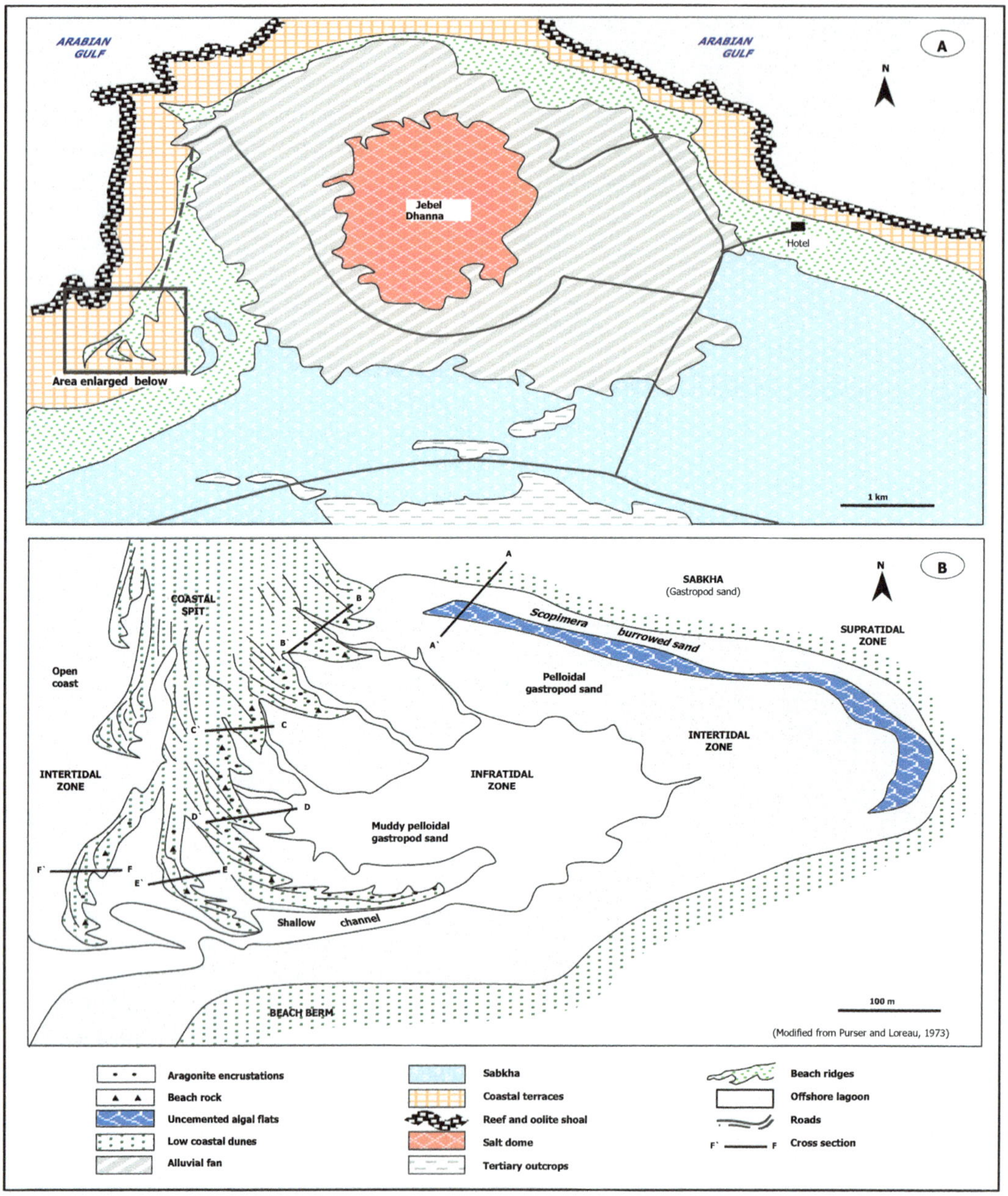

Fig. 11.13 (**a**) Geologic map of Jebel Dhanna (western Abu Dhabi) and (**b**) distribution of different sedimentary facies at southwest Jebel Dhanna

usually marks the base of the beach-face slope. As can be seen at Jebel Dhanna above the beach face, there is usually at least one berm backed by a line of hummocky dunes. In cross section, the landward-facing slope of a spit is convex and curves down to the approximate level of the old intertidal flat upon which it rests. The sediments show the sheetlike geometry, described by Imbrie and Buchanan (1965), for similar features from the Bahamas.

The spits increase in width seaward unless storm events generate another system of spits seaward of them. Spits appear to begin to form along lines of drifted seaweed that accumulate on the tidal flat. The inner edge of the coastal berms, along the Khor al Bazam, is fixed by *Salicornia*

Fig. 11.14 Annotated Google Earth image of coastal terraces in the vicinity of Al Tarif on the southern coast of the Khor Al Bazam

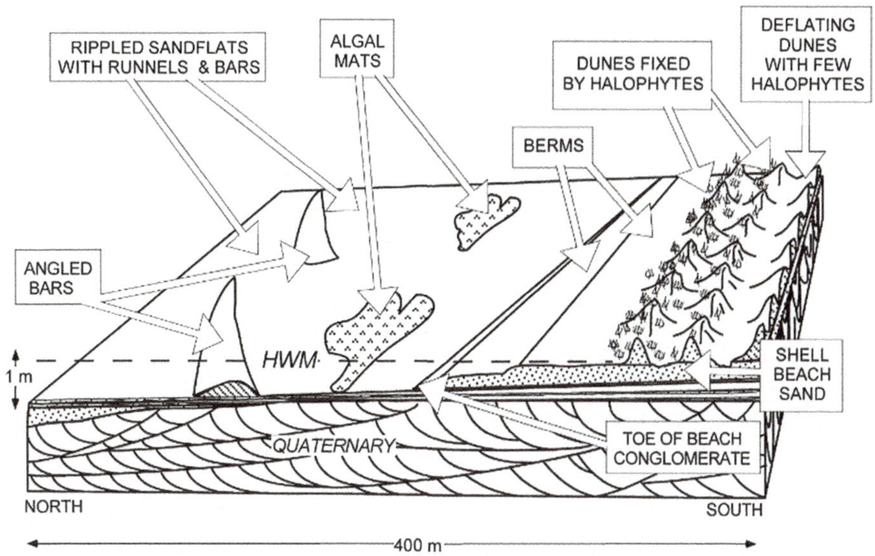

Fig. 11.15 Block diagram of morphology of angled bars and higher-energy beaches of the inner portion of the coastal terraces from Jebel Dhanna to Tarif-Al Mirfa areas of western Abu Dhabi

sp. This plant traps sediment and stabilizes the low dunes (Kendall and Skipwith 1969; Kirkham 1997). This vegetation is best developed at the seaward edge of the dune field and decreases in frequency and size landward. This change in character can be attributed to a combination of the daily dews that collect on these dunes and the presence of moisture-retaining bladder wrack, caught up in the sediments of the rapidly landward, as does the remnants of the bladder wrack. Ghost crabs with their characteristic burrows, and adjacent cone-like sand mounds, locally colonize the berm.

The entire coastline of the western Khor al Bazam is lined with an inland line similar compound forms of spits that are now stranded inland by coastal accretion, in the form of a series of beach ridges arc (Kendall and Skipwith 1969;

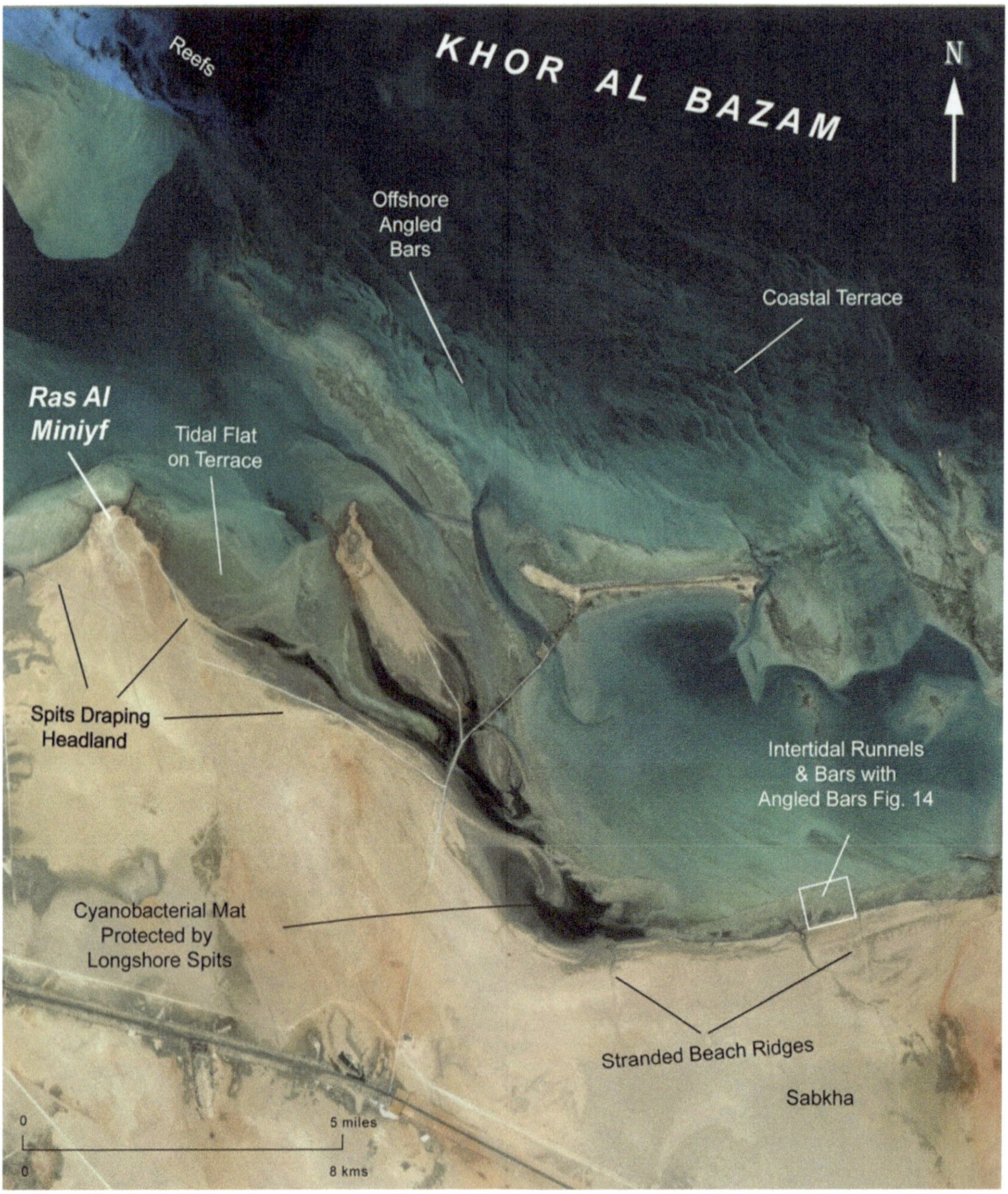

Fig. 11.16 Landsat image of coastal terraces, coralgal reef, and offshore angle bars and spits east of the Ras Al Miniyf peninsula

Kirkham 1997). Kirkham (1997) ascribes these to the early Holocene Flandrian transgression. Spits within each compound beach ridge are commonly hooked, sometimes forming into a looped bar (Evans 1942). Other spits drape headlands to form winged headlands (Kirkham 1997); examples include Ras Al Miniyf (Fig. 11.16) and west of Al Mirfa and the island of Marawwah (Kendall and Skipwith 1969).

West of Al Qanatir toward Jebel Dhanna appears to have had a history that suggests this portion of the Abu Dhabi Emirate coast has been exposed to constant wave action, while other areas have become more protected. This is sup-

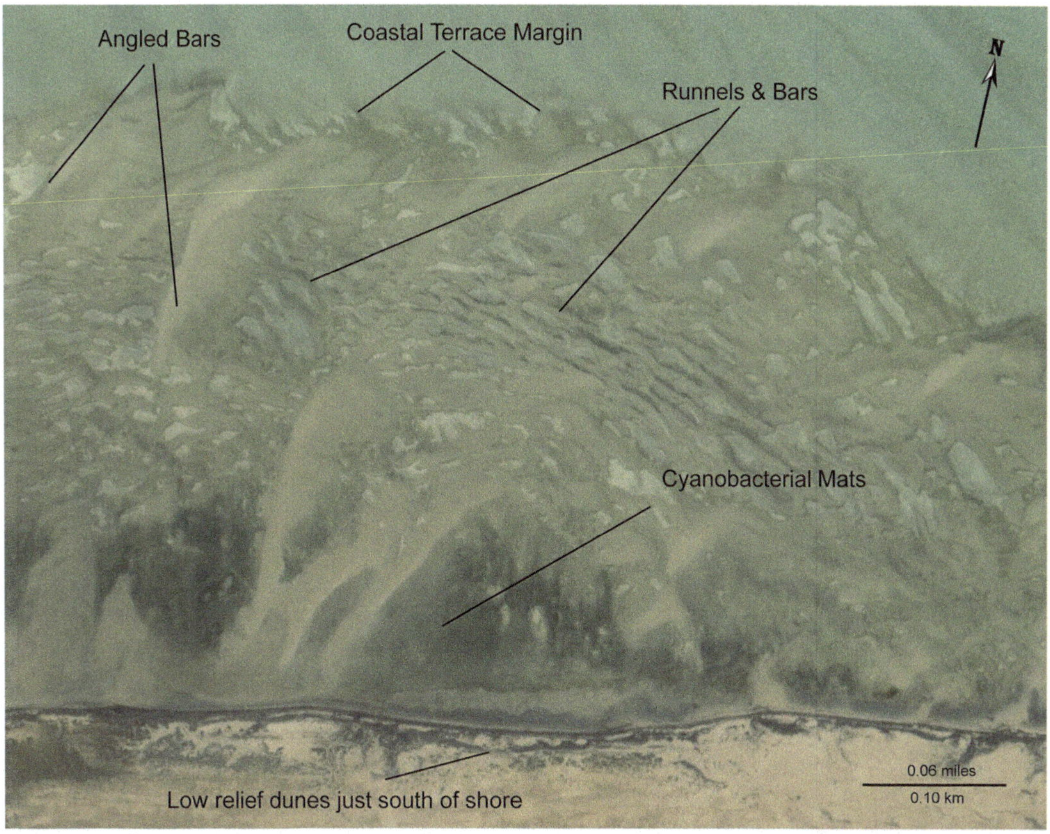

Fig. 11.17 Google Earth image (September 2008) east of the Ras Al Miniyf peninsula (located in white rectangle inset on Fig. 11.15). Note the tidal flats marked by runnels and bars, cyanobacterial mats protected by angled bars, and small dune covered beach ridges just south of the shoreline

ported by the abundance of the beach ridges that often delineate the upper margins of the tidal flat along the coast (Kirkham 1997). As in similar beach ridges, stranded in the sabkha to southwest of Abu Dhabi Island and in the area near Al Qanatir Island, these beach ridges are composed of accumulations of cerithid gastropod and bryozoan fragments. Locally, the ridges have isolated large embayments, between headlands of the Tertiary rock, which were probably filled and became isolated during the Flandrian transgression (Kirkham 1997). These embayments are now extensive sandy sabkhas, with the sediment derived from storm washover, aeolian materials, and outwash from the Neogene deposits. In these areas, halite is precipitated in the near subsurface and is underlain by carbonate sands and local cyanobacterial mats. Layering is disrupted by halite in the near surface of the sabkha, thus convoluting the bedding of these sediments.

The beach ridges are commonly uncemented; however, local beach rocks do occur. Thin aragonitic cement coatings are present in settings just seaward of the beach ridges and spits that mark the coast of the Khor al Bazam lagoon. For instance, near Ras Al Aish, surface cements coat marine indurated cemented carbonate crusts and beachrocks developing in the upper tidal flat sands situated seaward of the beach ridges. In fact, beachrock and indurated cemented crust surfaces are common in the middle tidal flat east of Jebel Dhanna (Fig. 11.12a). Most of these indurated cemented crust facies are not coated with carbonate cement crusts.

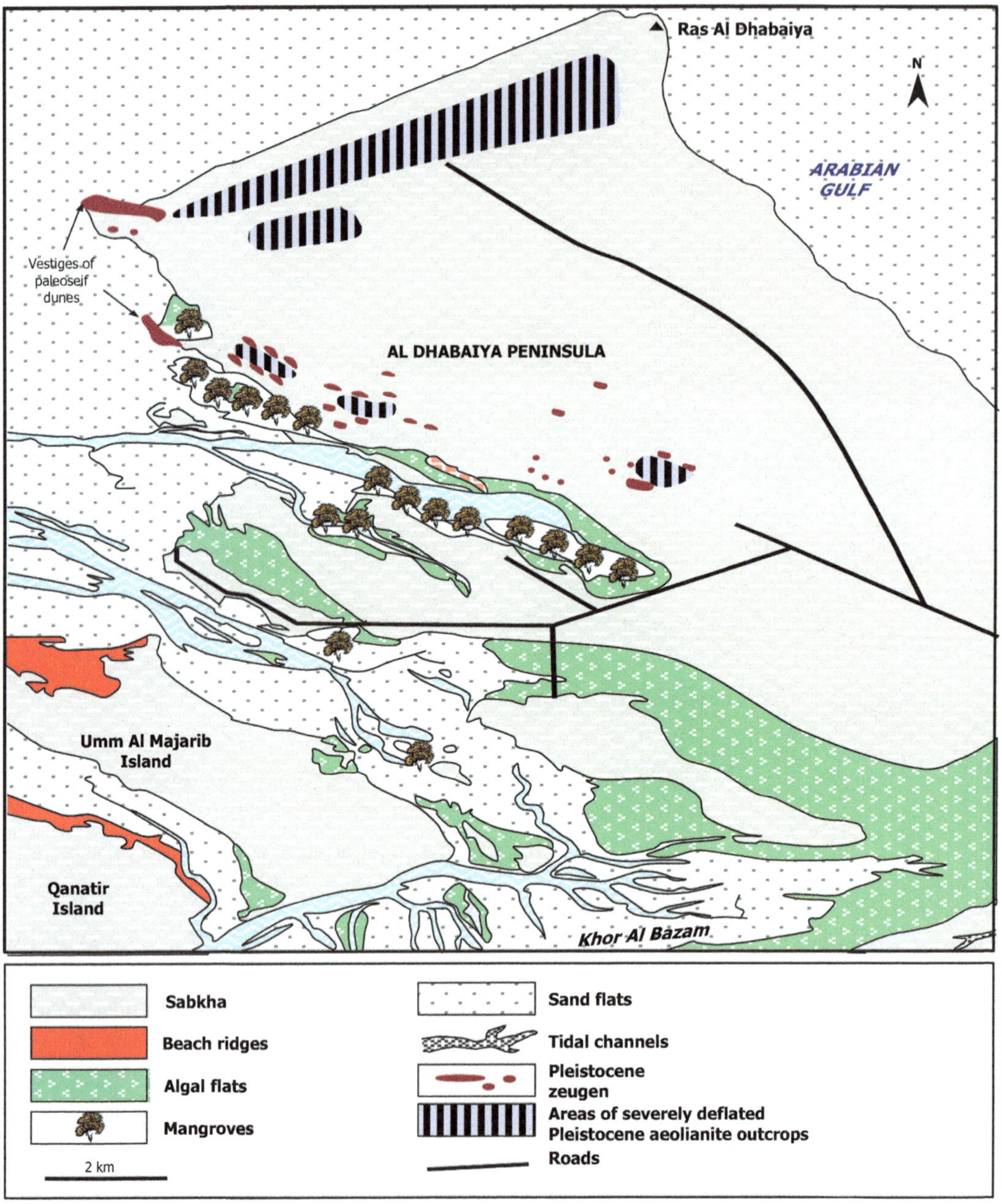

Fig. 11.18 Sedimentary facies map of Al Dhabaiya Peninsula, western Abu Dhabi

Mangrove Distribution

The southwestern coast of Al Dhabaiya Peninsula mangrove growth is common (Fig. 11.18). Here, a series of Pleistocene outcrops protect the west coast, and around and between them cyanobacterial flats and tidal creeks are lined by stands of the black mangrove (*Avicennia marina*). This mangrove has a characteristic system of protruding roots or "pneumatophores." It is the only mangrove that has been found in the Khor al Bazam, and it occurs as small bushes and trees in areas protected from turbulent wave action.

Fig. 11.19 (**a**) Mangroves occurring just seaward of intertidal beach rock; (**b**) crab burrowed tufted cyanobacterial mat, showing in the distance are Salicornia and black mangroves *Avicennia marina*; (**c**) cemented crab burrows along tidal creek; (**d**) mud cracks along the bank of a small tidal channel

Mangroves grow in many shallow protected areas in Abu Dhabi, usually lining the banks of creeks, draining algal flats or in narrow strips, parallel to the shore at the very top of the intertidal flats (Fig. 11.19a). These plants grow where they experience frequent tidal interchange, so that their roots and lower trunk are normally covered at high tide. Larger mangroves grow on the edges of channels, where there is better tidal interchange and they diminish in height and distribution inland. On the margins of the Khor al Bazam, as mangrove-lined channels become sediment-choked, the area becomes colonized by a crinkled cyanobacterial mat, and the mangroves die. These algal mats grow landward of the mangroves and are often underlain by a gypsum mush (Fig. 11.19b). The mats also grow on creek banks, are crinkled (Fig. 11.19c), and are occasionally tufted. The creek banks are often cemented to form a beachrock. In some cases dunes covered with halophytes, including *Salicornia* sp. and *Arthrocnemum glaucum* (Fig. 11.19c), enclose the cyanobacterial mat. Aragonite mud accumulates around the mangroves of the Khor al Bazam, as it does in Florida (Bathurst 1975) and in the Bahamas (Newell et al. 1951). In Abu Dhabi Emirate, this mud commonly is intensely burrowed by crabs (Fig. 11.19b). Some of this mud undoubtedly precipitates in situ, either in response to the evaporation of the seawater or local microbial activity.

There are three characteristic systems of coastal progradation, associated with the mangrove swamps of the southwestern coast of Al Dhabaiya peninsula. In one case are the mangrove-lined tidal creeks that pass laterally into a burrowed crab flat, with some *Salicornia* sp. colonizing local heights, into a gypsiferous crinkled cyanobacterial flat and finally sabkha flats. Penecontemporaneous cementation can be observed, forming indurated cemented crusts around the bushes of *Salicornia* sp. and extending across the burrowed crab flats. Mud-cracked zones are common here (Fig. 11.19d). In contrast, it is possible to traverse from mangrove-lined creeks, followed by burrowed crab flats, into *Salicornia* sp.-fixed gypsum dunes, across a narrow belt of algal flats (Fig. 11.19b). The third case occurs where *Salicornia* sp.-fixed dunes adjoin bioclastic beaches that are abruptly separated from adjacent sand flats by mangrove stands trapping fine-grained mud. In all three of these different mangrove settings, indurated cemented crusts and beachrocks are common.

Cyanobacterial Mats and Coastal Sabkhas

In the inner intertidal zone of the Abu Dhabi coast where there is maximum protection from the dominant north-northwest shamal winds, extensive microbial mats once dominated the tidal flats of this section of coast. These include three areas where wide-laminated mats of cyanobacteria once colonized the area before industrial development started along the coast. These include south and west of Abu Dhabi, the south east Khor Al Bazam, and south of the island of Khusaifa, also in the Khor Al Bazam but westward (Fig. 11.20).

There are numerous smaller mats in the western Khor al Bazam. These occur along the southern coast of exposed to wave action but in the shelter of islands, headlands, protected by small spits and angled swash bars on the upper positions of tidal flats. These incipient mates have crenulated surfaces and form ephemeral peats that are oxidized and eroded as the coast builds out. These small algal flats are prominent on the north side of the beach barrier on the shoals west of the island of Dagallah.

At the southeast end of the Khor Al Bazam lagoon, the largest of the Abu Dhabi algal flat parallels the coast for 42 km, while to the west, another smaller one parallels the coast for 9 km south of the island of Khusaifa. These flats are part of the seaward edge of a prograding coastal flat, with an average width of approximately 2 kilometers and cyanobacterial peat with a thickness of at least 30 cm. In some areas these peats extend landward in the subsurface for more than 2 km. They are capped by a thin cover of evaporites and wind-blown and storm washover sediments (Figs. 11.21 and 11.22).

The larger algal flats are divided on the basis of surface morphology, into four geographic belts. From the high-water mark seaward these are:

1. Cinder zone – a warty black algal surface, the color and size of the raised bumps resembling a weathered volcanic cinder layer (Fig. 11.23). ·These bumps, shaped like small pustules 2–3 centimeters in diameter, cap an unlaminated algal and sediment peat; this surface caps an unlaminated algal and sediment peat (Logan and others, 1964, code LLH).
2. Polygonal zone – algal mat separated into desiccation polygons a few centimeters to 2 meters in diameter, which cover laminated algal peat; carbonate sand and mud fills the cracks between the polygons (Fig. 11.24). The smaller of these polygons have raised edges between and under which sediment collects. Larger polygons grow in the small creeks that drain the lower portions of the flats. The growth of the laminae is limited by salinity. The size of the polygons is governed by the frequency of exposure to desiccation. Where the widths of the mats are reduced by sandy headlands, polygon mat zones are narrower, and in some cases, there are no polygonal and cinder zones.
3. Crinkle zone – leathery algal skin forming a blistered surface over gypsum and carbonate mush (Fig. 11.25). The origin of these blisters is unknown but probably is a combination of desiccation and the movement of trapped air with the rising tide. Wave scour is restricted, even during large storms

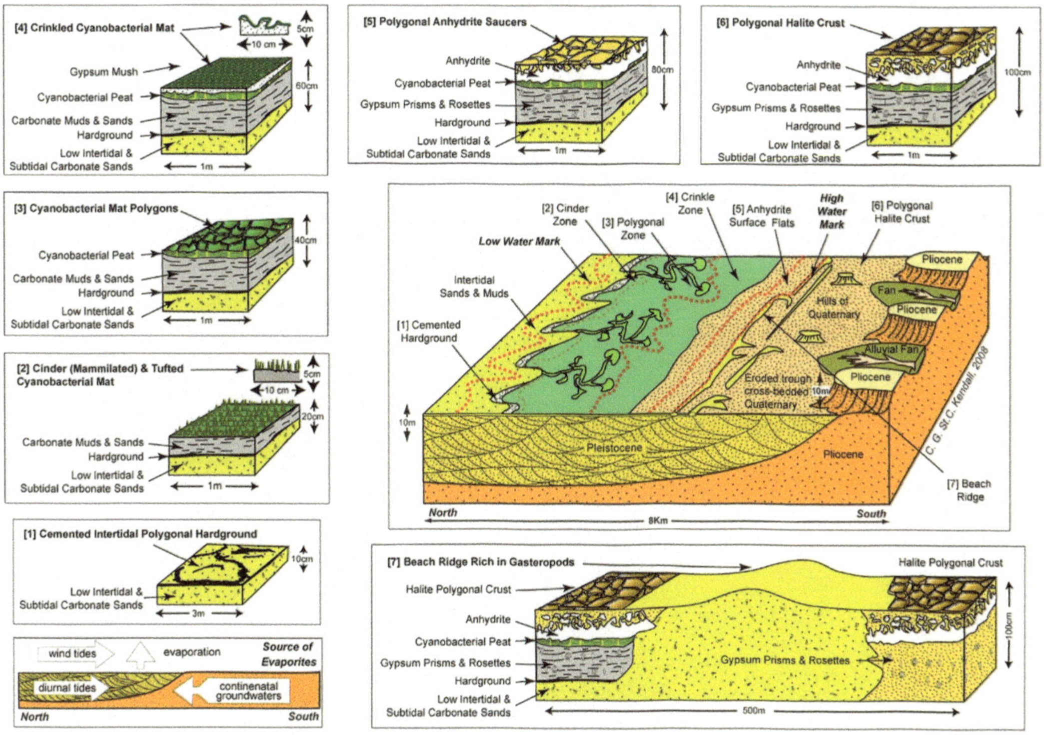

Fig. 11.20 Conceptual distribution of sabkha and cyanobacterial sediments of Abu Dhabi. (After Butler et al. 1982). (*1*) Seaward lagoonal carbonate sands and/or muds capped by carbonate hardgrounds; (*2*) poorly laminated lower tidal-flat cinder-like cyanobacterial peat; (*3*) upper tidal-flat cyanobacterial mat, commonly laminated and capped by desiccation polygons formed by growing cyanobacterial mat; (*4*) high intertidal to supratidal crenulated cyanobacterial mat capping a mush of gypsum crystals; (*5*) cap of supratidal anhydrite polygons with storm washover and wind-blown carbonate and quartz. Below, anhydrite is interlayered with storm washover carbonate and replaces the supratidal mush of gypsum crystals. Note the large gypsum crystals (prismatic or lenticular) that form within the cyanobacterial peat and carbonate and overlie and sometimes cement the hardground layers; (*6*) Halite crust formed into compressional polygons over the anhydrite layers, algal peat, hardgrounds, and intertidal and lagoonal carbonate; (*7*) cerithid-rich beach ridge flanked by cyanobacterial. (After Kendall and Alsharhan (2011))

4. Flat zone – firm, smooth algal mat, with no topographic relief, overlying a gypsum mush above algal peats of the polygonal zone. Landward carbonate sand and muds with dispersed quarts grains are incorporated in the surface, and anhydrite starts to replace the gypsum at the sediment surface.

The algal growth and structures of these different zones are related to the frequency and duration of subaerial exposure and the salinity of the tidal waters, while waves and tidal scour affect the edge of the cinder zone and along ebb channels. Of these agencies desiccation appears by far the most effective in controlling the surface morphology and salinity and hence the rate of algal growth. The algal mat morphology of the Khor al Bazam suggests that when you make an environmental interpretation of other algal stromatolites, allowance should be made for the effects of desiccation as well as wave and tidal scour. During normal diurnal flooding by tides, very little sediment is carried onto the mats, but during storm events lime mud and sands were transported onto their surface. The larger cyanobacterial flats pass landward into sabkha and landward a series of stranded high-energy small beach ridges located in what is now the supratidal zone (Kendall and Skipwith 1969; Kirkham 1997).

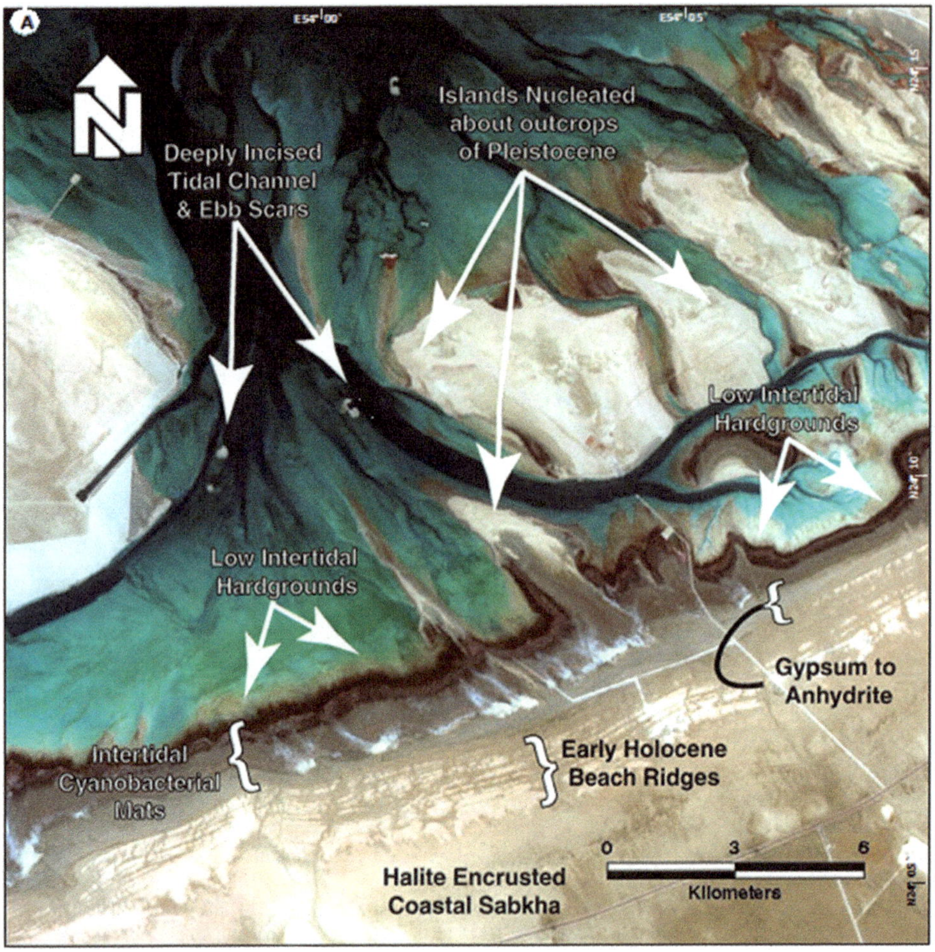

Fig. 11.21 Landsat image from between Al Dhabaiya and Abu Al Abyad Islands showing intertidal cyanobacterial mats, beach ridge, and coastal sabkha. In the offshore area deeply incised tidal channel, ebb scars, and low intertidal hardgrounds

Coastal Mainland Sabkha

Intertidal flats covered by cyanobacterial mats pass landward into a very flat supratidal zone of the coastal "mainland sabkha" plain without any break in slope (Evans et al. 1964a, b; Butler et al. 1982; Kirkham 1997). Its surface slopes gently seaward and lies above the level of normal spring tides and is covered by thin layer and surface of cyanobacteria intermixed with halite. Periodic storm induces flooding transport carbonate sand and lime mud onto its surface to form laminae. These are often disrupted by desiccation and evaporite mineral growth. These coastal sabkhas lie seaward of early Holocene Flandrian high-energy beaches composed of bioclastic sands rich in cerithid gastropods (Figs. 11.20 and 11.21) (Kirkham 1997; Kenig et al. 1991 and Strohmenger et al. 2011).

The "mainland sabkha" plain caps accreting Holocene supratidal, intertidal, and lagoonal carbonate sediments that extend over 320 km from Ras Ghanada to the Qatar peninsula to the west. Along the shore of the Khor al Bazam, the width of the sabkha varies, often as much as 32 km wide across embayments or nonexistent where peninsulas of the Neogene and Quaternary rocks outcrop, and in some places, the sabkha surface lies flush with eroded Quaternary rocks (Fig. 11.4d).

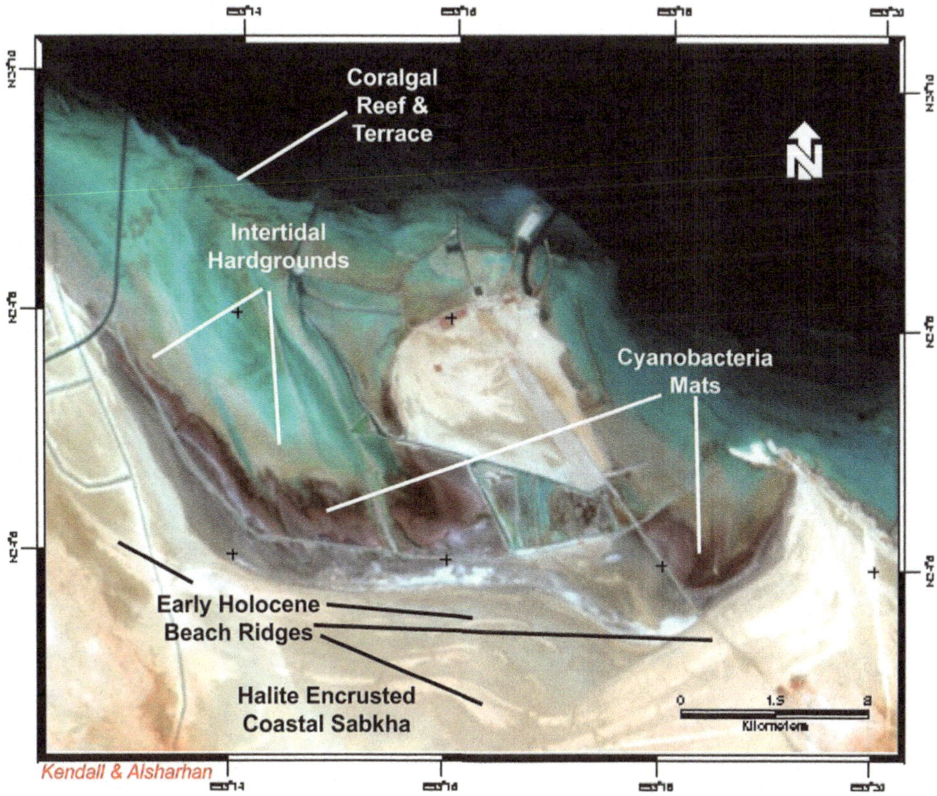

Fig. 11.22 Google image of Khusaifa intertidal cyanobacterial mats, beach ridge, and coastal sabkha. In the offshore-incised tidal channel, ebb scars, and low intertidal hardgrounds

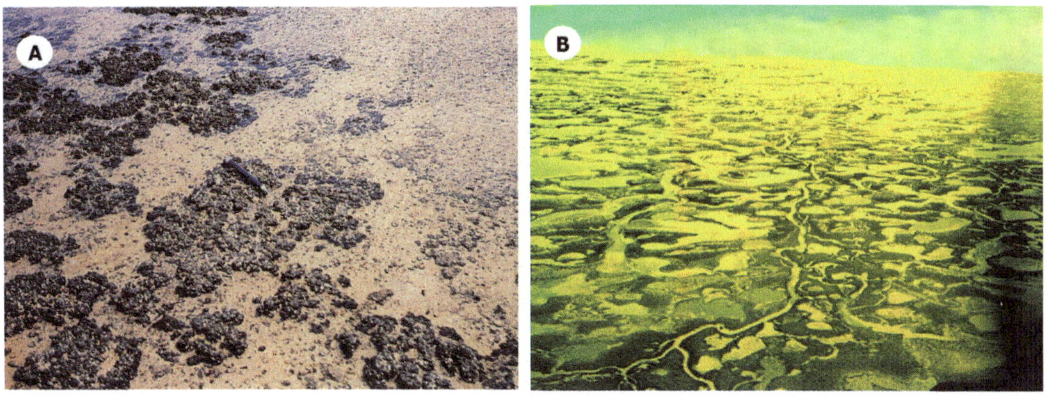

Fig. 11.23 (**a**) Cyanobacterial mats that look like coal cinders and mark the seaward edge of these flats along the coast of Al Qanatir Island. (**b**) Oblique aerial view of the Khor al Bazam cyanobacterial mat. Note that at this locality, this mat is the site of meandering tidal creeks and small round ponds. Most the cyanobacterial mat here is formed into the polygonal shapes that mark the cyanobacteria mat zone

Fig. 11.24 (**a**) Polygonal cyanobacterial mat surface in the middle of upper intertidal zone. (**b**) Large polygonal cyanobacterial mat overlying algal peat and underlying a poorly preserved algal surface. Both photos taken SE of Al Qanatir Island

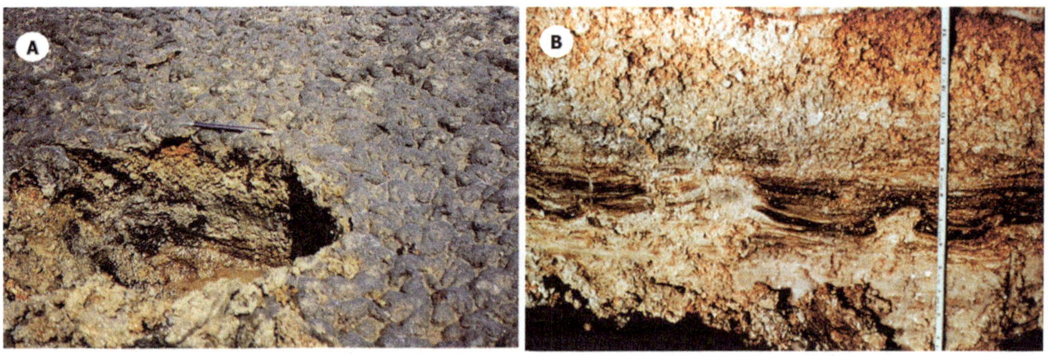

Fig. 11.25 (**a**) Crenulated or crinkle mat overlying the gypsum mush layer with the cyanobacterial peat below of the upper intertidal zone. (**b**) Surface trench located near the seaward margin of supratidal zone. Upper portion of the trench exposes bed of gypsum mush overlying algal peat and underlying a poorly preserved algal surface. Both photos taken from Al Qanatir Island

Trenches dug in the coastal sabkhas expose shoaling upward cycles of subtidal/intertidal carbonates capped by cyanobacterial peats overlain by supratidal evaporite facies (Fig. 11.13) (Kinsman and Park 1976; Butler et al. 1982; Kenig et al. 1989; and Kirkham 1997). At least 2 km inland from the present high-water mark (HWM), algal laminae are still preserved above the sandy low intertidal sediments and their capping thin beachrock crust. As could be seen in the walls of a canal cut at Mussafah, the cycles found in mainland coastal sabkha are complex (Kenig et al. 1991; Strohmenger et al. 2011). The cycle begins with a transgressive algal mat, followed by a prograding shoaling upward cycle of lagoonal muds and sands. Locally, these sands are underlain by indurated cemented crusts. Before the landfill from industrial development, cyanobacterial peats, anhydrite, and gypsum and halite capped the sands and muds that prograde seaward into the lagoons. The intense evaporation of the mixed continental groundwater and marine water table created high salt concentration in surface capillaries, also brought an upward movement of brines from the water table (Butler 1965; Wood et al. 2002). This caused the precipitation of gypsum and aragonite, while the Mg^{2+}/Ca^{2+} ratio rises, so local dolomitization of interstitial aragonite mud also occurred.

Fig. 11.26 At truncated core from the supratidal sabkha of the Al Qanatir traverse. The gypsum mush layer truncated from above A in the core. Note the gypsum rosettes in the laminated cyanobacterial peat of the polygonal zone. (From Butler et al. (1982))

As indicated at the seaward margin of the sabkha, its surface is an almost horizontal salt-encrusted flat algal flat. This thin algal mat covers the layer of gypsum mush that extends seaward beneath the crenulated mat of the crinkle zone and landward to be replaced by anhydrite. The mush is composed of lenticular gypsum crystals. Each crystal is about 0.5 mm in diameter and flattened perpendicular to its c-axis. Inland from the intertidal zone, this layer of gypsum crystals thickens to 20 cm (Fig. 11.25a, b). At the landward edge of flat algal flat, some of the surface gypsum mushes are replaced by small blebs of anhydrite (Fig. 11.25b). Some of the gypsum crystals in this zone show signs of solution, and calcium sulfate hemihydrate ($CaSO_4 \cdot O \cdot O_2H_2O$) may be present. Just seaward of the first occurrence of anhydrite, buried algal peat-capping lagoonal sediments occur. This peat contains larger flattened gypsum crystals (Fig. 11.26) up to 15 cm in diameter. The size of these prisms of gypsum is thought to be controlled by slow growth at depth within the water table. Where this gypsum is found in sandy lagoonal sediments, it contains many inclusions. However, in the algal sediments, gypsum characteristically contains fewer impurities.

Landward of the supratidal flat algal flat, anhydrite first forms as small nodules about 0.5–1 mm in diameter. This anhydrite lies within the sediment surface occurring as a soft blebs and nodules and layers with a white cheese-like texture and thixotropic. Like the gypsum mush, the anhydrite forms in the capillary zone. Traced into areas of higher salinity, the nodules become more abundant and larger, in places as much as 4–6 cm across. They form in and displace aeolian and storm washover sediments that accumulate on the upper algal flats and back onto inner portions of the sabkha. Landward of the initial nodules, a surface layer of anhydrite is precipitated and formed into a series of interlocking saucer-shaped structures that are polygonal in outline (Fig. 11.27a). As the washover sediments overlying the algal mat thicken in the Khor al Bazam and inland from Abu Dhabi Island, an anhydrite layer replaces the gypsum mush and in some cases is more than 20 cm thick and may become diapiric (Fig. 11.27b). Chicken wire nodules, and thin layers of anhydrite 2–3 cm thick, form concurrently in the same washover sediments. The layers show an enterolithic, contorted, crenulated, and tightly folded fabrics, similar to those of ptygmatic quartz veins (Fig. 11.27b). West of Tarif, antiform structures more than 20 m in diameter occur.

The sequence of anhydrite development in the Khor al Bazam ends with the influx of groundwater into the sabkha via outwash fans. Anhydrite is hydrated to gypsum, forming white, coarse, elongate, toothlike monoclinic crystals, about 2–3 cm long and 0.5 cm in diameter (Fig. 11.28a, b). The hydration process disturbs the surface of the sabkha, so much that it resembles a plowed field. The blistered surface contains large quantities of halite. Elongated, wispy crystals of rock salt are common in surface sediments of the sabkha, at the top of the capillary zone. After flash floods and marine incursions, surface waters evaporate and leave a thin white crust of halite,

Fig. 11.27 (**a**) Polygonal saucers of anhydrite exposed at salt flat surface (photo after Butler et al. 1982). (**b**) Stacked or "festooned" anhydrite polygons exposed in shallow trench. (**c**) Surface trench exposing gypsum in recycled aeolianite facies. The gypsum has almost completely replaced the anhydrite. All photos from south of Al Qanatir Island

Fig. 11.28 (**a**) Trench dug in the supratidal salt flats just landward of the stranded beach ridges marking the seaward edge of the southern coast of the Khor al Bazam on the Qanatir traverse. Note the gypsum layers that have almost completely replaced the anhydrite. (Photo from Butler et al. (1982)). (**b**) Trench dug in the supratidal salt flats just landward of the stranded beach ridges marking the seaward edge of the southern coast of the Khor al Bazam on the Qanatir traverse. Note the anhydrite layer

Fig. 11.29 (**a**) Sand-rich ridges of halite layer formed into polygons or tepeed surface that form in depressions behind the beach ridges marking the seaward edge of the southern coast of the Abu Dhabi sabkha. Similar depressions and features occur on the Qanatir traverse. (**b**) Halite polygonal crust that forms on the sabkha behind the beach ridges marking the seaward edge of the southern coast of the Khor al Bazam on the Qanatir Island. (From Butler et al. (1982))

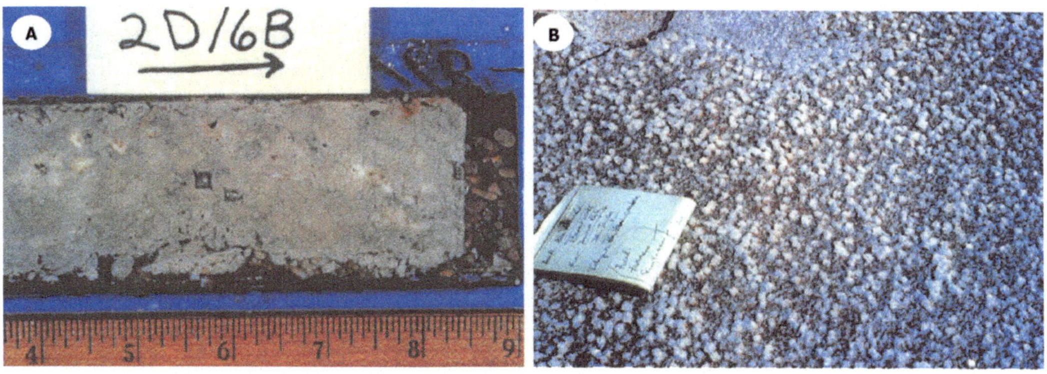

Fig. 11.30 (**a**) Core from the supratidal sabkha of Abu Dhabi. Note the halite hopper crystals that form just below the sediment surface. (**b**) Note the halite hopper crystals that form on the sabkha behind the beach

often expanded into polygonal saucers or, alternatively, sandy salt blisters (Fig. 11.29a, b). The salt is removed by wind and may be replaced by evaporation of capillary water. Where depressions in the sabkha retain pools of floodwater for any length of time, "hopper" crystals develop (Fig. 11.30a, b).

The most conspicuous features of the margin of the sabkha are the old beach ridges which are identified by their distinctive appearance in both the field and aerial photographs as a belt some 2–3 km wide (Figs. 11.7a and 11.8b). As with the "cheniers" Louisiana, they have linear shapes with smooth seaward margins and an irregular landward outline. Composed of well-sorted, coarse skeletal sands rich in cerithids, they mark stages of seaward accretion of the coast in the form of intertidal spits before the sabkha started its progradation representing barrier beaches that formed seaward of reworked earlier sedimentary fill and wadi outwash of the early Holocene (Wood et al. 2002). They drape headlands (as at Ras al Aish, Fig. 11.13) and cross embayments (as at Al Mirfa and Khusaifa, Figs. 11.14a, b and 11.15). These shoreline features formed some 4000 years B.P. (Kirkham 1997) when the Khor al Bazam was more open and deeper. Cross sections cut in some of the sandy sediment of the beach ridges reveal cross-bedding very similar to that shown by much of the miliolite, suggesting that they may be of Quaternary age. Other sediments of the sandy surface show crude horizontal laminae, with well-sorted horizons of foraminifera-rich sand that probably accumulated during wind-driven marine flooding.

Salt crusts occur seaward of the beach ridges or sandy surfaces between and behind beach ridges. These beds are commonly contorted, due to the crystallization of halite or the rotting of sargassum-like seaweed, washed inland with the foraminifera.

At some distance from the sea, large gypsum crystals commonly protrude through the sabkha surface. This occurs where the water table has fallen, and the surface sediments no longer bound together by capillary water are removed by deflation. The gypsum crystals are fragmented on exposure and are explained as the result of diurnal thermal expansion and contraction enhanced by the formation of small halite crystals within cleavage cracks.

The typical sabkha cycle is shown in Figs. 11.18b and 11.22. The surface of the sabkha is modified by both marine and aeolian erosion. For instance, on beach ridges, wind carries away only the finer grades of sand, leaving behind a lag deposit of gastropod shells. Aeolian erosion is limited to where sands are dampened, by capillary water from the water table, but marine erosion can extend deeper. In parts of the sabkha, where it has been particularly effective, marine erosion exposes the water table. A thin, dry, halite crust usually covers such areas, gypsum protrudes as vertical sand crystals, and the polygonal forms of anhydrite may be exposed. Marine flooding also breaches old beach lines, transporting their sediment onto the sabkha behind. The outgoing water re-breaches the beach ridges, to produce small deltas (e.g., west of Khusaifa and west of the Ras Al Miniyf peninsula). The sheets of floodwater are driven about the sabkha by strong winds.

Inland Sabkha Interpretation

The interior of Abu Dhabi Emirate is covered with aeolian sand and gravel plains, interrupted by the vast sabkhat Matti in the west and by a system of inland sabkhas. Along the coast a thin veneer of sand flanks the sea; southward, the sand cover becomes continuous with undulating sand sheets; barchan trains up to 20 m high; and plains covered by a veneer of sand and gravel form an interfingering patterns. The dune form gradually changes from undulating sheets and barchans to intersecting transverse dunes and sand mountains. South of Liwa, gravel plains and sabkhas form distinct hollows between the giant dunes and sand mountains. East of Liwa, undulating sand sheets and barchan systems extend to sabkhat Matti. Shamal wind-transported sand is incorporated in the sabkha or forms adhesion ripples in the western and central parts and accumulates on the eastern margin as small dunes or sand sheets. The surface of Abu Dhabi Emirate is covered by sand dunes of varying morphology.

Figure 11.31 demarks the distribution of main dune types and coastal and inland sabkha areas. Mega-dunes dominate the east, toward the north; both mega-dunes and interdune areas are crossed by small linear dunes that trend west-east (Glennie 1998). In contrast adjacent and parallel to the beaches lining the mainland sabkhas, foredunes form mounds of up to 2.5 m high; behind them there may be one or two beach ridges. For instance close to Al Tarif and Ras Al Miniyf no more than 30 m from the shore, the dunes begin to lose plant cover, probably because the amount of dew precipitated on the plants decreases rapidly with increasing distance from the sea (Figs. 11.14, 11.15, 11.16, and 11.17). Without the plant cover, the dunes are blown away, leaving behind beach ridges (Fig. 11.16).

The area extending from Jebel Baraka (west of Jebel Dhanna) to Al Sila encompasses the enormous inland sabkha, the "sabkhat Matti" (Fig. 11.32). This sabkha is a low-lying region that occurs on and across the western border of the United Arab Emirates with Saudi Arabia. It extends southward from the coast of Abu Dhabi,

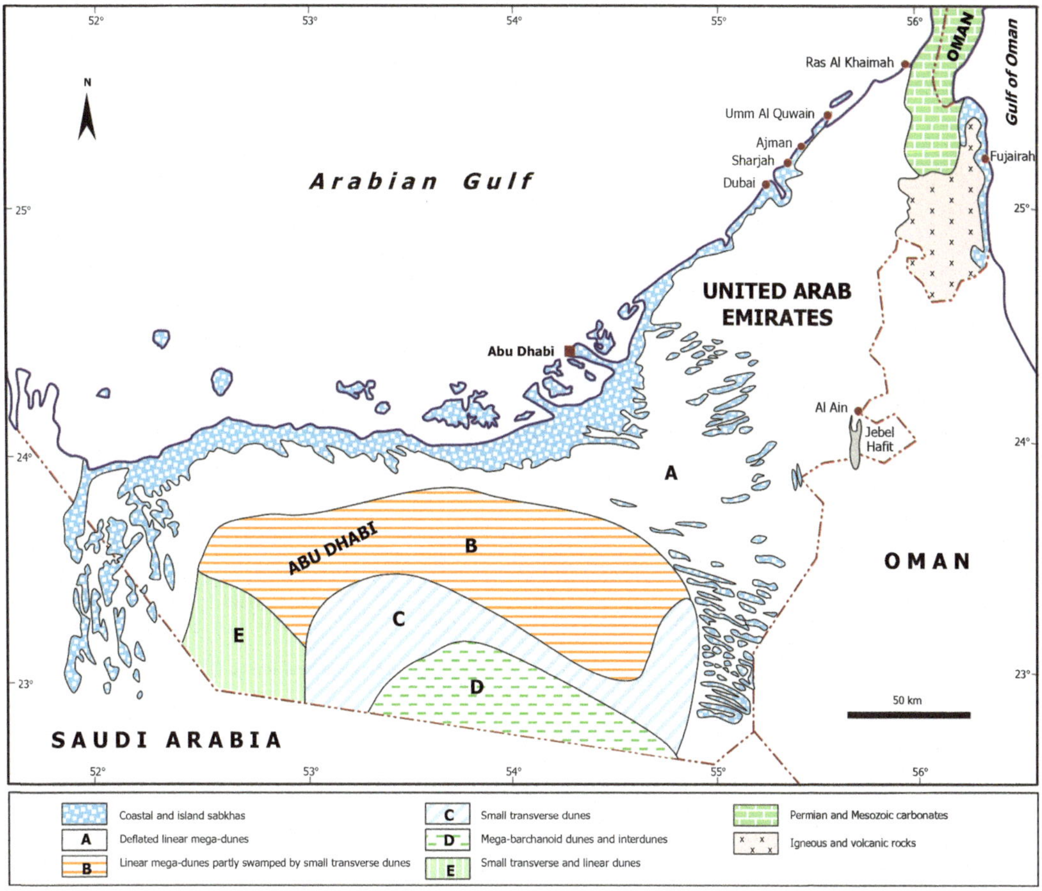

Fig. 11.31 Distribution of different sand dunes in Abu Dhabi Emirate. (Simplified from Glennie (1998))

for some 150 km in the desert, and it is characterized by a narrow strip of supratidal carbonate sands and evaporites that form a coastal sabkha (Fig. 11.32). Southward it grades into an area of inland siliciclastic sabkha. To the east, it passes into an erg-margin of aeolian deposits, while to the west, it is formed and underlain by the widespread fluvial gravels of a deflated gravel plain that probably formed during a pluvial phase of the Pleistocene (Glennie 1970). Late Pleistocene aeolian-rich sand dunes derived from the nearby marine terraces locally overlie the gravels.

The sabkhat Matti started forming some 6000 years ago, when sea level was higher and filled the Arabian Gulf. This raised the groundwater table of low-lying areas. Apart from the narrow coastal margin, the sabkhat Matti probably owes its origin to this rise in sea level (or local subsidence), which was followed by marine peneplanation and a gentle tilting to the north of this area (Glennie 1970). Sedimentation in sabkhat Matti area owes its character to postglacial alteration of a largely dune-rich desert that formed largely during the last glaciation. At this time the lower global sea level enabled the local sand dune to be transported from the north and northwest across a dry "Arabian Gulf" by the northern (shamal) winds. With the onset of postglacial flooding of the Arabian Gulf, this supply of aeolian sand was cutoff. Deflation of the near-coastal dunes sourced other dunes further south, removed sand down to the level of the rising water table, and caused the formation of both the coastal and inland sabkhas. In fact, during the winter months, the water table now coincides

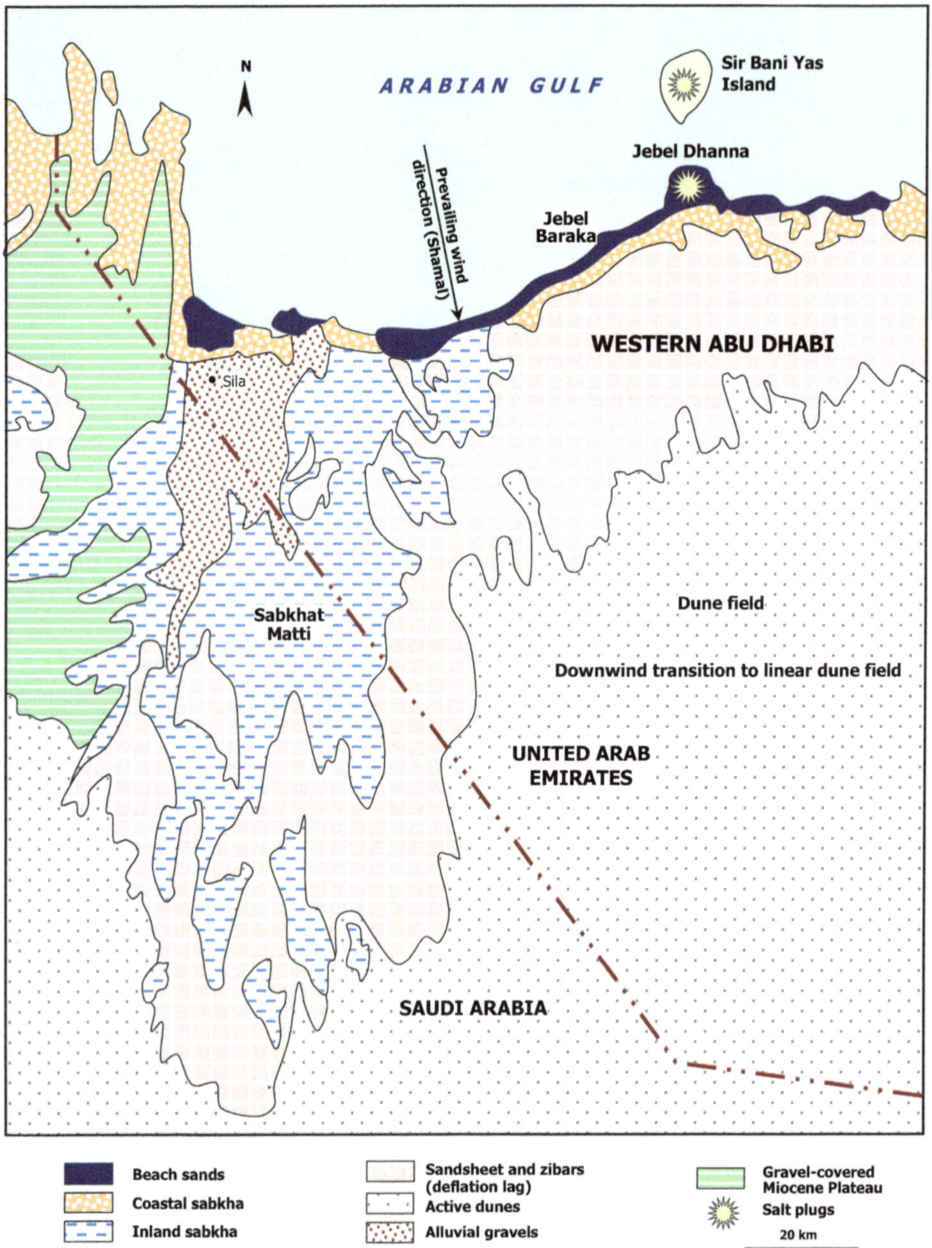

Fig. 11.32 Sedimentary facies distribution of sabkha Matti in western Abu Dhabi. (Modified from Hunting 1979, Glennie et al. 1994, Alsharhan and Kendall 2003)

with the present sabkha surface. In these areas evaporite growth is laterally extensive, a surface halite crust forms, while in the very shallow subsurface anhydrite and gypsum collects. Coupled with the high evaporation of these areas, the near-surface groundwaters cause the local sand dunes to be cemented by gypsum (Alsharhan and Kendall 2003; Glennie et al. 1994).

Conclusions

The sedimentary facies of the coastline of Abu Dhabi was mapped using remotely sourced satellite imagery linked to field descriptions and photography. This scheme identified ooid shoals on tidal deltas, beach ridges, sabkha flats, algal mats, mangroves, and sabkhas. Land and sea checking inferred other associated geological phenomenon including the distribution of the Holocene sedimentary facies of the coastal area of the Abu Dhabi Emirate including offshore pelecypod shell gravels, onshore coral reefs, coralgal sands, oolite shoals, inshore grapestones, and pelleted lime muds. The Holocene coastal facies complex onlaps Pleistocene cross-bedded carbonate sands rich in miliolids and shell fragments. Landward these, in turn, onlap exposures of Miocene mixed siliciclastics, carbonates. In western Abu Dhabi, the sabkhat Matti, a large inland lowland sabkha was mapped. This extends inland from the coast some 120 km, reaching a height of over 40 m above sea level at its southern tip.

This compilation of knowledge of the facies content and geometries of these bodies is intended for as analogs of similar features recognized in subsurface regional sequence stratigraphy. For this reason this paper is directed at the local and international geological and hydrocarbon exploration community, especially those starting their career.

References

Al-Asfour TA (1982) Changing sea-level along the North coast of Kuwait Bay. Published by Kegan Paul International, London. ISBN 0-7103-0010-7, 186Pp

Alsharhan AS, Kendall CGStC (2002) Holocene carbonate-evaporites of Abu Dhabi, and their Jurassic ancient analogs. In: Sabkha ecosystems, volume I: the Arabian Peninsula and adjacent countries, Hans-Jorg Barth and Benno Boer (Eds.), Kluwer Academic Publishers, pp 187–202

Alsharhan AS, Kendall CGSC (2003) Holocene coastal carbonates and evaporites of the southern Arabian Gulf and their ancient analogues. Earth-Sci Rev 61:191–243

Althausen J Jr, Christopher GSC, Kendall VL, Alsharhan AS, Whittle GL (2003) Using satellite imagery and GIS in the mapping of coastal landscapes in an arid environment: Khor Al Bazam, Western Abu Dhabi, United Arab Emirates. In: Sharhan AS, Wood WW, Goudie AS, Fowler A, Abdellatif EM (eds) The desertification in the third millennium. Swets & Zeitlinger Publishers (Balkema), Lisse, pp 415–422. ISBN: 90 5809 5711

Baltzer F, Kenig F, Boichard F, Plaziat JC, Purser BH (1994) Organic matter distribution, water circulation and dolomitization beneath the Abu Dhabi Sabkha (United Arab Emirates). In: Purser B, Tucker M, Zenger D (eds) Dolomites, international association of sedimentology, special publication, vol 21. Blackwell, Oxford, pp 409–428

Bathurst RGC (1975) Carbonate sediments and their diagenesis: developments in sedimentology, vol 12. Elsevier, Amsterdam, 658 pp

Berry RW, Brophy GP, Naqash A (1970) Mineralogy of the suspended sediments in the Tigris, Euphrates, and Shatt al-Arab rivers of Iraq, and the recent history of the Mesopotamian plain. J Sediment Petrol 40:131–139

Bush PR (1973) Some aspects of the diagenetic history of the sabkha in Abu Dhabi, Persian Gulf. In: Purser BH (ed) The Persian Gulf. Holocene carbonate sedimentation and diagenesis in a shallow Epicontenental Seas. Speringer-Verlag, Berlin, pp 395–408

Butler GP (1965) Early diagenesis in the Recent sediments of the Trucial Coast of the Persian Gulf: Unpublished M.Sc. Dissertation, University of London

Butler GP, Kendall CGStC, Harris PM (1982) Recent evaporites from the Abu Dhabi Coastal Flats: In Handford GR, Loucks RG, Davies GR (eds) Depositional and diagenetic spectra of evaporites. Society of Economic Paleontogists and Mineralogists Core Workshop 3, pp 33–64

Emery KO (1956) Sediments and water of Persian Gulf. Am Assoc Pet Geol Bull 40(10):2354–2383

Evans G (1942) The origin of spits, bars and related structures. J Geol 50:846–865

Evans G (1970) Coastal and nearshore sedimentation; a comparison of clastic and carbonate deposition. Proc Geol Assoc Lond 81:493–508

Evans G (1994) The Arabian Gulf; a modern carbonate-evaporite factory, a review. In: Arche A, Lopez-Gomez J (eds), Permico y Triasico de la peninsula Iberica; Permian and Triassic of the Iberian Peninsula, Cuadernos de Geologia Iberica, 19, pp 61–96

Evans G, Kendall CGSC, Skipwith P (1964a) Origin of the coastal flats, the sabkha, of the Trucial Coast. Persian Gulf. Nature 202:759–761

Evans G, Kinsman DJJ, Shearman DJ (1964b) A reconnaissance survey of the environment of recent carbonate sedimentation along the Trucial Coast, Persian Gulf. In: van Straaten LMJU (ed) Deltaic and shallow marine deposits. Elsevier Science, Amsterdam, pp 129–135

Evans G, Schmidt V, Bush P, Nelson H (1969) Stratigraphy and geologic history of the sabkha, Abu Dhabi, Persian Gulf. Sedimentology 12:145–159

Fairbridge RW (1961) Eustatic change in sea-level. In: Ahrens LH et al (eds) Physics and chemistry of the earth, vol 4. Perganmon Press, Oxford, pp 99–185

Ginsburg RN (1957) Early diagenesis and lithification of shallow water carbonate sediments in south Florida. In: Le Blanc RJ, Breeding JG (eds) Regional aspects of carbonate deposition, Society of economic paleontologists and mineralogists, special publication 5. Society of Economic Paleontologists and Mineralogists, Tulsa, pp 80–100

Glennie KW (1970) Desert sedimentary environments. Elsevier Publishing Company, Amsterdam

Glennie KW (1998) The desert of southeast Arabia: a product of Quaternary climatic change. In: Alsharhan AS, Glennie KW, Whittle GL, Kendall CGSC (eds) Quaternary deserts and climatic change. Balkema, Rotterdam, pp 279–291

Glennie KW, Pugh JM, Goodall TM (1994) Late quaternary Arabia desert models of Permian Rotliegend reservoirs. Shell Explor Bull 274:1–19

Godwin H, Willis EH (1959) Cambridge university Natural Radiocarbon measurements I. Am J Sci Radiocar, Suppl 1:63–75

Houbolt JJHC (1957) Surface sediments of the Persian Gulf near the Qatar Peninsula. Unpublished PhD thesis Utrecht University. The Hague, Mouton and Co., 113pp

Imbrie H, Buchanan H (1965) Sedimentary structures in modern carbonate sands of the Bahamas. Soc Sediment Geol (SEPM) Spl Publ 12:149–172

Kassler P (1973) The structure and geomorphic evolution of the Persian Gulf. In: Purser BH (ed) The Persian Gulf - Holocene carbonate sedimentation and diagenesis in a shallow epicontinental sea. Springer Verlag, New York, pp 11–32

Kendall CGStC, Alsharhan AS (2011) Quaternary carbonate and evaporite sedimentary facies and their ancient analogues. Special Publication 43, International Association of Sedimentologists, The Netherands, 478pp

Kendall CGSC, Skipwith PA (1969) Holocene shallow water carbonate and evaporite sediments of Khor Al Bazam, Abu Dhabi, Southwest Persian Gulf. Am Assoc Pet Geol Bull 53(4):841–869

Kendall CGSC, Lake P, Weathers III, Dallon H, Lakshmi V, Althausen J, Alsharhan AS (2003a) Evidence of rain shadow in the geologic record: repeated evaporite accumulation at extensional and compressional plate margins. In: Sharhan AS, Wood WW, Goudie AS, Fowler A, Abdellatif EM (eds) The desertification in the third millennium. Swets & Zeitlinger Publishers (Balkema), Lisse, pp 45–52. ISBN: 90 5809 5711

Kendall CGSC, Lakshmi V, Althausen J, Alsharhan AS (2003b) Changes in microclimate tracked by the evolving vegetation cover of the holocene beach ridges of the United Arab Emirate. In: Sharhan AS, Wood WW, Goudie AS, Fowler A, Abdellatif EM (eds) The desertification in the third millennium. Swets & Zeitlinger Publishers (Balkema), Lisse, pp 91–98. ISBN: 90 5809 5711

Kenig F, Huc AY, Purser BH, Oudin JL (1989) Sedimentation, distribution and diagenesis of organic matter in a recent carbonate environment Abu Dhabi, UAE. Org Geochem 16:735–745

Kenig F, Boichard R, Purser BH (1991) Holocene organosedimentary sequence of the Abu Dhabi coastal carbonate environment, United Arab Emirates. In: Bosellini A, Brander R, Flügel E, Purser BH, Schalger W, Tucker M, Zenger, D (eds), Dolomieu conference on carbonate platforms and dolomitization, International Association of Sedimentology, Oxford, UK (Abstract book), pp 134–140

King CAM (1959) Beaches and coasts. Edward Arnold, London, p 403

Kinsman DJJ (1964) Reef coral tolerance of high temperatures and salinities. Nature:1280–1282

Kinsman DJJ, Park RK (1976) Algal belt and coastal sabkha evolution, Trucial Coast, Persian Gulf. In: Walter MR (ed) Stromatolites. Elsevier, Amsterdam, pp 421–433

Kirkham A (1997) Shoreline evolution, aeolian deflation and anhydrite distribution of the Holocene, Abu Dhabi. GeoArabia 4:403–416

Kirkham A (1998a) A quaternary proximal foreland ramp and its continental fringe, Arabian Gulf, U.A.E. In: Carbonate R, Wright VP, Burchette TP (eds) Geological Society Special Publication, 149, pp 15–42

Kirkham A (1998b) Pleistocene Seif Dunes and their role in the development of complex past and present coastlines of the U.A.E. GeoArabia 3(1):19–31

Lakshmi V, Kendall CGSC, Althausen J, Alsharhan AS (2003) Studies of local climate change in United Arab Emirates using satellite data. In: Sharhan AS, Wood WW, Goudie AS, Fowler A, Abdellatif EM (eds) The desertification in the third millennium. Swets & Zeitlinger Publishers (Balkema), Lisse, pp 61–66. ISBN: 90 5809 5711

Larsen CE (1975) The mesopotamian delta region-a reconsideration of Lees and Falcon. J Am Orient Soc 95:43–57

Newell ND, Rigby JK, Whiteman AJ, Bradley JS (1951) Shoal water geology and environments Eastern Andros Island, Bahamas: American Museum Natural History Bulletin, pp 7–26

Newell ND, Purdy EG, Imbrie J (1960) Bahamian oolitic sand. J Geol 68:481–497

Pray, L.C. (1966) Hurricane Betsy (1965) and near shore carbonate sediments of the Florida Keys: Geological Society of America Annual Meeting, San Francisco, program with abstracts, pp 168–169

Privett DW (1959) Monthly charts of evaporation from the Indian Ocean (including the Red Sea and the Persian Gulf). J Quat Res of Meterol Soc 85:424–478

Purser BH, Seibold E (1973) The principle environmental factors influencing holocene sedimentation and diagenesis. In: Purser BH (ed) The Persian Gulf-Holocene carbonate sedimentation and diagenesis in a shallow epicontinental sea. Springer Verlag, New York, pp 1–9

Sabock DV, Gurr JA (1969) Persian Gulf fisheries. Commer Fish Rev 31(12):66–72

Sarnthein M (1972) Sediments and history of the postglacial transgression in the Persian Gulf and northwest Gulf of Oman. Mar Geol 12:245–266

Seibold E, Vollbrecht K (1969) Die bondengestalt des Persischen Golfs: Structural aspects of the Persian

Gulf (in German). "meteor" For Schungsergennisse, Reihe (Meteor for research result) 2(C), pp 29–56

Sheppard C, Price A, Roberts C (1992) Marine ecology of the Arabian region, patterns and processes in extreme tropical environments. Academic, London, p 359

Strohmenger, Christian J., Hesham Shebl., Abdulla Al-Mansoori, Khalil Al-Mehsin, Omar Al-Jeelani, Ismail Al-Hosani, Ali Al-Shamry And Salem Al-Baker, 2011 Facies stacking patterns in a modern arid environment: case study of the Abu Dhabi sabkha in the vicinity of Al Rafeq Island, United Arab Emirates, In: Christopher G. St C. Kendall, and Abdulrahman Alsharhan (eds) Quaternary carbonate and evaporite sedimentary facies and their ancient analogues: a tribute to Douglas James Shearman (Special Publication 43 of the IAS)., ISBN: 978-1-4443-3910-9, Oxford, Wiley-Blackwell, International Association. Sedimentology special publication 43, 149–183

Sugden W (1963) Some aspects of sedimentation in the Persian Gulf. J Sediment Petrol 33:355–364

Wagner CW, Van der Togt C (1973) Holocene sediment types and their distribution in the southern Persian Gulf. In: Purser BH (ed) The Persian Gulf. Springer, New York/Heidelberg/Berlin, pp 123–156

Weijermars R (1999) Quaternary evolution of Dawahat Zulum (Half Moon Bay) region of eastern province, Saudi Arabia. GeoArabia 4(1):71–90

Williams WW (1960) Coastal changes. Routledge/Kegan Paul, London, 220 pp

Williams AH, Walkden GM 2001 Carbonate eolianites from a eustatically influenced ramp-like setting: the quaternary of the Southern Arabian Gulf. In: Modern and ancient carbonate eolianites: sedimentology, sequence startigraphy and diagenesis. . SEPM Special Publication 71, pp 77–92

Williams AH, Walkden GM (2002) Late Quaternary highstand deposits of the southern Arabian Gulf: a record of sea-level and climatic change. In: Clift PD, Kroon D, Gaedicke C, Craig J (eds) The tectonic and climatic evolution of the Arabian Sea region, Geological Society London Special Publication 195. Geological Society, London, pp 371–386

Wood WW, Ward ES, Al Habshi AS (2002) Source of solutes to the coastal sabkha of Abu Dhabi. Geol Soc Am Bull 114(3):259–268

Climate Change Impacts on Salt Marsh Vegetation Ecophysiology

12

Isabel Caçador, Bernardo Duarte, and João Carlos Marques

Abstract

Ecosystems worldwide are experiencing the effects of climate change, and estuaries and salt marshes are no exception. The plant community will be one of the most affected elements by these climatic shifts, both in terms of structure and dynamics, with undeniable effects on its productivity. This thesis aims to study the effects of climate change on the marsh community dynamics, structure and productivity but also on its biogeochemical cycles and implications at the ecosystem level. Although special attention was given to primary productivity and plant physiology, a multidisciplinary approach was undertaken using both field assessments and mesocosmos trials. The results point out that these physically connected climatic changes are not only interconnected in a physical way but also at an ecological level. Mediterranean marshes will be more severely affected on their foundations, putting the entire ecosystem at risk and prone to climate change side effects and/or to synergistic events. The majority of the evaluated climatic changes have more negative impacts on the marsh pioneer species, affecting inevitably marsh establishment and expansion. Although to a lesser extent, the upper and middle marsh halophytes will also suffer from these climate-driven negative impacts. In addition, the appearance of resistant nonindigenous species (NIS) will add an increased threat to the marsh. The reduction of the pioneer zone in a large extension together with a middle marsh fragilization will open new ecological niches for the colonization of resistant NIS, imposing serious shifts in the marsh structure, dynamics and services provided to the estuarine ecosystem. This is even more evident when a holistic approach is undertaken focusing not only on plant physiology but also the marsh biogeochemistry and estuarine hydrological features. These evidences point out the need to adopt new management efforts, highlighting the desired marsh attributes and ecosystem services in the face of human activities that threaten salt marsh ecosystems.

Keywords

Climate change · Halophyte ecophysiology · Biogeochemistry · Ecosystem services

I. Caçador (✉) · B. Duarte
MARE – Marine and Environmental Sciences Centre, Faculty of Sciences, University of Lisbon, Lisbon, Portugal
e-mail: micacador@fc.ul.pt

J. C. Marques
MARE – Marine and Environmental Sciences Centre, c/o Department of Zoology, Faculty of Sciences and Technology, University of Coimbra, Coimbra, Portugal

B. Gul et al. (eds.), *Sabkha Ecosystems*, Tasks for Vegetation Science VI,
https://doi.org/10.1007/978-3-030-04417-6_12

Introduction

Our planet is undergoing an undeniable change in its climatic shift. This includes continuous increases in greenhouse gas (GHG) emissions from anthropogenic sources. Since 2007, annual global GHG emissions have continued to grow and reached 49.5 Gt of carbon dioxide equivalents (CO_2eq) in the year 2010, an unprecedented value (IPCC 2014). In addition, pollutants such as carbon monoxide (CO), volatile organic compounds (VOC), nitrogen oxides (NO_x) and sulphur dioxide (SO_2), which by themselves are negligible GHGs, have an indirect effect on the greenhouse effect by altering the abundance of important gases, such as methane (CH_4) and ozone (O_3), and/or by acting as precursors of secondary aerosols and also the amount of outgoing low wave radiation. These changes in the radiation emission have inevitable consequences on Earth's thermal budget. The radiative energy budget of the Earth is almost in balance, but the latest ocean heat content and satellite measurements indicate a small positive imbalance (Murphy et al. 2009; Trenberth et al. 2009; Hansen et al. 2011), consistent with the rapid changes in the atmospheric composition. These changes are in the genesis of all the climatic alterations our planet is facing. Global warming promoted by the increase in GHGs leads to changes in atmospheric and ocean circulation, season succession, storm and precipitation patterns, drought periods and the cascade of extreme thermal events observable in recent years with increasing frequency and severity (IPCC 2012, 2013).

Estuarine systems represent important ecosystems in the coastal landscape, providing essential ecosystem services such as water quality improvement, fisheries resources, habitat and food for migratory and resident animals and recreational areas for humans. In the past 40 years, estuarine conservation was recognized as a priority at national and international levels through several acts including the Ramsar Convention, 1977 (Munari and Mistri 2008), and Water Framework Directive, 2000. Despite all the conservation attempts on a worldwide scale, coastal, estuarine and transitional waters have been affected by human activities. Historically, developing human civilizations have often been concentrated in coastal areas where access to water promoted trade and commerce. As a consequence, human alteration of natural ecosystems is profound in coastal areas. This is a central concern in terms of environmental management in order to develop policies to balance socio-economic growth and environmental protection (Borja and Dauer 2008). Salt marshes are found fringing many of the world's coasts exposed to relatively low-energy hydrodynamic environments. They are characterized by a wide variety of herbaceous, succulent and woody vascular plants. Salt marshes have their upper limit occurrence approximately that of the highest astronomical tide (HAT), while the lower limit is rarely below mean high water neap (MHWN) tide level. These ecosystems appear mostly in estuarine ecosystem but can also be found in barrier islands, spits, embayments and open shores exposed to low wave energy (Allen 2000), as well as fringing coastal lagoons. Salt marshes can be found within all tidal regime ranges, from microtidal to macrotidal, presenting a great plasticity to water-level fluctuations which can be seasonal, as in the Baltic or the estuaries of southwest Western Australia (Hodgkin and Hamilton 1998), or unpredictable, as in some lagoons where water level is determined by rainfall patterns and lagoon opening. Salt marshes have a great ecological value for the estuarine ecosystem, namely, in terms of nutrient regeneration, primary production and wildlife habitat and as shoreline stabilizers. These ecosystems occupy the transition zone between terrestrial and marine biotopes and are characterized by a high productivity, which is considered essential in maintaining the detritus-based food chain of estuarine and coastal ecosystems (Marinucci 1982). Salt marshes are key areas for the estuarine system, namely, for primary production and nutrient regeneration (Caçador et al. 2009), becoming this way one of the most productive ecosystems on the planet (Lefeuvre et al. 2003). Estuarine wetlands, as salt marshes, constitute good carbon sinks having simultaneously reduced rates of GHGs emissions (Magenheimer et al. 1996), with a carbon sequestration capacity per unit area of about one order of magnitude higher than other wetland systems (Bridgham et al. 2006). Salt marshes are usually located in estuarine systems, and their primary production allows for a greater reduction of CO_2 in the atmosphere and incorporation in organic tissues through photosynthesis (Sousa et al. 2010). Wetlands represent the largest carbon pool with a capacity of 770 Gt of carbon, overweighing the total carbon storage of farms and rain forests (Han et al. 2005).

Being located in coastal systems, salt marshes are not only prone to the effects of altered atmospheric conditions (e.g. increased CO_2 and altered thermal environment) but also to indirect effects driven from changes in the marine system itself (e.g. sea level rise (SLR), altered dissolved carbon chemistry, salinity changes). Wetland systems are particularly vulnerable and susceptible to changes in quantity and quality of the water supply. Aside from the evident atmospheric alterations, climate change may have its most pronounced effects on salt marshes through changes in hydrological regimes, specifically, the nature and variability of the hydroperiod and the number and severity of extreme events. However, other factors related to climate may play important roles, including increased temperature and altered evapotranspiration, changes in the estuarine biogeochemistry, altered sedimentation and erosion patterns, fire, oxidation of organic sediments and the physical effects of the wave environment (IPCC 1998; Burkett and Kusler 2000; USGCRP 2000).

Climate change could alter the geographical distribution of salt marshes, which currently span temperate and arctic latitudes (Chapman 1977), or salt marsh plant species affecting ecosystem productivity. Turner (1976) suggests that the productivity of salt marshes is tied to latitude and climate with greater productivity at lower latitudes. Since all gas exchanges (plant photosynthesis or sediment respiration) are physically dependent on the thermal environment, it becomes inevitable to consider the effects of an altered thermal regime in the marsh productivity and biogeochemistry, namely, in terms of carbon stocks. Nowadays and with an increasing concern about climatic changes on our planet, several studies aroused focusing the effects of global warming on respiration in several ecosystems (e.g. Florides and Christodoulides 2009; Shakun et al. 2012). These studies point out an important role of increasing temperatures as major contributors to enhance respiration and therefore the CO_2 fluxes in the atmosphere (Kirschbaum 1995; Cox et al. 2000; Bond-Lamberty and Thomson 2010).

With climate change, the intensity of hurricanes and extreme rain events are expected to increase (IPCC 2007). As the first line of coastal defence, salt marshes are greatly affected by storms. In a long-term perspective, salt marshes are the ecosystems, which suffer less with the ecological disturbance caused by storms (Michener et al. 1997). In Cape Cod, for example, salt marshes were the only habitat with little reported damage from Hurricane Bob in 1991, mostly because unlike other upland habitats, salt marsh vegetation is highly adapted to salt stress and inundation (Valiela et al. 1998). Coastal storms do, however, have short-lived impacts in salt marsh communities. Storm surges push salt water up estuarine gradients, raising salinities in brackish and freshwater tidal marshes and temporarily shifting plant distributions. On the other hand, floodwaters can introduce large inputs of freshwater into salt marshes allowing the establishment of non-halophytic species in the salt marshes.

Alongside these climatic changes, the altered marsh environment along with the changes in interspecific relationships can provide conditions for the establishment of NIS that will impose an additional stress factor to the native marsh community.

Sea Level Rise

One particular aspect that acquired great meaning in recent years is the vulnerability of coastal areas to SLR, especially salt marshes. Climate-related changes to the carbon cycle are likely to alter the sequestration service provided by salt marshes, as well as affect long-term rates of salt marsh accretion and the ability of marshes to keep pace with SLR in ways that are still unclear. Climate change can affect salt marshes in a number of ways, including through SLR, particularly when sea walls and other anthropogenic structures squeeze marsh vegetation, preventing its movement upward and inland. However, some evidence indicates that SLR does not necessarily lead to the loss of marsh area, due to the ability of some marshes to accrete vertically, maintaining their elevation with respect to sea level. On the other hand, in organogenic marshes and in areas where sediment may be a limiting factor, these ecosystems may become prone to coastal squeeze, if some extreme predictions of accelerated rates of SLR are realized (Hughes 2004). McKee et al. (2004) suggest that temperature increases and decreases in rainfall driven by climate change may dramatically affect tidal marshes. Increased temperature may interact with other stressors damaging coastal marshes. For example, during the spring to fall period of 2000 in the Mississippi delta, there were large areas of salt marsh that were stressed and dying (Day et al. 2005). This resulted from a combination of effects driven from a strong La Niña event, resulting in prolonged low water levels, extreme drought and high air temperatures. This combination of factors apparently raised soil salinities to stressful toxic levels. Another example points to a replacement of the Gulf of Mexico salt marshes by mangroves due to its migration northward as a result of the increasing temperatures these systems have been experiencing year after year, allowing mangrove species to colonize new warmer environments.

Chen and Twilley (1998) developed a model of mangrove response to freeze frequency. These authors found that when freezes occurred more often than once every 8 years, mangrove forests could not survive. At a freeze frequency of once every 12 years, mangroves replaced salt marshes. In other locations like along the Louisiana coast, freezes historically occurred about every 4 years. By 2004, however, a killing freeze had not occurred for 15 years, and small mangroves started to appear over a large area near the coast. If this trend continues, mangroves will probably spread over much of the Northern Gulf and part of the South Atlantic coast. In fact, mangroves are already establishing in a wider range due to warming (Day et al. 2005).

Historically, SLR has been interconnected with sediment delivery and geologic development of salt marshes. As the rates of SLR have increased, rates of salt marsh accretion have also accelerated, sometimes exceeding SLR rates (Nixon 1980; Roman et al. 1997). However, as SLR continues to accelerate, there is the concern that its rate will outpace the rate of accretion, drowning salt marshes. Sea level rise effects are manifested in salt marshes in two different ways: (1) landward migration of salt marsh vegetation zones and submergence at lower elevations and (2) interior ponding and marsh drowning. The landward migration of low-marsh and high-marsh vegetation zones has been attributed to SLR in New England. Donnelly and Bertness (2001) observed a rapid landward expansion of *Spartina alterniflora* over the past 200 years, coincident with the rate of SLR. Warren and Niering (1993) compared vegetation cover data and peat profiles from 1947 to 1987 and found an increase in pioneers and dwarfs of *S. alterniflora* with a simultaneous decrease in *Spartina patens* and *Juncus gerardii* in the upper marsh. Since dwarf *S. alterniflora* and salt marsh pioneers generally occur in more waterlogged soil conditions, these results are also consistent with SLR. This rising phenomenon has also eroded the seaward edge of European salt marshes (e.g. Day et al. 1998). On the other side of the planet, SLR is contributing to the replacement of salt marsh by mangroves in Australia (Rogers et al. 2005). Rising water levels are of particular concern, since in several areas, the response of plant communities is constrained by sharp shorelines and anthropogenic structures. High marshes may be particularly vulnerable to drowning. On the other hand, productivity of low-marsh *S. alterniflora* increases with inundation (to a point). This led Morris et al. (2002) to suggest that increases in accretion linked to plant productivity act as a stabilizing agent allowing these marshes to counteract SLR. However, in the high marsh, several plants are less productive under inundation with slower accretion rates. This combined feedback from the whole marsh community can quickly transform high marsh into open water. This was already observed in Chesapeake Bay and Mississippi River delta marshes (Barras et al. 2004; Kearney et al. 1988). Despite the uncertainty about many of the potential effects of climate change on salt marshes, it seems clear that SLR will profoundly alter salt marshes by causing migrations of plant species and their associated fauna as well as marsh drowning.

Gathering a multidisciplinary approach from geochemistry to hydrodynamics and ecophysiology, a holistic point of view on this problem can be attained. Although some salt marshes maintain positive sedimentation rates in comparison to the mean SLR, this might not be true in the future due to increased anthropogenic pressures in the estuarine systems. In several estuaries around the globe, mean sea level (MSL) follows a well-identified trend, with increasing sea heights and higher inundation frequencies (Blum and Roberts 2009; Rybczyk et al. 1998). With this higher inundation, the sediment accumulation rate (SAR) in some salt marshes has an inverse pattern of behaviour, showing a decrease in the amount of sediments deposited as the MSL increases. Along with this decrease in sediment supply, the physical disturbances and chemical weathering can also affect salt marsh elevation (Duarte et al. 2013a, b). Still, these marshes apparently can keep up the pace with SLR, maintaining a positive feedback between SAR and MSL rise (Table 12.1).

Nevertheless, additional studies (Valentim et al. 2013) showed that a SLR scenario could

Table 12.1 Mean SLR and salt marsh sedimentation from 1963 to 2001 in two ecosystems differing in their maturity degree

	Mean SLR (cm year^{-1})	Young marsh sedimentation (cm year^{-1})	Mature marsh sedimentation (cm year^{-1})
1963–1986	−0.10	0.65	1.52
1986–2001	0.18	0.53	1.33
1963–2001	0.07	0.61	1.45

lead to changes in nutrient and sediment patterns around the salt marshes, and thus vegetation coverage percentage would be affected. Additionally, as a consequence of flood duration increase, sediment moisture will increase causing a stress condition to plants. Hence, the ratio of below-/aboveground biomass might increase, becoming critical to plant survival under conditions of accelerated SLR. Accordingly, both SLR and expected changes in vegetation coverage percentage in controlling salt marsh evolution have important implications in their stability and consequently in coastal management. This was lately confirmed by the findings presented in this chapter. Utilizing a multidisciplinary approach from geochemistry to hydrodynamics and ecophysiology, a more holistic point of view on this problem could be addressed.

The pioneer *S. maritima* undergoes periods of stress when exposed to prolonged tidal submersion (Fig. 12.1, Duarte et al. 2014b). Thus, this will have serious implications on its primary production, corroborating the hypothesis advanced by Valentim et al. (2013). This has serious implications not only on the ecosystem services provided on land by salt marsh halophytes but also for the entire estuarine system, as these are important areas with several functions for the whole ecosystem. Some examples of these functions are the contaminant remediative capacity of the salt marshes and its nutrient recycling role and food sources for secondary production (Duarte et al. 2014c). Due to the increasing stress conditions to which the halophytes are exposed under SLR, the senescence mechanisms will be more evident. Results from these studies point out an increasing number of necromass particles exported to the oceanic waters adjacent to the estuarine areas, as a combined result of the increased senescence and altered hydrodynamic features. If on one hand this will increase the fuelling of the secondary production of the coastal shelf by supplying higher amounts of particulate organic C and N (Fig. 12.2, Duarte et al. 2014c), on the other hand it will also contribute to an increasing contamination of these oceanic waters due to a higher number of contaminated

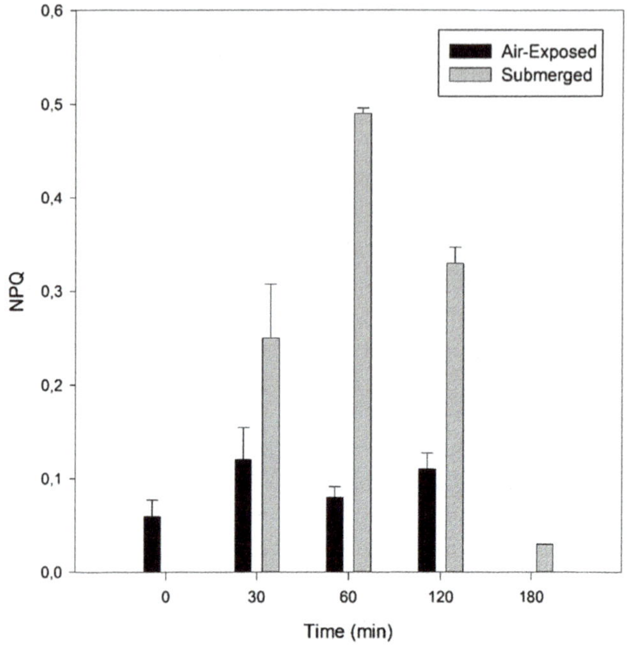

Fig. 12.1 Non-photochemical quenching (NPQ) in submerged and air-exposed *S. maritima* individuals (average ± standard deviation). (Figure from Duarte et al. 2014b)

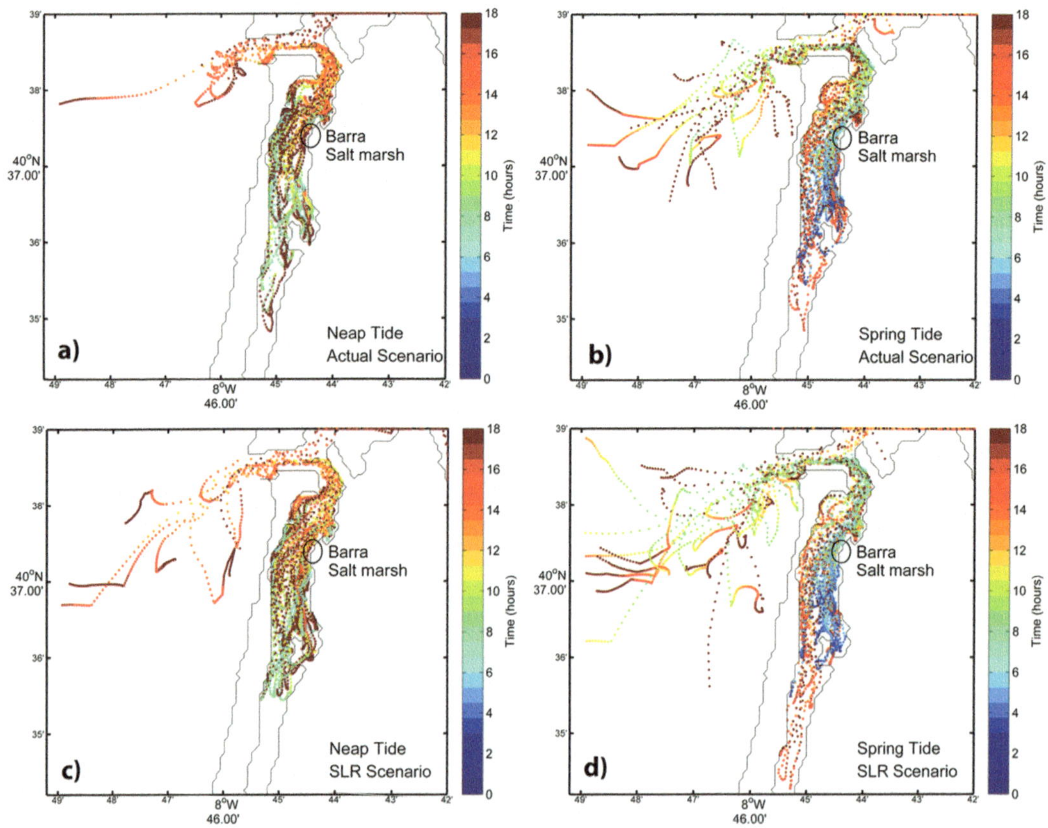

Fig. 12.2 Lagrangean transport in a salt marsh for neap tide in present (**a**) and SLR scenario (**c**) and for spring tides in present (**b**) and SLR scenario (**d**). (Figure from Duarte et al. 2014c)

detritus that will not be retained within the estuarine remediative area (Duarte et al. 2017).

Thus it becomes important to address SLR from a multidisciplinary approach, as impacts of these processes on the halophyte vegetation will condition the whole system at several different levels. Being trapped between the sea and urbanized lands, this data points out to a reduction of the salt marsh foundations (by reduction of its pioneer species) with enormous impacts on coastal erosion, possible eutrophication events and reduced remediative capacity of the ecosystem, making it prone to a possible collapse.

Atmospheric Carbon Dioxide Increase

Adding to the background level of several climatic changes and consequences is the greenhouse effect promoted by the increasing atmospheric GHGs concentration, such as CO_2. Changes in the atmospheric CO_2 concentration will have direct and indirect effects on the primary producers, due to its carbon fixation-based metabolism. As described for terrestrial ecosystems, in coastal ecosystems, the changes will mainly be dependent on plant metabolism. Due to metabolic differences, C_3 plants are expected to respond more positively to CO_2 increases than C_4 plants, a hypothesis that has been supported by field experiments that manipulate CO_2 concentrations (Ainsworth and Long 2005). Some experiments conducted in a brackish tidal marsh in the Chesapeake Bay showed that the experimental doubling of CO_2 concentration elicited a sustained biomass increase in *Scirpus olneyi* (C_3 sedge) (Erickson et al. 2007). Salt marshes dominated by C_4 grasses, such as *Spartina* and *Phragmites* spp., may respond differently. In the same Chesapeake Bay study, C_4 grasses *Spartina patens* and *Distichlis spicata* did not grow more rapidly under elevated CO_2 conditions. The same trend was found in growth chamber studies using European C_4 salt marsh species exposed to elevated CO_2 conditions (Lenssen et al. 1993; Rozema et al. 1991). This evidence suggests that the response of salt marshes to elevated CO_2 will be dependent on plant composition, with higher CO_2 concentrations favouring compositional shifts towards C_3 plants, as C_4 plants gradually reduce their area of colonization by competition.

In sum, C_3 plant growth will be promoted by the increasing concentration of this atmospheric carbon source, while C_4 species have their metabolism working at near maximum rates and thus won't see their primary productivity enhanced. In fact, C_4 grasses show evident signs of photochemical stress under elevated CO_2 concentrations, reducing its global photosynthetic ability (Fig. 12.3, Duarte et al. 2014e).

On the other hand, these estuarine environments divide their diel cycles between emersion and submersion conditions. Taking into account the present projections made by the Intergovernmental Panel for Climate Change (IPCC), it is expected that the increased levels of atmospheric CO_2 will lead to an inevitable increase in the dissolved CO_2 in water bodies, altering this way the availability of CO_2 underwater. Under submersion halophytes have their photosynthesis under non-saturated levels, expected to result in an increasing primary production along the increasing dissolved CO_2 concentrations. At this level, the estuarine halophyte communities act as a counterbalancing force, decreasing the dissolved CO_2, buffering estuarine water and increasing the oxygenation of the water column, an important service as a life support system (Table 12.2, Duarte et al. 2014d).

On the opposite side of the equation in the estuarine energy web, the decomposition processes will also be affected by changes in atmospheric CO_2. There is an evident shift in the biogeochemical function of salt marshes as organic matter recyclers, providing unbalanced amounts of nutrients to the nearby primary producers. All this points to evident changes in the whole ecosystem. The reduction of the colonization area of the ecosystem engineer *S. maritima* and expansion of C_3 halophytes will not only reduce the carbon livestocks of the salt marshes as effective sinks but also reduce the counteracting force of these halophytes as buffers of the estuarine waters. Aside from this, the deficient supply of inorganic nutrients due to altered biogeochemical cycles will also impact the primary

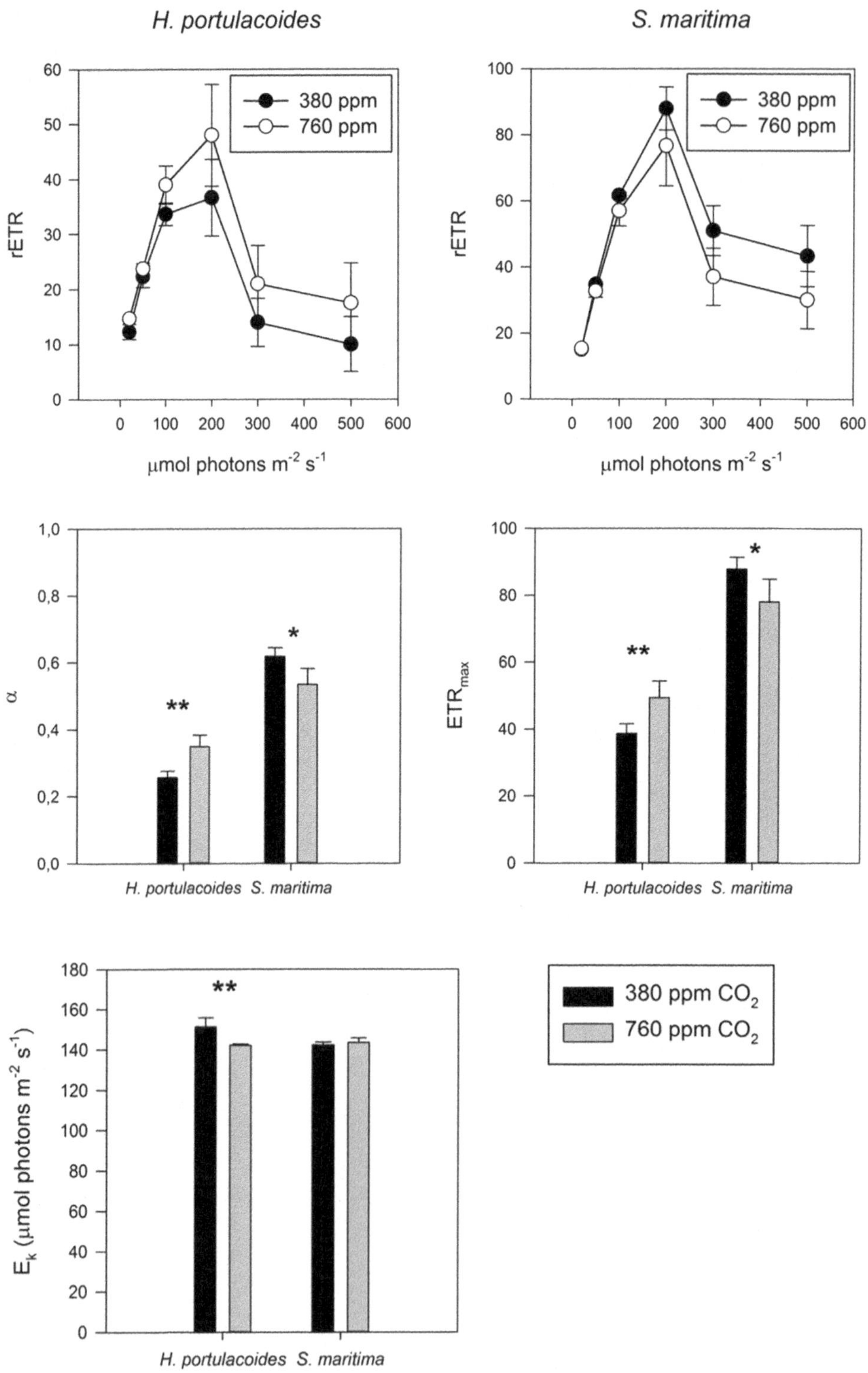

Fig. 12.3 Rapid light curves (RLCs), photosynthetic efficiency (α), maximum ETR (observed in the RLCs) and onset of light saturation (half-saturation light constant) in *H. portulacoides* and *S. maritima* individuals at both CO_2 tested concentrations (average ± standard deviation, $n = 5$; * $p < 0.05$, ** $p < 0.01$). (Figure from Duarte et al. 2014e)

Table 12.2 O_2 (mol) produced (+)/consumed (−) by each halophyte (considering all the coverage area in the Tagus estuary) at the four different considered scenarios per day, during the day- and night-time and considering all the estuarine halophyte community (Tagus Estuary), composed by these three species

	Dissolved CO_2 concentration (mM)			
Daytime	**0.05**	**0.5**	**1**	**2**
H. portulacoides	3952 ± 1397	12,050 ± 1397	8816 ± 2140	7990 ± 1301
S. maritima	−2039 ± 135	−8 79 ± 368	−590 ± 124	801 ± 224
Night	**0.05**	**0.5**	**1**	**2**
H. portulacoides	−5872 ± 196	−3882 ± 193	−316 ± 224	−297 ± 210
S. maritima	−6379 ± 465	−5885 ± 214	−6557 ± 207	−5840 ± 449
Daily budget	**0.05**	**0.5**	**1**	**2**
	−10,338	1404	1353	2654

productivity of the estuarine water column (Fig. 12.4).

Altered Thermal Environments

Warming of the climate system is unequivocal, and since the 1950s, many of the observed changes are unprecedented over past decades to millennia. The atmosphere and ocean have warmed, the amounts of snow and ice have diminished, sea level has risen, and the concentrations of greenhouse gases have increased (IPCC 2013). Additionally, the frequency and intensity of extreme thermal events are expected to increase (IPCC 2012). The same working group developed five possible scenarios of rising temperature of increasing thermal severity: the B1 scenario (+1.8 °C), the B2 scenario (+2.4 °C), the A1B scenario (+2.6 °C), the A2 scenario (+3.2 °C) and, the worst scenario, A1F1 (+4.0 °C). One of the salt marsh ecosystem services most dependent on the thermal environment is its carbon retention and emissions throughout sediment heterotrophic respiration. With a snowball effect, if carbon emissions increase, the greenhouse effect follows the same pattern leading to an increased atmospheric temperature. Nevertheless, salt marshes once again act as a buffer, counteracting global warming. In all projected scenarios, salt marshes tend to decrease their carbon emissions with the temperature increase, leading to higher carbon stocks and counteracting the greenhouse effect (Duarte et al. 2014a). This will certainly be an important service performed by wetlands in a future altered thermal environment. Additionally, to this projected global warming, some episodic extreme events will also affect inevitably all communities. Halophytes will be no exception. Another metabolic process highly dependent on temperature is photosynthesis. Namely, C_3 photosynthesis is expected to be more negatively affected by the occurrence of heat waves (Fig. 12.5). This will have impacts not only on the marsh foundations, due to a decreased productivity of species like *A. tripolium*, but also on the sediment stabilization and marsh build-up, due to its negative effects on upper marsh species productivity, like *H. portulacoides* (Duarte et al. 2015a, b). Cold spells will also have negative impacts on these species, although at a smaller scale. Additionally, these punctual events will also have impacts on the invasion ecology of some nonindigenous species, altering inevitably the marsh structure and function. The invader *S. patens*, for example, is more fit during heat spells and, along with its C_4 metabolism, will be able to replace the less-adapted C_3 species from the upper marsh. Its congener, the C_4 *S. maritima*, will also suffer dramatically from the increase of frequency and intensity of heat wave events. Again, the foundations and spreading of new marshes will be threatened, with another pioneer species having its primary productivity reduced (Duarte et al. 2016).

In contrast to *S. maritima*, under the present climate reality, the NIS *S. versicolor* proved to have a lower photosynthetic fitness under the Mediterranean climate, which seems to be slowing down its spread but not arresting it.

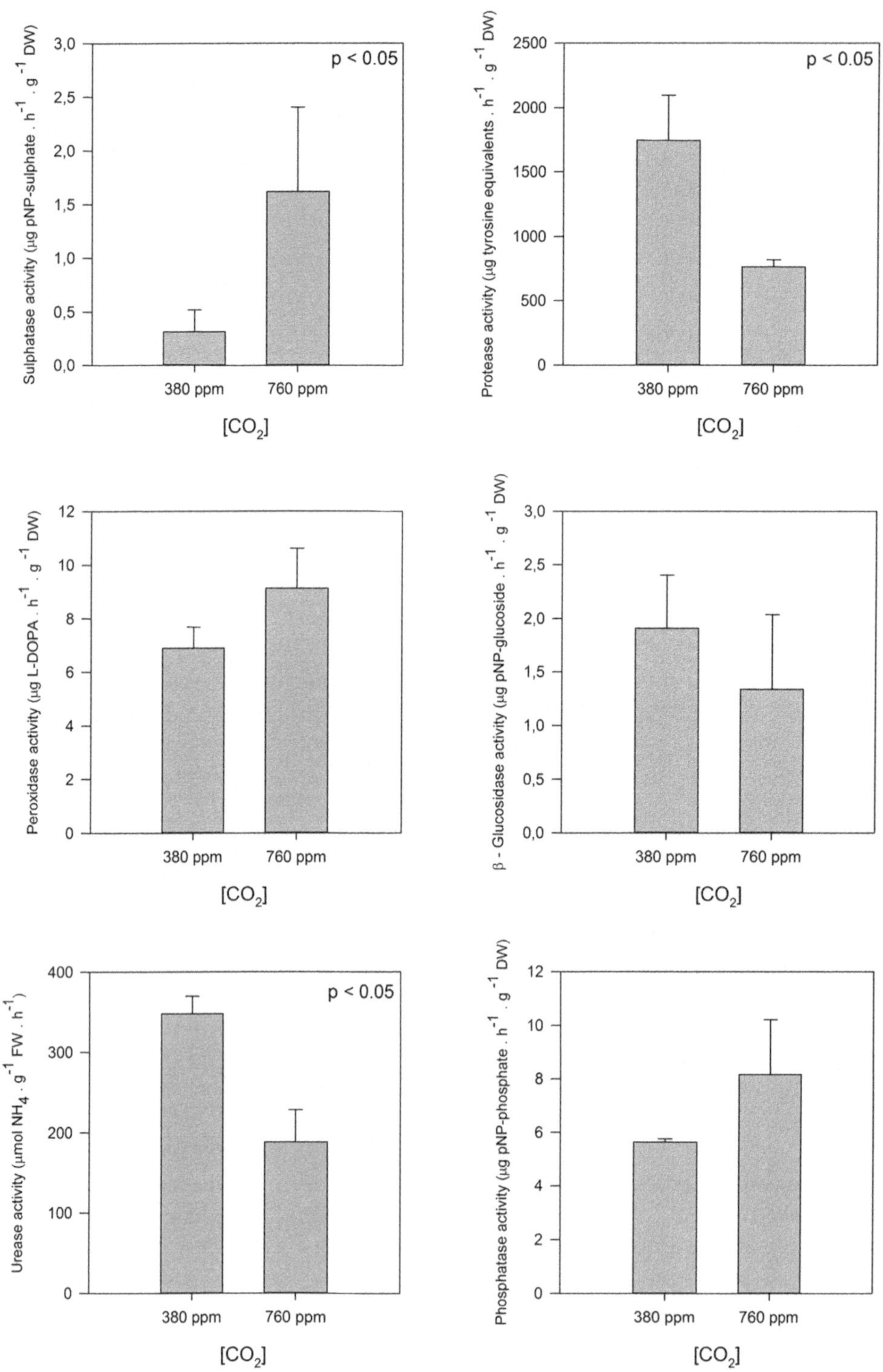

Fig. 12.4 Extracellular enzymatic activities in the rhizosediment of *S. maritima* exposed to present (380 ppm) and future (760 ppm) of CO_2 (average ± standard deviation, $n = 5$)

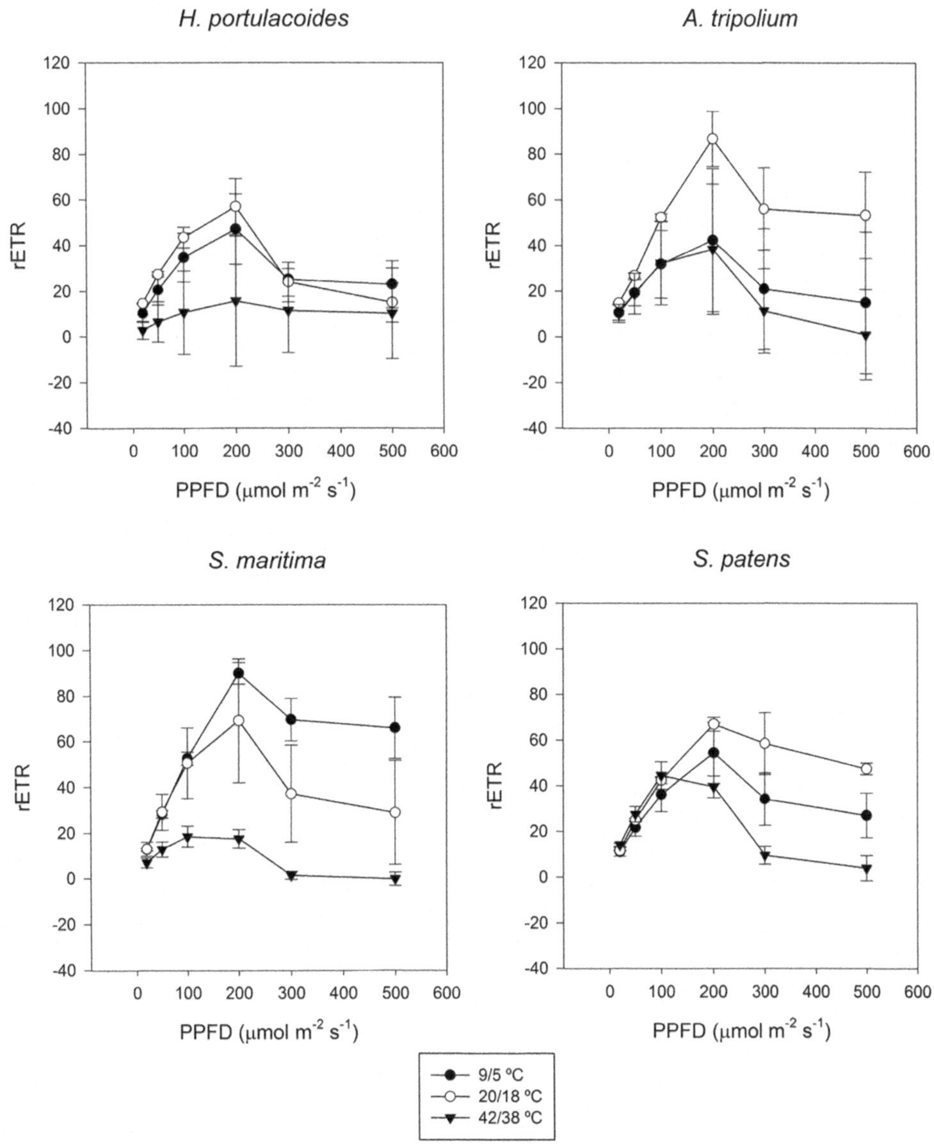

Fig. 12.5 Relative electron transport rates (rETR) in dark-adapted leaves of the four tested halophyte species under normal thermal regimes and after exposure to heat and cold waves (average ± standard error, $n = 5$)

Also, this NIS N-nutrition appears to be one of the factors impairing it spreading at higher rates (Fig. 12.6, Duarte et al. 2015c). Thus, as the ongoing climatic changes can be favouring drivers for a faster expansion of this species, taking the part of less productive Mediterranean species in a climate change scenario, the need arises to monitor this NIS and consider its spreading as a potential threat to the ecosystem biodiversity.

Overall, although being punctual events, extreme thermal conditions will leave their signature on the marsh structure. Although with more organic and cohesive sediments due to a higher organic carbon accumulation, marshes will tend to reduce in area as their spreading and accretion abilities will be decreased due to a progressive reduction of pioneer species abundance. At the same time, NIS will pose an increased threat by substituting for native halophytes in the already

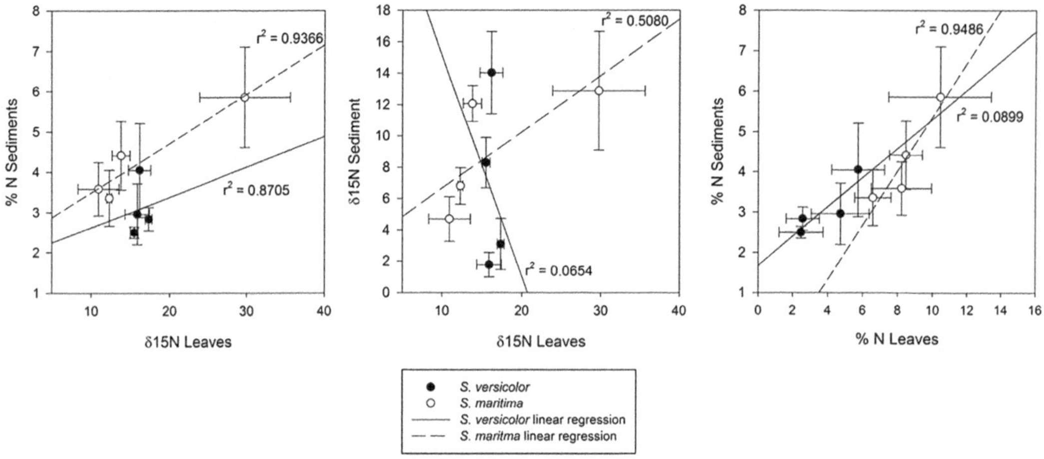

Fig. 12.6 Linear correlations between average seasonal foliar and sediment $\delta^{15}N$ and N contents in *S. versicolor* and *S. maritima* (average ± standard deviation, $n = 5$. * indicates significant differences between species at $p < 0.05$). (Figure from Duarte et al. 2015c)

mature marshes, changing completely the functions and services provided to the estuarine system.

Overview of the Impacts of Climate Change on Halophitic Vegetation

Back in 1977, the resilience of salt marshes to human activities was already recognized (Niering 1977). These are in fact resilient ecosystems, but as for all resilience mechanisms, it implies that the adaptation and recovery of the ecosystem must keep up the pace with the applied stressor. The climatic conditions nowadays impose a shift in the resilience capacity of our marshes. The speed at which changes are likely to occur in the future outpaces by far the ability of the marshes to adapt and thus the resilience of these ecosystems may be at stake. Niering (1977) unveiled the human signatures on New England salt marshes and pioneered the study of human impacts on coastal wetlands. Over the past several decades, however, eutrophication, overfishing and climate change have become global drivers of catastrophic impacts in coastal ecosystems (Jackson et al. 2001; Lotze et al. 2006), beyond Niering's predictions.

Human impacts are concentrated in coastal ecosystems due to a synergism of nearshore human activities, spillover from terrestrial impacts and the concentration of human settlement along shorelines (UNEP 2006). Despite this anthropogenic degradation of the world coastal areas, human populations continue to rely heavily on coastal ecosystem services. Estuaries and salt marshes provide more services than any other ecosystems (UNEP 2006). As stated several times before, these are important sources of storm protection (Costanza et al. 2008), nutrient regeneration (Valiela and Cole 2002), nurseries for commercially important fish and shellfish (Boesch and Turner 1984) and shelter for migratory birds (Teixeira et al. 2014). To preserve these functions and ecosystem services along with aesthetic and recreational value, salt marshes have become important conservation priorities in several countries under the shelter of the Ramsar Convention and other important directives (e.g. EU Water Framework Directive, EU Habitats Directive, South Africa Water Act and the US Clean Water Act).

Ecologists developed founding studies of plant succession (Clements 1916), assemblages (Chapman 1940), salt marsh development (Redfield 1965) and advances in community (Bertness 1991) and ecosystem ecology (Valiela and Teal 1979). The concern with the Portuguese marsh vegetation goes back to Esteves-de-Sousa (1950), who provided one of the first systematic surveys of the Tagus estuary marsh vegetation. A systematic understanding of these systems should provide tools to understand marsh dynamics and predict how these ecosystems will respond to human impacts associated with ongoing global changes. In fact, these are already impacted ecosystems, which can be observed from the data presented in Chap. 2 (Sect. 4), with an intense contamination record from industrial sources back to the 1960s (Duarte et al. 2013a, b). A similar record can be found which concerns the eutrophication of some marshes (Caçador and Duarte 2012, 2014). Going back to the resilience subject, it is admissible that in already disturbed ecosystems, any increasing stressor can overcome the resilience of its species. Nevertheless, to understand this, one must understand the physiological and biogeochemical mechanisms underlying the plasticity and resistance mechanisms of the marsh and its community.

Like other marshes around the world, the sediment supply in the Tagus estuary between 2001 and 1961 still allowed the marshes to keep pace with sea level rise, accreting vertically more than the sea level increased. Nevertheless, the most recent predictions impose an increased problem. If in the last decades sea level rise was mainly due to the melting of the polar ice caps, nowadays this was added by a decrease in the radiative energy dissipation of our planet, leading to increasing temperatures and inevitable expansion of the water bodies (IPCC 2013). This is likely to increase the rate of sea level rise by 2100 to values that can outpace the accretion rates of some slow growing marshes, like the ones located in the south side of the Tagus estuary, where in fact most of the marsh extensions are located. This will inevitably affect the tidal-regulated circadian rhythms of the plants, with adding prolonged flooding times and stress due to higher submersion heights, especially in the lower marsh. Although the low marsh isn't the most productive area of these systems (Caçador et al. 2009), the typical species found in the lowest part of the marsh have an extremely important role as "founding fathers", with an occupation that can go from 20 to 90% of the marsh (Caçador et al. 2013). The productivity of these species (e.g. *S. maritima*) will be inevitably affected, although to a smaller extent than would be expected. *Spartina maritima* exhibited a great plasticity and adaptation capacity, reducing the impacts in its primary productivity due to prolonged submersion periods (Duarte et al. 2014b). Nevertheless, a reduction is evident, which in a longer term can have significant effects on the population of these pioneer species.

The initiation and expansion of ponds in waterlogged marshes are considered to be an additional mechanism of marsh loss attributed to SLR in mid-Atlantic salt marshes (Hartig et al. 2002). Although there are few quantitative estimates of the expected extent of marsh loss as sea levels rise, most scientists agree that loss will be severe. Using current IPCC SLR scenarios and a salt marsh accretion model, Craft et al. (2009) predicted that 20–45% of salt marsh area in a Georgia estuary will be converted to low salinity marsh, tidal flat or open water by the year 2100. All these predictions are intrinsically connected with the physiological tolerance ranges exhibited by the different species, as observed for the pioneer *S. maritima*. Coupling these physiological evidences with robust hydrodynamic models (Valentim et al. 2013), the insights gathered from the physiological evidences are reinforced. The sea level rise scenario will inevitably lead to changes in nutrient and sediment patterns around the salt marshes, and thus vegetation coverage percentage would be affected. Additionally, as a consequence of flood duration increase, sediment moisture will increase imposing a stress condition to plants. Hence, the ratio of below-/aboveground biomass might increase, becoming critical to plant survival under conditions of accelerated SLR (Valentim et al. 2013). In the lower marsh, the main constraint impairing primary productivity and pioneer species establishment is the prolonged submersion time that plants will be exposed under increased sea level. On the other hand, in the upper marsh, this feature will have an additional stressor associated: increased saltwater intrusion. Combined with increased temperature and evapotranspiration, this will increase the salinity in the upper and middle marsh affecting the species hyaline habitat. This has consequences at the primary productivity level of, namely, the succulent shrubs, reducing it beyond its physiological optimum (Duarte et al. 2013b, 2014c). Again this will decrease the survival of species like *H. portulacoides* and *S. fruticosa*. Additionally, this opens space to the spreading of other species, namely, invasive ones like *S. patens*. This species is highly resistant to salinity changes tolerating very high salinity values (Duarte et al. 2015a). Nowadays the expansion of this species is constrained by a physiological misfit to the present climatic conditions observed in the Mediterranean marshes (Duarte et al. 2015b), preventing its rapid expansion. Thus, the ongoing climatic changes can be favouring drivers for a faster expansion of this species, taking part of less productive Mediterranean species in a climate change scenario, by direct competition.

In the areas where salt marsh is converted to vegetated mudflat or open water, the provision of ecosystem services is expected to decrease (Craft et al. 2009). This was also evident in one of the most important ecosystem services provided by Mediterranean marshes. These systems are often pointed out as sinks of contaminants and excessive nutrients. If this is true under present conditions, under SLR scenarios, this will be seriously affected. In the present conditions, the marsh detritus exports return to the marsh area after some tidal cycles or are maintained within the vicinity of their point of origin (Duarte et al. 2014d, 2017), maintaining the remediative service provided by these areas. In the future, a large part of these exports will reach the outer estuary and even the adjacent ocean shelf, transporting contaminants and organic nutrients trapped within these detritus (Duarte et al. 2014c, 2017). This is one of the more intrinsic evidences of the effects of climate change on the halophyte physiology and on the ecosystem services provided, with impacts beyond the estuarine system. The increased reach of the marsh-generated detritus has inevitable consequences on the coastal contamination levels but can also produce shifts in the coastal communities by enhancing the fuelling of secondary production in the coastal shelf or due to possible eutrophication processes promoted by excessive N exports (Duarte et al. 2014c, 2017).

As referred before, the main driver of all climatic changes is the increasing greenhouse effect promoted by the increasing atmospheric concentrations of GHGs, namely, CO_2. While SLR apparently affects the marsh vegetation by similar mechanisms (prolonged submersion, drowning and increased salinity), the increase in atmospheric CO_2 concentrations has a differential impact on the species depending on their own inherent characteristics. While an atmospheric increase of CO_2 concentration will favour C_3 primary production, C_4 photosynthesis apparently suffers no enhancement, even showing some signs of distress (Duarte et al. 2014e). Once again this will have its more negative effects on the marsh foundation and establishment, since as above-mentioned, the most abundant pioneer species, *S. maritima*, has a C_4 photosynthesis profile. In fact, the expansion of C_3 plants can also be at risk, if the C_4 pioneer species do not provide the conditions (for e.g. marsh elevation, sediment oxidation, hydrodynamic protection) for this increased productivity. This way the marsh will have its area reduced by this negative impact of the increased CO_2 on the sensitive and essential pioneer species. On the other hand, photosynthetic enhancement due to increased dissolved CO_2 was found for C_3 and C_4 halophytes, probably due to an alleviation of the stressful conditions imposed by submersion (as above-mentioned) by an increased CO_2 availability (Duarte et al. 2014e). Transposing these findings to the ecosystem and assuming increased dissolved CO_2 concentration scenarios, the halophyte community displays a new ecosystem function, in terms of water column oxygenation and as buffers of its acidification by withdrawing excess CO_2. As it will be further discussed for temperature rise, this appears to be a counteractive measure of the ecosystem towards the changes in the water chemistry. This buffer function allows reducing the excessive CO_2 levels in the water column and thus increasing its potential to again reabsorb higher amounts of atmospheric CO_2 while providing increased O_2 concentrations for the heterotrophic life forms inhabiting nearby. At a biogeochemical level, a similar buffer effect could be observed. Most of the carbon-related enzymes will be impaired during CO_2 increase (Caçador et al. 2015). If by one side this will impose serious constraints on the essential recycling processes provided by the marsh sediments, on the other hand, it apparently also counteracts the excessive CO_2 concentrations by reducing the mineralization of organic carbon sources to CO_2. Nevertheless, this will impose in the future a significant unbalance of the marsh biogeochemical cycles, shifting the ecosystem functions with inevitable changes in the ecosystem services provided, namely, at the nutrient regeneration level.

As above-mentioned, this self-buffering capacity of the ecosystem is not exclusive of its feedback towards increased atmospheric CO_2. A similar counteractive measure can be found while examining the effects of increased temperature on marsh CO_2 effluxes (Duarte et al. 2014a). Going through the IPCC storyline scenarios, it is evident that the sediments act as buffers of the temperature increase, by decreasing the CO_2 efflux and thus minimizing the greenhouse gas effect. This has inevitable consequences at the ecosystem services level, increasing in this way the sediment organic carbon pool and thus shifting the biogeochemical functions of these systems as above-mentioned and observed under CO_2 increase scenarios (Caçador et al. 2015). While permanent and long-term thermal disruptions, like warming, can impose long-term medium-scale effects on plant productivity, punctual but abrupt and extreme thermal events can induce severe damage on plant productivity on a shorter term (IPCC 2012). At this level, C_3 halophytes will be more severely damaged by heat waves (Duarte et al. 2015b, c) events which are expected to be more intense and more frequent in the near future (IPCC 2012). However, significant damage will be again felt by the pioneer species, either C_3 (e.g. *A. tripolium*) or C_4 (e.g. *S. maritima*). As previously referred, this will have more serious effects on marsh establishment and growth, becoming in this way more prone, for example, to SLR, also inherent to this global warming increase. Moreover, the reduction of the primary productivity of the most abundant native species will also open space for the colonization by NIS, changing completely the marsh landscape and, of course, its ecosystem services to the estuarine community (Duarte et al. 2015d).

It is far from evident that all these climatic changes are not only interconnected in a physical way but also at an ecological level. Mediterranean marshes will be more severely affected in their foundations, putting the entire ecosystem at risk and prone to other climate change side effects or to synergistic events. Accelerating human impacts are overwhelming salt marsh development and recovery by altering inundation regimes and the presence, identity and productivity of salt marsh foundation species. Humanity alongside with marshes faces a new challenge until the end of the century. Management efforts must highlight desired marsh attributes and ecosystem services in the face of human activities that threaten salt marsh ecosystems. This requires abandoning old-fashioned restoration goals, aiming to achieve "pristine" or "reference" conditions, replacing them by objectives aiming for enduring ecosystem subsistence and preservation of essential ecosystem services – contaminant sink, nursery habitat, nutrient regulation and shoreline protection.

References

Ainsworth EA, Long SP (2005) What have we learned from 15 years of free-air CO2 enrichment (FACE)? A meta-analytic review of the responses of photosynthesis, canopy properties and plant production to rising CO2. New Phytol 165:351–372

Allen JRL (2000) Morphodynamics of Holocene salt marshes: a review sketch from the Atlantic and Southern North Sea coasts of Europe. Quat Sci Rev 19:1155–1231

Barras J, Beville S, Britsch D, Hartley S, Hawes S, Johnston J, Kemp P, Kinler Q, Martucci A, Porthouse J, Reed D, Roy K, Sapkota S, Suhayda J (2004) Historical and projected coastal Louisiana land changes: 1978-2050. USGS Open-File Rep 03-334, 39p

Bertness MD (1991) Interspecific interactions among high marsh perennials in a New England salt marsh. Ecology 72:125–137

Blum M, Roberts H (2009) Drowning of the Mississippi Delta due to insufficient sediment supply and global sea-level rise. Nat Geosci 2:488–491

Boesch DF, Turner RE (1984) Dependence of fishery species on salt marshes: the role of food and refuge. Estuaries 7:460–468
Bond-Lamberty B, Thomson A (2010) Temperature-associated increases in the global soil respiration record. Nature 464:579–583
Borja A, Dauer DM (2008) Assessing the environmental quality status in estuarine and coastal systems: comparing methodologies and indices. Ecol Indic 8:331–337
Bridgham SD, Megonigal JP, Keller JK, Bliss NB, Trettin C (2006) The carbon balance of North American wetlands. Wetlands 26:889–916
Burkett V, Kusler J (2000) Climate change: potential impacts and interactions in wetlands of the United States. J Am Water Resour Assoc 36:313–320
Caçador I, Duarte B (2012) Tagus Estuary salt marshes: a historical perspective. In: Jordan JJ (ed) Estuaries: classification, ecology and human impacts. Nova Science Publishers, Inc
Caçador I, Duarte B (2014) Eutrophication impacts on salt marshes natural metal remediation. In: Ansari AA, Gill SS (eds) Eutrophication: causes, consequences and control, vol 2. Springer, Dordrecht
Caçador I, Caetano M, Duarte B, Vale C (2009) Stock and losses of trace metals from salt marsh plants. Mar Environ Res 67:75–82
Caçador I, Neto JM, Duarte B, Barroso DV, Pinto M, Marques JC (2013) Development of an angiosperm quality assessment index (AQuA – Index) for ecological quality evaluation of Portuguese water bodies – a multi-metric approach. Ecol Indic 25:141–148
Caçador I, Duarte B, Marques JC, Sleimi N (2015) Carbon mitigation: a salt marsh ecosystem service in times of change. In: Khan MA, Ozturk M, Gul B, Ahmed MZ (eds) Halophytes for food security in dry lands. Elsevier, Amsterdam
Chapman VJ (1940) Succession on the New England salt marshes. Ecology 21:279–282
Chapman VJ (1977) Introduction. In: Chapman VJ (ed) Wet coastal ecosystems. Elsevier, New York, pp 1–30
Chen RH, Twilley RR (1998) A gap dynamic model of mangrove forest development along gradients of soil salinity and nutrient resources. J Ecol 86:37–51
Clements FE (1916) Plant succession: an analysis of the development of vegetation. Carnegie Institution of Washington, Washington, DC
Costanza R, Perez-Maqueo O, Martinez ML, Sutton P, Anderson SJ, Mulder K (2008) The value of coastal wetlands for hurricane protection. Ambio 37:241–248
Cox PM, Betts RA, Jones CD, Spall SA, Totterdell IJ (2000) Acceleration of global warming due to carbon-cycle feedbacks in a coupled climate model. Nature 408:184–187
Craft C, Clough J, Ehman J, Joye S, Park R, Pennings S, Guo H, Machmuller M (2009) Forecasting the effects of accelerated sea-level rise on tidal marsh ecosystem services. Front Ecol Environ 7:73–78
Day JW Jr, Scarton F, Rismondo A, Are D (1998) Rapid deterioration of a salt marsh in Venice Lagoon, Italy. J Coast Res 14:583–590
Day JW, Narras J, Clairain E, Johnston J, Justic D, Kemp GP, Ko JY, Land R, Mitsch WJ, Steyer G, Templet P, Yanez-Arancibia A (2005) Implications of global climatic change and energy cost and availability for the restoration of the Mississippi delta. Ecol Eng 24:253–265
Donnelly JP, Bertness MD (2001) Rapid shoreward encroachment of salt marsh cordgrass in response to accelerated sea-level rise. Proc Natl Acad Sci 98:14218–14223
Duarte B, Caçador I, Marques JC, Croudace I (2013a) Tagus Estuary salt marshes feedback to sea level rise over a 40-year period: insights from the application of geochemical indices. Ecol Indic 34:268–276
Duarte B, Santos D, Marques JC, Caçador I (2013b) Ecophysiological adaptations of two halophytes to salt stress: photosynthesis, PS II photochemistry and anti-oxidant feedback – implications for resilience in climate change. Plant Physiol Biochem 67:178–188
Duarte B, Freitas J, Valentim J, Medeiros JP, Costa JL, Silva H, Dias JM, Marques JC, Caçador I (2014a) Modelling abiotic control of salt marsh sediments respiratory CO2 fluxes: new insights for climate change scenarios. Ecol Indic 46:110–118
Duarte B, Santos D, Marques JC, Caçador I (2014b) Biophysical probing of *Spartina maritima* Photosystem II changes during increased submersion periods: possible adaptation to sea level rise. Plant Physiol Biochem 77:122–132
Duarte B, Valentim JM, Dias JM, Silva H, Marques JC, Caçador I (2014c) Modelling sea level rise (SLR) impacts on salt marsh detrital outwelling C and N exports from an estuarine coastal lagoon to the ocean (Ria de Aveiro, Portugal). Ecol Model 289:36–44
Duarte B, Silva H, Marques JC, Caçador I, Sleimi N (2014d) Light-dark O_2 dynamics in submerged leaves of C_3 and C_4 halophytes under increased dissolved CO_2: clues for saltmarsh response to climate change. Ann Bot Plants 6:plu067
Duarte B, Santos D, Silva H, Marques JC, Caçador I (2014e) Photochemical and biophysical feedbacks of C_3 and C_4 mediterranean halophytes to atmospheric CO_2 enrichment confirmed by their stable isotope signatures. Plant Physiol Biochem 80:10–22
Duarte B, Santos D, Marques JC, Caçador I (2015a) Impact of extreme heat and cold events on the energetic metabolism of the C_3 halophyte *Halimione portulacoides*. Estuar Coast Shelf Sci 167:166–177
Duarte B, Goessling JW, Marques JC, Caçador I (2015b) Ecophysiological constrains of *Aster tripolium* under extreme thermal events impacts: merging biophysical, biochemical and genetic insights. Plant Physiol Biochem 97:217–228
Duarte B, Baeta A, Rousseau-Gueutin M, Ainouche M, Marques JC, Caçador I (2015c) A tale of two *Spartinas*: climatic, photobiological and isotopic insights on the

fitness of invasive versus native species. Estuar Coast Shelf Sci 167:178–190

Duarte B, Marques JC, and Caçador I (2015d) Ecophysiological responses of native and invasive Spartinaspecies to extreme temperature events in Mediterranean marshes. Biol Invasions 18, 2189–2205

Duarte B, Marques JC, Caçador I (2016) Ecophysiological responses of native and invasive *Spartina* species to extreme temperature events in mediterranean marshes. Biol Invasions 18:2189–2205

Duarte B, Vaz N, Valentim JM, Dias JM, Silva H, Marques JC, Caçador I (2017) Revisiting the outwelling hypothesis: modelling salt marsh detrital metal exports under extreme climatic events. Mar Chem 191:24–33

Erickson JE, Megonigal JP, Peresta G, Drake BG (2007) Salinity and sea level mediate elevated CO_2 effects on C_3-C_4 plant interactions and tissue nitrogen in a Chesapeake Bay tidal wetland. Glob Chang Biol 13:202–215

Esteves-de-Sousa A (1950) Acerca da sub-halosérie da região salgadiça litoral, entre Corroios e Talaminho. Revista da Faculdade de Ciências (Lisboa), Revista de Ciências Naturais

Florides GA, Christodoulides P (2009) Global warming and carbon dioxide through sciences. Environ Int 35:390–401

Han B, Wang XK, Ouyang ZY (2005) Saturation levels and carbon sequestration potentials of soil carbon pools in farmland ecosystems of China. Rural Eco-Environ 21:6–11

Hansen J, Sato M, Kharecha P, von Schuckmann K (2011) Earth's energy imbalance and implications. Atmos Chem Phys 11:13421–13449

Hartig EK, Gornitz V, Kolker A, Mushacke F, Fallon D (2002) Anthropogenic and climate-change impacts on salt marshes of Jamaica Bay, New York City. Wetlands 22:71–89

Hodgkin EP, Hamilton B (1998) Changing estuarine wetlands: a long term perspective for management. In: McComb AJ, Davis JA (eds) Wetlands for the future. Gleneagles Publishing, Adelaide, pp 243–255

Hughes RG (2004) Climate change and loss of saltmarshes: consequences for birds. Ibis 146(suppl 1):21–28

IPCC (1998) The regional impacts of climate change: an assessment of vulnerability. In: Watson RT, Zinyowera MC, Moss RH (eds) A special report of IPCC working group II. Cambridge University Press, Cambridge

IPCC (2007) Climate change 2007: the physical science basis. Cambridge University Press, New York

IPCC (2012) In: Field CB, Barros V, Stocker TF, Qin D, Dokken DJ, Ebi KL, Mastrandrea MD, Mach KJ, Plattner G-K, Allen SK, Tignor M, Midgley PM (eds) Managing the risks of extreme events and disasters to advance climate change adaptation. A special report of working groups I and II of the intergovernmental panel on climate change. Cambridge University Press, Cambridge/New York. 582 pp

IPCC (2013) In: Stocker TF, Qin D, Plattner G-K, Tignor M, Allen SK, Boschung J, Nauels A, Xia Y, Bex V, Midgley PM (eds) Climate change 2013: the physical science basis. Contribution of working group I to the fifth assessment report of the intergovernmental panel on climate change. Cambridge University Press, Cambridge/New York. 1535 pp

IPCC (2014) In: Edenhofer O, Pichs-Madruga R, Sokona Y, Farahani E, Kadner S, Seyboth K, Adler A, Baum I, Brunner S, Eickemeier P, Kriemann B, Savolainen J, Schlömer S, von Stechow C, Zwickel T, Minx JC (eds) Climate change 2014: mitigation of climate change. Contribution of working group III to the fifth assessment report of the intergovernmental panel on climate change. Cambridge University Press, Cambridge/New York

Jackson JB, Kirby MX, Berger WH, Bjorndal KA, Botsford LW, Bourque BJ, Bradbury RH, Cooke R, Erlandson J, Estes JA, Hughes TP, Kidwell S, Lange CB, Lenihan HS, Pandolfi JM, Peterson CH, Steneck RS, Tegner MJ, Warner RR (2001) Historical overfishing and the recent collapse of coastal ecosystems. Science 293:629–638

Kearney MS, Grace RE, Stevenson JC (1988) Marsh loss in Nanticoke Estuary, Chesapeake Bay. Geogr Rev 78:205–220

Kirschbaum MUF (1995) The temperature dependence of soil organic matter decomposition, and the effect of global warming on soil organic C storage. Soil Biol Biochem 27:753–760

Lefeuvre JC, Laffaille P, Feunteun E, Bouchard V, Radureau A (2003) Biodiversity in salt marshes: from patrimonial value to ecosystem functioning. The case study of the Mont-Saint-Michel bay. C R Biol 326:125–131

Lenssen GM, Lamers J, Stroetenga M, Rozema J (1993) Interactive effects of atmospheric CO_2 enrichment, salinity and flooding on growth of C_3 (*Elymus athericus*) and C_4 (*Spartina anglica*) salt marsh species. Vegetatio 104(105):379–388

Lotze HK, Lenihan HS, Bourque BJ, Bradbury RH, Cooke RG, Kay MC, Kidwell SM, Kirby MX, Peterson CH, Jackson JBC (2006) Depletion, degradation, and recovery potential of estuaries and coastal seas. Science 312:1806–1809

Magenheimer JF, Moore TR, Chmura GL, Daoust RJ (1996) Methane and carbon dioxide flux from a macrotidal salt marsh, Bay of Fundy, New Brunswick. Estuaries 19:139–145

Marinucci AC (1982) Trophic importance of *Spartina alterniflora* production and decomposition to the marsh estuarine ecosystem. Biol Conserv 22:35–58

McKee K, Mendelssohn IA, Materne MD (2004) Acute salt marsh dieback in the Mississippi River deltaic plain: a drought induced phenomenon? Glob Ecol Biogeogr 13:65–73

Michener WK, Blood ER, Bildstein KL, Brinson MM, Gardner LR (1997) Climate change, hurricanes and tropical storms, and rising sea level in coastal wetlands. Ecol Appl 7:770–801

Morris JT, Sundareshwar PV, Nietch CT, Kjerfve B, Cahoon DR (2002) Responses of coastal wetlands to rising sea level. Ecology 83:2869–2877
Munari C, Mistri M (2008) The performance of benthic indicators of ecological change in Adriatic coastal lagoons: throwing the baby with the water? Mar Pollut Bull 56:95–105
Murphy D, Solomon S, Portmann R, Rosenlof K, Forster P, Wong T (2009) An observationally based energy balance for the Earth since 1950. J Geophys Res Atmos 114:D17107
Niering WA (1977) Our dynamic tidal marshes: vegetation changes as revealed by peat analysis. Conn Arboretum Bull 22:2–12
Nixon SW (1980) Between coastal marshes and coastal waters--a review of twenty years of speculation and research on the role of salt marshes in estuarine productivity and water chemistry. In: Hamilton P, MacDonald K (eds) Estuarine and wetland processes. Plenum Publishing Corporation, New York, pp 437–523
Redfield AC (1965) Ontogeny of a salt marsh estuary. Science 147:50–55
Rogers K, Saintilan N, Heijnis H (2005) Mangrove encroachment of salt marsh in Western Port Bay, Victoria: the role of sedimentation, subsidence and sea level rise. Estuaries 28:551–559
Roman CT, Peck JA, Allen JR, King JW, Appleby PG (1997) Accretion of a New England (U.S.A.) salt marsh in response to inlet migration, storms, and sea-level rise. Estuar Coast Shelf Sci 45:717–727
Rozema J, Dorel F, Janissen R, Lenssen G, Broekman R, Arp W, Drake BG (1991) Effect of elevated atmospheric CO_2 on growth, photosynthesis and water relations of salt marsh grass species. Aquat Bot 39:45–55
Rybczyk JM, Callaway JC, Day JW (1998) A relative elevation model for a subsiding coastal forested wetland receiving wastewater effluent. Ecol Model 112:23–44
Shakun JD, Clark PU, He F, Marcott SA, Mix AC, Liu Z, Otto-Bliesner B, Schmittner A, Bard E (2012) Global warming preceded by increasing carbon dioxide concentrations during the last deglaciation. Nature 484:49–55
Sousa AI, Lillebo AI, Pardal MA, Caçador I (2010) The influence of *S. maritima* on carbon retention capacity in salt marshes from warm-temperate estuaries. Mar Pollut Bull 61:215–223
Teixeira A, Duarte B, Caçador I (2014) Salt marshes and Biodiversity. In: Khan MA, Böer B, Öztürk M, Al Abdessalaam TZ, Clüsener-Godt M, Gul B (eds) Sabkha ecosystems: volume IV: cash crop halophytes and biodiversity conservation, Tasks for vegetation science Vol. 47. Springer, Dordrecht
Trenberth KE, Fasullo JT, Kiehl J (2009) Earth's global energy budget. Bull Am Meteorol Soc 90:311–323
Turner RE (1976) Geographic variations in salt marsh macrophyte production: a review. Contrib Mar Sci 20:47–68
UNEP (2006) Marine and coastal ecosystems and human well-being: a synthesis report based on the findings of the Millennium ecosystem assessment. UNEP, Nairobi
USGCRP (2000) Climate change and America: overview document. In: A report of the national assessment synthesis team. US global change research program, Washington, DC
Valentim JM, Vaz L, Vaz N, Silva H, Duarte B, Caçador I, Dias JM (2013) Sea level rise impact in residual circulation in Tagus estuary and Ria de Aveiro lagoon. J Coast Res SI 65:1981–1986
Valiela I, Cole ML (2002) Comparative evidence that salt marshes and mangroves may protect seagrass meadows from land-derived nitrogen loads. Ecosystems 5:92–102
Valiela I, Teal JM (1979) The nitrogen budget of a salt marsh ecosystem. Nature 280:652–656
Valiela I, Peckol P, D'Avanzo C, Kremer J, Hersh D, Foreman K, Lajtha K, Seely B, Geyer WR, Isaji T, Crawford R (1998) Ecological effects of major storms on coastal watersheds and coastal waters: Hurricane Bob on Cape Cod. J Coast Res 14:218–238
Warren RS, Niering WA (1993) Vegetation change on a northeast tidal marsh: interaction of sea-level rise and marsh accretion. Ecology 74:96–103

Morphology and Betalain Characterization of 'Iceplants' (Aizoaceae) from the Coast of Wellington, New Zealand

13

Sumedha Madawala

Abstract

Exotic and native "iceplant" species (family: Aizoaceae) are found in coastal habitats in New Zealand. Iceplants produce hybrids through cross-pollination. Their identification in the field is difficult. A field survey was carried out to morphologically characterize some iceplant species and a "hybrid" commonly found along the coast of Wellington, New Zealand. Iceplants contain betalains in place of anthocyanins. The color variations also make their identification challenging. In this study, the betalains were quantified and characterized using spectrophotometric and high-performance liquid chromatographic (HPLC) methods, respectively. The results revealed that the selected iceplant species, namely *Carpobrotus edulis* (L.) N. E. Br., *C. chilensis* (Molina) N. E. Br., red and green morphotypes of *Disphyma australe* (W. T. Aiton) N. E. Br., and a putative "hybrid" between *C. edulis* and *Disphyma australe* (red morphotype), possess distinct morphological characteristics and choice of micro-habitats. The native *Disphyma australe* (green) and *C. chilensis* showed restricted distribution compared to the other species studied. *C. edulis* and the putative "hybrid" *C. edulis* × *D. australe* showed the most extensive distribution along the coast of Wellington. The robust growth of *C. edulis* and its ability to form thick mats indicate its ability to suppress the growth and establishment of other plants, thereby reiterating its success as an invasive species. The hybrid also showed invasive potential due to its robust growth and ability to establish on different substrates. Betalain levels corresponded to plants' red-magenta coloration with the lowest levels recorded in the exotic invasive *C. edulis* and the highest in *C. chilensis*. The HPLC analysis revealed six betacyanins in selected iceplant species, in different proportions and quantities. The study concludes that the iceplant species are easy to differentiate with the help of morphological characteristics. The results also support the idea that the putative hybrid has resulted from the cross-pollination between *C. edulis* and *D. australe*. The green morphotype of *D. australe* has shown restricted distribution, raising concerns for its conservation.

Keywords

Betalains · Dune vegetation · Hybrid · Invasion

S. Madawala (✉)
Department of Botany, Faculty of Science, University of Peradeniya, Peradeniya, Sri Lanka

B. Gul et al. (eds.), *Sabkha Ecosystems*, Tasks for Vegetation Science VI,
https://doi.org/10.1007/978-3-030-04417-6_13

Introduction

Iceplants and their putative hybrids occur naturally in coastal habitats throughout New Zealand (Chinnock 1972). Iceplant species such as *Carpobrotus edulis* (L.) N. E. Br. (commonly known as "hottentot-fig" or "highway iceplant") and *C. chilensis* (Mol.) N. E. Br. ("sea fig") are adventives, while *Disphyma australe* (Aiton) N. E. Br. ("horokaka") is endemic to New Zealand. According to Chinnock (1972), the native and adventive iceplant species are known to form hybrids through cross-pollination. These hybrids of iceplant species often overlap with the populations of adventives and endemics, exhibiting characteristics common to their parental species. In spite of many records of iceplant species and their putative hybrids along the coast of New Zealand, field identification is difficult due to morphological similarities.

C. edulis was introduced into New Zealand for use as an effective soil stabilizer in dune restoration (Albert 1995). Later, it successfully naturalized coastal habitats, mixing well with native communities (D'Antonio 1993; D'Antonio et al. 1993; Albert 1995). The genus *Carpobrotus* is well-known for its ability to hybridize among genotypes and with other co-occurring species. *Disphyma australe,* an endemic iceplant species in New Zealand, is known to exist in two different morphotypes (red and green), which can be distinguished by the color of their vegetative parts (Chinnock 1971). Studies have observed hybridization between *C. edulis* and *C. chilensis* (Schierenbeck et al. 2005; Vilá et al. 1998; Suehs et al. 2004; Jakobsson et al. 2008). Hybridization between *C. edulis* and *D. australe* was first recorded in New Zealand by Healy (1959) and later by Chinnock (1972). According to Chinnock (1972), the most common hybrids along the coast of New Zealand are from cross-pollination between *C. edulis* and *D. australe*, but not between *C. edulis* and *C. chilensis*. Heenan and Sykes (2010) termed hybrids of *C. edulis* and *D. australe* × *Carpophyma mutabilis* and hybrids of *C. chilensis* and *D. australe* × *Carpophyma pallida*. They noted that the hybrids were sterile and identified × *Carpophyma mutabilis* as the more common hybrid out of the two. Implications of hybridization have become more important, particularly in island habitats, as hybrids can become invasive, causing detrimental ecological and genetic impacts on native plant species (Vilá et al. 2000).

Iceplants contain betalains instead of anthocyanins. Betalains, which include the red-violet betacyanins and the yellow-orange betaxanthins, are water-soluble nitrogen-containing pigments present in different parts of plants in the order Caryophyllales, but excluding the families Caryophyllaceae and Molluginaceae (Moreno et al. 2008; Tanaka et al. 2008). Due to a restricted presence in higher plants, betalains play a role in chemotaxonomy (Waterman 2007). Major advances have been made in recent years in understanding the biosynthetic pathways of betalains (Tanaka et al. 2008; Han et al. 2009).

As with all other pigments, betalains play a crucial role in plants. The function of betalains in fruits and flowers is to attract pollinators, while their role in vegetative parts is still largely unknown (Ferruzzi et al. 2002; Fraser et al. 2004). Some researchers have observed betalain production in wounded tissues, highlighting its role in defense mechanisms against viral and fungal infections (Piattelli 1981; Stafford 1994). It is also known for its function as a UV-protector in iceplants (Vogt et al. 1999). Jain and Gould (2015) noted its role in ameliorating salinity stress in *D. australe*. Few attempts have been made to identify and characterize betalains in ornamental (*Bougainvillea, Celosia, Gomphrena, Portulaca, Mirabilis*), edible (*Beta vulgaris, Amaranthus* sp., *Opuntia* sp., *Hylocereus* sp., *Ullucus tuberosus*), and wild species (Cai et al. 1998, 2001; Stintzing et al. 2005; Kugler et al. 2007; Herrero et al. 2005; Azeredo 2009; Svenson et al. 2008; Wang et al. 2007; Jain and Gould 2015).

The present study attempts to distinguish the morphological characteristics of some selected iceplant species, namely *C. edulis, C. chilensis,* and *Disphyma australe* and a putative hybrid C. *edulis* × *D. australe*, found along the coast of Wellington, New Zealand. The study investigates the diversity of betalains in selected iceplant species and the putative hybrid using spectrophotometry and high-performance liquid chromatography (HPLC) coupled with mass spectrometry. The tissue distribution patterns of betalains were also observed.

Materials and Methods

Study Sites

The study was carried out along the coast of Wellington between September 2009 and March 2010. Specimens were collected from different coastal localities in and around Wellington (the area covers coastal suburbs including Lyall Bay, Island Bay, Days Bay, Seatoun, Eastbourne, and Paraparaumu) to obtain a good representation for selected species (Fig. 13.1). Due to limited distribution of *C. chilensis* and the green morphotype of *D. australe*, the specimens were collected from one colony each, near Eastbourne and Island Bay, respectively.

Species Descriptions

C. edulis and *C. chilensis* are perennial succulents with trailing habits and triangular-shaped leaves. Unlike *C. edulis*, the origin of *C. chilensis* and its reason for introduction into New Zealand are unknown (Vilá et al. 1998; Albert et al. 1997). *D. australe* is a native of New Zealand with two distinctively different morphotypes – one with shorter, light green leaves and the other with comparatively longer, dark green leaves often pigmented with red/magenta coloration (hereafter referred to as the green and red morphotypes, respectively). The red morphotype has a much wider distribution than the green morphotype. The putative hybrid *C. edulis* × *D. australe* (red

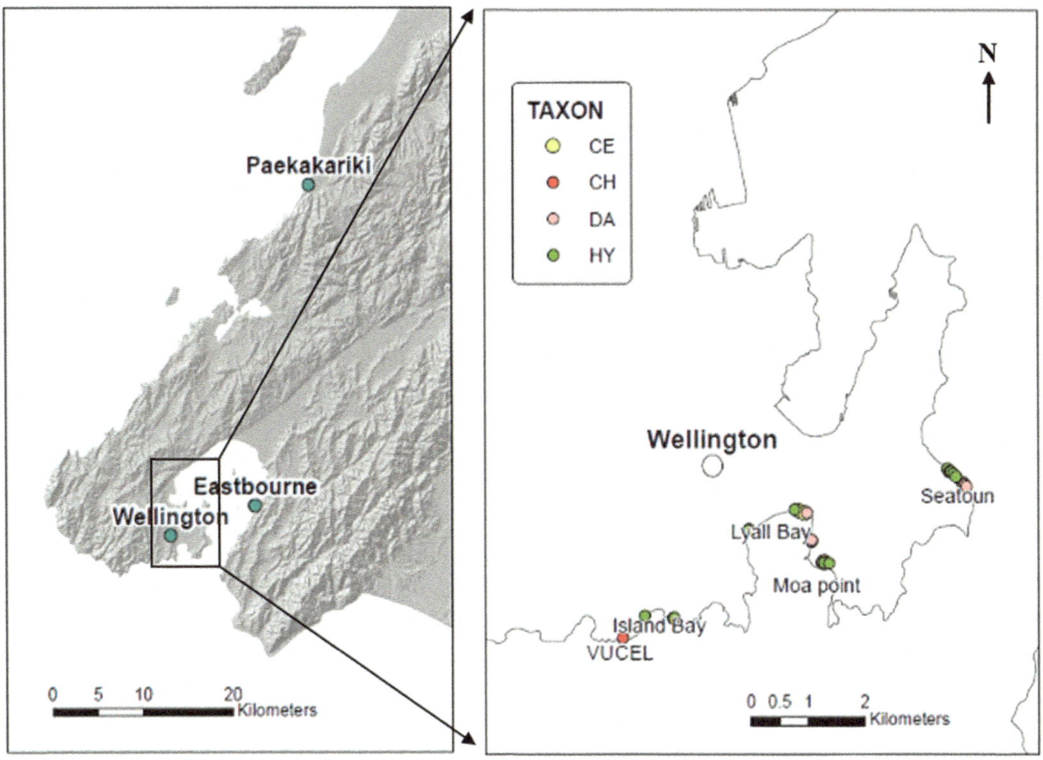

Fig. 13.1 Map showing the sampling locations of the selected iceplant species, *Carpobrotus edulis* (CE), *C. chilensis* (CH), *Disphyma australe* (DA), and the putative hybrid *C. edulis* × *D. australe* (HY) along the coast of Wellington, New Zealand

morphotype) has been identified in locations along the coast of Wellington, where parental species are most dominant. The hybrid *C. edulis* × *D. australe* bears some intermediate characters to its parental species with more resemblance to red morphotype of *D. australe.*

Collection of Specimens

In each sampling location, specimens of *C. edulis* and *D. australe* subsp. *australe* (red morphotype) and their putative hybrid were collected from ten different colonies. Due to limited distribution, *C. chilensis* and *D. australe* (green morphotype) were collected from single colonies found near Eastbourne and Island Bay, respectively. During plant collections, their preferred habitats and substrates were also noted. In the field, plants were identified by visual features such as leaf, flower, fruit, and clonal characteristics based on published descriptions of *C. edulis, C. chilensis,* and *D. australe* (Munz and Keck 1968; Vivrette 1993; Chinnock 1972).

Morphological Characterization and Microscopic Observations

Morphological characteristics such as leaf length, width, intermodal distance, leaf margins, tips, and flower color and diameter were measured in all fresh specimens (10 measurements were taken randomly from each colony). Epidermal peels and transverse sections of leaves and stems were observed under a confocal microscope (Olympus FV1000 spectral imaging microscope) to assess pigment distribution in tissues.

Characterization of Betalains

Randomly collected plant samples (n = 3) of selected iceplant species and the putative hybrid *C. edulis* × *D. australe* were quantified and chemically characterized for betalains using spectrophotometric and HPLC methods, respectively.

Spectrophotometric Analyses

From each taxon, five randomly selected specimens were used in the quantification of betalains using spectrophotometric analyses. Approximately 3–5 g of shoot samples were washed, blotted dry, and homogenized in 80% aqueous methanol containing 50 mM sodium ascorbate in a 1:5 tissue (g) to solvent (ml) ratio. After centrifugation at 15,000 g for 10 min, the supernatant was measured for absorbance at 480 nm (for betaxanthins), 538 nm (for betacyanins), and 650 nm (values of which were subtracted from A_{480} and from A_{530} to compensate for overlap in absorbance by extracted chlorophylls) using a Shimadzu 2,500 dual beam UV-vis spectrophotometer (Shimadzu, Kyoto, Japan). Betalain levels were calculated on a fresh weight basis using molar extinction coefficients of 60,000 mol^{-1} cm^{-1} for betacyanin and 48,000 mol^{-1} cm^{-1} for betaxanthin (Girod and Zryd 1991; Kugler et al. 2004).

HPLC Characterization

Three randomly chosen plant samples from selected iceplant taxa were used for the HPLC quantification and characterization of betalains. In preparation, 10 g of fresh plant sample (washed and cut into pieces) was freeze-dried in liquid nitrogen before homogenizing in 50 ml of 60% aqueous methanol containing 50 mM sodium ascorbate [in 1:5 ratio of tissue (g) to solvent (ml)]. The sample was homogenized and centrifuged to remove plant debris. Samples were then vacuum-filtered to obtain a clear suspension and stored at −25 °C until samples were transported to the Plant and Food Research Laboratories in Palmerston North, New Zealand for HPLC analyses.

Before HPLC analysis, the samples were reduced in volume using a rotary evaporator. Extracts were analyzed by HPLC to separate and identify individual pigments and spectrophotometrically to quantify pigments, as per Kugler et al. (2004, 2007) and Svenson et al. (2008). HPLC analysis was carried out using a

Dionex™ Ultimate™ 3000 solvent delivery system with a Phenomenex Synergi-Hydro RP (4 μm, 150 × 3.0 mm) column (column temperature 25 °C) and a Dionex™ 3000 PDA detector. Betaxanthins were detected at 470 nm and betacyanins at 538 nm. HPLC gradients were as per Kugler et al. (2007) but the flow rate was reduced to 0.5 ml min^{-1}. The injection volume was 20 L. Peaks on the HPLC chromatograms were assessed against known retention and spectral maxima for betalain pigments.

Data Analysis

Morphological parameters (n = 10) were presented as ranges (minimum to maximum). Betalain concentrations (n = 3) were compared between taxa using one-way analysis of variance (ANOVA) at a probability level of 5%. The mean separation was performed using Tukey's test. The data analysis was carried out using Minitab 16.0.

Results

Habitat Preferences and Morphological Characteristics of the Selected Iceplants

The selected iceplant taxa showed variations in their respective preferences in habitats and also their morphological characteristics (Fig. 13.2; Table 13.1). *C. edulis* was widely distributed and found predominantly on sandy, back dune habitats interspersed with other native plant communities. *C. edulis* also showed vigorous growth, forming thick mats. In contrast, *C. chilensis* showed fewer occurrences, but shared similar substrates to *C. edulis*. However, *C. chilensis* was often observed on sandy foredunes, sprawling across pebbles, and occasionally in back dune habitats. During the field survey, co-occurrence of *C. edulis* and *C. chilensis* was observed only very rarely. Furthermore, in contrast to *C. edulis*, *C. chilensis* showed more open growth architecture forming sparse and thin mats. In all study locations, *C. edulis* was more abundant than *C. chilensis*. The two *Carpobrotus* species showed some distinct morphological characteristics (in addition to their contrasting flower colors, which have been reported previously) enabling us to differentiate between the two species. Discrete morphological characteristics in leaves and consistent pigmentation patterns that were observed during the study may assist in more easily identifying the two species in the field. *C. edulis* bore pale green to dark green foliage with red/magenta pigmentation mainly along leaf margins and occasionally on leaf surfaces (especially on mature leaves). Young leaves of *C. edulis* were mostly green in color with pigmentation restricted to leaf margins. Leaf margins were sharp with serrated keels. Leaves were three-angled, succulent, and sabre-shaped. *C. edulis* bore large, distinctive lemon-yellow flowers, which turned pink as they matured. *C. chilensis,* on the other hand, had bluish-green leaves with red to magenta coloration starting from the tip and spreading toward the leaf base. Pigmentation also caused localized spots (as speckles) spreading over mature leaf surfaces. The leaves of *C. chilensis* were shorter in length than in *C. edulis* and serration was confined mainly to the upper part of the keel. Less frequent branching in *C. chilensis* resulted in a sparse growth pattern and thinly spread mats. *C. chilensis* bore striking rose-magenta colored flowers slightly smaller in size than those of *C. edulis* (see Table 13.1).

The two morphotypes of *D. australe* are distinguished by their size, shape, color, and leaf pigmentation (Fig. 13.3). The green morphotype was found exclusively on sea-facing rocky outcrops. The red morphotype, which carries red-magenta pigmented shoots, was observed on a variety of substrates, including rocky outcrops as well as gravelly/sandy substrates. The green morphotype bears fleshy, pale-green colored leaves with no pigmentation, while the red morphotype often has heavily pigmented leaves and stems. However, pigmentation varies among the individuals within a colony, with some individuals having largely dark blue-green leaves, while others had red-magenta colored leaves and stems. Both the morphotypes produced much

Fig. 13.2 The selected iceplant species and their varying habitats, (**a**) *C. edulis,* (**b**) *C. chilensis* (**c**) *D. australe* (green morphotype), (**d**) *D. australe* (red morphotype), and (**e**) putative hybrid *C. edulis* × *D. australe* (red morphotype)

smaller flowers than *C. edulis* and *C. chilensis* and the flower color varied from purple to pink to white. The leaves were closely obovate in shape with acute tips in both morphotypes. The leaf arrangement was crowded with short internodes.

The red morphotype showed variations in the level of pigmentation among colonies, suggesting phenotypic plasticity in relation to micro-environmental conditions. Plants on rocky and highly exposed habitats seem to reduce their leaf size, internodal distance, and flower size. The extent of pigmentation also showed fluctuations from one colony to another. The green morphotype was found in limited locations along the coast of Wellington and was restricted to rocky coastal habitats when compared to the more widespread distribution of the red morphotype.

Table 13.1 Morphological and habitat characteristics of iceplant species: *D. australe* (red and green morphotypes), *C. edulis*, the potential hybrid between red morphotype of *D. australe* and *C. edulis*, and *C. chilensis*

	C. edulis	*C. chilensis*	*D. australe* Red morphotype	*D. australe* Green morphotype	Hybrid (*D. australe* × *C. edulis*)
Habitat	Sandy back dunes	Gravelly and sandy substrates; mostly found in foredunes	Rocky, gravelly, and sandy substrates	Rocky outcrops facing sea	Sandy, rocky, gravelly substrates. Widespread along roads, roundabouts etc.
Habit	Form thick mats due to frequent branching; growing tip always directed upwards away from the substrate	Less crowded; form thin mats due to less frequent branching; growing tip stays close to the substrate	Grow close to the surface, less crowded; pigmented leaves and stems	Grow close to the substrate; leaves are crowded with short internodes; no red pigmentation	Leaves are linear and longer than the parental species; less crowded mats
Leaf characteristics					
Arrangement	Opposite	Opposite	Opposite	Opposite	Opposite
Color	Light to dark green	Bluish green	Bright to dark green	Light green	Dark green
Length (cm) [a]	7.0–10.0	6.0–7.2	1.5–2.3	1.2–1.8	3.0–4.5
Width (cm) [a]	1.0	1.0	0.6–0.8	0.6	0.4–0.6
Margin	Sharp and serrated mostly throughout the margin	Sharp and serrated, confine to the apex of the keel	Entire	Entire	Entire
Leaf tip	Acute	Acute	Bristle-tipped	Bristle-tipped	Blunt
Leaf cross-section	Sharply triangular	Sharply triangular	Roundly triangular	Roundly triangular	Roundly triangular
Betalain pigmentation	Along leaf margins and tips; also on mature leaves	Pigmentation on leaf surfaces and young stems	Leaves and young stems	No pigmentation	Leaves and young stems
Stem characteristics					
Color	Green (young), woody on mature	Pigmented	Frequently pigmented, woody on mature	Green (young) and woody on mature	Green and sometimes pigmented
Internodal distance (cm) [a]	4.5–5.0	2.0–4.0	3.0–4.5	1.0	6.0–6.7
Floral and fruit characteristics					
Color	Lemon-yellow	Magenta	Light purple to pink	White to light pink	Pink
Diameter (cm)[a]	8.0–12.0	4.0–7.5	2.0–4.0	2.0–2.5	3.5–3.8
Fruit	Fleshy, large, indehiscent	Fleshy, medium, indehiscent	Fleshy, indehiscent, red berry	Fleshy, dehiscent, green berry	Dry and not properly developed
Outer wall of the fruit	Thick (green)	Medium (red)	Thin (red)	Thin (green)	Thin

[a]Morphological parameters are presented as ranges (minimum– maximum). Ten measurements were taken from each colony. For each species, ten colonies were sampled

Fig. 13.3 The green and red morphotypes of *Disphyma australe* plants collected from Lyall Bay, Wellington, New Zealand

Fig. 13.4 Habit and the inflorescence of the putative hybrid between *C. edulis* and *D. australe* (red morphotype)

During the field survey, I came across the putative hybrid between *C. edulis* and *D. australe* (red type) in locations where the exotic and the native iceplant species were freely co-occurring (Fig. 13.4). The hybrids were commonly found on diverse substrates, including rocky, sandy, and/or gravelly, as well as in foredunes and back dunes. The hybrid was also observed in some restoration trials established by the local council. The hybrid exhibits intermediate characteristics of its parent species, and resembles more closely *D. australe* (in size and form) than *C. edulis*. Hybrids bear relatively longer, slender dark green leaves with entire margins and purple-pink flowers. Leaves are often pigmented with red-magenta coloration. The hybrid × *Carpophyma mutabilis* described by Heenan and Sykes (2010) was recorded with longer leaves and larger flowers compared to the putative hybrid that we observed along the coast of Wellington.

Tissue Distribution of Betalains

Microscopic observations revealed that the leaf epidermal cells contained varying levels of betalains in *C. chilensis* (Fig. 13.5). In contrast, the epidermal cells of *C. edulis* lacked betalains as pigmentation is mostly confined to the margins and tips of leaves (Fig. 13.5b). Stem sections of *D. australe* (red) demonstrated the presence of betalains in the immediate, outer cell layers

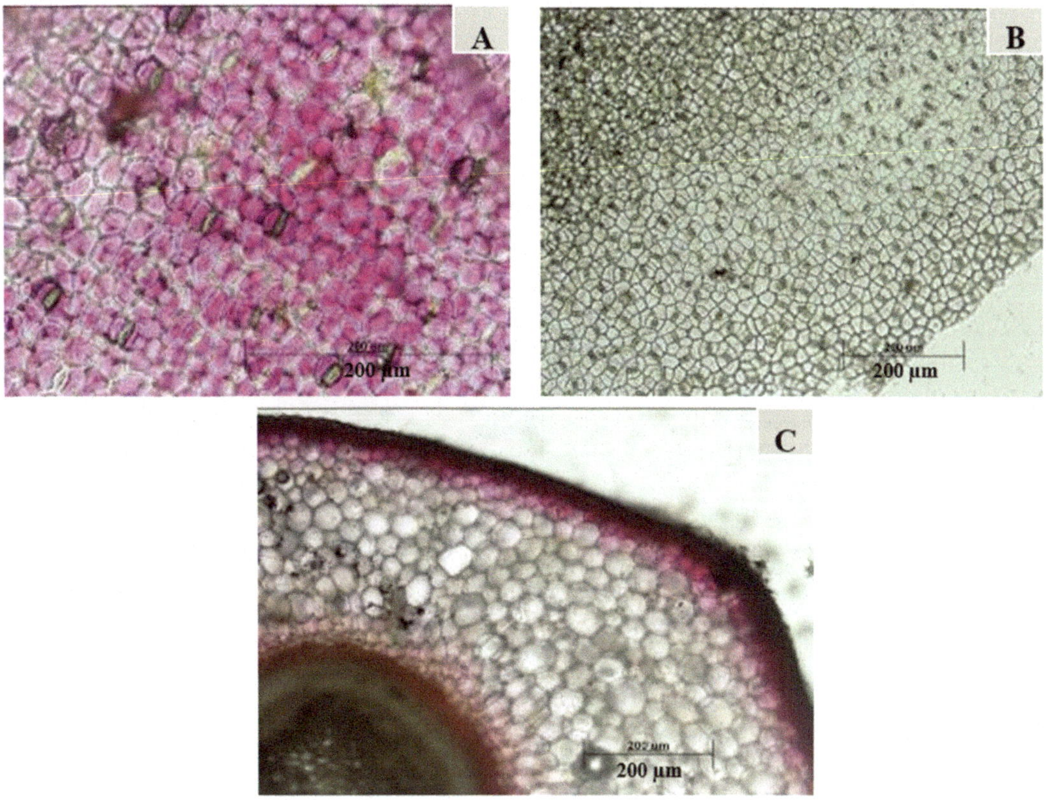

Fig. 13.5 Microscopic sections and peels showing the distribution of betalains: (**a**) epidermal peel showing betalain-containing epidermal cells in *C. chilensis*; (**b**) epidermal cells of *C. edulis*; and (**c**) transverse section of the young stem of *D. australe* (red)

surrounding the vascular bundle, in addition to their presence in cells just beneath the epidermal layer (Fig. 13.5c). In the present study, we also noted a higher and more widespread pigmentation in mature, senescing leaves of *C. edulis* than that of younger leaves.

Quantification of Betalains

Compared to the levels of foliar pigmentation, betalain concentrations were significantly varied among iceplant species and the hybrid *C. edulis* × *D. australe* (Fig. 13.6). Betacyanins were not detected in the green morphotype of *D. australe*. *C. chilensis* showed the highest total betalain content and the lowest was found in *C. edulis*. In all species, except *C. edulis*, betacyanins contributed more to the plants' total betalain content than that of betaxanthins. The *D. australe* (red) and the hybrid showed similar levels of betacyanin and betaxanthins in their foliage. *C. chilensis* was frequently highly pigmented under field conditions and had significantly higher betacyanin and total betalain concentrations than those of the other species tested.

Characterization of Betalains

The HPLC analysis of fresh plant samples detected the presence of six betacyanin pigments in the leaf extracts of iceplants. They were identified as betanin, isobetanin, betanidin, isobetanidin, lampranthin-II, and isolampanthrin-II (Table 13.2).

The six betacyanins detected in foliar samples were observed across all the selected ice-

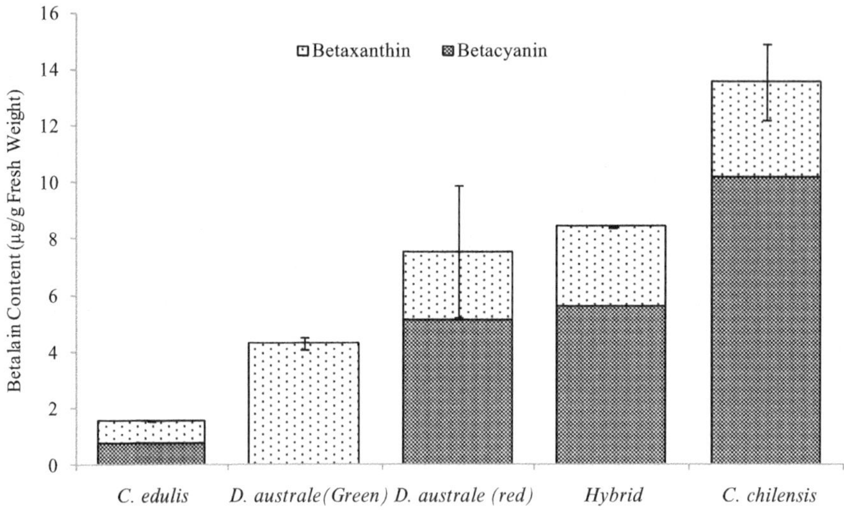

Fig. 13.6 Concentrations of betaxanthins and betacyanins (µg/g fresh weight; mean ±standard error) of the iceplant species, *Carpobrotus edulis, Disphyma australe* (green), *D. australe* (red), putative hybrid, and *C. chilensis* under field conditions, quantified using spectrophotometry

Table 13.2 Betalain peaks detected in extracts of *C. edulis, C. chilensis, D. australe* (red), and the hybrid leaf and stem tissues

Retention time (min) [a]		Betalain	Common name	Spectral max.
Betacyanins				
1	9.7[b]	Betanidin 5-O-β-glucoside	Betanin	534
2	10.9[b]	Isobetanidin 5-O-β-glucoside	Isobetanin	533
3	12.8	Betanidin	Betanidin	540
4	14.3	Isobetanidin	Isobetanidin	540
5	23.1[c]	Betanidin 5-O-[6′-O-(E)-feruloyl-β-glucoside]	Lampranthin-II	534
6	23.5	Isobetanidin 5-O-[6′-O-(E)-feruloyl-β-glucoside]	Isolampranthin-II	534

No betacyanins were detected in *D. australe* (green). Spectral maximum (nm) and molecular mass are reported for each peak detected at a particular retention time. Beetroot extract was used to provide some reference standards
[a]The structure for lampranthin-II as described by Strack et al. (1988)
[b] Betalain pigments extracted from beetroot as standard pigments
[c] Retention time is recorded from the betacyanin gradient

plant species except for the green morphotype of *D. australe*. However, their relative proportions varied (Table 13.3). Lampranthin II and isolampranthin II predominate in *C. edulis*, *D. australe* (red), and the hybrid, although higher proportions of betanin and isobetanin were detected in *D. australe* compared to the other two species. In contrast, betanin and isobetanin predominated in *C. chilensis*. The hybrid *C. edulis* × *D. australe* subsp. *australe* showed comparatively higher relative proportions of lampranthin II and isolampranthin II. However, other betacyanins were present in trace levels in the hybrid.

Discussion and Conclusion

The study indicated that the selected iceplant species and the hybrid showed contrasting morphological features and habitat preferences. Data also suggested that phenotypic characteristics are relatively stable within a species and morphotype. *Disphyma australe* (red) and the hybrid were

Table 13.3 Average relative proportions (%) of individual betalains within the total betalain profile detected in leaf tissue samples in the selected iceplant taxa (n = 3)

	Relative proportions (%) of individual betalains					
	Betanin	Isobetanin	Betanidin	Isobetanidin	Lampranthin-II	Isolampranthin-II
C. edulis	5.8	4.8	10.6	6.6	38.4	33.8
C. chilensis	43.1	21.5	7.9	6.3	11.4	11.8
D. australe (red)[a]	23.6	8.8	7.2	–	35.5	24.9
C. edulis x D. australe	5.3	2.8	7.7	4.9	47.5	42.1

[a]The green morphotype of *D. australe* did not undergo HPLC analysis

difficult to identify due to similarities in the shoot pigmentation patterns. However, the hybrid leaves are longer and less crowded than those of *D. australe* (red). According to Weber and D'Antonio (1997), leaf length and shape (leaf length:width ratio), fruit size, and flower color are considered to be important characteristics for differentiating iceplant species and their hybrids. They also suggest that the length and shape of leaves in iceplants show low phenotypic plasticity within a species. hence leading to strong inter-specific differences. The close vegetative resemblance of *C. edulis* and *C. chilensis* has prompted some scientists to claim that they are not different species but varieties (Akeroyd and Preston 1990, 1993), or perhaps a hybrid swarm (Vilá et al. 1998). However, the present study observed sufficiently distinct morphological characteristics between them, in addition to their distinct flower colors. Albert et al. (1997) have shown that *C. edulis* and *C. chilensis* are clustered into two distinct morphotypes, while their hybrids overlapped both species. Chinnock (1972) has found phenotypic differences among native and exotic iceplant species in a study carried out in coastal localities in New Zealand, and present results are comparable with Chinnock's findings. The field observations of the present study also suggested that *C. edulis* can hinder the growth and establishment of other native species due to their wide habitat preferences, vigorous growth, and their ability to form thick mats over substrates. In contrast, *C. chilensis* with its low growth rate and open, sparse colonies may cause less hindrance to establishing to other species.

The study also observed two morphotypes of *D. australe* with distinct morphological characteristics and habitat preferences. Chinnock (1971) acknowledged the presence of two distinct morphotypes with contrasting pigmentation in a study carried out along the coast of Wellington. In addition to their significant morphological differences, the present study noted their distinct micro-habitat preferences. The green morphotype, which has a restricted distribution, was observed on rocky outcrops along the coast facing direct sea spray. In contrast, the red morphotype is more widespread and was observed on diverse substrates and micro-habitats, but away from direct sea spray. Atkinson (1972) observed a strong correlation between *D. australe* and its choice of habitat (on cliffs with shallow rocky soils) in a study carried out in a few islands off the Hauraki Gulf in north New Zealand. However, previous studies did not record a highly localized occurrence of the green morphotype of *D. australe.*

Morphological and field evidence favors the idea that the putative hybrid that we came across along the coast of Wellington may have resulted from an inter-generic hybridization between *C. edulis* and *D. australe* (red). Field observations also showed that *C. edulis* and *D. australe* (red) are frequently co-occurring and overlap similar habitats where the putative hybrid is found. Further, *C. chilensis* has a restricted distribution compared to that of *C. edulis*, and the putative hybrid showed morphological characteristics more closely comparable to *C. edulis* and *D. australe* (red) than to *C. chilensis*. Field observations also suggested differences between the micro-habitat preferences of *C. edulis* and *C. chilensis.* They have already been identified as back dune and foredune species, respectively, by other workers (Bicknell and Mackey 1988; Schierenbeck et al. 2005). *C. edulis* showed widely distributed sympatric populations along the coast of Wellington. Differing habitat preferences and restricted distribution of *C. chilensis* may have resulted in less or no cross-pollination events between *C. edulis* and *C. chilensis* along the coast of Wellington, in spite of records of putative hybrids between *C. edulis* and *C. chilensis* in other parts of the world (Albert et al. 1997; Gallagher et al. 1997). Gallagher et al. (1997) noted that *C. chilensis* and *C. chilensis*-like hybrid morphotypes (as a result of a likely cross between *C. chilensis* and *C. edulis*) are morphologically and genetically difficult to differentiate. The field survey also noted that the putative hybrid *C. edulis* × *D. australe* has a widespread distribution along coastal locations in Wellington. However, the hybrid between *C. edulis* and *D. australe* (× *Carpophyma mutabilis*) identified by Heenan and Sykes in 2010 has longer leaves and larger flowers in comparison to the putative hybrid that we noted in the present study. The hybrid showed a more vigorous growth than one of its parental counterparts, *D. australe* (red). The hybrid's ability to spread on a variety of micro-habitats and substrates, and also its robust growth, can make it an invader similar to *C. edulis*. These growth differences between iceplant taxa were also noted by Guyo (2012). Arnold and Hodges (1995) suggest that hybrids with high fitness can lead to more abundance than their parental populations.

The morphological characteristics and pigmentation patterns showed more variations among populations of the hybrid and *D. australe* (red) compared to other species studied. In some populations, the level of pigmentation fluctuated among individuals within the same colony. Such variations may have been caused by different environmental conditions that the plants are exposed to in their respective micro-habitats. Previous studies observed similar morphological and pigmentation variations in *Carpobrotus* species and their hybrid populations (Schierenbeck et al. 2005). Stintzing et al. (2005) reported a broad range of betalain levels in different *Opuntia* colonies, confirming high variability among populations. However, due to less variability observed in *C. chilensis*, it has been considered as a species indicative of a genetic bottleneck (Gallagher et al. 1997). In the present study, we observed less morphological and pigmentation variability among the few colonies of *C. chilensis* along the coast of Wellington.

In the current study, there were higher levels of betalains in leaf epidermal cells of *C. chilensis*, while *C. edulis* lacked betalains in epidermal cells as the pigmentation was mostly confined to leaf margins. Jain and Gould (2015) suggested that the presence of foliar betalains may ameliorate responses to salinity stress in *D. australe*. The present study observed the presence of betalains in the immediate outer cell layers that encapsulate the vascular bundle in the red morphotype of *D. australe*. In spite of exposing the green morphotype to salinity stress, plants did not accumulate betalains in a study carried out by Jain and Gould (2015). Lee and Collins (2001) revealed the presence of betalains in spongy mesophyll tissue in species belonging to the family Amaranthaceae. Vogt et al. (1999) observed an accumulation of betacyanins in epidermal bladder cells in mature leaves of *Mesembryanthemum crystallinum* when plants were subjected to light. Similar observations were also reported by Ibdah et al. (2002) confirming spectral dependence of betacyanin accumulation in plants. In contrast, Wang et al. (2007) observed an increase of betacyanins in *Suaeda salsa* when in the dark, while light promoted the degradation of betacyanins. Field observations from the present study showed that the leaves and stems of *Carpobrotus* species possess more pigmentation on the upper surfaces that are exposed to the direct sunlight compared to the surfaces facing the substrate. However, this was not evident in either *D. australe* (red) or the putative hybrid, where pigmentation showed more or less an even distribution. Therefore, the present observations support the idea of light-induced betalain production, at least in *Carpobrotus* species. In addition to carotenoids, betalains also have a photoprotective function in plants (Mabry 1980). In terrestrial plants, the bulk of the photoprotective pigments are localized in their superficial structures, including the cuticle, epidermis, and their derivative structures such as trichomes and hairs (Steyn et al. 2002; Caldwell et al. 2003; Page and Towers 2002). The present results also indicate an accumulation of betalains in epidermal cells and cells immediately beneath this layer, though their presence in the cells surrounding the vascular bundle is not clear.

The present study noted a higher and more widespread betalain production in mature leaves of iceplants compared to younger leaves, and this was more marked in the two *Carpobrotus* species. On the contrary, Ibdah et al. (2002) noted that there is no production of betacyanins during senescence in *Mesembryanthemum crystallinium*. Similar results were also observed by Lee and Collins (2001). However, Hoch et al. (2003) provided experimental evidence to support the theory that photoprotective pigments, including anthocyanins, facilitate the recovery of foliar nutrients by protecting senescing leaves from excess light. With its photoprotective role, the enhanced betalain production during senescence in iceplants seems to have a similar function to anthocyanins. The different distributions of betalains in cells and tissues indicate that the pigments may have different functions including photoprotection and free radical scavenging.

The betalain concentrations also revealed variations between iceplant species and their hybrid. Among the iceplant species studied, total betalain contents ranged from 1.75 to 14.0 μg/g fresh plant material. The betacyanin levels, which contributed more to the total betalains in all iceplants excluding *D. australe* (green), were comparable to the field red-magenta coloration levels among species. A similar correlation between field coloration and betacyanin concentration was also reported in the cactus pear (*Opuntia* sp.) in a study conducted by Stintzing et al. (2005). Kugler et al. (2007) suggest that the relative betaxanthin:betacyanin ratios determine the broad color palette in *Gomphrena* and *Bougainvillea* flowers. However, when compared to other species, such as ulluco (*Ullucus tuberosus*), swiss chard (*Beta vulgaris* ssp. *cicla*), and red beetroot (*B. vulgaris*), the betalain concentrations in iceplants are relatively low (Svenson et al. 2008; Georgiev et al. 2010). The HPLC data suggested that the retention times and spectral data are consistent with data reported previously in *Carpobrotus* species (Piattelli and Impellizzeri 1970). In contrast to spectrophotometric analysis, the HPLC analysis could not detect betaxanthins in the studied iceplants. In a previous study, betaxanthins have been detected in *C. edulis* flowers (Herrero et al. 2005). However, it is also known that betaxanthins can form artificially during handling (such as during cell extraction, storage, etc.) through spontaneous reactions between betalamic acid and amino acids (Terradas and Wyler 1991; Hempel and Böhm 1997; Schliemann et al. 1999). Though *C. edulis* and *C. chilensis* are congeneric, the HPLC data indicated that the species had different betacyanin types. *C. edulis* had betacyanins such as lampranthin II and iso-lampranthin II, while *C. chilensis* had betanin and isobetanin. The putative hybrid *C. edulis* × *D. australe* showed more similar betacyanin types to *C. edulis* than that of *D. australe*, in spite of its closer morphological resemblance to *D. australe* (red).

In conclusion, the selected iceplant species from the coast of Wellington are relatively easy to distinguish due to their distinct morphological differences as well as their preferred substrates. This study also supports the idea that the commonly found putative hybrid may have arisen from cross-pollination between *C. edulis* and *D. australe*. The presence of the red- and green-colored morphotypes of *D. australe* was also confirmed during the study. Types of betacyanins, their relative proportions, and tissue distribution showed variations between the selected iceplants, suggesting different roles of betalains in iceplant species.

Acknowledgements I would like to acknowledge the numerous contributions made by Prof. Kevin Gould and Dr. Stephen Hartley of the Department of Biological Sciences, Victoria University of Wellington, New Zealand during the conduct of this study.

References

Akeroyd JR, Preston CD (1990) Notes on some Aizoaceae naturalized in Europe. Bot J Linn Soc 103:197–200

Akeroyd JR, Preston CD (1993) Carpobrotus N.E. Br. In: Tutin TG, Burges NA, Chater AO, Edmonson JR, Heywood VH, Moore DM, Valentine DH, Walters SM, Webb DA (eds) Flora Europaea Volume 1: Psilotaceae to Platanaceae. Cambridge University Press, Cambridge, p 135

Albert ME (1995) Morphological variation and habitat associations within the *Carpobrotus* species complex in coastal California. Master's thesis, University of California at Berkeley, Berkeley, CA

Albert ME, D'Antonio CM, Schierenbeck KA (1997) Hybridization and introgression in *Carpobrotus* spp. (Aizoaceae) in California. I. Morphological evidence. Am J Bot 84:896–904

Arnold ML, Hodges SA (1995) Are natural hybrids fit or unfit relative to their parents? Trends Ecol Evol 10:67–71

Atkinson IAE (1972) Vegetation and flora of Sail Rock, Hen and Chicken Islands. N Z J Bot 10:545–558

Azeredo HMC (2009) Betalains: properties, sources, applications, and stability – a review. Int J Food Sci Technol 44:2365–2376

Bicknell SH, Mackey EM (1988) Evidence of early occurrence of Carpobrotus aequilaterus N.E. Br. At Marina beach State Park, CA. Final report No. 05-10-065, California Department of Parks and Recreation, Rancho Santa Ana, California

Cai YZ, Sun M, Corke H (1998) Colourant properties and stability of Amaranthus betacyanin pigments. J Agric Food Chem 46:4491–4495

Cai YZ, Sun M, Corke H (2001) Identification and distribution of simple acylated betacyanins in the Ameranthecеae. J Agric Food Chem 49:1971–1978

Caldwell MM, Ballare CVL, Bomman JF, Flint SD, Bjorn LO, Teramura AH, Kulandaivelu G, Tevini M (2003) Terrestrial ecosystems, increased solar untraviolet radiation and interactions with other climatic change factors. J Photochem Photobiol 2:29–38

Chinnock RJ (1971) Studies in *Disphyma-a* genus related to *Mesernbryanthemum.* I. A revision of *Disphyma australe* (Ait.) *J. M.* Black. N Z J Bot 9:331–344

Chinnock RJ (1972) Natural hybrids between *Disphyma* and *Carpobrotus* (Aizoaceae) in New Zealand. N Z J Bot 10:615–626

D'Antonio CM (1993) Mechanisms controlling invasion of coastal plant communities by the alien succulent *Carpobrotus edulis*. Ecology 74:83–95

D'Antonio CM, Odion DC, Tyler CM (1993) Invasion of maritime chaparral by the introduced succulent *Carpobrotus edulis*. Oecologia 95:14–21

Ferruzzi MG, Böhm V, Courtney PD, Schwartz SJ (2002) Antioxidant and antimutagenic activity of dietary chlorophyll derivatives determined by radical scavenging and bacterial reverse mutagenesis assays. J Food Sci 67:2589–2595

Fraser LG, Harvey CF, Crowhurst RN, De Silva HN (2004) EST derived microsatellites from *Actinidia* species and their potential for mapping. Theor Appl Genet 108:1010–1016

Gallagher KG, Schierenbeck KA, D'Antonio CM (1997) Hybridization and Introgression in *Carpobrotus* spp. (Aizoaceae) in California II. AllozymeEvidence. Am J Bot 87:905–911

Georgiev VG, Weber J, Kneschke EM, Denev PN, Bley T, Pavlov AI (2010) Antioxidant activity and phenolic content of betalain extracts from intact plants and hairy root cultures of the red beetroot *Beta vulgaris* cv. Detroit Dark Red. Plant Foods Hum Nutr 65:105–111

Girod PA, Zryd JP (1991) Secondary metabolism in cultured red beet (*Beta vulgaris* L.) cells: differemial regulatian of betaxanthin and betacyanin biosynthesis. Plant Cell Tissue Organ Cult 25:l–12

Guyo DG (2012) Functional ecology of iceplant taxa and their role in eco-restoration of New Zealand sand dunes. M.Sc. thesis, Victoria University of Wellington, New Zealand

Han XH, Gao ZJ, Xiao XG (2009) Enzymes and genes involved in the betalain biosynthesis in higher plants. Afr J Biotechnol 8:6735–6744

Healy AJ (1959) Contributions to knowledge of the adventive flora of New Zealand. VIII. The "succulent" element of the adventive flora. Trans R Soc N Z 87:229–234

Heenan PB, Sykes WR (2010) Taxonomic notes on the New Zealand flora: × *Carpophyma mutabilis* and *Carpophyma pallida* (Aizoaceae), new names for two wild intergenric hybrids. N Z J Bot 48:225–230

Hempel J, Bo¨hm H (1997) Betaxanthin pattern of hairy roots from *Beta vulgaris* var *lutea* and its alteration by feeding of amino acids. Phytochemistry 44:847–852

Herrero FG, Escribano J, Carmona FG (2005) Betaxanthins as pigments responsible for visible fluorescence in flowers. Planta 222:586–593

Hoch WA, Singsaas EL, McCown BH (2003) Resorption protection. Anthocyanin facilitate nutrient recovery in autumn by shielding leaves from potentially damaging light levels. J Plant Physiol 133:1296–1305

Ibdah M, Krins A, Seidlitz HK, Heller W, Strack D, Vogt T (2002) Spectral dependence of flavanol and betacyanin accumulation in *Mesembryanthemum crystallinium* under enhanced ultraviolet radiation. J Plant Cell Environ 25:1145–1154

Jain G, Gould KS (2015) Functional significance of betalain biosynthesis in leaves of *Disphyma australe* under salinity stress. Environ Exp Bot 109:131–140

Jakobsson A, Padron B, Traveset A (2008) Pollen transfer from invasive *Carpobrotus* spp. To natives – a study of pollinator behavior and reproduction success. Biol Conserv 141:136–145

Kugler F, Stintzing FC, Carle R (2004) Identification of betalains from petioles of differently coloured swiss chard (*Beta vulgaris* L. ssp. Cicla [L.] Alef. Bright Lights) by high-performance liquid chromatography-electrospray ionization mass spectrometry. J Agric Food Chem 52:2975–2981

Kugler F, Stintzing F, Carle R (2007) Characterisation of betalain patterns of differently coloured inflorescences from *Gomphrena globosa* L. and *Bougainvillea* sp. by HPLC–DAD–ESI–MS. Anal Bioanal Chem 387:637–648

Lee DW, Collins TM (2001) Phylogenetic and ontogenetic influences on the distribution of anthocyanins and betacyanins in leaves of tropical plants. Int J Plant Sci 162:1141–1153

Mabry TJ (1980) Betalains. In: Bell EA, Harlwood BV (eds) Encyclopedia of plant physiology, New series, 8. Springer, Berlin, pp 513–533. Secondary products

Moreno DA, García-Viguera GJI, Gil-Izquierdo A (2008) Betalains in the era of global agrifood science, technology and nutritional health. J Phytochem Rev 7:261–280

Munz P, Keck D (1968) A California flora with supplement. University of California Press, Berkeley

Page JE, Towers GH (2002) Anthocyanins protect light-sensitive thiarubrine phototoxins. Planta 215:478–484

Piattelli M (1981) The betalains: structure, biosynthesis, and chemical taxonomy. In: Conn EE (ed) The biochemistry of plants. Academic, New York, pp 557–575

Piattelli M, Impellizzeri G (1970) 2-descarboxybetanidin, a minor betacyanin from *Carpobrotus acinaciformis*. Phytochemistry 9:2553–2556

Schierenbeck KA, Symonds VV, Gallagher KG, Bell J (2005) Genetial variation and phylogeographic analyses of two species of *Carpobrotus* and their hybrids in California. J Mol Ecol 14:539–547

Schliemann W, Kobayashi N, Strack D (1999) The decisive step in betaxanthin biosynthesis is a spontaneous reaction. Plant Physiol 119:1217–1232

Stafford HA (1994) Anthocyanins and betalains: evolution of the mutually exclusive pathways. Plant Sci 101:91–98

Steyn WJ, Wand SJE, Holcroft DM, Jacobs G (2002) Anthocyanins in vegetative tissues: a proposed unified function in photoprotection. New Phytol 155:349–361

Stintzing FC, Herbach KM, Mosshammer MR, Carle R, Yi WG, Sellappan S, Akoh CC, Bunch R, Felker P (2005) Colour, betalain pattern and antioxidant properties of cactus pear (*Opuntia* spp.) clones. J Agric Food Chem 53:442–451

Strack D, Bokern M, Marxen N, Wray V (1988) Feruloylbetanin from petals of *Lampranthus* and feruloylamaranthin from cell suspension cultures of *Chenopodium rubrum*. Phytochemistry 27:3529–3531

Suehs CM, Affre L, Medail F (2004) Invasion dynamics of two alien Carpobrotus (Aizoaceae) taxa on a Mediterranean island: 1. Genetic diversity and introgression. Heredity 92:31–40

Svenson J, Smallfield BM, Joyce NI, Sansom CE, Perry NB (2008) Betalains in red and yellow varieties of the andean tuber crop ulluco (*Ullucus tuberosus*). J Agric Food Chem 56:7730–7737

Tanaka Y, Sasaki N, Ohmiya A (2008) Biosynthesis of plant pigments: anthocyanins, betalains and carotenoids. J Plants 54:733–749

Terradas F, Wyler H (1991) 2,3- and 4,5-Secodopas, the biosynthetic intermediates generated from L-dopa by an enzyme system extracted from the fly agaric, *Amanita muscaria* L., and their spontaneous conversion to muscaflavin and betalamie acid, respectively, and betalains. Helv Chim Acta 74:124–140

Vilá M, Weber E, D'Antonio CM (1998) Flowering and mating system in hybridizing *Carpobrotus* (Aizoaceae) in coastal California. Can J Bot 76:1165–1169

Vilá M, Weber E, D'Antonio CM (2000) Conservation implications of invasion by plant hybridization. Biol Invasions 2:207–217

Vivrette N (1993) *Carpobrotus* (Aizoaceae). In: Hickman J (ed) The Jepson manual: higher plants of California. University of California Press, Berkeley

Vogt T, Ibdah M, Schmidt J, Wray V, Nimtz M, Strak D (1999) Light-induced betacyanin and flavonol accumulation in bladder cells of *Mesembryanthemum crystallinum*. Phytochemistry 52:583–592

Wang CQ, Zhao JQ, Chen M, Wang BS (2007) Involvement of hydrogen peroxide in betacyanin accumulation induced by dark in leaves of *Suaeda salsa*. J Plant Ecol 31:748–752

Waterman PG (2007) The current status of chemical systematics. J Phytochem 68:2896–2903

Weber E, D'Antonio CM (1997) Phenotypic plasticity in hybridizing *Carpobrotus* spp. (Aizoaceae). Am J Bot 84:93

14 Salt Tolerance and Potential Uses for Saline Agriculture of Halophytes from the Poaceae

Abdallah Atia, Ahmed Debez, Mokded Rabhi, Zouhaier Barhoumi, Chiraz Chaffei Haouari, Houda Gouia, Chedly Abdelly, and Abderrazak Smaoui

Abstract

Halophytes are mainly dicotyledonous plants, whereas most of the economically important crops are monocotyledonous and belong especially to the Poaceae family. Therefore, investigating these species is of major significance. Indeed, halophytes from the Poaceae have considerable forage value and economic uses, especially in saline and arid areas, and interestingly salt tolerance strategy of these species involves a wide range of adaptive mechanisms. These include high capacity of ion compartmentation, production of compatible osmolytes, high K^+ retention, and toxic ion secretion via salt glands, which leads to maintaining ion homeostasis. Understanding the salt tolerance mechanisms of the Poaceae halophytes would help in improving the salt tolerance of cultivated salt-sensitive species. Furthermore, the economic potential of these species is poorly reviewed. This review addresses mainly salt tolerance and the uses of these species for economic and ecological benefit. We also show that using these species to establish productive ecosystems and re-greening natural saline soils could be a useful approach to valorize and reclaim salt-affected areas and overcome water shortage issue.

Keywords

Halophytes · Poaceae · K^+ retention · Osmotic adjustment · Ions homeostasis · Sustainable development · Water shortage · Saline soil remediation

A. Atia
Research Unit, Nutrition and Nitrogen Metabolism and Stress Protein, Department of Biology, Faculty of Sciences of Tunis, Campus Universitaire El Manar I, Tunis, Tunisia

Biology Department, King Khalid University, Abha, Saudi Arabia

A. Debez (✉) · M. Rabhi · Z. Barhoumi · C. Abdelly · A. Smaoui
Laboratory of Extremophile Plants, Centre of Biotechnology of Borj-Cedria, Hammam-Lif, Tunisia

C. C. Haouari · H. Gouia
Research Unit, Nutrition and Nitrogen Metabolism and Stress Protein, Department of Biology, Faculty of Sciences of Tunis, Campus Universitaire El Manar I, Tunis, Tunisia

Introduction

Soil salinity is a major environmental constraint and poses a serious threat to crop production. Salinity affects ~45 Mha of irrigated land and costs agriculture an estimated US$ 27.3 billion per annum in lost revenue (Shabala et al. 2015). Most crops, including graminaceous crops, are very sensitive to ionic and osmotic stress. However, halophytes have evolved various mechanisms to overcome salt stress through long-term natural selection. Most halophytes are

B. Gul et al. (eds.), *Sabkha Ecosystems*, Tasks for Vegetation Science VI,
https://doi.org/10.1007/978-3-030-04417-6_14

dicotyledonous, whereas most of the economically important crops are monocotyledonous. Understanding the salt tolerance mechanism(s) of monocotyledonous halophytes will aid in improving the salt tolerance of cereals. Many halophytes species have developed several physiological strategies to tolerate extreme conditions in challenging environments. They have evolved various mechanisms that have enabled them to survive and reproduce in these environments (Flowers and Colmer 2008).

The adaptations by which these species respond to salinity are intensively studied. They include high germination capacity under salinity and osmotic stress and high germination recovery (Atia et al. 2011), selective uptake and transport of ions, ion compartmentation (Shabala 2013), production of compatible osmolytes (Slama et al. (2015), succulence, and salt secretion (Barhoumi et al. 2007). Salt tolerance in halophytes from the Poaceae is a complex trait that is under genetic control and involves various biochemical and physiological mechanisms. These species must release their osmotic adjustment to maintain water uptake and must maintain ion homeostasis, i.e., limit K^+ loss and avoid Na^+ toxicity. Hence, they accumulate and compartmentalize Na^+ and Cl^- ions in the vacuole and/or exclude them in the apoplast. Several halophyte genes associated with cation/proton antiporters on the plasma membrane (SOS1) and vacuolar membrane (NHX) and other genes coding vacuolar H^+-ATPases, vacuolar H^+-pyrophosphatase, high-affinity K^+ transporter (HAK), and high-affinity K^+ transporter (HKT) were implicated in ion homeostasis. The synthesis of osmolytes and the induction of antioxidant enzymes were also implicated in salinity tolerance of these species.

The halophytes from Poaceae are of considerable forage value as it is grazed by many herbivores, especially on the marginal area where the problem of water shortage is mentioned (Qasim et al. 2010 and Al-Shorepy et al. 2010). Thus, they are mainly used as forage species. However, other uses of these species, including the biomolecules production, the fiber production, and soil management, were also reported in literature. In this chapter we briefly review salt tolerance and the potentiality of these species in biosaline agriculture.

Salinity Tolerance

Salinity Effects on Growth, Photosynthesis, and Antioxidant Systems

Growth of halophytes is commonly stimulated by low salinity (Khan et al. 2000). In the most halophytes of Poaceae, growth is not affected at low salinities and decreases at higher salinities, and there is persistent growth under moderate salinities; however the range of salinity resistance is narrow. Glenn et al. (1999) assumed that although the growth of halophytes of Poaceae was not stimulated by NaCl, the more tolerant Poaceae species grew as well as the more tolerant dicots on higher salt solutions. This is the case of *Phragmites australis* and *Odyssea paucinervis*, which show their optimal growth in salt-free medium (Table 14.1). However, in some Poaceae halophytes, growth activity may be maximal under moderate salinity (Flowers and Colmer 2008). For instance, Al-Khateeb (2006) showed that growth stimulation in early seedlings of *Panicum antidotale* under low salinity (<100 mM NaCl). Koyro et al. (2013) reported that at 125 mM NaCl, biomass production of the same species was similar to the nonsaline control, and only the number of juvenile leaves was increased at this concentration (Table 14.1). Abideen et al. (2014) observed that in *Phragmites karka*, salinity stimulates root and shoot growth at 100 mM NaCl. However, in plants cultivated under 300 mM NaCl, the growth was significantly decreased in comparison to those grown at 100 mM NaCl and to the control. Recently, Sanadhya et al. (2015) reported that in the C4 halophyte *Aeluropus lagopoides*, the maximal growth occurred in the 100–300 mM NaCl range (Table 14.1).

Biomass production under saline conditions depends mainly on the ability to maintain high net photosynthesis at minimal water loss and energy consumption (Naidoo et al. 2012).

Table 14.1 Photosynthetic pathway (C3/C4), optimal growth conditions, osmolytes accumulated, and K^+ retention in some halophytes of Poaceae

Species	Photosynthetic pathway (C3/C4)	Optimal growth conditions	Osmolytes accumulated	K^+ retention	References
Spartina spp.	–	–	Dimethylsulfoniopropionate (DMSP)	–	Slama et al. (2015)
Spartina alterniflora	C4	H_2O	Proline and soluble sugars	+++	Vasquez et al. (2006)
Phragmites australis	C3–C4	H_2O	Sucrose and proline	+	Pagter et al. (2009)
Panicum antidotale	C 4	125 mM	Proline, soluble sugars	+++	Koyro et al. (2013)
Phragmites karka	C3	100 mM	Proline, soluble sugars	+++	Abideen et al. (2014)
Aeluropus lagopoides	C4	100–300 mM	Glycine betaine, proline, soluble sugars, and amino acids	+++	Sanadhya et al. (2015) and Moinuddin et al. (2014)
Odyssea paucinervis	C4	H_2O	Proline	++	Naidoo et al. (2008)
Thinopyrum ponticum	C3	100 mM	Glycine betaine, proline	+	Bhuiyan et al. (2015)
Puccinellia tenuiflora	C3	H_2O	Proline, soluble sugars, glycine betaine	+++	Yu et al. (2011)

Excessive NaCl concentrations leading to the production of ROS, in chloroplast, could be inhibitory to photosynthesis (Debez et al. 2008; Koyro et al. 2013). Thus, salt tolerance in plants is related to a plant's aptitude to maintain adequate net photosynthetic rate, high chlorophyll content, and stomatal conductance under saline conditions (Redondo-Gomez et al. 2006; Debez et al. 2008). Gas exchange rate and growth vary with plant species and abiotic factors such as salinity and drought (Geissler et al. 2009). Salt-induced photosynthesis reduction can be attributed to a limitation in stomatal conductance (*gs*) and/or a non-stomatal limitation, i.e., caused by damage in the photosynthetic apparatus (Rabhi et al. 2012). In the halophytes of Poaceae, the limitations of photosynthesis activity correlated to the observed restriction of plant growth (Koyro et al. 2013). The same authors also documented that the growth of *P. turgidum* was correlated to chlorophyll content, net photosynthetic rate, transpiration, and water-use efficiency. In this species, relative chlorophyll content was highest at moderate salinity (125 mM NaCl) and decreased significantly at high salinity (500 mM NaCl). There was also a transient increase in net photosynthetic rate (*A*) at 125 mM NaCl and a decline at higher salinity. Furthermore, the increase of salinity led to a transient increase of stomata number, their size, and maximum opening area which mean an adjustment to a particular demand of CO_2 in *P. turgidum*. However, the internal carbon dioxide concentration (Ci) was not influenced by increase in salinity (Koyro et al. 2013). This means that, in this species, high salinity that induced photosynthesis reduction can be attributed to a non-stomatal limitation (Rabhi et al. 2012). On the other hand, the increase in salinity was accompanied by an increase in H_2O_2 production and a decrease in Rubisco activity (Koyro et al. 2013). Therefore, high salinity can lead to structural alteration of both membranes and photosynthetic apparatus due to the production of reactive oxygen species (ROS) such as O_2^-, OH^-, or H_2O_2. It's well known that the latter molecule is toxic; however, it plays a very important role in signaling (Hussain et al. 2015) and is an important intermediate step in ROS defense in the halophytes of Poaceae, namely, at high salinity. Hence, in these species the production of H_2O_2

under high salinity (~ 400–500 mM) generates the activation of the enzymatic, i.e., catalase (CAT), superoxide dismutase (SOD), ascorbate peroxidase (APX), and glutathione reductase (GR), and/or nonenzymatic, i.e., ascorbate, glutathione, tocopherol, carotenoids, etc., antioxidant system (Muscolo et al. 2013 and Hussain et al. 2015). This was observed in some halophytes of Poaceae such as *Echinochloa crus-galli* (Abogadallah et al. 2010), *Puccinellia tenuiflora* (Yu et al. 2011), *Spartina alterniflora* (Subudhi and Baisakh 2011), *Pennisetum clandestinum* (Muscolo et al. 2013), and *Panicum antidotale* (Hussain et al. 2015). For instance, in the latter species, the enhanced SOD activity, at high salinity, corresponds well with the about twofold increase in the concentration of H_2O_2 (Hussain et al. 2015). Muscolo et al. (2013) observed that in *Pennisetum clandestinum* CAT activity increased significantly with an increase in salinity. The same observation was also recorded in *Panicum antidotale* (Hussain et al. 2015). Both APx and GR enzymes play a critical role in the removal of indigenous H_2O_2 (Duarte et al. 2013). In *Panicum antidotale*, the activities of the two latter enzymes were increased up to four- and fivefold, respectively, under salt treatment (Hussain et al. 2015). The increased GR activity in plants under saline conditions lead to maintain a high $NADP^+$/NADPH ratio, ensuring the ability of $NADP^+$ to accept electrons from photosynthetic electron transport chain and minimizing harmful effect of ROS on chloroplasts (Muscolo et al. 2013). APX is one of the enzymes capable of detoxifying ROS in the cell and utilizing the reducing power of ascorbic acid to eliminate the detrimental effect of H_2O_2. Thus, we can conclude that the halophytes of Poaceae have the capacity to eliminate the toxic effect of salt even at high salinity.

In conclusion, although the growth of the most halophytes of Poaceae was not stimulated by salinity, their performance under salinity is well beyond the capability of any known conventional crop species, making these species ideal for "saline agriculture" to the benefit of population in arid and semiarid regions around the world.

Salinity Effects on Osmotic Adjustment

All halophytes must meet the challenge of osmotic adjustment to a low external water potential (Flowers and Colmer 2008). For example, among 32 species of the Chenopodiaceae, inorganic ions, i.e., Na^+ and Cl^-, contributed 67% of the solute concentration (molar in shoot water), whereas in the Poaceae, the contribution of the same ions averaged only 32% of the solutes in 17 species; sugars were 19% in the Poaceae, but just 1% in the Chenopodiaceae (Flowers & Colmer, 2008). In fact, the halophytes of Poaceae accumulate carbon (C)-rich organic osmolytes, including sugars and sugar alcohols, and/or nitrogen (N)-rich organic osmolytes, including proline, glycinebetaine, and trigonelline (Moinuddin et al. 2014). Furthermore, the relative use of glycinebetaine and proline for osmotic adjustment varies between species, but generally the Poaceae accumulate proline and sugars by contrast to Chenopodiaceae which accumulate glycinebetaine in the most cases. For example, with *Halimione portulacoides* (Chenopodiaceae), the ratio of glycinebetaine to proline in leaves was ~85:1, whereas with *Puccinellia maritima* (Poaceae), this ratio was ~1:48 (reviewed by Norman et al. 2013). On the other hand, in the Poaceae, osmoticum used for osmotic adjustment was species dependent (Table 14.1). For instance, dimethylsulfoniopropionate (DMSP) is synthesized in many algae and in a few *Spartina* species, in sugar canes, and in *Wollastonia biflora* (Otte et al. 2004). In *Spartina alterniflora*, DMSP is produced in high concentration in leaves independently of the external salinity. However, Otte et al. (2004) suggest that this osmoticum could be involved in osmotic adjustment, without changes in its concentrations at the tissue level, by movement between the cytoplasm and the vacuoles within the cells, depending on the osmotic potential of the cytoplasm. In *Panicum antidotale* carbohydrates seem to play an important role in osmotic adjustment under salinity. In fact, in this species, the concentration of total soluble sugars was increased up to fourfold in leaves in plant cultivated under NaCl treatment (Hussain et al. 2012). Osmotic adjustment in many halophytes of Poaceae is achieved by accumulating both inorganic and organic solutes. For instance, Abideen et al. (2014) recorded that in *Phragmites karka*, Na^+ and K^+ collectively contributed to about 30% of the osmotic adjustment. However, proline and sugars play little role in osmotic adjustment (≤15%). They could be involved in osmo-protection. The salt-induced decrease in leaf water potential by the accumulation of Na^+ was also recorded in *Sporobolus virginicus* (Bell and O'Leary 2003), *Odyssea paucinervis* (Naidoo et al. 2008), *Panicum antidotale* (Hussain et al. 2012), and *Aeluropus lagopoides* (Ahmed et al. 2013). This may imply that some Poaceae prefers inexpensive means for osmotic adjustment as was observed in many Chenopodiaceae and Aizoaceae.

Salinity Effects on Ions Homeostasis

In saline environments, high salt concentration in the external medium favors Na^+ entry down the electrochemical gradient across the plasma membrane (Ardie et al. 2009). In the cytoplasm, Na^+ toxicity is thought not to differ between halophytes and glycophytes (*Flowers* and *Colmer,* 2008). Na^+ accumulation in the cytoplasm disturbs intracellular ion homeostasis and causes cell membrane dysfunction, metabolism perturbation, K^+ efflux, and ROS accumulation up to toxic levels, which ultimately leads to growth reduction and/or inhibition and ultimately to plant death. Hence, the difference in plants tolerance to salt is related to their capacity to exclude the toxic ions, like Na^+ and Cl^-, away from sensitive metabolic pathways and the retention of K^+ (Shabala et al. 2015). The halophytes of Poaceae have the capacity to maintain high K^+/Na^+ ratio (Table 14.1).

Complete exclusion of Na^+ and vacuolar compartmentalization of internal Na^+ are some of the mechanisms characteristic of salt-tolerant plants (Shabala 2013). For instance, when compared to wheat, the halophyte *Puccinellia tenuiflora* exhibits stronger selective absorption and transport for K^+ over Na^+ (Wang et al. 2009a, b). *P. tenuiflora* shows also more restricted unidirectional Na^+ influx that may contribute to its salinity tolerance (Wang et al. 2009a, b). By using X-ray microanalysis, Peng et al. (2004) recorded that in this species, the selective transport of K^+ over Na^+ occurs at the endodermis where the Casparian bands exist, which results in high K^+ accumulation in the shoot and large Na^+ retention in the root (Table 14.2).

Plasma membrane Na^+/H^+ antiporters, also known as SOS1, or NHA-type transporters, are responsible for extruding Na^+ out of the salt-stressed cell, while vacuolar membrane Na^+/H^+ antiporters, also known as NHX-type antiporters, are responsible for the sequestration of Na^+ from the cytosol into vacuoles under salt stress (Munns and Tester 2008; Ardie et al. 2009; Hasegawa 2013). The use of molecular tools and the transformation of some glycophytes give new insights about the implication SOS1-type transporters and NHX-type antiporters in salinity tolerance in the halophytes that belong to Poaceae. Thus, here, we discuss the transgenic expression of halophytic genes, from some halophytes of Poaceae, associated with salt tolerance mechanism in some glycophytes, for instance, increasing salt and drought stress tolerance in transformed plant (Table 14.3). The transformation of *Nicotiana tabacum* by HbNHX1 gene, coding a vacuolar Na^+/H^+ antiporter, from *Leptochloa fusca*, enhanced germination and root growth in transformed plant (Rauf et al. 2014) (Table 14.3). Zhang et al. (2008) recorded that the transformation of *Nicotiana tabacum* by vacuolar Na^+/H^+ antiporter (*NHX1*) from *Aeluropus littoralis* ameliorates salt tolerance by increasing Na^+ retention in roots and by increasing K^+/Na^+ ratio in leaves (Table 14.3). In *P. tenuiflora* the expression of PtNHA1 gene in the roots that encodes a plasma membrane Na^+/H^+ antiporter, also known as SOS1 transporters, plays an important role in Na^+ extrusion from the roots (Wang et al. 2011) and K^+ retention. In fact, the overexpression of this gene in *Arabidopsis thaliana* improved the salt tolerance of the transformed plant, together with less Na^+ content and more K^+ retention and enhanced ROS-scavenging enzyme activity. This

Table 14.2 Some example of salt resistance genes identified by functional analysis in the halophytes of Poaceae

Salt-resistance genes	Function	Species sources	Species transformed	Physiological response of transformed plant under salinity	References
PtNHA1	Plasma membrane Na(+)/H(+) antiporter	*Puccinellia tenuiflora*	*Arabidopsis thaliana*	Less Na (+) and more K(+) were accumulated	Wang et al. (2011)
				Overexpression of PtNHA1 results in enhanced ROS-scavenging enzymes activity	
HbNHX1	Vacuolar Na^+/H^+ antiporter (*NHX1*)	*Hordeum brevisubulatum*		Transgenic plants showed more tolerance to salt and drought stress	Lu et al. (2005)
LfNHX1	Vacuolar Na^+/H^+ antiporter (*NHX1*)	*Leptochloa fusca*	Tobacco	The overexpression of LfNHX1 gene enhanced germination and root growth	Rauf et al. (2014)
PutAKT1	AKT-1-type-K^+ channel	*Puccinellia tenuiflora*	*Arabidopsis thaliana*	Increased the K(+) content under normal, K(+)-starvation, and NaCl stress conditions and decrease in Na(+) accumulation	Ardie et al. (2010)
SaVHAc1	Vacuolar ATPase subunit	*Spartina alterniflora*	*Oryza sativa*	Enhanced salt tolerance by increasing the accumulation of messages of native genes involved in cation transport and ABA signaling	Baisakh et al. (2012)
				Increased K^+/Na^+ ratio	
				Reduced of stomata and increased water content	
SaSce9	SUMO conjugating enzyme	*Spartina alterniflora*	*Arabidopsis thaliana*	Improved salinity and drought tolerance of *Arabidopsis*	Karan and Subudhi (2012)
				Increasing chlorophyll and proline content	
				Decreased the levels of reactive oxygen species	
				Increased the expression of antioxidant genes, AtSOD and AtCAT, ion antiporter genes, AtNHX1 and AtSOS1, a gene involved in proline biosynthesis, AtP5CS, and a gene involved in ABA-dependent signaling pathway, AtRD22	
AlNHX1	Vacuolar Na^+/K^+ antiporters	*Aeluropus littoralis*	*Nicotiana tabacum*	Increase Na^+ retention in roots	Zhang et al. (2008)
				Increased K^+/Na^+ ratio in leaves	

might be due to efficient role of PtNHA1 in ions homoeostasis under NaCl salinity. On the other hand, the Na^+ sequestration in the vacuole by tonoplast Na^+/H^+ antiporters is energized by both vacuolar H^+ pumps: ATPase and pyrophosphatase (Bonales-Alatorre et al. 2013). Therefore, it is conceivable that an increase of proton gradient by the overexpression of the vacuolar H^+-ATPase and/or H^+-pyrophosphatase should enhance Na^+ sequestration through the NHX exchangers (Zhang and Shi 2013). For instance, the transformation of *Oryza sativa* by SaVHAc1 gene from *Spartina alterniflora* encoding vacuolar ATPase subunit enhanced salt tolerance, upregulated genes involved in cation transport and ABA signaling, increased K^+/Na^+ ratio, and increased water content (Baisakh et al. 2012) (Table 14.3).

High-affinity potassium transporters (HKT) play an important role in K^+ uptake by plant. HKT genes have been isolated from several halo-

Table 14.3 Halophytes of Poaceae for economic uses

Species	Uses	References
Polypogon monspeliensis	Forage and ornamental	Atia et al. (2011)
Diplachne fusca	Forage	Khan (2009)
Brachypodium distachyum	Forage	Barhoumi et al. 2010
Sporobolus virginicus	Forage and metals phytoremediation	Al-Shorepy et al. (2010) and Eid and Eisa (2010)
Halopyrum mucronatum	Oils production, ethanol production	Weber et al. (2007) and Abideen et al. (2011)
Panicum turgidum	Forage, ethanol production	Khan et al. (2009) and Abideen et al. (2011)
Spartina alterniflora	Metals phytoremediation and biomass and bioenergy production, soil fixation, and landscaping	Li et al. (2009), Meudec et al. (2007), and Manousaki et al. (2008)
Phragmites australis	Metals phytoremediation and fiber	Weis and Weis (2004) and Kefu et al., (2002)
Paspalum vaginatum	Ornamental	Choukr-Allah et al. (1996)
Thinopyrum ponticum	Soil fixation and landscaping	El Shaer (2006)
Puccinellia ciliate	Soil fixation and landscaping	El Shaer (2006)
Imperata cylindrica	Soil fixation and landscaping	Chaib and Boukhris (1998)
Sporobolus madraspatanus	Forage	Joshi et al. (2005)
Hordeum maritimum	Forage	Hafsi et al. (2010)
Odyssea paucinervis	Forage	Naidoo et al. (2008)
Aeluropus littoralis	Forage	Abdelly et al. (2006)
Spartina anglica	Metals phytoremediation	Williams et al. (1994)
Spartina densiflora	Metals phytoremediation	Luque et al. (1999)
Spartina patens	Metals phytoremediation, soil fixation, and landscaping	Eid and Eisa (2010)
Panicum antidotale	Forage	Ahmed et al. (2010)
Sporobolus airoides	Forage	Weber and Hanks (2008)
Sporobolus helvolus	Forage	Qasim et al. (2010)
Urochondra setulosa	Forage, ethanol production	Qasim et al. (2010), Abideen et al. (2011)
Puccinella scleroides	Metals phytoremediation	Toderich et al. (2010)
Leptochloa uninervia	Forage	Weber and Hanks (2008)
Leptochloa fusca	Soil desalination and biomass and bioenergy	Qadir et al. (1996); Lal (2008)
Agropyron sibiricum	Forage	Shamsutdinov and Shamsutdinov (2008)
Cymbopogon jwarancusa	Medicine	Qasim et al. (2010)
Leymus racemosus	Soil fixation and landscaping	Shamsutdinov and Shamsutdinov (2008)
Leymus chinensis	Forage and soil fixation and landscaping	Sun et al. (2008)

(continued)

Table 14.3 (continued)

Species	Uses	References
Catapodium rigidum	Forage	Barhoumi et al. (2010)
Phragmites karka	Forage, biofuel, and fiber production	Qasim et al. (2010), Abideen et al. (2011)
Puccinellia tenuiflora	Forage and soil fixation and landscaping	Wang et al. (2009a, b) and Koyro and Huchzermeyer (2004)
Spartina townsendii	Metals phytoremediation, soil fixation, and landscaping	
Lygeum spartum	Forage	Bouzid (2008)
Agropyron desertorum	Forage	Toderich et al. (2008)
Distichlis spicata	Forage	Al-Shorepy et al. (2010)
Desmostachya bipinnata	Medicine, ethanol production	Ahmad et al. (2009), Abideen et al. (2011)
Heteropogon contortus	Medicine	Katewa et al. (2001)
Eleusine indica	Medicine and ethanol production	Katewa et al. (2001)
Cenchrus ciliaris	Medicine and ethanol production	Katewa et al. (2001) and Abideen et al. (2011)
Aeluropus lagopoides	Forage and ethanol production	Abideen et al. (2011)
Dactyloctenium scindicum	Medicine	Katewa et al. (2001)
Cynodon dactylon	Medicine, ethanol production, and forage	Chaib and Boukhris (1998), Weber and Hanks (2008), and Abideen et al. (2011)
Vetiveria zizanioides	Medicine and metals phytoremediation	Ahmad et al. (2009); Xia (2004)
Setaria italica	Medicine	
Dichanthium annulatum	Forage, ethanol production	Ahmad et al. (2009)
Dactylis glomerata	Forage and metals phytoremediation	Chaib and Boukhris (1998); Toderich et al. (2008)
Paspalum paspaloides	Ethanol production	Abideen et al. (2011); Toderich et al. (2008)
Lasiurus scindicus	Ethanol production	Abideen et al. (2011)
Sporobolus ioclados	Ethanol production	
Chloris gayana	Forage production	Pasternak et al. (1993)

phytes (Ardie et al. 2009). In *Aeluropus lagopoides* shoots, AlHKT2;1 transcript upregulation by salinity was observed at 12 and 24 h, whereas in roots, maximum induction was observed at 48 h under salinity and K^+ starvation (Sanadhya et al. 2015). In *P. tenuiflora*, expression of PutHKT2;1 was induced by both high NaCl concentration and K^+-starvation stress in roots but only slightly regulated in shoots. PutHKT2;1 transcript levels were doubled under salinity by the depletion of K^+ (Ardie et al. 2009). This may imply that this protein facilitates K^+ uptake for maintaining a constant K^+/Na^+ ratio, even under low K^+ concentration and high Na^+ concentration (Ardie et al. 2009).

In *Puccinellia tenuiflora*, Guo et al. (2012) show that a significant positive correlation was found between root *PtSOS1* expression levels and selective transport capacity for K^+ over Na^+ by roots under NaCl salinity. Authors propose that PtSOS1 play a significant role in the plasma membrane of the xylem parenchyma cells in controlling K^+ transport into and Na^+ retrieval from the xylem.

More than 30 halophytes species that belong to Poaceae, mostly in the tribes Chlorideae, Sporoboleae, and Aeluropodeae, are known to remove salt from the leaf surface by means of bicellular epidermal salt glands (Moinuddin et al. 2014). Salt excretion from leaves is one of the most efficient mechanisms that prevent excessive accumulation of toxic ions in photosynthetic tissues and contribute significantly to maintain ion homeostasis and salinity tolerance (Barhoumi et al. 2007). For instance, in *Sporobolus spicatus* (Vahl) Kunth, an increase in salt secretion during the night resulted in the removal of salts that were accumulated during the day; thus, the plant was ready for the photosynthetic growth under saline environments in the following day (Ramadan 2001). Increasing salt concentration in the substrate increases secretion rates up to an optimal level, and then rates decline. Salinity levels at which maximum secretion is species dependent. Maximum rates are observed between 150 and 200 mM NaCl in moderately tolerant *Chloridoid* species, such as *Cynodon*, *C. gayana*, and *Eleusine*, at 200 mM NaCl in *Distichlis* and *Spartina*, and at 300 mM NaCl in *Sporobolus* species (reviewed by Céccoli et al. 2015). High rates of salt secretion contribute to K^+ retention under severe salinity, which resulted in high salt tolerance salt-excreting species (Naidoo and Naidoo 1998). Recently, Moinuddin et al. (2014) show that salt-excreting Poaceae (*Aeluropus lagopoides* and *Sporobolus tremulus*) had a higher shoot/root Na^+ ratio than non-excreting Poaceae (*Paspalum paspaloides* and *Paspalidium geminatum*). However, salt-excreting species have the capacity to increase their leaf succulence, have decreased life water potential to maintain their capacity to absorb water, and have the capacity to accumulate more compatible osmolyte, i.e., proline and glycinebetaine, with increasing salinity when compared to non-excreting Poaceae.

Halophytes of Poaceae for Economic Uses

Forage Halophytes

One of the most promising uses of halophytes of Poaceae may be as forage. The most known forage Poaceae were *Polypogon monspeliensis* (Atia et al. 2011), *Diplachne fusca* (Khan 2009), *Aeluropus littoralis*, *Brachypodium distachyum*, *Catapodium rigidum* (Barhoumi et al. 2010), *Sporobolus virginicus* (Eid and Eisa 2010), *Panicum turgidum* (Khan et al. 2009), *Sporobolus madraspatanus* (Joshi et al. 2005), *Hordeum maritimum* (Hafsi et al. 2010), *Odyssea paucinervis* (Naidoo et al. 2008), *Panicum antidotale* (Al-Khateeb 2006), *Sporobolus airoides* (Weber and Hanks 2008), *Sporobolus helvolus*, *Urochondra setulosa*, *Phragmites karka* (Qasim et al. 2010), *Leptochloa uninervia* (Weber and Hanks 2008), *Agropyron sibiricum* (Shamsutdinov and Shamsutdinov 2008), *Leymus chinensis* (Sun et al. 2008), *Puccinellia tenuiflora* (Wang et al. 2009a, b), *Lygeum spartum* (Bouzid 2008), *Agropyron desertorum* (Toderich et al. 2010), *Distichlis spicata* (Al-Shorepy et al. 2010), *Aeluropus lagopoides* (Ashraf et al. 2008), *Cynodon dactylon* (Chaib and Boukhris 1998; Weber and Hanks 2008), *Dichanthium annulatum* (Khan and Qaiser 2006), and *Dactylis glomerata* (Chaib and Boukhris 1998) (Table 14.3).

The diversity and the high percentage of halophytic species of Poaceae that can be used as forage constitute an opportunity for populations that survive in arid and semiarid regions. In fact, the domestication and the cultivation of these species may enhance livestock production since they can produce high consumable biomass in saline areas where productivity of non-halophytic species significantly decreased (Koyro and Lieth 2008; El Shaer 2010). For instance, the productivity of some *Sporobolus* species reached 29 t/ha under saline conditions (El Shaer 2010). Furthermore, the annual biomass production is 40 t DM/ha for *Distichlis spicata* when irrigated with saline water (Norman et al. 2013).

The forage products of some halophytes can occasionally contain moderate energy, usually 50–60% of the energy value of common fodder and low protein content (5–8%). For example, Al-Shorepy et al. (2010) report CP of 8.7% and 9.8% DM in *Sporobolus virginicus* and *Distichlis spicata*, respectively. However, the CP concentrations in grasses can reach acceptable values. For instance, in *Chloris gayana* and *Cynodon dactylon*, these values were 11.5% and 16% DM, respectively (Pasternak et al. 1993). Low CP in some halophytic species of Poaceae may be improved through agronomic means; for example, fertilizing halophytic grasses prior to harvest with nitrogen fertilizers and regular cutting (Bustan et al. 2005) have both increased CP in these species. Furthermore, saline water (12.5% seawater to 50% seawater) increased CP of *Sporobolus virginicus* from 6.8% to 9.0% (Ashour et al. 1997).

Forage products of salt-tolerant Poaceae must be used mixed with other products particularly those rich in energy feed resources. A field evaluation of forage potential of halophytes of Poaceae shows good results. For instance, in the case of *Sporobolus virginicus* and *Distichlis spicata* irrigated with saline water, results showed that their inclusion in silage up to 100% of the forge component in sheep and goat diets showed normal reproduction and growth. A mixed diet composed of *Atriplex* shrubs and *Sporobolus virginicus* resulted in better performances on sheep and goat compared with other diets containing either 100% *Sporobolus* grass or *Atriplex* shrub (Al-Shorepy et al. 2010). Thus, when managed properly, forage Poaceae have the potential to be used in integrated animal diets or completely replace the conventional ingredients in animal feeding system.

Medicinal Halophytes

Although halophytic species of Poaceae are mainly used for forage, they have been utilized, in the recent years, in other interesting domains. They were used as medicinal plants and as sources of bioactive substances. For instance, *Cymbopogon jwarancusa* (Qasim et al. 2010), *Desmostachya bipinnata* (Ahmad et al. 2009), *Heteropogon contortus*, *Cenchrus ciliaris*, *Dactyloctenium scindicum*, *Eleusine indica* (Katewa et al. 2001), *Cynodon dactylon* (Chaib and Boukhris 1998; Weber and Hanks 2008), *Vetiveria zizanioides* (Ahmad et al. 2009; Xia 2004), and *Setaria italica* (Ahmad et al. 2009) were used as source biomolecules and as medicinal plant (Table 14.3).

Halophytes grow in harsh salt environments; such conditions may incite them to produce several kinds of biologically active molecules they can use to respond to environmental stresses (Ksouri et al. 2007; Ksouri et al., 2008). For instance, the leaves of some edible halophytes are rich in carotenes (Guil-Guerrero and Rodríguez-García 1999). Polyphenols and other bioactive substances like α-tocopherol, ascorbic acid, carotenoids, and glutathione were also found in halophytes. These compounds show high antioxidant, antimicrobial, and anticancer activities (Trabelsi et al. 2010). They are also anti-inflammatory, anti-allergenic, antithrombotic, hepatoprotective, cardioprotective, and vasodilatory (Ksouri et al. 2008; Kong et al. 2010; Qasim et al. 2014). For instance, the leaves of the halophyte of Poaceae *Cymbopogon jwarancusa* were used as expectorant (Qasim et al. 2010). The leaves of *Eragrostis ciliaris* were used for healing infected wound in livestock (Qasim et al. 2010). *Cynodon dactylon* was used as diuretic (Qasim et al. 2010).

Halophytes Used in Soil Management and Phytoremediation

Land degradation and desertification constitute major constraints that affect agricultural areas and decrease crop productivity. They affect soil biological and physicochemical properties and lower its fertility. But interestingly, some halophytes showed high effectiveness in soil fixation and landscaping. Indeed, some Poaceae such as *Spartina alterniflora* (Li et al. 2009; Meudec et al. 2007; Manousaki et al. 2008), *Thinopyrum ponticum, Puccinellia ciliate* (El Shaer 2006),

Imperata cylindrica (Chaib and Boukhris 1998), *Spartina patens* (Eid and Eisa 2010), *Leymus racemosus* (Shamsutdinov and Shamsutdinov 2008), *Leymus chinensis* (Sun et al. 2008), *Puccinellia tenuiflora* (Wang et al. 2009a, b), and *Spartina townsendii* (Koyro and Huchzermeyer 2004) were used for soil fixation (Table 14.3). In fact, the cultivation of these species ameliorates the biological and the physicochemical properties of degraded soils. This is beneficial to increase soil organic matter and nutrient contents (Ashraf et al. 2010).

Some salt-tolerant species are found to be used in large-scale environmental cleanup efforts of metal-contaminated soils which increased with the increase of anthropogenic activity in the past years (Toderich et al. 2008; Hameed et al. 2012). In this study, we discovered that some Poaceae such as *Sporobolus virginicus* (Eid and Eisa 2010), *Spartina alterniflora* (Li et al. 2009; Meudec et al. 2007; Manousaki et al. 2008), *Phragmites australis* (Weis and Weis 2004; Kefu et al. 2002), *Spartina anglica* (Williams et al. 1994), *Spartina densiflora* (Luque et al. 1999), *Spartina patens* (Eid and Eisa 2010), *Puccinella scleroides* (Toderich et al. 2008), *Spartina townsendii* (Koyro and Huchzermeyer 2004), *Vetiveria zizanioides* (Ahmad et al. 2009; Xia 2004), and *Dactylis glomerata* (Toderich et al. 2008) were used for phytoremediation of metal-contaminated soils (Table 14.3).

There are two main mechanisms by which halophytes remediate metal-contaminated soils: (i) some halophytes have the ability to immobilize heavy metals through adsorption onto roots and/or precipitation within the rhizosphere (Weis and Weis 2004). This mechanism called is "metal phytostabilization." (ii) Some other halophytes have the ability to allow metal accumulation in their aboveground tissues, mainly shoots. Thereafter, metals are removed through shoot harvest. This option is called phytoextraction (Weis and Weis 2004). Several studies demonstrated that some tolerance mechanisms operating at the whole-plant level are not forever specific to sodium or chloride (Zaier et al. 2010). For instance, copper, zinc, and cadmium may accumulate in salt glands or trichomes of *Tamarix aphylla* and *Armeria maritima* (Toderich et al. 2008). Thus, significant ability for heavy metal removal has been noted for halophytes that accumulate large amounts of metals in their aerial parts (Caçador et al. 1996; Redondo-Gómez et al. 2010).

Other Uses of Halophytes of Poaceae

The two Poaceae, *Leptochloa fusca* (Lal 2008) and *Spartina alterniflora* (Li et al. 2009; Meudec et al. 2007; Manousaki et al. 2008), were used for biomass energy production. Biomass provides around 13–15% of the global energy demand (Lu et al. 2009). *Phragmites australis* (Kefu et al. 2002), *Urochondra setulosa, Cenchrus ciliaris, Aeluropus lagopoides, Cynodon dactylon, Sporobolus ioclados, Lasiurus scindicus, Panicum turgidum, Paspalum paspaloides, Phragmites karka, Desmostachya bipinnata, Dichanthium annulatum, Eleusine indica*, and *Halopyrum mucronatum* were used in ethanol production (Table 14.1) (Abideen et al. 2011).

Conclusion

The present contribution tried to compile the most recent data in order to further highlight both the fundamental and applied potentials of halophytes belonging to Poaceae, a botanical group including several strategic crops like cereals. This contribution although modest provides interesting data and should be placed in the context of global climate changes and the increasing salinization of lands and water resources, which are major challenges for developing countries, in particular.

References

Abdelly C., Barhoumi Z., Ghanya T., Debez A., Ben Hamed K., Sleimi N., Ouerghi Z., Smaoui A., Huchzermeyer B., Grignon C. (2006) Potential utilisation of halophytes for the rehabilitation and valorisation of salt-affected areas in Tunisia. In: Öztürk M., Waisel Y., Khan M.A., Görk G. (Eds), Biosaline agriculture and salinity tolerance in plants. Birkhäuser Verlag, Switzerland, pp 163–172

Abideen Z, Ansari R, Khan MA (2011) Halophyte: potential source of lignocellulosic biomass for ethanol production. Biomass Bioenergy 35:1818–1822

Abideen Z, Koyro H-W, Huchzermeyer B, Ahmed MZ, Gul B, Khan MA (2014) Moderate salinity stimulates growth and photosynthesis of *Phragmites karka* by water relations and tissue specific ion regulation. Environ Exp Bot 105:70–76

Abogadallah GM, Serag MM, Quick WP (2010) Fine and coarse regulation of reactive oxygen species in the salt tolerant mutants of barnyard grass and their wild-type parents under salt stress. Physiol Plant 138:60–73

Ahmad F, Khan MA, Ahmad M, Zafar M, Nazir A, Marwat SK (2009) Taxonomic studies of grasses and their indigenous uses in Salt Range area of Pakistan. Afr J Biotechnol 8(2):231–249

Ahmad MSA, Ashraf M, Ali Q (2010) Soil salinity as a selection pressure is a key determinant for the evolution of salt tolerance in blue Panicgrass (*Panicum antidotale* Retz.). Flora 205:37–45

Ahmed MZ, Shimazaki T, Gulzar S, Kikuchi A, Gul B, Khan MA, Koyro H-W, Huchzermeyer B, Watanabe KN (2013) The influence of genes regulating transmembrane transport of Na^+ on the salt resistance of *Aeluropus lagopoides*. Funct Plant Biol 40:860–871

Al-Khateeb S (2006) Effect of salinity and temperature on germination: growth and ion relations of *Panicum turgidum* Forssk. Bioresour Technol 97:292–298

Al-Shorepy SA, Alhadrami GA, El Awad AI (2010) Development of sheep and goat production system based on the use of salt-tolerant plants and marginal resources in the United Arab Emirates. Small Rumin Res 91:39–46

Ardie SW, Xie L, Takahashi R, Liu S, Takano T (2009) Cloning of a high-affinity K^+ transporter gene PutHKT2;1 from *Puccinellia tenuiflora* and its functional comparison with OsHKT2;1 from rice in yeast and Arabidopsis. J Exp Bot 60:3491–3502. https://doi.org/10.1093/jxb/erp184

Ardie SW, Liu S, Takano T (2010) Expression of the AKT1-type K(+) channel gene from *Puccinellia tenuiflora*, PutAKT1, enhances salt tolerance in Arabidopsis. Plant Cell Rep 29:865–874. https://doi.org/10.1007/s00299-010-0872-2

Ashour NI, Serag MS, Abd El-Haleem AK, Mekki BB (1997) Forage production from three grass species under saline irrigation in Egypt. J Arid Environ 37:299–307

Ashraf M, Hamed M, Arshad M, Ashraf Y, Akhtar K (2008) Salt tolerance of some potential forage grasses from cholistan desert of Pakistan. In: Khan MA, Weber DJ (eds) Ecophysiology of high salinity tolerant plants. Springer, Dordrecht, pp 31–54

Ashraf MY, Ashraf M, Mahmood K, Akhter J, Hussain F, Arshad M (2010) Phytoremediation of saline soils for sustainable agricultural productivity. In: Ashraf M et al (eds) Plant adaptation and phytoremediation. Springer Science+Business Media B.V, Dordrecht. https://doi.org/10.1007/978-90-481-9370-7-15

Atia A, Smaoui A, Barhoumi Z, Abdelly C, Debez A (2011) Differential response to salinity and water deficit stress in *Polypogon monspeliensis* (L.) Desf. provenances during germination. Plant Biol 13:541–545

Baisakh N, Ramana RMV, Rajasekaran K, Subudhi P, Janda J, Galbraith D, Vanier C, Pereira A (2012) Enhanced salt stress tolerance of rice plants expressing a vacuolar H^+-ATPase subunit c1 (SaVHAc1) gene from the halophyte grass *Spartina alterniflora* Löisel. Plant Biotechnol J 10:453–464. https://doi.org/10.1111/j.1467-7652.2012.00678.x

Barhoumi Z, Djebali W, Smaoui A, Chaïbi W, Abdelly C (2007) Contribution of NaCl excretion to salt resistance of *Aeluropus littoralis* (Willd) Parl. J Plant Physiol 164:842–850

Barhoumi Z, Atia A, Rabhi M, Djebali W, Abdelly C, Smaoui A (2010) Nitrogen and NaCl salinity effects on the growth and nutrient acquisition of the grasses *Aeluropus littoralis, Catapodium rigidum,* and *Brachypodium distachyum*. J Plant Nutr Soil Sci 173:149–157

Bell HL, O'Leary JW (2003) Effects of salinity on growth and cation accumulation of *Sporobolus virginicus* (*Poaceae*). Am J Bot 90:1416–1424

Bhuiyan MSI, Raman A, Hodgkins DS, Mitchell D, Nicol HI (2015) Physiological response and ion accumulation in two grasses, one legume and one saltbush under soil water and salinity stress. Ecohydrology 8:1547–1559. https://doi.org/10.1002/eco.1603

Bonales-Alatorre E, Shabala S, Chen ZH, Pottosin I (2013) Reduced tonoplast fast-activating and slow-activating channel activity is essential for conferring salinity tolerance in a facultative halophyte, *Chenopodium quinoa*. Plant Physiol 162:940–952

Bouzid N (2008) Salt tolerance strategies of *Lygeum spartum* L.: a new fodder crop for Algerian saline steppes. Flora 204:747–754

Bustan A, Pasternak D, Pirogova I, Durikov M, Devries TT, El-Meccawi S, Degen AA (2005) Evaluation of saltgrass as a fodder crop for livestock. J Sci Food Agric 85:2077–2084

Caçador I, Vale C, Catarino F (1996) Accumulation of Zn, Pb, Cu and Ni in sediments between roots of the Tagus estuary salt marshes, Portugal. Estuar Coast Shelf Sci 42:393–403

Céccoli G, Ramos J, Pilatti V, Dellaferrera I, Tivano JC, Taleisnik E, Vegetti AC (2015) Salt glands in the *Poaceae* family and their relationship to salinity tolerance. Bot Rev 81:162–178

Chaib M, Boukhris M (1998) Flore succinte et illustré des zones arides et sahariennes de Tunisie, p 290

Choukr-Allah R, Malcolm CV, Hamdy A (eds) (1996) Halophytes and biosaline agriculture. Marcel Dekker, New York

Debez A, Koyro HW, Grignon C, Abdelly C, Huchzermeyer B (2008) Relationship between the photosynthetic activity and the performance of *Cakile maritima* after long term salt treatment. Physiol Plant 133:373–385

Duarte B, Santos D, Caçador I (2013) Halophyte antioxidant feedback seasonality in two salt marshes with different degrees of metal contamination: search for an efficient biomarker. Funct Plant Biol 40:922–930

Eid MA, Eisa SS (2010) The use of some halophytic plants to reduce Zn, Cu and Ni in soil. Aust J Basic Appl Sci 4:1590–1596

El Shaer HM (2006) Halophytes as cash crops for animal feeds in arid and semi-arid regions. In: Biosaline agriculture and salinity tolerance in plants. Section III, pp 117–128. https://doi.org/10.1007/3-7643-7610-4-13

El Shaer HM (2010) Halophytes and salt-tolerant plants as potential forage for ruminants in the Near East region. Small Rumin Res 91:3–12

Flowers TJ, Colmer TD (2008) Salinity tolerance in halophytes. New Phytol 179:945–963

Geissler N, Hussin S, Koyro HW (2009) Interactive effects of NaCl salinity and elevated atmospheric CO2 concentration on growth, photosynthesis, water relations and chemical composition of the potential cash crop halophyte Aster tripolium L. Environ Exp Bot 65:220–231

Glenn EP, Brown JJ, Blumwald E (1999) Salt tolerance and crop potential of halophytes. Crit Rev Plant Sci 18:227–255

Guil-Guerrero JL, Rodríguez-García I (1999) Lipids classes, fatty acids and carotenes of the leaves of six edible wild plants. Eur Food Res Technol 209:313–316

Guo Q, Wang P, Ma Q, Zhang JL, Bao AK, Wang SM (2012) Selective transport capacity for K^+ over Na^+ is linked to the expression levels of PtSOS1 in halophyte *Puccinellia tenuiflora*. Funct Plant Biol 39:1047–1057

Hafsi C, Romero-Puertas MC, Gupta DK, Del Río LA, Sandalio LM, Abdelly C (2010) Moderate salinity enhances the antioxidative response in the halophyte *Hordeum maritimum* L. under potassium deficiency. Environ Exp Bot 69:129–136

Hameed A, Hussain T, Gulzar S, Aziz I, Gul B, Khan MA (2012) Salt tolerance of a cash crop halophyte *Suaeda fruticosa*: biochemical responses to salt and exogenous chemical treatments. Acta Physiol Plant 34:2331–2340

Hasegawa PM (2013) Sodium (Na^+) homeostasis and salt tolerance of plants. Environ Exp Bot 92:19–31

Hussain T, Koyro HW, Huchzermeyer B, Khan MA (2015) Eco-physiological adaptations of *Panicum antidotale* to hyperosmotic salinity: water and ion relations and anti-oxidant feedback. Flora 212:30–37

Joshi AJ, Mali BS, Hinglajia H (2005) Salt tolerance at germination and early growth of two forage grasses growing in marshy habitats. Environ Exp Bot 54:267–274

Karan R, Subudhi PK (2012) A stress inducible SUMO conjugating enzyme gene (*SaSce9*) from a grass halophyte *Spartina alterniflora* enhances salinity and drought stress tolerance in Arabidopsis. BMC Plant Biol 12:187. https://doi.org/10.1186/1471-2229-12-187

Katewa SS, Guria BD, Jain A (2001) Ethnomedicinal and obnoxious grasses of Rajasthan, India. J Ethnopharmacol 76:293–297

Kefu Z, Hai F, Ungar IA (2002) Survey of halophyte species in China. Plant Sci 163:491–498

Khan MA, Qaiser M (2006) Halophytes of Pakistan: characteristics, distribution and economic usages. In: Khan MA, Böer B, Kust GS, Barth HJ (eds) Sabkha ecosystems. Volume II: West and Central Asia. Springer, Dordrecht, pp 129–153

Khan MA, Ungar IA, Showalter AM (2000) Effects of salinity on growth, water relations and ion accumulation of the subtropical perennial halophyte, *Atriplex griffithii* var. stocksii. Ann Bot 85:225–232

Khan MA, Ansari R, Ali H, Gul B, Nielsen BL (2009) *Panicum turgidum*, a potentially sustainable cattle feed alternative to maize for saline areas. Agric Ecosyst Environ 129:542–546

Kong C, Um YR, Lee JI, Kim YA, Yea SS, Seo Y (2010) Constituents isolated from *Glehnia littoralis* suppress proliferations of human cancer cells and MMP expression in HT1080 cells. Food Chem 120:385–394

Koyro HW, Huchzermeyer B (2004) Ecophysiological needs of the potential biomass crop *Spartina townsendii* GROV. Trop Ecol 45:123–139

Koyro HW, Lieth H (2008) Global water crisis: the potential of cash crop halophytes to reduce the dilemma. In: Lieth H, Garciá-Sucre M (eds) Mangroves and halophytes: restoration and utilisation. Springer Science + Business Media B.V, Dordrecht, pp 7–19

Koyro HW, Hussain T, Huchzermeyer B, Khan MA (2013) Photosynthetic and growth responses of a perennial halophytic grass *Panicum turgidum* to increasing NaCl concentrations. Environ Exp Bot 91:22–29

Ksouri R, Megdiche W, Debez A, Falleh H, Grignon C, Abdelly C (2007) Salinity effects on polyphenol content and antioxidant activities in leaves of the halophyte *Cakile maritime*. Plant Physiol Biochem 45:244–249

Ksouri R, Megdiche W, Falleh H, Trabelsi N, Boulaaba M, Smaoui A, Abdelly C (2008) Influence of biological, environmental and technical factors on phenolic content and antioxidant activities of Tunisian halophytes. C R Biol 331:865–873

Lal R (2008) Crop residues as soil amendments and feedstock for bioethanol production. Waste Manag 28:747–755

Li B, Liao CH, Zhang XD, Chen HL, Wang Q, Chen ZI, Gan XJ, Wu JH, Zhao B, Ma ZJ, Cheng XL, Jiang LF, Chen JK (2009) *Spartina alterniflora* invasions in the Yangtze River estuary, China: an overview of current status and ecosystem effects. Ecol Eng 35:511–520

Lu SY, Jing YX, Shen SH, Zhao HY, Ma LQ, Zhou XJ, Ren Q, Li YF (2005) Antiporter gene from *Hordeum brevisubulatum* (Trin.) link and its overexpression in transgenic tobaccos. J Integr Plant Biol 47:343–349

Lu L, Tang Y, Xie J, Yuan Y (2009) The role of marginal agricultural land-based mulberry planting in biomass energy production. Renew Energy 34:1789–1794

Luque CJ, Castellanos EM, Castillo JM, Gonzales M, Gonzales-Vilches MC, Figuero ME (1999) Metals in halophytes of a contaminated estuary (Odiel Saltmarshes, SW Spain). Mar Pollut Bull 38:49–51

Manousaki E, Kadukova J, Papadantonakis N, Kalogerakis N (2008) Phytoextraction and phytoexcretion of Cd by the leaves of *Tamarix smyrnensis* growing on contaminated non-saline and saline soils. Environ Res 106:326–332

Meudec A, Poupart N, Dussauze J, Deslandes E (2007) Relationship between heavy fuel oil phytotoxicity and polycyclic aromatic hydrocarbon contamination in *Salicornia fragilis*. Sci Total Environ 381:146–156

Moinuddin M, Gulzar S, Ahmed MZ, Gul B, Koyro HW, Khan MA (2014) Excreting and non-excreting grasses exhibit different salt resistance strategies. AoB PLANTS 6:plu038. https://doi.org/10.1093/aobpla/plu038

Munns R, Tester M (2008) Mechanisms of salinity tolerance. Annu Rev Plant Biol 59:651–681

Muscolo A, Panuccio MR, Eshel A (2013) Ecophysiology of *Pennisetum clandestinum*: a valuable salt tolerant grass. Environ Exp Bot 92:55–63

Naidoo G, Naidoo Y (1998) Salt tolerance in *Sporobolus virginicus*: the importance of ion relations and salt secretion. Flora 193:334–337

Naidoo G, Somaru R, Achar P (2008) Morphological and physiological responses of the halophyte *Odyssea paucinervis* Staph *Poaceae*, to salinity. Flora 203:437–447

Naidoo G, Naidoo Y, Achar P (2012) Ecophysiological responses of the salt marsh grass *Spartina maritima* to salinity. Afr J Aquat Sci 37(1):81–88

Norman HC, Masters DG, Barrett-Lennard EG (2013) Halophytes as forages in saline landscapes: interactions between plant genotype and their environment change their feeding value to ruminants. Env and Exp Bot 92:96–109

Otte ML, Wilson G, Morris JT, Moran BM (2004) Dimethylsulphoniopropionate (DMSP) and related compounds in higher plants. J Exp Bot 55:1919–1925

Pagter M, Bragato C, Malagoli M, Brix H (2009) Osmotic and ionic effects of NaCl and Na_2SO_4 salinity on *Phragmites australis*. Aquat Bot 90:43–51. https://doi.org/10.1016/j.aquabot.2008.05.005

Pasternak D, Nerd A, Demalach Y (1993) Irrigation with brackish-water under desert conditions. The salt tolerance of 6 forage crops. Agric Water Manag 24:321–334

Peng YH, Zhu YF, Mao YQ, Wang SM, Su WA, Tang ZC (2004) Alkali grass resists salt stress through high K^+ and an endodermis barrier to Na^+. J Exp Bot 55:939–949

Qadir M, Qureshi RH, Ahmad N (1996) Reclamation of a saline- sodic soil by gypsum and *Leptochloa fusca*. Geoderma 74:207–217

Qasim M, Gulzar S, Shinwari ZK, Aziz I, Khan MA (2010) Traditional ethnobotanic uses of halophytes from Hub Balochistan. Pak J Bot 42(3):1543–1551

Qasim M, Abideen Z, Adnan MY, Ansari R, Gul B, Khan MA (2014) Traditional ethno-botanical uses of medicinal plants from coastal areas of Pakistan. J Coast Life Med 2(1):22–30

Rabhi M, Castagna A, Remorini D, Scattino C, Smaoui A, Ranieri A, Abdelly C (2012) Photosynthetic responses to salinity in two obligate halophytes: *Sesuvium portulacastrum* and *Tecticornia indica*. S Afr J Bot 79:39–47

Ramadan T (2001) Dynamics of salt secretion by *Sporobolus spicatus* (Vahl) Kunth from sites of differing salinity. Ann Bot 87:259–266

Rauf M, Shahzad K, Ali R, Ahmad M, Habib I, Mansoor S, Berkowitz G, Saeed N (2014) Cloning and characterization of Na^+/H^+ antiporter (LfNHX1) gene from a halophyte grass *Leptochloa fusca* for drought and salt tolerance. Mol Biol Rep 41:1669–1682

Redondo-Gomez S, Wharmby C, Castillo JM, Mateos-Naranjo E, Luque CJ, Cires A, Luque T, Davy AJ, Figueroa ME (2006) Growth and photosynthetic responses to salinity in an extreme halophyte, *Sarcocornia fruticosa*. Physiol Plant 128:116–124

Redondo-Gómez S, Mateos-Naranjo E, Andrades-Moreno L (2010) Accumulation and tolerance characteristic of cadmium in a halophytic Cd-hyperaccumulator, *Arthrocnemum macrostachyum*. J Hazard Mater 184:299–307

Sanadhya P, Agarwal P, Agarwal PK (2015) Ion homeostasis in a salt-secreting halophytic grass. AoB PLANTS 7:plv055. https://doi.org/10.1093/aobpla/plv055

Shabala S (2013) Learning from halophytes: physiological basis and strategies to improve abiotic stress tolerance in crops. Ann Bot 112:1209–1221

Shabala S, Bose J, Fuglsang AT, Pottosin I (2015) On a quest for stress tolerance genes: membrane transporters in sensing and adapting to hostile soils. J Exp Bot 67:1015–1031. https://doi.org/10.1093/jxb/erv465

Shamsutdinov N, Shamsutdinov Z (2008) Halophyte utilization for biodiversity and productivity of degraded pasture restoration in arid regions of Central Asia and Russia. In: Abdelly C, Öztürk M, Ashraf Mand Grignon C (eds) Biosaline agriculture and high salinity tolerance. Birkhäuser Verlag, Basel, pp 233–240

Slama I, Abdelly C, Bouchereau A, Flowers T, Savoure A (2015) Diversity, distribution and roles of osmoprotective compounds accumulated in halophytes under abiotic stress. Ann Bot 239. https://doi.org/10.1093/aob/mcu239

Subudhi P, Baisakh N (2011) *Spartina alterniflora* Loisel., a halophyte grass model to dissect salt stress tolerance. In Vitro Cell Dev Biol Plant 47:441–457

Sun Z, Zhou D, Ferreira LMM, Zhong Q, Lou Y (2008) Diet composition, herbage intake and digestibility in Inner Mongolian Cashmere goats grazing on native *Leymus chinensis* plant communities. Livest Sci 116:146–155

Toderich KN, Ismail S, Juylova EA, Rabbimov AA, Bekchanov BB, Shyuskaya LEV, Gismatullina G, Osamu K, Radjabov TB (2008) New approaches for biosaline agriculture development, management and conservation of sandy desert ecosystems. In: Abdelly C, Öztürk M, Ashraf M, Grignon C (eds) Biosaline agriculture and high salinity tolerance. Birkhäuser Verlag, Basel, pp 247–264

Trabelsi N, Megdiche W, Ksouri R, Falleh H, Oueslati S, Soumaya B, Hajlaoui H, Abdelly C (2010) Solvent effects on phenolic contents and biological activities of the halophyte *Limoniastrum monopetalum* leaves. LWT–Food Sci Technol 43:632–639

Vasquez EA, Gleen EP, Guntenspergen GR, Brown JJ, Nelson SG (2006) Salt tolerance and osmotic adjustment of *Spartina alterniflora* (*Poaceae*) and the invasive haplotype of *Phragmites australis* (*Poaceae*) along a salinity gradient. Am J Bot 93(12):1784–1790

Wang F, Xiong XR, Liu CZ (2009a) Biofuels in China: opportunities and challenges. In Vitro Cell Dev Biol Plant 45:342–349

Wang CM, Zhang JL, Liu XS, Li Z, Wu GQ, Cai JY, Flowers TJ, Wang SM (2009b) *Puccinellia tenuiflora* maintains a low Na^+ level under salinity by limiting unidirectional Na+ influx resulting in a high selectivity for K^+ over Na^+. Plant Cell Environ 32:486–496

Wang X, Yang R, Wang B, Liu G, Yang C, Cheng Y (2011) Functional characterization of a plasma membrane Na^+/H^+ antiporter from alkali grass (*Puccinellia tenuiflora*). Mol Biol Rep 38:4122–4813. https://doi.org/10.1007/s11033-010-0624-y

Weber DJ, Hanks DJ (2008) Salt tolerant plants from the great basin region of the United States. In: Khan MA, Weber DJ (eds) Ecophysiology of high salinity tolerant plants. Springer Science + Business Media B.V, Dordrecht, pp 69–106

Weber DJ, Ansari R, Gul B, Khan MA (2007) Potential of halophytes as source of edible oil. J Arid Environ 68:315–321

Weis JS, Weis P (2004) Metal uptake, transport and release by wetland plants implications for phytoremediation and restoration. Environ Int 30:685–700

Williams TP, Bubb JM, Lester JN (1994) The occurrence and distribution of trace metals in halophytes. Chemosphere 28:1189–1199

Xia HP (2004) Ecological rehabilitation and phytoremediation with four grasses in oil shale mined land. Chemosphere 54:345–353

Yu J, Chen S, Zhao Q, Wang T, Yang C, Diaz C, Sun G, Dai S (2011) Physiological and proteomic analysis of salinity tolerance in *Puccinellia tenuiflora*. J Proteome Res 10:3852–3870

Zaier H, Ghnaya T, Lakhdar A, Baioui R, Ghabriche R, Mnasri M, Sghair S, Lutts S, Abdelly C (2010) Comparative study of Pb-phytoextraction potential in *Sesuvium portulacastrum* and *Brassica juncea*: tolerance and accumulation. J Hazard Mater 183:609–615

Zhang JL, Shi H (2013) Physiological and molecular mechanisms of plant salt tolerance. Photosynth Res 115:1–22

Zhang GH, Su Q, An LJ, Wu S (2008) Characterization and expression of a vacuolar Na^+/H^+ antiporter gene from the monocot halophyte *Aeluropus littoralis*. Plant Physiol Biochem 46:117–126. https://doi.org/10.1016/j.plaphy.2007.10.022

15 Phenotyping Through Infrared Thermography in Stress Environment

Zamin Shaheed Siddiqui, Muhammad Umar,
Taek-Ryoun Kwon, and Soo Chul Park

Abstract

Abiotic stress like drought and salinity is the major environmental constraints that limit agricultural production. Physiological mechanisms explaining plant tolerance offer valuable insights for the development of genetically modified crops. Plant water status and alteration in photosynthetic capacity are some common physiological depictions which are induced by abiotic stress like drought and salinity. Chiefly it is because both stresses caused cellular dehydration in the plants, predominantly, during the initial phase of stress imposition. In water stress CO_2 availability is greatly reduced due to stomatal limitation; subsequently leaf temperature is elevated. So that studies on plant water status and stomatal regulation are important aspects in abiotic stress environment stabilizing the temperature inside plant/leaf. Therefore phenotyping using infrared thermography (heat sensitive sensor) could be a useful tool in the selection of tolerant genotypes. Generally infrared thermography is sensitive, less time-consuming, and nondestructive methodology which detects heat produced or generated by leaf under the influence of external factors. In general, temperature display pattern on IR images is inversely proportional to leaf water status. It was observed that infrared images are significantly correlated with some of the physiological traits indicating tolerance grading among genotypes.

Keywords

IR thermography · Phenotyping · Physiology · Salt stress · Drought stress · Stomatal conductance

Z. S. Siddiqui (✉) · M. Umar
Stress Physiology Phenomic Lab., Department of Botany, University of Karachi, Karachi, Pakistan
e-mail: zaminss@uok.edu.pk

T.-R. Kwon · S. C. Park
National Academy of Agricultural Sciences, Rural Development Administration, Suwon, Republic of Korea

Introduction

Drought and salinity are major environmental constraints antagonizing global agricultural production. Both abiotic stresses distress over more than 10 percent arable land in the world (Bray et al. 2000). In nature, abiotic stress like salinity and drought causes substantial decline in agricultural productivity showing more than 50% decrease of daily yield in most consumable crops (Kwon et al. 2015). Consequently, the improvement of salt and drought tolerance in crop plants is an imperative step toward crop breeding.

It was observed that the physiological nature of the tolerance in plant is a prerequisite step in

B. Gul et al. (eds.), *Sabkha Ecosystems*, Tasks for Vegetation Science VI,
https://doi.org/10.1007/978-3-030-04417-6_15

the development of crop plants for salt and drought tolerance. In plant tolerance can be attained against these stresses adopting numerous physiological characteristics and some metabolic adjustment (Flowers 2004). Water stress in terms of salt or drought stimulates numerous collective responses in plants such as osmotic potential, relative water content, stomatal conductance, quantum yield efficiency, etc. (Munns 2002; Bartels and Sunkar 2005; Chaves et al. 2009). During the early phase of stress imposition in plant, both stresses lead to cellular dehydration which results in osmotic imbalance and cytoplasmic movement of water into apoplastic space. Additionally metabolic changes, rising plant hormone like abscisic acid, and decreasing photosynthesis activity were also common in plants when subject to salt and drought stress (Davies et al. 2005).

Plants acclimatize to adversity of salt and drought stress by few physiological adjustments such as relative water in leaf, osmotic adjustment, oxidative protection, photosynthetic control, molecular signaling, and gene regulation (Silveira et al. 2012). Though, some adaptive responses of plants to those stresses are stress-dependent. Salt tolerance in plants activates precise mechanisms such as ion-specific signaling, ionic homeostasis, nutritional balance, and gene expression. Similarly, plants prompt gene expression and specific signaling to drought condition.

During photosynthesis, quantum efficiency of photosystem II is a main physiological response to abiotic stress (Chaves 1991; Munns et al. 2006; Siddiqui et al. 2014b). In water stress condition, CO_2 availability for photosynthesis via stomatal limitation (Flexas et al. 2007) and photosynthetic efficiency (Lawlor and Cornic 2002) via non-stomatal limitation is greatly influenced. Photosynthetic performance of stress-imposed plants is varied on positions of leaves and developmental stages of plant. Intensity and stress duration are also affected on the photosynthetic activity of plants (Garrity and O'Toole 1994; Tavakkoli et al. 2012).Therefore, it is important to comprehend the physiological mechanisms of stress tolerance at whole plant level (Munns 1993).

Soon after the initial phase, the growth of plant is mainly followed by ionic imbalance or toxicity, which seems to be excessive toxic ion accumulation particularly in the mature leaves causing less photosynthetic ability and triggering premature senescence. Manifestation regarding plant tolerance against the stresses at whole plant level is the maintaining ability of the plant to grow via adjusted metabolic process (Munns and Tester 2008; Siddiqui et al. 2014b). In last few decades, there were extensive improvements for salt tolerance in crop plants like maize, pear millet, barley, rice, and so on, mostly through pictorial assessments of stressed plants in breeding program (Ashraf and Wu 1994). Nevertheless, the biochemical/physiological mechanism underlying the phenotypic tolerance was not clear exclusively in the consideration as selection criteria. Consequently special efforts have been made to establish reliable biochemical/physiological parameters for evaluating the salt and drought tolerance. Recently, high-throughput phenotyping tactics constructed on the parameters of the tolerance-related biochemistry and physiology have been developed as most powerful tools for the screening of salt and drought tolerance. Phenomics is considerably a recent approach which originates from a theoretical background which can roughly be called as physiological and biochemical trait assessment with camera's images taken at several light spectrums including infrared, near-infrared, hyper-spectral spectrum RGB (red, green, and blue) and chlorophyll fluorescence (Lee et al. 2011). Scientists expanded the application of infrared thermography as a high-throughput phenotyping tool in field, demonstrating a possible screening for not only abiotic stress tolerance but also biotic stress tolerance (Prashar and Jones 2014). Screening of salt and drought tolerance in plants using infrared thermography together with a few physiological traits would be helpful in large-scale experimental setup. Bioassay regarding salt or drought tolerance in the field is always a problem in particular. In this situation, plant assessments with infrared camera can be a useful tool to select stress-tolerant genotypes in less time with key physiological traits.

Basis of IR Images

Infrared thermography is working with heat-sensitive sensor that detects heat produced in stressed plant due to limited water supply (Munns et al. 2010). In research, working with IR, it is likely to monitor the advancement of stressed plant over time through combination of vigilant image capture, color classification, and image analysis using a software. For the stressed plants, IR thermal sensing can deliver vast, effective, and rapid phenotyping, but it has to be adopted certain principles. For example, infrared thermography is a fragment of the electromagnetic range which releases a definite quantity of radiation as a function of their temperatures.

Mostly, plants display extreme temperature when they have less water; the more infrared radiation is discharged. It has specialty to sense the radiation in a way similar to the way an ordinary camera detects visible light. IR camera works well in complete darkness since ambient light level does not matter. IR thermal images are likely to have a unicolor channel since the camera generally uses an image sensor that does not distinguish wavelengths of infrared radiation. For temperature measurement, the brightest parts of the images are normally colored white, intermediate temperatures reds and yellows, and the dimmest (the coolest) parts blue or black (Fig. 15.1). Differences in color describe the temperature of plant or part of plant. It is completely relied on the color scheme adjustment. Leaf temperature can be correlated with RWC (Fig. 15.1). A scale should be displayed subsequent to a false color image to corresponding colors to a certain temperature (Fig. 15.2). Their resolution is varied considerably lower than that of optical cameras, frequently in between the range of 160 × 120 and 320 × 240 pixels, though some advanced cameras can also attain an enhanced resolution of 1280 × 1024 pixels. In uncooled detectors, the temperature differences at the sensor pixels are 1 °C. The pixel response time is also fairly slow, at the range of tens of milliseconds. Camera sensitivity is as fine as <0.02 °C for outstanding image quality and thermal patterns. Temperature range adjustment of IR camera is nearly up to 2000 °C for measuring the hottest objectives.

IR images give better results under environment in which temperature and light intensity can be controlled. This is because IR images of plant and leaves need a uniform background to recognize the color pattern. Figures 15.3, 15.4, and 15.5 have uniform background color that is why plants/leaves color is easily recognizable. The images captured in field environments do not have uniform background color and hence creating little bit confusion to the viewer (Figs. 15.2, 15.6, 15.7, and 15.8).

Generally, IR camera allows the detection of early heat generation in stressed parts of a plant. Under controlled environmental conditions, leaf temperature can be sensed by IR camera using

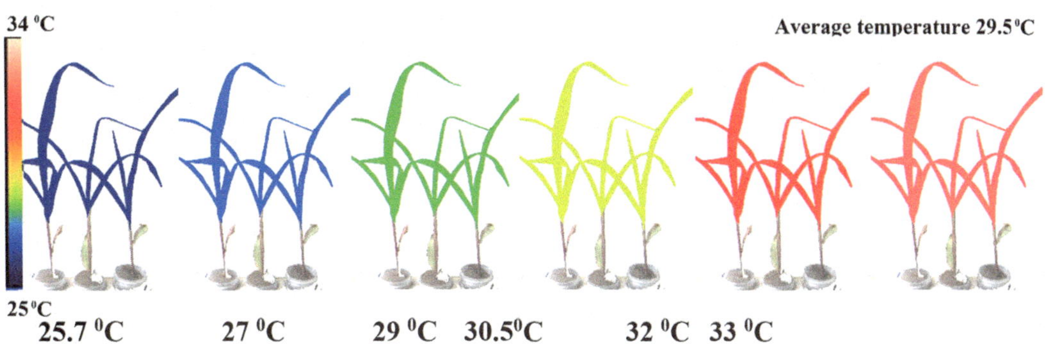

Fig. 15.1 Model showing relationship of IR image color with temperature using IR thermography. From left to right colors reflecting the increase in temperature. Blue color is demonstrating the lowest temperature while red and pink color showing highest temperature in leaves. Temperature ranging from 25 to 34 °C and RWC can vary from 80% to 30% depending on the genotypes and given environmental conditions

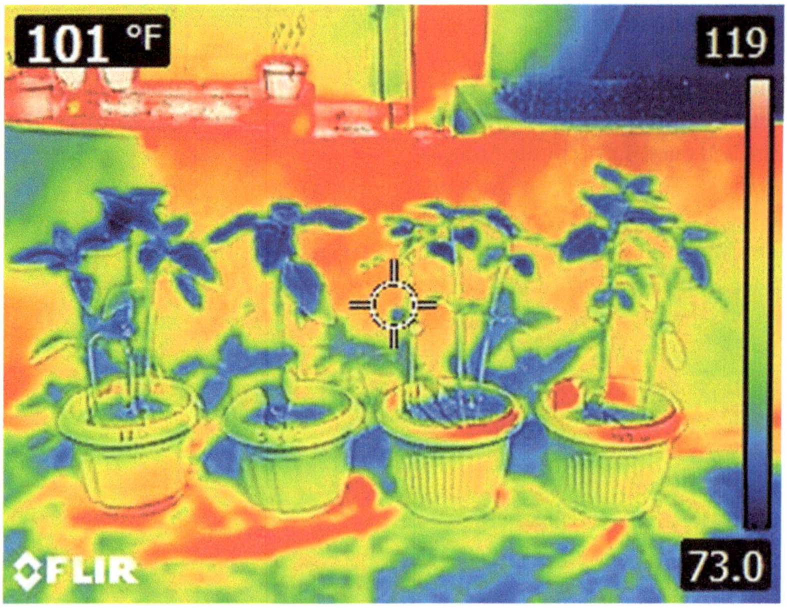

Fig. 15.2 Thermal images of Sunflower genotypes under controlled environments blue leaves showing the low temperature. Soil present inpots also has lower temperature. Report is generated using FLIR software

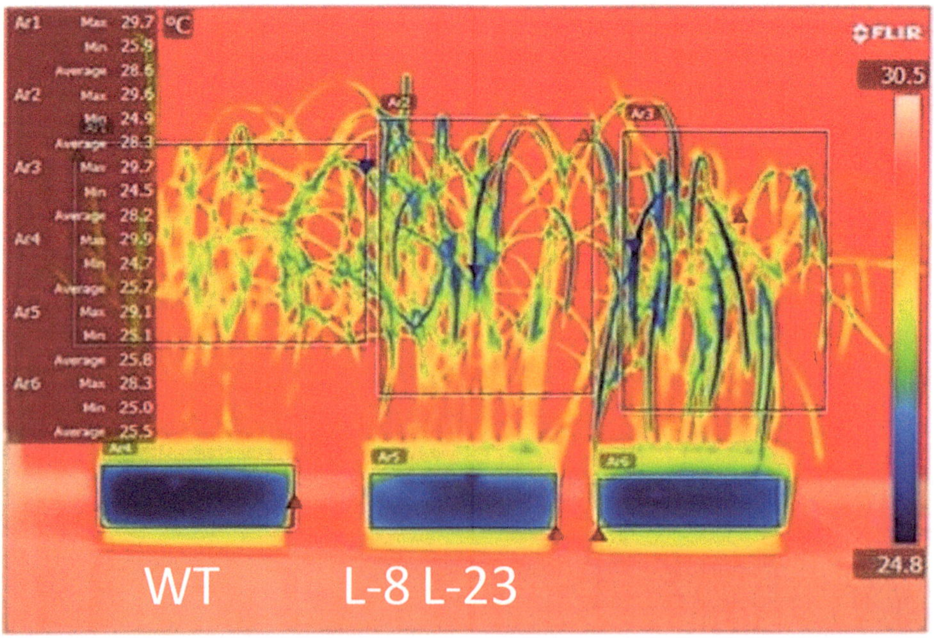

Fig. 15.3 IR thermography of transgenic rice plants under drought stresss environments. High-tech IR images of CaMsrB2 transgenic plants in drought stress environment. WT wild type, L-8 line 8 carrying single-copy T-DNA insertion, L-23 line 23 carrying two-copy T-DNA insertions

software. IR camera generates the images with distinct colors, and software defines the color pattern and average temperature of the whole plant or different plant parts (Fig. 15.3). Based on IR thermography, Siddiqui et al. (2014a) showed that transgenic rice plants showed better tolerance against the drought stress and software-developed images of leaves of transgenic plants

Fig. 15.4 IR images of sunflower plants under controlled and drought stress environments

Fig. 15.5 Four sunflower genotypes under salt stress environments. Leaves in blue color have lower temperature. Scale showing temperature in Fahrenheit (°F). Hysun-33 and Hysun-39 have higher temperature and less RWC compared to the genotypes S.28111 and SF0049

have different color compared to their wild type (Fig. 15.3).

Genotypes of same plant species may have different responses, depending on the growing environment. There is need to identify those genotypes that are tolerant to the stressful environments. In this regard, screening of resistant plant varieties is vital importance. Conventional screening on the basis of physiological trait analysis is hard, time-consuming, and destructive in nature and hence discouraged in plant physiological studies. In the last few decades, high-throughput thermal imaging systems and their computation allow nondestructive and early phenotyping in plant physiological studies. Thermal imaging is useful (1) to identify resistant genotypes against stressful environmental conditions, (2) to save time for screening, and (3) to provide better coverage and most importantly (4) fast and nondestructive in nature. Modern plant physiologists used various traits like biomass allocation, chlorophyll fluorescence, relative water contents, stomatal conductance, and osmotic potential for the selection of resistant or tolerant genotypes (Burke et al. 2006; Siddiqui et al. 2008; Munns et al. 2010; Siddiqui 2013).

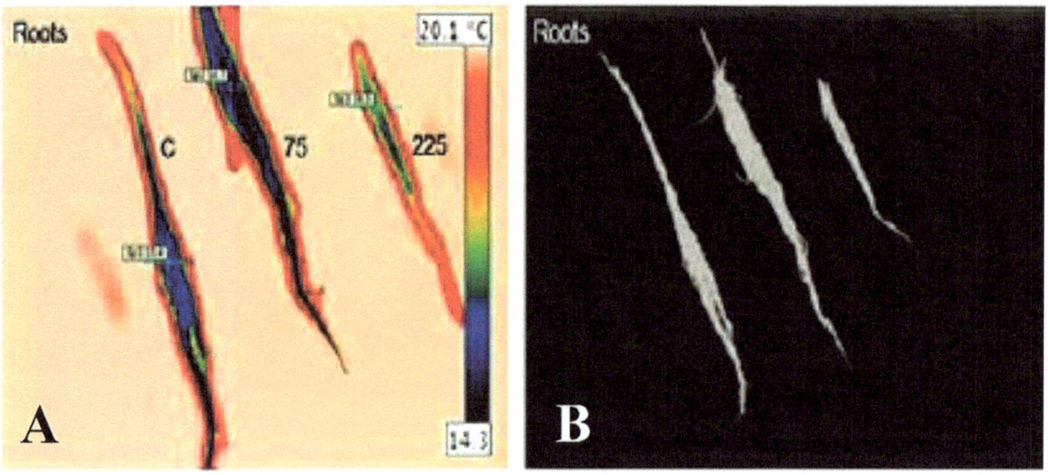

Fig. 15.6 Roots of rice plants under control and two salt concentrations (75 and 225 mM NaCl). Higher salinity increased the root temperature compared to control

Fig. 15.7 Infra-red images of sunflower plants under control (c) salt (S), drought (D) and combined salt and drought stress. Pots temperature also showing that in drought environment temperature increases and lower temperature in control and salt stressed pot

IR Thermography of Transgenic Plants Under Drought Stress

The area where agriculture output depends on rainfall sometimes faces serious problems. Cyclic and unpredictable drought is hampering the ecosystem and agricultural productivity. However, crop plants have to face these constraints and adapted to maintain their homeostasis through water use efficiency and other physiological adjustments (Siddiqui et al. 2014a). Several reports described the tolerance and responsive mechanisms of the plants under drought stress. Modern screening methods showed that many of the transgenic plants are drought resistant and can be of potential use in the drought-affected regions of the world (Kumar and Kumar 1996; Shinozaki and Yamaguchi-Shinozaki 2000; Munns 2002; Siddiqui et al. 2014a). The complexity of the environment and multiple characters in the field always cause difficulties in the bioassay of transgenic plant under drought

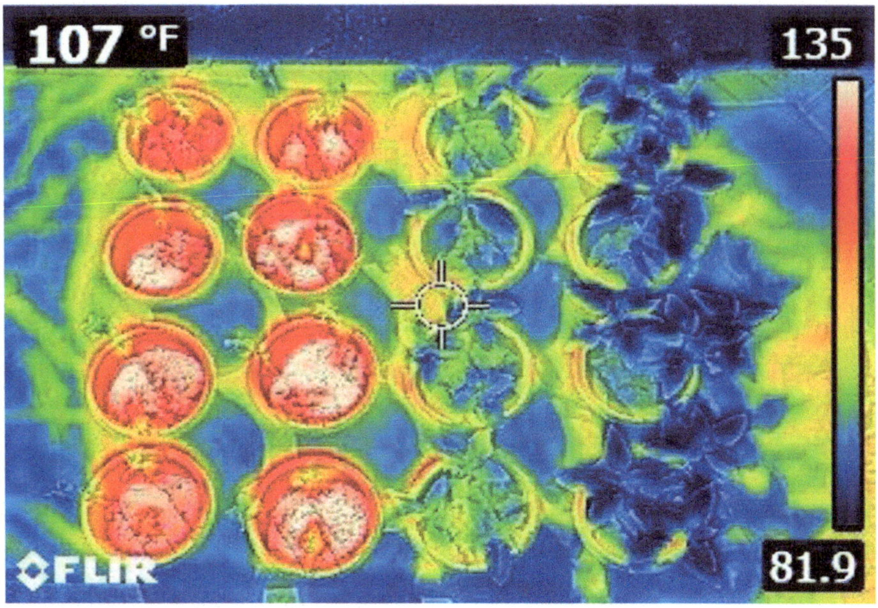

Fig. 15.8 Top view of Infra-red images of four sunflower genotypes (G1, G2, G3, G4) under control (c) salt (S), drought (D) and combined salt and drought stress. Soil temperature also indicating the less water and higher temperature

condition. In this scenario, IR camera can be a useful tool to identify the tolerant or resistant plant genotype in a short run with considerable importance. IR thermography works explain some of the physiological aspects of the plant under drought because it works on the heat production and explains the temperature differences in tested genotypes based on water relation of the plant. Siddiqui et al. (2014a) showed the variation in the color pattern of IR images in CaMsrB2 transgenic with their wild-type rice under drought stress environment (Fig. 15.3). The color pattern explained that transgenic rice plants had lower leaf temperature compared to the wild plant leaf, and the average plant temperature also showed the higher temperature in wild-type rice plant. The high-throughput phenotyping significantly correlated with many physiological traits like that of osmotic potential, relative water contents, and stomatal regulations. Table 15.1 showed that IR images of transgenic rice plants can also describe the status of RWC and stomatal conductance in drought-affected plants. Stomatal conductance was significantly correlated with relative water contents. However highly significant correlation was observed in relative water content with IR images. Some of the non-transgenic plants also showed the considerable differences in plant temperatures under drought environments (Fig. 15.4). IR thermal images also showed the temperature of soil in which plant is placed. In plants, the leaf temperature gives quick indication of plant water status under specific environmental conditions. Thermal sensing of drought-affected plants gives quick answer of plant temperature under drought stress (Munns 2002; Sirault et al. 2009). Average leaf temperature and their correlation with some physiological traits in CaMsrB2 transgenic rice lines were reported (Siddiqui et al. 2014a).

Measurement of Leaf Temperature Using FLIR Software

Phenotyping through IR thermography has greater efficiency to identify the drought-tolerant plant traits. In drought-sensitive plants, the leaf temperature increases with the deficiency of water in leaf. However, drought-tolerant plants maintain the thermoregulation of the plant

Table 15.1 Changes in temperature, RWC and stomatal conductance of CaMsrB2 transgenic lines in drought stress environments

	Pots temperature (°C)			Plant temperature (°C)			RWC %	
Genotypes	Min	Max	Avg	Min	Max	Avg		Stomatal conductance (mmol $m^{-2}S^{-1}$)
WT	24.72^{a}	29.94^{a}	27.33^{a}	25.94^{a}	29.66^{a}	27.80^{a}	40a	10a
L-8	25.11^{b}	29.11^{b}	27.11^{b}	24.94^{b}	27.61^{b}	27.27^{b}	54b	40b
L-23	25.05^{b}	28.55^{c}	26.80^{b}	24.50^{b}	29.61^{a}	27.07^{c}	74c	120c

Describe the relationship between average leaf temperature and Relative water contents in rice plants
WT wild type, L-8 line carrying single-copy T-DNA insertion (CaMsrB2), L-23 line 23 carrying two-copy T-DNA insertion
Similar alphabets on the graphs show non-significant difference at $P = 0.05$ (paired t test)

through stomatal regulation that plays a vital role in accumulation of water contents inside the leaf (Munns 2002; Jones et al. 2009). Hence the RWC and stomatal conductance are significantly correlated and consistent with leaf thermography in water-deficit conditions. Several reports indicate the potential use of IR thermography in plant breeding programs including wheat rice and sunflower (Fischer et al. 1998; Reynolds et al. 1998; Brennan et al. 2007). Data for the temperature of pots and plants presented in Table 15.1 were generated using a computer software "ThermaCAM Researcher Pro 2.10."

IR Thermography for Phenotyping of Salt Tolerance

Low-throughput physiological approaches are time-consuming, laborious, and destructive. There is need to correlate the conventional assessment and modern high-throughput phenotyping approaches and to integrate the protocol to establish adequate phenotypes under those environmental conditions that can control plant temperature (Merlot et al. 2002; Jones et al. 2009; Collins et al. 2010; Munns et al. 2010). Relative water contents and stomatal conductance are important characteristics in the salt and drought stress-related studies and have function to stabilize the temperature of plant organ or whole plant. IR camera can detect this change in temperature.

In rice genotypes IR thermal images showed different color pattern as the salinity increased (Siddiqui et al. 2014b). Increasing salinity in rice plants changed the color pattern of plant leaf from blue to green and then yellow and red (Fig. 15.1). Blue color expresses low temperature and higher relative water contents, while red color showed the elevated temperature and lower RWC. Siddiqui et al. (2014b) observed the pattern of IR thermal images of plants when compared to the physiological parameters had shown significant correlation with RWC and stomatal conductance in salt-induced changes. Table 15.2 explained the relationship of IR images with RWC and stomatal conductance. RWC of plant leaves is inversely proportional to the temperature of leaves. The higher RWC ensure the lower leaf temperature under salt stress. Stomatal conductance also decreased as the temperature increased. Sunflower genotypes when exposed to thermal imaging camera have shown variable responses. Genotype S.28111 and SF0049 were observed to have lowest temperature and highest RWC compared to the Hysun-33 and Hysun-39. The IR thermography based on sensing heat generation of plant is linked with plant's water status rather than other traits. In this regard, reports have shown that some of the physiological traits such as performance index and dark adapted quantum yield did not correlate with IR thermal images in rice plants. Several researchers used Fv/Fm ratio for the indication of salt stress tolerance and sensitivity in crop plants (Penuelas and Boada 2003). IR thermal images sense any variation in leaf temperature, and images should be taken under controlled environmental condition. Some reports indicated the effect of surrounding environments on the IR thermography of target plant. Transpiration through stomata has great effect on the leaf temperature. Before evaporation, water needs substantial amount of heat to convert from liquid to vapor state. This phenomenon explains the cooling effect of leaf because the hottest molecule in the leaf escapes first and then the cooler one (Jones et al. 2009). Hence the given environmental condition, stomatal transpiration, is key factor in maintaining leaf temperature. Osmotic stress due to salinity and drought causes reduction in transpiration of leaves to prevent the loss of water, resulting an increase in leaf temperature (Woo et al. 2008; Munns et al. 2010). IR thermal images of salt- and drought-stressed plants determine indirect water status of plant. The relationship between physiological traits and IR thermography is dependent on the type of genotype intensity and nature of the stress. It was observed that tolerant plant or genotype can uptake soil water effectively and maintains better stomatal regulation and their leaves have shown lower temperature (Jones et al. 2009; Berger et al. 2010; Lu et al. 2011). Furthermore, the leaf temperature is an indicator of stomatal conductance (Sirault et al. 2009). Plant temperature

Table 15.2 Temperatures and RWC of salt-treated and control plants and leaves calculated based on the IR thermal images

NaCl	Plants temperature (°C)			Leaf temperature (°C)			RWC %	Stomatal conductance (mmol $m^{-2}S^{-1}$)
	Min	Max	Avg	Min	Max	Avg		
Control	23.8a	32.9a	28.4a	27.6a	28.2a	27.9a	79%	135
75 mM	22.9b	32.8a	27.9b	27.7a	29.3b	28.5b	76%	107
150 mM	25.1c	33.7b	29.4c	29.4b	30.4c	29.9c	58%	52
225 mM	26.1d	32.8a	29.5c	29.4b	31.4d	30.4d	40%	35

Student t-test was done to compare control and salt treated samples. Different letters present significantly different values at $p < 0.05$

Student t-test was done to compare control and salt treated samples. Similar alphabets are non-significantly differed at $p < 0.05$. *Avg* average increased with the increasing salt intensity. Screening based on the growth and IR thermography was consistent (Sirault et al. 2009; Munns et al. 2010). The water contents inside the leaf are more sensitive during salt stress, and therefore it is main parameter to find out the stress responses of plants (Stepien and Klobus 2006).

The Fv/Fm ratio response against to salt stress is slow and can be observed at seedling stage (Woo et al. 2008; Jansen et al. 2009). Some reports showed that dark adapted quantum yield did not express committed relationship with leaf temperature and it could not be detected by IR thermography. Consequently, IR thermography may not be related to the photosynthetic performance of a plant under saline conditions. Meanwhile, leaf temperature allows the indication of the degree of stress in a plant with clear relationship with stomatal conductance and RWC. Plant temperature and water status are linked with water availability, leaf water potential, and stomatal conductance. IR images not only measure the IR thermography but are proved to be related to soil- and plant-based measures of water stress. IR thermography not only used to describe the plant temperature but also detect the soil water temperature. High-throughput phenotyping can be potentially used for identifying the differences among various genotypes under different irrigations and environmental stresses (Romano et al. 2011; Zia et al. 2011). Advance technology in the field of plant physiology needs to grow in order to identify the best possible protocol to optimize the data accurately. It is suggested that correlation analysis between modern high-throughput phenotyping and conventional low-throughput phenotyping for salt stress-induced changes of plant temperature could be a useful technique to screen the genotypes with target tolerance. Phenotyping through modern IR thermography offers less time-consuming, noninvasive, and mass bioassay for plant salt tolerance-related studies.

Conclusions

The biggest problems of the agricultural world are drought and salinity that badly affect the agricultural output. To tackle the global agricultural productivity under environmental extremes, crop breeding for the improvements of salt and drought resistance is an economically realistic approach. To develop the stress-resistant crop plants, it is necessary to understand the ultimate mechanism working on for the stress tolerance and based on physiological and biochemical characteristics. Conventional and modern plant breeding tactics have been applied to discover the tolerance of plant against desertification and salinization. A great concern of the researchers in the exploration of salt- and drought-tolerant genotypes was to create relation between target gene that is functioning for the tolerance and appropriate phenotype. Previous work on the screening of the plant genotypes was based on destructive and time-consuming approach to find the physiological and biochemical mechanism under stressful environment. Therefore the practical approach was difficult to develop the stress-tolerant genotypes in field environments. Current advancements in agricultural research such as IR thermography, digital camera images, plant-related software, automated robotics, and biotechnologies develop exclusive approaches to help the screening process. Modern approach provides noninvasive measurement of large amount of physiological data for the tolerance in a short time.

This chapter described the possible application of IR thermography as a potential tool to screen the transgenic and non-transgenic plants for the salt and drought stress tolerance. Recently some researcher reported the successful use of infrared images to distinguish between the tolerant and sensitive lines with significant correlation with stomatal conductance and RWC. IR images distinguish the genotypes by indicating the lower temperature and higher relative water contents in tolerant genotypes as compared to sensitive genotypes. Hence, it is suggested that modern IR thermography in agricultural sciences is nondestructive and mass measurement of drought and salt tolerance in lab as well as in field environments.

References

Ashraf M, Wu L (1994) Breeding for salinity tolerance in plants. Crit Rev Plant Sci 13:17–42

Bartels D, Sunkar R (2005) Drought and salt tolerance in plants. Crit Rev Plant Sci 24:23–58

Berger B, Parent B, Tester M (2010) High-throughput shoot imaging to study drought response. J Exp Bot 61:3519–3528

Bray EA, Bailey-Serres J, Weretilnyk E (2000) Responses to abiotic stresses. In: Gruissem W, Buchnnan B, Jones R (eds) Biochemistry and molecular biology of plants. American Society of Plant Physiologists, Rockville, pp 1158–1249

Brennan JP, Condon AG, Van-Ginkel M, Reynolds MP (2007) An economic assessment of the use of physiological selection for stomatal aperture-related traits in the CIMMYT wheat breeding program. J Agric Sci 145:187–194

Burke EJ, Brown SJ, Christidis N (2006) Modeling the recent evolution of global drought and projections for the twenty-first century with the Hadley Centre climate model. J Hydrometeorol 7:1113–1125

Chaves MM (1991) Effects of water deficits on carbon assimilation. J Exp Bot 42:1–16

Chaves MM, Flexas J, Pinheiro C (2009) Photosynthesis under drought and salt stress: regulation mechanisms from whole plant to cell. Ann Bot 103:551–560

Collins M, Fuentes S, Barlow EWR (2010) Partial root zone drying and deficit irrigation increase stomatal sensitivity to vapour pressure deficit in anisohydric grapevines. Funct Plant Biol 37:128–138

Davies WJ, Kudoyarova G, Hartung W (2005) Long-distance ABA signaling and its relation to other signaling pathways in the detection of soil drying and the mediation of the plant's response to drought. J Plant Growth Regul 24:285–295

Fischer RA, Rees D, Sayre KD, Lu ZM, Cordon AG, Saavedra AL (1998) Wheat yield progress associated with higher stomatal conductance and photosynthetic rate and cooler canopies. Crop Sci 38:1467–1475

Flexas J, Diaz-Espejo A, Galme's J, Kaldenhoff R, Medrano H, Ribas-Carbo M (2007) Rapid variations of mesophyll conductance in response to changes in CO_2 concentration around leaves. Plant Cell Environ 30:1284–1298

Flowers TJ (2004) Improving crop salt tolerance. J Exp Bot 55:307–319

Garrity DP, O'Toole JC (1994) Screening rice for drought resistance at the reproduc-tive phase. Field Crop Res 39:99–110

Jansen M, Gilmer F, Biskup B, Nagel KA, Rascher U, Fischbach A, Briem S, Dreissen G, Tittmann S, Braun S (2009) Simultaneous measurement of leaf growth and chlorophyll fluorescence via GROWSCREEN FLUORO allows detection of stress tolerance in *Arabidopsis thaliana* and other rosette plants. Funct Plant Biol 36:902–914

Jones HG, Serraj R, Loveys BR, Xiong L, Wheaton A, Price AH (2009) Thermal infrared imaging of crop canopies for the remote diagnosis and quantification of plant responses to water stress in the field. Funct Plant Biol 36:978–989

Kumar V, Kumar D (1996) Response of Indian mustard to saline water application at different growth stages. Trans Indian Soc Desert Technol 15:121–125

Kwon TR, Kim KH, Yoon HJ, Lee SK, Kim BK, Siddiqui ZS (2015) Phenotyping of plants for drought and salt tolerance using infra-red thermography. Plant Breed Biotechnol 3:299–307

Lawlor DW, Cornic G (2002) Photosynthetic carbon assimilation and associated metabolism in relation to water deficits in higher plants. Plant Cell Environ 25:275–294

Lee SK, Kwon TR, Seo EJ, Bae SC (2011) Current statues of phenomics and its application for crop improvement: imaging systems for high-throughput screening. Korean Plant Breed J 43:165–172

Lu Y, Hao Z, Xie C, Crossa J, Araus JL, Gao S, Vivek BS, Magorokosho C, Mugo S, Makumbi D, Taba S, Pan G, Li X, Rong T, Zhang S, Xua Y (2011) Large-scale screening for maize drought resistance using multiple selection criteria evaluated under water stressed and well-watered environments. Field Crop Res 124:37–45

Merlot S, Mustilli AC, Genty B, North H, Lefevre V, Sotta B, Vavasseur A, Giraudat J (2002) Use of infrared thermal imaging to isolate Arabidopsis mutants defective in stomatal regulation. Plant J 30:601–609

Munns R (1993) Physiological processes limiting plant growth in saline soil: some dogmas and hypotheses. Plant Cell Environ 16:15–24

Munns R (2002) Comparative physiology of salt and water stress. Plant Cell Environ 25:239–250

Munns R, Tester M (2008) Mechanisms of salinity tolerance. Annu Rev Plant Biol 59:651–681

Munns R, James RA, Lauchli A (2006) Approaches to increasing the salt tolerance of wheat and other cereals. J Exp Bot 57:1025–1043

Munns R, James RA, Sirault XRR, Furbank RT, Jones HG (2010) New phenotyping methods for screening wheat and barley for beneficial responses to water deficit. J Exp Bot 61:3499–3507

Penuelas J, Boada M (2003) A global change-induced biome shift in the Montseny mountains (NE Spain). Glob Chang Biol 9:131–140

Prashar A, Jones HG (2014) Infra-red thermography as a high-throughput tool for field phenotyping. Agronomy 4:397–417

Reynolds MP, Singh RP, Ibrahim A, Ageeb OAA, Larque-Saavenra A, Quick JS (1998) Evaluating physiological tools to complement empirical selection for wheat in warm environments. Euphytica 100:84–95

Romano G, Zia S, Spreer W, Sanchez C, Cairns J, Aeaus JL, Muller J (2011) Use of thermography for screening genotypic water stress adaptation in tropical maize. Comput Electron Agric 79:61–74

Shinozaki K, Yamaguchi-Shinozaki K (2000) Molecular response to dehydration and low temperature: differences and cross-talk between two stress signaling pathways. Curr Opin Plant Biol 3:217–223

Siddiqui ZS (2013) Effects of double stress on antioxidant enzyme activity in *Vigna radiata* (L.) Wilczek. Acta Bot Croat 72:145–156

Siddiqui ZS, Khan MA, Kim BG, Huang JS, Kwon TR (2008) Physiological response of *Brassica napus* genotypes in combined stress. Plant Stress 2:78–83

Siddiqui ZS, Cho JI, Kwon TR, Ahn BO, Lee KS, Jeong MJ, Ryu TH, Lee SK, Park SC, Park SH (2014a) Physiological mechanism of drought tolerance in transgenic rice plants expressing *Capsicum annuum* methionine sulfoxide reductase B2 (CaMsrB2) gene. Acta Physiol Plant 36:1143–1153

Siddiqui ZS, Cho JI, Park SH, Kwon TR, Ahn BO, Lee GS, Jeong MJ, Kim KW, Lee SK, Park SC (2014b) Phenotyping of rice in salt stress environment using high-throughput infrared imaging. Acta Bot Croat 73:149–158

Silveira JAG, Silva EN, Ferreira-Silva SL, Viégas RA (2012) Physiological mechanisms involved with salt and drought tolerance in *Jatropha curcas* plants. In: Carels N et al (eds) Jatropha, challenges for a new energy crop: volume 1: 125 Farming, Economics and Biofuel. https://doi.org/10.1007/978-1-4614-4806-8_7

Sirault XRR, James RA, Furbank RT (2009) A new screening method for osmotic component of salinity tolerance in cereals using infrared thermography. Funct Plant Biol 36:970–977

Stepien P, Klobus G (2006) Water relations and photosynthesis in *Cucumis sativus* L. leaves under salt stress. Biol Plant 50:610–616

Tavakkoli E, Fatehi F, Rengasamy P, McDonald GK (2012) A comparison of hydroponic and soil-based screening methods to identify salt tolerance in the field in barley. J Exp Bot 63:3853–3867. https://doi.org/10.1093/jxb/ers085

Woo N, Badger MR, Pogoson B (2008) A rapid non-invasive procedure for quantitative assessment of drought survival using chlorophyll fluorescence. Plant Methods 4:27

Zia S, Spohrer K, Wenyoung D, Spreer W, Romano G, Xiongkui H, Mller J (2011) Monitoring physiological responses to water stress in two maize varieties by infrared thermography. Int J Agric Biol Eng 4:7–15

The Mangroves of Myanmar

16

Christoph Zöckler and Cherry Aung

Abstract

This manuscript describes the biogeography and composition of mangroves in Myanmar. It underlines the significant and ongoing decline of mangrove coverage, the importance to protect these ecosystems, and their roles for critically endangered species of flora and fauna. Reasons for mangrove habitat loss are analysed. The monetary value of mangrove ecosystem services for coastal people has been estimated. UNESCO Biosphere Reserves and Ramsar sites have been suggested as best options for the future conservation and management of Myanmar's mangroves. Developing community-based protected areas and forest management within the framework of the Man and the Biosphere Programme would cater for most of the management options that also involve local communities.

Keywords

Mangrove biodiversity · Ramsar sites · UNESCO Biosphere Reserves

C. Zöckler (✉)
Manfred Hermsen Foundation, Bremen, Germany
e-mail: christoph.zoeckler@m-h-s.org

C. Aung
Department of Marine Science, Pathein University, Pathein, Myanmar

Introduction

Myanmar is the largest country in mainland Southeast Asia, with a continuous coastline of almost 3000 km, extending along the Bay of Bengal and Andaman Sea. The coastal habitats consist largely of soft marshes with mudflats and mangroves, interspersed with a few rocky outcasts and sandy beaches. In 2007 it held an estimated 437,000 ha of mangroves. It holds the 8th largest mangrove forest in the world (4% of the world mangroves) and stands 3rd largest in Southeast Asia (8.8% of Southeast Asia mangroves; Spalding et al. 2010) but is losing its mangroves fastest of all countries (Richards and Friess 2016), due to encroaching agriculture and aquaculture (Plate 16.1).

Distribution and Trends

Myanmar is the eighth largest mangrove nation with less than 500,000 ha mangroves left (Fig. 16.1).

Mangroves and adjacent mudflats are found in all coastal provinces and are widely distributed across the entire coastline (Figs. 16.2 and 16.3). The Central Delta region has been the most important for mangroves, which are concentrated along the southernmost parts of the Ayeyarwady Delta. However, in recent decades, the mangroves have been seriously depleted. Two other principal

B. Gul et al. (eds.), *Sabkha Ecosystems*, Tasks for Vegetation Science VI,
https://doi.org/10.1007/978-3-030-04417-6_16

Plate 16.1 Mangroves near Myeik. (C. Zöckler)

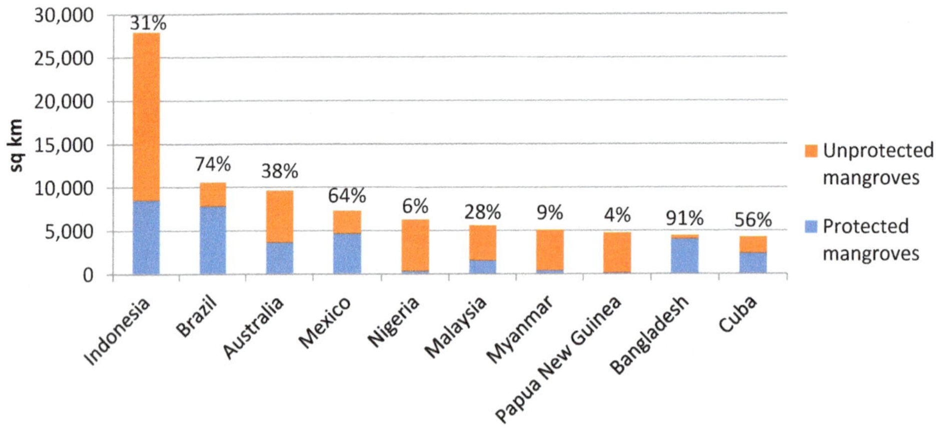

Fig. 16.1 Mangrove areas under protection for the ten largest mangrove nations. The estimates are based on the Global Distribution of Mangroves USGS (2011). Labels indicate the percentage of mangroves under protection

formations, along sheltered coasts in the Rakhine and Tanintharyi regions, are important. Figure 16.3 shows the current distribution of mangroves based on WCMC 2011 and updates from various sources (Zöckler et al. 2013; Harris et al. 2016; Bhagwat et al. 2016). The FAO (2010) report still mentions 437,931 ha, but recent data from Webb et al. (2014) show continued and strong decline in the Ayeyarwady Delta. All other regions suggest a strong decline since the 1980s in line with older assessments (FREDA and Actmang 2012). Richards and Friess (2016) estimated the overall loss from 2000–2012 to only 5.53%. While these identified losses continue through the regions, the losses in acreage seem less pronounced in the Tanintharyi Region. However, at closer look the level of destruction within the remaining mangrove stands is obvious and seems to continue even in protected sites, such as Meinmahla Kyun Wildlife Sanctuary (Yong 2016, Moses and Zöckler 2015, Zöckler 2016a). The degradation has been observed

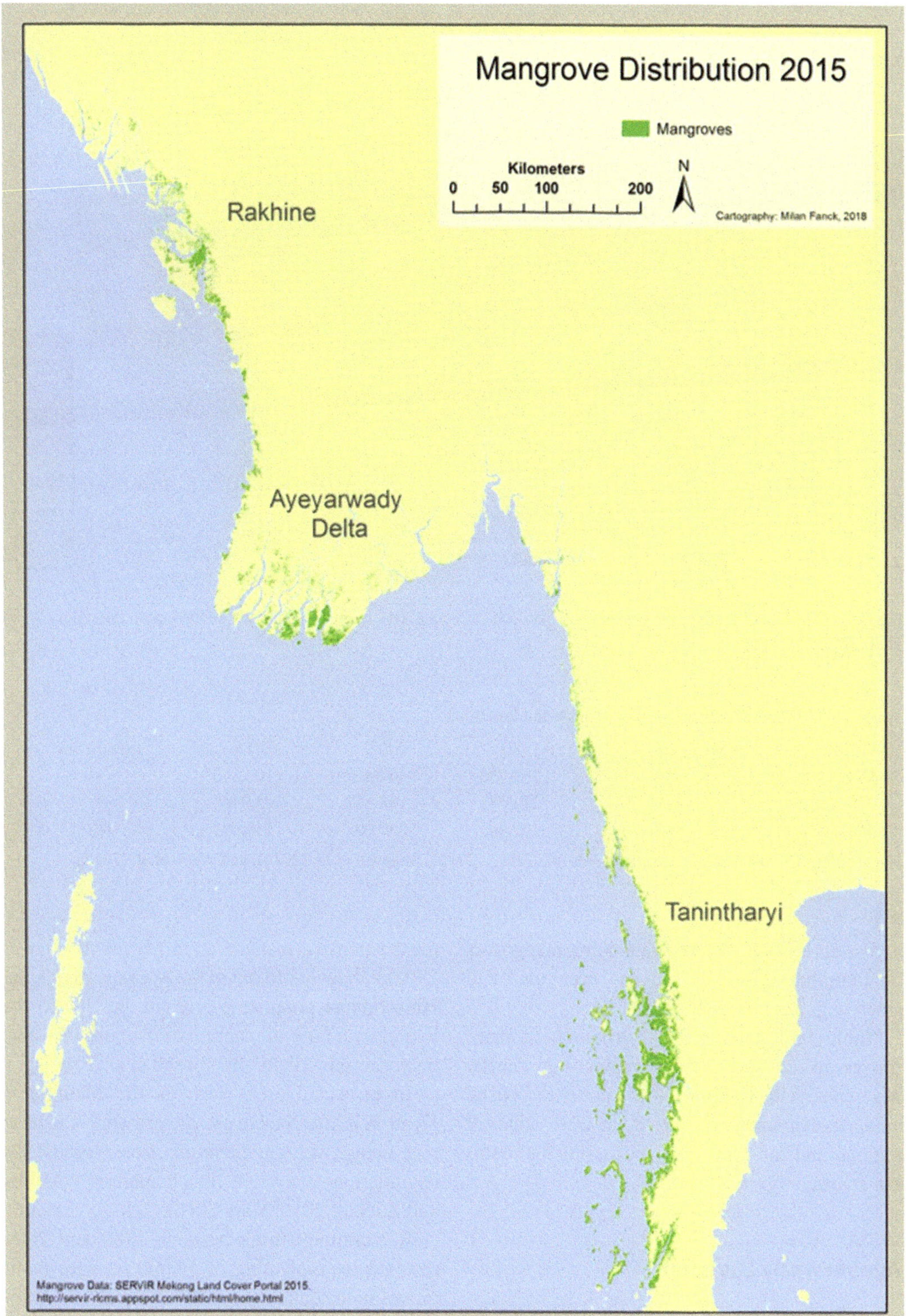

Fig. 16.2 Mangrove distribution by 2015 based on SERVIR-Mekong Land cover data. For updated information, see detailed regional maps

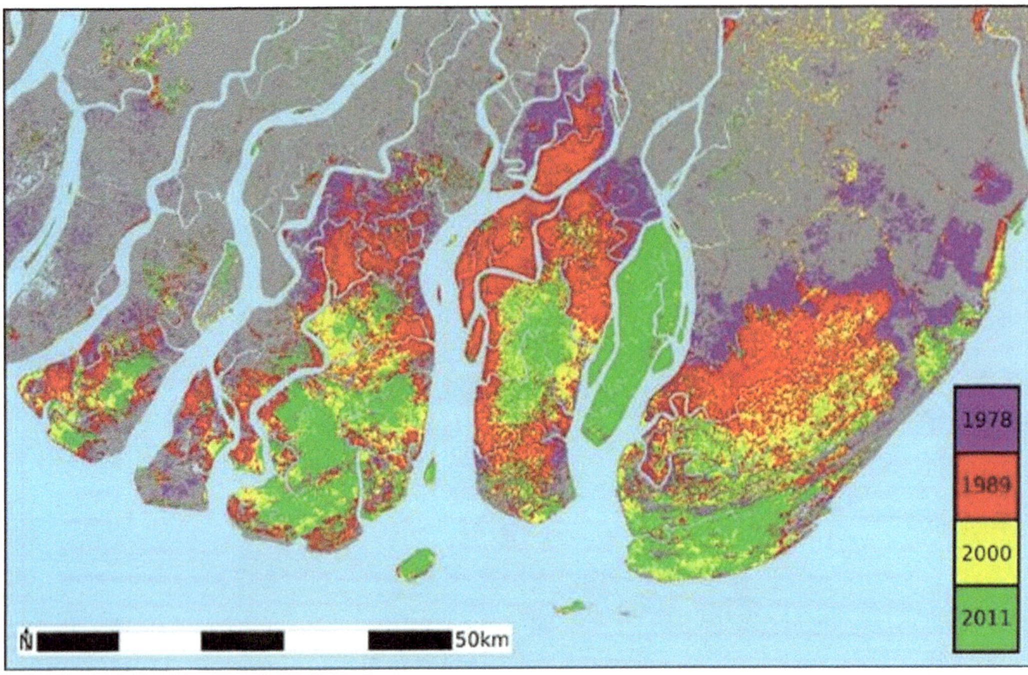

Fig. 16.3 Changes in mangrove forest cover between 1978 and 2011 in the Ayeyarwady Delta region. (Based on Webb et al. 2014)

Table 16.1 Mangrove distribution by region and loss in ha

Division	No. of tree spp.	1980	2002	2013	Loss in %
Rakhine	32	167,730	no data	102,840	>30
Ayeyarwady	29	274,781	138,341	45,048	>80
Tanintharyi	43	262,063	250,00	151,001	>40

FAO (2010), Cherry Aung (2016), Richards and Friess (2016), Bhagwat et al. (2016), and Webb et al. (2014)

across all areas in the vast remaining mangroves in Tanintharyi Region and has been unabated since (e.g. Zöckler 2016b).

Table 16.1 depicts the distribution of mangroves in the three main regions of coastal Myanmar, with Rakhine in the northern coastal area, Ayeyarwady in the delta and adjacent regions, and Tanintharyi encompassing the southern coastal strip.

Ayeyarwady Division

Mangroves have been covering the vast majority of the delta area of the Ayeyarwady Basin. They can reach as far as 60 km inland. Originally, the most extensive mangroves in Myanmar of over 270,000 ha were found in the Ayeyarwady Delta. High human population pressure has led to the loss of over 64% of mangrove cover over the past 35 years, more than 80% (Webb et al. 2014).

In protected areas such as the Meinmahla Kyun Wildlife Sanctuary, groundwork suggests that mangroves are suffering from precipitous degradation and rarely reach maturity (Moses and Zöckler 2016; Yong 2016).

Restoration efforts between 1980 and 2004 resulted in re-planting of 14,000 ha with mangroves in the delta region, mostly on old rice paddies and abandoned agricultural sites. The damage caused by the tropical storm Nargis has been estimated at 35,000 ha of mangrove forest

Plate 16.2 Mangroves at Meinmahla Kyun Wildlife Sanctuary. (C. Zöckler)

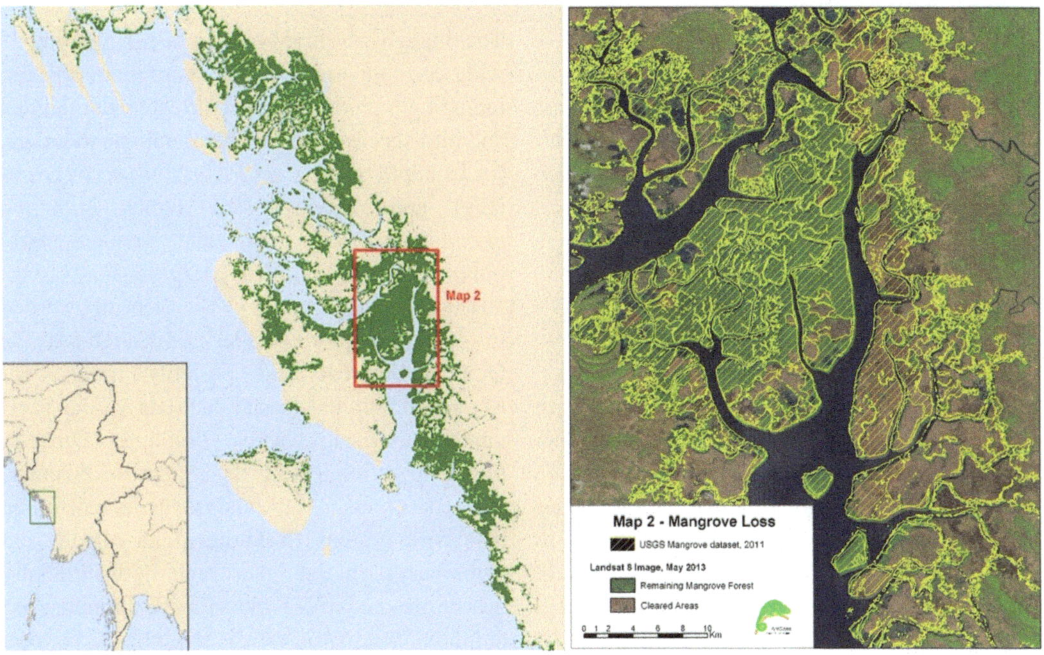

Fig. 16.4 Mangrove loss in the Wanbike area near Ramree Island in Rakhine region between 2001 and 2013. (Zöckler et al. 2013)

in two divisions of the delta region (FREDA and ACTMANG 2012). Comparison of satellite images between 1974 and February 1995 in the Ayeyarwady Delta area indicates mangrove losses from 32.2% to 5.8% of the forest area in Laputta and from 51.9% to 19.5% in Bogalay (Plate 16.2).

Rakhine State

Pressure on the Rakhine mangrove forest habitat is still increasing according to our own research on the ground, with new developments planned for deep sea ports, gas pipelines, infrastructure projects, and hotels (see Fig. 16.4). Adverse anthropogenic impacts on mangroves are expected to become even more pronounced as Myanmar opens its doors to increased foreign investment in agriculture and development (Zöckler et al. 2013).

In other parts of Rakhine, an additional threat is posed by shrimp farming. The mangrove degradation in the Rakhine region between 1974 and 1995 has been documented by Myint Pe (2002), showing a loss of ca. 30%; however, more recent data are not available. An analysis of the Wanbike area illustrates the loss of mangroves over the last 10–12 years (see Fig. 16.4), due to land conversion into rice paddies and aquaculture farms.

Tanintharyi Region

The Tanintharyi Region still holds large areas of mature mangroves, including the largest remaining mangrove area in Myanmar with >150,000 ha. However, this area has also decreased from 262,000 ha in the 1980s.

Decline in Mangrove Quality

In addition to the actual loss in mangrove area, there is also a deterioration of the remaining mangroves. The rapid assessment tool was also applied in the southern Tanintharyi Region (see Fig. 16.5). The rapid assessment method is being described elsewhere (Zöckler and Wodehouse in prep.), including factors such as logging, cutting, clear-cutting, fire, and the intensity of the impact.

Mangroves as Habitats

Plants and Vegetation

Myanmar hosts 44 species of mangrove trees of which species of the genus *Rhizophora*, *Sonneratia*, *Avicennia*, *Bruguiera*, *Ceriops*, *and Xylocarpus* spp. (Table 16.2) are dominant. This is within the range of expected 60 species in SE Asia (Yong 2016). Eight of these species are considered as globally threatened (Polidoro et al. 2010), and two are critically endangered (*Sonneratia griffithii* and *Bruguiera hainesii*). The former has been found with few individuals (<200) in Meinmahla Kyun (Yong 2016) and also a few in Kyaikkhami area, Mon State. The latter still remains with a few (<1000) in the Gwa region (Rakhine). Six plants of *Bruguiera hainesii* were found in the Magyi (Shwe Thaung Yan) mangrove area (Htoo Lwin Aung 2016). *Pemphis acidula* is one of the rare species, and it only occurred near Kyaikkhami, Mon State (Cherry Aung 1999), and in Lampi Island (San Tha Tun et al. 2008) (Plate 16.3).

Animal Biodiversity

The mangroves in Myanmar are supporting a wide range of vertebrate and invertebrate species, including several globally threatened mammal and bird species. Among the mammals these are, for example, fishing cat *Prionailurus viverrinus* (EN) and smooth-coated otters *Lutrogale perspicillata* (VU) that have been regularly observed in the Tanintharyi Region (Plate 16.4). Also dhole *Cuon alpinus* (VU) has been recorded in the delta region (Zöckler and Kottelat 2017). Quite prominent and possibly unique for Myanmar and its coastal habitats is the good numbers of Irrawaddy dolphins *Orcaella brevirostris* frequenting the mangrove channels and coasts near the mangroves in all three mangrove regions (Rakhine, delta region, and Tanintharv). In the latter region, the globally near-threatened (NT) Indo-Pacific humpbacked dolphin, *Sousa chinensis*, is still present in good numbers (Moses and Zöckler 2016) (Plate 16.4).

A total of over 230 species of birds were observed in Myanmar's mangroves (Zöckler et al. 2018). The lesser adjutant stork *Leptoptilos javanicus* (VU), mangrove pitta *Pitta megarhyncha* (NT), and brown-winged kingfisher *Pelargopsis amauroptera* (NT) are characteristic flagship species and still present in good numbers in the southern mangroves in Tanintharyi but scarce in the delta area and largely missing in the Rakhine region. Mangroves and associated

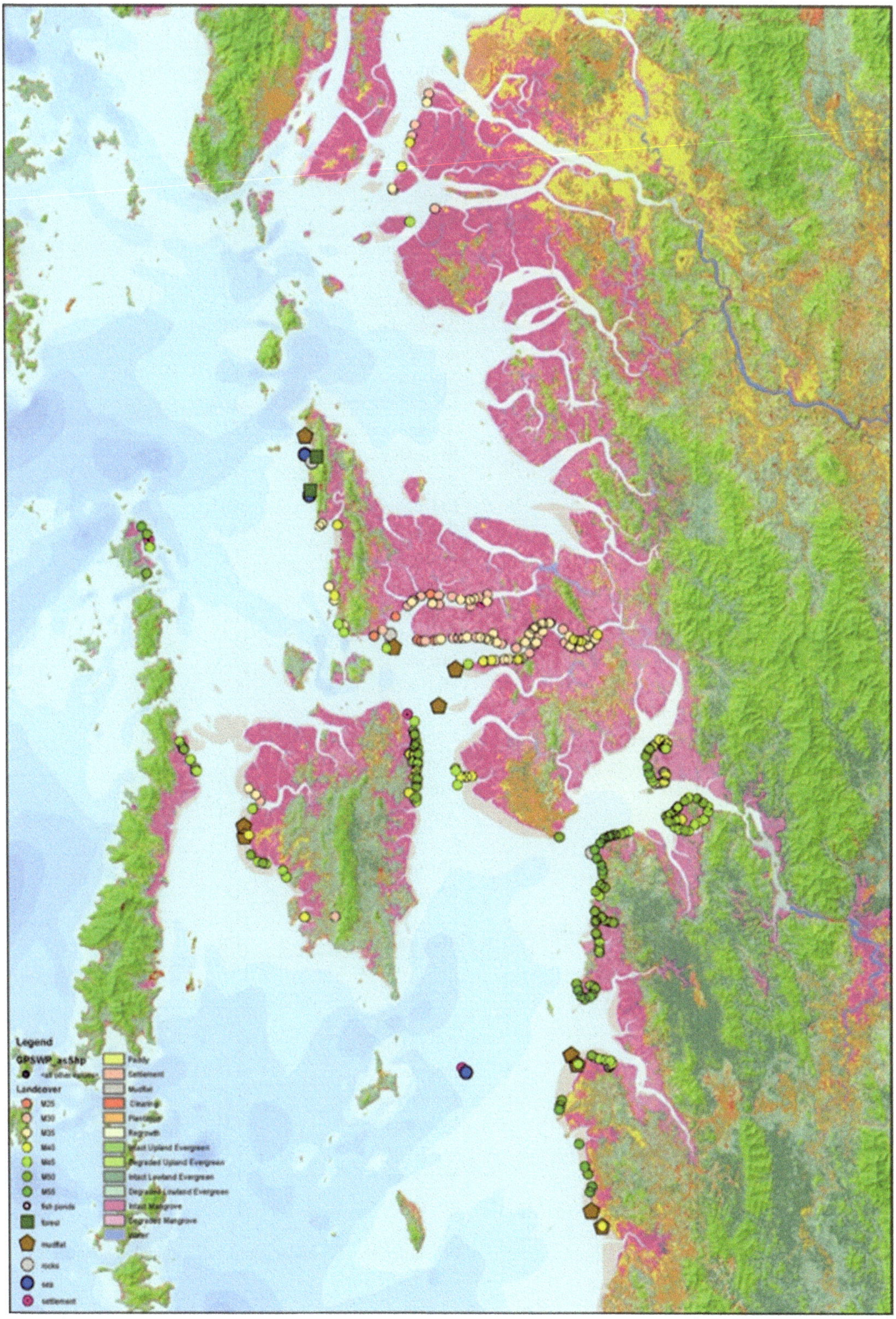

Fig. 16.5 Coastal habitats in the southern Tanintharyi Region, mangroves in pink, mudflats in grey, depicting the level of mangrove deterioration on a scale of M25 (heavy deterioration)–M55 (intact mangrove stands). (Zöckler 2016b)

Table 16.2 A list including the conservation status of the "true" mangrove species and hybrids observed during these studies in Myanmar

	Species	Family	Conservation status
1	*Acanthus ilicifolius*	Acanthaceae	LC
2	*Acanthus ebracteatus*	Acanthaceae	LC
3	*Acanthus volubilis*	Acanthaceae	LC
4	*Acrostichum aureum*	Pteridaceae	LC
5	*Acrostichum speciosum*	Pteridaceae	LC
6	*Aegiceras corniculatum*	Myrsinaceae	LC
7	*Aegialitis rotundifolia*	Plumbaginaceae	NT
8	*Aglaia cucullata*	Meliaceae	DD
9	*Avicennia alba*	Acanthaceae	LC
10	*Avicennia marina*	Acanthaceae	LC
11	*Avicennia officinalis*	Acanthaceae	LC
12	*Brownlowia tersa*	Malvaceae	LC
13	*Bruguiera cylindrica*	Rhizophoraceae	LC
14	*Bruguiera gymnorhiza*	Rhizophoraceae	LC
15	*Bruguiera parviflora*	Rhizophoraceae	LC
16	*Bruguiera hainesii*	Rhizophoraceae	CR
17	*Bruguiera sexangula*	Rhizophoraceae	DD
18	*Ceriops decandra*	Rhizophoraceae	NT
19	*Ceriops tagal*	Rhizophoraceae	LC
20	*Cynometra iripa*	Fabaceae	LC
21	*Dolichandrone spathacea*	Bignoniaceae	LC
22	*Excoecaria agallocha*	Euphorbiaceae	LC
23	*Heritiera fomes*	Malvaceae	EN
24	*Heritiera littoralis*	Malvaceae	LC
25	*Kandelia candel*	Rhizophoraceae	LC
26	*Lumnitzera littorea*	Combretaceae	LC
27	*Lumnitzera racemosa*	Combretaceae	LC
28	*Nypa fruticans*	Arecaceae	LC
29	*Phoenix paludosa*	Arecaceae	NT
30	*Pemphis acidula*	Lythraceae	LC
31	*Rhizophora apiculata*	Rhizophoraceae	LC
32	*Rhizophora mucronata*	Rhizophoraceae	LC
33	*Rhizophora stylosa*	Rhizophoraceae	LC
34	#*R* x *annamalayana*	Rhizophoraceae	EN
35	#*R* x *lamarckii*	Rhizophoraceae	VU
36	*Scyphiphora hydrophyllacea*	Rubiaceae	LC
37	*Sonneratia alba*	Lythraceae	LC
38	*Sonneratia apetala*	Lythraceae	LC
39	*Sonneratia caseolaris*	Lythraceae	LC
40	*Sonneratia griffithii*	Lythraceae	CR
41	*Sonneratia ovata*	Lythraceae	LC
42	#*Sonneratia* hybrids/spnov	Lythraceae	EN
43	*Xylocarpus granatum*	Meliaceae	LC
44	*Xylocarpus moluccensis*	Meliaceae	LC

Based on Cherry Aung (2016), and Yong (2016)

LC least concern, *DD* data deficient, *NT* near-threatened, *VU* vulnerable, *EN* endangered, *CR* critically endangered

Plate 16.3 Flowering mangrove tree *Lumnitzera racemosa* (Myeik). (C. Zöckler)

Plate 16.4 Smooth-coated otters *Lutrogale perspicillata* near Kan Maw (Myeik). (C. Zöckler)

mudflats are also home to a number of migratory waterbirds. A total of more than 20,000 migratory waterbirds have been counted regularly in winter in the southern Tanintharyi mangroves and mudflats alone. Among these are several globally threatened waterbirds (Zöckler et al. 2014, in prep.).

Among the reptiles such as snakes, crocodiles, and lizards, the most prominent examples are the estuarine crocodile *Crocodylus porosus* and the mangrove monitor lizard *Varanus indicus* (Thorbjarnarson et al. 2000). Several species of marine and freshwater turtles live exclusively in the mangroves, for example, the mangrove terrapin *Batagur baska*. However, this has no longer been observed in Myanmar's mangroves since early 2000 (Platt et al. 2008).

Fish, Crustaceans and other Invertebartes

Mangroves serve as nurseries for many marine and estuarine fish species, shrimps, and crabs. Several species that are associated with mangroves are listed as vulnerable, near-threatened,

Plate 16.5 Mudskipper Periophthalmus spp. Are prominent fish species in all mangrove ecosystems. Photo Stefan Pfuetzke

or (critically) endangered on the IUCN red list, such as the shovelnose ray, lemon shark, smalltooth sawfish, goliath grouper, and rainbow parrotfish.

Mudskipper fish *Pterophthalmus spec* are very common in Myanmar's mangroves (Plate 16.5). More details on fish species in the Ayeyarwady Delta are listed in Zöckler and Kottelat (2017).

Brachyuran crabs are common mangrove epifauna, of which the *Sesarmidae* and ghost crab *Ocypode* are dominating. Some Sesarmid crab species are specific to host plants. For example, *Sesarma intermedia* prefers to live among the *Nypa* palm trees. The gastropod *Littorina scabra* is a prominent species on the mangrove trees, especially on *Avicennia* species. Many different species of *Sipunculids*, *Echiurids*, worms, bivalves, and some crustaceans are dwelling in the mudflats. The peanut worm (phylum: *Sipunculida*), the prominent and mysterious *Sabellid polychaete* (Plate 16.6), the ancient Cambrian Brachiopod *Lingula anatina*, the small crab *Dotilla*, the slug *Onchidium* sp., and the insect larvae (Fly Fam: Dolichopidae) are just a few examples of the rich diversity of the mangrove and mudflat benthos. They also provide an important food sources for fish, crabs, shrimps, and many birds. Fireflies (Lampyridae) are well known to prefer the *Sonneratia caseolaris* trees. Many different bee species are visiting flowering mangroves, and mangrove honey is highly valued.

Mangrove Ecosystem Services

The Nargis storm in 2008 highlighted the need to emphasise the critical ecosystem services of high-quality mangroves, mainly in protecting coastal communities from storm surges. Moreover, Myanmar's mangroves are highly important for storing carbon and providing vital fish, crab, and shellfish nurseries.

Mature mangroves provide wood for boatbuilding, construction, firewood, charcoal, furniture, and fish traps, as well as non-timber forest products such as mangrove honey, fruit, medicine, and even wine, and nypa palm thatch for roofing and are also used to make syrup, juice, wine, and vinegar. Mangrove honey – with its subtle sweetness and tinge of saltiness – is becoming a valuable commodity in coastal communities.

The density and rot resistance of some types of mangrove wood and a corresponding ability to withstand exposure to saltwater are the reasons this timber is used in boat-building and often for construction of docks, fences, and fish traps. The wood of *Rhizophora*, *Bruguiera*, and *Ceriops* species is a heavy hardwood with tannin-rich barks and is often used as building material for homes, for railroad ties, and in the leather industry. Mangrove leaves are used as animal fodder for buffaloes, sheep, goats, and camels. In some locations, certain mangrove species are used in traditional medicine to treat a range of ailments

Plate 16.6 Sabellid polychaete with "bouquet" using mangrove leaves and sticks. (C. Zöckler)

from toothache to leprosy and tuberculosis (Govindasamy and Kannan 2012).

In the past, Ayeyarwady mangrove was famous for its *Ceriops* and *Bruguiera* trees. After World War II, a lot of *Ceriops* and *Bruguiera* mangroves were produced for charcoal. *Ceriops* charcoal is of excellent quality and exports from Ayeyarwady are famous. The palm *Phoenix paludosa* has been used for bridge and house construction (Cherry Aung 1999).

The mangrove seaweed *Catenella nipae* is another mangrove product. It grows at the trunks and pneumatophores of the plants in Kyaikkhami, Mon State. It provides good income for the local people that collect the seaweed and sell it for human consumption in the form of a salad.

Supporting and regulating services that mangrove habitats provide include nutrient cycling, carbon storage, flood control, water quality maintenance, shoreline stabilisation, sediment balance, land accretion, and the buffering of both land and marine habitats from the ravaging impacts of storms and rising sea levels (Nellemann et al. 2009; McIvor et al. 2012). As human habitation along the coasts has increased, the rich ecosystem services provided by mangroves have also grown.

The value of mangroves has been estimated in the range of 200,000–900,000 USD annually per km^2. Considering a remaining mangrove forest in Myanmar of about 300,000 ha (see Table 16.1), this would mean a total value of mangroves of 150–400 million USD per year, depending on the ecological functioning of the mangrove systems. Values include coastal protection, fisheries, coral reefs, and tourism (Costanza et al. 2008).

Causes of Mangrove Forest Degradation

Mangrove ecosystems, which make up less than 0.4% of the world's forests (Spalding et al. 2010), are being lost at the rate of about 1% per year (FAO 2010); in Myanmar the rate is as high as 5% per year (Richards and Friess 2016). Between 20% and 35% of the world's mangrove area has been lost since 1980 (FAO 2010). The rates of loss are highest in developing countries where mangroves are cleared for agriculture, coastal development, aquaculture, and timber and fuel production (Polidoro et al. 2010). Webb et al. (2014) listed agriculture expansion for rice production as the main driver for loss in mangrove area. Conversion for paddy cultivation of rice is a major threat to mangrove conservation, particularly in the Ayeyarwady Delta area. Agricultural expansion into mangrove areas to meet the requirements of regional food security is also common in the other two coastal regions, especially in the Rakhine region. Figure 16.5 showing mangrove loss between 2001 and 2013 overall depicts agricultural land replacing former mangrove areas (Zöckler et al. 2013).

Over-exploitation for fuelwood and timber production has degraded about 26% of mangrove forests around the world (Valiela et al. 2001), and similar figures are likely to apply with mangrove loss in Myanmar. Shrimp aquaculture has contributed to about 38% of global mangrove loss, and other types of aquaculture account for approximately another 14% (Gilman et al. 2008). For example, the Setse mangrove in Mon State was totally converted into shrimp farm in 2003 (Cherry Aung 2016). A similar scenario was observed in the Ayeyarwady and Rakhine region while the southern mangroves in Tanintharyi Region so far remained unaffected by aquaculture.

While direct anthropogenic impacts including coastal development comprise the biggest threat to mangrove ecosystems at present, climate change is another factor of concern regarding changes in mangrove ecosystems (Gilman et al. 2008).

Over-exploitation of Firewood and Charcoal Production

For at least a hundred years, the Ayeyarwady mangrove forests had provided firewood and charcoal to Yangon, the capital city of Myanmar, and other towns in the delta area. The annual firewood requirement for Yangon is about 700,000 tons, and this demand is increasing steadily due to dynamic population growth (Myint Pe 2002). However, the Ayeyarwady Delta firewood production, of about 432,200 tons per year, is no longer sustainable since the 1980s, and the mangrove forest is diminishing fast, due to over-exploitation of the forest for charcoal production.

A similar development is taking place in the Tanintharyi Region. Especially in the past 5–8 years, an acceleration in charcoal production and mangrove degradation has been observed. This is largely due to the increasing demand and export to Thailand and the cheap import of cheap chain saws (Zöckler 2016b).

Aquaculture Development

Throughout their biogeography, mangroves and intertidal mudflats are exposed to aquaculture development. In Myanmar, aquaculture in mangroves was introduced in 1980 as a pilot scheme in the northern part of Rakhine (see photos). Later, in the southwestern part of the Ayeyarwady Delta area, mangroves have been converted into brackish shrimp ponds, and this development is still continuing (FREDA and ACTMANG 2012) and adding to the loss from agricultural conversion.

The culture of tiger prawns (*Penaeus monodon*) using traditional methods was initiated in 1975 and increased from about 28,000 ha in 2000 up to about 49,000 ha in 2002 (Myint Pe 2002). There has been no further increase after 2003 because there was a severe disease problem affecting shrimp ponds such as white spot virus around 2001–2002. The shrimp business declined in those days because of decreasing market demand. Cyclone Nargis in 2008 destroyed a lot of shrimp farm in Yangon and Ayeyarwady Region, and many shrimp farmers left the job, but recently they started to rebuild with foreign aid support from, e.g. Germany and Japan (FREDA and ACTMANG 2012).

Coastal Development

Myanmar is one of the least developed countries in Asia and only recently opened for foreign investment. This, in turn, leads to coastal development activities in some coastal areas. In Rakhine, oil and gas terminals have been constructed alongside with roads. These have heavily impacted the existing coastal mangrove areas on the Ramree Island. Similar plans exist for the northern Rakhine Province near Dawei, where a major coastal development project encroaches into mangrove habitats. Further south in the Myeik region the industrial zone has been expanding north into mangrove areas.

Plastic and Fishing Nets

Another problem for mangroves in particular is the pollution with macro-plastic, including ghost nets, damaging and sometimes uprooting young mangroves plants.

Mangrove Protection and Restoration

Protected Areas

With an ever-increasing pressure on the remaining mangroves, there is an urgent need to safeguard the remaining mangrove forests of Myanmar. However, only a small fraction of the mangroves in Myanmar are currently protected inside the Meinmahla Kyun Reserve and in the Lampi Marine National Park covering a total of <1% of Myanmar's overall mangrove stands. No further extensions are planned at present (NBSAP 2015). A priority task should be to expand the existing protected area network in Myanmar. Strengthening the legislative framework applicable to protected area management will enhance law enforcement as well as providing incentives for restoration of degraded mangrove areas.

Meinmahla Kyun

Meinmahla Kyun Wildlife Sanctuary, 13,000 ha in size, was established in 1993, protected and managed by the Forest Department. No villages or settlements are within the area of the sanctuary, but villagers are still allowed to go fishing and also cutting mangroves. A new management plan has been drafted that will reduce fishing and harvest of timber. In 2017 the area was extended to include adjacent intertidal mud- and sandflats and islands, increasing the total protected area to about 50,000 ha in total, and the site was designated as Myanmar's third Ramsar site.

Lampi Marine National Park

The Lampi Island Marine National Park (MNP) is a demarcated marine national park located in Myeik Archipelago. It was designated in 1996 and comprises Lampi, the biggest island of the park and the core of the site, and 20 smaller islands in its surroundings. It protects a rich diversity of habitats including mangrove, beaches, dunes, coral reefs, and sea-grass beds.

Ramsar Sites

In 2017 Myanmar designated for the first time two coastal wetlands. In addition to the above mentioned Meinmahla Kyun site, parts of the Gulf of Mottama were designated as Ramsar site adding another 90,000 ha of intertidal mudflats as coastal protected areas. The Ramsar protection tool is particular valuable to cover intertidal mudflats that host important numbers of migratory waterbirds.

UNESCO Man and Biosphere Reserves

Developing community-based protected areas and forest management within the framework of the Man and the Biosphere Programme would cater for most of the management options that also involve local communities. It could create a vital platform for the long-term protection and restoration of mangroves along the Myanmar coast. A good showcase can be the southern Myeik Archipelago, mangroves, and mudflats in the Tanintharyi Region (see Fig. 16.6). Here we still have a large area of mature mangrove stands, raising pressure on the existing natural resources and an open-minded community, interested in jointly developing scenarios and to redress further mangrove loss. Establishing a large

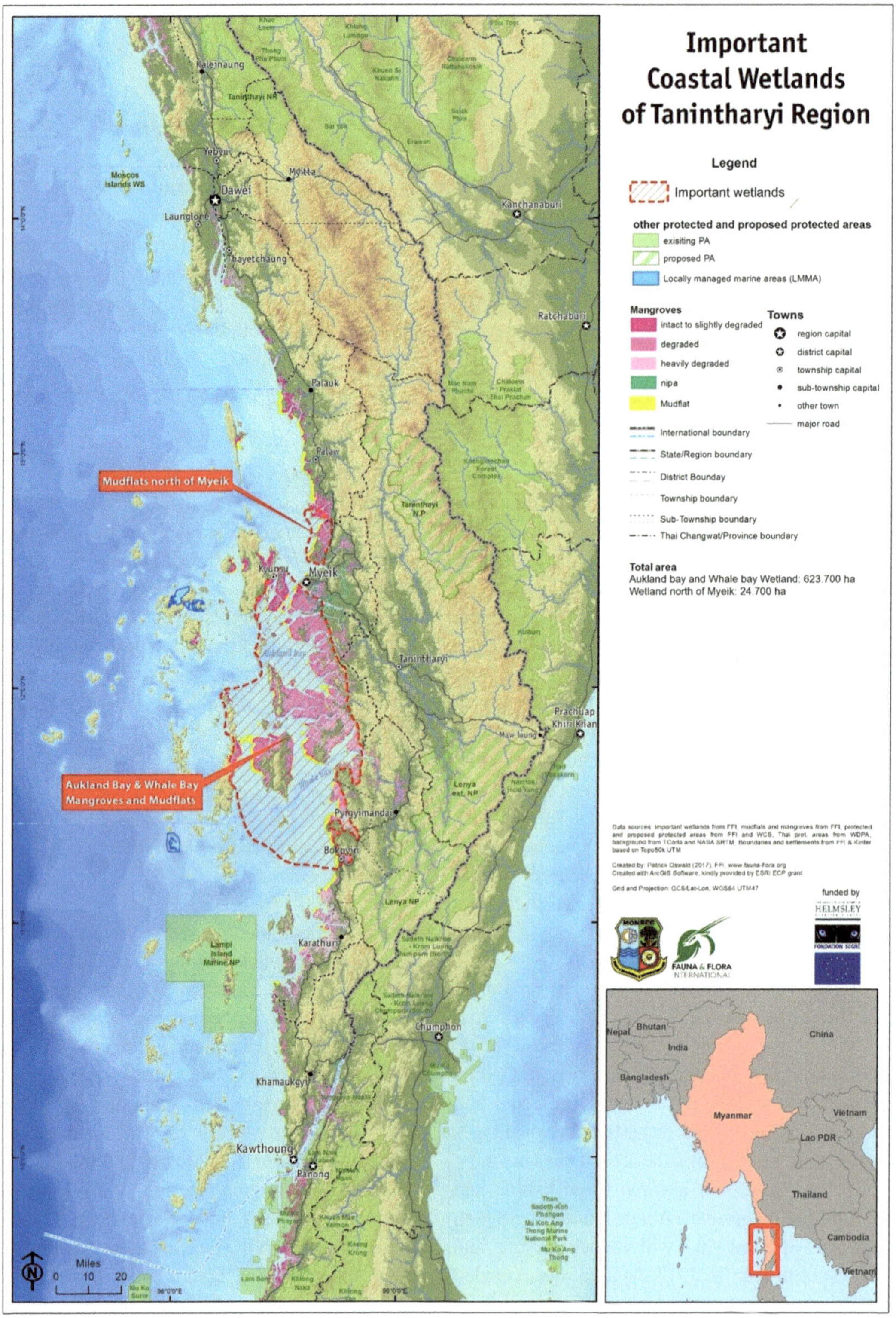

Fig. 16.6 Key coastal biodiversity areas that could be protected as biosphere reserve and as Ramsar sites. (Zöckler et al. 2018)

biosphere reserve with several core areas as protected sanctuaries, combined with Ramsar sites, buffer, and development zones, where nature conservation in reconciliation with sustainable human development and good land use practices can be tested and applied, is being proposed.

Restoration

Recent restoration activities involving local forest user groups pointed the way towards successful government-NGO collaborations to curb further mangrove degradation. Restoration efforts have been taking place in the Ayeyarwady Delta since the 1980s, and a total of 13,000 ha has been restored.

Curbing mangrove deforestation is an urgent conservation priority. A study in Thailand found that the cost of restoring mangroves was US $946 per hectare, while the cost for protecting existing mangroves was only US$ 189 per hectare (Ramsar Secretariat 2001).

Some areas of the mangroves in the Ayeyarwady Delta have been improved by the Forestry Department's rehabilitation programmes, including reforestation and the protection of natural mangroves in Ayeyarwady and Tanintharyi regions.

Establishment of a mangrove plantation programme with suitable mangrove and other forest-growing species has been launched in degraded and abandoned lands.

The Ministry of Natural Resources and Environmental Conservation is cooperating with Denmark on a mangrove conservation project in Rakhine and Tanintharyi regions. The project started in August 2018.

It would be important to coordinate these laudable mangrove conservation and restoration efforts, including with the establishment of a combination of UNESCO Biosphere Reserves and Ramsar Wetlands. This will be major contributions to redress the ongoing mangrove habitat loss and involve the local communities to safeguard the continuous provision of essential ecosystem services provided by intertidal ecosystems.

Acknowledgements We would like to thank Fauna and Flora International (FFI), the Lighthouse Foundation, and the Manfred Hermsen Stiftung for support of surveys and assessments over the years from 2008 to 2018 and the Forest Department of the Ministry of the Environment and Natural Resources for granting permits in remote coastal areas. We would like to thank Patrick Oswald and Milan Fanck who assisted with GIS support.

References

Bhagwat T, Hess A, Horning N, Khaing T, Thein ZM, Aung KM, Aung KH, Phyo P, Ye Lin Tun, Aung Htat Oo, Neil A, Win Myo Thu, Songer M, LaJeunesse Connette K, Bernd A, Connette G, Leimgruber P (2016) Losing a jewel—rapid declines in Myanmar's intact forests from 2002–2014. Unpublished report

Cherry Aung (1999) Study on some ecology of mangrove at Setse, Mon State. Unpublished Master thesis submitted to Mawlamayne University

Cherry Aung (2016) The status of mangroves in Ayeyarwady Region. Paper presented at symposium proceeding of third Japan-Myanmar joint symposium, Pathein University

Costanza R, Pérez-Maqueo O, Martinez M, Sutton P, Anderson S, Mulder K (2008) The value of coastal wetlands for hurricane protection. Ambio 37(4):241–248

Ellison AM (2008) Managing mangroves with benthic biodiversity in mind: moving beyond roving banditry. J Sea Res 59:2–15

FAO (2010) The world's mangroves 1980–2005, FAO forestry paper 153. Forest Resources Division, FAO. The Food and Agriculture Organisation of the United Nations, Rome, 77 pp

FREDA & ACTMANG (2012) Ten years in Pyindaye-Yangon. Thin Publishing House. 164 pp

Gilman E, Ellison J, Duke N, Field C (2008) Threats to mangroves from climate change and adaptation options: a review. Aquat Bot 89(2):237–250

Govindasamy C, Kannan R (2012) Pharmacognosy of mangrove plants in the system of unani medicine. Asian Pac J Trop Dis:S38–S41

(PDF) (Mangroves: A Rich Source of Natural Bioactive Compounds). Available from: https://www.researchgate.net/publication/290929348_Mangroves_A_Rich_Source_of_Natural_Bioactive_Compounds. Accessed 2 Sept 2018

Harris C, Lorenz K, Zöckler C (2016) Land cover classification Mawdin Coast, Ayeyarwady Region, Myanmar Unpubl Rep for FFI Cambridge, 28pp

Htoo Lwin Aung (2016) Species diversity and community structure of mangrove of Magyi, Shwe Thaung Yan Sub-township. Unpublished master of research thesis submitted to Pathein University

McIvor AL, Spencer T, Möller I, Spalding M (2012) Storm surge reduction by mangroves. Natural coastal protection series: report 2. Cambridge coastal research unit working paper 41. Published by The Nature Conservancy and Wetlands

International. 35 pages. ISSN2050-7941. URL: http://www.naturalcoastalprotection.org/documents/storm-surge-reduction-by-mangroves

Moses S, Zöckler C (2015) Bird survey Report Ayeyarwaddy Delta November – December 2015. Unpublished report for Fauna and Flora International. Fauna and Flora International

Moses S, Zöckler C (2016) Bird survey report Ayeyarwady Delta, Myanmar 29 November – 13 December 2015. Unpublished report for FFI. Yangon 2016

Myint Pe (2002) National Report of Myanmar- on the sustainable management of the Bay of Bengal Large Marine Ecosystem (BOBLME) GCP/RAS/179/WBG

Nellemann C, Corcoran E, Duarte CM, Valdés L, DeYoung, C, Fonseca L, Grimsditch G (eds) (2009) Blue carbon: a rapid response assessment. United Nations Environment Programme, GRID-Arendal

Platt K, Platt SG, Thirakhupt K, Rainwater TR (2008) Recent records and conservation status of the critically endangered mangrove terrapin, *Batagur baska*, in Myanmar. Chelonian Conserv Biol 7:261–265

Polidoro B, Carpenter K, Collins L, Duke N, Ellison A (2010) The loss of species: mangrove extinction risk and geographic areas of global concern. PLoS ONE 5(4):e10095. https://doi.org/10.1371/journal.pone.0010095

Ramsar Secretariat (2001) www.ramsar.org

Richards DR, Friess DA (2016) Rates and drivers of mangrove deforestation in Southeast Asia, 200-2012. PNAS 113(2):344–349

San Tha Tun, Tint Swe, Tint Tun (2008) Preliminary study on the mangroves of Lampi Island and adjacent areas. 18pp. Published by Europe Conservation Switzerland (ECoSwiss) and Diversity and Nature Conservation Association (BANCA)

Spalding M, Kainuma M, Collins L (2010) World atlas of mangroves. Earthscan Publishing, London

The Republic of the Union of Myanmar (2015) National Biodiversity Strategy and Action Plan (NBSAP) Yangon, 121pp

Thorbjarnarson JB, Platt SG, Saw Tun Khaing (2000) A population survey of the estuarine crocodile in the Ayeyarwady Delta, Myanmar. Oryx 34:317–324

Valiela I, Bowen JL, York JK (2001) Mangrove forests: one of the world's threatened major tropical environments. Bioscience 51:807–815

Webb EL, Jachowski NRA, Phelps J, Fries DA, Than MM, Ziegler AD (2014) Deforestation in the Ayeyarwady Delta and the conservation implications of an internationally-engaged Myanmar. Glob Environ Chang. https://doi.org/10.1016/j.gloenvcha.2013.10.007

Yong JWH (2016) An ecological and plant biodiversity assessment of the Meinmahla Kyun Wildlife Sanctuary (MKWS) in relation to biodiversity conservation and restoration, and human livelihood. FFI report. 37 pp.

Zöckler C (2016a) Bird survey and training report Ayeyarwady Delta, Myanmar. Unpublished report for Fauna & Flora International

Zöckler C (2016b) The bird and mangrove survey of the inner Myeik Archipelago March 2016. Unpublished report for Fauna & Flora International

Zöckler C, Kottelat M (2017) Biodiversity of the Ayeyarwady Basin. Ayeyarwady State of the Basin Assessment (SOBA) Report 4.5. National Water Resources Committee (NWRC), Myanmar

Zöckler C, Wodehouse D (in prep) A new rapid assessment tool for determining mangrove degradation

Zöckler C, Delany S, Barber J (2013) Sustainable coastal zone management in Myanmar. ArcCona Ecological Consultants and Flora Fauna International. Cambridge, p 60

Zöckler C, Zaw Naing T, Moses S, Nou Soe R, Htin Hla T (2014) The importance of the Myanmar coast for water birds. Stilt 66:37–51

Zöckler C, Moses S, Thu Lwin (2018) The importance of the Myeik mangroves and mudflats, Taninthary, Myanmar for migratory waterbirds. (subm. *Wader Study*)

Potentially Domesticable Chenopodiaceae Halophytes of Iran

17

Mohammad Kafi and Masoumeh Salehi

Abstract

Climate change and water mismanagement have confronted Iran with water scarcity, desertification, and reduction of water quality. In 2020, Iran will need 165 billion m^3 of water annually for producing 120 million tons of foodstuffs. In 2012, the agriculture sector used more than 90% of the renewable water of Iran. Therefore, this country faces water crisis, and the extension of the agriculture sector based on fresh water is nearly impossible. Using saline water and soil for the cultivation of halophytes or high salt-tolerant crops could be considered as an alternative solution. Among native halophytes, Chenopodiaceae has the highest (44%) number of species with wide adaptability and different usage (forage, vegetable, food, and oil). Successful utilization of native and exotic chenopods in Iran like *Atriplex* and *Kochia scoparia* as forage and *Salicornia* as oil crop and quinoa as a food is promising. *Halocnemum strobilaceum*, *Seidlitzia rosmarinus*, *Halostachys caspica*, *Suaeda fruticosa*, and *Salsola* were evaluated as forage crops, and among them *Halocnemum strobilaceum* showed high salt tolerance and wide adaptability. However, the highest forage quality belonged to *Halostachys caspica.* Evaluation of forage quality showed that low metabolic energy is the main problem of using these species as feed. *Suaeda aegyptiaca*, *Salsola soda*, and *Salicornia persica* could be considered as vegetables. *Arthrocnemum macrostachyum*, *Suaeda fruticosa*, *Suaeda aegyptiaca, Halocnemum strobilaceum*, and *Salicornia persica* with 10 up to 30% edible oil could be promising oil crops in the future. In addition, chenopods are host of four species of *Cistanche* that are capable to use as medicinal plants.

Keywords

Chenopods · Forage · High saline water · Medicinal · Oil

Introduction

Iran covers an area of 1,648,195 km^2, 50% of which consists of mountains and deserts. The largest mountain chain of Iran is Zagros, which runs from the northwest of the country southward, first to the shores of the Persian Gulf and then continues eastward till the most southeastern province. Other mountain ranges (Alborz) run

M. Kafi (✉)
Department of Agrotechnology, Ferdowsi University of Mashhad, Mashhad, Iran
e-mail: m.kafi@um.ac.ir

M. Salehi
National Salinity Research Center, Agricultural Research, Education and Extension Organization (AREEO), Yazd, Iran

B. Gul et al. (eds.), *Sabkha Ecosystems*, Tasks for Vegetation Science VI,
https://doi.org/10.1007/978-3-030-04417-6_17

from the northwest to the east along the southern edge of the Caspian Sea. The Central or Interior Plateau is located in between these mountain chains and covers over 50% of the country. It is partly covered by a remarkable salt swamp (kavir) and partly by areas of loose sand or stones. Approximately 90% of the country is climatologically arid and semiarid. The summer is extremely hot with temperatures in the interior reaching as high as 55 °C. In winter, temperatures lower than freezing point are common in many places, reaching as low as −30 °C in the northwest (Cheraghi and Halim 2012).

Salinity is one of the main constraints of agricultural production. Most of the saline lands (55 million ha almost 34% of the country) are located in the central plateau, coastal area, and Khuzestan province (Table 17.1, Fig. 17.1). Of the total land area, some 90 million ha or 54.6% are rangelands; 12.4 million ha or 7.5% are forests; and 34 million ha equal to 20.6% are deserts (Table 17.1, Fig. 17.1).

The rate of population growth in Iran is around 1.5%. Hence, more water resources will be needed in order to produce enough food for this population (Shahrestani 2013). Annual precipitation of Iran is 250 mm which is almost 30% of world rainfall, while 70% of this precipitation is evaporated. The average annual potential evaporation of the country is ranging from less than 700 mm along the Caspian Sea shore to over 4000 mm in the deserts and the southwestern part of Khuzestan province. The country's total renewable water is around 120 billion m^3, and water consumption of Iran is estimated to be 96 billion m^3. Presently, Iran is using 80% of its water resources (Shahrestani 2013), but some parts of the country are of water resources, and the agriculture sector is the main water consumer (around 90%).

Table 17.1 Salt-affected areas of Iran (Momeni 2011)

Lands	Million ha	%
Lands with low to high salinity	43.0	26.1
Lands with rather high to high salinity	4.8	3.0
Lands with very high salinity to rather high	7.8	4.7
Central desert	20.0	12.1
Plains	1.2	0.7
Other lands	88.0	53.4
Total	164.8	100

One approach for reducing the current pressure on fresh water resources is using saline water for agricultural production. Biosaline agriculture defines as using halophytes, saline soil, and water for agricultural production. However, not all agricultural crops could produce economic yield while being irrigated by high saline water (above 12 dS/m). Introducing halophytes as new or alternative crops for developing countries like Iran with arid and semiarid climate and water scarcity problem could be a reliable solution. In Iran, there's a wide area that is prone to producing halophytic forage as livestock feed. The aim of this chapter is to introduce chenopods family and evaluate the current status of Chenopodiaceae cultivation in Iran and finally selection of promising species for cultivation with hypersaline irrigation water.

Forage and Fodder Production in Iran

Forage production of rangelands is estimated around 10.7 million tons per year, and fodder crops are mainly alfalfa, clover, and maize with a production rate of 18.44 million tons in 2012 (Ahmadi et al. 2014). The recorded animal unit of the country is 110,041,000. Each animal unit uses 511 kg of forage per year. Hence, 56.2 million tons of forage products are required for animal feed in Iran. The total forage production of the country (rangelands and forage in agriculture systems) is 29.1 million tons annually, and this is almost half of the forage requirement of the country. Therefore, the remaining amount of forage requirement should be imported or provided by overgrazing in rangelands which causes excessive pressure on them. High water requirement of forage crops, low water quality, and sensitivity of current forage crops to salinity and water stress restrict their extension. At the same time, large quantity of saline lands and saline water resources is available which could be allocated to the production of forage by halophyte farming.

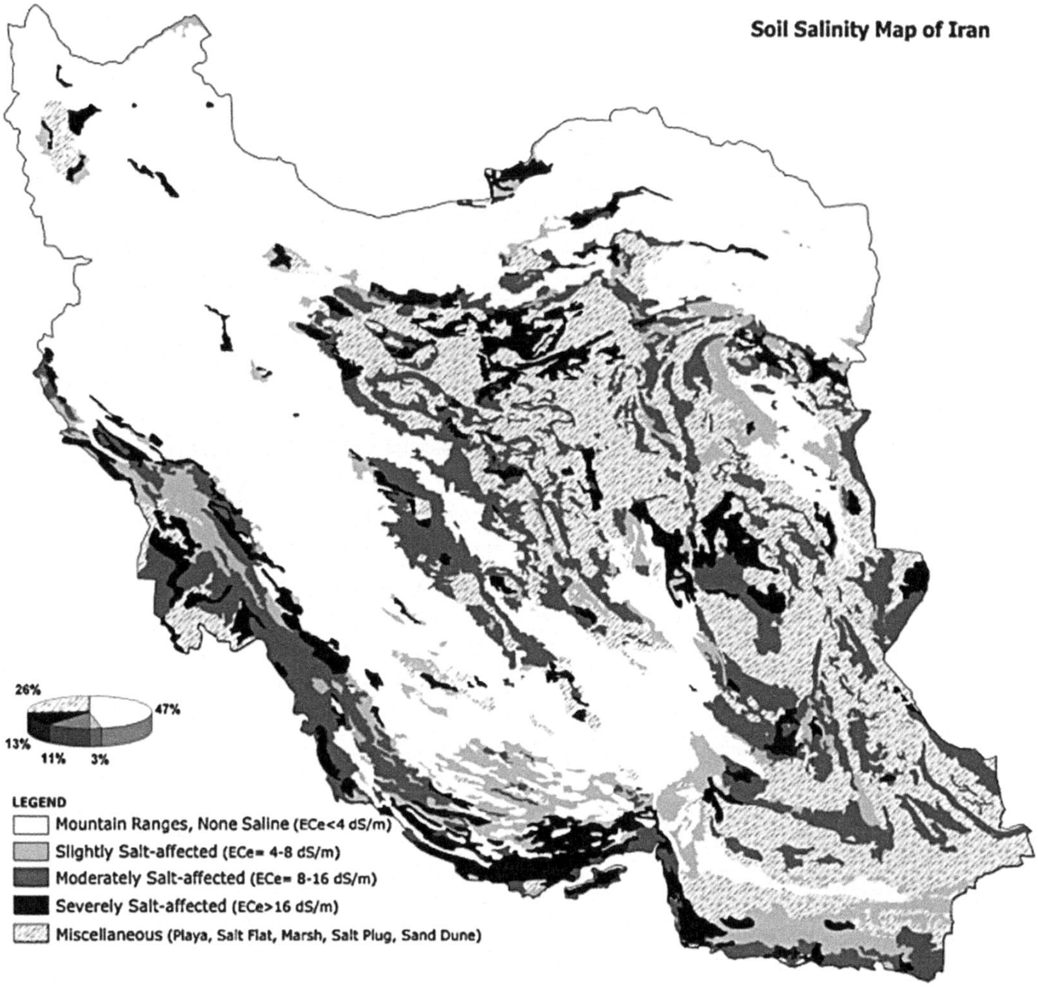

Fig. 17.1 Soil salinity map of Iran. (Banie 2001)

Distribution of Chenopodiaceae Family in Iran

Chenopodiaceae is the most important plant family in arid and semiarid areas and salt marshes. This family has 45 genera and approximately 204 species in Iran. The largest genera are *Salsola* with 39 species, *Atriplex* with 21 species, *Suaeda* with 16 species, and some genera with 1 species (Camphorosma) (Hatami and Khosravi 2013). This family has the highest number of C_4 species among other plant families. All five types of Kranz anatomy and two biochemical subtypes were observed in this family (Hatami and Khosravi 2013). Chenopods with C_4 photosynthesis pathway are widely distributed in Iran, especially in arid and hyperarid zones, and C_3 photosynthesis pathway species are mostly observed in areas with higher rate of precipitation (Hatami and Khosravi 2013). The ratio of C_4 to C_3 species of Chenopodiaceae in Iran is about 2:1. Salsoloideae comprises the highest number of C_4 species, and Salicornia is made up of the lowest number of C_4 species (only one) (Akhani et al. 1997). *Haloxylon ammodendron* and *H. persicum* are C_4 trees of this family. They have been cultivated in salt lands and sand dunes for decades by the Iranian government, but because of unsuitable ecological habitat, most of them have been destroyed (Akani 2006). Three species

(*Bienertia cycloptera, Bienertia sinuspersici, and Borszczowia aralocaspica*) have single-cell C_4 photosynthesis pathway (Akani 2006). Due to the different ecological habitats of Iran, halophyte chenopods represent high biodiversity. Out of more than 200 chenopod halophyte species, 76% are observed in the central plateau, 44% in the northeast, 35% in northwest, and less than 30% in the other parts of the country. Most of the chenopods grow in the harsh environment of central plateau in which the annual rainfall is less than 200 mm. Among chenopod halophytes, *Salsola* genus with 17 species, after that *Atriplex* and *Suaeda* with 9, and *Salicornia* with 8 species are the most abundant chenopod halophytes in the Iranian flora (Table 17.2).

Chenopodiaceae Family Cultivation in Iran

Atriplex canescens was introduced to Iran first in 1961, and this species along with the other members of this genus including *Atriplex lentiformis* and *A. halimus* could produce high biomass in the harsh environment of many parts of the country. They are recommended for forage production in the rangelands of Iran. Exotic *Atriplex* can produce high biomass yield and reproduces in saline rangelands, but natural regeneration of *A. canescens* has not been reported in Iran, and it has been removed after vital cycle. Seed germination of this species needs enough water and a temperature above 15 °C. These conditions rarely occur in Iranian rangelands, simultaneously. *Atriplex* increased the surface (0–10 cm) soil salinity and changed the density and composition of native species. The effect of increasing soil salinity in exotic *Atriplex* was more than that of native *Atriplex* (158% in *A. halimus* versus 4% in *A. leucoclada*) (Heshmati et al. 2007). Khatirnamani (2004) concluded that *Atriplex* sowing in an arid condition with 200 mm rainfall increased soil moisture and establishment of annual species, but in semiarid condition with 250–350 mm rainfall, it increased surface soil salinity and negatively affected domestic species. The allopathic effect of *Atriplex* and its competition for water and nutrition absorption was the other reason for the elimination of local species. *Atriplex* sowing caused the replacement of poor annual forage quality species, for example, replacing of *Artemisia sieberi* with *Launaea spinosa* and *Astragalus squarrosus* with *Hulthemia persica*. Low forage quality and anti-nutritional effect of *Atriplex* caused lower grazing by livestock; then *Atriplex* produced high amount of biomass and leaves. Based on the results of 50 years of *Atriplex* farming in Iran, researchers now express that it would be more rational to use native *Atriplex* and limit *Atriplex* farming to arid and degraded rangelands. Although exotic *Atriplex* caused ecological problems in the rangelands of Iran, this species is the main forage source of camels in very harsh environments.

Among exotic species, *Chenopodium quinoa* is considered as a promising crop for food production from halophytes irrigated with saline water. It has a high nutritive value compared to other cereals and the gluten-free diet with high price in European market (Peiretti et al. 2013). It could be a reliable alternative crop in saline soils and water with limited access to nonsaline water resources. This species is also considered in countries not affected by salinity. Seed yield of quinoa (Triticaca cv.) was 2.4 t ha^{-1} by using saline water with 14–17 dS m^{-1} in Yazd (Salehi unpublished data) (Fig. 17.2). Quinoa can tolerate ECe up to 40 dS/m (Adolf et al. 2012). Quinoa water requirement was half of wheat, and it produced economic yield with saline water above 12 dS m^{-1} and completed its growth cycle in 108 days. Based on the climate condition, early mature quinoa cultivars could be cultivated two times a year in the central plateau of Iran. It could replace rice and wheat in the Iranian diet, cure some of the micronutrient deficiencies in children, and replace common cereal in the diet of individuals suffering from diabetes and celiac. The detailed results of this work will be presented in another chapter of this book.

There are different evidence of successful cultivation of *Salicornia* irrigated with seawater in Eritrea, Mexico, and the UAE (Glenn et al. 2013). Because of high input demand of edible oil production in Iran, *Salicornia* is considered

Table 17.2 List of halophytes and salt-tolerant Chenopodiaceae family of Iran

Species	Life form	Distribution								Halotype	Photosynthesis
		NW	N	NE	W	C and E	SW	S	SE		
Anabasis aphylla L	CH	+	+	+		+				E	C 4
A. calcarea	H			+		+				XH	
A. eriopoda	H			+		+				XH	
A. haussknechtii	CH					+				E	C 4
A. salsa	H					+				E	
A. setifera Moq.	H, T					+	+	+	+	XH	C 4
Arthrocnemum macrostachyum	CH								+	E	C 3
Atriplex dimorphostegia	T			+		+				E	
A. flabellum Bge.	T		+	+						E	C 4
A. griffithii Moq.	CH					+				XH	
A. hastata L	T	+		+		+				E	
A. leucoclada	H	+	+	+	+	+	+	+	+	E	C 4
A. micrantha	T	+								E	
A. moneta Bge.	T			+		+				E	
A. nitens	T	+				+				E	
A. tatarica L.	T	+				+				E	C 4?
Bassia eriantha	T			+		+				E	
Bassia eriophora	T					+	+	+	+	E	
B. hyssopifolia	T	+				+				E	C 4
Bienertia cycloptera	T	+		+		+	+	+	+	E	Single-cell C4
Bienertia sinuspersici								+			Single-cell C4
Borszczowia aralocaspica								+			Single-cell C4
Beta maritime L.	T							+	+	E?	C 3
Camphorosma monspeliaca L.	H, CH	+	+	+	+	+				XH	C 4
Climacoptera brachiata	T	+		+		+				E	
C. crassa	T	+								E	C 4
C. lanata	T		+	+		+				E	C 4
C. turcomanica	T									E	
Cornulaca leucacantha	T					+	+	+	+	PH	C 4
C. monacantha Del.	CH					+	+	+	+	XH	C 4
Gamanthus gamocarpus	T			+		+				E	
G. pilosus	T	+								E	
Halanthium rariflorum	T	+	+	+		+				E	
Halimione verrucifera	H	+		+		+				E	
Halimocnemis mollissima Bge.	T			+		+				E	
Halocharis hispida	T		+	+		+				E	
H. sulphurea	T				+	+	+	+	+	E	C 4
H. violacea	T					+			+	E	
Halocnemum strobilaceum	CH	+	+	+	+	+	+	+	+	E	C 3
Halopeplis perfoliata	CH							+		E	C 3
H. pygmaea	T	+				+				E	C 3
Halostachys caspica	CH	+	+	+		+				E	C 3
Halothamnus auriculus	H, CH	+		+		+				XH	

(continued)

Table 17.2 (continued)

Species	Life form	Distribution								Halotype	Photosynthesis
H. glaucus	CH	+	+	+		+				XH	C 4
H. hierochunticus	T		+			+	+		+	E, XH	
H. subaphyllus	CH			+		+				XH	C 4
Halotis pilifera	T			+		+				E	
Haloxylon aphyllum	P		+	+		+				E, XH	C 4
H. recurvum	CH								+	E	
H. salicornicum	CH				+		+	+	+	E, XH	C 4
Kalidium caspicum	CH	+	+	+		+				E	C 3
Kochia iranica	T	+		+		+				XH	C 4
K. odontoptera	T					+				XH	C 4
Microcnemum coralloides	T	+				+				E	C 3
Panderia pilosa	T				+	+				E	C 4
Petrosimonia brachiata	T	+	+	+		+				E	C 4
P. glauca	T	+	+	+		+				E	
Piptoptera turkestana Bge.	T					+				E, PH	
Salicornia europaea L.	T	+	+							E	C 3
S. iranica	T					+				E	C 3
S. sinus-persica	T							+		E	C 3
S. persica Akhani subsp. *persica*	T					+				E	C 3
S. persica subsp. *rudshurensis*	T	+								E	C 3
S. perspolitana	T					+				E	C 3
S. x tashkensis	T					+				E	C 3
S. ramosissima	T					+				E	
Salsola abarghuensis	CH			+		+				E	
S. arbuscula	CH			+		+				E	
S. baryosma	SS						+	+	+	E	C 4
S. chorassanica	T					+				E	
S. dendroides	H	+	+	+		+				E	
S. drummondii	CH							+	+	E	
S. gossypina Bge.	T			+						E	
S. incanescens	T		+	+		+				E	
S. jordanicola	T				+					E	
S. kerneri	CH					+				XH	
S. leptoclada	T			+		+				E	C 4
S. nitraria	T	+				+				E	C 4
S. nodulosa	CH	+								E	
S. orientalis	CH			+		+				XH	
S. sclerantha	T			+		+				E	
S. soda L.	T	+	+			+				E	
S. tomentosa	CH		+	+		+				XH	
Seidlitzia florida	T	+	+		+	+		+	+	E	C 4
S. rosmarinus	CH			+		+		+	+	E	C 4
Suaeda acuminata	T	+			+	+	+	+		E	
S. aegyptiaca	T				+		+	+	+	E	C 4
S. altissima	T	+	+	+	+	+				E	C 4
S. arcuata	T			+		+				E	
S. crassifolia	T	+	+							E	

(continued)

Table 17.2 (continued)

Species	Life form	Distribution								Halotype	Photosynthesis
S. fruticosa	CH					+	+	+	+	E	C 4
S. maritima	T	+	+			+		+		E	
S. microphylla	CH	+	+	+		+				E	C 4
S. microsperma	T		+	+		+				E	
S. physophora	CH	+	+							E	

Life form: *CH* chamaephyte, *H* hemicryptophyte, *P* phanerophyte, *T* therophyte. Halophytic type: *E* euhalophyte, *XH* xerohalophyte, *PH* psammohalophyte
Akhani and Ghorbanli (1993), and Akani (2006)

Fig. 17.2 Field trail of quinoa in Yazd and whole plant with produced seed (**a**) and cleaned quinoa seeds (**b**) (irrigated with 14 dS/m saline water)

for producing oil using hypersaline water in coastal areas. Members of Salicornia are found in soils with saturated salinity from 17 up to 150 dS/m. The highest salinity tolerance was observed in *S. europaea* in Urmia lake (Zare and Keshavarzi 2007). The plant of this tribe is found in a wide range of altitudes (0–1600) except for *Arthrocnemum macrostachyum* which is limited to Persian Gulf coast (Zare and Keshavarzi 2007). Eight different species of salicornia have been identified in Iran, six species in the center and south and two species in the north of Iran, from which photosynthesis pathway of *S. bigelovii* and *S. persica* is C_4 and C_3, respectively (Akhani 2008). The results of a 2-year experiment on salicornia farming showed that the main problem of salicornia cultivation is low germination at seawater salinity. Among the salicornia species, *S. persica Akhani* subsp. *persica* which is observed in the central plateau of Iran represented the highest tolerance to salinity during germination such that 87% of the seed germinated at 600 mM NaCl salinity (Fig. 17.2). Germination response of *S. sinus-persica* which grows naturally in

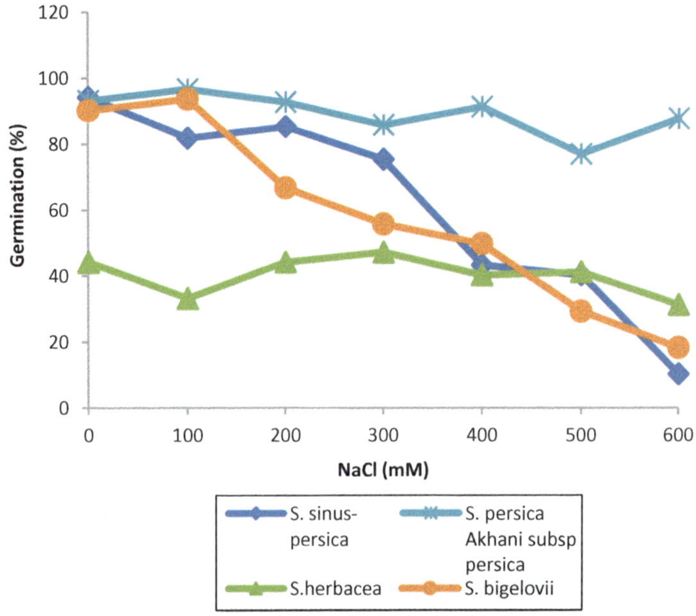

Fig. 17.3 Germination response of four species of *Salicornia* in the presence of different levels of salinity (NaCl)

Fig. 17.4 Transplanting of *S. bigelovii* seedling (40 days after sowing) to saline soil and irrigating with seawater (25–55 dS/m) after 30 days

coastal area in the south of Iran and *S. bigelovii* was the same, and 50% of germination occurred at 380–400 mM NaCl (Salehi unpublished data) (Fig. 17.3).

Therefore, seedling establishment would be a critical step for growing salicornia irrigated with seawater. Results of the transplanting method showed that seedling established sufficiently when directly irrigated with seawater after transplanting (Fig. 17.4). Shoot biomass and plant height of *Salicornia bigelovii* reduced significantly after 30 days of irrigation using 25, 35, 45, and 55 dS/m salinity (Fig. 17.5) (Salehi et al. 2016).

Direct seed sowing of salicornia demonstrated that *S. persica Akhani* subsp. *persica* has a reliable establishment in the field and it emerged after 7 days, but *S. bigelovii* seeds needed 20 days for 50% emergence with 15 dS/m saline water. Flowering time of *Salicornia* species was sensitive to photoperiod, so *S. bigelovii* and *S. persica* needed 13- and 12-h day length to start flowering, respectively. In our experiment, *Salicornia bigelovii* and *persica* produced 1.9 and 1.5 t ha^{-1} seed (215 days after planting), respectively (Fig. 17.5) (Salehi et al. 2016). Biomass production at flowering stage (140 days after planting) was 32 and 15 t FW ha^{-1}, respectively (Salehi

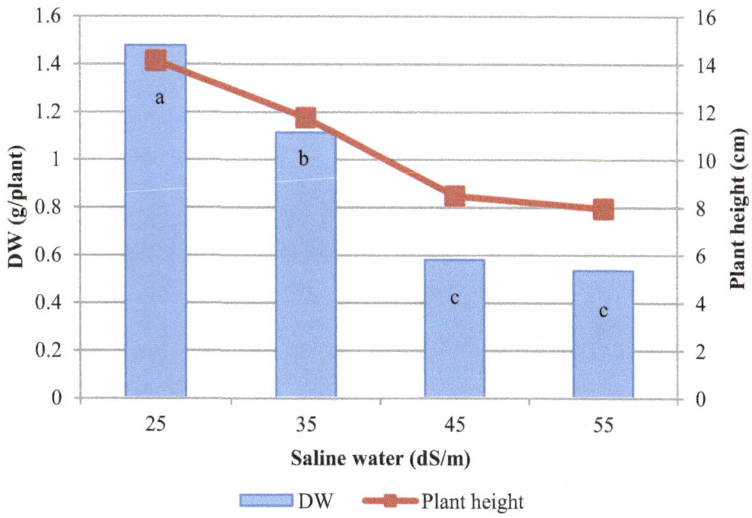

Fig. 17.5 Effect of saline water on dry weight (DW) and plant height (cm) of *Salicornia bigelovii* after 30 days

Fig. 17.6 Field trail of two species of *Salicornia bigelovii* (**a**) and *persica* (**b**) in Yazd, Iran

et al. 2016). Plant shape, color, and seed shape of these two species were completely different (Figs. 17.5 and 17.6).

Salicornia farming as an oil crop still has major problems, and finding the best solution for their cultivation increases their feasibility for farmers. Some problems of salicornia farming include seed falling especially in *S. persica* and significant yield loss during the mechanical harvesting of its small seeds. Salicornia needs high volume of water during its growing period (Glenn et al. 1997). High water requirement of salicornia causes nitrate leaching and algae bloom in coastal areas (Bailis and Yu 2012) and also high cost of water pumping. Sowing method is very important because tiny seeds could not germinate in deep sowing, surface sowing causes insect damage, and water moves the seeds away from their primary location.

Kochia scoparia is another species of Chenopodiaceae family which is considered for forage production with saline water in Iran (Kafi et al. 2014). Its cultivation is one of the success stories of halophytes farming.

Selection Criteria of Forage Halophyte Species

Biomass production and salinity tolerance are the main criteria for selecting halophyte as forage crop. After that, agronomic factors such as germination and establishment with saline water irrigation, fertilizer, forage quality, and cutting frequency are very important. In this section, seed germination, forage quality, and the range of soil salinity in which some of the chenopods can grow will be presented.

Seed Germination and Establishment Under Saline Condition

Salt tolerance in halophytes varies with the stage of development. Some halophytes are not so much tolerant to salinity during germination and seedling growth stage but are tolerant at later growth stages. Seed germination is usually delayed with increasing salinity. The seed of halophytes may either germinate or remain viable for recovery during more favorable conditions (Khan et al. 2006). Evaluation of germination under salinity of Chenopodiaceae family in Iran showed that *S. persica Akhani* subsp. *persica*, *Salsola arbuscula*, *S. abrghuensis*, *S. dendroides*, *Kochia scoparia*, and *S. yazdiana* seeds have high salt tolerance during germination stage (Table 17.3). Almost all species had a high germination rate under nonsaline conditions. Fifty percent of germination of *Suaeda aegyptiaca*, *S. richteri*, *Haloxylon aphyllum*, *Hammada salicornia*, and *Halostachys caspica* occurred at 200 mM or lower NaCl, although some of these species have high salt tolerance during later growth stages. It was reported that seed germination of some halophytes occurred after rainfall in arid and semiarid conditions (Khan and Gul 2006).

Native Chenopodiaceae for Farming with Hypersaline Water as Forage Crop

Biomass production and salinity tolerance of halophytes are very important criteria, and information about soil salinity of the habitat may help to select high salt-tolerant species. One of the most tolerant species in Meyghan habitat in central Iran is *Halocnemum strobilaceum*. This species could produce 15.5% cover at 123 dS m^{-1} and survive at 250 dS m^{-1} soil salinity in the south of Iran (Najafi 2004) (Table 17.4). It can be found in all saline lands of Iran with different climate conditions (Fig. 17.7). The main restriction for the distribution of this species is water table level. If the groundwater level falls below 5 m, the growth of *Halocnemum strobilaceum* will be restricted (Najafi 2004). Among the evaluated chenopod species, *Atriplex leucoclada*, *Seidlitzia rosmarinus*, and *Salsola baryosma* are not able to grow at soil salinity above 50 dS/m, while other chenopod halophyte species can survive at higher saline conditions (Table 17.4). *Salsola tomentosa* and *Suaeda acuminata* can survive at 279 dS/m soil salinity. There is not so much data on biomass production and salt tolerance of native chenopods species using high saline water. Kafi et al. (2010) evaluated biomass production of four species of the Chenopodiaceae family (*Suaeda dimorphestegia*, *Atriplex arcuata*, *Salsola crassa*, and *Kochia scoparia*) under saline water (26 dS/m). The results indicated that *S. dimorphestegia* produced the highest biomass 12 t DM ha^{-1}, and *Salsola*, *Atriplex*, and *Kochia* produced 7, 4, and 3 t DM ha^{-1}, respectively.

Table 17.3 Salt tolerance of some of the chenopods species

Species	Germination under nonsaline (%)	50% germination	25% germination	References
Salsola arbuscula	90	500	800	Mosleh et al. (2011)
S. yazdiana	55	200	700	
Suaeda fruticosa	75	150	200	Pouresmaeil et al. (2003)
Salsola rigida	100	350	500	Teimouri et al. (2005)
S. dendroides	100	450	650	
S. richteri	80	100	170	
Salsola arbuscula	95	260	350	Amini and Abedini (2012)
Sedlitzia rosmarinus	100	270	430	Rasouli et al. (2012)
Sedlitzia rosmarinus	88	200	270	Hadi et al. (2007)
Sedlitzia rosmarinus	62	120	210	Azarnivand et al. 2006
Hammada salicornia	98	120	160	
Haloxylon aphyllum	80	100	150	
Arthrocnemum macrostachyum	80	350	550	Khan and Gul, (2002)
Arthrocnemum macrostachyum	70	170	370	Khan (1999)
S. fruticosa	85	250	340	
Cressa cretica	90	170	350	
Salicornia sinus-persica	98	380	500	Salehi (unpublished data)
S. persica Akhani subsp. *persica*	93	87% germination at 600		
S. herbacea	45	500	Above 600	
S. bigelovii	90	400	550	
Kochia scoparia	100	600–800	1100	Khan et al. (2001)
Suaeda salsa (Brown seed)	98	700		Li et al. (2005)
Suaeda salsa (Black seed)	90	550	700	
Halostachys caspica	58	50	150	Jalali et al. (2013)
Halocnemum strobilaceum	82	120	180	
Halocnemum strobilaceum	89	220	350	Qu et al. (2008)
Suaeda aegyptiaca	55	50	160	El-Keblawy et al. (2016)
Salsola vermiculata L.	90	300	450	Guma et al. (2010)

Chenopods' Forage Quality

Forage quality of some chenopod species which have high relative palatability and salt tolerance is reported in this section. Differences in palatability of halophytes help agronomists to select the best species for forage production (Norman et al. 2013).

Halocnemum strobilaceum has a high salt tolerance and wide range of adaptability to different climate conditions. Metabolic energy (ME), dry matter digestibility (DMD), and crud protein (CP) at different regions and growth stages were higher than critical values (Table 17.5). It is highly salt-tolerant, and 25% of seed germination of this species occurs at 350 mM (Qu et al. 2008). This species has a very important role in forage supply during winter and autumn, and it also protects high palatable species (*Puccinellia distance*) (Hosseini and Shahmoradi 2011). The autecological evaluation of this species indicated that it has high water requirements. All these factors show that this species could be considered for seawater farming (Figs. 17.8 and 17.9).

Table 17.4 Minimum, maximum, and average of soil salinity (0–100 cm) of some Chenopodiaceae species in natural saline rangelands of Iran

Species	Min	Max	Mean	References
Seidlitzia rosmarinus	19.16	48.09	33.62	Khodahami (1995)
Halocnemum strobilaceum	5.42	250	70.92	Najafi (2004), Khodahami (1995)
Hammada salicornia	2.74	160.15	61.47	Khodahami (1995)
Suaeda fruticosa	16.01	106.85	56.34	Najafi (2004), Khodahami (1995)
Halocnemum strobilaceum	16.01	150.95	64.59	Najafi (2004)
Seidlitzia rosmarinus	7.64	52.20	22.70	
Salsola baryosma	4.36	52.65	37.12	
Atriplex leucoclada	2.58	40.35	26.35	
Halocnemum strobilaceum	30.00	110.00	70.80	Hosseini and Shahmoradi (2011)
Desmostachya bipinnata			250	Najafi (2004)
Suaeda acuminata			279	Khodahami (1995)
Salsola tomentosa			279	
Salsola crassa			224	
Halocharis sulphurea				
Cressa cretica			81	
Suaeda vermiculata			83	
Atriplex leucoclada			173	
Salsola turcomanica	36	97	97	Akbarpour (2005)
Halostachys caspica	36	97	97	
Halostachys belangeriana	35	79	79	
Microcnemum coralloides			52.6	Zare and Keshavarzi (2007)
Salicornia persica			96	
Salicornia europaea	17	150	150	
Bienertia cycloptera	21.7	89.6		Akhani et al. (2003)

Fig. 17.7 Produced seeds of *Salicornia bigelovii* (a) and *persica* (**b**)

Seidlitzia rosmarinus grows fairly well in saline and alkaline soils. Its seeds have high viability, and 50% germination occurs at 270 mM salinity. This halophyte is suitable for camel grazing and has high preference value in June, when the other forage plants are not highly available in Iranian deserts. Annual biomass production of this species in natural ecosystems of the central desert of Iran (Ardakan) is 594 kg ha^{-1}. Apart from its forage application, Na_2CO_3 is also extracted after burning of *S. rosmarinus*, and this chemical substance is used for soap making,

Table 17.5 Forage quality of *Halocnemum strobilaceum*

Province	Growth stage	DMD	CP (%)	ME	References
Golestan	Vegetative	66.45	12.09	5.27	Pasandi (2015)
	Ripening	49.28	9.04	4.39	
Qom	Vegetative	72.97	13.76	1.41	Ahmadi and Sanadgol (2010)
	Ripening	69.46	8.09	9.80	
Uromieh		69.2	9.7	9.80	
Qom		73.6	9.1	10.50	
Yazd		75.7	11.7	10.90	
Gomishan	Vegetative	62.89	11.26	8.67	Mirzaali et al. (2006)
	Flowering	62.7	12.01	8.7	
	Ripening	53.96	18.64	7.17	
Critical value		50	7	8	

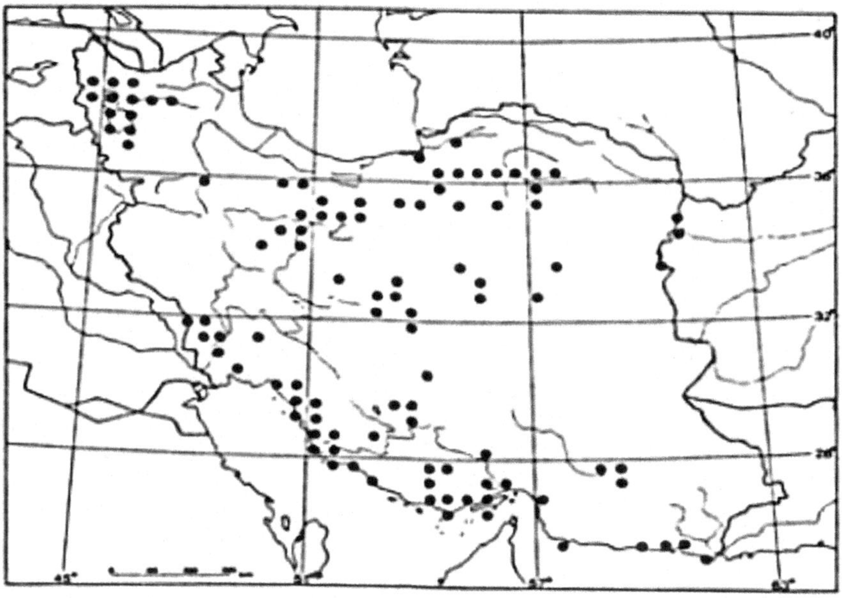

Fig. 17.8 *Halocnemum strobilaceum* distribution in Iran (Assadi et al. 1989)

detergent, glass work, paper making, dyeing, and tile industry. Its forage quality is higher than *Salsola arbuscula*, *Suaeda fruticosa*, and *Hammada salicornia* (Towhidi and Zhandi 2007). Autecological evaluation of *Seidlitzia rosmarinus* showed that it cannot tolerate high water table and its salinity tolerance is lower than *Halocnemum strobilaceum.* This species could be considered for halophyte farming with drainage water below 30 dS/m salinity in the central plateau or Khuzestan Province (Table 17.6).

Evaluation of the forage quality of *Halostachys caspica*, a perennial halophytes, showed that its metabolic energy, dry matter digestibility, and crud protein are higher than critical value (Table 17.7), but 50% reduction in germination occurred at 150 mM NaCl (Jalali et al. 2013). It grows in saline rangelands with lower than 33 dS m^{-1} soil salinity and soil with higher P and K content.

Salsola species are the largest chenopod group in Chenopodiaceae family and have an important

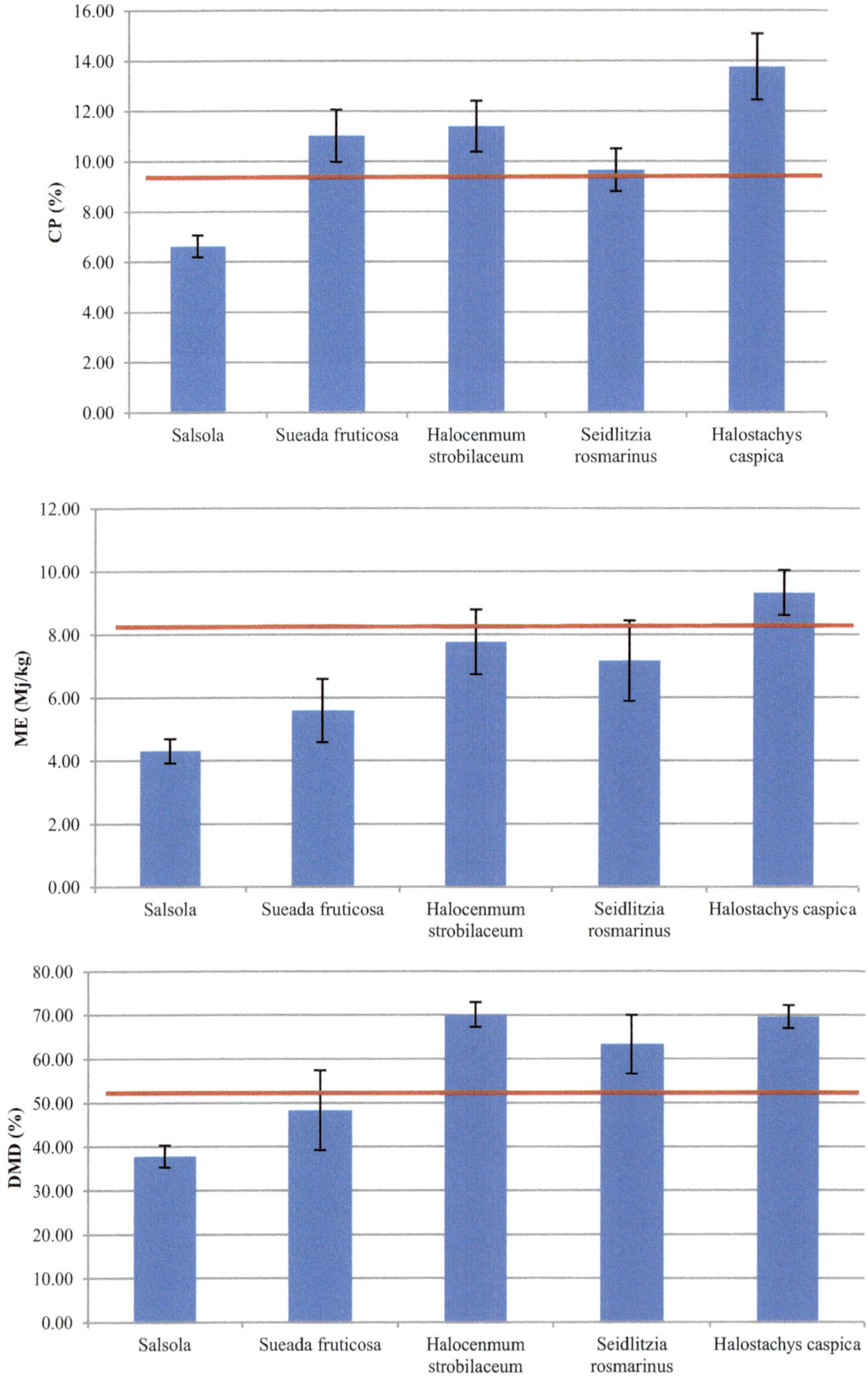

Fig. 17.9 Crude protein (CP) (%), metabolizable energy (ME) (Mj/kg), and dry matter digestibility (DMD) (%) of five selected halophytes (*Suaeda fruticosa*, *Halocnemum strobilaceum*, *Seidlitzia rosmarinus*, *Halostachys caspica*, and *Salsola*) from Chenopodiaceae family. Red lines show the critical value and bars show the standard error

Table 17.6 Forage quality of *Seidlitzia rosmarinus*

Province	Growth stage	DMD	CP (%)	ME	References
Qom and Yazd			9.50	4.80	Towhidi et al. (2011)
Semnan				4.50	Towhidi and Zhandi (2007)
Sistan				6.91	
Yazd		46.26	9.50		
Khuzestan		68.30	11.71	4.80	Abarghani et al. (2014)
Sistan		48.33	11.95	6.98	Yousef Elahi et al. (2014)
Qom	Vegetative	75.68	7.84	10.87	Ahmadi and Sanadgol (2010)
	Ripening	77.92	7.39	11.25	
Critical value		50	7	8	

Table 17.7 Forage quality of *Halostachys caspica*

Region	Growth stage	DMD	CP (%)	ME	References
Golestan	Vegetative	66.53	12.66	5.87	Pasandi (2015)
	Ripening	48.43	7.92	4.82	
Uromieh		75.77	17	10.88	Rasouli et al. (2011)
Qom		72.33	10.82	10.28	
Yazd		72.1	11.6	10.24	
	Early growth stage	74.15	13.9	10.60	
	Vegetative	73	15.9	10.41	
	Ripening	72.9	9.6	10.39	
Gomishan	Vegetative	71.86	19.87	10	Mirzaali et al. (2006)
	Flowering	68.68	18.35	9.67	
Critical value		50	7	8	

role in the dry and saline rangelands of Iran. This genus has annual and perennial species and produces high forage quantity during winter. *Salsola rigida*, *S. kali*, and *S. dendroides* are among palatable species in saline rangelands. *Salsola rigida* is a perennial species with high water-use efficiency and strong root, and it has a major role in soil conservation. The comparison of forage quality of *Halocnemum strobilaceum*, *Halostachys caspica*, *Seidlitzia rosmarinus*, *Salicornia herbacea*, *Alhagi persarum*, *Salsola rigida*, and *Atriplex leucoclada* indicated that *S. rigida* has the highest, while *A. leucoclada* has the lowest forage quality specifically at flowering stage (Rasouli and Amiri 2015).

Salsola arbuscula has a lower preference value in comparison to other *Salsola* species. *Salsola yazdiana* is a native species and grows in the central desert of Iran. DMD and ME of this species are lower than critical value (Yousef Elahi et al. 2014). Panahi et al. (2012) evaluated the forage quality of three species of *Salsola* (*Salsola arbuscula*, *S. orientalis*, and *S. tomentosa*). *Salsola arbuscula* demonstrated the lowest quality, and its vegetative stage had better quality than the other growth stages (Table 17.8).

CP and ME of *Salsola turcomanica* during vegetative stage are lower than normal, and during ripening stage, its DMD is lower than critical value. Ranchers of Golestan province collect *Salsola turcomanica* during seed ripening and use it as a winter forage (Pasandi 2015).

Suaeda fruticosa is an obligate and leaf succulent perennial halophyte that produces a lot of seed under saline condition. Its seeds contain high-quality edible oil with 73% unsaturated fatty acid (Weber et al. 2007). Its seed germination reduced to 75% at 200 mM NaCl (Pouresmaeil et al. 2003), but after emergence it can tolerate 106 dS m^{-1} soil salinity (Najafi 2004). Dry matter digestibility (DMD) and metabolizable energy (ME) of this species are

Table 17.8 Forage quality of *Salsola species*

Species	Region	Growth stage	DMD	CP (%)	ME	References
Salsola arbuscula	Semnan	Vegetative	18	7.8	1.06	Panahi et al. (2012)
		Flowering	25	7.7	2.25	
		Ripening	18	7	1.06	
S. orientalis		Vegetative	33	4.8	3.61	
		Flowering	34	6.2	3.78	
		Ripening	25	6.7	2.25	
S. tomentosa		Vegetative	37	6.0	4.29	
		Flowering	30	8.0	3.1	
		Ripening	27	8.0	2.59	
Salsola arbuscula	Markazi	Vegetative	17	5.8	0.89	
		Flowering	22	5.5	1.74	
		Ripening	20	6.8	1.4	
S. orientalis		Vegetative	32	5.9	3.44	
		Flowering	37	5.0	4.29	
		Ripening	40	10	4.8	
S. tomentosa		Vegetative	32	10.2	3.44	
		Flowering	42	10	5.14	
		Ripening	37	7.9	4.29	
Salsola arbuscula	Isfahan	Vegetative	17	5.0	0.89	
		Flowering	24	7.5	2.08	
		Ripening	14	5.8	0.38	
S. orientalis		Vegetative	40	3.0	4.8	
		Flowering	35	2.0	3.95	
		Ripening	33	2.0	3.61	
S. tomentosa		Vegetative	35	4	3.95	
		Flowering	33	1.9	3.61	
		Ripening	32	4.0	3.44	
Salsola turcomanica	Golestan	Vegetative	78	5.4	5.89	Pasandi (2015)
		Ripening	32	7.1	4.96	
Salsola vermiculata	Sistan	Ripening	34	6.0	4.81	Yousef Elahi et al. (2014)
Salsola yazdiana			47	11.5	6.91	
Salsola richteri	Sabzevar	Vegetative	68	14.7	9.55	Dianati et al. (2012)
		Flowering	65	9.4	9.19	
		Ripening	64	8.6	9.01	
S. arbuscular		Vegetative	62	6.3	8.64	
		Flowering	54	4.3	7.26	
		Ripening	54	3.7	7.18	
Salsola orientalis	Isfahan	Vegetative	62	8.1	8.5	Saeedfar et al. (2006)
		Flowering	58	8.7	7.8	
		Ripening	44	6.9	5.5	
Salsola tomentosa	Yazd		37	13.3	4.2	Towhidi and Zhandi (2007)
Salsola yazdiana			23	7.3	1.9	
Mean		Vegetative	41	7.0	4.5	
		Flowering	38	6.3	4.5	
		Ripening	34	6.5	3.9	
Critical value			50	7	8	

Table 17.9 Forage quality of *Suaeda fruticosa*

Region	DMD	CP (%)	ME	References
Sistan	37.47	9.60	5.35	Yousef Elahi et al. (2014)
Qom and Yazd	68.62	13.50	5.00	Towhidi and Zhandi (2007)
Semnan	18.47	7.90	2.80	
Yazd	68.62	13.50	9.67	Towhidi and Zhandi (2007)
Khuzestan	59.21	12.00	5.37	Abarghani et al. (2014)
Critical value	50	7	8	

lower than critical value (Table 17.9). Its forage quality was also lower than *Seidlitzia rosmarinus* and *Salsola arbuscula* in Semnan province and had the same quality as *Salsola tomentosa* and *S. rosmarinus* in Yazd province.

> There are interactions between species and environment on forage quality. Differences in soil type, rainfall, growth stage, and soil salinity affect feeding value of halophytes. (Norman et al., 2013)

Average reported data on the forage quality of four halophyte species and *Salsola* spp. showed that *Halostachys caspica* had the highest DMD, ME, and CP. DMD and CP of *Halocnemum strobilaceum*, *Seidlitzia rosmarinus*, and *Suaeda fruticosa* were higher than the critical value of *Salsola* species. ME of *Halocnemum strobilaceum* was almost equal to critical value of *Salsola* species, but other species had lower values. Comparing the data with critical value indicated that the ME of all species except *Halostachys caspica* was lower than *Salsola* genus. Increasing ME of halophytes or animals which are fed with chenopods is an important factor. Chenopods have a high amount of CP, sulfur, and minerals, which are very critical for ruminants. However, oxalate, mineral toxicities induced deficiencies and have adverse effects on animals. Antioxidants of halophytes detoxify reactive oxygen, alleviate deficiency, and improve meat quality (Norman et al. 2013). Chenopods accumulate high amounts of potassium, chloride, calcium, and magnesium, and this high level of minerals is above the maximum tolerable levels for livestock feeding (Masters et al. 2007). High concentrations of sodium will cause depressed feed intake and under some conditions will compromise animal health (Masters et al. 2007). Anti-nutritional factor (nitrate, oxalate, coumarin, and tannin) is another restriction factor of using chenopod forage. Intensive livestock production is a good opportunity for using halophytes as part of animal feed. For instance, *Kochia scoparia* can comprise 20–40% of the beef cattle diet (Cohen et al. 1989). Atriplex can be used 75% with *Panicum antidotale* (25%) in Baluchi sheep feed (Kashki and Tavakoli 2005).

Other Utilizations of Chenopodiaceae Family

Suaeda aegyptiaca grows naturally in the south of Iran and local people use it as a vegetable. It can produce the highest biomass at 10 and 20 dS/m salinity. It could produce 20 mg vitamin C in 100 g biomass at salinity lower than 40 dS/m (Zakery-Asl et al. 2014), but its germination is sensitive to salinity, and a 75% loss in germination occurred at 150 mM NaCl (El-Keblawy et al. 2016). *Salsola soda* is also observed in Rud Shour, Gomishan, and Agh Gol in Hamedan (Akhani et al. 2007). It is more commonly known as "agretti" and can be irrigated with seawater. It is cultivated in Italy and consumed as a vegetable (Centofanti and Bañuelos 2015). Salicornia is used as a sea vegetable, and in addition to being used as green salad, it has medicinal attributes such as lipid-lowering, immune modulatory, and hypoglycemic render (Patel 2016).

Arthrocnemum macrostachyum, *Suaeda fruticosa*, *Suaeda aegyptiaca*, *Halocnemum strobila-*

Table 17.10 Oil quantity and quality of some species of Chenopodiaceae

Species	Oil (%)	Unsaturated fatty acid (%)	Saturated fatty acid (%)	References
Arthrocnemum macrostachyum	25	64.7	35.5	Weber et al. (2007)
Suaeda fruticosa	30.7	73.6	25.7	Weber et al. (2007), Shahi (2013)
Suaeda aegyptiaca	32.99	87.71	9.87	Firouzabadi et al. (2014)
Halocnemum strobilaceum	17.76	83.24	11.61	
Salicornia persica Akhani subsp. *rudshurensis Akhani*	9.3	70.00	30.00	Ahmadi et al. (2016)
Salicornia herbacea	13.8	65.99	34.01	Shahi (2013), Choi et al. (2014)

ceum, and Salicornia persica seeds are the source of edible oil with 70–80% unsaturated fatty acids (Firouzabadi et al. 2014, Weber et al. 2007) (Table 17.10).

Chenopodiaceae is the host of four species of *Cistanche*. *Seidlitzia rosmarinus* is the host of *Cistanche laxiflora*, *Suaeda* is the host of *C. flava*, and *Haloxylon* and *Arthrocnemum macrostachyum* are the host of *C. tubulosa* (Saeidi Mehrvarz and Shahi Shavvon 2008). All the species of *Cistanche* and the entire plant family belong to Orobanchaceae, and they are parasitic. *Cistanche* is a parasite on the roots of *Halostachys*, *Halocnemum*, and *Sueada*. The stubby rhizome of *Cistanche* is attached to the fine roots of the host plants. The stems of *Cistanche* are sliced to produce the pharmaceutical materials (Dharmananda 2004). *Cistanche* products are available in the market, and cultivation of this species besides other Chenopodiaceae family species increases the farmers' income and encourages them to protect and extent the rangeland cover.

References

Abarghani A, Ghaji M, Mansori H, Mamouei M, Mirzadeh K (2014) Chemical composition, metabolizable energy and in vitro digestibility of three camel browsed halophytes species. Anim Sci J 106:29–42

Adolf VI, Jacobsen SE, Shabala S (2012) Salt tolerance mechanisms in quinoa (Chenopodium quinoa Willd.). Environ Exp Bot 92:43–54

Ahmadi A, Sanadgol A (2010) Nutritive value of Zandi sheep diets (halophyte plants) grazing in Abbasabad desert rangelands of Qom. J For Wood Prod 63

Ahmadi H, Noroozy J, Farhoodi M, Mohammad M, Rahmatzadeh B (2016) Extraction and physicochemical properties of salicornia (Salicornia persica Akhani sub sp. Rudshurensis Akhani) oil. Iran J Nutr Sci Food Technol 11:67–74

Ahmadi K, Gholizadeh H, Ebadzadeh HR, Hosseini Pour R, Hatami F, Fazli B, Kazemiyan A, Rafiee M (2014) Agricultural statistics of Iran Ministry of Agriculture

Akani H (2006) Biodiversity of halophytic and sabkha ecosystems in Iran. In: Khan MA, Boer B, Stanislavovich Kust G, Barth HJ (eds) Sabkha ecosystems volume II: west and central Asia. Springer, Dordrecht

Akbarpour H (2005) Identification of saline reagions and halophyte plants in Golestan Province. Final report of research plan, Agriculture research, education and extention organization, Agriculture and Natural Resources Research Center of Golestan, grant no. 79-031-0213909-01

Akhani H (2008) Taxonomic revision of the genus Salicornia L.(Chenopodiaceae) in Central and Southern Iran. Pak J Bot 40:1635–1655

Akhani H, Edwards G, Roalson EH (2007) Diversification of the old world Salsoleae sl (Chenopodiaceae): molecular phylogenetic analysis of nuclear and chloroplast data sets and a revised classification. Int J Plant Sci 168:931–956

Akhani H, Ghobadnejhad M, Hashemi S (2003) Ecology, biogeography and pollen morphology of *Bienertia cycloptera* Bunge ex Boiss.(Chenopodiaceae), an enigmatic C4 plant without Kranz anatomy. Plant Biol 5:167–178

Akhani H, Ghorbanli M (1993) A contribution to the halophytic vegetation and flora of Iran. In Towards the rational use of high salinity tolerant plants. Springer, Dordrecht

Akhani H, Trimborn P, Ziegler H (1997) Photosynthetic pathways inChenopodiaceae from Africa, Asia and Europe with their ecological, phytogeographical and taxonomical importance. Plant Syst Evol 206:187–221

Amini F, Abedini E (2012) Evaluation of seed germination, growth and plant anatomy of Salsola arbuscula Pall. in salt stress in In vitro

Assadi M, Maassoumi AR, Khatamsaz M, Mozaffarian V (1989) Flora of Iran. Ministry of Agriculture, Islamic Republic of Iran, Tehran

Azarnivand H, Zandii Esfahan E, Shahriari E (2006) Effect of salinity on germination of Haloxylon aphyllum, Sedlitzia rosmarinus and Hammada salicornica. Desert 11:187–196

Bailis R, Yu E (2012) Environmental and social implications of integrated seawater agriculture systems producing Salicornia bigelovii for biofuel. Biofuels 3:555–574

Banie MH (2001) Soils map of Iran: land resources and potentialities. Soil and Water Research Institute (SWRI), Tehran

Centofanti T, Bañuelos G (2015) Evaluation of the halophyte Salsola soda as an alternative crop for saline soils high in selenium and boron. J Environ Manag 157:96–102

Cheraghi SAM, Halim MK (2012) Status and new developments on the use of brackish water for agricultural production in the near East: Iran country report. FAO, Cario

Choi D, Lim GS, Piao YL, Choi OY, Cho KA, Park CB, Chang YC, Song YI, Lee MK, Cho H (2014) Characterization, stability, and antioxidant activity of Salicornia herbaciea seed oil. Korean J Chem Eng 31:2221–2228

Cohen RDH, Iwaasa AD, Mann ME, Coxwontu E, Kernan JA (1989) Studies on the feeding valte of *Kochia scoparia* (L.) Schrad. hay for beef cattle. Can J Anim Sci 69:735–743

Dharmananda S (2004) CISTANCHE and endangered species issues affecting the herb supply, ITM

Dianati TGA, Haidarian Aghakhani M, Filehkesh I, Naghipour Borji AA (2012) Investigation on the effects of phenological stages on forage quality and soluble carbohydrates in salsola arbuscula and salsola richteri species in saline rangelands of sabzevar. Iran J Range Desert Res 18:652–661

El-Keblawy A, Gairola S, Bhatt A, Mahmoud T (2016) Effects of maternal salinity on salt tolerance during germination of Suaeda aegyptiaca, a facultative halophyte in the Arab Gulf desert. Plant Species Biol

Firouzabadi A, Jafari M, Assareh M, Arzani H, Javadi S (2014) Investigation on the potential of halophytes as a source of edible oil (case study: Suaeda aegyptiaca and Halocnemum strobilaceum). Int J Biosci 5:87–93

Glenn E, Miyamoto S, Moore D, Brown JJ, Thompson TL, Brown P (1997) Water requirements for cultivatingSalicornia bigeloviiTorr. with seawater on sand in a coastal desert environment. J Arid Environ 36:711–730

Glenn EP, Anday T, Chaturvedi R, Martinez-Garcia R, Pearlstein S, Soliz D, Nelson SG, Felger RS (2013) Three halophytes for saline-water agriculture: An oilseed, a forage and a grain crop. Environ Exp Bot 92:110–121

Guma I, Padrón-Mederos M, Santos-Guerra A, Reyes-Betancort J (2010) Effect of temperature and salinity on germination of Salsola vermiculata L.(Chenopodiaceae) from Canary Islands. J Arid Environ 74:708–711

Hadi MR, Taheri R, Sharif MA (2007) Study effect of salinity on the seed germination of Seidlitzia rosmarinus. Pajuhesh Sazandegi 76:151–157

Hatami E, Khosravi A (2013) Mapping of geographic distribution of C3 and C4 species of the family Chenopodiaceae in Iran. Iran J Bot 19:263–276

Heshmati G, Naseri K, Ghanbarian G (2007) A critique on Atriplex canescens planting in rangelands of Iran from ecological view. J Agric Sci Nat Resour 13:186–198

Hosseini SA, Shahmoradi AA (2011) Autecology of Halocnemum strobilaceum (Pall.) M. Bieb. in saline and alkaline rangelads of Golestan province. J Plant Sci Res 22:18–30

Jalali M, Shahraki S, Khajavi M (2013) Evluation effect of salt stress on three species *Frankenia hirsuta, Halocnemum strobilaceum and Halostachys caspica*. Irani J Plant sci 7:26–31

Kafi M, Gul B, Salehi M (2014) Kochia (Kochia scoparia (L.) Schrad) unwanted or wanted plant for forage production in harsh environments. In Sabkha ecosystems: volume IV: cash crop halophyte and biodiversity conservation. Springer, Dordrecht

Kafi M, Zamani G, Pouyan M (2010) Study the domestication possibility of four halophyte species using brakish and saline irrigation water. Iran J Range Desert Res 17:276–291

Kashki V, Tavakoli H (2005) Nutritive value and proper level of mixed feeding of Atriplex cancsens and Panicum antidotale in Balouchi sheep. J Dairy Sci. Amer Dairy Science Assoc 1111 N Dunlap Ave, Savoy, IL 61874 USA, pp 273–273

Khan MA. Comparative influence of salinity and temperature on the germination of subtropical perennial halophytes. Halophyte uses in different climates I: ecological and ecophysiological studies: proceedings of the 3rd seminar of the EU Concerted Action Group IC 18CT 96-0055, 20 July, 1998, Backhuys Publishers, Florence, Italy, p 77

Khan MA, Ansari R, Gul B, Qadir M (2006) Crop diversification through halophyte production on salt-prone land resources. CAB Rev Perspect Agric Vet Sci Nutr Nat Resour 1:048

Khan MA, Gul B (2002) Salt tolerant plants of coastal sabkhas of Pakistan. In: Barth H, Boer B (eds) Sabkha Ecosystems, vol 1. Kluwer Academic Publisher, Dordrecht, pp 123–140

Khan MA, Gul B (2006) Halophyte seed germination. Ecophysiology of high salinity tolerant plants. Springer, Dordrecht

Khan MA, Gul B, Weber DJ (2001) Influence of salinity and temperature on the germination of Kochia scoparia. Wetl Ecol Manag 9:483–489

Khatirnamani J (2004). Evaluation the effect of Atriplex on rangeland soil of Golestan provincce. Final report, Management Planning Country

Khodahami G (1995). Identification of saline areas and halophytes, study of tolerance or avoidance mechanism and introduce suitable salt resistance plants in Fars Province. Final Report of Research Plan, Agriculture Research, Education and extention Organization,

Agriculture and Natural Resources Research Cneter of Fars, Grant No. 73-310217100-632
Li W, Liu X, Khan MA, Yamaguchi S (2005) The effect of plant growth regulators, nitric oxide, nitrate, nitrite and light on the germination of dimorphic seeds of Suaeda salsa under saline conditions. J Plant Res 118:207–214
Masters DG, Benes SE, Norman HC (2007) Biosaline agriculture for forage and livestock production. Agric Ecosyst Environ 119:234–248
Mirzaali E, Erfanzadeh R, Mesdaghi M (2006) The study of effects of exclosure on vegetation and soil surface in saline ranges of Gomishan, Golestan province. J Agric Sci Nat Resour 78:79–84
Momeni A (2011) Geographical distribution and levels of soil salinity. Iran Soil Res 24:203–215
Mosleh AA, Bakhshi KG, Nemati N, Soltani M (2011) Investigation on the effect of salinity stress on seed germination of Salsola abarghuensis, Salsola arbuscula and Salsola yazdiana 18:267–279
Najafi, K. (2004). Identification of saline areas and halophytes, study of tolerance or avoidance mechanism and introduce suitable salt resistance plants in Hormozgan Province. Final report of Research Plan, Agriculture Research, Education and extention Organization, Agriculture and Natural Resources Research Cneter of Hormozgan
Norman HC, Masters DG, Barrett-Lennard EG (2013) Halophytes as forages in saline landscapes: interactions between plant genotype and environment change their feeding value to ruminants. Environ Exp Bot 92:96–109
Panahi F, Assareh MH, Jafari M, Jafari A, Arzani H, Tavili A, Zandi Esfahan E (2012) Phenological effects on forage quality of Salsola arbuscula, Salsola orientalis and Salsola tomentosa in three habitats in the central part of Iran. Middle-East J Sci Res 11:800–807
Pasandi M (2015). Determination of Chemical compositon and digestibility (In vitro) of important species in saline and alkaline rangelands of Golestan province. Final report of Research Plan, Agriculture Research, Education and extention Organization, Agriculture and Natural Resources Research Cneter of Golestan, Grant No. 2-57-13-88020
Patel S (2016) Salicornia: evaluating the halophytic extremophile as a food and a pharmaceutical candidate. Biotech 6(3):1–10
Peiretti PG, Gai F, Tassone S (2013) Fatty acid profile and nutritive value of quinoa (Chenopodium quinoa Willd.) seeds and plants at different growth stages. Anim Feed Sci Technol 183:56–61
Pouresmaeil M, Ghorbanli M, Khavarinejhad R (2003) Effet of salinity on germination, fresh and dry mass, ion content, proline, soluble sugar and starch content in Suaeda fruticosa. Desert 10:257–265
Qu X-X, Huang Z-Y, Baskin JM, Baskin CC (2008) Effect of temperature, light and salinity on seed germination and radicle growth of the geographically widespread halophyte shrub Halocnemum strobilaceum. Ann Bot 101:293–299
Rasouli B, Amiri B (2015) Assessment of new forage sources in saline areas of Iran. Environ Resour Res 3:75–83
Rasouli B, Amiri B, Assareh MH, Jafari M (2011) Nutritive value of a halophyte species, Halostachys caspica in three different phenological stages and three different sites. Iran J Range Desert Res 18:32–41
Rasouli B, Amiri B, Jafari M, Assareh MH (2012) Effect of NaCl and Na2SO4 on germination and intial growth of Seidlitzia rosmarinus. Iran J Rangel Desert Res 18:505–514
Saeedfar M, Feyzi MT, Shahmoradi A (2006) Autecology of *Salsola orienthalis* in steppe rangelands of Isfahan province. Iran Desert Rangel Res 13:116–126
Saeidi Mehrvarz S, Shahi Shavvon R (2008). Notes on the genus Cistanche (Orobanchaceae) in Iran
Salehi M, Dehghani F, Ebrahimi NG (2016) Successful story of Salicornia seed production with saline water. J Water Sustain Dev, In press
Shahi M (2013) Amount oil content of two halophytes Suaeda fruticosa and Salicronia gerbacea. MSc thesis, Birjand UniversityBrijand
Shahrestani H (2013) Organization and management optimize water use in agriculture. J Agric Nat Resour Eng 12:37–41
Teimouri A, Moghaddam M, Heidari Sharif Abad H, Jafari M, Azarnivand H (2005) Effect of salinity on seed germination in three Salsola species (*S rigida, S dendroides, S richteri*). Iran J Nat Res 58:609–710
Towhidi A, Saberifar T, Dirandeh E (2011) Nutritive value of some herbage for dromedary camels in the central arid zone of Iran. Trop Anim Health Prod 43:617–622
Towhidi A, Zhandi M (2007) Chemical composition, in vitro digestibility and palatability of nine plant species for dromedary camels in the province of Semnan, Iran. Egypt J Biol:9
Weber DJ, Ansari R, Gul B, Ajmal Khan M (2007) Potential of halophytes as source of edible oil. J Arid Environ 68:315–321
Yousef Elahi M, Peyravi M, Mirzaei HR (2014) Determination of nutritive value of five species of halophyte plants of Sistan by In vitro and In situ techniques. Res Anim Prod 5:50–68
Zakery-Asl MA, Bolandnazar S, Oustan S (2014) Effect of salinity and nitrogen on growth, sodium, potassium accumulation, and osmotic adjustment of halophyte Suaeda aegyptiaca (Hasselq.) Zoh. Arch Agron Soil Sci 60:785–792
Zare G, Keshavarzi M (2007) Morphological study of Salicornieae (Chenopodiaceae) native to Iran. Pak J Biol Sci 10:852–860

18 Plant Growth-Promoting Bacteria Associated to the Halophyte *Suaeda maritima* (L.) in Abbas, Iran

Edgar Omar Rueda-Puente, Oscar Bianciotto, Saifolah Farmohammadi, Omid Zakeri, Jesús López Elías, Luis Guillermo Hernández-Montiel, and Murillo Amador Bernardo

Abstract

Suaeda maritima (L.), regarded as a promising halophyte, is widely distributed along the coastal plains of Abbas, Iran. *Suaeda maritima* has been highly incorporated with the traditional agriculture to support the Iranian agricultural economy. However, its productivity is limited by a lack of available nitrogen. Application of halotolerant plant growth-promoting rhizobacteria (PGPR) suggested being an alternative biological fertilizer. Increasing the knowledge of halotolerant PGPR associated to the native crops remains important. Nine endemic colonies were isolated from the roots of *S. maritima*. Those isolates were cultured in different salinity conditions (0, 0.25, 0.5, and 0.75 M NaCl) and maintained at different temperature regimes (30 and 55 °C). The nitrogen fixation ability of the isolated endemic colonies was screened by acetylene reduction assay. Among them, only one showed high acetylene reduction activity and capacity to solubilize phosphates. This bacterium was identified as *Bacillus amyloliquefaciens*. Seeds inoculated with *Bacillus amyloliquefaciens*, in conjunction with *Azospirillum halopraeferens* as a biological control, were tested for seed germination and early growth promotions of *S. maritima* exposed to high salinities under in vitro conditions. *Bacillus amyloliquefaciens* showed a high specificity for the wild *S. maritima*. This is the first report of *Bacillus amyloliquefaciens* as nitrogen-fixing bacterium associated with the oilseed *S. maritima* a novel halophyte crop. Through this work, a reliable biological method was found, based on beneficial bacteria, to contribute to maintain or improve the fertility of soils sustaining *Suaeda* fields.

Keywords

Arid desert · Halophytes · Nitrogen fixation

E. O. Rueda-Puente (✉) · J. L. Elías
Agriculture Department, Sonora University, Hermosillo, Sonora, Mexico
e-mail: erueda04@santana.uson.mx; lopez_eliasj@guayacan.uson.mx

O. Bianciotto
Universidad Nacional de Tierra del Fuego A.I.A.S, Ushuaia, Tierra del Fuego, Argentina

S. Farmohammadi · O. Zakeri
Academic Center for Education Culture and Research (ACECR) of Hormozgan, Iran and Province Department of Natural Resources, Plantation Office, Tehran, Hormozgan, Iran

L. G. Hernández-Montiel · M. A. Bernardo
Center of Biological Researchers of Norwest, Instituto Politécnico Nacional, La Paz, Baja California Sur, México
e-mail: lhernandez@cibnor.mx

B. Gul et al. (eds.), *Sabkha Ecosystems*, Tasks for Vegetation Science VI,
https://doi.org/10.1007/978-3-030-04417-6_18

Introduction

Over 800 million hectares of land surface are adversely affected by salinity throughout the world (Munns 2005). Like other agricultural areas, Abbas, Iran, agricultural practices are dependent on groundwater from wells. Unfortunately, water extraction in excess amount, saline intrusion, and inappropriate use of synthetic fertilizers may increase salinity in salt-affected coastal regions (Rueda et al. 2004, 2009a, b). Production alternatives include establishing improvement programs for salt tolerance on traditional crops. Other option is to evaluate, select, and improve salt-tolerant plants that already are adapted to salt flat areas, focusing on those that might make desirable crops (Ungar 2000).

Actually there is an increasing interest about the remediation effects of saltbush on a range of salt levels in soils (Song et al. 2008). *Suaeda maritima* is an annual halophyte, tolerant of high salinity and alkaline environments, and is therefore considered as a promising halophyte for forage and edible oil in dry arid zones and coastal saline wetlands (Wang et al. 2004). It seems feasible to incorporate some of these important plants as part of the crops which support the agricultural economy of those areas affected by salinity (Khan and Gul 2002; Towhidi and Zhandi 2007; Rueda et al. 2011). In Iran country, *S. maritima* has a wide distribution along the coasts. This plant was identified as a novel crop for possible domestication due to its high-quality oilseed resource and so marked as an agro-industrial commodity with high potential (Khan and Gul 2002). However, the productivity of this halophyte is limited by a lack of available nitrogen, a condition that affects its growth potential and establishment (Rueda et al. 2011). Traditionally, farmers apply synthetic fertilizers to compensate for soil nitrogen deficiency. However, indiscriminate use of these fertilizers might increase salinity and severely damage the soil microbial structure and composition (Banwari and Rao 1990).

Few studies related to bacterial diversity within the rhizosphere of salt marsh plants have been published. The most studied plants are *Salicornia bigelovii* (Rueda et al. 2004) and *Spartina alterniflora* (Lovell et al. 2000; Nielsen et al. 2001). Recent molecular studies suggested that numerous microbial genera associated with the halophytes *Salicornia bigelovii*, *Spartina*, and *Leptochloa fusca* were unclassified (Reinhold et al. 1987; Munns 2005). It is important to increase the knowledge on known salt-tolerant nitrogen-fixing bacteria associated with the native crops (Hamdi 1999; Whipps 2000) as a potential bio-fertilizer in a salt production areas like Abbas, Iran. Several halophyte plants, including *S. maritima*, which already is considered as a new resource for arid regional agriculture, can benefit with soil-enriched biological N2 fixers (Bashan et al. 1992).

The work presented here focused on the composition of natural bacteria associated with the root system of the halophyte *S. maritima* to isolate and collect endemic bacteria able to fix atmospheric nitrogen and to measure the effect of isolated bacteria on germination of *Suaeda*. The hypothesis reveals the inoculation of identified endemic halotolerant plant growth-promoting bacteria associated with the halophyte *S. maritima* enhances seed germination and early seedling growth.

Materials and Methods

Sample Collections

Soil samples (1 kg) and root system (50 g) were taken from the natural population of *S. maritima*, along the coastal area of Abbas, Iran (27°12′06.61″N-56°24′51.40″E), with an average annual precipitation of 251 mm and temperature of 49 °C. Four young and mature plants and moist soil samples were collected in four different locations: (a) 27°12′42.57″N–56°25′54.39″E, elevation 5 masl; (b) 27°12′39.83″N–56°25′55.51″E, elevation 5 masl; (c) 27°12′37.21″N–56°25′54.42″E, elevation 5 masl; and (d) 27°12′41.81″ N–56°26′0.04″E, elevation 5 masl. Samples were placed in black plastic bags and labeled with date and location of collection. During collection, plants were kept on an ice-cold recipient for 5 h and immediately processed at laboratory arrival.

Isolation of Endemic Plant Growth-Promoting Bacteria from Roots and Soil

The experiments were conducted at Academic Center for Education Culture and Research of Hormozgan, Iran. The root samples were carefully taken from the soil surface and were cut into pieces of 2 cm pieces. Two sets of roots were inoculated in different liquid N-free media (1:100, roots/water) of OAB (Okon and Hadar 1987) and Rennie's medium (Rennie 1981) and incubated for 48 h at 30 °C and 55 °C with 150 rpm agitation. The resulting solutions were diluted six times 10^{-1}, in 0.85% saline solution, to achieve a total dilution of 10^{-6}. Five replicates of dilution from both media were dispersed in four NaCl concentrations each (0.0, 0.25, 0.50, and 0.75 M) by spreading in plates containing the same OAB and Rennie's solid N-free media. The plates were incubated at 30 °C and 55 °C for 1 to 4 days (at 30 °C = 40 plates per sample = 320 plates/location; at 55 °C = 40 plates per sample = 320 plates/location =640 plates/4 locations in total).

From each root sample, 4 g of soil, obtained from each root sample, in sterile distilled water (1:100, soil/water, by volume) were diluted. After five 10–1 dilutions to achieve total dilution of 10–5, a volume of 0.1 mL to each N-free medium, in five replicates each, was then applied in plates with four NaCl concentrations (0.0, 0.25, 0.50, and 0.75 M). The inoculated petri dishes were incubated at 30 °C and 55 °C (at 30 °C = 40 plates per sample = 320 plates/location; at 55 °C = 40 plates per sample = 320 plates/location = 640 plates/4 locations in total).

Vigorous colony growth was observed in all plates containing 0, 0.25, 0.5, and 0.75 M NaCl at 30 and 55 °C. Colonies from these incubated petri dishes were re-isolated, plated, and purified separately on both N-free mediums (OAB and Rennie's) containing 0.5 M NaCl. Fit in mention that all isolates, including diazotrophs, were stored in 15% glycerol at −70 °C (Carrillo et al. 1998).

Phosphate Solubilization Assay

In order to determine the phosphate-solubilizing ability of isolated halotolerant endemic plant growth-promoting bacteria, colonies were cultured in a solid medium SRSM1 and liquid medium SRSM2. The solid medium SRSM1 containing glucose, 10 g/l; tribasic calcium phosphate, 5 g/l; ammonium sulfate, 0.5 g/l; potassium chloride, 0.2 g/l; magnesium sulfate heptahydrate, 0.3 g/l; manganese sulfate, 0.002 g/l; yeast extract (Difco), 0.5 g/l; bromocresol purple, 0.1 g/l with pH 7.2 (Sundara-Rao and Sinha 1963; Vázquez et al. 2000); and four different concentrations of NaCl (0, 0.25, 0.5, and 0.75 M) was incubated at 30 °C and 55 °C.

Phosphate-solubilizing activity was considered as positive, when the medium appeared transparent to the eye. To validate the results, five replicates were performed with *Azospirillum halopraeferens* as a control which is a nitrogen-fixing bacterium with the capacity to solubilize phosphate and grow under salinity conditions and is considered as growth-promoting bacteria (De Troch and Vaderleyden 1996). Colonies of morpho-types showing clear halos and large solubilization were measured with a digital caliper (GENERAL No. 143, General Tools Manufacturing Co., Inc., New York) and an illuminator (Laminator portable-APCO ISSUE N. BJ-84,161). Later, the determination of phosphate concentration according to the standard method of Strickl and Parsons (1972) was determined.

The liquid medium SRSM2 containing as like as SRSM1 medium, but where tri-basic calcium phosphate is replaced by 1.1 g/l of di-potassium phosphate tri-hydrate in 250-ml Erlenmeyer flasks at 32 ± 2 °C with 120 rpm constant agitation. The bacterial cultures were washed three times by 20,000 g centrifugation for 10 s with 2% sodium chloride at 27 ± 2 °C. The bacterial culture was adjusted to 1.0 ± 0.1 at 540 nm absorbance by using spectrophotometer (Milton Roy Company Spectronic 20D). An aliquot of 250 µl of each bacterial suspension was added in a 125 ml flask containing 25 ml of SRSM2 liquid medium containing insoluble calcium phosphate, non-inoculated bacterial flask containing SRSM2 liquid medium considered as control, incubated for 24 h. After incubation, two replicates (500 µl) were taken from each culture and centrifuged at 20,000 g for 10 s, and the pellet was discarded. Twenty to 50 milliliters (depending on the predetermined level of phosphate solubilization) of each isolates were sampled and diluted in 50 ml of distilled water (Vazquez et al. 2000), in quadruplicate. Because it has been reported that after successive transfers isolates of inorganic phosphate-solubilizing bacteria (IPBS) lose their phosphate-solubilizing capacity (Craven and Hayasaka 1982), the positive isolates were stored at −20 ±3 °C in SRSM2 liquid medium supplemented with 30% glycerol.

Acetylene Reduction Assay

Isolated bacteria were grown in appropriate bottles with OAB and Rennie's N-free media containing 0.5 M NaCl. The bottles were sealed and incubated at 30 °C for 5 days. After incubation, 1 ml of air was extracted with a syringe and 1 ml of acetylene was injected. Culture stocks were incubated for 48 h at 30 °C after this procedure. Five replicates were used to validate the results, and the biological control was a strain (*Azospirillum halopraeferens* AU10), which is a nitrogen-fixing bacterium with the capacity to develop in roots (Reinhold et al. 1987, De Troch and Vaderleyden 1996). Subsequently, nitrogen activity was assayed by gas chromatography. Ethylene analysis was performed with a gas chromatograph (Varian 6000, Vary Instrument Group, Atlanta, Georgia) equipped with a hydrogen flame ion detector (FID). A different set of five bottles was used for each reading. C2H2 reduction was evaluated after 120 h; the amount of C2H4 produced was expressed in nanomole C2H4 per culture (Holguin et al. 1992).

Identification of N_2-Fixing Bacteria

Molecular bacteria characteristics were analyzed by gas chromatography of cell fatty acid methyl esters (FAME) (Sasser 1990) and 16S rRNA sequencing (ACCULAB, Newark, DE). The bacterial identifications assigned in this report are based on percent genetic distances (%GD), which are defined as the number of nucleotide differences between two sequences, expressed as a percent. All DNA sequences were generated using PE Biosystems' MicroSeq 16S rRNA Gene Kit or MicroSeq 500 rDNA Bacterial Sequencing Kit. Samples were then identified using the MicroSeq Microbial Identification software and databases. Phylogenetic tree was generated using the neighbor-joining algorithm (Saitou and Nei 1987). All isolates, including diazotrophs, were stored at 4 °C. Both the ATCC stock and the nitrogen-fixing isolate were stored in 15% glycerol at −70 °C (Carrillo et al. 1998).

Evaluation of Bacillus amyloliquefaciens as Inoculant on the Germination and Seedling Growth of S. maritima Under Salinity Conditions

On germination stage, *S. maritima* seeds were inoculated with *Bacillus amyloliquefaciens* and with *A. halopraeferens* AU10 as a control. The following set of methodologies was used to test the ability of *Bacillus amyloliquefaciens* to enhance the germination percentage and rate, fresh and dry weight, root length, plant height, and CFU.ml^{-1} (colony-forming bacterial units) and to adhere efficiently to the root system of *S. maritima*. Seeds were obtained from mature, dried plants of a natural population in the rural coastal community of Bandar Abbas near Hormozgan, Iran. Seed collection was carried out during June 2012, when plants naturally liberate seeds. Plants were sifted to separate mature seeds. The seeds were cleaned, and the largest seeds with a uniform color and without apparent damage were selected.

Then, seeds were disinfected by immersion in sodium hypochlorite (3% active chlorine) during 30 s and washed three times with sterilized distilled water. Disinfected seeds were inoculated separately with bacteria (*Bacillus amyloliquefaciens* and *Azospirillum halopraeferens*) treatments according to Carrillo et al. (1998), at a concentration of 108 cells ml^{-1}. Germination tests were performed in sterilized petri dishes, each with a cloth layer substrate (150 × 15 mm) covering the bottom of the dish and moistened with uniform amounts of NaCl solution (0, 0.25, 0.5, and 0.75 M). Germination tests were done in a growth chamber at 27 ± 0.5 °C, 35 ± 1% RH and with continuous white light. Twenty milliliters of the appropriate solution was changed and added every 4 days to each dish. Seeds were considered germinated when the radical was at least 2 mm long. The number of germinated seeds was recorded daily (germination rate), and the final percentage of germination was determined after 30 days. Germination rate was calculated using the formula described by Maguire (1962):

M = n1/t1+n2/t2+…n20/t20where n1, n2,… n20 are the numbers of germinated seeds at times t1, t2,…t20 in days, respectively. The selected hierarchic experiment of a randomized design considering two factors, each one with five replicates of ten seeds each, was used. The first factor (inoculants) had three levels: no inoculation, inoculation with the bacterium *Bacillus amyloliquefaciens*, and inoculation with *Azospirillum halopraeferens*. The second factor had four levels: concentrations of NaCl 0, 0.25, 0.5, and 0.75 M. The two factors of three and four levels, respectively, resulted in 12 treatments (600 experimental units). The data for percentage germination was analyzed after applying an arcsine transformation (Snedecor 1956; Sokal and Rohlf 1988), in order to apply two-way analyses of variance (ANOVA). Germination rates, which consisted in the sums of germinations per day, were transformed (percent to arcsine) before analysis.

Thirty-five plants from each treatment were chosen randomly. Dry and fresh weights were measured on the 60th day. Root length and height were measured with a digital caliper (GENERAL No. 143, General Tools Manufacturing Co., Inc., New York). Dry weight was determined after drying each organ in a forced air dryer at 110 °C for 36 h. In addition, protein, total lipid, and ash contents were assayed. Proteins were evaluated by the micro-Kjeldahl method; ash by weight difference after burning for 24 h at 500 °C and total lipids were measured according to Barnes and Blackstock (1973).

Quantification of bacteria adhering to the root system of *S. maritima* was carried out at the conclusion of the study 60 d later. Seven plants of each pot of each treatment (420 seedlings) were washed with sterile distilled water and submerged during 1 min into Eppendorf tubes with sterile water. The tubes were agitated during 1 min to detach bacteria from the roots. Three samples of 100 μl were taken from the bacterial solution of each tube and sowed by dispersion on petri dishes with N-free media OAB and incubated 24 h at 30 °C for colony-forming unit determination (CFU).

Statistical Analysis

Results were analyzed by two-way ANOVA, and the F-test was applied to determine statistically different values (Snedecor 1956). Least significant differences between means of treatments were separated by Duncan's multiple range test at $P < 0.05$. All statistical tests were done with SAS (2001).

Results

Isolation of Nitrogen-Fixing Bacteria from S. maritima Rhizosphere

To investigate the bacterial abundance and morphological characteristics of halotolerant endemic nitrogen-fixing PGPR associated with the root system of *Suaeda*, nine colonies were selected. These nine morpho-types were present at the four sampled sites. When grown in different concentrations of NaCl, the nine selected colonies of nitrogen-fixing bacteria grew most quickly in 0.25 M NaCl at 30 °C and 55 °C. At this concentration, the maximum CFU ml^{-1} of bacteria occurred at 13 to 24 h, while at 0, 0.5, and 0.75 M NaCl, the growth was slower considering the two temperatures. ·

Phosphate Solubilization Assay

After 26 h of incubation, all isolates were capable of dissolving insoluble phosphate. However, considering *Azospirillum halopraeferens* as a control, one morpho-type named BACTSUAEDA showed significant differences ($P < 0.05$) at 0 and 0.25 M NaCl at 30 °C and 55 °C (Table 18.1). At 0.5 and 0.75 M NaCl with two temperatures were not significantly different ($P < 0.05$). In liquid medium, bacterial species in culture medium containing insoluble calcium phosphate showed solubilization after 21 h of incubation. BACTSUAEDA showed significant differences ($P < 0.05$) compared with other bacterial isolates. BACTSUAEDA showed maximum solubilization after 21 h while *A. halopraeferens* at 16 h (140 ± 13 mg ml^{-1} at 21 h; 145 ± 11 mg ml^{-1} at 16 h), respectively.

Acetylene Reduction Assay and 16S rRNA Characterization

According the acetylene reduction assays, only one of the nine isolated bacteria showed significant ($P < 0.05$) acetylene reduction activity; this nitrogen-fixing bacterium with highest acetylene reduction ability coincided with the same isolated bacterium which showed significant values in dissolving insoluble phosphate (BACTSUAEDA); the value of acetylene reduction assay of BACTSUAEDA was 7.13/0.28 nmole $culture^{-1}$ h^{-1}, in contrast to the control *Azospirillum halopraeferens* with 7.11/1.9 nmole $culture^{-1}$ h^{-1}, while in other isolated bacteria, the values were not more than 4.45 nmole $culture^{-1}$ h^{-1}. The nitrogen-fixing bacterium with highest acetylene reduction and dissolving insoluble phosphate ability according to molecular bacteria characteristics was related with *Bacillus amyloliquefaciens* (Fig. 18.1), considering as a reference the strain as *Bacillus* sp. by Liu et al. (2012).

Evaluation of B. amyloliquefaciens as Inoculant on the Germination and Seedling Growth of *Suaeda* spp. Under Salinity Conditions

Seeds of *S. maritima* were inoculated with the nitrogen-fixing bacterium with highest acetylene reduction and dissolving insoluble phosphate ability, which was related to *Bacillus amyloliquefaciens* at the germination stage. Compared with *Azospirillum halopraeferens* as a biological control, the results showed significant differences ($P < 0.05$) among the treatments (Table 18.2). Also, it was observed that when salinity was higher, inhibition of germination was greater for *Suaeda*. Seeds inoculated with *Azospirillum halopraeferens* and *Bacillus amyloliquefaciens* germinated about 87% more than the control without inoculated bacteria at 0 M NaCl. Treatments with higher salinity affected negatively the germination percentage, independent of the presence of bacteria.

On the other hand, for two parameters (plant height and root length), halotolerant nitrogen-fixing bacteria influenced plant growth, with significant differences ($P < 0.05$) between treatments (Table 18.2). Treatment with *Bacillus amyloliquefaciens* promoted seedling growth significantly ($P < 0.05$), and the higher number of bacterial cells adhered was observed to the roots with both studied bacteria (Table 18.2). Moreover, results confirm the declining trend of colony-forming bacterial units (CFU) with increasing salinity. *Bacillus amyloliquefaciens* and *Azospirillum halopraeferens* were unaffected below 0.25, 0.50, and 0.75 M NaCl, respectively, when assayed on *Suaeda*. However, the length of root system was significantly ($P < 0.05$) increased with *Bacillus amyloliquefaciens* with significant difference, compared to *Azospirillum halopraeferens*.

Seeds inoculated with *Bacillus amyloliquefaciens* and *Azospirillum halopraeferens* showed impact in proteins and ash contents with

Table 18.1 Phosphate solubilization determined by halo size (mm) under different concentrations of salinity and temperature

	Salinity (NaCl)							
	0 M		0.25 M		0.5 M		0.75 M	
	Temperature and size of halos (mm)							
Bacterium	30 °C	50 °C	30 °C	50 °C	30 °C	50 °C	30 °C	50 °C
BACTSUAEDA	3.43 ± 0.22a	2.77 ± 0.34 a	2.77 ± 0.15 a	2.03 ± 0.06 a	0.99 ± 0.05 b	1.13 ± 0.19a	0.43 ± 0.13a	0.13 ± 0.23a
Azospirillum halopraeferens	1.43 ± 0.19b	1.33 ± 0.30 b	1.77 ± 0.10 b	1.77 ± 0.05 b	1.43 ± 0.10 a	1.17 ± 0.22a	0.43 ± 0.19 a	0.11 ± 0.21a

Means followed by the same letter are not significantly different at $P < 0.05$. Comparisons were made within columns using Duncan's multiple range tests. Values represent means of five replications

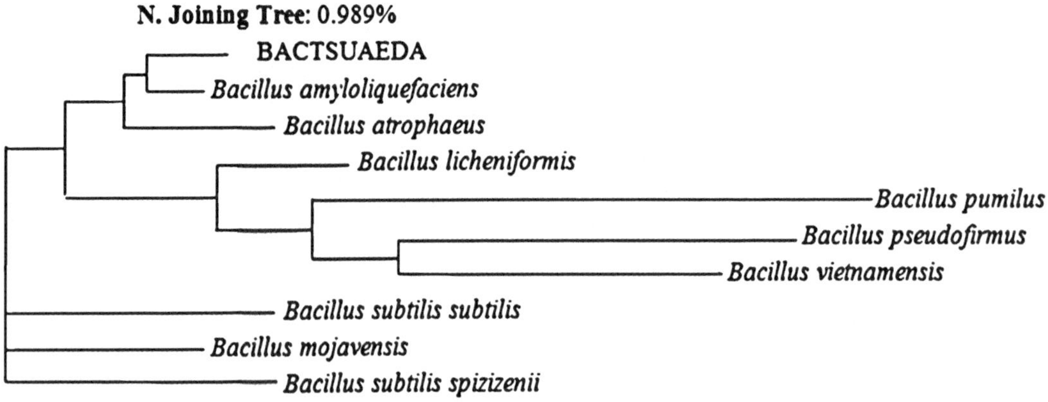

Fig. 18.1 Phylogenetic tree of 16 rRNA sequences of isolated halotolerant bacteria (*Bacillus amyloliquefaciens*) from *S. maritima* rhizosphere in Abbas, Iran

significant differences ($P < 0.05$) (Table 18.3). There is no significant difference observed in lipids ($P < 0.05$); however, the lipid content reached the maximum value with *Bacillus amyloliquefaciens* at 0.50 and 0.75 M NaCl (18%) and with *Azospirillum halopraeferens* at 0.5 and 0.75 M NaCl (16 and 17%), respectively. There is no significant difference observed in proteins ($P < 0.05$) at 0 and 0.25 M NaCl with inoculated and non-inoculated seeds.

Discussion

Because *Suaeda* has been improved for use in the food and agricultural industries and to reforest disturbed zones under salty conditions (FAO 1998), it is important to consider strategies for its reproduction and cultivation. According to sustainable strategies and taking into account the bio-fertilization, it has been suggested that *S. maritima* microorganisms are associated with *Suaeda maritima* (Goodfriend et al. 2000; Bagwell et al. 2001). It was observed that eight of the nine colonies that grew in N2-free media in our study were unable to reduce acetylene in large quantities. This could be attributable to dependence on other bacteria, a phenomenon that is quite common among microorganisms (Holguin et al. 1992; Rodelas et al. 1996; Rojas et al. 2001). The results showed the ability of these bacteria can grow in a dry arid saline environment, which salinity exceeds 0.3 M NaCl concentrations (Velarde et al. 2003).

The ability of *Bacillus* spp. to solubilize phosphate is known for seawater and marine sediments. Terrestrial IPSB like *Bacillus amyloliquefaciens* have been used together with ectomycorrhizal fungi for inoculation of Douglas fir seedlings to promote their growth, and the weathering of calcium phosphate rocks has been attributed to ectomycorrhizal fungi and bacteria and *Bacillus licheniformis* isolated from Brazilian soils (Akhavan et al. 1991; Puente 2004). Previous results were agreed with our isolated bacterium *Bacillus amyloliquefaciens* (Liu et al. 2012), and other reports indicated that it has an ability to produce phytohormones (Rodelas et al. 1996; Zexun and Wei 2000). *Bacillus amyloliquefaciens* has been detected in soil and water, associated with plants, and has a significant nitrogen-fixing ability (Vázquez et al. 2000; Puente 2004; Villegas et al. 2010).

According to the evaluation of seed germination, several experiments were carried out exposed to high salinity (Felker et al. 1981). Considering our evaluation, it was observed that both *Bacillus amyloliquefaciens* and *Azospirillum halopraeferens* are good nitrogen fixers and plant growth promoters for *Suaeda* on germination and early seedling stage. Compared with other plants and beneficial microorganisms, similar results have been obtained according to Goodfriend et al. (2000) and Villegas et al. (2010).

Table 18.2 Effects of *Bacillus amyloliquefaciens* and *Azospirillum halopraeferens* on mean values of root length, plant height, and fresh and dry weights of *S. maritima* seedlings under four concentrations of NaCl

Bacterial inoculant[a]	Salinity (M)	Germination (%)	Plant height (cm)	Root length (cm)	Fresh weight (mg)	Dry weigh (mg)	CFU·ml
Control	0.00	45 ± 4b	1.88 ± 0.20ab	1.37 ± 0.23b	10.20 ± 3.5b	1.04 ± 0.12a	0
Control	0.25	38 ± 5bc	1.26 ± 0.31c	1.35 ± 0.20b	10.22 ± 4.3b	1.05 ± 0.45a	0
Control	0.50	15 ± 12d	0.28 ± 0.10e	1.34 ± 0.15b	20.06 ± 5.4a	1.00 ± 0.56a	0
Control	0.75	4 ± 4e	0.09 ± 0.04e	0.02 ± 0.04d	20.00 ± 6.1a	1.08 ± 0.65a	0
B. amyloliquefaciens	0.00	89 ± 11a	2.10 ± 0.15a	1.38 ± 0.19b	10.21 ± 3.8 b	1.06 ± 0.32a	3689 ± 467a
B. amyloliquefaciens	0.25	87 ± 13a	1.48 ± 0.23c	1.87 ± 0.11a	10.24 ± 2.8b	1.05 ± 0.21a	3456 ± 654a
B. amyloliquefaciens	0.50	50 ± 9b	0.30 ± 0.09e	1.81 ± 0.19a	20.06 ± 2.9a	1.01 ± 0.13a	2313 ± 453b
B. amyloliquefaciens	0.75	35 ± 12bc	0.35 ± 0.04e	0.10 ± 0.04d	20.03 ± 6.7a	1.09 ± 0.11a	2130 ± 340b
A. halopraeferens	0.00	82 ± 13a	2.03 ± 0.15a	1.58 ± 0.16ab	10.22 ± 3.9b	1.05 ± 0.24a	3645 ± 342a
A. halopraeferens	0.25	82 ± 18a	1.32 ± 0.19c	1.77 ± 0.12a	10.21 ± 2.9b	1.04 ± 0.23a	2476 ± 645ab
A. halopraeferens	0.50	53 ± 9b	1.06 ± 0.12 cd	1.23 ± 0.10bc	20.16 ± 3.7a	1.01 ± 0.19a	2108 ± 342b
A. halopraeferens	0.75	33 ± 7c	0.20 ± 0.21e	0.04 ± 0.02d	20.01 ± 3.9a	1.08 ± 0.14a	1075 ± 547bc

Means followed by the same letter are not significantly different at $P < 0.05$. Comparisons were made within columns using Duncan's multiple range test. Values represent means of five replications

[a]Bacterium (1 × 108 CFU ml^{-1})

Table 18.3 Effect of *Bacillus amyloliquefaciens* and *Azospirillum halopraeferens* on the contents of total lipid (mg g^{-1}), protein (%), and ash (%) in seedlings of *S. maritima* under four concentrations of NaCl (0, 0.25, 0.5, and 0.75 M)

Bacterial inoculant[a]	Salinity (M)	Lipid (%)	Protein (%)	Ash (%)
Control	0	0.11 ± 0.03a	11.33 ± 1.21a	0.93 ± 0.12bc
Control	0.25	0.14 ± 0.04a	11.13 ± 1.29a	0.89 ± 0.11b
Control	0.5	0.15 ± 0.03a	5.34 ± 1.34b	0.51 ± 0.02c
Control	0.75	0.13 ± 0.06a	3.125 ± 1.02c	0.24 ± 0.02d
B. amyloliquefaciens	0	0.14 ± 0.03a	11.36 ± 1.20a	1.00 ± 0.12a
B. amyloliquefaciens	0.25	0.17 ± 0.03a	11.35 ± 1.92a	0.93 ± 0.10ab
B. amyloliquefaciens	0.5	0.18 ± 0.04a	5.09 ± 1.21b	0.55 ± 0.09c
B. amyloliquefaciens	0.75	0.18 ± 0.03a	3.24 ± 1.09c	0.31 ± 0.10d
A. halopraeferens	0	0.15 ± 0.04a	11.48 ± 1.82a	1.01 ± 0.11a
A. halopraeferens	0.25	0.16 ± 0.05a	11.36 ± 1.21a	0.95 ± 0.10ab
A. halopraeferens	0.5	0.17 ± 0.02a	6.14 ± 1.33b	0.93 ± 0.12ab
A. halopraeferens	0.75	0.17 ± 0.03a	3.51 ± 1.02c	0.33 ± 0.13d

Means followed by the same letter are not significantly different at $P < 0.05$. Comparisons were made within columns using Duncan's multiple range test. Values are represented by means of five replications

[a]Bacterium (1 × 108 CFU ml^{-1})

Those results are agreed with our results, and the positive effects of bacteria on *Suaeda* plant possibly suggest that the production of plant growth-promoting substances was often reported as responsible for the enhancement of plant growth promotion (Goodfriend et al. 2000; Vázquez et al. 2000; Bagwell et al. 2001). Moreover our obtained results indicate a specificity of plant-bacterium interaction, according to Baldani and Dobereiner (1980), which is controversial with Okon and Hadar (1987). Arsac et al. (1990) stated that the effect of PGPR is not strain-dependent among different plant species.

Other studies with inoculants on plants indicate some inhibitory effects on germination (Díaz et al. 2001). Similar studies have been reported on the development of germination rate in *Arthrocnemum indicum* (Khan and Gul 1998; Khan et al. 2000; Khan and Gul 2002); *Salicornia rubra*, *Suaeda corniculata*, and *Suaeda salsa* (Little 1950); and *Suaeda japonica* (Yokoishi and Tanimoto 1994). The importance of these studies is focused because *Suaeda* spp. seeds could be classified as one of the most salt-tolerant species during germination. *Suaeda* spp. can also be recommended for silviculture practices, principally where there is a possible investment for sustainable forest management. At the same time, halophytes such as *Suaeda* cultivation would help to balance the carbon cycle and reduce global warming, as this plant thrives in dry zones (Robles et al. 2009). Moreover the increase in soil salinity may inhibit the growth of common economic plants, *Suaeda* can able to grow in extreme salty soils, and it is more helpful to arid-zone management (Velarde et al. 2003).

Conclusion

This study is the first step to enhance the growth of halophytes of *Suaeda* by using halotolerant plant growth-promoting bacteria as an efficient and reliable biological product, as well as in the extension of the range of host plants for *A. halopraeferens*. Furthermore, studies on *Suaeda* spp. interaction with *Bacillus* as a bio-fertilizer are recommended to determine the extent to which these observations can be reproduced under field conditions. In this work, a reliable biological method based on beneficial bacteria in germination and seedling stage was assayed to contribute to maintain or improve the fertility of soils sustaining *Suaeda* fields. Also it was observed that both *Azospirillum halopraeferens* and *Bacillus amyloliquefaciens* are good nitrogen fixers and plant growth promoters for *Suaeda*.

Acknowledgments This work was supported by Academic Center for Education Culture and Research

(ACECR) of Hormozgan, Iran. Thanks to Universidad de Sonora and Project of CONACYT (Apoyoscomplementarios para la consolidacióninstitucional de grupos de investigación) 2007; code: 74592 of Dr. Bernardo Murillo Amador, CIBNOR.

References

Akhavan K, Campbell W, Jurinak J, Dudley L (1991) Effects of $CaSO_4$, $CaCl_2$, and NaCl on leaf nitrogen, nodule weight, and acetylene reduction activity in *Phaseolus vulgaris* L. Arid Soil Res Rehabil 5:97–103

Arsac J, Lamothe C, Mulard D, Fages J (1990) Growth enhancement of maize (Zea mays L) through *Azospirillum lipoferum* inoculation: effect of plant genotype and bacterial concentration. Agronomie 10:640–654

Bashan Y, Holguin G, Puente M (1992) Alternativaagrícola regional por fertilizantesbacterianosenuso y manejo de los recursos naturales en la Sierra de la Laguna Baja California Sur. In: Ortega R (ed) Uso y Manejo de los Recursos Naturales en la Sierra de la Laguna, B.C.S. Centro de InvestigacionesBiológicas del Noroeste, La Paz, pp 46–67

Bagwell C, Dantzler M, Bergholz P, Llovell C (2001) Host-specific ecotype diversity of rhizoplane diazotrophs of the perennial glasswort *Salicornia virginica* and selected salt mash grasses. J Aqu Microbiol Ecol 23:293–300

Baldani V, Dobereiner I (1980) Host-plant specificity in the infection of cereals with *Azospirillum* spp. Soil Biol Biochem 12:443–439

Banwari I, Rao V (1990) Effect of *Azospirillum brasilense* on growth and nitrogen content of *Cynodon dactylon* under different moisture regimens. Ind. J Plant Physiol 33:210–213

Barnes H, Blackstock J (1973) Estimation of lipids in marine animal and tissues: detailed investigation of sulphophosphovanil method for 'total' lipids. J Exp Mar Biol Ecol 12:103–118

Carrillo A; Puente ME, Castellanos E, Bashan Y (1998) Aplicaciones Biotecnológicas de EcologíaMicrobiana. Pontificia Universidad Javeriana, Santa Fe de Bogotá, Colombia and Centro de InvestigacionesBiológicas del Noroeste Manual de Laboratorio, Manual de Laboratorio, La Paz, B.C.S., México, pp 15–20

Craven PA, Hayasaka S (1982) Inorganic phosphate solubilization by rhizosphere bacteria in a *Zostera marina* community. Can J Microbiol 28:605–610

De Troch P, Vaderleyden J (1996) Surface properties and motility of rhizobium and *Azospirillum* in relation to plant root attachment. Microb Ecol 32:149–169

Díaz V, Ferrera C, Almaraz S, Alcántar G (2001) Inoculation of plant growth-promoting bacteria in lettuce. Terrain 19:327–335

Food and Agriculture Organization, FAO (1998) Red Latinoamericana de CooperaciónTécnicaen SistemasAgroforestales. EspeciesArbóreas y Arbustivas para las zonas Áridas y Semiáridas de América Latina, p 320

Felker P, Clark PR, Laag AE, Pratt P (1981) Salinity tolerance of the tree legumes mesquite (*Prosopis glandulosa var torreyana*, *P. velutina*, and *P. articulate*) algarrobo (*P. chilensis*), Kiawe (*P. pallida*) and tamarugo (*P. tamarugo*) grown in sand culture on nitrogen free media. Plant Soil 61:311–317

Goodfriend W, Olsenm M, Frye R (2000) Soil microfloral and microfaunal response to *Salicornia bigelovii* planting density and soil residue amendment. Plant Soil 1:23–32

Hamdi H (1999) Rhizobium-legume symbiosis and nitrogen fixation under severe conditions and in arid climate. Microb. Mol Biol Rev 63:968–989

Holguin G, Guzman M, Bashan Y (1992) Two new nitrogen-fixing bacteria from the rhizosphere of mangrove trees: isolation, identification and in vitro interaction with rhizosphere *Staphylococcus* sp. Federation of European Microbiological Societies. Microbiol Ecol 101:207–216

Khan MA, Gul B (1998) High salt tolerance in germinating dimorphic seeds of *Arthrocnemum indicum*. Int J Plant Sci 159:826–832

Khan MA, Gul B (2002) Salt tolerant plants of coastal Sabkhas of Pakistan. In: Sabkha A, Barth H, Boer B (eds) Ecosystems. Kluwer Academic Press, Dordrecht

Khan MA, Gul B, Weber DJ (2000) Germination response of *Salicornia rubra* to temperature and salinity. J Arid Environ 45:207–214

Little EL (1950) Southwestern trees – a guide to the native species of New Mexico and Arizona. USD, Handbook No. 9, Government Printing Office, Washington, DC, 560 p

Liu W, Wang X, Wu L, Chen M, Tu C, Luo Y, Christie P (2012) Isolation, identification and characterization of *Bacillus amyloliquefaciens* BZ-6, a bacterial isolate for enhancing oil recovery from oily sludge. Chemosphere 87(10):1105–1110

Lovell C, Piceno Y, Bagwell C (2000) Molecular analysis of diazotroph diversity in the rhizosphere of the smooth cordgrass, *Spartina alterniflora*. Appl Environ Microbiol 66:3814–3822

Maguire J (1962) Speed of germination-aid in selection and evaluation for seedling emergence and vigour. Crop Sci 2:176–177

Munns R (2005) Genes and salt tolerance: bringing them together. New Phytol 167:645–663

Nielsen I, Finster K, Welsch A, Donelly R, Herbert R, De Wit L, Lomstein B (2001) Sulphate reduction and nitrogen fixation rates associated with roots, rhizomes and sediments from *Zostera noltii* and *Spartina maritima* meadows. Environ Microbiol 3:63–71

Okon Y, Hadar Y (1987) Microbial inoculants as crop-yield enhancers. CRC Crit Rev Biotechnol 6:6–85

Puente M (2004) Poblacionesbacterianasendófitas y del rizoplano de plantas del desiertodegradadotas de roca y suefectosobre el crecimiento del cardón (Pachycereuspringlei [S. WATS] BRITT. and ROSS).

Doctoral thesis, Centro de Investigaciones Biológicas del Noroeste, La Paz, B.C.S., México, pp 1–166
Rennie R (1981) A single medium for the isolation of acetylene reducing (dinitrogen-fixing bacteria from soil). Can J Microbiol 27:8–14
Reinhold B, Hurek T, Fendrik I, Pot B, Gillis M, Kersters K, Thielmans S, De Ley J (1987) *Azospirillum halopraeferens* sp. novo a nitrogen-fixing organism associated with roots of Kallar grass (*Leptochloa fusca* L. Kunth). Int J Syst Bacteriol 37:43–51
Rodelas B, González I, Salmeron V, Pozo C, Martinez T (1996) Enhancement of nodulation, N-2 fixation and growth of faba bean (Viciafaba L.) by combined inoculation with *Rhizobium leguminosarum* by Viceaesp and *Azospirillum brasilense*. Symbiosis 21:175–186
Rojas A, Holguin G, Glick B, Bashan Y (2001) Synergism between Phyllobacterium sp (N_2-fixer) and *Bacillus licheniformis* (P-solubilizer), both from a semiarid mangrove rhizosphere. Microb Ecol 35:181–187
Rueda PEO, Barrón H, Jojanes H (2009a) BacteriasPromotoras Del Crecimiento Vegetal. Editorial Plaza y Valdes, México, p 112
Rueda PEO, Barrón H, Tarazón H, Preciado R (2009b) La Salinidad: Un Problema o Una Opción Para La Agricultura? Editorial Plaza y Valdes, México City, p 264
Rueda PEO, Castellanos T, Troyo E, De León J (2004) Effect of *Klebsiella pneumoniae* and *Azospirillum halopraeferens* on the growth and development of two *Salicornia bigelovii* genotypes. Aust J Exp Agric 44:65–74
Rueda PEO, Félix A, Beltrán M, Ruíz H, Valdez C, García H, Ávila N, Partida L, Murillo B (2011) Sustainable options for soil management in arid zones: uses of the halophyte *Salicornia bigelovii* (Torr.) and biofertilizers in the modern agriculture. Trop Subtrop Agroecosyst 13:157–167
Robles AB, Ruiz M, Ramos ME, González R (2009) Role of livestock grazing in sustainable use, naturalness promotion in naturalization of marginal ecosystems of southeastern Spain (Andalusia). In: Rigueiro-Rodriguez A, McAdam J, Mosquera-Losada MR (eds) Agroforesty in Europe. Editorial Springer, Lugo, p 445
Saitou N, Nei M (1987) The neighbour-joining method: a new method for reconstructing phylogenetic trees. Mol Biol Evol 4:406–425
SAS, Institute (2001) SAS/STAT User's Guide. Version 6.12 SAS Institute, Cary, NC
Sasser M (1990) Identification of bacteria through fatty acid analysis. In: Clement Z (ed) Methods in Phytobacteriology, vol 565. Akadamiai RU Kiado, Budapest
Snedecor G (1956) In: Freeman SR (ed) Statistical methods applied to experiments in agriculture and biology. The Iowa State College Press, Ames
Sokal R, Rohlf F (1988) Biometry. In: Freeman SR (ed) The principles and practice of statistics in biological research, San Francisco, p 650
Song J, Fan H, Zhao YY, Jia YH, Du XH, Wang BS (2008) Effect of salinity on germination, seedling emergence, seedling growth and ion accumulation of a euhalophyte *Suaeda salsa* in an intertidal zone and on saline inland. Aquat Bot 88:331–337
Strickl J, Parsons T (1972) A practical handbook of sea water analysis. Bull Fish Res Can 167:49–52
Sundara-Rao W, Sinha M (1963) Phosphate dissolving micro-organisms in the soil and rhizosphere. Indian J Microbiol 41:999–1011
Towhidi A, Zhandi M (2007) Chemical composition, in vitro digestibility and palatability of nine plant species for dromedary camels in the province of Semnan. Iran Egypt J Biol 9:47–52
Ungar I (2000) Ecophysiology of vascular halophytes. Department of Botany, CRC Press, Ohio University, Athens, p 209
Vázquez P, Holguin G, Puente M, López-Cortes A, Bashan Y (2000) Phosphate-solubilizing microorganisms associated with the rhizosphere of mangroves in a semiarid coastal lagoon. Biol Fertil Soils 30:460–468
Velarde M, Felker P, Degano C (2003) Evaluation of argentine and Peruvian Prosopis germplasm for growth at seawater salinities. J Arid Environ 55:515–531
Villegas E, Rueda PEO, Puente ME, Muñiz SR, Avilés MS, Grimaldo JO, Murillo A, Preciado R (2010) First report of plant growth promoting bacteria from mesquite (*Prosopis glandulosa*) rhizosphere on volcans of Sonora desert. 12 International symposium on microbial ecology, Cairns, Australia, pp 15–25
Wang BS, Luttge U, Ratajczak R (2004) Specific regulation of SOD isoforms by NaCl and osmotic stress in leaves of the C_3 halophyte *Suaeda salsa* L. J Plant Physiol 161:285–293
Whipps J (2000) Microbial interactions and biocontrol in the rhizosphere. J Exp Bot 52:487–511
Yokoishi T, Tanimoto S (1994) Seed germination of the halophyte *Suaeda japonica* under salt stress. J Plant Res 107:385–388
Zexun I, Wei S (2000) Effect of cultural conditions on IAA biosynthesis by *Klebsiella oxytoca* SG-11. Chinese J Appl Environ Biol 6:66–69

Quinoa: A New Crop for Harsh Environments

19

Kameswara Rao Nanduri, Abdelaziz Hirich, Masoumeh Salehi, Saeed Saadat, and Sven Erick Jacobsen

Abstract

Extraordinary adaptations of quinoa caused expansion into different geographic areas with different soil and climate conditions. Water scarcity, soil salinity, and low water quality are main reasons of low food production in the Middle East and North Africa region. Quinoa was considered for food production using saline water and soil in this region. Field experiments in saline area of Iran showed that quinoa (Titicaca cv.) could produce 2.4 t ha^{-1} seed yield in 14 dS/m saline water and 2.3–3 t ha^{-1} in 20 dS/m saline water in Turkey and Morocco. Almost 7–10 t ha^{-1} seed yield was obtained with 16–18 dS/m saline water in UAE. Among the genotype, Titicaca had high yield stability in different climate conditions. The studies in the region showed that quinoa has high adaptation to the agroclimatic conditions and, therefore, has excellent potential as an alternative crop to rehabilitate salt-affected farms which have become uneconomical for the cultivation of the traditionally grown crops. All countries in the region worked on adaptability and agronomic practices. Scaling up the project needs agro-climatologically zoning and selecting appropriate areas with saline water and soil, which are not suitable for conventional crop production and seed processing, marketing, mechanization, and national government policy, for quinoa extension.

Keywords

Chenopodium quinoa · Saline water · Halophyte

K. R. Nanduri · A. Hirich
International Center for Biosaline Agriculture, Dubai, United Arab Emirates
e-mail: n.rao@biosaline.org.ae

M. Salehi (✉)
National Salinity Research Center, Agricultural Research, Education and Extension Organization (AREEO), Yazd, Iran

S. Saadat
Soil and Water Research Institute, Education and Extension Organization (AREEO), Karaj, Iran

S. E. Jacobsen
Department of Agriculture and Ecology, Faculty of Life Sciences, University of Copenhagen, Tastrup, Denmark
e-mail: seja@plen.ku.dk

Introduction

The Middle East and North Africa region (MENA) is the most water scarce region in the world. The World Bank defines a region as water stressed if it has less than 1700 m^3 per capita. The MENA region on average has 1274 m^3 per capita. Sixteen of the 23 nations within the MENA region are already suffering from extreme water stress (less than 1000 m^3 per capita) (Verdier

B. Gul et al. (eds.), *Sabkha Ecosystems*, Tasks for Vegetation Science VI,
https://doi.org/10.1007/978-3-030-04417-6_19

2011). All the countries in MENA will experience an increase in their water gap up to 2050, especially countries such as Iran, Iraq, and Egypt. Agriculture uses more than 80% of water in the MENA region, partly due to low irrigation efficiency of 30% compared to the world average of 45% (Negewo 2012). In addition to water scarcity, low water quality and soil salinity are other important constraints for agricultural production. Most of the crop and forage species are salt-sensitive and cannot produce economic yield with saline water above 12 dS/m. The use of halophytes may be a viable alternative to reduce pressure on freshwater resources and land for conventional cropping systems and the utilization of land degraded by salinity.

Halophytes can complete their life cycle in saline habitats when the salt concentration of the soil solution is above 5 g/L. They grow in habitat ranging from sea coasts to areas with saline water and soil that are still not suitable for agriculture (Khan and Gul 2006). Flowers et al. (1986) defined halophytes as plants that have an ability to complete their life cycles in salt concentrations of at least 200 mM NaCl. From an agricultural point of view, halophytes are plants that can produce economical yield in a range of salinities in which conventional crops cannot be produced. For example, in Iran saline water above 12 dS/m is considered only for halophyte production.

Of more than 50,000 edible plant species in the world, only a few hundred contribute significantly to food supplies. Just 15 crops provide 90% of the world's food energy intake, with rice, maize, and wheat comprising up to two-thirds of this (Loftas and Ross 1995). Quinoa is a facultative halophyte plant species, and some varieties are able to complete their life cycles at 400 mM salinity (Hariadi et al. 2011a; Razzaghi et al. 2011). Quinoa growth was reported to be stimulated by moderate salinity (10 dS/m) (Brakez et al. 2013; Jacobsen et al. 2003; Koyro et al. 2008). Salt-tolerance traits important to quinoa are Na^+ sequestration in vacuoles, xylem Na^+ loading, high ROS tolerance, K^+ retention, stomata regulation and density, and salt bladders (Adolf et al. 2012b).

Quinoa has been cultivated in the Bolivian and Peruvian Andean region for 7000 years (Garcia et al. 2015). Worldwide interest on this crop is increasing because of its high nutritional quality (Nowak et al. 2015; Jacobsen 2003; Repo-Carrasco et al. 2003). Quinoa has a high protein content and an optimal balance of amino acids with high amounts of lysine and methionine and high amount of fiber and minerals, such as calcium and iron (Abugoch James 2009). It contains antioxidants such as polyphenols (Nsimba et al. 2008). Finally, quinoa is gluten-free and suitable for celiac patients. Whole grain consumption also prevents type 2 diabetes, and because of its low glycemic index, quinoa can be used to replace common cereals in a diabetic diet (De Munter et al. 2007; Zevallos et al. 2015).

Soil and Water Distribution of the Near East and North Africa Region

Salt-affected soils are defined as soils that contain high concentrations of soluble salts, which adversely affect the growth of most plants. Salt-affected soils are widely spread in arid and semi-arid areas compared to the humid regions. The scarcity and salinity of water are one of the most important factors for the extension of agriculture in these areas. Distribution of saline soils is closely related to some factors such as climate, agricultural operation, geochemical and hydrological conditions, etc. Salt-affected soils occur in all continents and under almost all climatic conditions. Therefore, it is necessary to understand the origin of salt-affected soils, physical and chemical characteristics, and their formation processes. The most dominant types of ions in salt-affected soils are Na^+, Ca^{2+}, Mg^{2+}, Cl^-, SO_4^{2-}, CO_3^{2-}, and HCO_3^-.

Salt-affected soils are extensive in all the continents, but their extent and distribution have not been studied in detail in some countries. A first attempt to study the extent of salt-affected soils on a worldwide basis was made by the FAO/UNESCO Soil Map of the World (Massoud and Girard 1978) (Table 19.1).

Table 19.1 Salt-affected soils in some countries

Continent	Country	Area, 1000 ha		
		Saline/solonchaks	Sodic/solonetz	Total
Africa	**Afars and Issas**	1741	–	1741
	Algeria	3021	129	3150
	Angola	440	86	526
	Botswana	5009	670	5679
	Chad	2417	5850	8267
	Egypt	7360	–	7360
	Ethiopia	10,608	425	11,033
	Gambia	150	–	150
	Ghana	200	118	318
	Guinea	525	–	525
	Guinea-Bissau	194	–	194
	Kenya	4410	448	4858
	Liberia	362	44	406
	Libyan Arab Jamahiriya	2457	–	2457
	Madagascar	37	1287	1324
	Mali	2770	–	2770
	Mauritania	640	–	640
	Morocco	1148	–	1148
	Namibia	562	1751	2313
	Niger	–	1389	1389
	Nigeria	665	5837	6502
	Rhodesia	–	26	26
	Senegal	765	–	765
	Sierra Leone	307	–	307
	Somalia	1569	4033	5602
	Sudan	2138	2736	4874
	Tunisia	990	–	990
	Union of the Comoros	–	671	671
	Cameroon			
	United Rep. of Tanzania	2954	583	3537
	Zaire	53	–	53
	Zambia	-	863	863
South Asia	**Afghanistan**	3103	–	3101
	Bangladesh	2479	538	3017
	Burma	634	–	634
	India	23,222	574	23,796
	Iran	26,399	686	27,085
	Iraq	6726	–	6726
	Israel	28	–	28
	Jordan	180	–	180
	Kuwait	209	–	209
	Muscat and Oman	290	–	290
	Pakistan	10,456	–	10,456
	Qatar	225	–	225
	Sarawak	1538	–	1538
	Saudi Arabia	6002	–	6002
	Sri Lanka	200	–	200
	Syria	532	–	532
	United Arab Emirates	1089	–	1089

(continued)

Table 19.1 (continued)

Continent	Country	Area, 1000 ha		
		Saline/solonchaks	Sodic/solonetz	Total
North and Central Asia	**China**	36,221	437	36,658
	Mongolia	4070	–	4070
	USSR	51,092	119,628	170,720
Southeast Asia	**Democratic Kampuchea**	1291	–	1291
	Indonesia	13,213	–	13,213
	Malaysia	3040	–	3040
	Socialist Rep. of Vietnam	983	–	983
	Thailand	1456	–	1456

Massoud and Girard (1978)

Table 19.2 Area (million ha) and percent of saline and sodic soils of some regions

Regions	Total area	Saline soils	(%)	Sodic soils	(%)
Africa	1899.1	38.7	2.0	33.5	1.8
Asia and the Pacific and Australia	3107.2	195.1	6.3	248.6	8.0
Europe	2010.8	6.7	0.3	72.7	3.6
Latin America	2038.6	60.5	3.0	50.9	2.5
Near East	1801.9	91.5	5.1	14.1	0.8
North America	1923.7	4.6	0.2	14.5	0.8
Total	12781.3	397.1	3.1	434.3	3.4

Several authors have attempted to estimate the extent of salt-affected soils in the last 30 years, for example, using the FAO/UNESCO Soil Map of the World (1970–1980). FAO estimated that globally the total area of saline and sodic soils were 397 and 434 million ha, respectively. From 230 million ha of irrigated lands, 45 million ha (19.5%) and, from almost 1500 million ha of rainfed lands, 32 million ha (2.1%) are salt-affected soils as indicated in Table 19.2.

Oldeman et al. (1991) estimated that the total area affected by waterlogging was over ten million ha and that affected by salinity was over 76 million ha. They did not distinguish between irrigated and rainfed areas. Dregne et al. (1991) estimated that about 43 million ha of irrigated lands in dry lands were affected by various processes of degradation, mainly waterlogging, salinization, and sodication. Umali (1993) estimated that 1–1.5 million ha of soils are lost by salinization, annually. Ghassemi et al. (1995) estimated that salinization of irrigated lands caused losses of annual income about 12 billion USD globally. Nelson and Maredia (2001) estimated that about 12 million ha of irrigated lands may have gone out of production as a result of salinization. Data from FAO's database Aquastat shows that in some countries, the areas affected by salinity can be as high as 50% of the areas fully equipped for irrigation. Some of these estimations are the result of expert judgment or the aggregation of statistics which have been collected by different methods; therefore they are difficult to compare. There is still the need for data on the rate of change in areas affected by salinization, sodication, and waterlogging at a regional and global level.

Distribution of salt-affected soils in the Near East and North Africa region varies geographically with climate, agricultural activities, irrigation methods, and policies related to land management. These soils are mainly confined to irrigated farming systems in the arid and semiarid zones. The salts present are either of intrinsic origin (typical of Egypt, Sudan, and Iran) or are the result of seawater intrusion in coastal regions or of irrigation with brackish or saline groundwater. In the irrigated zones of Morocco, continuous irrigation has resulted in soil salinization. Secondary salinization due to irrigation with saline water is also reported in the NENA region. In Libya, Sudan, Iran, Iraq, and United Arab Emirates, large tracts of lands have been degraded due to heavy irrigation with groundwater. Salinity, sodicity, or the combination of both is seriously affecting productive areas like the Nile Delta of Egypt and the Euphrates Valley in Iraq and Syria. The situation is further complicated by association with problems of waterlogging and high $CaCO_3$. In Kuwait and the United Arab Emirates, soil salinization is mainly confined to coastal areas but also occurs on irrigated farms.

Iran

Most of the soils of the area are lithosols due to heavy erosion which does not allow profile development. Other soils are alluvial-colluvial with steady rejuvenation of the profile and which occur in a variety of forms not always well differentiated by the pedologist but readily distinguished from one another by vegetation. Climatically, the soils of Iran can be classified as humid, semi-humid, and arid.

Out of the 165 million ha that comprise the country's area, about 37 million ha are suitable for irrigated and dry-land agriculture, of which 20 million ha are irrigated and 17 million ha are dry land. Of the 37 million ha of agricultural lands, currently 18.5 million ha are devoted to horticulture and field crop production. Of these, 6.4 million ha are under annual irrigated crops, two million ha are under horticultural crops, and about 6.2 million ha are under five annual dry-land crops, while the remaining 3.9 million ha are fallow. Roughly one-third of Iran's total surface area is suited for farmland, but because of poor soil and lack of adequate water distribution in many areas, most of it is not under cultivation. Only 12% of the total land area is under cultivation (arable land, orchards, and vineyards), but less than one-third of the cultivated area is irrigated; the rest is devoted to dry farming. Some 92% of agro products depend on water. The western and northwestern parts of the country have the most fertile soils.

Egypt

The total agricultural lands of Egypt are about 7.8 million feddans, which are almost entirely dependent on irrigation. About 0.9 Mha suffer from salinization in cultivated irrigated areas: 6% of the Northern Delta region, 20% of the Southern Delta and Middle region, and 25% of the Upper region. About 1.26 Mha are salt-affected, as well as about 50% of the irrigated soils affected. The majority of salt-affected soils are located in the northern central part of the Nile Delta and on its eastern and western sides. Other areas are found in Wadi El-Natroun, Tell El Kebir, the Oases, as well as in many parts of the Nile Delta and Valley and El-Fayoum province. Sodium chloride is the dominant salt and it is the main source of salinization. Inundating the soils with sea and lake salty water for a long time is the most important factor for salinization in the mentioned soils, besides the tidal effect and the salty soil solution. Shalma and El-Hamoul South Burullus Lake represent this type. Magnesium chloride and magnesium sulfate are the dominant salts and main source of salinity in the Manzala Lake area, while sodium chloride is the dominant salt representing the main source of salinization in the area of Abees South Mariut Lake. Sodium chloride and sodium sulfate are the dominant salts of Mariut and Tell El Kebir areas, respectively. The main factor responsible for the deterioration of these soils is seepage from irrigation canals in Mariut and from Ismailia canals in Tell El Kebir. Sodium carbonate and sulfate are the dominant salts in Ferhash area. The main factor for alkalinization is the effect of biological sulfate reduction. Sodium chloride is the dominant salt, and sodium saline irrigation water (5000 ppm) is the main factor for salinity formation in this area.

Pakistan

From a variety of sources, about 60% of Pakistan's total land area is classified as unusable for forestry or agriculture because it consists of deserts and mountains. Pakistan's soils are mostly dry and have high concentrations of calcium carbonate and a low content of organic matter. The major soil groupings are Indus basin soils, mountain soils, and sandy desert soils. In most rainfed areas of Pakistan, the soils have developed from wind- and water-transported materials and consist of loess, old alluvial deposits, mountain outwash, and recent stream valley deposits. Some are derived from shale and sandstones, while the soils in central irrigated Punjab, NWFP, and Sind provinces vary from clay loam to silty loam (Baig et al. 1985). The soils of the huge Thal desert, Cholistan desert, and Tharparker belt are alluvial with sandy textured sand dunes covering 50–60% of the area (Khan and Anwar 1968).

Turkey

Turkish soils are diverse, and several soil groups have been recognized in different geographical zones. Major groups by region can be outlined as follows: Black Sea coasts, podzolic soils; Aegean and Mediterranean coasts, Mediterranean soils; mountains and high-elevation areas with sufficient rainfall, forest and rendzina soils; arid areas, brown and reddish-brown soils; semiarid areas, lime-free brown, chestnut, and reddish-chestnut soils; plains and adjacent slopes, alluvial and colluvial soils; and finally volcanic areas, volcanic soils. This is quite a generalized outline and it does not cover the other minor soil groups. Land use capability classes and their share in total land area are given in Table 19.3 (General Directorate of Soil and Water 1978). Over two-thirds of the country consists of classes VI and above.

Jordan

In the Valley, 6500 ha salt affected, 1400 ha slightly, 1600 ha moderately, the rest severely. In the Jordanian desert, soils contain an amount of salt content ranging from 1% to 10%.

Table 19.3 Land use capability classes and percent of land area of Turkey

Soil use classes	Total area (ha)	Percent
I	4,973,162	6.53
II	6,705,943	8.81
III	7,532,049	9.89
IV	7,163,439	9.41
V	165,076	0.22
VI	10,189,857	13.39
VII	36,232,151	47.59
VIII	3,163,654	4.16

Libya

Salt-affected soils are about 700,700 ha: 199,300 slight, 174,400 moderate, and 327,000 severe sabkhas and sodic soils. Area affected by salinity and waterlogging is about 250,000 ha.

Algeria

Irrigated area is about 350,000 ha, of which 25% are salt-affected soils. About 8% of the irrigation waters are very saline and 21% have moderate salinity. Salinity and water logging affect some areas in the northern west and the south.

Yemen

The salt-affected soils in this country total 483,467 ha. The principal soil types in the various physiographic regions are as follows:

- The soils of the coastal plains are either alluvial fans or coarse inter-wadi soils. In wadis and flood plains, the soils are loamy to silt and clay which is considered good agricultural land. The inter-wadi areas are dominated by dune formations and coarse skeletal sandy soils subject to wind erosion. The coastal fringes of the plains consist of very saline tidal flats and are known as "sabkhas."
- The soils of the western slopes range from bare rock and very shallow soils near the mountain peaks, while stony and very stony calcareous soils with pH around 8 and low organic matter occur in the middle slopes. The lower slopes have generally deep silty and loamy soils. This region has relatively extensive alluvial loams and silt loams which make good agricultural land. Around southern highland, thick loess deposits occur which have developed deep silty soils. The south of the midlands is occupied by rock outcrops with pockets of shallow soils.
- The highlands have large stretches of plains between the mountains which constitute extensive loamy, silty, and fine silty soils on

level surfaces, one-third of which bear organic matter within the surface layer. Associated with these soils is a minor component of clay soils, which also have a dark layer rich in humus. These constitute very productive agricultural lands. On the lower slopes of the highlands, silt loams and silty clay loams prevail, while the flat basins comprise silty and loamy soils. The eastern slopes' region comprises mainly rock outcrops, with some shallow soils confined to pockets. Deep loamy soils are only encountered within local depressions and wadis.

- In the eastern plains, wadi flood plains have deep alluvial soils which are medium textured, while the restricted areas where flooding takes place regularly have stratified sandy loams and silt loams.

North Africa

North Africa includes roughly 10 countries, depending on where one draws its borders. By far, the most dominant feature of the region is the Sahara desert, which extends across 3.3 million square miles – an area not much smaller than that of the United States. Climate and landforms are strongly linked here, since it is the presence of a specific landform – the Atlas Mountains – that creates the desert conditions farther east. Most of the soils are formed through flooding, soil erosion, and layering.

Arenosols Soils Arenosols soils contain particles that are transported or deposited by the wind, known as eolian. These soils are formed when the wind causes particles to "sandblast" or grind against the surface of sandstone or other rocks, leaving behind minute particles called psamments. Psamments can be fine or gravel-like and uneven. Arenosols soils contain fine, coarse, or medium grains and blow around easily with the wind because they contain virtually no water to bind particles together. Arenosols soils in areas such as the Sinai Peninsula contain sand along with clay or loam. These soils lack natural nutrients and organic matter, which is why few plants can grow in them easily.

Alluvial Soils Derived from the Latin word meaning "to wash against," these soils are created by fine deposits of sediments that are formed when mountainous rocks erode and end up in lakes, rivers, and seas by means of flowing water. Alluvial soils in Egypt are in the Delta and Nile Valley due to the Nile's historical seasonal flooding every year. Alluvial soils formed by deposits of river sediment are known as fluvial. The Nile Valley's alluvial fluvial soil has a fine texture and is extremely fertile because of the organic matter and minerals. Farmers cultivate the land in and around the Delta and Nile Valley due to the organic matter, which is beneficial for a healthy and bountiful produce.

Marine Soils Marine soils exist near the edge or coastal areas of the Mediterranean and Red seas. This type of soil along coastal areas is formed when eroded material and rock fragments travel toward the sea through rivers and streams until they end up on the shore.

Calcareous Soils Like the marine soils, calcareous soils are also found along shores of Mediterranean and Red seas but are deposited at the foot of slopes known as colluvial. These soils are formed by a mixture of eroded material and rock fragments that naturally collect at the base of mountains or steep hills.

Morocco

The total irrigated area is 1 Mha, of which about 21% is salt-affected and 57% of the Gharb is irrigated. Salt-affected soils include 350,000 ha. In the north and northwest (Gharb and Loukkos), soils are affected by waterlogging induced by excess rainfall and irrigation water. In the south and east (Tadla, Moulouya, Ouarzazat, and Tafilalet), soils are affected by waterlogging and salinization induced by rising water tables due to irrigation. The major soil types include the following: (i) yermosols, lithosols, regosols, and sierozems, which are the dominant types because of mountainous topography and the large area of the Saharan and arid zones, (ii) rendzinas, (iii) xerosols, (iv) vertisols, (v) gleysols, and (vi) saline soils.

Tunisia

Salt-affected soils approximate 1.5 Mha, 10% of the total area. Salinization of irrigated areas results from the climatic deficit and the use of marginal water and soil resources. Irrigated areas cover about 375,000 ha. Salinization and waterlogging affected about 50% of the areas, 10% severely.

Based on the saline soil resources in the region and freshwater scarcity, there is a huge opportunity for salt-loving and salt-tolerant crop production. It may be possible to produce food with halophyte farming in this region. The use of halophytes as cash crops is growing even in countries with non-saline soils (Panta et al. 2014). Among salt-tolerant crops, quinoa is considered a staple food which can be produced in saline soil.

Adaptation of Quinoa to Saline Agroclimatic Condition of UAE

The United Arab Emirates (UAE) has an arid climate, characterized by poor and erratic rainfall, high daytime summer temperatures, and poorly developed soils. In the past two decades, salinity has emerged as a major problem for agriculture in the country. Expansion of the agricultural area and over-pumping of groundwater have increased the salt concentration of the aquifers to the extent that groundwater is no longer suitable for growing salt-sensitive crops. An analysis of the agricultural sector in 2013 to define the main problems showed that 75% of the farms surveyed had very high salinity of 2–45 dS m^{-1}, limiting the choice of crops that can be successfully cultivated.

Preliminary Evaluation

The International Center for Biosaline Agriculture (ICBA) in Dubai (UAE) has been evaluating a range of salt-tolerant crops for their ability to grow and produce economic yields, ultimately to introduce them to the farmers. Among the half dozen or so new crops tested, quinoa seemed highly promising. During the cropping season, November 2006–March 2007, 120 germplasm accessions acquired from the US Department of Agriculture (USDA) were evaluated for local adaptation and growth performance at ICBA's research station, where the soil was fine sand, moderately alkaline (pH 8.2), and strongly calcareous (up to 53% $CaCO_3$) with very low organic matter (<0.5%). The plants were irrigated with low-salinity water with an electrical conductivity (EC_w) of 2–3 dS m^{-1} using the drip system. Seeds were harvested from 73 accessions. The top 20 accessions selected on the basis of seed yield were further tested for performance during the subsequent cropping seasons (2007–2008 and 2008–2009). The seed yield among the accessions varied between 53.86 and 359.86 g m^{-2} in 2007–2008 and between 3.32 and 258.42 g m^{-2} in 2008–2009, with the average yield over the two seasons ranging between 0.3 and 2.4 t ha^{-1} (Rao and Shahid 2012). The reasonably high yields obtained in some accessions showed that quinoa has good adaptation and can be successfully cultivated in the UAE. In the follow-up study during the 2009–2010 cropping season (November–April), the five top-ranking accessions selected on the basis of average seed yield in the previous years were further evaluated in a replicated trial using low-salinity irrigation water (EC_w 2.8 dS m^{-1}). The seed yield among the five accessions ranged between 374.4 g m^{-2} (Ames 13,757) and 533.6 g m^{-2} (Ames 13,761) with an average of 456.6 g m^{-2} over the accessions. The dry matter yield, averaged over the accessions, was 1464 g m^{-2}, with accession Ames 13,742 producing 1624 g m^{-2} seed yield (Rao and Shahid 2012). During the course of time, mass selection to eliminate inferior plant types and enhance the yield potential of some selected accessions, namely, Ames 13,727, Ames 13,742, Ames

13,761, Ames 22,157, and NSL 10639, has resulted in five improved lines/genotypes, Q1–Q5, respectively.

Multiplication Trials

The results from the preliminary trials undertaken at ICBA research station indicated that quinoa has a significant potential as a new crop for the UAE. To validate the results, further studies were undertaken at Ghayathi in the Western Region of Abu Dhabi Emirate in 2012–2013 with two of the five selected lines (Q3 and Q5) and NSL 106399 and subsequently in five locations, three in the Northern Emirates (Dibba, Hamraniah, and Al Dhaid) and two in the Western Region of Abu Dhabi (Madinat Zayed and Ghayathi) in 2013–2014, with four of the five selected quinoa lines (Q1, Q2, Q3, and Q5). While the two locations in the Western Region were agricultural farms, those in the Northern Emirates were the federal government's regional research stations. In the 2012–2013 growing season, each of three genotypes was sown in a plot of size 30 × 4 m, while in 2013–2014, a randomized complete block design (RCBD) with three replications was used to evaluate the performance of the four genotypes. The plot size was four rows of 7 m with a spacing of 50 cm between rows and 1 m between two plots. In all cases, the plant-to-plant distance within the rows was maintained at 25 cm. The plants were irrigated with drip system, and standard recommended agronomic practices for sandy soils were adopted to ensure proper crop growth and development. The data recorded in the trials included plant height, days to flowering and maturity, and dry biomass and seed yields. At maturity, the panicles were harvested from 1 m^2 quadrats – three from each plot which served as within-plot replicates in the 2012–2013 study and one per plot in the 2013–2014.

In the 2012–2013 studies, the irrigation water at Ghayathi had a salinity of 14–15 dS m^{-1}. Despite the high salinity, a mean seed yield of 750 g m^{-2} was obtained of the three genotypes. The dry biomass yield was also high, the mean of the three cultivars being 0.72 kg m^{-2}, indicating the potential of quinoa as an alternative food and forage crop for saline areas (Rao et al. 2013).

The results of the 2013–2014 multi-location yield trial, along with the soil and water characteristic of the sites, are presented in Table 19.4. As can be seen, salinity of the irrigation water at the five locations ranged between 2.3 and 18.9 dS m^{-1}. Analysis of variance showed significant effects of the location and genotype for plant height, days to flowering and maturity, and dry weight ($p < 0.05$). Seed yields among the locations were significantly different ($p < 0.001$), but differences among the four genotypes were found to be insignificant ($p > 0.05$). The interaction effect of location and genotypes was significant for days to flowering and maturity and dry weight, though not for plant height and seed yield. Seed yields in the Northern Emirates correlated negatively with the salinity of irrigation water, which was 6.1 dS m^{-1} in Dibba, 4.5 dS m^{-1} in Hamraniah, and 2.3 dS m^{-1} in Al Dhaid. Thus, averaged over lines, the seed yield was highest in Al Dhaid (541 g m^2), followed by Hamraniah (398 g m^{-2}) and Dibba (190 g m^2) (Fig. 19.1). In the Western Region, in spite of the high salinity of the irrigation water (15–18 dS/m), the seed and biomass yields were considerably higher compared to the Northern Emirates (Fig. 19.1). In Ghayathi, the mean seed yields of the four lines were 1044 g m^{-2} – much higher than the yields recorded in the previous year (750 g m^{-2}) from the same location. In Madinat Zayed, the mean seed yield of the four lines was 714 g m^{-2}, which was similar to the yield obtained from Ghayathi in the previous year. A perusal of the climatic data revealed no major differences among the locations (data not presented); therefore the differences in yields could be due to the differences in chemical properties of the soils and waters other than the EC and pH that were measured. Also, being agricultural farms, the two sites in the Western Region possibly had a more favorable soil microclimate compared the research stations in Northern Emirates. Nevertheless, the lowest mean seed yields of 190 g m^{-2} obtained in these studies are close to the average yields of quinoa from favorable environments in the Andes

Table 19.4 The soil and water quality at the five locations and morpho-agronomic variation in four improved quinoa lines evaluated at five locations in the United Arab Emirates

	Soil		Water		Days to flowering		Days to maturity		Plant height (cm)		Dry weight (g m^2)		Seed yield (g m^2)	
Location	Texture	pH	Salinity (dS m^{-1})	pH	Range	Mean	Range	Mean	Range	Mean	Range	Mean	Range	Mean
Ghayathi	Sand	7.79	16.3	7.32	45–57	52	80–92	90	134.7–140.6	137.2	1881–2686	2233	890–1197	1044
Madinat Zayed	Sand	7.43	18.9	6.92	54–68	63	89–103	98	142.7–160.5	152.2	1641–3441	2519	620–843	714
Dibba	Sand	7.71	6.1	7.31	50–53	51	87–94	91	52.5–77.8	67.1	455–1219	852	133–239	190
Hamraniah	S. loam	8.22	4.5	7.70	39–45	43	78–90	84	107.8–146.9	123.6	1250–2920	2081	258–545	398
Al Dhaid	Sand	8.25	2.3	7.76	40–44	42	75–85	80	93.5–128.3	118.6	2282–3723	3069	442–634	541

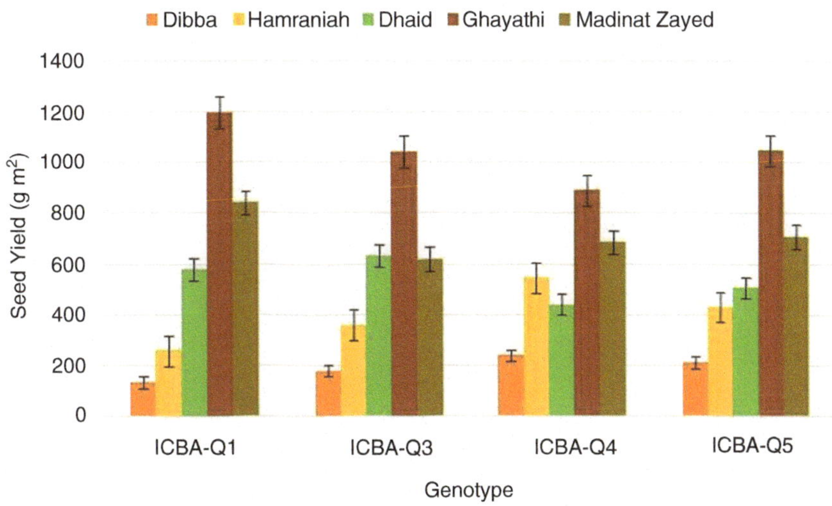

Fig. 19.1 Seed yields of four quinoa lines grown at five locations in the UAE during 2003–2014. The error bars represent LSD of the means (0.05)

region. Thus, the results from current studies besides confirming quinoa's ability to withstand high salinity also demonstrated its good adaptation to hyperarid desert conditions.

Constraints and Opportunities of Quinoa Production in the UAE

The high yields obtained with saline irrigation water in the Western Region showed that quinoa, which is a facultative halophyte, not only endures salinity but also that some of its cultivars prosper under saline conditions. Therefore it has huge potential as an alternative food and feed crop when traditional crops become uneconomical due to increased groundwater salinity. In the trials conducted in the UAE, significantly high variability was observed in yields across locations. Therefore, understanding of the influence of edaphic and climatic factors on productivity is important. Knowledge on quinoa pests and diseases, in areas outside its traditional growing regions, is missing, although no major problems were encountered in the current studies. A close relative such as *Chenopodium murale* is a common weed in the UAE, which may harbor biotic stressors which can migrate to quinoa. Introduction and scaling-up of novel crops such as quinoa ton on traditional environments also require the study and establishment of the entire value chain and building capacity of researchers and farmers in production, harvesting, storage, and processing technologies.

Potential of Quinoa Production with Saline Water in Iran

Climate Condition

Based on rainfall, 28 different agroclimatic zones are distinguished in Iran (Ghaffari et al. 2014). Saline water resources and saline soils are located mainly in the central plateau, in south Iran, and a small part in the north by the Caspian sea with dry-land salinity. About 54% of the land has arid climate with very warm and warm summer, high evaporation, and high water requirement (Fig. 19.1). In aridity index of 0.03–0.2, rainfall is 100 up to 350 mm and in 0.2–0.5 rainfall range is 250–500 mm. Saline lands with different degrees of salinity are separated in four different climate conditions (Fig. 19.2).

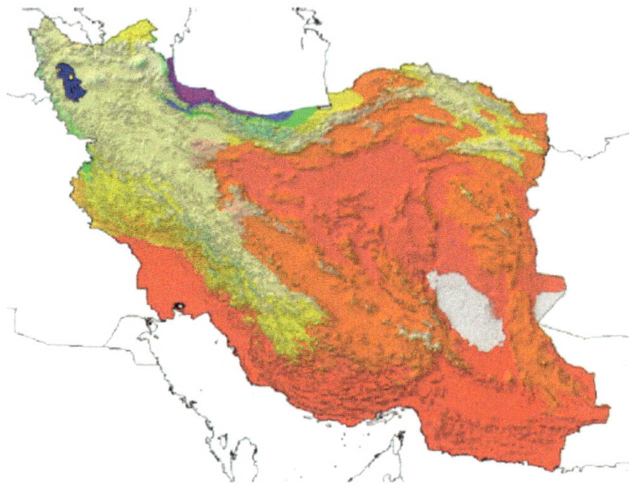

Colour	Moisture regime	Aridity	Temperature regime winter	Range winter	Temperature regime Summer	Range summer	% of country	Approx area Km²	Quinoa cropping system
	Arid	0.03-0.2	Mild	10-20	Very warm	>30	16.7	286822	Rainfed or irrigated winter crop
	Arid	0.03-0.2	Cool	0-10	Very warm	>30	18.7	305814	Sowing in July
	Semiarid	0.2-0.5	Cool	0-10	Very warm	>30	1.6	26454	Rainfed winter crop or spring crop*
	Arid	0.03-0.2	Cool	0-10	warm	20-30	18.7	305814	Spring crop or July sowing

* Sowing date depend on genotype.

Fig. 19.2 Agroclimate zone of Iran based on UNESCO method with legend of selected (saline) area. (Ghaffari et al. (2014))

Adaptation and Seed Production of Quinoa

Because of the climatic diversity of Iran, sowing date is an important agronomic factor to secure emergence and achieve high seed yield. Sowing date is depending on climate and cultivar. For selection of the best sowing date and evaluation of the temperature effect of different sowing dates on quinoa (cv. Titicaca), eight sowing dates were tested (22 August, 6 September, 26 September, 7 October, 24 October, 24 February, 7 March, and 30 March).

Emergence was not sensitive to high temperature. Titicaca cv. emerged 5–8 days after sowing at a temperature of 19–32 °C, with max temperature of 42 °C (Fig. 19.3). Jacobsen and Bach (1998) reported that base, optimum, and maximum temperatures of quinoa for germination were 3, 30–35, and 50 °C, respectively. Bois et al. (2006) reported that base temperature of quinoa is between −1.9 and +0.2. Jacobsen et al. (1999) recommended sowing quinoa when the soil temperature is 8–10 °C. GDD requirement for emergence of quinoa in the field was 25–50 degree-day. Jacobsen and Bach (1998) reported 30 degree-day requirement for germination of quinoa.

Early maturing cultivars of quinoa are suitable for sowing in August and February in the central plateau of Iran. Sowing date depends on climate conditions. Titicaca started flowering 50–60 days after sowing with a mean temperature during

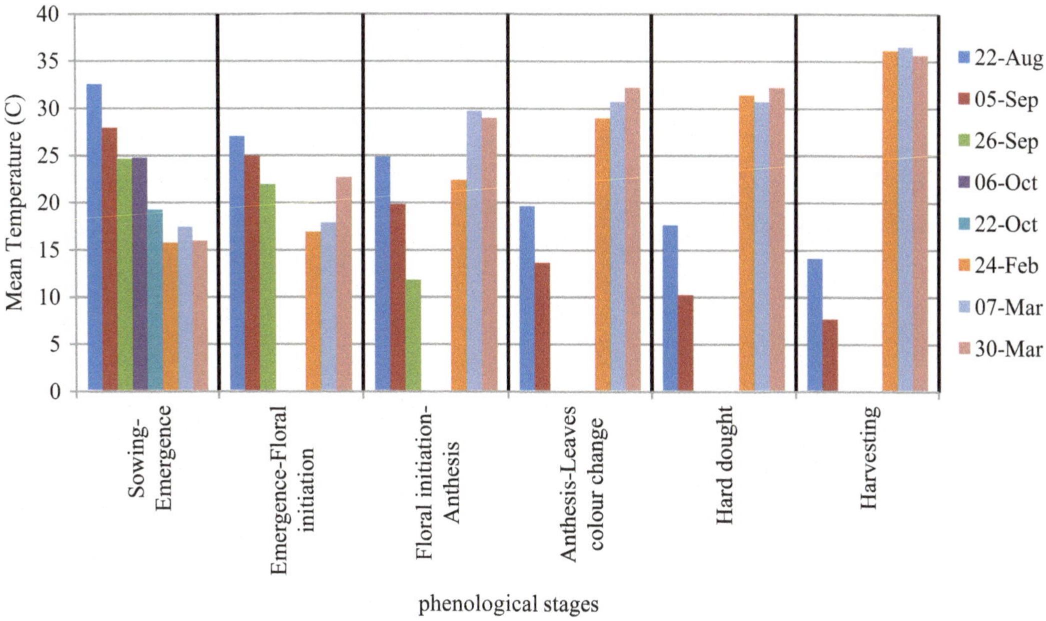

Fig. 19.3 Mean temperature at different growth stages of quinoa (Titicaca), Yazd, Iran

flowering and seed filling of 13–32 °C, with minimum and maximum temperatures of 13 and 35 °C, respectively (Fig. 19.3). Titicaca seed yield at temperature 13–19 °C on 5 September sowing date was 0.56 g m^{-2}. In 22 August sowing date, 560 mm saline water (14–17 dS/m) was applied in Yazd. The amount of applied water and rainfall during the growth period for this genotype was almost the same in Turkey, where 2.3 t ha^{-1} seed yield with 20 dS/m saline water was obtained (Yazar et al. 2015).

Changing the sowing date to July increased the water requirement because of high evaporation. In this sowing date, water would be available using minimum water requirements to produce optimum seed yield. For 22 August sowing date, flowering time is very important, and the optimal genotype should reach anthesis before mean temperatures fall below 20 °C. Among the tested genotypes, only Titicaca needs as little as 60 days for flowering, and it was selected for this sowing date (Fig. 19.4).

Very late flowering genotypes such as NSRCQ2, NSRCQ8, and NSRCQ12 may suffer from drought and heat stress during grain filling to ripening in spring cropping. They could be considered for forage production in different saline regions. Blanco (2015) reported that quinoa is used as fodder for ruminants especially in conditions where other crops cannot grow due to the saline conditions. Quinoa by-products, such as stalk, bran, dry leaves, and forage silage, can be used for animal feed.

Genotypes with longer vegetative phases are suitable for winter cropping because of their longer time to flowering and during winter remain in vegetative phase and in suitable condition start flowering. These genotypes are suitable for dry-land agriculture in semiarid areas with cool winter conditions of Iran, where they use winter and spring rainfall to germinate. Late maturing genotype (NSRCQ10) at November sowing date could tolerate winter temperature and start growing after temperature increases and eventually produce seed (Fig. 19.5). Among the evaluated genotypes, NSRCQ5, NSRCQ6, NSRCQ7, NSRCQ10, and NSRCQ13 are suitable for dry-land salinity agriculture system in the north and south of Iran with 250–350 mm rainfall as a winter crop. Quinoa (Titicaca) produces 1.7 t $ha^{-1}$of seed in dry-land farming in Turkey with 224 mm rainfall (Lavini et al. 2014).

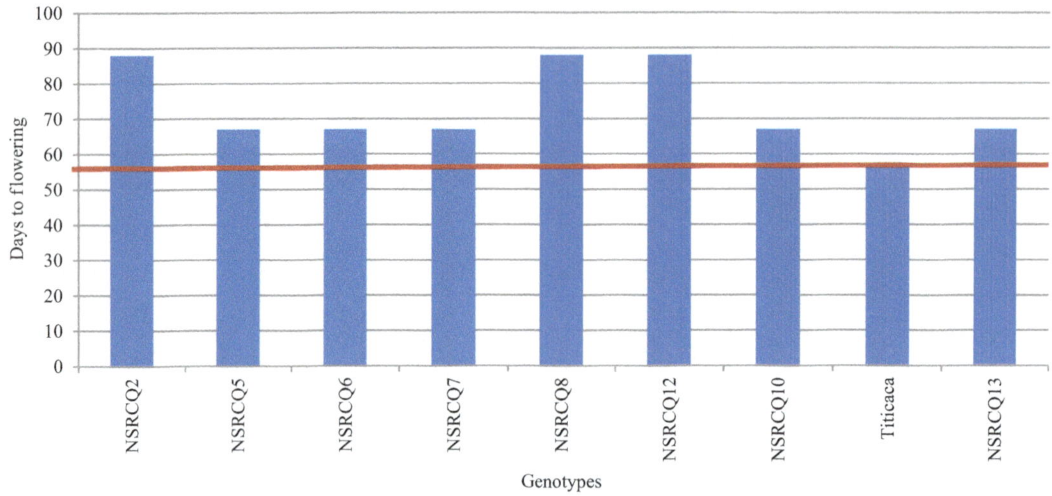

Fig. 19.4 Days to flowering of different evaluated genotypes in Yazd, Iran, as spring cropping

Fig. 19.5 Late mature cultivar with longer vegetative phase as winter crop in cool winter condition (NSRCQ10)

Quinoa is one of the few crops that can tolerate frost to a certain extent, but this depends largely on the duration of the frost, the quinoa variety, the phenological stage of the plant when frost occurs, the relative humidity, and the microlocation of the fields (Garcia et al. 2015). At −4.5 °C, quinoa had frost damage at the floral initiation stage 77 and 66 days after sowing on 23 September and 7 October, respectively, but at 10–12 leaves growth stage, plants can tolerate −4.5 °C (Fig. 19.6). Quinoa seedlings are resistant to freezing up to −6 °C for 5 h (Bois et al. 2006). Phenological stage is very important for frost damage; plants are affected more at the 12-leaf stage and anthesis when compared to the 2-leaf stage, which can tolerate −8 °C for 2–4 h (Jacobsen et al. 2005). For the 23 October sowing date, early mature cultivar (Titicaca) at 12-leaf stage can tolerate −4.5 °C but after winter could not produce seed (Fig. 19.7).

Developmental scale is very important for evaluating the effects of environment on yield,

Fig. 19.6 Frost damage at two growth stages (floral initiation 77 days after sowing (**a**) and 10–12 leaves growth sage (**b**))

frost damage, drought stress, pest, diseases, weed control, or fertilizer application (Bazile et al. 2015). Quinoa has a short-day quantitative response to photoperiod. This shows that longer days extend the duration of some growth stages, but plants reach flowering in all ranges of photoperiod, and the timing of development is also sensitive to temperature (Bazile et al. 2015).

Sowing at different dates showed that grain filling is sensitive either to low (lower 20 °C) or high temperatures. The differences between mean temperatures of August and February sowing dates during grain filling were 6 °C with 33 °C and 38 °C mean of max temperature, respectively. One thousand seed weights and seed yields in the February sowing date were 25% and 37.5% lower than the August sowing date, respectively (Table 19.2). Bertero et al. (1999) showed that the most limiting factor to phenological adaptation in nontropical environ-

Fig. 19.7 Titicaca cultivar in 23 October sowing date (140 days after sowing)

Table 19.5 Effect of sowing date on growth stage duration, seed yield, and 1000 seed weight of quinoa (Titicaca genotype) under saline condition (14 dS/m saline water)

	Sowing date							
Phenological stage	22 August	6 September	26 September	7 October	24 October	24 February	7 March	30 March
Emergence (days)	5	6	6	7	9	7	6	7
Floral initiation (days)	32	36	37	Frost damage		41	41	39
Flowering (days)	56	56	77			75	61	74
Color change (days)	68	70	Frost damage			112	105	94
Harvesting (days)	108	123			146	132	126	110
Seed yield (Kg ha^{-1})	2343 a	0.56	No seed	No seed	No seed	1637 ab	1478 ab	680 b
1000 seed weight (g)	2.93 a	–	–	–	–	1.67 b	1.36 bc	1.16 c

ments is photoperiod and temperature during grain filling. The most limiting factor between different sowing dates is high or low temperature during grain filling, and the optimum temperature during this stage is 20 °C (Table 19.5).

One of the main pests of quinoa is *Spodoptera exigua* which feed on leaves during quinoa's vegetative growth stage. It originated from Southern Asia (Wilson 1932) and is the main pest of sugar beet, corn, cotton, and potato in Iran. This larva was observed in the February and May sowing dates, and insecticide was applied two times (Fig. 19.8).

There is a wide variation on salt tolerance in quinoa; this represents important resources for selection and breeding for even higher tolerance and for cultivar adapted to different climate conditions and soils (Bendevis et al. 2014). Titicaca is selected cultivar in Denmark from material originated in southern chili, Panicle is yellow to orange, early mature with yellow seed and day length netural (Adolf et al. 2012a). Seed yield of this genotype is evaluated in the field trail in Italy, Morocco, Turkey, and Iran; it was from 2.3–3 t ha^{-1} with saline water (14–20 dS/m) (Table 19.6). Titicaca genotype adapted to various agroclimate conditions and had high yield stability (Dost 2015; Razzaghi et al. 2011; Yazar et al. 2015; Lavini et al. 2014). At 400 mM NaCl, concentration treatment biomass and plant height of this genotype decreased to 43.9% and 26.9%, respectively (Adolf et al. 2012a). Adolf et al. (2012a) compared 14 quinoa varieties and reported that two varieties belonging to real type (Utusaya and Pandela rosada) adapted to the

Fig. 19.8 *Spodoptera exigua* larva in May, October, and March

Table 19.6 Seed yield of quinoa in irrigated saline water

Saline water (dS/m)	Seed yield (t ha^{-1})	Cultivar	Country	References
Freshwater	3.3	Titicaca	Italy	(Lavini et al. (2014
22	2.3–2.7			
0.26	2.9–3.1	Titicaca	Turkey	(Yazar et al. (2015
10	2.7–2.8			
20	2.3–3			
30	2.2–2.7			
1	3.9	Titicaca	Morocco	(Lavini et al. (2014
10	3.5			
20	2.9			
30	2.5			
14–17	2–3	Promising genotypes	Yazd, Iran	Salehi (Unpublished data)
14–17	2.4	Titicaca		
Freshwater	3.3	Titicaca	Iranshahr, Iran	(Dost (2015
Freshwater	1.0–1.8	Sajma, Santamaria	South of Iran	(Sepahvand (2012
	No seed	Sajma, Santamaria	North of Iran	
16–18.9	8–10	ICBA Q1,3,4,5	UAE	(Choukr-Allah et al. 2016)
6.6	1.6	Chill-2011	Iraq	

extremely harsh climate condition. Utusaya genotype biomass and plant height decreased to 23.8 and 7.7%, respectively, at 400 mM NaCl. These genotypes are originated from the *Salares* region of Bolivia, and it was a short-day genotype and could not be successful at all sowing dates.

Constraints and Opportunities of Quinoa Production in Iran

Quinoa may adapt to the Mediterranean region, which is under the effect of climate changes such as drought and high temperatures (Jacobsen 2014). Iran has various climate conditions, and with the diversity to be found in quinoa with more than 3000 accessions in gene banks, it should be possible to find suitable genotypes for production. However, quinoa material is not freely available, which is a constraint to increasing quinoa production outside South America.

Quinoa is tolerant to soil salinity, but for germination and emergence, irrigation with freshwater or rainfall is needed. Farmers rarely have access to freshwater in saline areas of Iran. An alternative method should be used at early growth stages. Munir and Basra (2009) evaluated different priming methods on seed germination, and results showed that hormonal priming with benzyl aminopurine (BAP) reduced germination time by 50% at EC 10 and 20 dS/m. Our investigation showed that transplanting and irrigating with saline water (17 dS/m) are other ways for establishment under saline condition.

Iran is in water crisis, so introducing a new crop with low water requirement and salinity tolerance would be a benefit. Wheat production systems in arid conditions need 800–900 mm fresh or brackish water, and rice production systems in Iran consume high amount of freshwater (3–3.5 m). Quinoa production in this climate needs only 560 mm water, which may even be saline (14 dS/m). Quinoa seed yield with freshwater is lower than wheat, but using saline water resources (above 12 dS/m), which is not possible to produce common crops, is a potential resource for quinoa production. Quinoa could replace cereals (wheat and rice) for children, diabetes, celiac, and those with other diets. Quinoa seeds are highly nutritious with high protein content and are excellent sources of vitamins and minerals.

Saline Water for Quinoa Production in North Africa

Adaptation of Quinoa to the Agroclimatic Conditions of North Africa

Climate change and population increase have created an urgent need in North Africa to introduce new crops in order to achieve food security, especially for the rural poor populations. According to literature and a recent publication from FAO Morocco, a North African country was the first in North Africa to introduce quinoa (Bazile 2014). The first trials on quinoa were carried out within the IAV-BYU project conducted jointly by Hassan II Institute of Agronomy and Veterinary Medicine (Morocco) and Brigham Young University (USA) in 1999 (Benlhabib et al. 2014). The objective of this project was to improve the food security of Moroccan subsistence farmers in high land areas (Atlas Mountains) by introducing new crops because quinoa has superior nutritional value and is abiotic stress tolerant. Quinoa was evaluated in two locations at Khenifra (mountain region) and Rabat (coastal region). Results showed that among the tested 14 lines, G205-95DK performed the best under high-altitude conditions due to its origin in the high altitudes of Altiplano (Andean Plateau), while most of the other lines showed adaptation to the coastal region, probably because of provenance effect (Choukr-Allah et al. 2016).

Later in the 2006–2007 cropping season, 30 selected lines were tested with two sowing dates (November and February) in two localities (IAV Rabat and Tnine Ait Boukhayou at My Bouazza). Testing of quinoa began again in 2008 in the semiarid region of Rhamna under the EU project SWUP-MED (sustainable water use securing food production in dry areas of the Mediterranean region). In a field experiment conducted in Rhamna, five selected quinoa lines (L11, L119, L123, L142, L143) along with two commercial varieties Titicaca and Puno were evaluated for phenological and agro-morphological characteristics (Filali 2011). Titicaca and L143 recorded the highest seed yield (1.5 t ha^{-1}), while L11 showed the lowest seed yield (0.46 t ha^{-1}).

Within the same project (SWUP-MED) Hirich et al. (2012b) have conducted a series of experiments in the south of Morocco (Agadir region) in order to evaluate the quinoa productivity under deficit irrigation (Hirich 2013; Hirich et al. 2012a, 2013), salinity (El Youssfi et al. 2012; Hirich et al. 2014c, 2014d), organic amendment (Hirich et al. 2013, 2014a), nitrogen fertilization (Choukr-Allah et al. 2016), and sowing date (Hirich et al. 2014b) conditions. Results from these experiments showed that applying deficit irrigation during vegetative growth stage using 50% of irrigation requirement stabilized yield and maximized the crop water productivity (Fig. 19.9).

Organic amendment was applied for quinoa combined with water deficit. Obtained results indicated that organic amendment of 10 and 5 t ha^{-1} increased grain yield by 16 and 3%, respectively, under full irrigation conditions, and by 18 and 13% under deficit irrigation, respectively. The highest seed yield (66.3 g/plant) was recorded when quinoa was subjected to

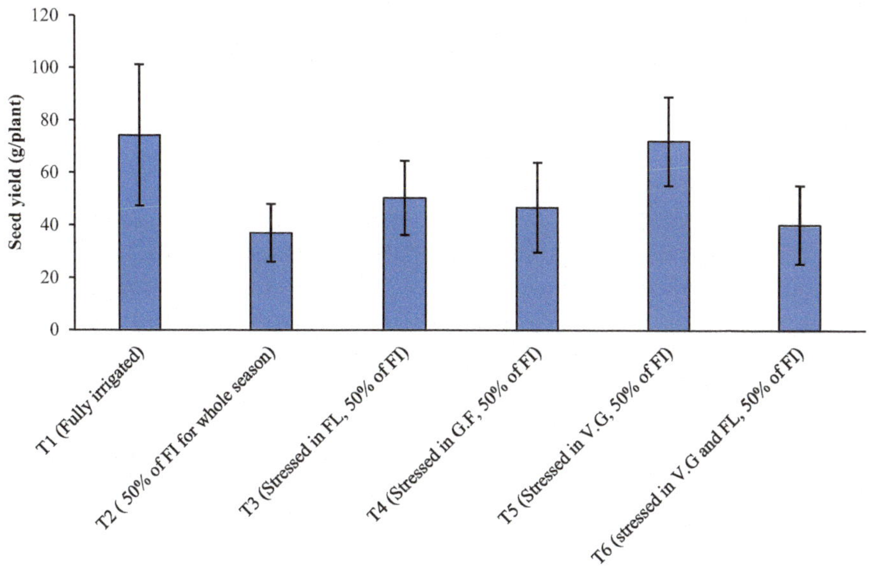

Fig. 19.9 Effect of deficit irrigation applied during different growth stages of quinoa (Hirich et al. 2012a)

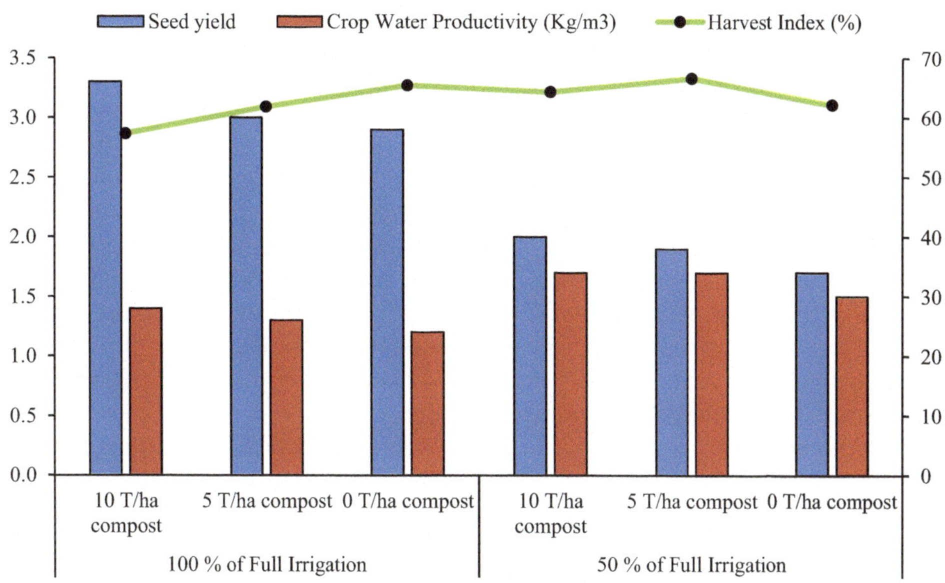

Fig. 19.10 Combined effect of organic amendment and deficit irrigation on quinoa (Hirich et al. 2014a)

full irrigation and received 10 t ha^{-1}of compost (Fig. 19.10).

Results from field trials conducted in sandy soil during two growing seasons (2011 and 2012) in Marrakech region showed that deficit irrigation (50 and 33% of full irrigation) affected quinoa performance and resulted in seed yield reductions of 15.8 and 30.1%, respectively, in the first season (2011), and by 15.2 and 41.5%, respectively, in the second season (2012) when compared to full irrigation (Fghire et al. 2015).

Research conducted on quinoa in the south of Morocco (Agadir region) focused only on testing seven nitrogen doses (0, 40, 80, 120, 160, 200, and 240 Kg ha^{-1}) under four irrigation regimes (25, 50, 75, and 100% of full irrigation). Results suggested that seed yield increased with increasing nitrogen supply, though the response varied with the level of water stress (Choukr-Allah et al. 2016). Quinoa has been sown at 10 sowing dates (from 1st of November to 30th of March with increments of 15 days). The study showed that sowing dates affected growth and productivity due to differences in temperature, precipitation, and radiation over time. Seed and dry matter yields were the highest when quinoa was sown in November and early December (Hirich et al. 2014b).

An experiment in the Rabat region was carried out on sandy–loam soil to compare three tillage systems: one-pass using a tine cultivator followed by a roller (CD-R), one-pass using a disk harrow (CC), and no-till seeding or direct seeding (DS) (Oussible et al. 2013). The primary objective was to develop an automated strategy for quinoa crop management. Results related to quinoa yield revealed the advantages of a direct-seeding system (DS) which produced the highest plant population at harvest (55,208 plants/ha), largest grain yield (0.64 t ha^{-1}), and highest harvest index (0.46).

Recently, FAO collaborated with several partners worldwide to access quinoa seeds and to access their expertise in the field of crop cultivation. Various research centers, universities, and seed firms were mobilized to find quinoa seeds of different varieties to be tested. Twenty different genotypes were eventually made available to countries for the international trial program (Bazile et al. 2016).

In Algeria two sites were analyzed and thirteen genotypes were tested. The yield average obtained was 1.65 t ha^{-1} at the first site (AL1) and 0.26 t ha^{-1} at the second site (AL2) with an average of 0.96 t ha^{-1}. The three genotypes Q26 (2.62 t ha^{-1}), Q18 (2.27 t ha^{-1}), and Q27 (2.17 t ha^{-1}) recorded the highest yields among other genotypes in the first site.

In Egypt, thirteen genotypes were tested in two sites. The average yield representing the country was equal to 1.89 t ha^{-1}at the first site (EG1) and 2.35 t ha^{-1}at the second site (EG2). The highest yields were obtained with Q12 (3.87 t ha^{-1}), Q18 (3.17 t ha^{-1}), and Q29 (3.41 t ha^{-1}) among other tested genotypes. In Mauritania, three sites were analyzed for this study, and nine genotypes were tested with an average yield equal to 0.08 t ha^{-1} at the first site (MA1), 0.08 t ha^{-1} at the second site (MA2), and 0.05 t ha^{-1} at the third site (MA3). This very low average was explained by low yields recorded for tested genotypes in the three experimental sites.

A series of studies were conducted on quinoa toward the goal of including these new crops into the crop structure of the desert area of Egypt (Shams 2011a). Quinoa seemed to be very important to grow in the Egyptian desert on newly reclaimed sandy soil in order to combat degradation and diversification. Thirteen varieties of quinoa were tested under two soil systems (till and no-till systems). Quinoa yield under supplementary irrigation was better than under rainfall only (rainfall amounts ranged from 100–150 mm/year). Seeding on flat beds recorded higher yields than sowing on ridges. The yield of quinoa significantly increased with diminished plant spacing from 20 cm to 15 cm. There were also gradual reductions in seed yield/field with delaying the time of seeding starting from the middle of November until the mid of February.

In a field trial conducted in Egypt during 2008–2009 and 2009–2010 winter seasons, five concentrations of nitrogen fertilizer (0, 90, 180, 270, and 360 Kg N/ha) were tested for improvement of growth and yield in sandy soils. Results revealed that fertilizing quinoa with 360 Kg N/ha resulted in maximum grain yield of 1203 and 1088 Kg/ha in the first and second seasons, respectively (Shams 2011b).

Quinoa Production Under Saline Conditions in North Africa

Many studies have been carried out on quinoa in order to evaluate its responses to salinity, and all showed that quinoa as a halophyte tolerates to high levels of salinity and can survive even under seawater salinity (Adolf et al. 2012a; Eisa et al. 2012; Hariadi et al. 2011b; Jacobsen et al. 2000;

Pulvento et al. 2012; Razzaghi et al. 2011, 2012; Ruiz-Carrasco et al. 2011; Koyro et al. 2008). Quinoa has been subjected to several trials on salinity in North Africa. Most of the published studies have been conducted in Morocco.

Impact of Salinity on Quinoa Germination and Seedling Emergence

Seed germination and growth are key factors in the life of a plant and are especially affected in the presence of limiting factors, such as salt. Germination tests were conducted by Brakez et al. (2014) in order to test the effect of several saline treatments of 50, 100, 150, and 200 mMNaCl solutions and 10, 20, 30, and 40% seawater (SW) dilutions in petri dishes. All seeds germinated in all NaCl and seawater treatments and the control. However saline treatments delayed germination when compared to the control causing more delay of germination than NaCl. The hypocotyl and radicle lengths and dry weights were affected more by NaCl than by seawater. The radicles were the most affected by salt stress when compared to hypocotyls.

A seedling test was carried out in order to evaluate the response of seedling establishment to irrigation water salinity (Hirich et al. 2014c, 2014d). The test was conducted in seedling trays filled with peat, three seeds sown per tray. Eight irrigation water salinity treatments were applied, 1, 2, 4, 6, 8, 10, 12, and 14 dS/m for chickpea and 1, 2, 4, 8, 12, 16, 20, and 24 dS/m for quinoa. There were a total of four replications per treatment. The number of seeds established was counted during the first 12 days since they were sown.

The effect of increasing levels of irrigation water salinity on the seedling establishment of quinoa is presented in Fig. 19.11. A trend of decreasing seedling emergence with increasing water salinity concentrations was found. Fifty percent of seedlings emerged in treatments with an EC value equal to 16 dS/m. When applying saline water with an EC value exceeding 20 dS/m, quinoa demonstrated only 10% of seedling establishment. More than 80% of seedlings emerged when irrigating with saline water with an EC value up to 12 dS/m.

Such data clearly indicate that the use of saline water with an EC value up to 12 dS/m for quinoa is the one to be recommended to obtain high germination percentages without any notable reduction in the seedling number. Salinity treatments caused a delay in germination compared to freshwater treatments, and increased salinity levels led to delayed seedling establishment. The control (1 dS/m) needed only 2 days to reach maximum

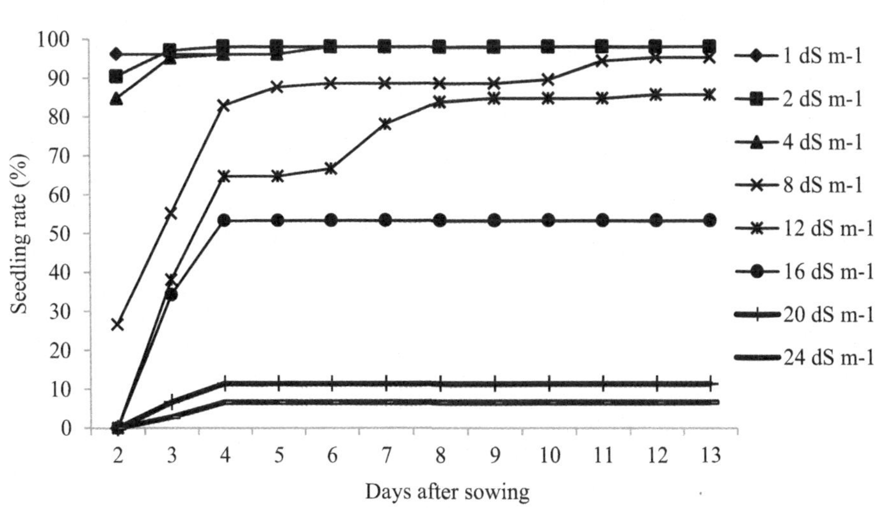

Fig. 19.11 Seedling establishment of quinoa under increasing irrigation water salinity (Hirich et al. 2014d)

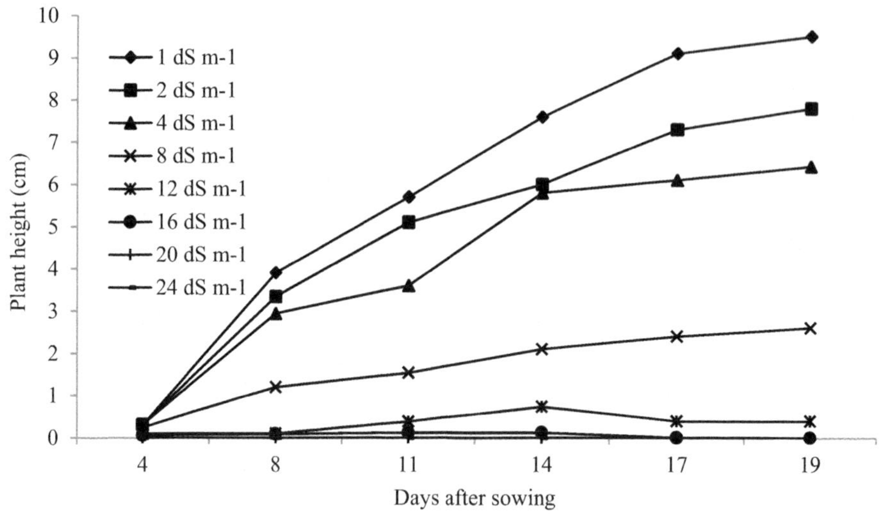

Fig. 19.12 Variation of seedling height under increasing irrigation water salinity (Hirich et al. 2014d)

seedling establishment while treatments with an EC value equal to 8 dS/m needed about 11 days.

Figure 19.12 shows the plant height during the seedling period 20 days after sowing. Increasing irrigation water salinity negatively affected seedling height and led to a severe reduction, especially when irrigating with saline water that exceeded 8 dS/m for quinoa. Increasing salinity up to 16 dS/m led to inhibition of plant growth after germination.

The relationship between seedling fresh matter and irrigation water salinity is presented in Fig. 19.13. Results clearly indicate that seedling biomass accumulation was inhibited under salinity with an EC value equal to 12 dS/m. The seedlings of quinoa are shown to be more resistant to salinity as the slope of the relationship equation is −0.09.

Priming Effect on Quinoa Seedlings Under Saline Conditions

Applying priming on quinoa seeds could improve seedling emergence under saline conditions (Hirich 2014). Figure 19.14 shows the changes in seedling rate and seedling half time T50 (the time in days required to reach the seedling emergence of 50% of sown seeds). Statistical analysis revealed significant differences in terms of both seedling rate and T50. Hormonal priming applying IAA (indoleacetic acid), GA_3 (gibberellic acid), or AA (ascorbic acid) leads to high seedling emergence rate, while KCl priming resulted in the lowest rate. The highest T50 was obtained using AA, distilled water (H_2O), and control (without priming), while the lowest T50 was obtained using IAA. These findings indicate that using IAA as a seed primer leads to high seedling emergence rate and simultaneously reduces the seedling half time. It is the most recommended based on results of this study to use in order to improve and enhance seedling emergence of quinoa.

Quinoa Production Under Salinity

The irrigation water salinity level had a negative impact on the dry matter production of quinoa ($p < 0.01$). Figure 19.15 shows the variation of dry matter during the growing period (Hirich 2014). The highest dry matter was obtained by the control treatment irrigated with freshwater. An increase in irrigation water salinity caused a significant decline in dry matter

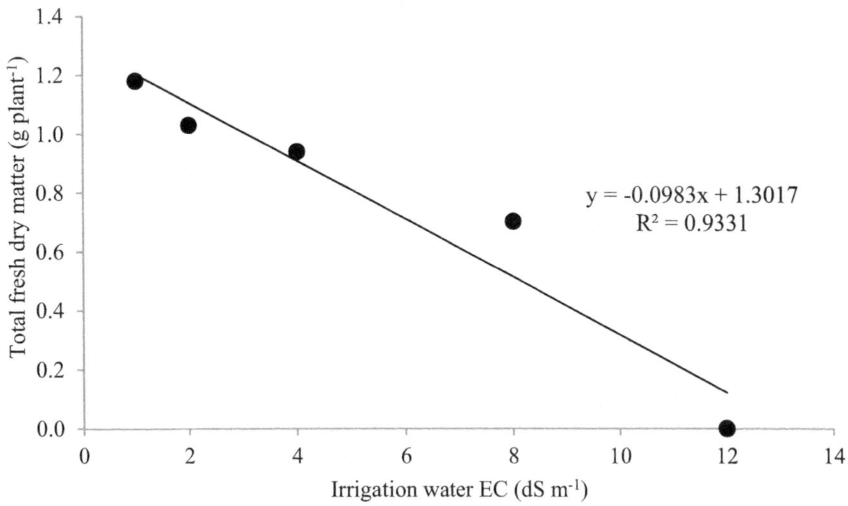

Fig. 19.13 Relationship between irrigation water salinity and seedling fresh matter (Hirich et al. 2014d)

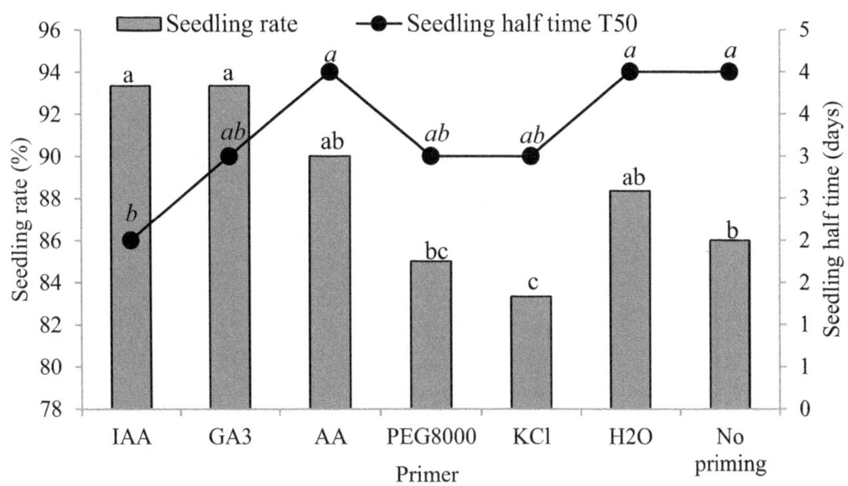

Fig. 19.14 Changes in seedling rate under irrigation water salinity of 15 dS/m and different priming treatments (Hirich 2014)

weight. Differences in dry matter weight between the treatments increased and became more pronounced at later stages of crop growth. There was no significant difference in terms of dry matter of quinoa obtained under irrigation with salinity level up to 10 dS/m. There was a reduction of 22 and 53% in terms of dry matter of quinoa when applying saline water with an EC value equal to 20 and 30 dS/m, respectively.

The effect of irrigation water salinity on final dry biomass, root volume, grain yield, harvest index (HI), and crop water productivity (CWP) of both chickpea and quinoa is reported in Table 19.7 (Hirich et al. 2014c, 2014d).

The results show that the highest dry biomass weight has been recorded under the control treatment for both crops. The gradual increments in the salinity level of irrigation water resulted in

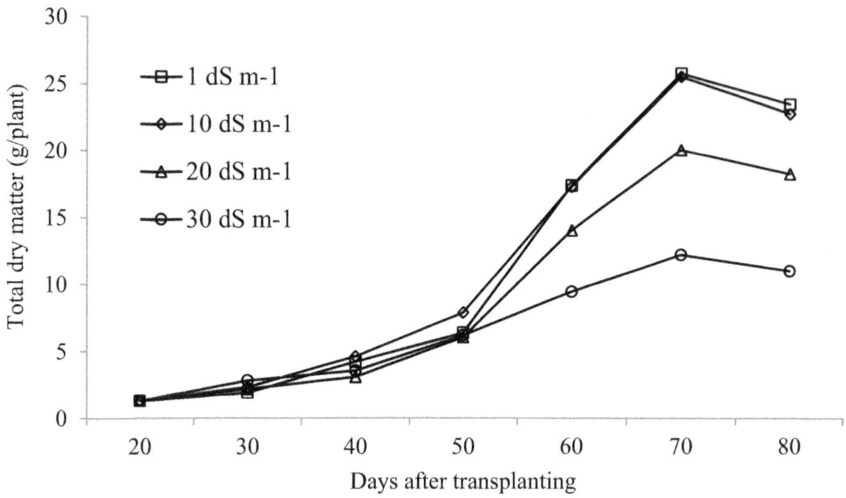

Fig. 19.15 Variation of dry matter including root, stem, and leaf weight during growing period (Hirich 2014)

Table 19.7 Dry biomass, root volume, grain yield, harvest index, and CWP for tested saline treatments

Treatments	1 dS/m (control)	10 dS/m	20 dS/m	30 dS/m
Dry biomass (g/pot)	70.2 a	68.1 a	54.6 b	32.9 c
Root volume (cm^3)	6.7 a	5.8 b	4.8 c	4.3 c
Grain yield (g/pot)	39.1 a	35.5 a	29.7 b	25.7 b
Harvest index (%)	55.7 b	52.1 b	54.4 b	77.9 a
Water productivity (Kg/m)	0.6 b	0.7 ab	0.7 ab	0.8 a

Hirich et al. (2014d)
Different letters (a, b, and c) indicate statistically significant at $P < 0.05$

notable reduction in the biomass production of quinoa, with values 3, 22, and 53% lower than that where irrigation was with freshwater for the irrigation salinity levels of 10, 20, and 30 dS/m, respectively. Root volume was also negatively affected by irrigation water salinity. The reduction in root volume under the highest salinity level (30 dS/m) was equal to only 35% of the control. Grain yield was affected by irrigation water salinity. It was less affected by salinity level when compared to the biomass. Irrigation with saline water up to 10 dS/m did not affect grain yield of quinoa, and yield loss was only 9% when compared to freshwater. An increase of salinity level to 20 and then to 30 dS/m resulted in reduction in grain production by only 24 and 34%, respectively, compared to that obtained under freshwater (1 dS/m).

The harvest index (HI) was calculated as the ratio between grain yield and dry biomass. The HI varied slightly with increasing irrigation water salinity up to 20 dS/m. The highest HI was obtained under the most stressed treatment (30 dS/m). The lowest harvest index was obtained under treatment that received saline water with an EC value equal to 10 dS/m. Crop water productivity (CWP) was calculated by dividing grain yield by total evapotranspiration. CWP was affected positively by irrigation water salinity; the most efficient treatment according to statistical analysis ($p < 0.001$) was the most stressed treatment (30 dS/m) which produced 0.8 Kg for

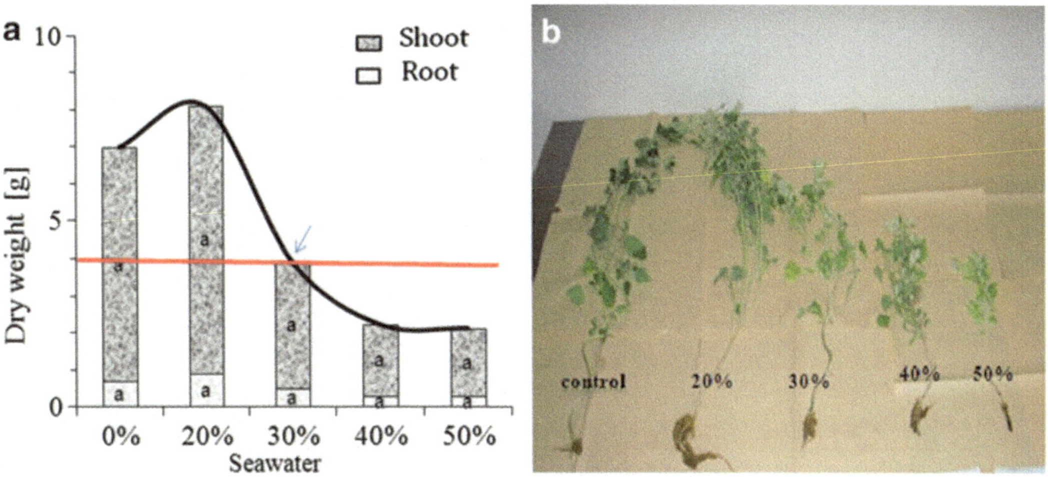

Fig. 19.16 Biomass production of *C. quinoa* in different seawater treatments (Brakez et al. 2013)

Table 19.8 Biochemical composition of quinoa seeds under non-saline and saline conditions

Compound	Unit	Control	125 mM NaCl
Protein	(%)	21.63	23.88
Starch	(%)	46.28	49.02
Carbohydrates	(%)	47.12	44.98
Vitamin C	(mg/g)	0.03	0.03
Polyphenols	(mg/g)	1.22	1.12
Flavonoids	(mg/g)	0.12	0.13
Lipids	(%)	6.57	7.98

Brakez et al. (2015)

1 m^3 of water. The lowest CWP (0.6 Kg m^{-3}) was obtained under control treatment.

Three lines of quinoa (QM1113, QS0938, D0708) cultivated in open fields were irrigated with treated wastewater with different levels of salinity in order to assess the impact on yield (El Youssfi et al. 2012). Significant differences were found in the performance of the lines with QM1113 being the most productive with an average grain yield of 6.92 t ha^{-1} at 6 dS/m salinity and was followed by D0708 with an average yield of 5.65 t ha^{-1} at 3 dS/m.

Five-week-old seedlings were irrigated with four seawater (SW) dilutions (20, 30, 40, and 50%) in comparison with control plants of the same age irrigated with freshwater (Brakez et al. 2013). Maximum biomass was noted in 20% seawater treatment (EC 12 dS/m) and decreased with the increase of salinity. Fifty percent of biomass reduction was observed under 40% of seawater (Fig. 19.16).

An experiment was carried out in an open field during the 2011–2012 growing season in Agadir, Morocco. The study was conducted using the Hualhuas cultivar of quinoa, and two treatments were investigated: 0 mMNaCl as a control and 125 mMNaCl, which gave the maximum growth of quinoa in controlled conditions (Brakez et al. 2015). The yield of quinoa seeds per plant increased in 125 mMNaCl treatment when compared to the control treatment without salinity. The biochemical analysis of quinoa seeds showed that the 125 mMNaCl treatment induced higher and/or similar concentration for the analyzed elements (Table 19.8).

Identification of Quinoa Germplasm with Potential Traits for Marginal Environments in the North Africa

The first quinoa germplasm evaluation and screening in North Africa are carried out in 2000 in Morocco under high-altitude conditions

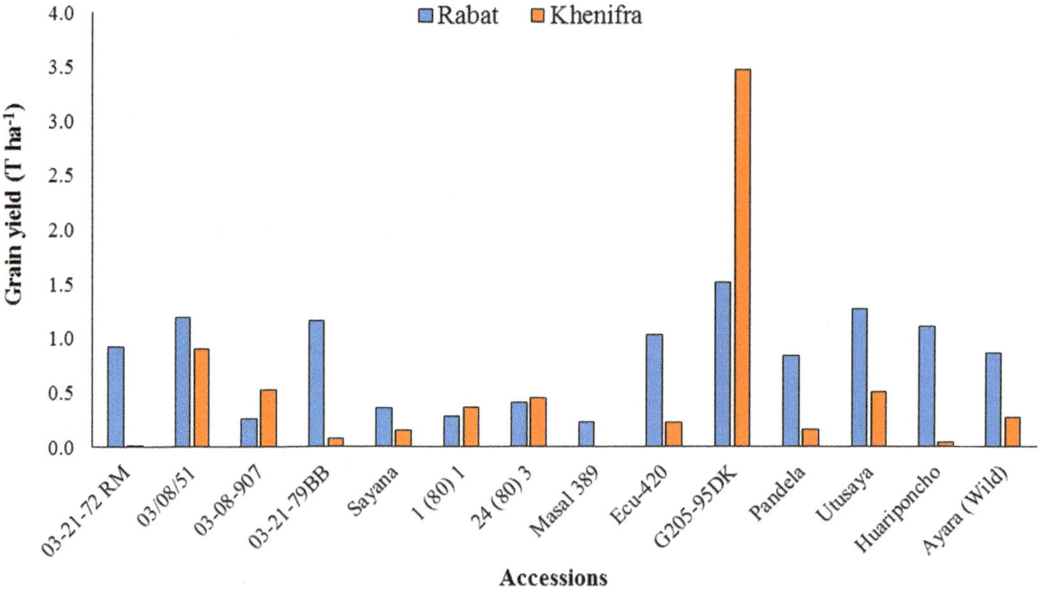

Fig. 19.17 Quinoa grain yields (t ha^{-1}) under high-altitude (Khenifra) and coastal (Rabat) conditions in 2000–2001 season

in the Khenifra region and coastal conditions in the Rabat region. The results for yield of 14 genotypes are presented in Fig. 19.17 (Benlhabib et al. 2014; Choukr-Allah et al. 2016). The results clearly indicate that accession G205-95DK had the highest yields in both locations. However, a large variation within the same accession was observed in terms of location. Most of the accessions shown were more adapted to coastal conditions.

Filali et al. (2016) conducted a screening experiment of five quinoa accessions under four irrigation treatments. Results showed that yield and dry matter of all accessions were affected by water stress, while crop water productivity (CWP) increased with decreased applied water. The highest yield was obtained by accession (L119) and the lowest by L11 (Table 19.9).

An investigation has been carried out to evaluate the genetic diversity of 78 quinoa accessions developed through recurrent selection from Andean germplasm introduced to Morocco in the winter of 2000. Twenty-three quantitative and qualitative characters were used for the evaluation of genetic diversity and the relationship between the accessions and also for the establishment of a core collection in Morocco (Mhada et al. 2014). The phylogenic tree displayed in Fig. 19.18 reveals the presence of four distinct clusters and ten sub-clusters. Cluster I represents 5% of the germplasm-containing accessions with tall plant height, low harvest indexes, and high resistance to downy mildew. Cluster II gathered late-maturing accessions with high levels of resistance to downy mildew. Cluster III includes 37% of the accessions that have an early maturing and short plant height. The last cluster, VI, included early maturing, high harvest index, and high-yielding accessions. This group represented 28% of the quinoa germplasm.

Table 19.9 Yield, dry matter, harvest index, water supply, and CWP of five quinoa accessions under several irrigation levels

Accessions	Irrigation treatments	Yield (t/ha)	Dry matter (t/ha)	Harvest index	Water supply (mm)	Crop water productivity (Kg/m^3)
L11	100% FI	3.4	9.6	35%	335	1
	75% FI	2.8	8.6	33%	251	1.1
	50% FI	2.1	6.7	31%	167	1.3
	25% FI	2	5.8	34%	83	2.4
L119	100% FI	4.6	16.1	29%	335	1.4
	75% FI	2.8	8	35%	251	1.1
	50% FI	2.5	7.8	32%	167	1.5
	25% FI	2.6	8	33%	83	3.1
L123	100% FI	3.5	11.1	32%	335	1
	75% FI	2.9	11.2	26%	251	1.2
	50% FI	3.2	9.6	33%	167	1.9
	25% FI	2.3	6	38%	83	2.8
L142	100% FI	3.8	12.6	30%	335	1.1
	75% FI	3.6	14.3	25%	251	1.4
	50% FI	2.8	9.5	29%	167	1.7
	25% FI	1.9	6.2	31%	83	2.3
L143	100% FI	4.3	15.2	28%	335	1.3
	75% FI	3.1	11.5	27%	251	1.2
	50% FI	2.2	7.8	28%	167	1.3
	25% FI	1.6	4.7	34%	83	1.9

Filali et al. (2016)

Constraints and Opportunity of Quinoa Production in the North Africa

North African agriculture is relatively advanced with a supply chain capable of meeting the requirements and standards needed to export to the EU market for a number of agricultural products, which the region has cultivated for a long time. This institutional set-up provides a basis for introducing a new crop that has some benefits related to problems faced by farmers and potentially could contribute a higher-value product for the region to export as well as a protein-rich commodity for domestic consumers (Pedersen et al. 2013).

Several farmers in Morocco have previous knowledge about quinoa from their neighbors. Many farmers indicate that when they consider the introduction of a new crop in their production system, they make some sort of comparison between the crops that they already produce and the new crop. They look at the market availability, product prices, yield, production costs, and ease of production. A number of farmers perceive quinoa as a likely crop to be included in their crop rotation if the current prices can be obtained on the market. On the consumers' side, quinoa is especially good for people who are intolerant to gluten, which could be an important market segment for national markets and for exports to Europe (Table 19.10).

Conclusions

Introducing quinoa in this new region was a three-step process (Fig. 19.19). The first step was crop adaptability, which includes agronomic practices (sowing date, sowing density, fertilizer,

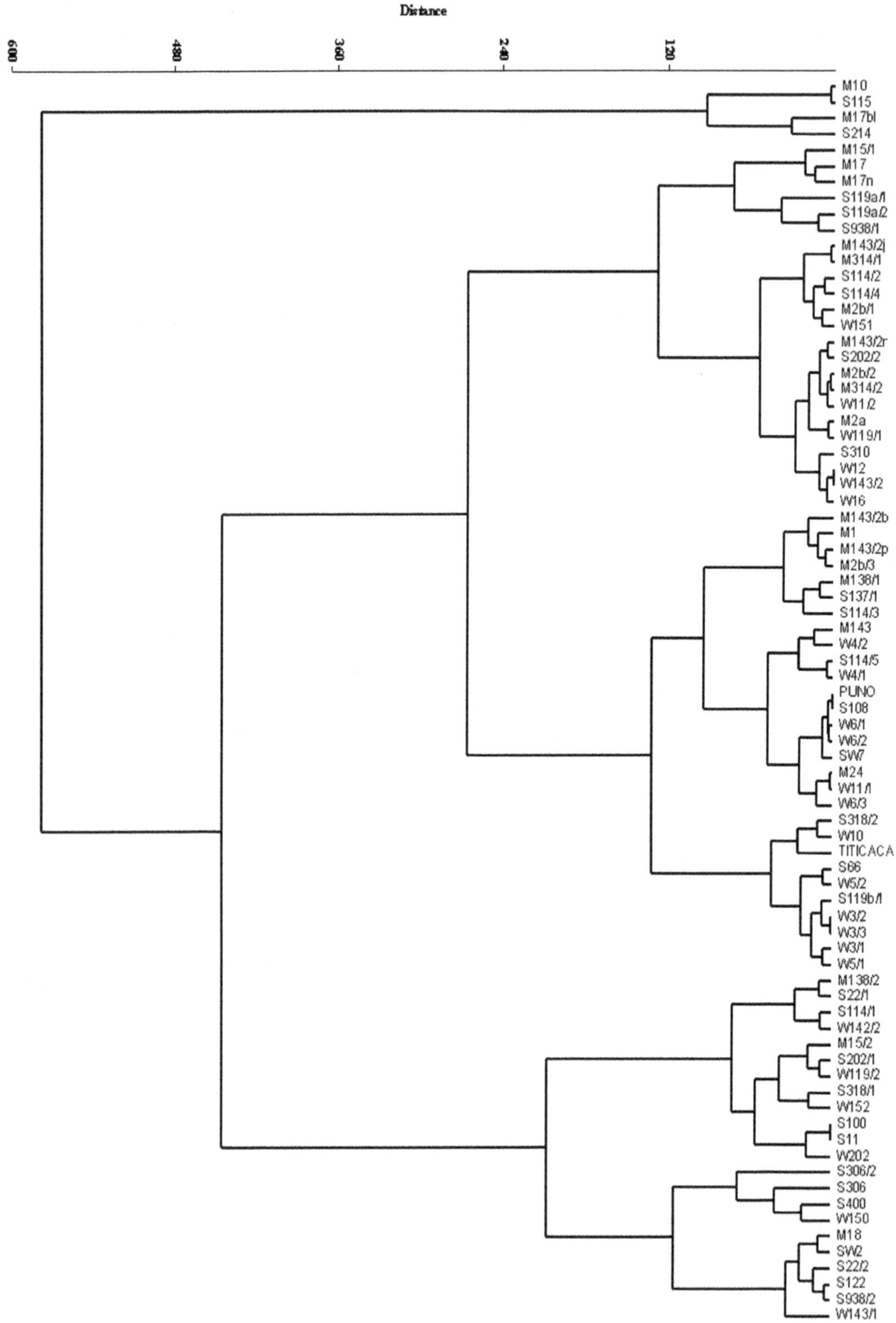

Fig. 19.18 Dendrogram based on agronomic and morphological traits using UPGMA (Mhada et al. 2014)

Table 19.10 Overview of advantages and barriers for implementing quinoa as a new crop among farmers in North Africa

Advantages	Barriers
General	***General***
Drought resistant	Low yields compared with cereals
Salt resistant	Difficult to get seed
High price	Seed germination and emergence problems
Gluten free (added value)	
High durability	Competition with other high-yielding crops
	High content of saponine – quinoa must be cleaned before consumption
Specific to North Africa	***Specific to North Africa***
EU market proximity encourage farmers to adopt quinoa in North Africa	Farmers have little if any knowledge about quinoa
	Low yield is a problem: when a significant share of farmers use farm products for local/home consumption
Availability of salt-affected soils in North Africa where quinoa can be introduced	
	Irrigation scheme is about to be implemented
A significant share of rainfed areas	
Cleaning	
Initiatives made by women's cooperatives to clean quinoa	
A high value in restaurants in Morocco	
Can be used for local dishes (e.g., couscous, bread)	

and water requirement) and genotype selection. Quinoa experiments began in 2000 and 2006 in the region by Morocco and ICBA. Some farmers then started growing quinoa in Morocco and Egypt. Adaptability of the crop was evaluated in Iran, Iraq, UAE, Turkey, Egypt, Lebanon, Yemen, Sudan, and Mauritania in regional projects by FAO (Dost 2015). Based on these results agroclimatology zoning can be done for selecting potential areas for quinoa production in saline lands in order to select the best genotypes, appropriate sowing dates, and other agronomic practices. The first step took 10–16 years in the region, and there are still some obstacles for implementing the next step of development.

Low seed germination and seedling establishment are the main constraints in most of the countries in the region. In the entire region, grain filling of quinoa occurred during summer or late of spring. Temperature and photoperiod of the mother plant also influence seed dormancy. Spring sowing, in which grain fill occurs in the summer, promotes dormancy in quinoa seed. However, autumn ripening reduces dormancy (Ceccato et al. 2015). In Iran, sowing dates in which the seed ripened in autumn produced the high viable seeds (96% germination during 3 days). In the new region, a private seed company should produce the high seed quality in the best climate condition.

Phase two includes winnowing, post-harvest processing, sowing, and threshing. Along with marketing, the infrastructure of quinoa is scaling up, and the national government should support marketing in these first years. Most of the farmers in developing countries use wheat and rice straw for animal feed. Using by-product of quinoa and training of farmers are very important for acceptance of new crops by farmers. Almost all the countries in the region are just starting the second step. For progress to the next step and large-scale production, national governments' and international institutes' supporting strategies are very important.

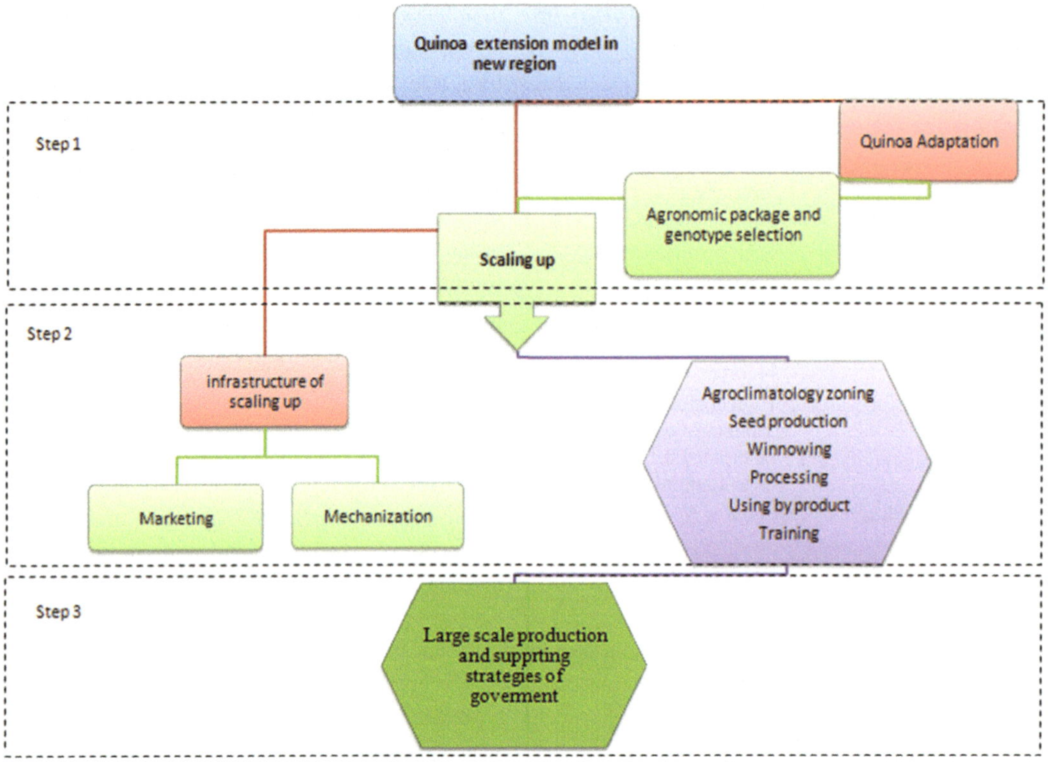

Fig. 19.19 Flow chart of quinoa extension in the new region

Acknowledgments Thanks to Mohammad Shahid and Khalil Ur Rahman Butt for assisting in the fieldwork and in the data collection in UAE. Thanks to Dr. Basra, Dr. Shahid, and Dr. Jalal Kamli for sending germplasm to Iran and Vali Soltani, Hassan Zare, and Seyyed Jalil Hasheminasab for fieldwork and data collection in Iran.

References

Abugoch James LE (2009) Chapter 1 Quinoa (*Chenopodium Quinoa* Willd.): composition, chemistry, nutritional, and functional properties. In: Steve LT (ed) Advances in food and nutrition research. Academic, London

Adolf V, Shabala S, Andersen M, Razzaghi F, Jacobsen S-E (2012a) Varietal Differences Of Quinoa's Tolerance To Saline Conditions. Plant Soil 357:117–129

Adolf VI, Jacobsen SE, Shabala S (2012b) Salt tolerance mechanisms in quinoa (*Chenopodium Quinoa* Willd.). Environ Exp Bot 92:43–54

Baig A, Baig M, Ali Q, Khan C (1985) Agro-ecological zonation of pothwar. A. Wheat. B. Maize. Pakistan Agricultural Research Council, Islamabad, 68

Bazile D (2014) State of the art report on quinoa around the world in 2013. In: Bazile D (ed) Food and agriculture organization of the United Nations (FAO), Santiago

Bazile D, Bertero H, Nieto C (2015) State of the art report on quinoa around the world in 2013, FAO

Bazile D, Pulvento C, Verniau A, Al-Nusairi MS, Ba D, Breidy J, Hassan L, Mohammed MI, Mambetov O, Otambekova M, Sepahvand NA, Shams A, Souici D, Miri K, Padulosi S (2016) Worldwide evaluations of quinoa: preliminary results from post international year of quinoa Fao projects in 9 countries. Front Plant Sci 7:850

Bendevis MA, Sun Y, Shabala S, Rosenqvist E, Liu F, Jacobsen S-E (2014) Differentiation of photoperiodinduced ABA and soluble sugar responses of two quinoa (*Chenopodium quinoa* Willd.) cultivars. J Plant Growth Regul 33:562–570

Benlhabib O, Jacobsen SE, Jellen EN, Maughan PJ, Choukr-Allah R (2014) Status of quinoa production and research in morocco. In: Bazile D (ed) State of the art report on quinoa around the world in 2013. Food and Agriculture Organization of the United Nations (FAO), Santiago

Bertero H, King R, Hall A (1999) Modelling photoperiod and temperature responses of flowering in quinoa (*Chenopodium Quinoa* Willd.). Field Crop Res 63:19–34

Blanco JA (2015) Fodder and animal feed. In: Bazile D, Bertero H, Nieto C (eds) State of the art report on quinoa around the world in 2013. Fao

Bois J-F, Winkel T, Lhomme J-P, Raffaillac J-P, Rocheteau A (2006) Response of some andean cultivars of quinoa (*Chenopodium Quinoa* Willd.) to temperature: effects on germination, phenology, growth and freezing. Eur J Agron 25:299–308

Brakez M, El Brik K, Daoud S, Harrouni MC (2013) Performance of chenopodium quinoa under salt stress. In: Developments in soil salinity assessment and reclamation. Springer, Dordrecht

Brakez M, Harrouni M, Tachbibi N, Daoud S (2014) Comparative effect of nacl and seawater on germination of quinoa seed (*Chenopodium Quinoa* Willd). Emirates J Food Agric 26:1091

Brakez M, Daoud S, Harrouni MC, Tachbibi N, Brakez Z (2015) Nutritional value of chenopodium quinoa seeds obtained from an open field culture under saline conditions. In: Halophytes for food security in dry lands. Elsevier, Amsterdam, p 37

Ceccato D, Bertero D, Batlla D, Galati B (2015) Structural aspects of dormancy in quinoa (*Chenopodium Quinoa*): importance and possible action mechanisms of the seed coat. Seed Sci Res 25:267–275

Choukr-Allah R, Nanduri KR, Hirich A, Shahid M, Alshankiti A, Toderich K, Gill S, Butt KUR (2016) Quinoa for marginal environments: towards future food and nutritional security in MENA and central Asia regions. Front Plant Sci 7:1–11

De Munter JS, Hu FB, Spiegelman D, Franz M, Van Dam RM (2007) Whole grain, bran, and germ intake and risk of type 2 diabetes: a prospective cohort study and systematic review. PLoS Med 4:E261

Dost M (2015) Field evaluation results across locations and identification of suitable quinoa varieties. Project (Tcp/Rab/3403–Fao)

Dregne H, Kassas M, Rozanov B (1991) A new assessment of the world status of desertification. Desertification Control Bull 20:6–18

Eisa S, Hussin S, Geissler N, Koyro HW (2012) Effect of Nacl salinity on water relations, photosynthesis and chemical composition of quinoa (*Chenopodium Quinoa* Willd.) As a potential cash crop halophyte. Aust J Crop Sci 6:357–368

El Youssfi L, Choukr-Allah R, Zaafrani M, Mediouni T, Ba Samba M, Hirich A (2012) Effect of domestic treated wastewater use on three varieties of quinoa (*Chenopodium Quinoa*) under semi arid conditions. World Acad Sci Eng Technol 68:306–309

Fghire R, Wahbi S, Anaya F, Issa Ali O, Benlhabib O, Ragab R (2015) Response of quinoa to different water management strategies: field experiments and saltmed model application results. Irrig Drain 64:29–40

Filali K (2011) Caractérisation Et Évaluation Du Rendement De Lignées De Quinoa (Chenopodium Quinoa) Dans La Région De Rhamna. Diplôme De Master En Biotechnologies Et Amélioration Génétique Des Productions Agricoles, Institut Agronomique Et Vétérinaire Hassan Ii

Filali K, Hirich A, Benlhabib O, Choukr-Allah R, Ragab R (2016) Yield and dry matter simulation using satlmed model for five quinoa (Chenopodium Quinoa) accessions under deficit irrigation in South Morocco. Irrigation and Drainage, In Press

Flowers T, Hajibagheri M, Clipson N (1986) Halophytes. Q Rev Biol 61:313–337

Garcia M, Condori B, Castillo CD (2015) Agroecological and agronomic cultural practices of quinoa in South America. In: Quinoa: improvement and sustainable production. Wiley, Hoboken, pp 25–46

General Directorate Of Soil And Water (1978) Soil resources of Turkey. General Directorate of Soil and Water, Ankara

Ghaffari A, Ghasemi VR, De Pauw E (2014) Agro-climatically zoning of Iran by Unesco approach. Iran Dryland Agron J 4:63–95

Ghassemi F, Jackman AJ, Nix AH (1995) Salinization of land and water resources. Cab International, Wallingford

Hariadi Y, Marandon K, Tian Y, Jacobsen S-E, Shabala S (2011a) Ionic and osmotic relations in quinoa (Chenopodium Quinoa Willd.) Plants grown at various salinity levels. J Exp Bot 62:185–193

Hariadi Y, Marandon K, Tian Y, Jacobsen SE, Shabala S (2011b) Ionic and osmotic relations in quinoa (*Chenopodium Quinoa* Willd.) Plants grown at various salinity levels. J Exp Bot 62:185–193

Hirich A (2013) Using deficit irrigation with treated wastewater to improve crop water productivity of sweet corn, chickpea, faba bean and quinoa. Revue Marocaine Des Sciences Agronomiques Et Vétérinaires 2:15–22

Hirich A (2014) Effects of deficit irrigation using treated wastewater and irrigation with saline water on legumes, corn and quinoa crops. Phd, Hassan Ii Institue of Agronomy and Veterinary Medicine

Hirich A, Choukr-Allah R, Jacobsen S-E, El Yousfi L, El Omari H (2012a) Using Deficit Irrigation With Treated wastewater in the production of quinoa (*Chenopodium Quinoa* Willd.) in Morocco. Revista Científica Udo Agrícola 12:570–583

Hirich A, Choukr-Allah R, Jacobsen S-E, Fahmi H, Rami A, Laajaj K, El Youssfi L, El Omari H (2012b) Etude De L'effet De L'irrigation Déficitaire Et De La Salinité Sur Céréales, Légumineuses Et Quinoa. 1ère Journées Doctoriales De L'institut Agronomique Et Vétérinaire Hassan Ii. Rabat, Maroc

Hirich A, Choukr-Allah R, Jacobsen S-E (2013) The combined effect of deficit irrigation by treated wastewater and organic amendment on quinoa (Chenopodium Quinoa Willd.) Productivity. Desalin Water Treat 52:2208–2213

Hirich A, Choukr-Allah R, Jacobsen SE (2014a) Deficit irrigation and organic compost improve growth and yield of quinoa and pea. J Agron Crop Sci 200:390–398

Hirich A, Choukr-Allah R, Jacobsen SE (2014b) Quinoa in Morocco – effect of sowing dates on development and yield. J Agron Crop Sci 200:371–377

Hirich A, Choukr-Allah R, Jelloul A, Jacobsen S-E (2014c) Quinoa (*Chenopodium Quinoa* Willd.)

Seedling, water uptake and yield responses to irrigation water salinity. Acta Hortic 1054:145–152
Hirich A, Jelloul A, Choukr-Allah R, Jacobsen SE (2014d) Saline water irrigation of quinoa and chickpea: seedling rate, stomatal conductance and yield responses. J Agron Crop Sci 200:378–389
Jacobsen SE (2003) The worldwide potential for quinoa (*Chenopodium Quinoa* Willd.). Food Rev Int 19:167–177
Jacobsen SE (2014) New climate-proof cropping systems in dry areas of the mediterranean region. J Agron Crop Sci 200:399–401
Jacobsen S, Bach A (1998) The influence of temperature on seed germination rate in quinoa (Chenopodium Quinoa Willd). Seed Sci Technol (Switzerland) 26:515–523
Jacobsen S, Quispe H, Mujica A (1999) Quinoa: an alternative crop for saline soils in the Andes. Scientist and farmer-partners in research for the 21st century. Cip Prog Rep 2000:403–408
Jacobsen S-E, Quispe H,Mujica A (2000) Quinoa: an alternative crop for saline soils in the Andes. Cip Program Report
Jacobsen SE, Mujica A, Jensen C (2003) The resistance of quinoa (*Chenopodium Quinoa* Willd.) To adverse abiotic factors. Food Rev Int 19:99–109
Jacobsen S-E, Monteros C, Christiansen J, Bravo L, Corcuera L, Mujica A (2005) Plant responses of quinoa (*Chenopodium Quinoa* Willd.) To frost at various phenological stages. Eur J Agron 22:131–139
Khan MA, Anwar CM (1968) Sand dune rehabilitation in Thal, Pakistan. J Range Manag 21:316–321
Khan MA, Gul B (2006) Halophyte Seed Germination. In: Ecophysiology of high salinity tolerant plants. Springer, Dordrecht
Koyro HW, Lieth H, Eisa SS (2008) Salt tolerance of *Chenopodium Quinoa* Willd., grains of the Andes: influence of salinity on biomass production, yield, composition of reserves in the seeds, water and solute relations. Mangroves Halophytes Rest Util:133–145
Lavini A, Pulvento C, D'andria R, Riccardi M, Choukr-Allah R, Belhabib O, Incekaya Ç, Metin Sezen S, Qadir M, Jacobsen SE (2014) Quinoa's potential in the mediterranean region. J Agron Crop Sci 200:344–360
Loftas T, Ross J (1995) Dimensions of need: an atlas of food and agriculture. Food & Agriculture Organisation, Rome
Massoud F, Girard M (1978) The use of satellite imagery in detecting and delineating a salt affected soils
Mhada M, Jellen E, Jacobsen S, Benlhabib O (2014) Diversity analysis of a quinoa (Chenopodium Quinoa Willd.) germplasm during two seasons. World Acad Sci Eng Technol Int J Biol Biomol Agric Food Biotechnol Eng 8:273–276
Munir H, Basra S (2009) Improved germination and seedling vigor enhancement of quinoa (*Chenopodium Quinoa* Willd.) Grown at different salinity levels. In: Sustainable management of saline waters and salt-affected soils for agriculture: proceedings of the second bridging workshop Aleppo, Syria 15–18 November, Icarda, p 33
Negewo BD (2012) Renewable energy desalination: an emerging solution to close the water gap in the middle East and North Africa. World Bank Publications, Washington, DC
Nelson M, Maredia M (2001) Environmental impacts of the cgiar: an assessment. A report from Tac's standing panel on impact assessment. Fao, Rome
Nowak V, Du J, Charrondiere UR (2015) Assessment of the nutritional composition of quinoa (Chenopodium Quinoa Willd.). Food Chem
Nsimba RY, Kikuzaki H, Konishi Y (2008) Antioxidant activity of various extracts and fractions of *Chenopodium Quinoa* And *Amaranthus* Spp. seeds. Food Chem 106:760–766
Oldeman L, Hakkeling R, Sombrock W (1991) Global assessment of soil degradation (Glasod). World map of the status of human–induced soil degradation, Isric Wageningen
Oussible M, Baamal L, Yousfi Y, Belcaid M, Benlhabib O (2013) The effect of different soil tillage systems on the stand establishment and grain yield of quinoa crop in Morocco. In: Choukr-Allah R (Ed) International conference on sustainable water use for securing food production in the mediterranean region under changing climate. Agadir, Morocco
Panta S, Flowers T, Lane P, Doyle R, Haros G, Shabala S (2014) Halophyte agriculture: success stories. Environ Exp Bot 107:71–83
Pedersen SM, Benlhabib O, Xhoxhi O, Lind KMH, Ørum JE (2013) Farmers perception of quinoa-experience from bouchane region in Morocco and Adana in Turkey. Proceeding: sustainable water use for securing food production in the mediterranean region under chainging climate, Agadir, Morocco
Pulvento C, Riccardi M, Lavini A, Iafelice G, Marconi E, D'andria R (2012) Yield and quality characteristics of quinoa grown in open field under different saline and non-saline irrigation regimes. J Agron Crop Sci 198:254–263
Rao N, Shahid M (2012) Quinoa–a promising new crop for the Arabian Peninsula. Am Eurasian J Agric Environ Sci 12:1350–1355
Rao NK, Rahman K, Ismail S (2013) Quinoa: prospects as an alternative crop for salt-affected areas. In: Hall R, Rudebjer P, Padulosi S (eds) 3rd international conference on neglected and under-utilized species (Nus) For Food Secure Africa In Book Of Abstracts
Razzaghi F, Ahmadi SH, Adolf VI, Jensen CR, Jacobsen SE, Andersen MN (2011) Water relations and transpiration of quinoa (*Chenopodium Quinoa* Willd.) under salinity and soil drying. J Agron Crop Sci 197:348–360
Razzaghi F, Ahmadi SH, Jacobsen SE, Jensen CR, Andersen MN (2012) Effects of salinity and soil–drying on radiation use efficiency, water productivity and yield of quinoa (*Chenopodium Quinoa* Willd.). J Agron Crop Sci 198:173–184

Repo-Carrasco R, Espinoza C, Jacobsen S-E (2003) Nutritional value and use of the andean crops quinoa (*Chenopodium Quinoa*) and Kañiwa (*Chenopodium Pallidicaule*). Food Rev Int 19:179–189

Ruiz-Carrasco K, Antognoni F, Coulibaly AK, Lizardi S, Covarrubias A, Martãnez EA, Molina-Montenegro MA, Biondi S, Zurita-Silva AS (2011) Variation in salinity tolerance of four lowland genotypes of quinoa (*Chenopodium Quinoa* Willd.) As assessed by growth, physiological traits, and sodium transporter gene expression. Plant Physiol Biochem 49:1333–1341

Sepahvand NA (2012) Investigation on adaptation, phonological and agronomical characters and quality of quinoa in Iran. Final report, agricultural research, education & extension organization, Project Grant Number 2-03-03-86275

Shams A (2011a) Combat degradation in rain fed areas by introducing: new drought tolerant crops in Egypt. Int J Water Res Arid Environ 1:318–325

Shams A (2011b) Response of quinoa to nitrogen fertilizer rates under sandy soil conditions. Int J Water Res Arid Environ 1:318–325

Umali DL (1993) Irrigation-induced salinity: a growing problem for development and the environment. World Bank Publications, Washington, DC

Verdier F (2011) Mena regional water outlook part II desalination using renewable energy task 1–desalination potential. Fichtner Stuttgart

Wilson J (1932) Notes on the biology of *Laphygma Exigua* Hübner. Fla Entomol 16:33–39

Yazar A, Incekaya Ç, Sezen SM, Jacobsen S-E (2015) Saline water irrigation of quinoa (*Chenopodium Quinoa*) under mediterranean conditions. Crop Pasture Sci 66:993–1002

Zevallos V, Herencia L, Ciclitira P (2015) Quinoa, coeliac disease and gluten-free diet. In: Bazile D, Bertero H, Nieto C (eds) State of the art report on quinoa around the world in 2013. FAO, Rome

Different Antioxidant Defense Systems in Halophytes and Glycophytes to Overcome Salinity Stress

20

Prabhakaran Soundararajan,
Abinaya Manivannan, and Byoung Ryong Jeong

Abstract

Metabolic processes, such as photosynthesis and respiration, lead to the generation of reactive oxygen species (ROS) as a side product. Chloroplasts, mitochondria, peroxisomes, glycosomes, and plasma membranes are the predominant metabolically active cell organelles which release ROS. Plants possess enzymatic and non-enzymatic antioxidant defense systems to maintain the ROS level. Enzymatic antioxidants include superoxide dismutase (SOD), ascorbate peroxidase (APX), guaiacol/glutathione peroxidase (POD), catalase (CAT), monodehydroascorbate reductase (MDAR), dehydroascorbate reductase (DHAR), aldehyde dehydrogenases (ALDH), and glutathione reductase (GR). Non-enzymatic antioxidants include ascorbate (AsA), glutathione (GSH), proline, and phenolic compounds. Under stress conditions, ROS are excessively generated in the plant. Based on the tolerance level to the salinity plants are divided into two categories: glycophytes and halophytes. Glycophytes are salt-sensitive plants and halophytes are salt-resistant plants. To adapt to the saline environment, halophytes have evolved varied anatomical features such as a salt gland or bladder, vacuolar compartmentalization, and stomata closure timing. In glycophytes under salt stress, higher lipid peroxidation, impairment of photosynthesis, osmotic stress, and ionic imbalance cause excessive generation of ROS. Perhaps higher accumulation and uncontrollable level of ROS leads to cross-reaction with other vital metabolic pathways and damages macromolecules such as lipids, proteins, and nuclei acids. Whereas in halophytes, ROS are spatial and temporal in nature. Plants with an efficient antioxidant system generally have a higher tolerance against stress. In this chapter, antioxidant defense mechanisms present in glycophytes and halophytes are described using model plants such as *Arabidopsis thaliana* (glycophyte) and *Cakile maritima, Suaeda salsa* L., and *Thellungiella halophila* (halophytes).

P. Soundararajan
Horticulture Major, Division of Applied Life Science (BK21 Plus), Graduate School, Gyeongsang National University, Jinju, South Korea

A. Manivannan
Institute of Agriculture and Life Science, Gyeongsang National University, Jinju, South Korea

B. R. Jeong (✉)
Horticulture Major, Division of Applied Life Science (BK21 Plus), Graduate School, Gyeongsang National University, Jinju, South Korea

Institute of Agriculture and Life Science, Gyeongsang National University, Jinju, South Korea

Research Institute of Life Science, Gyeongsang National University, Jinju, South Korea
e-mail: brjeong@gnu.ac.kr

B. Gul et al. (eds.), *Sabkha Ecosystems*, Tasks for Vegetation Science VI,
https://doi.org/10.1007/978-3-030-04417-6_20

Keywords
Antioxidant · Glycophyte · Halophyte · Reactive oxygen species · Salinity · Stress

Introduction

Salinity is one of the major constraints that affect growth and development of plants. Pitman and Läuchli (2002) mentioned that at least 20% of irrigated lands are affected by salt and in some cases it could be even higher than 50%. He further added losses on agriculture is difficult to estimated as it is substantially increasing. Nonetheless, recently Shrivastava and Kumar (2015) reported that 20–50% of loss in major agricultural crops is due to drought and higher level of salts in soil. Even some river basins become unsuitable for the direct consumption of water by humans due to the higher level of salt (Vengosh 2003). Excessive salt in soil affects its physical and chemical properties in three ways: (i) lower osmotic potential of soil reduces water available for plant, (ii) soil aeration and water permeability is decreased, and (iii) accretion of other toxic ions or a deficiency of essential nutrients for plant. These inhibitory effects disrupt the ecological system. Plant species growing in arid or semi-arid regions are able to use excessive salt as an osmoticum. Rainfall in arid or semi-arid regions is also insufficient to leach salt to a deeper layer in soil (i.e., away from the root zone) (Zhao et al. 2007). High salt content in marshland habitats facilitates the growth of halophytes. Many halophytes require a high level of salt in environment for its growth and development (Wiciarz et al. 2015; Pardo and Quintero 2002).

Generally, high salinity hinders growth and development of plants. During salt stress, damage such as chlorosis and necrosis occurs in the plant (Wiciarz et al. 2015). Major reason for decrease in growth is excessive accumulation of cytosolic sodium (Na^+). Higher Na^+ creates an ionic imbalance in cytoplasm (Zhu 2003; Marschner 1995; Munns and Tester 2008). Most importantly, this ionic and osmotic imbalance leads to oxidative stress, i.e., an enormous accumulation of ROS. Uncontrollable accumulation of ROS increases the peroxidation of lipids in the cell membrane and damages macromolecules such as proteins, lipids, and nucleic acids (Foyer and Noctor 2003, 2005). Survival of the fittest ensures evolutionary changes occur in organisms to develop specific system needed to improve growth and survival under extreme conditions such as cold, heat, drought, and floods. Salt tolerance in plants is possible due to the following mechanisms: delays in germination or growth, exclusion of salt in the root zone, excretion of salt via glands or hairs, compartmentalization, and/or an efficient antioxidant system to alleviate oxidative stress (Hasegawa et al. 2000).

Plants that grow under high salinity conditions are generally classified as halophytes and plants that are sensitive to salt are termed glycophytes. Halophytes are found in the ocean, in coastal areas, mangrove forests, and desert regions as their traits are evolutionarily tailored to resist high salinity. Apart from the physiological adaptation, improved biochemical strategies such as enhanced antioxidants and transporters determine the tolerance against oxidative stress caused under high salinity conditions. In this chapter, we discuss the antioxidant defense systems of glycophytes and halophytes used to scavenge ROS.

Reactive Oxygen Species

Reactive oxygen species (ROS) play an important role in all growth and development stages of plants, from germination to senescence and/or rejuvenation (Bailly 2004; Cheeseman 2007; Dietz et al. 2010; Soundararajan et al. 2016). ROS are generated from metabolically active sites such as mitochondria, chloroplasts, peroxisomes, glyoxysomes, and plasma membranes. Electron transfer chain (ETC) of mitochondria and chloroplast is considered as primary source for the generation of superoxide (O_2^-). In addition, the excitation and transfer of electrons in photosystems I and II releases often release ROS such as O_2^-, singlet oxygen ($^1O_2$), and hydrogen peroxide (H_2O_2). Breakdown of lipids by glycolate oxidase generates O_2^- and H_2O_2. In plants, ROS are produced from the following three most important sites during photosynthetic transport system: S_3 state of water splitting cycle, reducing site of PSII, and acceptor site of PSI (Cheeseman 2007; Dietz et al. 2010). According to Inze (1995) report, PSI is major site for production of O_2^-. In peroxisomes and glyoxysomes, O_2^- and H_2O_2 are rapidly released during catabolism of lipids and purines (Corpas et al. 2001). Various reactions that occur in peroxisomes lead to the generation of ROS as side products. When xanthine is catalyzed by xanthine oxidase into uric acid, this process releases O_2^-. Oxidation of fatty acids by flavin oxidase generates the hydroxyl radical (•OH) and NO. H_2O_2 is generated during the conversion of glycolate into glyoxylate by glyoxylate oxidase (Foyer and Noctor 2003, 2005). Specifically, involvement of cell-wall-dependent peroxidases, oxalate oxidases, and NADPH oxidases in transfer of electron makes plasma membrane as an important site for the synthesis of ROS. Therefore, plasma membrane is also an important site for ROS synthesis, as most cell signaling occurs on cell wall (Chandrakuntal et al. 2010). Bailly et al. (2004) mentioned that activation of NADPH oxidase, amine oxidase, cytochrome p450, cell-wall peroxidase, and germin-like oxalate involved the dispersion of H_2O_2 from cell-to-cell. Transient expression of ROS acts as signaling molecules in various aspects of plant growth (Soundararajan et al. 2016). Most metabolically active sites are a source of ROS. Though antioxidant enzyme activity can be contradictory and it does not always depend on the accumulation of ROS, beneficial (signaling) or deleterious (oxidative stress) actions of ROS are determined by its viscosity on cytoplasm. Role on the developmental stages are determined by localizations of ROS (Leymarie et al. 2012). During the initial period, ROS generated in cytoplasm moves to nucleus and to cell wall for its propagation (Chandrakuntal et al. 2010). Time of production and viscosity of ROS transfer defines its function. Soundararajan et al. (2016) recently mentioned that ROS are not completely quarantined as they have a vital function during each stage of the plant lifecycle. Under an oxidative stress condition, excessively generated ROS causes damage to proteins, lipids, and nucleic acids, and can lead to plant death. Higher plants possess enzymatic and non-enzymatic antioxidant mechanisms to detoxify ROS. This varies between the types of stress, species, and also stages. Therefore, antioxidant enzymes play an important role in maintaining ROS level in plants.

Antioxidant System

Antioxidants are enzymes or components involved in the elimination of free radicals released due to cellular reactions (Foyer and Noctor 2005). Higher antioxidant activity by plants reduces ROS accumulation (Desikan et al. 2001). Adaptation of plants to high salinity levels using ion homeostasis is associated with an adequate antioxidant system and osmolytes. Antioxidants can be divided into two types: enzymatic and non-enzymatic. In the enzymatic antioxidant, major enzymes such as superoxide dismutase (SOD), ascorbate peroxidase (APX), guaiacol/glutathione peroxidase (GPX), catalase (CAT), monodehydroascorbate reductase (MDAR), dehydroascorbate reductase (DHAR), aldehyde dehydrogenases (ALDH), and glutathione reductase (GR) detoxify the excessive levels of ROS groups such as O_2^-, H_2O_2, and •OH. Meanwhile, other low molecular non-enzymatic antioxidants such as proline, ascorbate (AsA), glutathione (GSH), phenolics, flavonoids, etc., can also aid the activity of antioxidant enzymes. Coming to osmolytes, proline, α-tocopherol, GSH, and AsA are the preliminary respondents. For example, in comparison with the salt-sensitive varieties, salt-tolerant varieties of cotton have threefold to fourfold higher levels of α-tocopherol, a lipophilic antioxidant (Gossett et al. 1994). Proline is an active osmolyte involved in balancing the osmotic potential when the Na^+ affects the potassium (K^+) flux. According to

Munns (2005), antioxidants play an important role in maintaining the physiology of plants during salt stress. Antioxidants are essential to uphold the leaves on its active state instead of the low-stomatal conductance seen under stress conditions. As a consequence of the protection of metabolic function, antioxidants prevent premature senescence (James et al. 2002). Therefore, the level of ROS is maintained by the antioxidant system and this is a vital process for metabolic pathways.

Generally, plants with a higher level of antioxidants possess greater resistance against oxidative stress. In most studies, the activity of antioxidants was higher in rice and pea plants (Nor'aini et al. 1997; Hernandez et al. 2000) compared to wheat (Meneguzzo et al. 1999). In all the metabolic process, ROS levels are maintained by the antioxidant system. Changes in the condition and level of ROS such excessive production, excessive accumulation, and cross-reaction with other metabolic pathways result in the deleterious effects of ROS. Dietz et al. (2010) mentioned that redox homeostasis is an important factor for protein oxidation (carbonylation and decarbonylation) and mRNA synthesis (Nystrom 2005). However, activity or expression of antioxidant enzymes are contradictory or not uniform. It is difficult to define the single hypothesis, since responses of antioxidant enzymes are reported differently in same species even on same treatment under different development stages (Shannon et al. 1994). Inconsistency is because the activity of the antioxidant enzymes is interchangeable and depends on its substrate or other enzymes in the metabolic cycle. Inhibition of CAT (3-aminotriazol) under oxidative stress increased activity of APX and glutathione reductase (Kang et al. 1999). In another study, suppression of APX and CAT by transformation of its antisense orientation activated monoascorbate reductase and also an alternative antioxidant system (Rizhsky et al. 2002). Constitutive expression of gene encoding for antioxidants and osmo-protectants determines the capacity of plant to alleviate deleterious effects of high salinity levels.

Glycophytes

Plants that are sensitive to even a slight increase in salinity are generally referred to as glycophytes. Generally, the growth of glycophytes will be retarded in the presence of salt. Most glycophytes lack specialized anatomical features as well as vacuolar compartmentalization. Therefore, concentration of salt ions in the glycophytes is generally lower. Even presence of Na^+ and Cl^- at lower level can cause deleterious effects on growth and development.

Arabidopsis thaliana

Arabidopsis thaliana is one of the most studied model plants due to its small genome and shorter lifecycle (Zhang et al. 2004). This completely sequenced plant is sensitive to high salinity. As the efficiency of transformation is higher in *Arabidopsis* spp., and characterization of most of the genes was carried out, several overexpression and mutant studies were performed on *A. thaliana for* the better understanding of plant model to various stress conditions. Under salinity stress, an enormous amount of ROS generated in plants cause oxidative stress. In addition, there was ionic and osmotic imbalance due to the increase in Na^+ uptake over the K^+ ion (Zhu 2003). Malfunction of electron excitation and recycling in photosynthetic process leak charged/excited electrons. Overexpression of Mn-SOD, an isoform of SOD prominently present in mitochondria, inhibits ROS and increases tolerance against NaCl stress (Wang et al. 2004). Leakage of electrons in the mitochondria occur at three important sites: complex I, complex III, and ubiquinone (Moller 2001). Additionally, transgenic overexpression of Mn-SOD also enhanced the activity and expression of other isomers such as CuZn-SOD in chloroplast and Fe-SOD in plastid (Bowler et al. 1989). SOD is primarily involved in the elimination of O_2^- and conversion to H_2O_2. Enhancement of Mn-SOD stimulated CAT and POD for the reduction of H_2O_2 into H_2O and O_2 (Giannopolitis and Ries 1977).

AsA is one of the most important low molecular weight antioxidants. It acts as the substrate for glutathione, α-tocopherol, and other antioxidant enzymes, specifically APX and GR (Asada 1999; Pignocchi and Foyer 2003). Plant lacking in the stimulated production of AsA failed to overcome oxidative stress (Smirnoff 1996; Smirnoff and Wheeler 2000). A lower CO_2 assimilation rate makes photosystem I sensitive to salt stress (Krall and Edwards 1992). Assimilation of CO_2 is important for production of ATP and NADPH. To reach steady state, NADPH is important for the coupling of electrons. Meanwhile, amplitude and feedback inhibition photosynthetic electron transport is regulated by the proton motive force coupling with electron transport. Dissipation in energy disrupts ATP and NADPH synthesis is unable to couple with CO_2 fixation. Therefore, chlorophyll fluorescence is affected (Krause et al. 1994). Activities of other important antioxidant enzymes such as MDAR and DHAR in the AsA-GSH cycle are also decreased in the *vtc-1* mutant (Huang et al. 2005).

Nicotinamide adenine dinucleotide phosphate-oxidase (NADPH-oxidase) is another important antioxidant enzyme involved in H_2O_2 generation (Miller et al. 2009; Torres et al. 2002). Plasma membrane-localized NADPH oxidase is also known as the respiratory burst oxidase homologue (Rbohs). In *A. thaliana* 10 NADPH oxidases have been identified and characterized as *AtRboh A* to *J* (Fluhr 2009). Transient increases in endogenous H_2O_2 level act as signaling component and early defense response for plant. This induces root hairs and stomatal closure (Foreman et al. 2003). Mutant of *AtRbohD* and *AtRbohF*, main isoforms involved in salt stress resistance, failed in the production of initial H_2O_2 for stress signaling followed by decreased activity of antioxidant enzymes such as SOD, CAT, APX, and GR (Rejeb et al. 2015). Alleviation of phytotoxic effects was disturbed in the NADPH oxidase mutant correlated with a lower expression of CuZn-SOD, an enzyme mainly localized in chloroplast for dismutation of O_2^- into H_2O_2. Similarly, CAT and APX enzymes are required for reduction of H_2O_2 into H_2O and O_2. In *A. thaliana* CAT is encoded by three genes: *CAT1*, *CAT2*, and *CAT3* (Frugoli et al. 1996). Ye et al. (2011) reported that abscisic acid (ABA) is involved in regulation of H_2O_2 as well as CAT for maintaining H_2O_2 lower than the cytotoxic level. Activity of important kinases such as mitogen-activated protein kinase (MAPK) and Ca^{2+}-dependent protein kinase (CPK) are reciprocal towards the antioxidant response for environment stress (Xing et al. 2007; Zhang et al. 2012). Meanwhile, APX is encoded by five genes: *APX1*, *APX2*, *APX3*, *APX4*, and *APX5* (Santos et al. 1996). Interestingly, *APX-1* mutant of *A. thaliana* failed to control higher generation of H_2O_2 (Pnueli et al. 2003). Activity of SOD, CAT, APX, and GR was blocked by the inhibitors of MAPK (Zhang et al. 2006). Increase in cytosolic SOD is correlated with higher activity of APX and GR. Lower *AtRbohD* and *AtRbohF* expression impairs the Ca^{2+} binding (Rejeb et al. 2015), whereas ABA-treatment induced *AtRbohD* and *AtRhohF* along with antioxidant enzyme activity. Reduction in the cytosolic Ca^{2+} ratio failed to activate the antioxidant defense system (Kader and Lindberg 2010). With long-term stress, rather than the higher activity of antioxidant enzymes in the initial stage, after 2 days SOD, CAT, APX, and GR activities were decreased. However, the accumulation of ROS was controlled by the non-enzymatic antioxidants α-tocopherol, glutathione, and AsA (Rejeb et al. 2015). Under salt stress, increased lipid peroxidation and impairment in photosynthetic process cause oxidative stress in glycophytes. Therefore, the level of antioxidant enzyme defense at the present level is not sufficient to prevent excessive accumulation of ROS in glycophytes. Hence, overexpression of antioxidant enzymes is necessary for glycophytes to alleviate the oxidative stress.

Halophytes

Halophytes are adapted to salty environments. Tolerance against salt stress is multifaceted. Most halophytes have specific characteristics enabling them to live in extreme conditions. There are two types of adaptation processes that halophytes use to cope with high salt levels: one is to accumulate the salt and the other is to exclude the salt. Salt glands or bladders in halophytes store the excess salts (Breckle 2002). Another way is to compartmentalize the salt ions into the vacuole and minimize its presence in the cytosol (Maathuis et al. 1992). Many excluders block uptake of Na^+ and Cl^- at the root endodermis. This selective permeability prevents uptake of Na^+ and Cl^-. Pumping out excessive ion is another strategy followed by blocking the entry of salt ions. In some cases, plants accumulate excessive salt in older leaves and defoliate when saturated with salt ions (Gorham 1995).

As mentioned earlier, during exposure to salt, excessive amounts of ROS will be generated. Higher amounts of ROS can lead to cross-reactions and this can affect the whole adaptation physiology. Hence, plants possess an antioxidant mechanism to increase tolerance against salinity (Amor et al. 2007). In this section, the antioxidant mechanism of three important halophytes, *Cakile maritima, Suaeda salsa* L., and *Thellungiella halophilla,* in comparison with *Arabidopsis thaliana,* is covered.

Cakile maritima

Cakile maritima, universally referred to as sea rocket, is a succulent halophyte. This plant grows along the coastal regions of Africa and Europe (Clausing et al. 2000). It also has a notable economic value. It contains about 40% oil with more erucic acid present in dried seeds, which is useful for industrial purposes (Zarrouk et al. 2003).

Debez et al. (2008) showed that *C. maritima* possessed a complex survival strategy for acclimation and adaptation in higher saline regions. They further described how *C. maritima* has adopted higher water and nitrogen use efficiency to keep the PSII active. Longer exposure of *C. maritima* to high saline environment showed its improved antioxidant system to prevent oxidative damage. Ellouzi et al. (2011) reported that in comparison with the *A. thaliana* (glycophyte), *C. maritima* had higher antioxidant activity and preserved ionic balance. Specifically, *A. thaliana* showed visible growth reduction and damage in 100 mM NaCl after 72 h. *C. maritima* grew at 400 mM NaCl even after several days. During initial period *C. maritima* maintained relative water content (RWC) above −0.7 MPa; however, the RWC of *A. thaliana* reached less than −1.6 MPa. This result showed decreased osmotic potential in glycophyte compared to halophyte. Although both plants had higher Na^+ content, *C. maritima* accumulated more Na^+ in shoot compared to root. This demonstrates the detoxification mechanism adopted by *C. maritima* to confine the Na^+ and lower the osmotic potential. Consequently, increased Na^+ is stored in vacuole to prevent accumulation of Na^+ in cytoplasm. This could be due to the activity of Na^+/H^+ antiporter, which catalyzes tonoplast exchange of Na^+ with H^+ (Shi et al. 2003). Induction of Na^+/H^+ is an efficient method for compartmentalization of Na^+ to vacuole from cytoplasm (Munns 2005). Meanwhile, *C. maritima* also increased the uptake of K^+ and Ca^{2+}. The blocking of the binding site of transport systems and the electrostatic binding of Na^+ with the K^+ and Ca^{2+} in the cell wall increased the uptake of potassium and calcium (Aleman et al. 2009; Hernandez et al. 2010). Former ion (K^+) maintains the osmotic potential while latter (Ca^+) protects the cell integrity (Marschner 1995).

Compared with the glycophytes, the accumulation of H_2O_2 and malondialdehyde (MDA) as well as activities of SOD, CAT, and POD were higher in halophytes. Though few studies (Shalata and Neumann 2001; Zhu et al. 2004; Munns and Tester 2008) stated that induction of antioxidant enzymes and resistance against stress are not directly correlated, increased antioxidant enzymes corresponds to the accumulation or transient localization of ROS. Accumulation of ROS leads to deleterious effects. Transiency of ROS has an intermediary role and induces stress-regulatory mechanisms (Foyer and Noctor 2005). Therefore, in *C. maritima* transient presence of H_2O_2 acts as a signaling molecule (Ellouzi et al. 2011). It enhances α-tocopherol, an efficient antioxidant that helps to maintain membrane stability. Tocopherol plays a major role in intracellular signaling and cyclic electron transport around PSII. In *A. thaliana* decreased α-tocopherol was directly linked with higher MDA. Therefore, inhibition of LPO by α-tocopherol maintains membrane integrity and was activated by transient increases in H_2O_2. Meanwhile, in *C. maritima*, an increase in H_2O_2 is directly correlated with SOD. Higher activity of SOD reduced O_2^- to H_2O_2 and O_2. Subsequently, CAT and POD activity was directly correlated with H_2O_2 content. This result illustrates that H_2O_2 acts as a signal for induction of antioxidants in halophytes. Nonetheless, accumulation of H_2O_2 cause irreversible damage in glycophytes.

From the long-term stress study conducted by Amor et al. (2007) on oleaginous halophytes, they stated that the increase in tolerance even after 20 days under 100–200 mM NaCl was directly correlated with the increase in activity of SOD, POD, and AsA-GSH cycle enzymes. However, decreased tolerance at the higher concentration of NaCl, i.e., 400 mM after 10 days, might be due to loss or diminished activities of CAT, POD, DHAR, and GR, which directly decreased AsA and GSH content. Decrease in water absorbance creates ionic imbalance and toxicity. It also leads to malfunction of CO_2 assimilation. Abnormal metabolic changes that occurred under salt stress failed to decompose excessively generated ROS in glycophytes (Ellouzi et al. 2011). An increase in the potential of antioxidants was reported to have a higher capacity to tolerate salt-stress in halophytes. Growth rate of *C. maritima* at 100 mM NaCl was almost equal to the control. However, growth was reduced by approximately 30% without showing any symptoms of necrosis or chlorosis at 200 or 400 mM NaCl. Similarly, MDA, electrolytic leakage, and H_2O_2 content remained constant at 100 mM NaCl. At 400 mM NaCl antioxidant activity decreased over two-fold when compared to control. Activity of SOD increased over time and intensity of NaCl (100–400 mM). However, no changes in CAT were observed until 10 days, and after that it increased. POD and GR activity increased until 10 days and started to decrease thereafter, which correlated with the DHAR. Both AsA and GSH increased with 100 mM and started to decrease with higher concentration of NaCl. Higher activities of enzymatic and non-enzymatic antioxidants were directly related to low solute leakage and membrane damage. Membrane functionalities of halophytes were directly correlated with higher salt tolerance (Chaparzadeh et al. 2004). Lipids in the membrane are considered as the first target of salt stress due to excessive H_2O_2 cause electrolyte leakage. Scavenging of H_2O_2 was brought about by activation of POD and APX in association with O_2^- radical scavenger, SOD. Amor et al. (2007) observed that SOD, APX, and MDHAR showed an immediate response against ROS accumulation. Whereas CAT, POD, DHAR, and GR failed to show such rapid reactions. This variation could be correlated with AsA-GSH cycle enzymes. Hernandez et al. (2000) reported that AsA-GSH enzyme activity was increased in short term whereas in long term their activities increased gradually, which illustrates the role of ROS in signal transduction and expression of antioxidants transcripts. Sometimes, expression of transcripts and activities of antioxidant enzymes can vary based on the turnover of enzymes. Increased GSH/GSSG ratio could indicate overcompensation for the recycling and active form of glutathione. Increase in ASC/DHA is directly correlated with reduction in H_2O_2. Quick response of antioxidant enzymes such as SOD, CAT, POD, APX, and GR and in cases of longer exposure, antioxidant enzyme activity, was compensated for by the low molecular antioxidants and proline component that maintain a low ROS level.

Suaeda salsa **L.**

Suaeda salsa L. generally grows in saline soil. The foliage of *S. salsa* can be used as animal feed. Compartmentalization of Na^+ and Cl^- is highly active in *S. salsa* (Maathuis et al. 1992). This plant is usually grown to remediate the growing medium by removing Na^+ and Cl^- (Maathuis et al. 1992; Zhao 1991). *S. salsa* possesses a high level of betalains, an important free radical scavenger and a natural additive for food, medicines, and cosmetics.

Under treatment of NaCl, decrease in O_2^- was correlated with higher expression of CuZn-SOD

I rather than CuZn-SOD and Fe-SOD I in *S. salsa* High salinity increased the CO_2 assimilation and photosystem II (PSII) phytochemistry (Lu et al. 2003). Qiu-Fang et al. (2005) reported that enhancement of thylakoid-bound SOD isozymes was correlated with the improved tolerance against salt stress. Fe-SOD is necessary for the proper function of PSI. Mutant of Fe-SOD leads to the failure of PSI. Overexpression of Fe-SOD protects chloroplast against oxidative stress. Mn-SOD is more important in protecting chloroplast than Fe-SOD. Excessive accumulation of O_2^- due to the malfunction of PSII electron transport systems alter the chlorophyll fluorescence properties of chloroplast (Asada 1999). Protection of chloroplast in *S. salsa* could be due to either avoidance of excessive O_2^- formation or an efficient antioxidant defense system. In *S. salsa*, Na^+ is present in the vacuole rather than in the cytosol (Ellouzi et al. 2011). After the treatment of NaCl, the Na^+ increased by factors of 4.0 and 10.9 in *S. salsa* without affecting the osmotic and ionic balance. Application of 200 mM NaCl to *S. salsa* increased growth to 93.8% and activities of antioxidant enzymes. No significant changes in Fv/Fm indicated that photo-inhibition had not occurred in 200 mM NaCl treatment. MDA and H_2O_2 levels decreased after 7 days' treatment of 200 mM NaCl. Reduction of oxidative stress parameters was directly correlated with the increased AsA and GSH content. The H_2O_2 is a potential oxidant that is able to diffuse across the membrane, cause oxidation of proteins, inhibit the Calvin cycle, and can lead to •OH generation. The •OH can possibly mutate the DNA strands and alter the synthesis of proteins. Antioxidants such as CAT, APX, and GR involved in scavenging of H_2O_2 are increased in *S. salsa* at 200 mM of NaCl. In particular, the S-APX (around 75% of total APX) and thylakoid bound APX (50% of total chloroplast APX) was higher under saline condition. Similarly, thylakoid-bound and S-GR also increased in the salt-treated *S. salsa* In *S. salsa* GSSG to GSH as a basic requirement for AsA and APX production was highly activated during stress and this increased the turnover of enzymes involved in the AsA-GSH cycle. Level of H_2O_2 present in chloroplast determined the metabolism of plant, and chloroplast-APX is necessary to alleviate the excessive H_2O_2 (Pang et al. 2005). Increased GSH content directly induced the APX.

Noctor and Foyer (1998) reported that APXs are actively involved in scavenging H_2O_2 into water using AsA as an electron donor. There are five distinct APXs present in plants, i.e., the cytosol (cAPX), the microbody (mAPX), the peroxisome (pAPX), the stroma (sAPX) of chloroplasts, and the thylakoid membranes (tAPX) of chloroplasts. Overexpression of *cAPX* and *tAPX* in *A. thaliana* increased tolerance against oxidative stress. Li et al. (2012) reported that *S. salsa* has an effective antioxidant defense system. Chloroplast is susceptible to excessively generated ROS. Therefore, activity of chloroplast APX is important for plants to act against oxidative stress. Transgenic expression of chloroplast APX from *S. salsa*, especially stroma APX (*Ss. sAPX*) for *A. thaliana,* increased tolerance against salt stress up to 120 mM NaCl. Growth, such as root length and chlorophyll content, was increased in the *Ss.sAPX* lines when compared to wild-type plants. Li et al. (2012) suggested that higher expression of sAPX was important for higher tolerance and adaptive response of plants against salt stress. As mentioned earlier, photosynthetic capacity of *Suaeda salsa* L. was not reduced by NaCl (Wang et al. 2004). However, impairment of electron transport in chloroplast and mitochondria increased the ROS. Along with enzymes such as SOD, CAT, and ASA-GSH, components such as carotenoids, AsA, glutathione, and α-tocopherol were also involved in the ROS scavenging. Antioxidant capacity of chloroplast is important in scavenging of ROS as chloroplast is an essential organelle for photosynthesis. In the chloroplast, to scavenge the O_2^-, function of CuZn-SOD is associated with thylakoid membrane of chloroplast, where Fe-SOD is associated with chloroplast stroma. Meanwhile, Mn-SOD played a more important role in alleviation of O_2^- in mitochondria. However, previous reports suggest that SOD isoforms have conflicting results in the reduction of O_2^- (Kwon et al. 2002). Increase in CO_2 assimilation under NaCl treatment in *S. salsa*, was correlated with higher

Fe-SOD and Mn-SOD activity. Oxidative stress leads to the induction of betacyanin. Betalains scavenge the excessive free radicals. Wang et al. (2007) reported that H_2O_2 generation in the root is directly proportional to the betacyanin in *S. salsa* In-built capacity or evolution of an effective enzymatic and non-enzymatic antioxidant system of *S. salsa* L. allows this plant to survive even at 600 mM NaCl.

Thellungiella halophila

Thellungiella halophila is an important halophyte used as a model plant to study salt stress. It is referred as the salt cress (Bressan et al. 2001). The other advantages of using this plant belong to the mustard family are, it has a shorter lifecycle, relatively small genome size, and is closely related to *A. thaliana*. *T. halophila* is generally referred as an extremophile as it can survive under various extreme conditions such as low temperature, drought, and high salt (Inan et al. 2004).

Growth and development of *T. halophila* increased at 100–200 mM NaCl (Inan et al. 2004). Accumulation of Na^+ and Cl^- in glycophytes decreased the physiological function either directly or indirectly. In comparison, *A. thaliana* growth was diminished at 35–50 mM NaCl. *A. thaliana* does not survive at 100 mM NaCl (Nublat et al. 2001). This might be due to the absence of efficient internalization and compartmentalization of salt ions. Accumulation of ions in the leaf apoplast causes leaf dehydration and membrane disruption. Failure of leaf compartmentalization builds ions to a toxic level in cytoplasm and affects essential metabolic pathways (Munns and Passioura 1984; Speer and Kaiser 1991). The ionic and osmotic imbalance collectively disturbs the stomata opening/closure, photosynthesis, and other essential metabolism pathways (Meloni et al. 2003). However, the function of halophytes remains modest in spite of an accumulation of Na^+; rather its growth was improved under Na^+ accumulation (M'rah et al. 2006). Leaf hydration remains constant, even with the addition of NaCl. Though in normal condition chlorophyll content of *T. halophila* is slightly higher than *A. thaliana*, under NaCl treatment chlorophyll content was very much higher than *A. thaliana*. Increase in chlorophyll was directly correlated with an efficient yield of the photosystem. Proper circulation and use of photosynthetic electrons promoted the photosynthetic process. Meanwhile inhibition of the photosystem and the efficiency decreased the net photosynthetic rate and CO_2 assimilation (Krall and Edwards 1992; Stallaert et al. 1995; M'rah et al. 2006). Therefore, higher oxidative stress in *A. thaliana* was associated with increased lipid peroxidation. This leads to cell wall damage and higher accumulation of ROS. Though enhanced antioxidant enzyme activity was reported to protect cell organelles from ROS, most of antioxidant activity was constitutive. In *T. halophila*, SOD activity and CuZn-SOD isoform expression was decreased at higher NaCl concentrations (Sabah et al. 2006), peroxidase especially guaiacol peroxidase was increased, and catalase activity was decreased at 200 mM NaCl (M'rah et al. 2006). In another study, SOD activity was increased twice as much in the halophytes (including *T. halophila*) than the glycophytes (Radyukina et al. 2007).

Kartashov et al. (2008) reported that the enzyme SOD was either stimulated or depressed under the stress condition depending on the plant species, types of stress intensity, stage, duration, and cumulative effect of other enzymes. Activation of SOD occurred normally based on the post-translation of metabolites, isoforms, or its transcripts. Baranenko (2005) suggested that activation of SOD is a prerequisite for the plant under stress conditions. Accumulation of Na^+ in the leaves was higher than the root, which could be efficiently compartmentalized. Meanwhile higher accumulation of Na^+ induced proline biosynthesis (Kartashov et al. 2008). Even under normal conditions, SOD in *T. halophila* is higher than in glycophytes. During salt treatment, activity of SOD, and guaiacol-dependent was induced in *T. halophila*. At the same time, no effect was found in the free peroxidase and covalently-bound peroxidase. As proline is principally involved in the maintenance of osmolytes as well as signaling, the extreme tolerance of *T. halophila* against salt, i.e., to grow in even more than 600 mM NaCl, is directly related to the higher concentration of proline (Radyukina et al. 2007; Kant et al. 2006). The increase in SOD is indirectly proportional to the proline in the glycophytes (Kartashov et al. 2008). However, in a few examples of halophytes, including *T. halophila,* increase in proline can be accompanied with a decrease in SOD activity. Ozturk and Demir (2002) suggested that proline could contribute as the parallel replacement of SOD for scavenging the SOD. According to Radyukina et al. (2007), accumulation of proline is substantially increased even when SOD activity starts to decrease after long-term treatment. This kind of compensation mechanism stimulates diverse and alternative pathways for the utilization of H_2O_2 generated from O_2^-. Increased POD is directly related to the detoxification of accumulated H_2O_2 (Kartashov et al. 2008). Salt tolerance of *T. halophila* largely depends on its biochemical adaptation mechanism. This effect was mainly cumulative and solicited based on the protecting system. As gene expression and the activity/isoforms are varied and complex, differential regulation can be changed according to the time and stage of growth. An efficient antioxidant and physiological protection along with stimulated Na^+/H^+ antiporter prevents plant from desiccation, oxidative damage, and function loss in the photosynthetic systems.

Conclusions

It can be concluded that halophytes have well-developed antioxidant response systems when compared to glycophytes. This is in addition to other unique halophyte features such as salt gland or bladder, compartmentalization of ions, and changes in the stomatal opening period. Evolutionary development of morphology, physiology, organelles, activation of defense-related pathways, and mainly compartmentalization of Na^+ ions of halophytes has created functional differences from glycophytes. Though the conclusion cannot be obtained from only higher or lower activity of antioxidant enzymes, timely or optimal regulation can be seen in halophytes compared to that of glycophytes. Several series of coupled reactions are activated in the halophytes. More efficient antioxidant system has evolved to avoid the oxidative, osmotic, and ionic stress from high salinity. Perhaps the salt treatment enhances growth with more co-operative regulation of H_2O_2 scavengers. Therefore, the well-developed supporting system and utilization of Na^+ ions for metabolism helps the halophytes to grow, complete their lifecycle, and disperse seeds to maintain the ecosystem.

Acknowledgements Prabhakaran Soundararajan and Abinaya Manivannan were supported by a scholarship from the BK21 Plus Program, the Ministry of Education, Republic of Korea.

References

Aleman F, Nieves-Cordones M, Martínez V, Rubio F (2009) Potassium/sodium steady-state homeostasis in *Thellungiella halophila* and *Arabidopsis thaliana* under long-term salinity conditions. Plant Sci 176(6):768–774

Amor NB, Jiménez A, Megdiche W, Lundqvist M, Sevilla F, Abdelly C (2007) Kinetics of the anti-oxidant response to salinity in the halophyte *Cakile maritima.* J Integr Plant Biol 49(7):982–992

Asada K (1999) The water-water cycle in chloroplasts: scavenging of active oxygens and dissipation of excess photons. Annu Rev Plant Biol 50(1):601–639

Bailly C (2004) Active oxygen species and antioxidants in seed biology. Seed Sci Res 14:93–107

Baranenko VV (2005) Superoxide dismutase in plant cells. Tsitologiia 48(6):465–474

Bowler C, Alliotte T, De Loose M, Van Montagu M, Inzé D (1989) The induction of manganese superoxide dismutase in response to stress in *Nicotiana plumbaginifolia.* EMBO J 8(1):31–38

Breckle SW (2002) Salinity, halophytes and salt affected natural ecosystems. In: Salinity: environment-plants-molecules. Springer, Dordrecht, pp 53–77

Bressan RA, Zhang C, Zhang H, Hasegawa PM, Bohnert HJ, Zhu JK (2001) Learning from the Arabidopsis experience. The next gene search paradigm. Plant Physiol 127(4):1354–1360

Chandrakuntal K, Shah AK, Thomas NM, Karthika V, Laloraya M, Kumar PG, Laloraya MM (2010) Blue light exposure targets NADPH oxidase to plasma membrane and nucleus in wheat coleoptiles. J Plant Growth Regul 29(2):232–241

Chaparzadeh N, D'Amico ML, Khavari-Nejad RA, Izzo R, Navari-Izzo F (2004) Antioxidative responses of Calendula officinalis under salinity conditions. Plant Physiol Biochem 42(9):695–701

Cheeseman JM (2007) Hydrogen peroxide and plant stress: a challenging relationship. Plant Stress 1(1):4–15

Clausing G, Vickers K, Kadereit JW (2000) Historical biogeography in a linear system: genetic variation of Sea Rocket (*Cakile maritima*) and Sea Holly (*Eryngium maritimum*) along European coasts. Mol Ecol 9(11):1823–1833

Corpas FJ, Barroso JB, del Rio LA (2001) Peroxisomes as a source of reactive oxygen species and nitric oxide signal molecules in plant cells. Trends Plant Sci 6:145–150

Debez A, Koyro HW, Grignon C, Abdelly C, Huchzermeyer B (2008) Relationship between the photosynthetic activity and the performance of *Cakile maritima* after long-term salt treatment. Physiol Plant 133(2):373–385

Desikan R, Soheila AH, Hancock JT, Neill SJ (2001) Regulation of the Arabidopsis transcriptome by oxidative stress. Plant Physiol 127(1):159–172

Dietz KJ, Jacquot JP, Harris G (2010) Hubs and bottlenecks in plant molecular signalling networks. New Phytol 188(4):919–938

Ellouzi H, Ben Hamed K, Cela J, Munné-Bosch S, Abdelly C (2011) Early effects of salt stress on the physiological and oxidative status of *Cakile maritima* (halophyte) and *Arabidopsis thaliana* (glycophyte). Physiol Plant 142(2):128–143

Fluhr R (2009) Reactive oxygen-generating NADPH oxidases in plants. In: Reactive oxygen species in plant signaling. Springer, Berlin/Heidelberg, pp 1–23

Foreman J, Demidchik V, Bothwell JH, Mylona P, Miedema H, Torres MA, Davies JM (2003) Reactive oxygen species produced by NADPH oxidase regulate plant cell growth. Nature 422(6930):442–446

Foyer CH, Noctor G (2003) Redox sensing and signalling associated with reactive oxygen in chloroplasts, peroxisomes and mitochondria. Physiol Plant 119(3):355–364

Foyer CH, Noctor G (2005) Redox homeostasis and antioxidant signaling: a metabolic interface between stress perception and physiological responses. Plant Cell 17(7):1866–1875

Frugoli JA, Zhong HH, Nuccio ML, McCourt P, McPeek MA, Thomas TL, McClung CR (1996) Catalase is encoded by a multigene family in *Arabidopsis thaliana* (L.) Heynh. Plant Physiol 112(1):327–336

Giannopolitis CN, Ries SK (1977) Superoxide dismutases I. Occurrence in higher plants. Plant Physiol 59(2):309–314

Gorham J (1995) Mechanism of salt tolerance of halophytes. Halophyt Biosal Agric 31

Gossett DR, Millhollon EP, Lucas M (1994) Antioxidant response to NaCl stress in salt-tolerant and saltsensitive cultivars of cotton. Crop Sci 34(3):706–714

Hasegawa PM, Bressan RA, Zhu JK, Bohnert HJ (2000) Plant cellular and molecular responses to high salinity. Annu Rev Plant Biol 51(1):463–499

Hernandez JA, Jimenez A, Mullineaux P, Sevilia F (2000) Tolerance of pea (*Pisum sativum* L.) to long-term salt stress is associated with induction of antioxidant defences. Plant Cell Environ 23(8):853–862

Hernandez M, Fernandez-Garcia N, Diaz-Vivancos P, Olmos E (2010) A different role for hydrogen peroxide and the antioxidative system under short and long salt stress in *Brassica oleracea* roots. J Exp Bot 61(2):521–535

Huang C, He W, Guo J, Chang X, Su P, Zhang L (2005) Increased sensitivity to salt stress in an ascorbate-deficient Arabidopsis mutant. J Exp Bot 56(422):3041–3049

Inan G, Zhang Q, Li P, Wang Z, Cao Z, Zhang H, Shi H (2004) Salt cress. A halophyte and cryophyte Arabidopsis relative model system and its applicability to molecular genetic analyses of growth and development of extremophiles. Plant Physiol 135(3):1718–1737

James RA, Rivelli AR, Munns R, von Caemmerer S (2002) Factors affecting CO2 assimilation, leaf injury and growth in salt-stressed durum wheat. Funct Plant Biol 29(12):1393–1403

Kader MA, Lindberg S (2010) Cytosolic calcium and pH signaling in plants under salinity stress. Plant Signal Behav 5(3):233–238

Kang KS, Lim CJ, Han TJ, Kim JC, Jin CD (1999) Changes in the isozyme composition of antioxidant enzymes in response to aminotriazole in leaves ofarabidopsis thaliana. J Plant Biol 42(3):187–193

Kant S, Kant P, Raveh E, Barak S (2006) Evidence that differential gene expression between the halophyte, *Thellungiella halophila*, and *Arabidopsis thaliana* is responsible for higher levels of the compatible osmolyte proline and tight control of Na^+ uptake in *T. halophila*. Plant Cell Environ 29(7):1220–1234

Kartashov AV, Radyukina NL, Ivanov YV, Pashkovskii PP, Shevyakova NI, Kuznetsov VV (2008) Role of antioxidant systems in wild plant adaptation to salt stress. Russ J Plant Physiol 55(4):463

Krall JP, Edwards GE (1992) Relationship between photosystem II activity and CO2 fixation in leaves. Physiol Plant 86(1):180–187

Krause GH, Foyer CH, Mullineaux PM (1994) The role of oxygen in photoinhibition of photosynthesis. In: Causes of photooxidative stress and amelioration of defense systems in plants. CRC Press, Boca Raton, pp 43–76

Kwon SY, Jeong YJ, Lee HS, Kim JS, Cho KY, Allen RD, Kwak SS (2002) Enhanced tolerances of transgenic tobacco plants expressing both superoxide dismutase and ascorbate peroxidase in chloroplasts against methyl viologen-mediated oxidative stress. Plant Cell Environ 25(7):873–882

Leymarie J, Vitkauskaité G, Hoang HH, Gendreau E, Chazoule V, Meimoun P, Bailly C (2012) Role of reactive oxygen species in the regulation of Arabidopsis seed dormancy. Plant Cell Physiol 53(1):96–106

Li K, Pang CH, Ding F, Sui N, Feng ZT, Wang BS (2012) Overexpression of *Suaeda salsa* stroma ascorbate peroxidase in Arabidopsis chloroplasts enhances salt tolerance of plants. S Afr J Bot 78:235–245

Lu C, Qiu N, Wang B, Zhang J (2003) Salinity treatment shows no effects on photosystem II photochemistry, but increases the resistance of photosystem II to heat stress in halophyte *Suaeda salsa*. J Exp Bot 54(383):851–860

M'rah S, Ouerghi Z, Berthomieu C, Havaux M, Jungas C, Hajji M, Lachaâl M (2006) Effects of NaCl on the growth, ion accumulation and photosynthetic parameters of *Thellungiella halophila*. J Plant Physiol 163(10):1022–1031

Maathuis FJM, Flowers TJ, Yeo AR (1992) Sodium chloride compartmentation in leaf vacuoles of the halophyte *Suaeda maritima* (L.) Dum. and its relation to tonoplast permeability. J Exp Bot 43(9):1219–1223

Marschner H (1995) Mineral nutrition of higher plants, 2nd edn. Academic, San Diego

Meloni DA, Oliva MA, Martinez CA, Cambraia J (2003) Photosynthesis and activity of superoxide dismutase, peroxidase and glutathione reductase in cotton under salt stress. Environ Exp Bot 49(1):69–76

Meneguzzo S, Navam-Izzo F, Izzo R (1999) Antioxidative responses of shoots and roots of wheat to increasing NaCI concentrations. J Plant Physiol 155(2):274–280

Miller G, Schlauch K, Tam R, Cortes D, Torres MA, Shulaev V, Mittler R (2009) The plant NADPH oxidase RBOHD mediates rapid systemic signaling in response to diverse stimuli. Sci Signal 2(84):1–r10

Moller IM (2001) Plant mitochondria and oxidative stress: electron transport, NADPH turnover, and metabolism of reactive oxygen species. Annu Rev Plant Biol 52(1):561–591

Munns R (2005) Genes and salt tolerance: bringing them together. New Phytol 167(3):645–663

Munns R, Passioura JB (1984) Hydraulic resistance of plants. III. Effects of NaCl in barley and lupin. Funct Plant Biol 11(5):351–359

Munns R, Tester M (2008) Mechanisms of salinity tolerance. Annu Rev Plant Biol 59:651–681

Noctor G, Foyer CH (1998) Ascorbate and glutathione: keeping active oxygen under control. Annu Rev Plant Biol 49(1):249–279

Nor'aini MF, Finch RP, Burdon RH (1997) Salinity, oxidative stress and antioxidant responses in shoot cultures of rice. J Exp Bot 48(2):325–331

Nublat A, Desplans J, Casse F, Berthomieu P (2001) sas1, an Arabidopsis mutant overaccumulating sodium in the shoot, shows deficiency in the control of the root radial transport of sodium. Plant Cell 13(1):125–137

Nystrom T (2005) Role of oxidative carbonylation in protein quality control and senescence. EMBO J 24(7):1311–1317

Ozturk L, Demir Y (2002) In vivo and in vitro protective role of proline. Plant Growth Regul 38(3):259–264

Pang CH, Zhang SJ, Gong ZZ, Wang BS (2005) NaCl treatment markedly enhances H_2O_2-scavenging system in leaves of halophyte *Suaeda salsa*. Physiol Plant 125(4):490–499

Pardo JM, Quintero FJ (2002) Plants and sodium ions: keeping company with the enemy. Genome Biol 3(6):1

Pignocchi C, Foyer CH (2003) Apoplastic ascorbate metabolism and its role in the regulation of cell signalling. Curr Opin Plant Biol 6(4):379–389

Pitman MG, Läuchli A (2002) Global impact of salinity and agricultural ecosystems. In: Salinity: environment-plants-molecules. Springer, Dordrecht, pp 3–20

Pnueli L, Liang H, Rozenberg M, Mittler R (2003) Growth suppression, altered stomatal responses, and augmented induction of heat shock proteins in cytosolic ascorbate peroxidase (Apx1) deficient Arabidopsis plants. Plant J 34(2):187–203

Qiu-Fang Z, Yuan-Yuan L, Cai-Hong P, Cong-Ming L, Bao-Shan W (2005) NaCl enhances thylakoid-bound SOD activity in the leaves of C3 halophyte *Suaeda salsa* L. Plant Sci 168(2):423–430

Radyukina NL, Ivanov YV, Kartashov AV, Shevyakova NI, Rakitin VY, Khryanin VN, Kuznetsov VV (2007) Inducible and constitutive mechanisms of salt stress resistance in *Geum urbanum* L. Russ J Plant Physiol 54(5):612–618

Rejeb KB, Benzarti M, Debez A, Bailly C, Savoure A, Abdelly C (2015) NADPH oxidase-dependent H2O2 production is required for salt-induced antioxidant defense in *Arabidopsis thaliana*. J Plant Physiol 174:5–15

Rizhsky L, Liang H, Mittler R (2002) The combined effect of drought stress and heat shock on gene expression in tobacco. Plant Physiol 130(3):1143–1151

Santos M, Gousseau H, Lister C, Foyer C, Creissen G, Mullineaux P (1996) Cytosolic ascorbate peroxidase from *Arabidopsis thaliana* L. is encoded by a small multigene family. Planta 198(1):64–69

Shalata A, Neumann PM (2001) Exogenous ascorbic acid (vitamin C) increases resistance to salt stress and reduces lipid peroxidation. J Exp Bot 52(364):2207–2211

Shannon MC, Grieve CM, Francois LE (1994) Whole-plant response to salinity. In: Plant-environment interactions. Dekker, New York, pp 199–244

Shi H, Lee BH, Wu SJ, Zhu JK (2003) Overexpression of a plasma membrane Na+/H+ antiporter gene improves salt tolerance in *Arabidopsis thaliana*. Nat Biotechnol 21(1):81–85

Shrivastava P, Kumar R (2015) Soil salinity: a serious environmental issue and plant growth promoting bacteria as one of the tools for its alleviation. Saudi J Biol Sci 22(2):123–131

Smirnoff N (1996) Botanical briefing: the function and metabolism of ascorbic acid in plants. Ann Bot 78(6):661–669

Smirnoff N, Wheeler GL (2000) Ascorbic acid in plants: biosynthesis and function. Crit Rev Biochem Mol Biol 35(4):291–314

Soundararajan P, Manivannan A, Jeong BR (2016) Signaling patterns of reactive oxygen species and phytohormones during transition period of quiescent seeds into metabolically active organisms. In: New challenges in seed biology-basic and translational research driving seed technology. InTech, Croatia, pp 75–95

Speer M, Kaiser WM (1991) Ion relations of symplastic and apoplastic space in leaves from *Spinacia olerace*a L. and Pisum sativum L. under salinity. Plant Physiol 97(3):990–997

Stallaert VM, Ducruet JM, Tavernier E, Blein JP (1995) Lipid peroxidation in tobacco leaves treated with the elicitor cryptogein: evaluation by high-temperature thermoluminescence emission and chlorophyll fluorescence. Biochim Biophys Acta Bioenerg 1229(2):290–295

Torres MA, Dangl JL, Jones JD (2002) Arabidopsis gp91phox homologues AtrbohD and AtrbohF are required for accumulation of reactive oxygen intermediates in the plant defense response. Proc Natl Acad Sci 99(1):517–522

Vengosh A (2003) Salinization and saline environments. Treat Geochem 9:333–365

Wang SM, Zhang JL, Flowers TJ (2007) Low-affinity Na+ uptake in the halophyte *Suaeda maritima*. Plant Physiol 145(2):559–571

Wang Y, Ying Y, Chen J, Wang X (2004) Transgenic Arabidopsis overexpressing Mn-SOD enhanced salt-tolerance. Plant Sci 167(4):671–677

Wiciarz M, Gubernator B, Kruk J, Niewiadomska E (2015) Enhanced chloroplastic generation of H2O2 in stress-resistant Thellungiella salsuginea in comparison to Arabidopsis thaliana. Physiol Plant 153(3):467–476

Xing Y, Jia W, Zhang J (2007) AtMEK1 mediates stress-induced gene expression of CAT1 catalase by triggering H2O2 production in Arabidopsis. J Exp Bot 58(11):2969–2981

Ye N, Zhu G, Liu Y, Li Y, Zhang J (2011) ABA controls H_2O_2 accumulation through the induction of OsCATB in rice leaves under water stress. Plant Cell Physiol 52(4):689–698

Zarrouk M, El Almi H, Youssef NB, Sleimi N, Smaoui A, Miled DB, Abdelly C (2003) Lipid composition of seeds of local halophytes: *Cakile maritima*, *Zygophyllum album* and *Crithmum maritimum*. In: Cash crop halophytes: recent studies. Springer, Dordrecht, pp 121–124

Zhang JZ, Creelman RA, Zhu JK (2004) From laboratory to field. Using information from Arabidopsis to engineer salt, cold, and drought tolerance in crops. Plant Physiol 135(2):615–621

Zhang A, Jiang M, Zhang J, Tan M, Hu X (2006) Mitogen-activated protein kinase is involved in abscisic acid-induced antioxidant defense and acts downstream of reactive oxygen species production in leaves of maize plants. Plant Physiol 141(2):475–487

Zhang H, Ni L, Liu Y, Wang Y, Zhang A, Tan M, Jiang M (2012) The C2H2-type zinc finger protein ZFP182 is involved in abscisic acid-induced antioxidant defense in Rice. J Integr Plant Biol 54(7):500–510

Zhao KF (1991) Desalination of saline soils by *Suaeda salsa*. Plant Soil 135:303–305

Zhao J, Ren W, Zhi D, Wang L, Xia G (2007) Arabidopsis DREB1A/CBF3 bestowed transgenic tall fescue increased tolerance to drought stress. Plant Cell Rep 26(9):1521–1528

Zhu JK (2003) Regulation of ion homeostasis under salt stress. Curr Opin Plant Biol 6(5):441–445

Zhu Z, Wei G, Li J, Qian Q, Yu J (2004) Silicon alleviates salt stress and increases antioxidant enzymes activity in leaves of salt-stressed cucumber (*Cucumis sativus* L.). Plant Sci 167(3):527–533

21 Diversity and Distribution of Salt-Tolerant Plants of the United Arab Emirates: Perspectives for Sustainable Utilization and Future Research

Sanjay Gairola, Tamer Mahmoud, and Hatem A. Shabana

Abstract

This chapter provides an overview of halophytic flora of the United Arab Emirates (UAE) and efforts to conserve these genetic resources through seed banking. The halophytic vegetation of UAE is comprised of 74 species, belonging to 19 families, and distributed along the coastal and nearby inland saline areas in the country. The majority of these species belong to Amaranthaceae, followed by Cyperaceae, Poaceae, Fabaceae, and Zygophyllaceae. More information about habitat distribution, salinity tolerance, and other ecophysiological details of these species is needed. The conservation of halophytic species and assessment of viability are underway in SSBH, and so far great progress has been made to conserve their germplasm. The conserved germplasm in seed banks can be used for various research and development purposes including combating desertification, habitat restoration, and promoting greenery in the coastal as well as inland habitats in the hyper-arid deserts. Halophytes of the UAE could be sustainably exploited for an array of uses including chemicals, native plant landscaping, biofuel, medicinal, soil rehabilitation, and a number of other useful purposes.

Keywords

Desert · Halophytes · Conservation · Sharjah Seed Bank and Herbarium · UAE

Introduction

Salt-tolerant or halophytic plants constitute about 2% of the world's flora (Flowers and Colmer 2008). These are highly evolved and specialized plants with well-adapted morphological and physiological characteristics allowing them to proliferate in soils possessing high-salt concentrations (Khan and Duke 2001). Among the 2500–3000 halophytic species present worldwide, 700 are reported from the Mediterranean region (Khan and Duke 2001), and recent new regular discoveries enrich the flora of this region. The largest numbers of halophytes are found in the Amaranthaceae family (Daoud et al. 2015). Saline coastal and inland soils of arid and semi-arid climates are mainly dominated by the halophytic plants, where evaporation greatly exceeds precipitation (Manousaki and Kalogerakis 2011). These plants can grow in a variety of habitats, including salt marshes, estuaries, saline

S. Gairola (✉) · T. Mahmoud · H. A. Shabana
Sharjah Seedbank and Herbarium, Sharjah Research Academy, Sharjah, United Arab Emirates
e-mail: sgairola@sra.ae

B. Gul et al. (eds.), *Sabkha Ecosystems*, Tasks for Vegetation Science VI,
https://doi.org/10.1007/978-3-030-04417-6_21

depressions (sabkhas), and saline inland deserts, and on coastal sand dunes or rocky coasts. It is well understood that the tolerance or response of plants to salt, which is a kind of ecological adaptations, varies between halophytic species.

The halophytic plants of the Arabian Peninsula have great ecological and socioeconomic importance. Several halophytic plants have traditionally been used for medical, nutritional, phytoremediation, rehabilitation of degraded salt marshes, production of oil as biodiesel, and many other purposes; therefore, there is an increasing interest toward these species. Phytochemical investigation of some halophytes of the UAE revealed that these plants are a source of unique active phytochemicals, potentially due to their tolerance to the extreme environmental conditions of the country (Cybulska et al. 2014). Species such as *Cressa cretica*, *Sesuvium* spp., *Suaeda* spp., *Arthrocnemum macrostachyum*, *Halocnemum strobilaceum*, *Limonium axillare*, *Salsola* spp., and *Tamarix aphylla* have great potential to be used for medicinal purposes (e.g., Khan and Qaiser 2006; Weber et al. 2007; Sakkir et al. 2012; Cybulska et al. 2014; Bidak et al. 2015) and *Salicornia* spp. as oilseed crop. So, these valuable biological resources are an integral part of natural heritage, which potentially can yield important ecological goods and services and need sustainable utilization and better conservation efforts.

In general, the coastal and sabkha ecosystems and salt-tolerant plants of the UAE received attention of few researchers (Embabi 1993; Deil and Müller-Hohenstein 1996; Deil 1998; Böer and Gliddon 1998; Evans et al. 2002; Lieth and Menzel 2002; Böer 2002; Brown and Boer 2005; Böer and Saenger 2006; Brown 2006; Karim and Dakheel 2006; Karim et al. 2013), and there is less information with regard to the coastal and inland sabkha vegetation of the northern emirates. The information about halophytic vegetation is necessary for research, coastal management and development, and restoration of degraded arid and saline habitats (Ghazanfar et al. 2014). This chapter deals with halophyte diversity and distribution pattern, germplasm conservation in SSBH, and seed dispersal strategies. Moreover, the importance and utilization potential of these plants and prospective for future research in the UAE are also discussed.

A Brief Overview on the Climate and Halophytic Vegetation of UAE

The UAE has a hot sunny climate characterized by low rainfall which is often sporadic in occurrence, long drought periods, and high temperatures and can be considered as an arid or hyper-arid region based on the geographical region. Due to these unpredictable conditions, such arid regions are particularly susceptible to soil and water salinization. The salinization of natural and agricultural lands is widespread in the country (Karim and Dakheel 2006). The landscape of the UAE is dominated by low-lying, sandy desert, but extensive salt flats (sabkhas) occur in coastal areas of the country (Brown 2006). Beside coastal areas, sabkhas are also found inland, where they are scattered in different areas throughout the country. In coastal sabkha, the tidal action of the sea provides saline water, while in the inland sabkha, the episodic flooding takes place after the infrequent rainfall events, which may sometimes form ephemeral lakes (Brown and Boer 2005). In areas where the water table lies close to the surface, the high concentration of salts on the sabkha surface sometimes prevents the growth of most plant species.

The extremely harsh climatic conditions prevailing in the UAE determine the nature of plant communities and the sequence of their occurrence. The spatial and temporal gradients of salinity and soil moisture can be considered extremely important for plant distribution in hyper-arid deserts of the UAE. Halophytic perennials predominate in the coastal vegetation. The characteristic halophytic vegetation of the UAE mainly consists of *Arthrocnemum macrostachyum*, *Halocnemum strobilaceum*, and *Halopeplis perfoliata*. These species are very common in salt marsh environments, and *Suaeda vermiculata* and *Limonium axillare* are found in more specialized conditions (Brown and Boer 2005). Locally, annual species such as *Suaeda aegyptiaca* and *Tetraena simplex* may be more conspicuous near coastal areas. Avicennia marina is the dominant mangrove, occurring in patches in coastline of the country. Also, *Juncus rigidus* and *Aeluropus* spp. occur predominantly in littoral zones where they mainly form dense, monospecific stands. A halophytic dwarf shrub, *Cornulaca monacantha*, usually grows on dunes close to the coast. Succulence is considered one of the important adaptive structural strategies and characterizes several important species of the UAE sabkhas, and these include *Arthrocnemum macrostachyum*, *Halocnemum strobilaceum*, *Halopeplis perfoliata*, *Haloxylon* spp., *Salicornia* spp., *Salsola* spp., *Suaeda* spp. and *Tetraena* spp. Succulence also occurs in other facultative halophytes, such as *Sesuvium portulacastrum* and *Mesembryanthemum nodiflorum*. Some halophytes have high ecological amplitude and can survive in wide range of habitats. For example, *Frankenia* spp. are found both in coastal salt marsh and inland saline habitats of the country.

Halophytic Plants of UAE

There are approximately 830 plant species in the UAE, and the halophytic vegetation comprises an important group among these plants. A list of halophytic plants was developed from Sharjah Seed Bank and Herbarium (SSBH) database of UAE plants, which is based on extensive field studies and other available literatures (e.g., Böer 2002; Jongbloed et al. 2003; Böer and Saenger 2006; Karim and Dakheel 2006; Brown 2006; Karim and Fawzi 2007; Karim et al. 2013; Ghazanfar et al. 2014). The present study has revealed that the halophytic vegetation of UAE is diverse with 74 species and constitutes about 9% of the flora of the country. These belong to 19 families (Table 21.1). Family Amaranthaceae (26), Cyperaceae (8), Poaceae (5), and Fabaceae (5) have the highest contribution to the total halophytic flora, while other families are represented by less species.

Following Khan and Qaiser (2006), we categorized halophytes into three categories based on their habitats: hydrohalophytes (present in salt marshes), psammophytes (sand loving plants found on littoral or inland sand dunes), and xerohalophytes (salt desert species). Our data showed that among the total halophytic plants, species could be classified as xerohalophytes, hydrohalophytes, and psammophytes, and the rest could be under one or more categories. However, further investigation on habitat distribution will give more insights on halophyte types in the UAE. Among the total halophytic species of the UAE, a total of 18 species can be classified as rare including four new records to the UAE. The genera with the highest number of halophytic species are *Salsola*, *Tetraena*, and *Limonium*. Generally, the inland salty ground supports associations dominated by species of *Salsola*, *Suaeda*, *Tetraena*, *Sesuvium*, and *Mesembryanthemum*.

Distribution Patterns of Halophytic plants

The distributional records of halophytic species in the northern emirates (i.e., Ajman, Fujairah, Ras Al Khaimah, Sharjah, and Umm Al Quwain) and Dubai are provided in Fig. 21.1. However, the future work will include the distribution of the halophytic vegetation throughout the country including the Emirate of Abu Dhabi. The present records are based on SSBH field explorations and available published literatures (e.g., Böer and Gliddon 1998; Böer 2002; Jongbloed et al. 2003; and Karim and Fawzi 2007). However, more surveys in other localities in the country may add wider distributional range for them. The map showed predominant distribution in coastal and inland saline area in the northwest of the area. The eastern coastal region of Sharjah and both inland and coastal regions of Ajman and Umm Al Quwain also showed high concentration of halophytic species due to large mangrove cover and sabkhas in both sides. Distribution maps of some of the potentially useful halophytic species are presented (Fig. 21.2). Populations of *A. macrostachyum* and *H. strobilaceum* are usually found close to the seashores and in the intertidal zones.

Table 21.1 An update list of salt tolerant plants of the UAE

Species	Plant type	Occurrence/ status	Growth form
Aizoaceae			
Mesembryanthemum nodiflorum L.	Xerohalophyte	Rare	Herb
Sesuvium verrucosum Raf.	Psammophyte	Rare	Herb
Amaranthaceae			
Agriophyllum minus Fisch & C.A.Mey	Hydrohalophyte	Common	Herb
Anabasis setifera Moq.	Xerohalophyte	Common	Shrub
Arthrocnemum macrostachyum (Moric.) K.Koch	Hydrohalophyte	Common	Shrub
Atriplex canescens James	Xerohalophyte	Not common	Shrub
Atriplex dimorphostegia Kar. & Kir.	Psammophyte	Not common	Herb
Atriplex leucoclada Boiss.	Xerohalophyte	Common	Shrub
Bassia muricata (L.) Aschers.	Xerohalophyte	Common	Herb
Beta vulgaris L. *subsp maritima* (L.) Arcang	Xerohalophyte	Rare	Herb
Bienertia cycloptera Bunge ex Boiss.	Xerohalophyte	Not common	Herb
Cornulaca aucheri Moq.	Xerohalophyte	Not common	Herb
Cornulaca monacantha Delile	Xerohalophyte	Common	Shrub
Halocnemum strobilaceum (Pall.) M.Bieb.	Hydrohalophyte	Common	Shrub
Halopeplis perfoliata (Forssk.) Bunge ex Asch. & Schweinf.	Hydrohalophyte	Common	Shrublet
Haloxylon persicum Bunge	Psammophyte	Not common	Shrub
Haloxylon salicornicum (Moq.) Boiss.	Xerohalophyte	Common	Shrub
Salicornia europaea L.	Hydrohalophyte	Rare	Herb
Salicornia sp.*	Hydrohalophyte	Rare	Shrublet
Salsola cyclophylla Baker	Xerohalophyte	Rare	Shrublet
Salsola drummondii Ulbr.	Hydrohalophyte	Common	Shrublet
Salsola imbricata Forssk.	Xerohalophyte	Common	Shrub
Salsola rubescens Franch.	Xerohalophyte	Not common	Shrub
Salsola schweinfurthii Solms	Xerohalophyte	Very rare	Shrublet
Seidlitzia rosmarinus Ehrenb. ex Bunge	Xerohalophyte/ Hydrohalophyte	Locally Common	Shrub
Suaeda aegyptiaca (Hasselq.) Zohary	Hydrohalophyte	Not common	Herb
Suaeda maritima (L.) Dumort.	Hydrohalophyte	Rare	Herb
Suaeda vermiculata Forssk. ex J.F.Gmel.	Hydrohalophyte	Common	Shrub
Apiaceae			
Anethum graveolens L.	Xerohalophyte	Rare	Herb
Acanthaceae			
Avicennia marina (Forssk.) Vierh.	Hydrohalophyte	Common	Tree
Caryophyllaceae			
Polycarpaea repens (Forssk.) Aschers & Schweinf.	Psammophyte	Not common	Shrublet
Sphaerocoma aucheri Boiss	Hydrohalophyte	Locally common	Shrublet
Spergularia diandra (Guss.) Heldr. & Sartori	Psammophyte	Rare	Herb
Spergularia marina (L.) Griseb.	Psammophyte	Not Common	Herb
Convolvulaceae			
Cressa cretica L.	Hydrohalophyte	Not common	Herb
Ipomoea pes-caprae (L.) R.Br.	Psammophyte	Rare	Vine
Cyperaceae			
Bolboschoenus maritimus (L.) Palla	Hydrohalophyte	Rare	Herb
Cyperus arenarius Retz.	Hydrohalophyte	Locally common	Sedge
Cyperus conglomeratus Rottb.	Hydrohalophyte	Common	Sedge

(continued)

Table 21.1 (continued)

Species	Plant type	Occurrence/ status	Growth form
Cyperus laevigatus L.	Hydrohalophyte	Locally common	Sedge
Cyperus rotundus L.	Hydrohalophyte	Common	Sedge
Fimbristylis cymosa R.Br.	Hydrohalophyte	Rare	Sedge
Fimbristylis ferruginea (L.) Vahl	Hydrohalophyte	Very rare	Sedge
Schoenus nigricans L.	Hydrohalophyte	Rare	shrub
Fabaceae			
Alhagi graecorum Boiss.	Xerohalophyte/ psammophyte	Not common	Shrub
Lotus garcinii DC.	Xerohalophyte/ psammophyte	Locally common	Shrublet
Lotus halophilus Boiss. & Spruner	Xerohalophyte	Common	Herb
Prosopis cineraria (L.) Druce	Xerohalophyte	Common	Tree
Prosopis juliflora (Sw.) DC.	Xerohalophyte	Common	Small tree
Frankeniaceae			
Frankenia pulverulenta L.	Psammophyte	Locally common	Herb
Hydrocharitaceae			
Halophila stipulacea (Forssk.) Asch.	Hydrohalophyte	Rare	Herb
Juncaceae			
Juncus rigidus Desf.	Hydrohalophyte	Not common	Shrub
Juncus socotranus (Buchenau) Snogerup	Hydrohalophyte	Not common	Shrub
Plumbaginaceae			
Limonium axillare (Forssk.) Kuntze	Hydrohalophyte	Common	Shrublet
Limonium carnosum Kuntze	Hydrohalophyte	Not common	Subshrub
Limonium stocksii Kuntze	Hydrohalophyte	Rare	Shrub
Poaceae			
Aeluropus lagopoides (L.) Thw.	Hydrohalophyte	Common	Grass
Halopyrum mucronatum (L.) Stapf	Psammophyte	Locally common	Grass
Phragmites australis (Cav.) Trin. ex Steud.	Hydrohalophyte	Locally common	Reed
Sporobolus ioclados (Trin.) Nees	Psammophyte	Common	Grass
Sporobolus spicatus (Vahl) Kunth	Psammophyte	Common	Grass
Portulacaceae			
Portulaca oleracea L.	Xerohalophyte	Common	Herb
Portulaca quadrifida L.	Xerohalophyte	Not common	Herb
Rhizophoraceae			
Rhizophora mucronata Lamk.	Hydrohalophytes	rare	Tree
Scrophulariaceae			
Cistanche tubulosa (Schenk) Hook.f. Parasite on halophytes	Psammophyte/ Xerohalophyte	Not common	Herb
Tamaricaceae			
Tamarix aphylla (L.) H.Karst.	Xerohalophyte	Not common	Small tree
Tamarix nilotica (Ehrenb.) Bunge	Xerohalophyte	Common	Small tree
Tamarix aucheriana (Decne.) Baum	Xerohalophyte	Common	Shrub
Verbenaceae			
Phyla nodiflora (L.) Greene	Hydrohalophyte	Not common	Herb

(continued)

Table 21.1 (continued)

Species	Plant type	Occurrence/ status	Growth form
Zygophyllaceae			
Tetraena boulosii A.Hosny	Chamaephyte	Rare	Shrublet
Tetraena mandavillei Hadidi	Xerohalophyte	Common	Shrub
Tetraena qatarense Hadidi	Xerohalophyte	Not common	Shrublet
Tetraena simplex L.	Xerohalophyte	Common	Herb

Asterisk represents new plant records for UAE collected by SSBH during 2016. Species name is under confirmation through DNA analysis

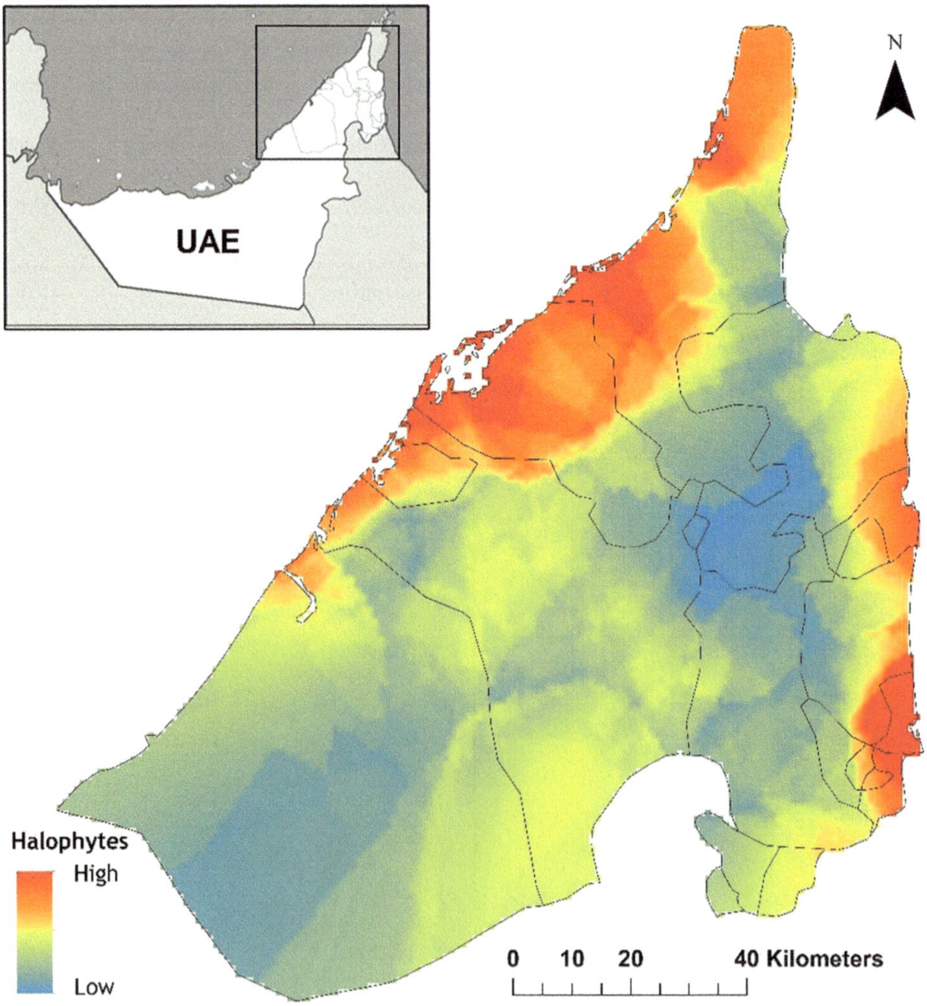

Fig. 21.1 Map showing distribution of halophytic plants in northern emirates

While *A. marina* and *Salicornia* sp. are distributed mainly in the tidal zones, *H. salicornicum* is widely distributed in different habitats in the region and shows wide ecological amplitude. *H. salicornicum* may have benefited somewhat over the recent decades as a result of its less palatability which plays an important role in the abundance of this species (author's personal

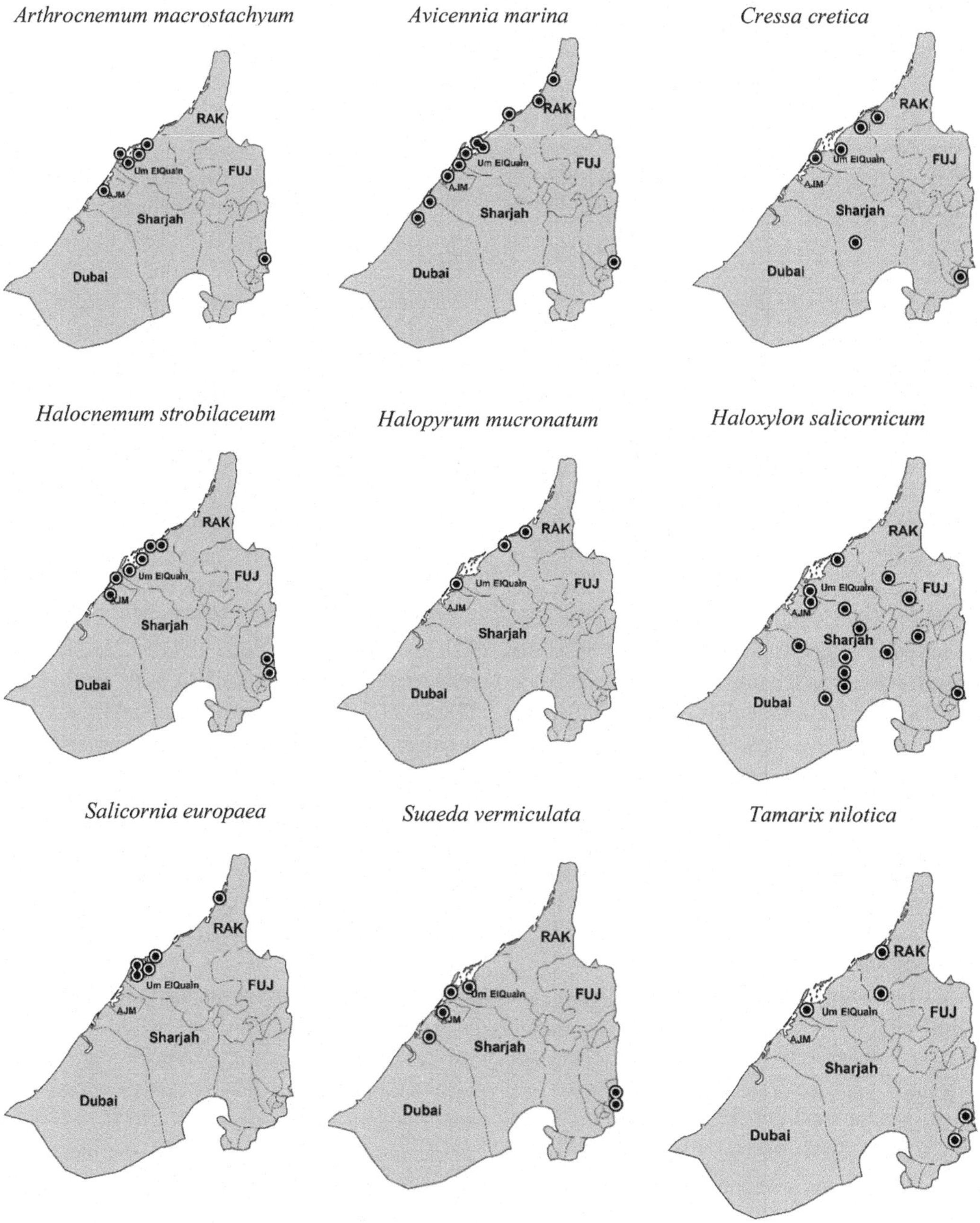

Fig. 21.2 Updated distributional records of some of the potentially useful halophytic species

observations). Inland sabkha develops frequently in the interdunes, often less vegetation, and sometimes without vegetation. In these habitats, species such as *Cressa cretica*, *Tamarix nilotica*, *Suaeda vermiculata*, *Halocnemum strobilaceum*, *Halopyrum mucronatum*, and other characteristic species are often found in various densities. The diversity and distributions of inland halophytic species are correlated, most probably, with the presence of salty soil and the vicinity of the salt marsh and in the interdunes. Such distributional records would form baseline for comparing past and future changes in the distribution of species in any region.

In arid and semiarid regions, salinity is closely linked to lowlands or depressions (Badreldin et al. 2015). In general, it is recognized that salinity, tidal range altitude, and topography are the main factors affecting species distribution in saline environments and the local environment could be decisive for vegetation distribution. In saline habitats, salt marshes are among the most abundant, fertile, and accessible habitats on earth and therefore are highly threatened by human activities (industrial pollution, urbanization, agriculture, etc.), which have damaged many existing salt marshes in the world (Gedan et al. 2009; Al Hassan et al. 2016). The main threats to the present state of biodiversity in the Arab Gulf region are habitat destruction and fragmentation and the emergent impact of climate change that forebodes unprecedented risk and rates of endangerment of important species (El-Keblawy 2014a). Besides, the problems of salinity have been exacerbated in recent decades due to a more widespread use of irrigation water and inappropriate land use. Many of the natural habitats in the Arabian Gulf region are threatened both naturally and anthropologically (Böer 1998; Tourenq and Launay 2008).

Seed Dispersal Strategies of Some Halophytes

In harsh and unpredictable condition of desert environment, plants have evolved different dispersal strategies Shabana et al. (2018). In this chapter, the characteristics of diaspore dispersal structure and dispersal mechanisms of some common halophytic species are presented (Table 21.2). The data was collected from the available literatures and includes authors' field observations. Many halophytic species produce an enormous amount of highly dispersible diaspores every year, most probably to ensure the regeneration in the harsh and stressful desert conditions. Well-developed dispersal structures are reported more common in perennial than in annual species (Werner 1979). We found that family Amaranthaceae and Zygophyllaceae have the maximum average value of diaspore size. Among 19 species, 84% are reported to be dispersed by fruits; some of them have predominant utricles, while 16% of the species have seed as a dispersal unit. Among 19 studied species, *Tamarix aphylla* has the smallest diaspore unit (0.05 cm), while the biggest unit was recorded for *Salsola drummondii* (0.7 cm). The size variation of diaspores may reflect differences in plant recruitment strategy. Studies show that the big healthy seeds of *A. setifera* that matured in a stress-free environment germinate faster than seeds that matured under stressful high-salinity and hyper-arid conditions (El-Keblawy et al. 2016). Also, Telenius and Torstensson (1991) reported that differences in seed size are always associated with differences in germination rate and establishment.

The variations in diaspores traits of studied species may help them to adapt to harsh unpredictable desert environments (Liu et al. 2014). Diaspores of 7 species (36.5%) had appendages like wings or spongy perianth/bracts and pappus; however, 12 species (63.5%) lack appendages (Table 21.2). The restricted dispersal was predominant (68.4%), and therefore semachory was the main dispersal mode (47.4%) followed by anemo-meteochory (31.6%). Similar patterns were observed by Shabana et al. (2018) for hyper-arid desert species of the UAE. Restricted dispersal of the desert plants could help seeds to stay near the mother plants, "safe sites" for seedling establishment (Shabana et al. (2018). It is reported that the dispersal ability of halophyte diaspores may enhance their ability in exploring suitable habitats or safe sites for germination and establishment in the unpredictable and stressful desert environments (El-Keblawy 2014b). Hence, more studies on dispersal ecology of desert plants will advance the understanding of plant population and community dynamics in such environments.

Germplasm Conservation Through Seed Banking

Seeds are convenient for storage in ex situ gene banks for long term. The Arabian Gulf region is known to have a comprehensive and rich endowment of unique and genetically diverse plant

Table 21.2 Diaspore characteristics and dispersal strategies of some halophytes of the UAE

Species	Plant type	Diaspore type	Spatial dispersal	Diaspore size (cm)	Appendages	Dispersal mode
Amaranthaceae						
Anabasis setifera	W	Utricle	DAV	0.5	Wings	Anemo-meteochory
Arthrocnemum macrostachyum	W	Utricle	RD	0.3	Absence	Semachory
Atriplex leucoclada	W	Utricle	RD	0.4	Bracts	Semachory
Cornulaca monacantha	W	Utricle	DAV	0.3	Absence	Anemo-meteochory
Halocnemum strobilaceum	W	Utricle	RD	0.15	Absence	Barochory
Halopeplis perfoliata	W	Fleshy capsule	RD	0.3	Absence	Barochory
Salsola drummondii	W	Utricle	DAV	0.7	Wings	Anemo-meteochory
Suaeda aegyptiaca	H	Utricle (fruiting spongy perianth)	RD	0.3	Spongy perianth	Barochory
Suaeda vermiculata	W	Fleshy capsule	DAV	0.3	Absence	Anemo-meteochory
Convolvulaceae						
Cressa cretica	W	Seed	RD	0.25	Absence	Semachory
Juncaceae						
Juncus rigidus	G	Seed	RD	0.1	Absence	Semachory
Molluginaceae						
Limeum indicum	H	Mericarp	RD	0.2	Absence	Semachory
Plumbaginaceae						
Limonium carnosum	W	Utricle	RD	0.5	Absence	Semachory
Poaceae						
Aeluropus lagopoides	G	Caryopsis without appendages	RD	0.12	Absence	Semachory
Halopyrum mucronatum	G	Caryopsis with appendages (short)	RD	0.1	Short appendages	Semachory
Sporobolus ioclados	G	Caryopsis without appendages	RD	0.08	Absence	Semachory
Tamaricaceae						
Tamarix aphylla	W	Seed	DAV	0.05	Pappus	Anemo-meteochory
Zygophyllaceae						
Tetraena mandavillei	W	Capsule	RD	0.5	Absence	Barochory
Tetraena simplex	H	Capsule	DAV	0.2	Wings	Anemo-meteochory

H herbaceous, *G* graminoids, *W* woody, *RD* restricted spatial dispersal, *DAV* developed abiotic vectors

genetic resources. At present, conservation of these genetic resources is a major concern. In the UAE, the conservation of plant genetic resources including the halophytic species is underway in the SSBH (Gairola et al. 2013). The stored germplasm of halophytic species can serve as a possible repository of useful stress tolerance traits and new traits for varietal improvement in future. Among the conserved seed accessions, ca. 215 belong to halophytes (Fig. 21.3), and some of them are of rare halophytes. In the SSBH, families with high-seed accessions include Amaranthaceae, Poaceae, Fabaceae, Cyperaceae, and Aizoaceae. It is important that the seeds collected and stored in seed banks must be of high quality and at maximum viability. So, periodic seed viability checks of the stored collections are important in order to improve the management of the seed bank.

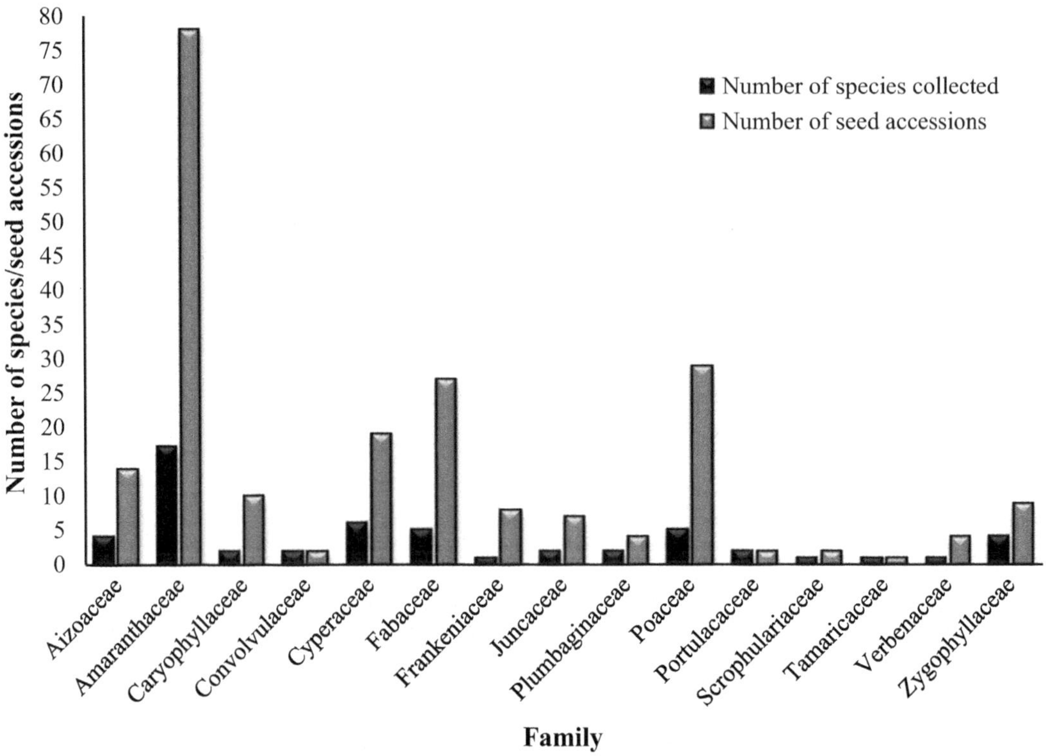

Fig. 21.3 Graph showing seed accessions of halophytes conserved at the Sharjah Seed Bank

We have tested the viability of stored seed accessions of *Anabasis setifera*, *A. macrostachyum*, *Bassia muricata*, *H. strobilaceum*, *H. perfoliata*, *Salsola drummondii*, *S. imbricata*, *S. rubescens*, *Suaeda aegyptiaca*, and *Suaeda vermiculata*. There is limited information on germination responses of seeds of these species after drying at 15 °C and 15% RH and storage at −18 °C in seed bank. The germination trials were carried out under controlled conditions of light and temperature. From the germination results obtained after 3 years in storage, some accessions showed increased germination percentages from initial values, some showed a decrease in percentage germination, and others show no significant change (manuscript under review). The halophytes seed stored in the SSBH maintained their viability after 3 years of storage at −18 °C, and some of the accessions had germination percentages of ~65% or higher. Further studies from representative accession of different species are underway which would provide optimal conditions for their long-term storage standardizing the seed bank protocol. In summary, testing the viability of stored samples through germination is essential for the maintenance of seed bank as the results obtained through germination trials may suggest the need for recollection or regeneration. The stored seed germplasm can potentially utilized for basic or applied research, restoration and species reintroduction program, native plants landscaping, and any other applications.

Prospective for Sustainable Utilization

Desert halophytes play an important role in providing a range of ecosystem services, support biodiversity, and have great utilization potentials. This promising group of plants has various applications in different environmental contexts due to their special physiological characteristics and biochemical composition (Buhmann and Papenbrock 2013). At present, the potential utilization and economic consideration of halophytes are receiving increased attention particularly in arid regions, where salinity problems are very crucial. Based on her extensive research experience of the Arabian plants, Ghazanfar et al. (2014) highlighted the utilization potential of halophytes for restoration of degraded saline locations, coastal stabilization, production of cheap biomass for renewable energy, cash crops for fodder, medicine, building materials, and sequestering CO_2. Therefore, there is a need to identify promising plants and assess their potentials.

The potential of UAE halophytes as a source of phytochemical and biofuel production was highlighted recently (Cybulska et al. 2014). In the UAE, current efforts are underway to develop halophytes as crops for seawater irrigation. Among other plants, *S. portulacastrum* is an important medicinal plant which exhibits remarkable antioxidant, antibacterial, and antifungal properties (Magwa et al. 2006). Similarly, the possible use of *S. vermiculata* in nutraceuticals can be due to edibility containing free radical scavenging anti-inflammatory compounds (Cybulska et al. 2014). The high oil content in the seeds of *Salicornia* spp., *Suaeda fruticosa*, and *Arthrocnemum* spp. makes these species a highly promising oilseed crop (Swingle et al. 1996; Weber et al. 2007; Glenn et al. 2013; Cybulska et al. 2014). In UAE, one of the latest developments in halophytes research is the aviation biofuel project of the Masdar Institute of Science and Technology. The project combines an integrated system of aquaculture, halo-agriculture, and mangrove silviculture to produce sustainable biofuels for aviation and other by-products such as seafood (https://sbrc.masdar.ac.ae/index.php/projects/seas/item/76-the-seawater-energy-and-agriculture-system).

Using native halophytes in landscaping is beneficial as these plants are adapted to the climate, geography, and hydrology and evolved to local conditions of a particular area over thousands of years (El-Keblawy et al. 2015). For example, species like *Halocnemum* sp., *Limonium* spp., and *Tamarix* spp. have potential to be used in embellishment and landscaping in UAE. These species will not require irrigation with freshwater and eventually will help in the sustainable use of water resources. The use of native plants in phytoremediation is always preferred because it provides advantages over other introduced species and helps re-establish the heritage of flora lost through disturbances. Further, the phytoextraction or phytostabilization potential of halophytic plants such as *Atriplex halimus* (Manousaki and Kalogerakis 2009; Nedjimi and Daoud 2009), *Mesembryanthemum crystallinum* (Ghnaya et al. 2007), *S. portulacastrum* (Ghnaya et al. 2007), and *Tamarix smyrnensis* (Kadukova et al. 2008) is well reported; however, they should be tested in the field conditions to determine such potential (Gairola et al. 2015).

Research Prospective

In the UAE, studies on habitat distribution, ecological amplitudes, and potential economic usages of halophytic flora need to be attempted in detail. Further, research on halophytes' responses to salt stress has indeed increased in recent years and requires information at the molecular, biochemical, and physiological bases of their tolerance to high soil salinity (Gil et al. 2014). It is also important to explore the morpho-anatomical adaptations (Grigore et al. 2014), structure and function of salt glands, nature of secreted material, mechanism(s) of ion secretion, and physiological significance of the halophytic species of the Arabian Peninsula. The halophytes that are reported as better salt/metal accumulators should be tested in the field to see if they are ideal for widespread use in the region. It has been emphasized that the halophytic plants present an emerging trend in phytoremediation research and are expected to receive considerable attention in the near future (Shabala 2013). It was observed that morphs with different foliage colors in some halophytes seems to be an adaptation strategy which increases the possibility of establishment in unpredictable environments by producing seeds and differs in germination requirements and salinity tolerance (El-Keblawy et al. 2013). So, the apparent flower color polymorphisms in wind-pollinated species such as *S. rubescens*, *H. salicornicum*, and *A. setifera* need to be studied to ascertain the role of color variation. The coastal ecosystems are one of the unique ecosystems and should be efficiently managed to conserve their threatened and unique biodiversity elements. Genetic resource conservation through seed banking is necessary to preserve the biodiversity of these important ecosystems and possible use in habitat restoration.

Conclusions

Diversity and distribution of the halophytic flora of the UAE and ongoing efforts to conserve these genetic resources through seed banking were presented. Further assessment and extensive information about the habitat distribution and salinity tolerance of these promising plants are crucial to monitor their populations and potential for sustainable utilization. Halophytes provide valuable ecosystem goods and services, and some species has great utilization potential for different purposes. The germplasm conservation of halophytes is underway in SSBH, and so far great progress has been made. The germination test/viability status results can also indicate how often a conservation collection should be replenished or regenerated. The seed dispersal strategies of some halophytes are highlighted in this chapter because the current interest in studying seed dispersal in spatial ecology is growing as these processes might determine the spatial structure of plant populations in unpredictable desert ecosystems. Future research on habitat ecology of halophytic plants, distribution pattern, potential uses, and their responses to salt stress is needed to generate information on the molecular, biochemical, and physiological bases of their tolerance to high soil salinity.

Acknowledgments We thank Dr. Amr Abdel-Hamid, Director General of the Sharjah Research Academy, for the encouragement and support. Thanks are also to Prof. Ali El-Keblawy (University of Sharjah) for his valuable comments/suggestions that greatly improved this book chapter. Mohammed Hasan and Mohammed Fiaz of SSBH are thanked for their help during field collections and seed separation. Reena Cordeiro provided valuable assistance at various stages of preparation of this chapter, and we sincerely thank her.

References

Al Hassan M, Chaura J, López-Gresa MP, Borsai O, Daniso E, Donat-Torres MP, Mayoral O, Vicente O, Boscaiu M (2016) Native-invasive plants vs. halophytes in Mediterranean salt marshes: stress tolerance mechanisms in two related species. Front Plant Sci 7:473

Badreldin N, Uria-Diez J, Mateu J, Youssef A, Stal C, El-Bana M, Magdy A, Goossens R (2015) A spatial pattern analysis of the halophytic species distribution in an arid coastal environment. Environ Monit Assess 187(5):1–5

Bidak ML, Kamal SA, Halmy MW, Heneidy SZ (2015) Goods and services provided by native plants in desert ecosystems: examples from the northwestern coastal desert of Egypt. Glob Ecol Conserv 3:433–447

Böer B (1998) Anthropogenic factors and their potential impacts on the sustainable development of Abu Dhabi's terrestrial biological resources. Int J Sust Dev World 5:125–135

Böer B (2002) The coastal and sabkha flora of the United Arab Emirates, Sabkha ecosystems 1. Springer, Dordrecht, pp 303–309

Böer B, Gliddon D (1998) Mapping of coastal ecosystems and halophytes (case study of Abu Dhabi, United Arab Emirates). Mar Freshw Res 49(4):297–301

Böer B, Saenger P (2006) The biogeography of the coastal vegetation of the Abu Dhabi gulf Coast. In: Khan MA, Böer B, Kust BS, Barth HJ (eds) Sabkha ecosystems: west and Central Asia. Springer, Dordrecht, pp 31–36

Brown G (2006) The sabkha vegetation of the United Arab Emirates. In: Khan MA, Böer B, Kust BS, Barth HJ (eds) Sabkha ecosystems: west and Central Asia. Springer, Dordrecht, pp 37–51

Brown G, Boer B (2005) Terrestrial habitats. In: Hellyer P, Aspinall S (eds) The Emirates: a natural history. Trident Press Limited, London, pp 95–107

Buhmann A, Papenbrock J (2013) Biofiltering of aquaculture effluents by halophytic plants: basic principles, current uses and future perspectives. Environ Exp Bot 92:122–133

Cybulska I, Brudecki G, Alassali A, Thomsen M, Brown JJ (2014) Phytochemical composition of some common coastal halophytes of the United Arab Emirates. Emir J Food Agric 26(12):1046–1056

Daoud S, Elbrik K, Tachbibi N, Bouqbis L, Brakez M, Harrouni MC (2015) The potential use of halophytes for the development of marginal dry areas in Morocco. In: Khan MA, Ozturk M, Gul B, Ahmed MZ (eds) Halophytes for food security in dray lands. Elsevier Inc, pp 141–156

Deil U (1998) Coastal and Sabkha vegetation. In: Ghazanfar SA, Fisher M (eds) Vegetation of the

Arabian Peninsula. Kluwer Academic Press, Dordrecht, pp 209–228

Deil U, Müller-Hohenstein K (1996) An outline of the vegetation of Dubai (UAE). Verhandlungen-Gesellschaft Fur Okologie 25:77–96

El-Keblawy A (2014a) Impact of climate change on biodiversity loss and extinction of endemic plants of arid land mountains. J Biodiver Endanger Species 22

El-Keblawy A (2014b) Effects of seed storage on germination of desert halophytes with transient seed bank. In: Khan MA, Böer B, Öztürk M, Al Abdessalaam TZ, Clüsener-Godt M, Gul B (eds) Sabkha ecosystems: cash crop halophyte and biodiversity conservation. Springer Netherlands, Dordrecht, pp 93–103

El-Keblawy AA, Bhatt A, Gairola S (2013) Perianth colour affects germination behaviour in wind-pollinated *Salsola rubescens* in Arabian deserts. Botany 92(1):69–75

El-Keblawy A, Bhatt A, Gairola S (2015) Developing sea-watered landscapes: a potential way to reduce stress on freshwater resources. Curr Sci 108(10):1773–1774

El-Keblawy A, Gairola S, Bhatt A (2016) Maternal salinity environment affects salt tolerance during germination in *Anabasis setifera*: a facultative desert halophyte. J Arid Land 8(2):254–263

Embabi NS (1993) Environmental aspects of geographical distribution of mangrove in the United Arab Emirates. In: Lieth H, Al Masoom AA (eds) Towards the rational use of high salinity tolerant plants. Springer, Dordrecht, pp 45–58

Evans G, Kirkham A, Carter RA (2002) Quaternary development of the United Arab Emirates coast: new evidence from Marawah Island, Abu Dhabi. Geoarabia-Manama 7:441–458

Flowers TJ, Colmer TD (2008) Salinity tolerance in halophytes. New Phytol 179(4):945–963

Gairola S, Mahmoud T, Bhatt A, El-Keblawy A (2013) Importance of seed banking and herbarium collections in biodiversity conservation and research: a new initiative in the United Arab Emirates. Curr Sci 105(8):1048–1050

Gairola S, Bhatt A, El-Keblawy A (2015) A perspective on potential use of halophytes for reclamation of salt-affected lands. Wulfenia J 22(1):88–97

Gedan KB, Silliman BR, Bertness MD (2009) Centuries of human-driven change in salt marsh ecosystems. Mar Sci 1:117–141

Ghazanfar SA, Altundag E, Yaprak AE, Osborne J, Tug GN, Vural M (2014) Halophytes of Southwest Asia. In: Khan MA, Böer B, Öztürk M, Al Abdessalaam TZ, Clüsener-Godt M, Gul B (eds) Sabkha ecosystems: cash crop halophyte and biodiversity conservation. Springer, Dordrecht, pp 105–133

Ghnaya T, Slama I, Messedi D, Grignon C, Ghorbel MH, Abdelly C (2007) Effects of Cd on K, Ca and N uptake in two halophytes *Sesuvium portulacastrum* and *Mesembryanthemum crystallinum*: consequences on growth. Chemosphere 67:72–79

Gil R, Bautista I, Boscaiu M, Lidón A, Wankhade S, Sánchez H, Llinares J, Vicente O (2014) Responses of five Mediterranean halophytes to seasonal changes in environmental conditions. AoB Plants 6:plu049

Glenn EP, Andaya T, Chaturvedi R, Martinez-Garciaa R, Pearlsteina S, Soliza D, Felgera RS (2013) Three halophytes for saline-water agriculture: an oilseed, a forage and a grain crop. Environ Exp Bot 92:110–121

Grigore MN, Ivanescu L, Toma C (2014) General morphological and anatomical adaptations in halophytes. In: Grigore MN, Ivanescu L, Toma C (eds) Halophytes: an integrative anatomical study. Springer, Cham, pp 33–37

Jongbloed M, Feulner G, Böer B, Western AR (2003) The comprehensive guide to the wild flowers of the United Arab Emirates. Environmental Research and Wildlife Development Agency, Abu Dhabi

Kadukova J, Manousaki E, Kalogerakis N (2008) Pb and Cd accumulation and phyto-excretion by salt cedar (*Tamarix smyrnensis* Bunge). Int J Phytorem 10(1):31–46

Karim FM, Dakheel AJ (2006) Salt-tolerant plants of the United Arab Emirates. ICBA, Dubai

Karim FM, Fawzi NM (2007) Flora of the United Arab Emirates. United Arab Emirates University, Al Ain

Karim FM, Dakheel AJ, Rao NK (2013) Salt-tolerant plants of the United Arab Emirates. ICBA, Dubai

Khan MA, Duke NC (2001) Halophytes – a resource for the future. Wetl Ecol Manag 9(6):455–456

Khan MA, Qaiser M (2006) Halophytes of Pakistan: characteristics, distribution and potential economic usages. In: Khan MA, Böer B, Kust BS, Barth HJ (eds) Sabkha ecosystems: west and Central Asia. Springer, Dordrecht, pp 129–153

Lieth H, Menzel U (2002) Sabkhat in coastal areas with special emphasis on the UAE. In: Barth HJ, Böer B (eds) Sabkha ecosystems: the Arabian Peninsula and adjacent countries. Kluwer Academic Publishing, Dordrecht, pp 203–214

Liu HL, Zhang DY, Duan SM, Wang XY, Song MF (2014) The relationship between diaspore characteristics with phylogeny, life history traits, and their ecological adaptation of 150 species from the cold desert of northwest China. Sci Wor J 2014: 149–168.

Magwa ML, Gundidza M, Gweru N, Humphrey G (2006) Chemical composition and biological activities of essential oil from the leaves of *Sesuvium portulacastrum*. J Ethnopharmacol 103(1):85–89

Manousaki E, Kalogerakis N (2009) Phytoextraction of Pb and Cd by the Mediterranean saltbush (Atriplex halimus L.): metal uptake in relation to salinity. Environ Sci Pollut Res 16:844–854

Manousaki E, Kalogerakis N (2011) Halophytes-an emerging trend in phytoremediation. Int J Phytoremediation 13(10):959–969

Nedjimi B, Daoud Y (2009) Cadmium accumulation in *Atriplex halimus* subsp. *schweinfurthii* and its influence on growth, proline, root hydraulic conductivity and nutrient uptake. Flora 204:316–324

Sakkir S, Kabshawi M, Mehairbi M (2012) Medicinal plants diversity and their conservation status in the United Arab Emirates (UAE). J Med Plants Res 6(7):1304–1322

Shabala S (2013) Learning from halophytes: physiological basis and strategies to improve abiotic stress tolerance in crops. Ann Bot 112:1209–1221

Shabana HA, Navarro T, El-Keblawy A (2018) Dispersal traits in the hyper-arid hot desert of the United Arab Emirates. Plant Ecol Evol 151:194–208

Swingle RS, Glenn EP, Squires V (1996) Growth performance of lambs fed mixed diets containing halophyte ingredients. Anim Feed Sci Technol 63(1):137–148

Telenius A, Torstensson P (1991) Seed wings in relation to seed size in the genus *Spergularia*. Oikos 1:216–222

Tourenq C, Launay F (2008) Challenges facing biodiversity in the United Arab Emirates. Manage Environ Qual Int J 19(3):283–304

Weber DJ, Ansari R, Gul B, Khan MA (2007) Potential of halophytes as source of edible oil. J Arid Environ 68:315–321

Werner PA (1979) Competition and coexistence of similar species. In: Solbrig OT, Jain S, Johnson GB, Raven PH (eds) Topics in plant population biology. Macmillan Education UK, London, pp 287–310

22 Framework for Rapid Evaluation of a Mangrove Restoration Site: A Case Study from Indian Sundarban

Sourabh Kumar Dubey, Udo Censkowsky, Malancha Roy, Bimal Kinkar Chand, and Ajanta Dey

Abstract

Documented knowledge on the evaluation of mangrove plantation/restoration site is very limited in India. In this effort, an attempt has been made to evaluate a mangrove plantation site managed by a local non-governmental organization (NGO) after 2 years of implementation. In this framework, the plantation project was assessed by (i) biodiversity richness, (ii) composition and vegetation structure and (iii) interaction of biotic and abiotic factors and (iv) societal aspects. The floristic diversity includes ten true mangrove species and nine mangrove associates in the patch. The natural regeneration potential has remarkably increased after plantation, and *Sonneratia apetala* is considered as 'champion species' in terms of natural regeneration (seedling density 3730 ha^{-1}). The colonization of the macro-benthos was significantly amplified after plantation, and three species of molluscs and fourteen crustacean species (brachyuran crabs, grapsid crabs, hermit crabs and scorpion mud lobster) with ample bioturbation dynamics were encountered during the study. The present short-term evaluation depicted that the implementing NGO is primarily successful in re-establishing the natural diversity and ecosystem functioning in an increasing anthropogenic interfered landscape, and local communities are managing the site. Thus, to achieve a successful mangrove plantation/restoration programme, formulation of a location-specific long-term monitoring plan and adoption of community-based co-management practices are recommended.

Keywords

Mangrove plantation · Mangrove monitoring · Natural regeneration · Sundarban

S. K. Dubey (✉)
Bluesensus – Sustainability & Seafood Consulting, Kolkata, West Bengal, India
e-mail: s.dubey@blue-sensus.com

U. Censkowsky
Bluesensus – Sustainability & Seafood Consulting, Weilheim, Germany

M. Roy
Nature Environment & Wildlife Society, Kolkata, West Bengal, India

National Centre for Sustainable Coastal Management (NCSCM-Kolkata), Salt Lake Sector V, Kolkata, West Bengal, India

B. K. Chand
West Bengal University of Animal and Fishery Sciences, Kolkata, West Bengal, India

A. Dey
Nature Environment & Wildlife Society, Kolkata, West Bengal, India

B. Gul et al. (eds.), *Sabkha Ecosystems*, Tasks for Vegetation Science VI,
https://doi.org/10.1007/978-3-030-04417-6_22

Introduction

Mangrove conservation through plantation, restoration, rehabilitation of degraded mangrove sites and regeneration of mangrove patches in mudflats throughout tropics is one of the prime areas of action research (Feller et al. 2017). The mangrove management strategy primarily aims to maintain the health of mangrove ecosystems, hence improving the resiliency (Schmitt and Duke 2015). Strategically situated at the confluence of land and sea interfaces, mangroves are among the most productive wetland ecosystems on earth (Kathiresan and Bingham 2001). Mangrove forests are the economic foundations of many tropical coastal regions as they provide ecosystem services to the adjoining systems and humankind (Walters et al. 2008) and considered as a multi-valued ecosystem (MEA 2005). The 'Ecosytem Services' encompass tangible goods like timber and non-timber forest produces and intangible services like carbon sequestration and provide shelter, breeding and nursery ground for aquatic and non-aquatic animal species, protection from natural hazards and erosion, land accretion and recreational services (MEA 2005; UNEP 2014).

Despite the immense socio-ecological importance, more than half of the world's mangrove forests had been degraded due to various anthropological forces (Giri et al. 2011, 2015). The rapid coastal development resulted in a global decrement in mangrove areas from 187,940 km^2 (1980) to about 152,310 km^2 (2005) with an estimated annual loss of 0.16% and 0.39%, respectively (FAO 2007; Hamilton and Casey 2016). In that context, successful implementation of mangrove plantation/restoration programmes becomes highly important to reconcile local livelihood and mangrove conservation.

General aspects of the mangrove plantation programme have been documented (Saenger 2002; Macintosh and Ashton 2003; Primavera and Esteban 2008; Macintosh et al. 2012). A mangrove restoration/plantation programme must be focused on suitable plantation site, distribution of background species, followed by selection of appropriate species, best-suited plantation technique and time for the selected site (Schmitt and Duke 2015). Finally, the most crucial step is the monitoring and evaluation with quantifiable parameters like ecological study of the ecosystem with supportive societal imperatives (Biswas et al. 2009).

Since the early 1980s, the mangrove conservation/restoration programme got momentum in India through various government and private sector agencies (Bhatt et al. 2011). Although scientific monitoring and evaluation programmes were very limited (Macintosh et al. 2012) and many plantation programmes in the Indian Subcontinent have been extremely disparaged due to failures in restoring the natural and ecological functionality (Datta et al. 2012; Datta and Deb 2017). In this effort, an attempt has been made to evaluate a mangrove plantation site managed by a local non-governmental organization (NGO) after 2 years of mangrove plantation. The immediate objectives taken by this NGO were to implement a mangrove plantation programme by the local women communities and to promote the concept of 'anthropogenic change resilience' through restoring and preserving the functions of the ecosystem. The prime objective of the present evaluation was to examine the composition of forest structure, biodiversity richness, biotic and abiotic factors and societal aspects under prevailing community-NGO co-management structure.

Study Site

Sundarban is the largest deltaic mangrove zone of the planet, partly spread across 9630 km^2 in India (Chaudhuri and Choudhury 1994). The Indian Sundarban delta lies between 21°40 ′N and 22°40 ′N latitudes and 88°03′E and 89°07′E longitudes. The United Nations Educational, Scientific and Cultural Organization (UNESCO) declared the Indian part of Sundarban as a 'World Heritage Site' in 1987, and the 'Sundarban Biosphere Reserve' (SBR) was designated under the UNESCO Man and the Biosphere (MAB) programme in 2001. The region is characterized by a subtropical humid climate with a dry season between November and April and a wet monsoonal period over the rest of the year. The average maximum temperature is 34 °C and the average minimum temperature is 13.7 °C. The relative humidity is 75–80%. The maximum temperature varies between 26 and 40 °C. The average annual rainfall ranges between 1500 and 2000 mm. Most of the rainfall (about 74% of the total) occurs during the period of June–September. Extreme weather with frequent cyclonic depressions occurs during mid-March to mid-September. The Indian Sundarban is polyhaline in nature with higher salinity near the coast, and the water is almost fresh on the inland side boundary of the Sundarban. The soils of Sundarban area are mostly silty clay to clay loam in nature. The soils are generally rich in potassium and have a moderate amount of available phosphorus but low in nitrogen.

This study was carried out in the Lakshmipur village (latitude 21°56′ 44.46″ N longitude 88° 9′ 42.76″ E) located outside protected area network of Sundarban (Fig. 22.1). The site is situated on the western banks of the River Baratala-Muriganga (Hooghly channel) that discharges into the Bay of Bengal. The environmental and biophysical setting of the afforestation site is depicted in Table 22.1.

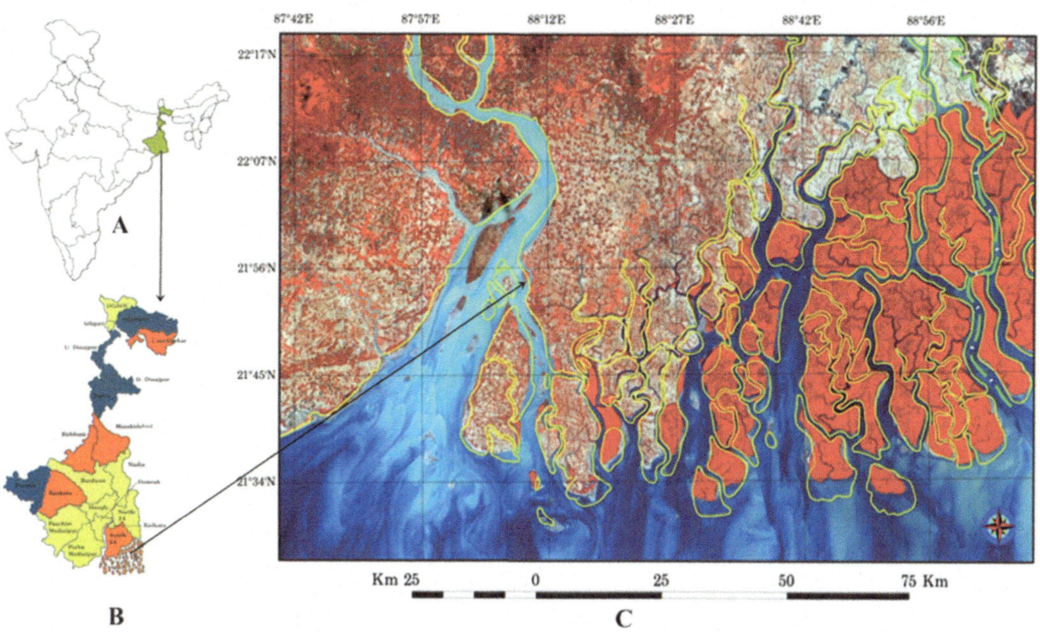

Fig. 22.1 Map of the study sites. A, B and C denote India, West Bengal and Indian Sundarban, respectively

Table 22.1 Environmental and biophysical setting of the study site before plantation

Name of site	Lakshmipur mudflat
Size of the planted area	1.2 ha
Main river course	Hooghly/Baratala
Type of substratum	Soft muddy appearance at the lower end of the substratum and the middle portion of the mudflat is harder and stable
Salinity	5–25 ppt throughout the seasons of the year
pH	8.02
Water inundation	Average 15 days throughout the month
Vegetation coverage	Nearly 20% area covered by sparse salt marsh grass vegetation
Existing mangroves	*Excoecaria agallocha* L. and *Acanthus ilicifolius* L.
Existing mangrove associates	*Derris trifoliata* Lour., *Porteresia coarctata* (Roxb.) Tateoka, *Crinum defixum* Ker Gawl.
Existing macro-benthos	*Telescopium telescopium* (Linnaeus, 1758), *Uca (Tubuca) rosea* (Tweedie, 1937), *Uca triangularis* (A. Milne-Edwards, 1873)

Note: Vegetation coverage, existing mangroves, mangrove associates and macro-benthos were assessed during the pre-plantation phase in August 2016.

Table 22.2 Habitat suitability of mangrove rehabilitation for the proposed sites

Biophysical	Yes	No	Not determined
A. Out-planting: River (sea) front			
Exposed to air during neap tide (low tide)	√		
Protected from wave action		√	
Substrate firm (foot does not sink above the ankle)		√	
No waterlogged		√	
Remaining mangroves	√		
B. Out-planting: Inner-intertidal			
Exposed to air during neap tide (low tide)	√		
Protected from wave action	√		
Substrate firm (foot does not sink above the ankle)	√		
No waterlogged	√		

Note: Criteria for the mangrove rehabilitation sites adopted from Primavera et al. (2012)

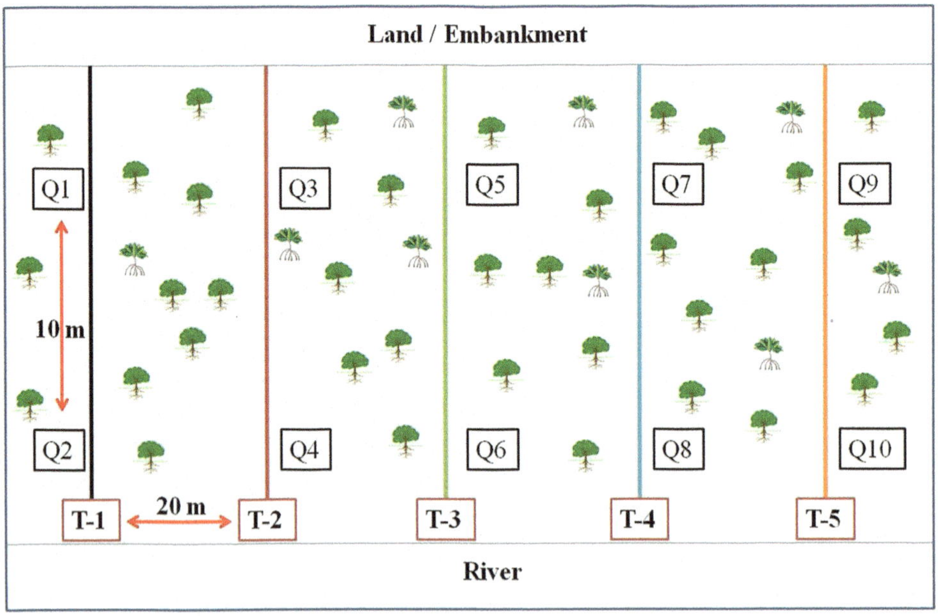

Fig. 22.2 Schematic representation of sampling design for mangrove structure assessment. Q and T denoted quadrates and transect lines, respectively

Habitat suitability of mangrove rehabilitation for the proposed sites is depicted in Table 22.2.

Methodologies

Data Collection

Data collection was performed following a 'stratified random sampling design' (Raffaelli and Hawkins 1996). A schematic representation of sampling design for mangrove structure assessment is given in Fig. 22.2. The river track was considered as the main transect line, and five vertical transects were established at a 20-m interval along the main transect. Rectangular plots (quadrates) of 10 × 10 m were established along the line transects to record necessary quantitative and biological parameters of mangroves, mangrove associates and prevailing macro-fauna. The natural regeneration or newly recruited mangrove species (which are distinct in comparison to the planted one) were also counted falls within each plot. In each line transect, two quadrates were laid at 10 m interval. Altogether 10 quadrates were sampled within the 1 ha plantation site during the study period. Plants having height > 4 m were recorded as trees, 1–4 m as saplings and < 1 m as seedlings. The procedure of the data collection was followed as described by Kathiresan (2000). In the present study, a mangrove species was defined according to Tomlinson's criteria (Tomlinson 1994), supplemented by a few additional species, supported by the expanded definition provided by Duke (1992).

Data Analysis

The data obtained from each quadrate such as frequency, density, abundance and their relative values were calculated (Curtis and Mcintosh 1950; Misra 1968; Muller- Dombois and Ellenberg 1974).

$$\text{Frequency} = \frac{\text{Total number of quadrats in which species occurred}}{\text{Total number of quadrats studied}} \times 100$$

$$\text{Density} = \frac{\text{Total number of individuals of the species}}{\text{Total number of quadrats studied}} \times 100$$

$$\text{Relative Frequency} = \frac{\text{Frequency of a species}}{\text{Sum of frequency of all species}} \times 100$$

$$\text{Relative Density} = \frac{\text{Density of a species}}{\text{Sum of density of all species}} \times 100$$

Apart from these, univariate measures or biological diversity, viz., Simpson Index, Dominance Index, Shannon Weiner Diversity Index, Pielou's evenness index, Brillouin's index, Margalef's richness index, Equitability Index and Berger-Parker dominance, were analyzed using PRIMER v7 programme (Plymouth Routines in Multivariate Ecological Research, Clarke and Gorley 2015) without any transformation of data.

Soil and Water Quality Analysis

Composite surface soil samples (from the uppermost soil layer of 0–15 cm) were collected from selected subplots of the plantation site. Soil pH, electrical conductivity (dS m^{-1}) and soil organic carbon (mg ha^{-1}) content were analyzed based on the standard methods (Page et al. 1982; Singh et al. 1999). Soil available nitrogen (N) (kg ha^{-1}) and phosphorus (P) (kg ha^{-1}) were estimated following Subbiah and Asija (1956) and Olsen et al. (1954). Soil textural analysis and calculation were done as per USDA method.

Tidal water temperature (°C), pH, dissolved oxygen (mg l^{-1}) and salinity (g l^{-1}) were determined directly with a digital water analysis instrument (HANNA, HI 9828, Germany), while ammonia-nitrogen, NH_3–N (mg l^{-1}); nitrate-nitrogen, NO_3–N (mg l^{-1}); nitrite-nitrogen, NO_2–N (mg l^{-1}); and phosphate-phosphorus, PO_4-P (mg l^{-1}) were measured using HACH spectrophotometer (DR 2800, Germany). Total alkalinity (mg $CaCO_3$ l^{-1}) and total hardness (mg $CaCO_3$ l^{-1}) were measured as per APHA (2012).

Results and Discussions

Mangrove Vegetation Structure and Composition

Seven planted true mangrove species belonging to six families and seven genera (Table 22.3) were identified from the present rapid evaluation

Table 22.3 Taxonomy and global conservation status of the observed mangroves in the plantation site

Sl	Family	Genera	Mangrove species	Local name	IUCN status	Global population trend[a]
1	Rhizophoraceae	*Rhizophora*	*Rhizophora apiculata* Blume.	Garjan	LC	D
2	Acanthaceae	*Avicennia*	*Avicennia marina* (Forsk.) Vierh.	Piara baen	LC	D
3			*Avicennia alba* Blume. [b]	Kalo baen	LC	D
4		*Acanthus*	*Acanthus ilicifolius* L. [b]	Horgonja	LC	U
5	Lythraceae	*Sonneratia*	*Sonneratia apetala* Buch.-Ham.	Keora	LC	D
6	Rhizophoraceae	*Bruguiera*	*Bruguiera gymnorhiza* (L.) Lam.	Kankra	LC	D
7		*Ceriops*	*Ceriops tagal* (Perr) CB. Rob.	Mat Garan	LC	D
8	Malvaceae	*Heritiera*	*Heritiera fomes* Buch. Ham.	Sundari	EN	D
9	Meliaceae	*Xylocarpus*	*Xylocarpus moluccensis* (Lam.)	Passur	LC	D
10	Euphorbiaceae	*Excoecaria*	*Excoecaria agallocha* L.[b]	Gneoa	LC	D

D decreasing, *U* unknown, *EN* endangered, *NT* near threatened, *LC* least concern
[a]IUCN global population trend
[b]Existing mangroves and or natural recruits

Table 22.4 Structural features of mangroves in the plantation site

Name of species	Frequency (%)	Density (Sp. ha^{-1})	Relative frequency (%)	Relative density (%)
Seedling				
Bruguiera gymnorhiza (L.) Lam.	60.00	1510.00	27.27	63.18
Rhizophora apiculata Blume.	50.00	450.00	22.73	18.83
Ceriops tagal (Perr) CB. Rob.	20.00	50.00	9.09	2.09
Xylocarpus moluccensis (Lam.) M. Roem.	20.00	130.00	9.09	5.44
Heritiera fomes Buch. -Ham.	30.00	140.00	13.64	5.86
Avicennia marina (Forsk.) Vierh.	20.00	40.00	9.09	1.67
Sonneratia apetala Buch.-Ham.	20.00	70.00	9.09	2.93
Sapling				
Rhizophora apiculata Blume.	50.00	690.00	22.73	28.87
Natural regeneration (Seedling)				
Avicennia marina (Forsk.) Vierh.	70.00	350.00	36.84	7.53
Sonneratia apetala Buch.-Ham.	70.00	3730.00	36.84	80.22
Acanthus ilicifolius L.	30.00	530.00	15.79	11.40
Excoecaria agallocha L.	20.00	40.00	10.53	0.86
Total		**7730**		

of reforested patch. *Heritiera fomes* is listed as an endangered species as per the IUCN Red List of Threatened Species. Among the non-planted mangrove species, *Acanthus ilicifolius* showed dominancy followed by *Avicennia alba* and *Excoecaria agallocha*. Finally, ten true mangrove species were documented from the regenerated forest strand.

Structural features of mangroves in the plantation site are given in Table 22.4. The total mangrove density in the evaluated site was found to be 7730 ha^{-1}. A total of 4650 ha^{-1} seed-

Fig. 22.3 Natural regeneration in the planted patch (**a**) *Sonneratia apetala* recruits at 2016, (**b**) colony of *Sonneratia apetala* (**c**) continued the growth of *Sonneratia apetala* at 2018, (**d**) *Sonneratia* with *Acanthus ilicifolius* in 2018. Grass species are *Cyperus rotundus* (**a**) and *Porteresia coarctata* (**b–d**). (Photo: S. K. Dubey)

lings of *Avicennia*, *Sonneratia*, *Acanthus* and *Excoecaria* had grown by natural regeneration. The presence of saplings and seedlings greater than 50% of the number of mature trees per hectare (Gan 1995) and seedling density over 2500 ha^{-1} (Srivastava and Bal 1984) indicates good regeneration potential of a mangrove patch. Consistent with the above, this planted mangrove patch showed a good regenerative potential. The density and abundance of saplings and seedlings depend on soil nutrient availability and vegetation inheritance of the forest (Twilley 1995). The analysis of data revealed that especially *Sonneratia* has good regeneration potential (density 3730 ha^{-1}) and is considered as 'champion species' followed by *Acanthus* (density 530 ha^{-1}), *Avicennia* sp. (density 350 ha^{-1}) and *Excoecaria* (density 40 ha^{-1}), respectively (Table 22.4) (Fig. 22.3).

Species-wise, the density of planted mangroves varied from 40 to 1510 ha^{-1}, and the highest seedling densities were recorded for the *Bruguiera gymnorrhiza* (Fig. 22.4), followed by *Rhizophora apiculata* (Fig. 22.5), and lowest for *Avicennia marina*. The same trend was observed in case of frequency (%) of individual species. The density represents the numerical strength of species in the community (Misra 1968). Frequency refers to the degree of dispersion of individual species in an area and is usually expressed in terms of percentage occurrence. The hierarchy of relative frequency (%) and relative

Fig. 22.4 Growth of planted *Bruguiera gymnorrhiza*. (**a**) 2017 (**b**) 2018. Grass species *Cyperus rotundus*. (Photo: S. K. Dubey)

density (%) for the planted mangrove species can be depicted as *Bruguiera gymnorrhiza* > *Rhizophora apiculata* > *Heritiera fomes* > *Xylocarpus moluccensis* > *Sonneratia apetala* > *Ceriops tagal* > *Avicennia marina* (Table 22.4). Relatively taller seedlings were planted in sea-front sites with high sediment load and in deeper water, so the higher leaves remain exposed and are less prone to gathering sediment and flooding, increasing the plant's survival chances.

Highest Simpson's index (0.66) of naturally regenerated mangroves in the present assessment showed that the site has good regeneration potential in presence of good management strategy. Shannon Weiner Diversity Index (> 1), Margalef's richness index (> 0.5), Pielou's evenness index (> 0.5), Brillouin's index (> 1) and Dominance Index (> 0.5) for the planted mangrove strand partially depicted heterogeneous forest structure and balanced survival rate.

Composition of Mangrove Associates and Mangrove-Dependent Macro-Faunal Community

Mangroves sometimes include additional plant species within the same environments that are generally considered as mangrove associates (Tomlinson 1994). Many mangrove-dependent biota establish interlinked community assemblages, with a number of unique plant-animal relationships (Saenger 1994). The present assessment provided nine mangrove associates belonging to eight families and eight genera. These are *Clerodendrum inerme* (L.) Gaertn. (ban juhi), *Crinum defixum* Ker Gawl. (sukha darsan), *Cryptocoryne ciliata* (Roxb.) Fisch. ex Wydler. (Kerali), *Heliotropium curassavicum* L. (Nona Hatisnur), *Derris trifoliata* Lour. (pan lota), *Derris indica* (Lam.) Benn. (karanja), *Ipomoea pes-caprae* (L.) R. Br. (chagal kunri), *Cyperus*

Fig. 22.5 *Rhizophora apiculata* plantation during (**a**) high tide and (**b**) low tide. Grass species are *Cyperus rotundus* and *Porteresia coarctata* (**b**). (Photo: S. K. Dubey)

rotundus L. (mutha ghash) and *Porteresia coarctata* (Roxb.) Tateoka (dhani ghash).

Although mangrove plantation can assist to re-establish other plant communities, its effectiveness to recover the macro-benthic community remains uncertain because the establishment of the macro-benthic community is a complex intermingled process of environmental conditions and life history of macro-benthos. In the present rapid assessment study, 3 species of molluscs belonging to 3 families and 14 crustacean species (sesarmid crabs, hermit crabs, scorpion mud lobster etc.) belonging to 7 families were encountered which was quite high in comparison with the pre-plantation baseline study (Table 22.5). Among numerous mangrove-dependent macro-benthic communities, the molluscs and crustaceans are one of the most conspicuous groups of the mangrove-dependent fauna that serve as a useful indicator for the restoration of habitat functionality (Kumar and Khan 2013).

During the assessment course, at the timing of low tide (within the middle margin of the intertidal zone to the waterfront), different bioturbation structures were observed within the newly mangrove planted area. Bioturbation structures like tracks and trails, footprints, burrows, mounds and pellets produced by the interaction of living organisms and soft sediments are evident of a wide range of features (Das 2016). The average bioturbation structure (especially brachyuran crab holes and thalassinid mound) was found to be 34.90 ± 16.05 for each quadrate plotted. Among the different bioturbatory structures, grazing trails of *Telescopium*, feeding and dwelling burrows of *Uca* sp., dwelling and escaping burrows of *Metaplax* sp. as well as other grapsid crabs, mounds or heaps of *Thalassina anomala* were observed. These macro-benthic faunas are considered as ecosystem engineer due to their different bioturbatory structures because they influence nutrient productivity and sediment functions that influence forest growth and productivity (Dubey et al. 2012, 2013).

Table 22.5 Lists of the observed mangrove associated macro-fauna in the plantation site

Sl	Phylum	Family	Scientific name
1	Mollusca	Assimineidae	*Assiminea beddomiana* (Nevill, 1881)
2		Potamididae	*Telescopium telescopium* (Linnaeus, 1758)
3		Onchidiidae	*Onchidium tigrinum* (Stoliczka, 1869)
4	Arthropoda (Crustacea)	Ocypodidae	*Uca (Tubuca) acuta* (Stimpson, 1858)
5			*Uca (Tubuca) rosea* (Tweedie, 1937)
6			*Austruca lactea* (De Haan, 1835)
7			*Uca (Austruca) triangularis* (A. Milne-Edwards, 1873)
8			*Uca (Tubuca) dussumieri* (H. Milne-Edwards, 1852)
9		Dotillidae	*Dotilla intermedia* de Man, 1888
10			*Dotilla blanfordi* Alcock, 1900
11		Thalassinidae	*Thalassina anomala* (Herbst, 1804)
12		Sesarmidae	*Perisesar mabidens* (De Haan, 1835)
13			*Sesarmoides longipes* (Krauss, 1843)
14			*Parasesarm aplicatum* (Latreille, 1803)
15		Varunidae	*Metaplax intermedia* de Man, 1888
16		Portunidae	*Scylla serrata* (Forskål, 1775)
17		Diogenidae	*Clibanarius padavensis* (de Man, 1888)
18	Chordata (Pisces)	Gobiidae	*Boleophthalmus boddaerti* (Pallas, 1770)
19			*Pseudapocryptes elongatus* (Cuvier, 1816)
20			*Odontamblyopus rubicundus* (Hamilton, 1822)
21		Eleotridae	*Eleotris fusca* (Forster, 1801)

Note: Taxonomy and nomenclature followed as per World Register of Marine Species (WoRMS 2017) and FishBase (Froese and Pauly 2017)

Primary Succession

Saltmarsh grass (*Porteresia coarctata*) was counted as a pioneer along the banks and constituted majorities of the total herbaceous populations, which is the preliminary stages of primary succession (Fig. 22.6). Stabilization and reclamation of the denuded mudflat by halotolerant grasses can increase the chances for successful mangrove restoration programmes (Begam et al. 2017). The cover of *Porteresia coarctata* in the lower intertidal zones acts as a protective barrier for mangrove species growing in the middle intertidal zones.

Soil Edaphic Factors and Physicochemical Properties of Water in the Plantation Site

Sub-surface soil samples collected from the mangrove plantation sites exhibited pH range from 6.85 to 7.35 (Table 22.6). The clay (%) of the soil varied between 78% and 80% thus depicted clay type of soil texture. The organic carbon value was found to be higher during post-plantation phases. The total organic carbon of the sediment influences fertility of the soil, thereby enhancing biological activity (Kumar 1996). During the study, higher soil available N and available P was observed after restoration and natural regeneration of mangroves. Decomposition of mangrove leaves and bioturbation increase the redox state of soil, which influence nutrient availability (Reef et al. 2010). As a result, high rates of denitrification deplete the nitrate and nitrite pools and produce the most common form of N in mangrove soils (Reef et al. 2010). Benthic microbial communities and chances of mycorrhizal colonization can be high in N-rich mangrove soil and thus influence uptake of immobile nutrients, especially P (Sengupta and Chaudhuri 2002).

Physicochemical parameters of tide-fed water source (Hooghly-Baratala channel) during high tide were analyzed (Table 22.7). The surface water temperature showed variation with respect to seasons. Dissolved oxygen, total alkalinity and

Fig. 22.6 Stabilization of mudflat through the settlement of Saltmarsh grass (*Porteresia coarctata*). (**a**) 2016 (**b**) 2018. (Photo: S. K. Dubey)

total hardness showed slight variation and increasing trend after mangrove plantation. The nitrogenous and phosphorous compound did not show any remarkable variations. It can be noted that soil edaphic factors and physicochemical properties of water in the plantation site are well suited for mangrove plantation and recolonization of macro-benthos (Kumar and Khan 2013).

Probable Hazard

Without any protection measures, a mangrove plantation is of little use or futile. During the first couple of years, the plantations are vulnerable to various anthropogenic and natural stressors. Therefore field-level monitoring (e.g. growth and survival), maintenance (e.g. removing debris, algae and other pests) and protection (e.g. repairing of fencing, restriction from livestock grazing, inhibiting human impact and protection from wave actions) are the most important management options of mangrove plantation. Apart from these, barnacles, crabs, insects and pest attacks (Schmitt and Duke 2015) can also damage young mangroves. Probable hazards in the mangrove plantation site accompanied by their degree of occurrence were portrayed in Table 22.8.

Community Involvement

To ensure success in mangrove management, the involvement of local communities is an effective way of maintaining and enhancing the protection function of the mangrove while providing livelihoods for local people and contributing to better assessment and governance of natural resources (Schmitt and Duke 2015). Primavera and Esteban (2008) also suggested incorporating such mechanisms, by which the communities can work as de facto mangrove managers, are granted tenurial rights. In this evaluation study, we presented community participation as a form of case studies. It

Table 22.6 Different soil parameters (mean ± standard deviation) of the mangrove plantation site

Edaphic factors	Pre-plantation[a]	Post-plantation[b]
Soil pH	6.85 ± 0.21	7.35 ± 0.25
EC (electrical conductivity) (dS m^{-1})	22.78 ± 1.31	24.55 ± 0.35
Soil organic carbon (mg ha^{-1})	15.25 ± 0.25	17.85 ± 0.68
Available nitrogen N (kg ha^{-1})	325.45 ± 1.20	345.75 ± 2.25
Available phosphorus P (kg ha^{-1})	14.75 ± 1.50	17.50 ± 1.05
Clay (%)	78.75 ± 0.25	79.50 ± 0.01
Silt (%)	10.55 ± 0.10	11.65 ± 0.02
Sand (%)	10.70 ± 0.05	8.85 ± 0.01

Note: Analysis was carried out at the Department of Aquatic Environment Management, Faculty of Fishery Sciences, West Bengal University of Animal and Fishery Sciences, Kolkata and School of Environmental Studies, Jadavpur University

[a]Samples were collected during monsoon period (August 2016)

[b]Samples were collected during pre-monsoon period (April 2017)

Table 22.7 Different physicochemical properties of water (mean ± standard deviation) of the mangrove plantation site

Physicochemical properties	Pre-plantation[a]	Post-plantation[b]
Water temperature (°C)	30.87 ± 1.81	31.96 ± 1.52
pH	8.17 ± 0.23	7.15 ± 0.32
Dissolved oxygen (mg l^{-1})	5.96 ± 0.30	6.98 ± 0.36
Total alkalinity (mg $CaCO_3$ l^{-1})	110.75 ± 1.61	115.14 ± 1.02
Total hardness (mg $CaCO_3$ l^{-1})	111.37 ± 1.03	113.10 ± 1.02
Ammonia-nitrogen NH_3-N (mg l^{-1})	0.21 ± 0.04	0.20 ± 0.03
Nitrate-nitrogen NO_3-N (mg l^{-1})	0.24 ± 0.03	0.27 ± 0.03
Nitrite-nitrogen NO_2-N (mg l^{-1})	0.02 ± 0.01	0.02 ± 0.01
Phosphate-phosphorus PO_4-P (mg l^{-1})	0.25 ± 0.02	0.25 ± 0.04

Note: Analysis was carried out at the Department of Aquatic Environment Management, Faculty of Fishery Sciences, West Bengal University of Animal and Fishery Sciences, Kolkata and School of Environmental Studies, Jadavpur University

[a]Samples were collected during monsoon period (August 2016)

[b]Samples were collected during pre-monsoon period (April 2017)

Table 22.8 Possible observed hazards in the mangrove plantation site

Attributes	Observation
Physical	
Wave action	General wave action and flushing due to tidal amplitude are observed. This is a particular problem where inundation and sedimentation rates are high, as in the lower intertidal to sub-tidal flats which caused damage to net fencing
Biological	
Infestation of filamentous algae	Not observed
Boring isopods	Not observed
Barnacle infestation	Not observed
Oysters infestation	Not observed
Leaf disease	Observed in *Heritiera fomes* (seasonal)
Anthropogenic	
Boat traffic and fishing gears	Usually, 6–7 country fishing boats were observed within the plantation site in a daily basis. They usually sail to offshore during high tide; however, resource user's right to the community cannot be shut down
Debris	Garbage and debris (fishing nets, plastic bags, etc.) were observed and that maximum was brought by the tidal action
Domestic animals	Generally not observed within the plantation site

was revealed during community interaction that the women groups participated during propagule collection, nursery building, plantation and post-plantation care. As the site is attached with landward embankment, the women group has initiated a monetary penalty (with the permission of local government) to the owner of livestock to stop grazing. This is a good community co-management approach to build people's conservation fund. Biswas et al. (2009) advocated that community participation is essential at least during three phases: (i) ecosystem synthesis/situation analysis, (ii) identification of priorities for interventions and development of restoration strategy and action plan and (iii) participatory implementation and monitoring. Biswas et al. (2009) also described the possible cause of failure in community participation, which depends on cultural and economic conditions, ownership in the community, transparency of the efforts and pluralism. When a mangrove restoration effort complies with their traditional cultures and generates economic returns and community ownership with the transparency of the entire system, then only community members will be accountable to their own roles, and the restoration may be successful (Biswas et al. 2009). Although considerable biological diversity and tree health are encountered in the present evaluation, long-term monitoring and protection should be required engaging local community to achieve a stable mangrove stands.

Case Study 22.1: Community Involvement in the Mangrove Restoration

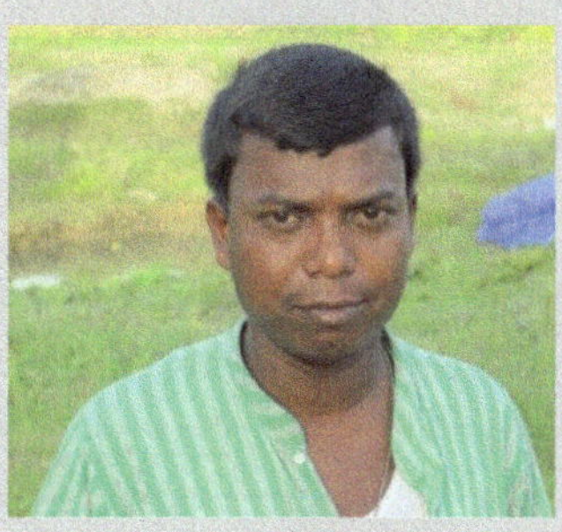

Prabir Mandal
Male, 33 years

Prabir is a permanent resident of Lakshmipur village. Previously, his father ran a food stall nearby the present plantation site which was in nascent state at that time. During 1995–1996, his family experienced heavy flood with severe erosions and lost their homestead belongings as well as the food stall. Subsequently, in 2009 (during severe tropical cyclone *Aila*), due to the breach of the embankment, they lost their house and was impelled to move towards upland which is 500 m away from the embankment. Since then Prabir understood the role of mangroves in shoreline protection. He performs the lead role in making and repairing river embankments and other community development in his village. Besides these activities, he also serves as 'mangrove steward' and actively participated in the mangrove plantation programme. 'Mangroves give us various services and I am sure that only mangroves can protect the embankment and resist coastal erosions', Prabir said during the interaction. He said that, due to fencing and protection of the plantation area, natural regenerations took place, which starts stabilizing the mudflat. Due to this plantation initiative, finfish and shellfish recruitment are increasing day by day. He said that *lal chengo mach* (*Odontamblyopus rubicundus* Hamilton, 1822) previously was not available in this mudflat and now it becomes available due to mangrove restoration.[1] Prabir wants to stop crab and shrimp seed collection activities, which can disturb the plantation area. He thinks that community mobilization and capacity building programme is essential for community development in larger context.

[1]Das (2017) also documented catch of mangrove-dependent fish have significantly increased after mangrove plantation which gave extra income to the artisanal fishermen.

Case Study 22.2: Community Involvement in the Mangrove Restoration

Nila Boidya
Female, 36 years

Nila is a homemaker and resides at Lakshmipur village. She lives in a mud-made thatched house situated just the outer layer of the embankment. She serves as 'mangrove steward' and participated in nursery bed preparation, in preparation of jute bag rearing the seedling and in the plantation process. She was also engaged in the watering of the planted trees and monitored the post-plantation process. As the site is attached with landward embankment, Nila took a lead role to initiate a monetary penalty to the owner of livestock to stop grazing. 'Many villagers are depended on the mangrove resources in that area and augmentation of mangrove cover would be beneficial for them'. She also added during the interaction, 'if there are no mangroves, the river has no meaning. It's like a tree without roots'. She also added that due to this plantation initiative, the availability of finfish and shellfish is increasing day by day. She told that formation of more number of mangrove nursery rearing group in her village would be directly beneficial for the women group for supplementary income and indirectly for protection of the environment.

Conclusion and Recommendation

Documented knowledge on the evaluation of mangrove restoration site is very limited in India. In this chapter, we have tried to develop an evaluation framework for mangrove plantation site. In reality, the success of mangrove plantation/restoration programme should be evaluated in terms of achieving ecological, socio-economic and institutional sustainability, but that requires a considerable long time-scale (at least of 30 years) of management and monitoring (Datta et al. 2012; Datta and Dev 2017). From the present short-term evaluation, it can be depicted that the implementing NGO is primarily successful in re-establishing the natural ecological stability in an increasingly man-made landscape. However, ensuring capacity for sustainable livelihood for the local villagers, a long-term development plan is required. Most mangrove plantation/restoration efforts in Indian Sundarban have followed a trial-and-error method without any explicit and integrated framework, baseline ecological information or proper consideration of community involvement. Majority of the projects are externally funded with a limited period (1–5 years), and the whole effort collapsed as soon as the external support is discontinued. Our evaluation framework is in line with Ruiz-Jaen and Aide (2005) and Biswas et al. (2009) who advocated three effective measures for assessing restoration success: (i) biodiversity richness, (ii) composition and vegetation structure and (iii) interaction of biotic and abiotic factors. Technical and financial resources are key driving forces for the assessment of simple to complex functional components of an ecosystem. Furthermore, to achieve a successful mangrove plantation/restoration programme, formulation of location-specific long-term monitoring plan consulting with standard documents (e.g. Lewis 2005; Lewis et al. 2006; Primavera et al. 2012; Lewis and Brown 2014) and community-based co-management are recommended.

Acknowledgements The study was financially supported by European retailer Alnatura (CSR fund). Germany-based consulting firm 'bluesensus-sustainability & seafood consulting' implemented this plantation project with its Indian partner Nature Environment and Wildlife Society (NEWS). Authors are thankful to the women group during the course of field visit and evaluation.

References

APHA (American Public Health Association) (2012) Standard methods for the examination of water and wastewater, 22nd edn. American Public Health Association, American Water Works Association, and Water Environment Federation, Washington, DC

Begam MM, Sutradhar T, Chowdhury R, Mukherjee C, Basak SK, Ray K (2017) Native salt-tolerant grass species for habitat restoration, their acclimation and contribution to improving edaphic conditions: a study from a degraded mangrove in the Indian Sundarbans. Hydrobiologia 803(1):373–387. https://doi.org/10.1007/s10750-017-3320-2

Bhatt JR, Macintosh JD, Nayar TS, Pandey CN, Nilratna BP (2011) Towards conservation and management of mangrove ecosystems in India. Ministry of Environment and Forest, New Delhi

Biswas SR, Mallik AU, Choudhury JK, Nishat A (2009) A unified framework for the restoration of southeast Asian mangroves – bridging ecology, society and economics. Wetl Ecol Manag 17(4):365–383. https://doi.org/10.1007/s11273-008-9113-7

Chaudhuri AB, Choudhury A (1994) Mangroves of the Sundarbans (vol I). India, The IUCN Wetlands Programme, IUCN, Bangkok, 247 pp

Clarke KR, Gorley RN (2015) PRIMER v7: user manual/tutorial. PRIMER-E, Plymouth, 296 pp

Curtis JT, Mcintosh RP (1950) The interrelations of certain analytic and synthetic phytosociological characters. Ecology 31(3):434–455. https://doi.org/10.2307/1931497

Das GK (2016) Bioturbation structures. In: Tidal sedimentation of the Sunderban's Thakuran Basin. Springer International Publishing, Cham, pp 123–140. https://doi.org/10.1007/978-3-319-44191-7_8

Das S (2017) Ecological restoration and livelihood: contribution of planted mangroves as nursery and habitat for artisanal and commercial fishery. World Dev 94:492–502. https://doi.org/10.1016/j.worlddev.2017.02.010

Datta D, Deb S (2017) Forest structure and soil properties of mangrove ecosystems under different management scenarios: experiences from the intensely humanized landscape of Indian Sunderbans. Ocean Coast Manag 140:22–33. https://doi.org/10.1016/j.ocecoaman.2017.02.022

Datta D, Chattopadhyay RN, Guha P (2012) Community based mangrove management: a review on status and sustainability. J Environ Manag 107:84–95. https://doi.org/10.1016/j.jenvman.2012.04.013

Dubey SK, Choudhury A, Chand BK, Trivedi RK (2012) Ecobiological study on burrowing mud lobster *Thalassina anomala* (Herbst, 1804) (Decapoda: Thalassinidea) in the intertidal mangrove mudflat of deltaic Sundarbans. Explor Anim Med Res 2:70–75

Dubey SK, Chakraborty DC, Chakraborty S, Choudhury A (2013) Burrow architecture of red ghost crab *Ocypode macrocera* (H. Milne-Edwards, 1852): a case study in Indian Sundarbans. Explor Anim Med Res 3:136–144

Duke NC (1992) Mangrove floristics and biogeography. In: Robertson AI, Alongi DM (eds) Tropical mangrove ecosystems. Coastal and marine estuarine series 41. American Geophysical Union, Washington, DC, pp 63–100

Feller IC, Friess DA, Krauss KW, Lewis RR (2017) The state of the world's mangroves in the 21st century under climate change. Hydrobiologia 803(1):1–12. https://doi.org/10.1007/s10750-017-3331-z

Food and Agriculture Organization (2007) The world's mangroves, 1980–2005: a thematic study in the framework of the Global Forest Resources Assessment 2005. FAO of United Nation Rome, Italy. Forestry Paper 153

Froese R, Pauly D (eds) (2017) Fish base. World Wide Web electronic publication. www.fishbase.org, version (02/2017)

Gan BK (1995) A working plan for the Matang Mangrove Forest Reserve (Fourth Revision). The State Forest Department of Perak Darul Ridzuan, Malaysia

Giri C, Ochieng E, Tieszen LL, Zhu Z, Singh A, Loveland T, Masek J, Duke N (2011) Status and distribution of mangrove forests of the world using earth observation satellite data. Glob Ecol Biogeogr 20(1):154–159. https://doi.org/10.1111/j.1466-8238.2010.00584.x

Giri C, Long J, Abbas S, Murali RM, Qamer FM, Pengra B, Thau D (2015) Distribution and dynamics of mangrove forests of South Asia. J Environ Manag 148:101–111. https://doi.org/10.1016/j.jenvman.2014.01.020

Hamilton SE, Casey D (2016) Creation of a high spatio-temporal resolution global database of continuous mangrove forest cover for the 21st century (CGMFC-21). Glob Ecol Biogeogr 25(6):729–738. https://doi.org/10.1111/geb.12449

Kathiresan K (2000) Flora and fauna in mangrove ecosystems: a manual for identification. Ministry of Environment and Forests, India: CAS in Marine Biology

Kathiresan K, Bingham BL (2001) Biology of mangroves and mangrove ecosystems. Adv Mar Biol 40:81–251. https://doi.org/10.1016/s0065-2881(01)40003-4

Kumar SR (1996) Distribution of organic carbon in the sediments of Cochin mangroves, south west coast of India. Indian J Mar Sci 25:56–61

Kumar P, Khan A (2013) The distribution and diversity of benthic macroinvertebrate fauna in Pondicherry mangroves, India. Aquat Biosyst 9(1):15. https://doi.org/10.1186/2046-9063-9-15

Lewis RR (2005) Ecological engineering for successful management and restoration of mangrove forests. Ecol Eng 24(4):403–418. https://doi.org/10.1016/j.ecoleng.2004.10.003

Lewis RR, Brown B (2014) Ecological mangrove rehabilitation – a field manual for practitioners. Mangrove Action Project, Canadian International Development Agency and OXFAM

Lewis RR, Quarto A, Enright J, Corets E, Primavera J, Ravishankar T, Djamaluddin R (2006) Five steps to successful ecological restoration of mangroves. Mangrove Action Project, Yogyakarta, Indonesia. Elsevier B.V

Macintosh DJ, Ashton EC (2003) Draft code of conduct for the sustainable management of mangrove ecosystems. World Bank, ISME, cenTER Aarhus

Macintosh DJ, Mahindapala R, Markopoulos M (eds) (2012) Sharing lessons on mangrove restoration. Bangkok, Thailand: mangroves for the future and gland. IUCN, Switzerland

Millennium Ecosystem Assessment (MEA) (2005) Ecosystems and human well-being: biodiversity synthesis. World Resources Institute, Island Press, Washington, DC

Misra R (1968) Ecology workbook. Oxford and IBH publishing company. Ltd, New Delhi

Muller-Dombois, Ellenberg H (1974) Aims and methods of vegetation ecology. Wiley, New York

Olsen SR, Cole CV, Watnabe FS, Dean LA (1954) Estimation of available phosphorus in soils by extraction with sodium bicarbonate. US Department of Agriculture. Circular 939

Page AL, Miller RH, Keeney DR (1982) Methods of soil analysis part 2: chemical and microbiological properties. American Society of Agronomy, Wisconsin

Primavera JH, Esteban JMA (2008) A review of mangrove rehabilitation in the Philippines: successes, failures and future prospects. Wetl Ecol Manag 16(5):345–358. https://doi.org/10.1007/s11273-008-9101-y

Primavera JH, Savaris JD, Bajoyo B, Coching JD, Curnick DJ, Golbeque R, Guzman AT, Henderin JQ, Joven RV, Loma RA, Koldewey HJ (2012) Manual on community-based mangrove rehabilitation – mangrove manual series no. 1. ZSL, London. 240 pp

Raffaelli D, Hawkins S (1996) Intertidal ecology. Chapman and Hall, London

Reef R, Feller IC, Lovelock CE (2010) Nutrition of mangroves. Tree Physiol 30(9):1148–1160. https://doi.org/10.1093/treephys/tpq048

Ruiz-Jaen MC, Mitchell Aide T (2005) Restoration success: how is it being measured? Restor Ecol 13(3):569–577. https://doi.org/10.1111/j.1526-100x.2005.00072.x

Saenger PE (1994) Mangroves and saltmarshes. In: Hammond L, Synnot RN (eds) Marine biology. Addison Wesley Longman, Sydney

Saenger P (2002) Mangrove ecology, silvicultureand conservation. Kluwer Academic Publishers, Dordrecht. 360pp

Schmitt K, Duke NC (2015) Mangrove management, assessment, and monitoring. Trop For Handb :1725–1759. https://doi.org/10.1007/978-3-642-54601-3_126

Sengupta A, Chaudhuri S (2002) Arbuscular mycorrhizal relations of mangrove plant community at the Ganges river estuary in India. Mycorrhiza 12(4):169–174. https://doi.org/10.1007/s00572-002-0164-y

Singh D, Chhonkar PK, Pandey RN (1999) Soil, plant & water analysis – a method manual. IARI, New Delhi

Srivastava PBL, Bal HS (1984) Composition and distribution pattern of natural regeneration after second thinning in Matang mangrove reserve. Perak Malaysia. In: Soepadimo E, Rao AN, Macintosh DJ (eds) Proceedings of the Asian symposium on mangrove environment: research managements. Universiti Malaya, Kuala Lumpur, pp 761–784

Subbiah BV, Asija GL (1956) A rapid procedure for the determination of available nitrogen in soils. Curr Sci 25:259–260

Tomlinson PB (1994) The botany of mangroves. Cambridge University Press, Cambridge

Twilley RR (1995) Properties of mangrove ecosystems related to the energy signature of coastal environments. In: Hall CAS (ed) Maximum power: the ideas and application of Odum HT. University of Colorado Press, Boulder/Colorado, pp 43–62

UNEP (2014) The importance of mangroves to people: a call to action. In: van Bochove J, Sullivan E, Nakamura T (eds) United Nations Environment Programme World Conservation Monitoring Centre, Cambridge, 128 pp

Walters BB, Rönnbäck P, Kovacs JM, Crona B, Hussain SA, Badola R, Primavera JH, Barbier E, Dahdouh-Guebas F (2008) Ethnobiology, socio-economics and management of mangrove forests: a review. Aquat Bot 89(2):220–236. https://doi.org/10.1016/j.aquabot.2008.02.009

WoRMS Editorial Board (2017). World register of marine species. Available from www.marinespecies.org. Accessed 30 May 2018. https://doi.org/10.14284/170

Mangrove Biogeography of the Indo-Pacific

23

P. Saenger, P. Ragavan, C.-R. Sheue, J. López-Portillo, J. W. H. Yong, and T. Mageswaran

Abstract

Studies on biogeography are useful to understand the present and past distribution patterns of biological diversity and their underlying environmental and historical causes. The mangrove realm, largely confined to sheltered tropical and subtropical coastlines within latitudes of around 32°N and 38°S, is divided longitudinally into an Atlantic realm and an Indo-Pacific realm. Because of the relatively recent closure of the Central American Isthmus, a small incursion of the Atlantic realm species has occurred into the Indo-Pacific realm along the tropical and subtropical American Pacific coast and spread north and south since the last glaciation. Although the biogeography of mangroves has been widely discussed, recent investigations of mangrove floristics and genetic diversity using molecular studies warrant a further comprehensive account of the biogeography of mangroves. Considering these facts, the biogeography of the Indo-West Pacific (IWP) mangroves has been examined in detail, and the Atlantic East Pacific mangrove region (AEP) has only been briefly discussed. The difference in species richness between the IWP and the AEP is evidence of the effectiveness of the African land mass and the East Pacific Ocean barriers. Within the IWP, two trends in relation to mangrove species richness can be identified: first, maximal species richness which occurs along the shorelines of Makassar Strait, between Borneo and Sulawesi in Indonesia, and second, the marked attenuation of species numbers with increasing latitude generally related to limiting temperatures. The contemporary and his-

P. Saenger (✉)
School of Environment, Science and Engineering, Southern Cross University, Lismore, NSW, Australia
e-mail: peter.saenger@scu.edu.au

P. Ragavan
SERB-National Post Doctoral Fellow, CSIR-National Botanical Research Fellow, Lucknow, Uttar Pradesh, India

C.-R. Sheue
Department of Life Sciences and Center of Global Change Biology, National Chung Hsing University, Taichung, Taiwan

J. López-Portillo
Instituto de Ecologia, A.C. (INECOL), Xalapa, Veracruz, Mexico

J. W. H. Yong
ARC Centre for Mine Site Restoration, School of Biological Sciences, University of Western Australia, Perth, Australia

T. Mageswaran
National Centre for Sustainable Coastal Management, MoEFCC, Chennai, Tamil Nadu, India

B. Gul et al. (eds.), *Sabkha Ecosystems*, Tasks for Vegetation Science VI,
https://doi.org/10.1007/978-3-030-04417-6_23

torical processes leading to these trends are summarized and discussed. Finally, the threats to the mangroves of the various regions within the IWP are subsequently briefly reviewed, and given the ecological and economic values of mangroves, the potential of mangrove restoration/rehabilitation to offset mangrove losses is also evaluated.

Keywords
Biogeography · Diversity · Mangrove · Rehabilitation · Restoration · Species richness

Introduction

Studies on biogeography are useful to understand the present and past distribution patterns of biological diversity and their underlying environmental and historical causes. The mangrove realm, largely confined to sheltered tropical and subtropical coastlines within latitudes of around 32°N and 38°S, is divided longitudinally into an Atlantic realm and an Indo-Pacific realm (Tomlinson 1986; Ricklefs and Latham 1993). Because of the relatively recent closure of the Central American Isthmus (about 3.5 million years ago), a small incursion of the Atlantic realm species has occurred into the Indo-Pacific realm along the tropical and subtropical American Pacific coast and spread north and south since the last glaciation (Sandoval-Castro et al. 2014). Although the biogeography of mangroves has been widely discussed, recent studies of mangrove floristics and genetic diversity using molecular studies warrant a further comprehensive account of the biogeography of mangroves. Considering these facts, the biogeography of the Indo-West Pacific (IWP) mangroves has been examined in detail, and the Atlantic East Pacific mangrove region (AEP) has only been briefly discussed. The threats to the mangroves of the various regions within the IWP are subsequently briefly reviewed, and given the ecological and economic values of mangroves, the potential of mangrove restoration/rehabilitation to offset mangrove losses is also evaluated.

Recent Taxonomic and Distributional Changes in the Mangrove Flora

Recent taxonomical studies led to an understanding of extended distribution of few extant mangrove species and the discovery of new entities. Some of the significant findings are as follows. The taxonomic relationships of the Australian *Excoecaria* species were examined by Maguire and Saenger (2000) using both leaf morphological data and DNA sequence data from the internal transcribed spacer (ITS) region of ribosomal genes. The nucleotide differences in the examined ITS1 region showed that *E. agallocha* consisted of two species – the widespread *E. agallocha* and the species confined to northern Australian mangroves, previously described as *E. ovalis*. The leaf morphological data also support the re-erection of *E. ovalis:* single factor analysis of variance consistently separated *E. ovalis* from *E. agallocha* on the basis of leaf width, leaf length, length of petiole and leaf margin.

A recent re-examination of the genus *Kandelia*, long considered monotypic, showed that two well-differentiated populations existed, separated by the South China Sea. These were established as two distinct species *Kandelia candel* and *K. obovata* (Sheue et al. 2003).

The genus *Ceriops* has also been revised recently (Sheue et al. 2009a, b, 2010) and has been found to consist of three species on the basis that the species formerly known as *C. decandra* consisted of a species complex: *Ceriops decandra* occurs in India, Bangladesh, Myanmar and Thailand; *Ceriops pseudodecandra* occurs in Australia, Papua New Guinea and in the Indonesian provinces of Irian Jaya and Seram; and *Ceriops zippeliana* occurs in Indonesia, Malaysia, Vietnam and the Philippines.

Sheue et al. (2009a) have re-examined the distinctiveness of *Ceriops tagal* and *Ceriops australis*. In the process, *Ceriops australis* was found to occur widely in Indonesia, constituting a major increase in its known distribution from Australia to the southern coast of New Guinea.

Sheue et al. (2005) recorded the following species of *Bruguiera* from Singapore:

B. sexangula, *B. gymnorhiza*, *B. parviflora* and *B. cylindrica*, as well as the new record of *Bruguiera hainesii*, which has also recently been recorded from northeastern Queensland (Cooper et al. 2016).

The distributional limits of *Rhizophora* species have been extended in the southwestern Pacific by Duke (2010), and new records of *Sonneratia ovata* and *S. lanceolata*, together with the hybrids *S.* × *urama*, *S.* × *gulngai* and *Rhizophora* × *annamalayana*, have been made from the Andaman and Nicobar Islands, India (Ragavan et al. 2016).

Considering the above taxonomical results, the mangrove flora of the world is updated and listed in Appendix 1. The mangrove niche is inhabited by around 81 tree and shrub species of 30 genera from 17 families worldwide. Eleven named and unnamed hybrids (Table 23.1) and several subspecies/varieties of mangroves have been described. Hybrids are not discussed further because they are usually of limited distribution and, as they occur within the range of the putative parental species, they do not offer a great deal of biogeographic information. The same goes for varieties and subspecies. For further information on mangrove hybrids, see the recent review by Ragavan et al. (2017). Recently Ono et al. (2016) reported on the hybrid origin of *B. hainesii* (*B. gymnorhiza* × *B. cylindrica*), although here it is considered as distinct species until further confirmation is available.

Biogeography of Indo-West Pacific Mangroves

The species diversity and distribution of mangroves are variable at different spatial scales, i.e. global, regional, estuarine and intertidal. On a global scale, mangrove distribution appears to be limited by temperature, while at a regional scale, various biotic and abiotic factors control their distribution. In the past, it was assumed that because of long distance dispersal by buoyant propagules, mangrove species were widely and continuously distributed in suitable habitats within the region of their occurrence (Tomlinson 1986; Duke and Kudo 2018). More recently it seemed that buoyant propagules are better adapted to longshore and interisland dispersal rather than intercontinental transmission by oceanic currents (Clarke 1993; McGuiness 1997). Mangrove dispersal is thus limited in different

Table 23.1 Indo-Pacific mangrove hybrids with putative parental species

Combretaceae
Lumnitzera × *rosea (L. littorea* × *L. racemosa)*
Lamiaceae
Unnamed hybrid *(Avicennia marina* × *A. rumphiana)*
Lythraceae
Sonneratia × *gulngai (S. alba* × *S. caseolaris)*
Sonneratia × *hainanensis (S. alba* × *S. ovata)*
Sonneratia × *urama (S. alba* × *S. lanceolata)*
Unnamed hybrid *(Sonneratia alba* × *S. griffithii*
Pteridaceae
Unnamed hybrid (*Acrostichum aureum* × *A. speciosum*
Rhizophoraceae
Bruguiera × *rhynchopetala (B. gymnorhiza* × *B. sexangula)*
Unnamed hybrid *(Ceriops tagal* × *C. australis)*
Rhizophora × *annamalayana (R. apiculata* × *R. mucronata)*
**Rhizophora* × *harrisonii (R. mangle* × *R. racemosa)*
Rhizophora × *lamarckii (R. apiculata* × *R. stylosa)*
Rhizophora × *mohanii (R. mucronata* × *R. stylosa)*
Rhizophora × *selala (R. stylosa* × *R. mangle)*
Rhizophora × *tomlinsonii (R. apiculata* × *R. mangle)*

Notes: The hybrid *Rhizophora* × *brevistyla* was reduced into synonymy with *R. mangle* by Hou (1960). Similarly, Hou (1960) reduced *Rhizophora samoensis* into synonymy with *R. mangle*. This has been followed by later authors, e.g. Breteler (1977), Tomlinson (1986) and Tyagi (2003), although Duke (2010) maintains its species status. There have been suggestions that *Rhizophora harrisonii* is a hybrid between *R. mangle* and *R. racemosa* (Keay 1953; Breteler 1969; Wilcox 1985), but data from foliar wax chemistry did not support any hybrid status in that *R. harrisonii* showed no intermediate characteristics from those of its putative parental species (Dodd et al. 1995). More recent molecular data, however, supported the hybrid nature of this taxon (Cerón-Souza et al. 2010). Recently Ono et al. (2016) reported on the hybrid origin of *Bruguiera* × *hainesii (B. gymnorhiza* × *B. cylindrica)*: here, because of its high fruit set, it is considered as a distinct species until more data are forthcoming

Hybrids confined to the Atlantic East Pacific only are indicated with an *

Finally, since this paper was prepared another hybrid mangrove has been described from north-east Australia (Duke and Kudo (2018)

ways by both land barriers and wide expanses of water, which in turn cause the disjunct pattern of contemporary mangrove distribution. As discussed below, recent phylogeographic studies, using molecular methods, also have shown the existence of disjunction in populations of the mangrove species throughout the Indo-West Pacific.

The well-known disjunction pattern of mangroves is the species-poor Atlantic East Pacific (AEP) and species-rich Indo-West Pacific (IWP). Despite the equivalent area of mangrove habitat in each region, mangroves of the IWP are three times more diverse than in the AEP. Mangroves of IWP consist of 69 mangrove species, whereas the AEP region has 13 species (Appendix 1). Only one species (*Acrostichum aureum*) and three genera (*Acrostichum*, *Avicennia* and *Rhizophora*) have a common distribution in both the IWP and the AEP. Except for the family Tetrameristaceae (formerly Pelliceriaceae), all mangrove families of the AEP are also represented in the IWP. Difference in species richness between the IWP and the AEP is evidence of the effectiveness of the African land mass and the East Pacific Ocean barriers.

On the whole, 20 out of 25 genera are common in the IWP. The exceptions are *Aglaia*, *Brownlowia*, *Camptostemon*, *Diospyros* and *Osbornia*. *Diospyros* is restricted to Australasia, and *Aglaia*, *Brownlowia* and *Phoenix* are restricted to South and South East Asia. *Camptostemon* and *Osbornia* are restricted to South East Asia and Australasia. The mangrove flora of South and South East Asia shares 42 species, of which 11 species (*Acanthus volubilis*, *Aglaia cucullata*, *Brownlowia tersa*, *Ceriops decandra*, *Excoecaria indica*, *Heritiera fomes*, *Kandelia candel*, *Phoenix paludosa*, *Sonneratia apetala*, *Sonneratia griffithii* and *Aegialitis rotundifolia*) are not found in Australasia. Australasia shares 40 species with South East Asia, of which 9 species (*Aegialitis annulata*, *Avicennia rumphiana*, *Bruguiera hainesii*, *Brownlowia argentata*, *Camptostemon schultzii*, *Ceriops australis*, *Ceriops pseudodecandra*, *Osbornia octodonta* and *Bruguiera exaristata*) are restricted to South East Asia and Australasia. Three species *Avicennia integra*, *Diospyros littorea* and *Excoecaria ovalis* are restricted to Australasia, and *Camptostemon philippinense*, *Ceriops zippeliana* and *Aegiceras floridum* are restricted to South East Asia. Both *Kandelia obovata* and *Acanthus xiamenensis* are found only in East Asia.

Regional Distribution of Mangrove Species Around the Indo-Pacific

Based on recent taxonomic and distributional records discussed above and the personal knowledge of the authors, the distribution pattern and gradients in five geographical regions of the IWP (East Africa, Middle East, South Asia, South East Asia, East Asia and Australasia) and of the East Pacific are summarized in Tables 23.2, 23.3, 23.4, 23.5, 23.6 and 23.7.

The mangroves of the East Pacific coast are listed in Table 23.2, which has been compiled from data given in Breteler (1969), Gentry (1982), Jiménez (1987), Cerón-Souza et al. (2005, 2010), Spalding et al. (2010), Barreto and Barreto-Pittol (2012), Castillo-Cárdenas et al. (2015) and Santamaria-Damián et al. (2016). Detailed data on the distribution of mangroves and associated species are given by Ellison (2004) for Central America, Sandoval-Castro et al. (2014) for Mexico and Gentry (1982) and Cerón-Souza et al. (2005) for Colombia.

No natural mangroves occur in the Hawaiian archipelago, but various species have been introduced including *Rhizophora mangle*, *R. mucronata*, *Bruguiera sexangula*, *B. parviflora*, *Ceriops tagal* and *Conocarpus erectus*. Of these species, only *R. mangle*, *B. sexangula* and *C. erectus* have established self-maintaining populations (Allen 1998). Other introduced species in Eastern Polynesia include *Rhizophora stylosa* on Mo'orea, Tahiti and Bora Bora, *Rhizophora mangle* on Enewetak Atoll and *Bruguiera gymnorhiza* on Bikini Atoll (Saenger 2002).

The mangroves of the East African region and the Middle East consist of 10 and 4 species, respectively. Species occurrences are listed in Tables 23.3 and 23.4, respectively. They have been compiled from data given in Béguinot (1918), Dale (1938), Koechlin et al. (1974),

Table 23.2 Mangrove species distribution in the East Pacific region

Species	Mex	Gua	ElS	Hon	Nic	CRi	Pan	Col	Ecu	Per
Acrostichum aureum	*	*		*	*	*	*	*	*	
Acrostichum danaeifolium	*	*	*	*	*		*	*	*	
Avicennia bicolor	*	*	*	*	*	*	*	*		
Avicennia germinans	*	*	*	*	*	*	*	*	*	*
Conocarpus erectus	*	*	*	*	*	*	*	*	*	*
Laguncularia racemosa	*	*	*	*	*	*	*	*	*	*
Mora oleifera							*	*		
Pelliciera rhizophorae					*	*	*	*	*	
Rhizophora mangle	*	*	*	*	*	*	*	*	*	*
Rhizophora racemosa			*	*	*	*	*	*	*	*
Tabebuia palustris						*	*	*		
Species totals	**7**	**7**	**7**	**8**	**9**	**9**	**11**	**11**	**8**	**5**

Mex Mexico, *Gua* Guatemala, *EIS* El Salvador, *Hon* Honduras, *Nic* Nicaragua, *CRi* Costa Rica, *Pan* Panama, *Col* Colombia, *Ecu* Ecuador, *Per* Peru

Table 23.3 Mangrove species distribution of the East African region

Species	SAf	Moz	Mad	Tan	IOI	Ken	Som	Dji
Acrostichum aureum	*	*		*	*			
Avicennia marina	*	*	*	*	*	*	*	*
Bruguiera gymnorhiza	*	*	*	*	*	*	*	
Ceriops tagal	*	*	*	*	*	*	*	*
Heritiera littoralis		*	*	*	*	*		
Lumnitzera racemosa	*	*	*	*	*	*	*	
Pemphis acidula		*	*	*	*			
Rhizophora mucronata	*	*	*	*	*	*	*	
Sonneratia alba		*	*	*	*	*	*	
Xylocarpus granatum[a]		*	*	*	*	*		
Species totals	**6**	**10**	**9**	**10**	**10**	**8**	**6**	**2**

SAf South Africa, *Moz* Mozambique, *Mad* Madagascar and Mauritius, *Tan* Tanzania, *IOI* Indian Ocean islands of Comoros, Mayotte and the Seychelles, *Ken* Kenya, *Som* Somalia including the island of Socotra, *Dji* Djibouti

[a]Only *X. granatum* is listed in Spalding et al. (2010), but Dale (1938) mentions 'the two species of *Xylocarpus*' in the Kenyan mangroves, listing both *X. granatum* (as *X. benadirensis*) and *X. moluccensis*. Other records of *X. moluccensis* are from Somalia (Ciferri 1939), Tanzania (Mangora 2011) and Madagascar (Koechlin et al. 1974; Hughes and Hughes 1992)

Zahran (1977), Ruwa (1993), Banyikwa (1986), Ormond et al. (1988), Cornes and Cornes (1989), Hughes and Hughes (1992), Faye (1993), Spalding et al. (2010), Kumar et al. (2010a, b, 2011), Moore et al. (2015) and Almahasheer (2018). Limited areas of suitable habitat and a prevalence of harsh environmental conditions limit extensive growth of mangroves in these regions. Mangroves of this region are considered as a subset of the highly diverse mangroves of other regions of the IWP (Duke et al. 1998), and no endemic species have been reported. Distributional gradients in this region appear relatively simple and feature a decline in species richness with increasing latitude associated with temperature and/or aridity.

South Asia (also known as Indian subcontinent) represents approximately 7% of the global mangrove area and hosts 42 species belonging to 22 genera. The mangroves of South Asia are listed in Table 23.5, which have been compiled from data given in Naskar and Mandal (1999), Spalding et al. (2010) and Ragavan et al. (2016). No endemic species of mangroves are known from South Asia. Maximum number of species occurs in mainland India (37 species), with a

Table 23.4 Mangrove species distribution in the Middle East

Species	Eri	Sud	Egy	SAr	Yem	Oma	UAE	Qat	Bah	Kuw	Ira
Avicennia marina	*	*	*	*	*	*	*	*	*	**	*
Bruguiera gymnorhiza	*	*	*		*		**				
Ceriops tagal											*
Rhizophora mucronata	*		*	*	*		**				*
Species totals	**3**	**2**	**3**	**2**	**3**	**1**	**1**	**1**	**1**	**0**	**3**

Eri Eritrea, *Sud* Sudan, *Egy* Egypt, *SAr* Saudi Arabia, *Yem* Yemen, *Oma* Oman, *UAE* United Arab Emirates, *Qat* Qatar, *Bah* Bahrain, *Kuw* Kuwait, *Ira* Iran
**Introduced

marked decline in species numbers with increasing distance from the continental land mass (e.g. Sri Lanka, the Maldives and the Chagos Archipelago). Different geomorphological settings between west and east coasts of India and absence or presence of river deltas are responsible for the greater species richness of the east coast. None of the species on the west coast is unique whereas several species (*Scyphiphora hydrophylacea*, *Acrostichum speciosum*, *Aegialitis rotundifolia*, *Aglaia cucullata*, *Heritiera fomes*, *Brownlowia tersa*, *Pemphis acidula*, *Nypa fruticans*, *Phoenix paludosa* and *Acanthus volubilis*) are restricted to east coast of India.

Another important mangrove region in South Asia comprises the Andaman and Nicobar Islands (ANI) that host 34 mangroves species. The east coast of these islands, in particular, has well-developed contiguous mangroves. *Rhizophora stylosa*, *Lumnitzera littorea*, *Sonneratia ovata*, *Sonneratia griffithii* and *Sonneratia lanceolata* are restricted to the ANI in South Asia. Earlier reports of the occurrence of *Rhizophora stylosa* and *Sonneratia griffithii* on the east coast of India are erroneous. It is often noted that floristically ANI has a close similarity with Southeast Asian countries due to its geographical proximity. However, mangrove floristics of ANI show a closer similarity with mainland India, sharing only five species with South East Asia (*Rhizophora stylosa*, *Lumnitzera littorea*, *Sonneratia ovata*, *Sonneratia griffithii* and *Sonneratia lanceolata*). In contrast, there is highest similarity between mainland India and the Southeast Asian countries like Thailand, Myanmar, Malaysia and Singapore. It indicates the possible dispersal route between these South Asian countries and mainland India through the Strait of Malacca to the shores of the Bay of Bengal. However, the disjunct distribution of *Lumnitzera littorea* in this region is difficult to explain: it is restricted to Sri Lanka and ANI but is not found on either coasts of mainland India.

The mangroves of the East Asia are listed in Table 23.6, which has been compiled from data given in Li and Lee (1997), Hsueh and Lee (2000) and Spalding et al. (2010). The most notable feature of this region is the high level of species richness across the region, with a rapid loss of species to the north-east (China and Japan) as a result of unfavourable temperature regimes (Li and Lee 1997; Chen et al. 2017). These trends are discussed further below.

The mangroves of the Pacific and Australasian region are listed in Table 23.7, which have been compiled from data given in Saenger et al. (1977), Maguire and Saenger (2000), Duke (2006), Ellison (2009), and Spalding et al. (2010). The continental land masses in this region (Australia and Papua New Guinea) show high levels of species richness and a degree of endemism (*Avicennia integra*, *Diospyros littorea* and *Excoecaria ovalis*), but the various island groups to the south and east show a rapid decline in species numbers, reflecting both dispersal limitations due to distance and limited availability of suitable mangrove habitat. This is particularly well illustrated by the tropical Christmas Island (10° 25′S; 105° 40′ E), lying some 360 km south of Java with its diverse mangroves and with an annual rainfall in excess of 2000 mm, which has four recorded mangrove species. Clearly, dispersal distance and adverse currents, together with

Table 23.5 Mangrove species distribution in South Asia

Species	Pak	Ind	Mal	ANI	Sri	Ban
Acanthus ebracteatus		*		*		
Acanthus ilicifolius		*		*	*	*
Acanthus volubilis		*		*		
Acrostichum aureum		*	*	*	*	*
Acrostichum speciosum		*		*		
Aegialitis rotundifolia		*				*
Aegiceras corniculatum	*	*		*	*	*
Aglaia cucullata		*				*
Avicennia alba		*				*
Avicennia marina	*	*	*	*	*	*
Avicennia officinalis		*		*	*	*
Brownlowia tersa		*		*		
Bruguiera cylindrica		*	*	*	*	
Bruguiera gymnorhiza	*	*	*	*	*	*
Bruguiera parviflora		*		*		
Bruguiera sexangula		*	*		*	*
Ceriops decandra	*	*				*
Ceriops tagal	*	*	*	*	*	*
Cynometra iripa		*		*	*	
Dolichandrone spathacea		*		*	*	
Exceocaria agallocha		*	*	*	*	*
Excoecaria indica		*		*	*	*
Heritiera fomes		*				*
Heritiera littoralis		*	*	*	*	*
Kandelia candel		*				*
Lumnitzera littorea				*	*	
Lumnitzera racemosa		*	*	*	*	
Nypa fruticans		*		*	*	*
Pemphis acidula		*	*	*	*	
Phoenix paludosa		*		*		
Rhizophora apiculata	*	*	*	*	*	*
Rhizophora mucronata	*	*	*	*	*	*
Rhizophora stylosa		*		*		
Scyphiphora hydrophylacea		*		*	*	
Sonneratia alba		*		*	*	
Sonneratia apetala		*				*
Sonneratia caseolaris	*	*	*	*	*	*
Sonneratia griffithii		*		*		
Sonneratia lanceolata				*		
Sonneratia ovata				*		
Xylocarpus granatum		*		*	*	*
Xylocarpus moluccensis		*		*		*
Species totals	**8**	**39**	**13**	**34**	**24**	**24**

Pak Pakistan, *Ind* India, *Mal* Maldives, *ANI* Andaman and Nicobar Islands, *Sri* Sri Lanka, *Ban* Bangladesh

restricted availability of mangrove habitats, are the major limiting factors.

Contemporary Biogeographical Processes Affecting Mangrove Distributions

A much-debated question concerning mangroves is where they first arose and how they got where they are today (Duke 1995; Ellison et al. 1999). Whatever the exact historical origin(s) and dispersal routes of mangroves, the present distributions of mangroves show several features reflecting some of the more modern processes described below (Saenger 1998). These modern constraints are manifested as a reduction of mangrove species with increasing latitude on the one hand and selective removal or loss of species by natural upheavals or human activity on the other.

The latitudinal limits of mangrove vegetation on the major land masses of the Indo-Pacific (Table 23.8) show that these limits are quite variable. They can, however, be broadly related to temperature and/or aridity (Saenger and Moverley 1985; Quisthoudt et al. 2012; Chen et al. 2017). It should also be noted that there is a gradual reduction of mangrove development (e.g. height and extent of mangrove vegetation) with latitude (Saenger and Snedaker 1993) and that an attenuation of species with increasing latitude precedes these latitudinal limits. By way of example, the southward attenuation of species on the non-arid east coasts of Africa and Australia is shown in Table 23.9. This comparison indicates that on continental coasts where mangroves are not limited by aridity, temperature is the major factor in reducing number of species with latitude. Clearly while there is some intraspecific variability between the two continents, the data support the suggestion that, in the presence of adequate rainfall, latitude as a proxy for temperature is related to the southern limits of mangrove species distribution. Similar limitations apply to northern distributional limits (Quisthoudt et al. 2012; Chen et al. 2017).

Table 23.6 Mangrove species distribution of East Asia

Species	Mya	Mal	Sin	Indo	Tha	Cam	Vie	Phi	Chi	Jap
Acanthus ebracteatus	*	*	*	*	*	*	*	*	*	
Acanthus ilicifolius	*	*	*	*	*	*	*	*	*	
Acanthus volubilis	*	*	*	*	*					
Acanthus xiamenensis[a]									*	
Acrostichum aureum	*	*	*	*	*	*	*	*	*	*
Acrostichum speciosum	*	*	*	*	*			*	*	
Aegialitis annulata				*						
Aegialitis rotundifolia	*			*	*					
Aegiceras corniculatum	*	*	*	*	*	*	*			
Aegiceras floridum		*		*			*	*		
Aglaia cucullata	*	*		*						
Avicennia alba	*	*	*	*	*	*	*	*		
Avicennia marina	*	*	*	*	*	*	*	*	*	*
Avicennia officinalis	*	*	*	*	*		*	*		
Avicennia rumphiana		*	*	*				*		
Brownlowia argentata	*	*		*				*		
Brownlowia tersa	*	*	*	*	*	*		*		
Bruguiera cylindrica	*	*	*	*	*		*	*	*	
Bruguiera exaristata				*						
Bruguiera gymnorhiza	*	*	*	*	*	*	*	*	*	*
Bruguiera hainesii	*	*	*	*	*		*			
Bruguiera parviflora	*	*	*	*	*		*	*		
Bruguiera sexangula	*	*	*	*	*	*	*	*	*	
Camptostemon philippinense				*				*		
Camptostemon schultzii				*				*		
Ceriops australis				*						
Ceriops decandra	*	*		*	*	*	*	*		
Ceriops tagal	*	*	*	*	*	*	*	*	*	
Ceriops pseudodecandra				*						
Ceriops zippeliana		*	*							
Cynometra iripa				*	*		*			
Dolichandrone spathacea		*	*	*			*			
Exceocaria agallocha	*	*	*	*	*	*	*	*	*	*
Excoecaria indica		*	*	*		*	*			
Heritiera fomes	*				*					
Heritiera littoralis	*	*	*	*	*	*	*	*	*	*
Kandelia candel	*	*	*	*	*		*	*		
Kandelia obovata							*		*	*
Lumnitzera littorea	*	*	*	*	*	*	*	*	*	
Lumnitzera racemosa	*	*	*	*	*	*	*	*	*	*
Nypa fruticans	*	*	*	*	*	*	*	*	*	*
Osbornia octodonta		*		*				*		
Pemphis acidula		*	*	*	*	*	*	*	*	*
Phoenix paludosa		*		*	*					
Rhizophora apiculata	*	*	*	*	*	*	*	*	*	*
Rhizophora mucronata	*	*	*	*	*	*	*	*		
Rhizophora stylosa		*	*	*			*	*	*	*
Scyphiphora hydrophylacea		*	*	*	*	*	*	*	*	
Sonneratia alba	*	*	*	*	*	*	*	*	*	*

(continued)

Table 23.6 (continued)

Species	Mya	Mal	Sin	Indo	Tha	Cam	Vie	Phi	Chi	Jap
Sonneratia apetala	*								**	
Sonneratia caseolaris	*	*	*	*	*	*	*	*	*	
Sonneratia griffithii	*	*			*					
Sonneratia lanceolata				*						
Sonneratia ovata	*	*	*	*	*	*	*	*	*	
Xylocarpus granatum	*	*	*	*	*	*	*	*	*	
Xylocarpus moluccensis	*	*	*	*	*			*		
Species totals	**36**	**43**	**36**	**50**	**37**	**25**	**34**	**38**	**25**	**11**

Mya Myanmar, *Mal* Malaysia, *Sin* Singapore, *Indo* Indonesia, *Tha* Thailand, *Cam* Cambodia, *Vie* Vietnam, *Phi* The Philippines, *Chi* China, *Jap* Japan

**Introduced

[a]It has been suggested that *Acanthus xiamenensis* is synonymous with *A. ilicifolius*, but it is retained here until further data becomes available

The distribution of mangrove species richness is shown in Fig. 23.1, based on data from Spalding et al. (2010) as contained at the portal site called TroCEP (Tropical Coastal Ecosystem Portal): http://www.nies.go.jp/TroCEP/index.html. While the nature of the data has some limitations (e.g. Western Australia has 20 mangrove species listed for the entire ~21,000 km of coastline, where mangrove species numbers range from 20 in the north to 1 in the south) and the algorithm has difficulties modelling mid-ocean rather than near-shore species numbers, Fig. 23.1 nevertheless shows two important biogeographical trends.

First, the distribution of mangrove species richness across the IWP clearly shows that maximal richness occurs along the shorelines of Makassar Strait, between Borneo and Sulawesi in Indonesia. This area has the greatest interdigitation of land and sea in the world, a condition stretching back to the Holocene. While some seaways may have closed due to tectonic or sea-level changes, the area has always maintained some direct connections between the tropical Pacific and tropical Indian oceans as Makassar Strait lies more or less seaward of the Sunda Shelf. High precipitation and relatively low salinities in this area also provided favourable conditions for the development and maintenance of the mangrove niche. Furthermore, cycles in sea level during the numerous late Pliocene and Pleistocene glacial-interglacial intervals have led to the appearance and disappearance of islands, the openings and closing of marine corridors and bridges and alternating higher and lower salinities (Voris 2000). In turn, these processes have led to bottleneck and founder effects on the abundant mangroves of the region and led to high rates of speciation.

The second identifiable trend in Fig. 23.1 is the attenuation of species numbers with increasing latitude generally related to limiting temperatures. Thus, the rapid loss of species on the east and west coast of Australia, in southern Africa and along the shores of the East China Sea can be appreciated. The rapid decline in species on the northern shores of East Africa and into the Red Sea, Gulf of Oman and Arabian Gulf is almost certainly due to limiting aridity rather than temperature. An attenuation of species numbers from continental land masses to offshore archipelagoes is also discernible although less marked.

Other contemporary biogeographical processes result from human activities such as pollution, water diversion, selective clearing and, most recently, major mangrove afforestation or rehabilitation programmes, with or without novel species introductions. These are briefly reviewed below.

Historical Biogeographical Processes Affecting Mangrove Distributions

Historical biogeographic processes involve changing the genetic diversity of populations directly or via altered or redirected gene flow

Table 23.7 Mangrove species distribution in the Pacific and Australasian region

Species	Mar	MIs	Kir	Sam	Fij	PNG	NCa	NZ	Au	CI
Acanthus ebracteatus						*			*	
Acanthus ilicifolius						*	*		*	
Acrostichum aureum						*	*			
Acrostichum speciosum						*	*		*	
Aegialitis annulata						*			*	
Aegiceras corniculatum						*			*	
Avicennia alba						*				
Avicennia integra									*	
Avicennia marina						*	*	*	*	
Avicennia officinalis						*				
Avicennia rumphiana						*				
Brownlowia argentata						*			*	
Bruguiera cylindrica						*			*	
Bruguiera exaristata						*			*	
Bruguiera gymnorhiza	*	*	*	*	*	*	*		*	*
Bruguiera hainesii						*			*	
Bruguiera parviflora						*			*	
Bruguiera sexangula						*			*	*
Camptostemon schultzii						*			*	
Ceriops australis									*	
Ceriops pseudodecandra						*			*	
Ceriops tagal						*	*		*	
Cynometra iripa						*			*	
Dolichandrone spathacea						*	*		*	
Diospyros littorea						*			*	
Exceocaria agallocha	*				*	*	*		*	
Excoecaria ovalis									*	
Heritiera littoralis					*	*	*		*	*
Lumnitzera littorea		*	*		*	*	*		*	
Lumnitzera racemosa						*	*		*	
Nypa fruticans						*			*	
Osbornia octodonta						*			*	
Pemphis acidula		*				*	*		*	*
Rhizophora apiculata						*	*		*	
Rhizophora mucronata			*			*			*	
Rhizophora stylosa	*			*			*		*	
Scyphiphora hydrophylacea						*	*		*	
Sonneratia alba		*	*			*	*		*	
Sonneratia caseolaris						*	*		*	
Sonneratia lanceolata						*			*	
Sonneratia ovata						*				
Xylocarpus granatum		*		*	*	*	*		*	
Xylocarpus moluccensis						*			*	
Species totals	**3**	**5**	**4**	**3**	**5**	**39**	**18**	**1**	**38**	**4**

Mar Marianas Islands, *MIs* Marshall Islands, *Kir* Kiribati, *Sam* Samoa, *Fij* Fiji, *PNG* Papua New Guinea, *NCa* New Caledonia, *NZ* New Zealand, *Au* Australia, *CI* Christmas Island

between populations. Thus, for example, loss or fragmentation of habitat is usually associated with a reduction in population size that is accompanied by a reduction in genetic diversity: Parani et al. (1997) found that *Avicennia marina* in India, under stress from intense

Table 23.8 Northern and southern limits of mangrove vegetation on major land masses of the Indo-Pacific

Continental land mass	Northern limit	Southern limit
Pacific America	30° 15′	5° 30′
Red Sea/Southern Africa	30° 31′	32° 59′
Western Australia	–	33° 16′
Eastern Australia	–	38° 45′
Pacific Asia	31° 21′	–

Table 23.9 Southern latitudinal limits of shared mangrove species on the eastern coasts of Australia and Africa

Species	E. African Southern limit	E. Australian Southern limit
Sonneratia alba	23° 55′	22° 30′
Pemphis acidula	24° 18′	24° 01′
Xylocarpus granatum	26° 01′	25° 30′
Heritiera littoralis	26° 33′	22° 15′
Ceriops tagal	26° 50′	28° 11′
Lumnitzera racemosa	26° 50′	27° 30′
Acrostichum aureum	28° 58′	26° 05′
Rhizophora mucronata	31° 42′	18° 25′
Bruguiera gymnorhiza	32° 14′	29° 25′
Avicennia marina	32° 59′	38° 45′

grazing and pollution, showed reduced levels of polymorphism.

Restricted gene flow between populations can be caused by such factors as changing sea levels, closing of marine seaways, alterations in current patterns or distance. Given that eight major glacial-interglacial intervals have occurred during the last 3 million years (Late Pliocene and Pleistocene) with sea-level changes ranging from −70 to 90 m (Rossi 1981), changing configuration of land masses has been significant (Voris 2000). The emergence during Pleistocene times of the Sunda and Sahul shelves would have constituted major barriers to gene flow both to the north-west and to the south-east.

From their study of three populations of *Ceriops zippeliana* (as *C. decandra*) on the west, east and southern coasts of Malaysia, Tan et al. (2005) found that the three populations showed a high level of genetic differentiation and, when grouped per geographic regions, there was a clear differentiation between populations from the eastern Indian Ocean (West Malaysia) from those of the western Pacific Ocean (East and Southern Malaysia). They suggested that this split resulted from the historical lowering of sea level near the Strait of Malacca during the recent Pleistocene glaciations. Similar findings were also reported for *Ceriops tagal* (Huang et al. 2012).

Similarly, Wee et al. (2017) and Yang et al. (2017) have shown that the widely distributed mangrove *Sonneratia alba* has very restricted gene flow due to various genetic barriers and geographic distance. Thus the northern Australian population is genetically differentiated from populations on New Caledonia and in South East Asia, while the populations of Mozambique, the Bay of Bengal and southern Japan are also distinctive. They concluded that the effects of oceanic barriers due to sea-level change were compounded by the fact that *Sonneratia alba* propagules have a limited dispersal distance.

Minobe et al. (2010) examined the genetic diversity of *Bruguiera gymnorhiza* and found that genetic diversity was low within any local population and that differentiation between the Pacific Ocean, Bay of Bengal and the Arabian Sea was the result of very low gene flow between each of these regional seas, coupled with frequent fluctuation of population size due to changes in sea level. In the West Central Pacific, the genetic diversity of *Bruguiera gymnorhiza* was significantly higher in specimens from Iriomote Island, Okinawa, than on the more remote Pohnpei Island, Micronesia (Sugaya et al. 2003). Dispersal limitations related to propagule viability over extended distances appear to be the major cause.

Fujimoto et al. (1996) have investigated what constitutes the critical rise of sea level affecting mangroves. They found that with gradual sea-level rise of 1–2 mm/year over the last 2000 years, extensive mangrove forests developed on Kosrae

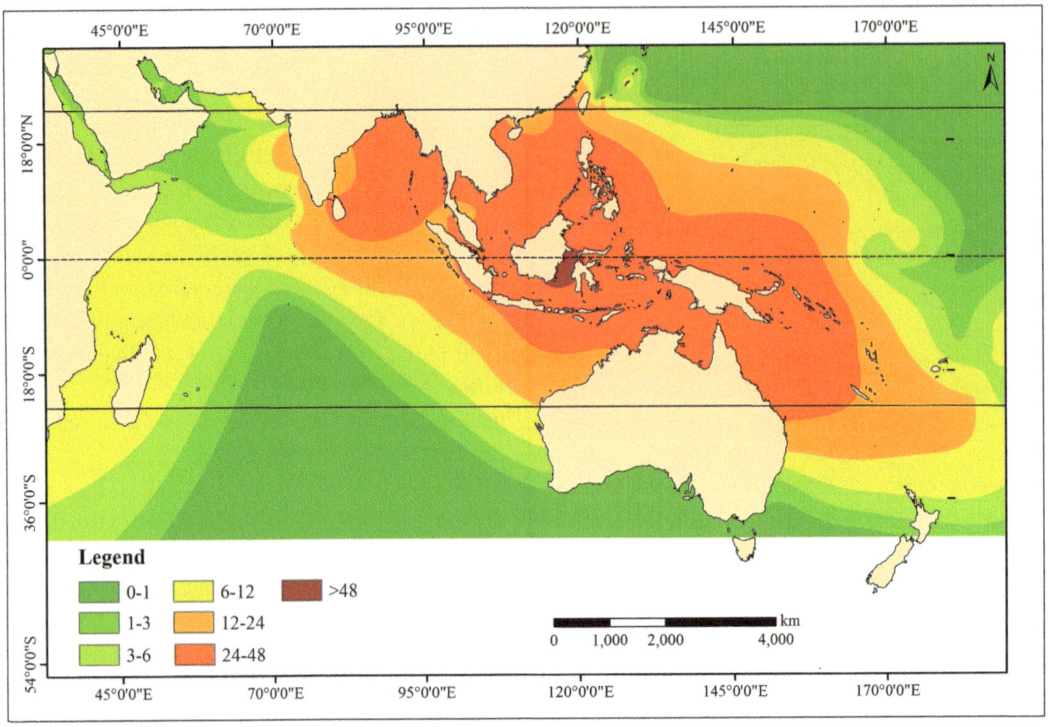

Fig. 23.1 Approximate distribution of mangrove species richness in the Indo-West Pacific. (Based on data from Spalding et al. (2010) in http://www.nies.go.jp/TroCEP/index.html. Species size classes are based on a geometrical progression)

Island, Micronesia, by accumulating mangrove peat (Fujimoto et al. 1996). However, during the period of rapid sea-level rise of around 10 mm/year that occurred between 4100 and 3700 year BP, the mangrove forest ceased peat accumulation and retreated landward, suggesting that the critical rate of sea-level rise enabling mangrove peat accretion on islands was between 2 and 10 mm/year.

The distributions of the two species of *Kandelia* offer a good example of vicariance, where two populations develop more or less allopatrically with no or very limited gene flow between them (Kado et al. 2004). However, because of the recent divergence times of many pantropical taxa (including mangroves, e.g. *Avicennia*; see Li et al. 2016), it is increasingly recognized that long distance dispersal might be the more important process determining the present distribution of many pantropical taxa.

Several mangrove species show strong inbreeding and self-fertilization, particularly at range edge, e.g. *Avicennia marina* (Maguire et al. 2000; Arnaud-Haond et al. 2006; De Ryck et al. 2016), *Rhizophora stylosa* (Islam et al. 2014), *Kandelia obovata* (Chen et al. 1996), *Bruguiera gymnorhiza* (Islam et al. 2006) and *Sonneratia alba* (Wee et al. 2017). Thus, the findings of De Ryck et al. (2016) on the east African coast highlight the genetically impoverished situation in peripheral populations due to the rarity of long distance dispersal, followed by inbreeding and dispersal limitation due to the coastal geomorphology and the availability of suitable habitats. Interestingly, in their study of *Sonneratia alba*, Wee et al. (2017) clearly demonstrated the range edge effect by regressing H_o (observed heterozygosity) and F_{IS} (inbreeding coefficient) against latitude: H_o decreased with increasing latitude, and F_{IS} increased with increasing latitude.

A summary of where some of the processes described above have been identified in mangrove species of the IWP is shown in Table 23.10 and Fig. 23.2. This summary indicates that the South China Sea and the Strait of Malacca, with their generally shallow depths and seasonally

Table 23.10 Areas of restricted gene flow in the Indo-West Pacific identified from molecular studies of mangroves

Location	Indicative lat/long	Species	Presumed cause	References
1. SWIO	30°S/33°E	*A. marina*	Dispersal limitations and habitat availability	De Ryck et al. (2016)
2. Arabian Sea	14°N/74°E	*B. gymnorhiza*	Bottlenecks during glaciation	Minobe et al. (2010)
3. WCIO	10°S/65°E	*S. alba*	Dispersal limitations	Wee et al. (2017)
4. Malacca Str.	1°N/103°E	*B. gymnorhiza*	Bottlenecks during glaciation	Minobe et al. (2010)
		C. tagal	Dispersal limitations	Huang et al. (2012)
		C. zippeliana	Glaciation	Tan et al. (2005)
		S. alba	Glaciation	Yang et al. (2017)
		S. alba	Dispersal limitations	Wee et al. (2017)
5. SCS	10°N/110°E	*C. tagal*	Bottlenecks during glaciation	Ge and Sun (2001)
		S. alba	Glaciation	Yang et al. (2017)
		A. marina	Dispersal limitations	Kado et al. (2004)
		K. candel	Dispersal limitations	Kado et al. (2004)
6. WCPO	20°N/140°E	*B. gymnorhiza*	Dispersal limitations	Sugaya et al. (2003)
7. Timor Sea	10°S/127°E	*S. alba*	Land barrier during glacial periods	Wee et al. (2017)
		C. tagal	Dispersal limitations	Huang et al. (2012)
8. Coral Sea	20°S/154°E	*S. alba*	Dispersal limitations	Wee et al. (2017)
		C. tagal	Dispersal limitations	Huang et al. (2012)

SCS South China Sea, *WCIO* Western Central Indian Ocean, *SWIO* South-West Indian Ocean, *WCPO* Western Central Pacific Ocean

reversing ocean currents, are significant areas of demonstrated restricted gene flow for several mangrove species and are worthy of further investigation for other IWP mangroves.

Contemporary Anthropogenic Factors Affecting Mangrove Distributions

From the brief descriptions below, it is obvious that the major anthropogenic factors affecting mangroves and their distribution are overexploitation, land use changes, hydrological alteration and pollution.

Overexploitation

Around the world, mangroves have been overexploited (Blasco et al. 2001; Zorini et al. 2004; Hauff et al. 2006; Mangora 2011; Richards and Friess 2016). For example, in eastern Africa, uncontrolled wood extraction, particularly of *Rhizophora*, *Bruguiera* and *Ceriops* – which, together with *Heritiera*, are used for construction – has adversely affected many mangrove areas. *Avicennia marina*, which is widely used for firewood and charcoal, has also been affected significantly (Ngoile and Shunula 1992; Rasolofo 1993). Along the Arabian peninsula, mangrove wood has been used for building huts and boats, for making bird traps and for firewood while the foliage has been grazed by camels and been used for fodder. Over-exploitation as camel fodder around the town of Djibouti has been documented (Faye 1993).

Giri et al. (2010) have shown that the total mangrove extent has in recent years declined to 137,760 km^2, some 12% below the latest estimates by the Food and Agriculture Organization. Of even more concern is that of much of the remaining areas of mangroves, a large proportion is degraded and that undisturbed, dense

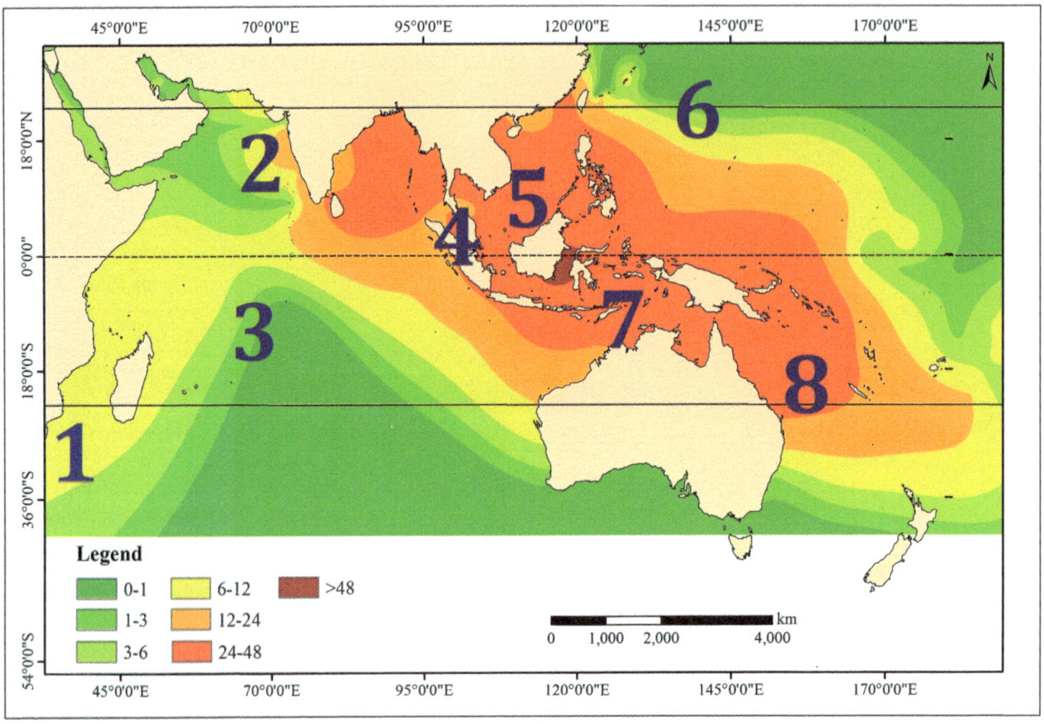

Fig. 23.2 Approximate distribution of mangrove species richness in the Indo-West Pacific with the numbered locations (as per Table 23.10) of those areas where restricted gene flow has been demonstrated for several mangrove species

mangroves have disappeared from many parts of the world (Blasco et al. 2001).

Land Use Changes

Conversion of mangroves for agricultural purposes is widespread in the Indo-West Pacific. For example, in the Rufiji Delta in Tanzania, mangrove conversion to rice commenced in the 1960s, which was ultimately prohibited by law in 1987. While initially good harvests were obtained, ultimately rice fields decreased productivity and were abandoned. Rice cultivation on the landward edge of the mangroves in Madagascar has also been attempted on a large scale (e.g. in the delta of the Tsiribihina), but it is now largely abandoned because of the salinization of the soil. Salt ponds also exist in the delta as well as shrimp culture, causing the loss of extensive mangrove areas (Rasowo 1992; Rasolofo 1993).

Saltworks commenced in the Rufiji Delta, Tanzania, in 1930. However, had spread in the 1970–80 period and by 1989, there were 3100 ha of saltponds in the mangroves, many of which were poorly constructed. The mangroves have been most affected in Bagamoyo, Dar es Salaam, Tanga, Mtwara and Lindi.

In parts of the Middle East, the principal concern has been coastal infill and construction work in and around mangroves, sometimes on a very large scale (Vousden and Price 1985). Infilling and reclamation works are still occurring in many parts of the world (e.g. Tarut Bay, Saudi Arabia).

Possibly the largest effect on mangroves has arisen from the construction of aquaculture facilities in the mangroves. In countries like Indonesia, India, Thailand, Vietnam and China, massive destruction of mangroves occurred in the 1980s and beyond. In 1978 the first shrimp mariculture attempts were made in Kenya supported by UN agencies. With average production levels of 525 kg/ha/year, the Ngomeni aquaculture pilot project caused the destruction of 60 ha of mangroves in Ngomeni Creek (Rasowo 1992). Similarly, shrimp farming in the Mekong Delta of Vietnam increased massively in the 1990s and became unsustainable due to the unplanned development, the self-pollution of farms, the outbreak of viral diseases and the destruction of mangroves (de Graaf and Xuan 1998).

Hydrological Alteration

Mangrove losses caused by changed hydrological conditions are widespread in the IWP (Ghosh et al. 2015; Haque and Reza 2017). On the other hand, localized losses have also been reported. For example, near the port on Khor Farasan, Saudi Arabia, a causeway has adversely affected a large area of *Avicennia marina*, and it appears that initially insufficient culverts were provided. As a result, some mangrove mortality resulted. More culverts were subsequently inserted, and the surviving mangroves now appear to be doing well. Mandura and Khafajii (1993) concluded that 63.3% of the trees died upstream of the causeway.

Pollution

Adverse effects of pollution on mangroves have been widely reported throughout the IWP (Vane et al. 2009; Chowdhury et al. 2017). Of least concern is pollution by sewage, which the mangroves appear to tolerate reasonably well. On the other hand, heavy metal pollution has been found to reduce genetic diversity in *Avicennia marina* (Manurung et al. 2017). They found that genetic diversity in this species was inversely proportional to pollution levels in three estuaries, ranging from non-polluted to moderately and highly polluted.

In contrast to the Red Sea, the mangroves of the Arabian Gulf coast are extremely limited in their distribution and were adversely affected by the large oil spill from the Gulf War (Pashaei et al. 2015). Smaller oil spills, e.g. at Umm-Al-Quwain Bay, United Arab Emirates, in January 1998, can still affect considerable mangrove areas – in this case some 20 km of coastline (Youssef et al. 2000).

Reversing the Anthropogenic Changes

Reducing Habitat Loss

With the current annual loss of mangrove area at 0.4% around the world (Hamilton and Casey 2016), there is increasing interest in implementing better management of existing healthy mangrove areas. Rogers et al. (2016) suggested several actions that could improve protection of coastal wetlands in Australia and the ecosystem services they provide. As these actions seem appropriate more widely, they are briefly listed and discussed. The suggested initiatives include benchmarking and improving coastal wetland extent and health through restoration and/or rehabilitation, reducing complexity and inconsistency in governance arrangements and facilitating wetland adaptation and ecosystem service delivery using a range of relevant mechanisms. They suggested that actions that build on the momentum to mitigate climate change by sequestering carbon could achieve multiple desirable objectives such as climate-change mitigation and adaptation, floodplain rehabilitation and habitat protection. Clearly for mangroves, such an approach should include the establishment or extension of mangrove green belts to provide a natural, self-repairing barrier for coastal protection, as well as the afforestation of suitable coastal areas with mangroves (Saenger 2011).

Reverse Trend of Mangrove Loss by Rehabilitation

López-Portillo et al. (2017) have recently reviewed mangrove restoration or rehabilitation projects around the world and concluded that, with detailed and site-specific project design, mangrove restoration/rehabilitation schemes can offset some mangrove losses and replace some of the ecosystem services that may have been reduced or lost during mangrove deterioration. They provided a framework for planning and implementation of restoration projects, as well as monitoring and reporting protocols. A similar framework, together with a case study of the restoration of the Chokoria Sundarbans of Bangladesh, was described by Biswas et al. (2009).

To date, insufficient research has focused on the establishment and maintenance of restoration projects, and accumulated practical experience has not been widely disseminated. In addition, innovative research in this area has only just commenced. For example, El-Tarabily and Youssef (2011) have recently shown that bacteria capable of producing high levels of 1-aminocyclopropane-1-carboxylic acid deaminase, such as *Pseudoalteromonas maricaloris,* have the potential as biological inoculants to promote growth of mangrove seedlings in rehabilitation projects in nutrient-poor sediments in hypersaline coastal areas.

Some Examples of Success Stories

Many mangrove restoration/rehabilitation projects have been undertaken, but there are a few studies that have evaluated the flow of ecosystem services from such regenerated ecosystems (see Field 1996; Primavera and Esteban 2008). It is only recently that attempts have been made to understand and quantify restoration trajectories by which success or otherwise of restoration/rehabilitation projects can be usefully compared (Twilley and Rivera-Monroy 2005; Aung et al. 2011; Salmo et al. 2013; López-Portillo et al. 2017).

Mangrove restoration around Paradip port of Orissa, India, was undertaken using ten mangrove species that had been nursery-raised. Planted over a 10 ha degraded wetland area in pure and mixed stands, survival rates greater than 70% were found for *S. apetala*, *A. officinalis*, *R. mucronata*, *K. candel* and *H. fomes*. *Sonneratia apetala* recorded maximum growth in height (3.0 m after 2 years), and generally, growth performance was better with plants in mixed stands than in monospecific ones (Das et al. 1997).

The deltaic Chokoria Sundarbans, situated south of Chittagong on the Bangladesh east coast, were badly degraded due to a proliferation of shrimp cultivation and salt farms. With the removal of mangroves, shrimp cultivation became unsustainable, and most farms closed, leaving the area vulnerable to frequent cyclones and tidal surges. Considering the run-down conditions of the site, an area-specific restoration and community plan was prepared by IUCN Bangladesh, with active participation of the local community. While some initial success was reported, the necessary database to assess pre- and post-restoration conditions was not realized due to funding shortages, and it was likely community interest would wane (Biswas et al. 2009). The project nevertheless assisted with developing a useful planning framework.

Das (2017) assessed the contribution of *Avicennia marina* plantations in the Indian state of Gujarat to the inshore and offshore fisheries' sectors and found that the mangrove plantations had significantly increased the catch of mangrove-dependent fish in both sectors. Despite these plantations being monospecific and sparse, they nevertheless contributed around US$ 0.57 billion to the fishery sector of the state.

Importance of Mangroves, Seagrasses and Saltmarshes

With annual mangrove losses of around 0.4% and other halophytes declining at similar rates, it seems timely to reconsider the importance of halophytes in general and mangroves in particular.

Trophic Significance

Early studies suggested that mangroves provided the primary carbon source fuelling detrital-based food webs, and therefore they represented the major route of energy flow in such systems. However, mangrove litter dynamics, however, have shown that all mangrove systems are highly variable and that the export of organic matter is at best about one-third of the total organic matter production. More importantly, stable isotope analyses have investigated the real pathways of energy flow in mangrove ecosystems and have progressively shown that the contribution of mangroves to many species in the estuarine food web was highly variable, but not as significant as previously believed (Loneragan et al. 1997; Thimdee et al. 2004, 2008; Taylor et al. 2018).

Mangrove foliage is nutritious and has been used as fodder. For example, mangrove propagules of *Avicennia marina* have been used as a substitute for corn in commercial feed for fingerlings of bluespot mullet (*Valamugil seheli*) at rates of up to 300 g kg^{-1} (Belal 2004). Use of mangrove propagules rather than dietary corn led to increased growth, body moisture, ash and protein and a reduction in body fat.

Ecosystem Services

Mangroves provide a range of ecosystem services of considerable economic value (Saenger 2002; Hussain and Badola 2008; Iftekhar and Takama 2008; Walters et al. 2008; Van Oudenhoven et al. 2014; Friess 2016). In the Sundarban mangroves of Bangladesh, Rahman et al. (2018) found that mangrove fodder accounted for 2% of the provisioning services obtained by 100 households in 3 villages and that fishery capture, fuel energy and honey accounted for 86%, 7% and 4.7%, respectively, of an annual total of US$1135 per ha.

The significant role of mangroves in carbon sequestration has recently been highlighted as part of the climate-change discussions. Mangrove ecosystems contain some of the largest carbon stocks quantified to date (Rahman et al. 2015).

Using an economic cost-benefit model, Tri et al. (1998) showed that mangrove rehabilitation undertaken, *inter alia*, to enhance sea defence systems in three coastal districts of northern Vietnam, is desirable from an economic perspective based solely on the direct use benefits by local communities. Such activities have even higher benefit-cost ratios with the inclusion of the indirect benefits resulting from the avoided maintenance cost for the sea dike system that the mangrove stands protect from coastal storm surges. Thus, Tri et al. (1998) concluded that a strong case for mangrove rehabilitation could be made as an important component of a sustainable coastal management strategy. It should also be noted that the correlation between those coastal communities not protected by mangrove forests and the resulting high loss of life and property as a result of the 2004 tsunami was significant and has led to enhanced restoration attempts of mangrove habitat along several Indian Ocean shorelines (Erwin 2009; Saenger 2011).

Biotechnology Potential of Halophytes

Mangroves have large potential as sources of novel agrochemicals and compounds of medicinal value (Bandareanayake 2002). Also, mangroves help with the immobilization of heavy metals in mangrove sediments (Chowdhury et al. 2017). Because of the ability of mangroves to tolerate and bioremediate the high levels of nutrients and pollutants found in sewage water, they have been proposed as a simple and effective alternative to conventional sewage treatment, particularly given their low cost and low maintenance (Lambs et al. 2011). Mangroves have also been found to be tolerant of, and to accumulate, high concentrations of trace metals and to thus minimize metal movement to adjacent ecosystems (Vane et al. 2009; Chowdhury et al. 2017).

Finally, mangroves and halophytes in general carry the genetic material that has given them their salt-tolerance capability. While the physiological and biochemical basis of salt tolerance is only gradually being elucidated, there is a growing appreciation of some of the genetic complexities involved. The possibility of using mangrove proteins or genes to enhance the salt resistance of our crop plants is increasingly promising.

References

Allen JA (1998) Mangroves as alien species: the case of Hawaii. Glob Ecol Biogeogr Lett 7:61–71

Almahasheer H (2018) Spatial coverage of mangrove communities in the Arabian Gulf. Environ Monit Assess 190:85. https://doi.org/10.1007/s10661-018-6472-2

Arnaud-Haond S, Teixeira S, Massa SI, Billot C, Saenger P, Coupland G, Duarte CM, Serrão EA (2006) Genetic structure at range edge: low diversity and high inbreeding in Southeast Asian mangrove *(Avicennia marina)* populations. Mol Ecol 15:3515–3525

Aung TT, Than MM, Katsuhiro O, Yukira M (2011) Assessing the status of three mangrove species restored by the local community in the cyclone-affected area of the Ayeyarwady Delta, Myanmar. Wetl Ecol Manag 19:195–208

Bandaranayake WM (2002) Bioactivities, bioactive compounds and chemical constituents of mangrove plants. Wetl Ecol Manag 10:421–452

Banyikwa FF (1986) The geographical distribution of mangrove forests along the East African coast. In: Status and utilisation of mangroves. Proceedings of a workshop on 'Save the mangrove ecosystems in Tanzania', 21–22 February 1986, Dar es Salaam. Published by the Faculty of Science, University of Dar es Salaam

Barreto MB, Barreto-Pittol E (2012) First report of *Rhizophora racemosa* Meyer (Rhizophoraceae) in the mangrove forests of the Venezuelan Caribbean Coast. Interciencia 37:133–137

Béguinot A (1918) Sulla constituzione dei boschi di Mangrovie nella Somalia italiana. Boll R Soc Geogr Ital 7:295–306

Belal IEH (2004) Replacement of corn with mangrove seeds in bluespot mullet *Valamugil seheli* diets. Aquac Nutr 10:25–30

Biswas SR, Malik AU, Choudhury JK, Nishat A (2009) A unified framework for the restoration of Southeast Asian mangroves – bridging ecology, society and economics. Wetl Ecol Manag 17:365–383

Blasco F, Aizpuru M, Gers C (2001) Depletion of mangroves of continental Asia. Wetl Ecol Manag 9:245–256

Breteler FJ (1969) The Atlantic species of *Rhizophora*. Acta Bot Neerl 18:434–441

Breteler FJ (1977) America's Pacific species of *Rhizophora*. Acta Bot Neerl 26:225–230

Castillo-Cárdenas MF, Díaz-Gonzales F, Cerón-Souza I, Sanjur O, Toro-Perea N (2015) Jumping a geographical barrier: diversification of the mangrove species *Pelliciera rhizophorae* (Tetrameristaceae) across the Central American Isthmus. Tree Genet Genomes 11:822. https://doi.org/10.1007/s11295-014-0822-1

Cerón-Souza I, Toro-Perea N, Cárdenas-Henao H (2005) Population genetic structure of neotropical mangrove species on the Colombian Pacific coast: *Avicennia germinans* (Avicenniaceae). Biotropica 37:258–265

Cerón-Souza I, Rivera-Ocasio E, Medina E, Jiménez JA, McMillan WO, Bermingham E (2010) Hybridization and introgression in New World red mangroves, *Rhizophora* (Rhizophoraceae). Amer J Bot 97:945–957

Chen X, Lin P, Lin Y (1996) Mating system and spontaneous mutation rates for chlorophyll deficiency in populations of the mangrove *Kandelia candel*. Hereditas 125:47–52

Chen L, Wang W, Li QQ, Zhang Y, Yang S, Osland MJ, Huang J, Peng C (2017) Mangrove species' responses to winter air temperature extremes in China. Ecosphere 8(6):e01865

Chowdhury R, Favas PJC, Jonathan MP, Venkatachalam P, Raja P, Sarkar SK (2017) Bioremoval of trace metals from rhizosediment by mangrove plants in Indian Sundarban wetland. Mar Pollut Bull 124:1078–1088

Ciferri R (1939) La associazioni del litorale marino della Somalia meridionale. Riv Biol Colon 2:5–42

Clarke PJ (1993) Dispersal of grey mangrove (*Avicennia marina*) propagules in south-eastern Australia. Aquat Bot 45:195–104

Cooper WE, Kudo H, Duke NC (2016) *Bruguiera hainesii* C.G.Rogers (Rhizophoraceae), an endangered species recently discovered in Australia. Austrobaileya 9:481–488

Cornes MD, Cornes CD (1989) The wild flowering plants of Bahrain. Immel Publishing, London

Dale IR (1938) Kenya mangroves. Zeitschr Weltforstwirtschaft 5:413–421

Das S (2017) Ecological restoration and livelihood: contribution of planted mangroves as nursery and habitat for artisanal and commercial fishery. World Dev 94:492–502

Das P, Basak UC, Das AB (1997) Restoration of the mangrove vegetation in the Mahanadi delta, Orissa, India. Mangrove Salt Marshes 1:155–161

de Graaf GJ, Xuan TT (1998) Extensive shrimp farming, mangrove clearance and marine fisheries in the southern provinces of Vietnam. Mangrove Salt Marshes 2:159–166

De Ryck DJR, Koedam N, Van der Stocken T, van der Ven R, Adams J, Triest L (2016) Dispersal limitation of the mangrove *Avicennia marina* at its South African range limit in strong contrast to connectivity in its core East African region. Mar Ecol Prog Ser 545:123–134

Dodd RS, Fromard F, Rafii ZAA, Blasco F (1995) Biodiversity among West African *Rhizophora*: foliar wax chemistry. Bichem Syst Ecol 23:859–868

Duke NC (1995) Genetic diversity, distributional barriers and rafting continents – more thoughts on the evolution of mangroves. Hydrobiologia 295:167–181

Duke NC (2006) Australia's mangroves: the authoritative guide to Australia's mangrove plants. University of Queensland, Brisbane

Duke NC (2010) Overlap of eastern and western mangroves in the South-western Pacific: hybridization of all three *Rhizophora* (Rhizophoraceae) combinations in New Caledonia. Blumea 55:171–188

Duke NC, Kudo H (2018) *Bruguiera* × *dungarra*, a new hybrid between mangrove species B. exaristata and *B.*

gymnorhiza (Rhizophoraceae) recently discovered in north-east Australia. Blumea 63:279–285

Duke NC, Ball MC, Ellison JC (1998) Factors influencing biodiversity and distributional gradients in mangroves. Glob Ecol Biogeogr 7:27–47

Ellison AM (2004) Wetlands of Central America. Wetl Ecol Manag 12:3–55

Ellison JC (2009) Wetlands of the Pacific Island region. Wetl Ecol Manag 17:169–206

Ellison AM, Farnsworth EJ, Merkt RE (1999) Origins of mangrove ecosystems and the mangrove biodiversity anomaly. Glob Ecol Biogeogr Lett 8:95–115

El-Tarabily KA, Youssef T (2011) Improved growth performance of the mangrove *Avicennia marina* seedlings using a 1-aminocyclopropane-1-carboxylic acid deaminase-producing isolate of *Pseudoalteromonas maricaloris*. Plant Growth Regul. https://doi.org/10.1007/s10725-011-9618-6

Erwin KL (2009) Wetlands and global climate change: the role of wetland restoration in a changing world. Wetl Ecol Wetl 17:71–84

Faye B (1993) Mangrove, sécheresse and dromadaire. Sécheresss 4:47–55

Field CD (ed) (1996) Restoration of mangrove ecosystems. International Society for Mangrove Ecosystems/ ITTO, Okinawa

Friess DA (2016) Ecosystem services and disservices of mangrove forests: insights from historical colonial observations. Forests 7:183. https://doi.org/10.3390/f7090183

Fujimoto K, Miyagi T, Kikuchi T, Kawana T (1996) Mangrove habitat formation and response to Holocene sea-level changes on Kosrae Island. Micronesia Mangroves Salt Marshes 1:47–57

Gentry AH (1982) Phytogeographic patterns as evidence for a Choco refuge. In: Prance GT (ed) Biological diversification in the tropics. Columbia University Press, New York, pp 112–136

Ghosh A, Schmidt S, Fickert T, Nüsser M (2015) The Indian Sundarban mangrove forests: history, utilization, conservation strategies and local perception. Diversity 7:149–169

Ge X-J, Sun M (2001) Population genetic structure of Ceriops tagal (Rhizophoraceae) in Thailand and China. Wetlands Ecology and Management 9:203–209

Giri C, Ochieng E, Tieszen LL, Zhu Z, Singh A, Loveland T, Masek J, Duke N (2010) Status and distribution of mangrove forests of the world using earth observation satellite data. Glob Ecol Biogeogr. https://doi.org/10.1111/j.1466-8238.2010.00584.x

Hamilton S, Casey D (2016) Creation of a high spatio-temporal resolution global database of continuous mangrove forest cover for the 21st century. Glob Ecol Biogeogr 25:729–738

Haque MZ, Reza MIH (2017) Salinity intrusion affecting the ecological integrity of Sundarbans mangrove forests, Bangladesh. Int J Conserv Sci 8:131–144

Hauff RD, Ewel KC, Jack J (2006) Tracking human disturbance in mangroves: estimating harvest rates on a Micronesian Island. Wetl Ecol Manag 14:95–105

Hou D (1960) A review of the genus *Rhizophora* with special reference to the Pacific species. Blumea 10:625–634

Hsueh M-L, Lee H-H (2000) Diversity and distribution of the mangrove forests in Taiwan. Wetl Ecol Manag 8:233–242

Huang Y, Zhu C, Li X, Hu L, Tan F, Zhou R, Shi S (2012) Differentiated population structure of a genetically depauperate mangrove species *Ceriops tagal* revealed by both Sanger and deep sequencing. Aquat Bot 101:46–54

Hughes RH, Hughes JS (1992) Répertoire des zones humides d'Afrique. IUCN/UNEP/WCMC, Cambridge. 808 pp

Hussain SA, Badola R (2008) Valuing mangrove ecosystem services: linking nutrient retention function of mangrove forests to enhanced agroecosystem production. Wetl Ecol Manag 16:441–450

Iftekhar MS, Takama T (2008) Perceptions of biodiversity, environmental services, and conservation of planted mangroves: a case study on Nijhum Dwip Island, Bangladesh. Wetl Ecol Manag 16:119–137

Islam MS, Lian CL, Kameyama N, Wu B, Hogetsu T (2006) Development and characterization of ten new microsatellite markers in a mangrove tree species *Bruguiera gymnorrhiza* (L.). Mol Ecol Notes 6:30–32

Islam MS, Lian C, Kameyama N, Hogetsu T (2014) Low genetic diversity and limited gene flow in a dominant mangrove tree species (*Rhizophora stylosa*) at its northern biogeographical limit across the chain of three Sakishima islands of the Japanese archipelago as revealed by chloroplast and nuclear SSR analysis. Plant Syst Evol 300:1123–1136

Jiménez JA (1987) A clarification on the existence of Rhizophora species along the Pacific coast of Central America. Brenesia 28:25–32

Kado T, Fujimoto A, Giang LH, Tuan M, Hong PN, Harada K, Tachida H (2004) Genetic structures of natural populations of three mangrove species, *Avicennia marina, Kandelia candel* and *Lumnitzera racemosa,* in Vietnam revealed by maturase sequences of plastid DNA. Plant Species Biol 19:91–99

Keay RWJ (1953) *Rhizophora* in West Africa. Kew Bull 8:121–127

Koechlin J, Guillamet J-L, Morat P (1974) Flore et végétation de Madagascar. Cramer, Vaduz, p 687

Kumar A, Khan M, Muqtadir A (2010a) Distribution of mangroves along the Red Sea Coast of the Arabian Peninsula: part 1: the Northern Coast of Western Saudi Arabia. Earth Sci India 3:28–42

Kumar A, Khan M, Muqtadir A (2010b) Distribution of mangroves along the Red Sea Coast of the Arabian Peninsula: part 2: the Southern Coast of Western Saudi Arabia. Earth Sci India 3:154–162

Kumar A, Khan M, Muqtadir A (2011) Distribution of mangroves along the Red Sea Coast of the Arabian Peninsula: part 3: Coast of Yemen. Earth Sci India 4:29–38

Lambs L, Léopold A, Zeller B, Herteman M, Fromard F (2011) Tracing sewage water by ^{15}N in a mangrove ecosystem to test its bioremediation ability. Rapid Commun Mass Spectrom 15:2777–2784

Li MS, Lee SY (1997) Mangroves of China: a brief review. Rorest Ecol Manag 96:241–259

Li X, Duke NC, Yang Y, Huang L, Zhu Y, Zhang Z, Zhou R, Zhong C, Huang Y, Shi S (2016) Re-evaluation of phylogenetic relationships among species of the mangrove genus *Avicennia* from Indo-West Pacific based on multilocus analyses. PLoS One 11(10):e0164453

Loneragan NR, Bunn SE, Kellaway DM (1997) Are mangroves and seagrasses sources of organic carbon for penaeid prawns in a tropical Australian estuary? A multiple stable-isotope study. Mar Biol 130:289–300

López-Portillo J, Lewis RR III, Saenger P, Rovai A, Koedam N, Dahdouh-Guebas F, Agraz-Hernández C, Rivera-Monroy VH (2017) Mangrove forest restoration and rehabilitation. In: Rivera-Monroy VH, Lee SY, Kristensen E, Twilley RR (eds) Mangrove ecosystems: a global biogeographic perspective. Springer, Berlin, pp 301–345

Maguire TL, Saenger P (2000) The taxonomic relationships within the genus *Excoecaria* L. (Euphorbiaceae) based on leaf morphology and rDNA sequence data. Wetl Ecol Manag 8:19–28

Maguire TL, Saenger P, Baverstock P, Henry R (2000) Microsatellite analysis of genetic structure in the mangrove species *Avicennia marina* (Forsk.) Vierh. (Avicenniaceae). Mol Evol 9:1853–1862

Mandura AS, Khafaji AK (1993) Human impact on the mangrove of Khor Farasan Island, Southern Red Sea coast of Saudi Arabia. In: Lieth H, Al Masoom A (eds) Towards the rational use of high salinity tolerant plants, vol I. Springer Netherlands, Dordrecht, pp 353–361

Mangora MM (2011) Poverty and institutional management stand-off: a restoration and conservation dilemma for mangrove forests of Tanzania. Wetl Ecol Manag 19:533–543

Manurung J, Siregar IZ, Kusmana C, Dwiyanti FG (2017) Genetic variation of the mangrove species *Avicennia marina* in heavy metal polluted estuaries of Cilegon Industrial Area, Indonesia. Biodiversitas 18:1109–1115

McGuinness KA (1997) Dispersal, establishment and survival of *Ceriops tagal* propagules in a north Australian mangrove forest. Oecologia 109:80–87

Minobe S, Fukui S, Saiki R, Kajita T, Changtragoon S, Ab Shukor NA, Latiff A, Ramesh BR, Koizumi O, Yamazaki T (2010) Highly differentiated population structure of a mangrove species, *Bruguiera gymnorhiza* (Rhizophoraceae) revealed by one nuclear *GapCp* and one chloroplast intergenic spacer *trnF-trnL*. Conserv Genet 11:301–310

Moore GE, Grizzle RE, Ward KM, Alshihi RM (2015) Distribution, pore-water chemistry, and stand characteristics of the mangroves of the United Arab Emirates. J Coast Res 31:957–963

Naskar K, Mandal R (1999) Ecology and diversity of Indian mangroves. 2 Vols., Daya Publishing House, Delhi

Ngoile MAK, Shunula JP (1992) Status and exploitation of the mangrove and associated fishery resources in Zanzibar. Hydrobiologia 247:229–234

Ono J, Yong JWH, Takayama K, Saleh MNB, Wee AKS, Asakawa T, Yllano OB, Salmo SG, Meenakshisundaram SH, Watano Y, Webb EL, Kajita T (2016) *Bruguiera hainesii*, a critically endangered mangrove species, is a hybrid between *B. cylindrica* and *B. gymnorhiza* (Rhizophoraceae). Conserv Genet 17:1137–1144. https://doi.org/10.1007/s10592-016-0849-y

Ormond RFG, Price ARG, Dawson-Shepherd AR (1988) The distribution and character of mangroves in the Red Sea, Arabian Gulf and Southern Arabia. In: Proceedings of the UNDP/UNESCO Regional Mangrove Project, Colombo, 11–14 November 1986, pp 125–130

Parani M, Lakshmi M, Elango S, Ram N, Anuratha CS, Parida A (1997) Molecular phylogeny of mangroves II. Intra and inter specific variation in *Avicennia* revealed by RAPD and RFLP markers. Genome 40:487–495

Pashaei R, Gholizadeh M, Iran KJ, Hanifi A (2015) The effects of oil spills on ecosystem at the Persian Gulf. Int J Rev Life Sci 5:82–89

Primavera JH, Esteban JMA (2008) A review of mangrove rehabilitation in the Philippines: successes, failures and future prospects. Wetl Ecol Manag 16:345–358

Quisthoudt K, Schmitz N, Randin CF, Dahdouh-Guebs F, Robert EMR, Koedam N (2012) Temperature variation among mangrove latitudinal range limits worldwide. Trees 26:1919–1931

Ragavan R, Saxena A, Jayaraj RSC, Mohan PM, Pavichandran K, Saravanan S, Vijayaraghavan A (2016) A review of the mangrove floristics of India. Taiwania 61:224–242

Ragavan R, Zhou R, Ng WL, Rana TS, Mageswaran T, Mohan PM, Saxena A (2017) Natural hybridization in mangroves – an overview. Bot J Linn Soc 185:208–224

Rahman MM, Khan MNI, Hoque AKF, Ahmed I (2015) Carbon stock in the Sundarbans mangrove forest: spatial variations in vegetation types and salinity zones. Wetl Ecol Manag 23:269–283

Rahman MM, Jiang Y, Irvine K (2018) Assessing wetland services for improved development decision-making: a case study of mangroves in coastal Bangladesh. Wetl Ecol Manag 26(4):563–580. https://doi.org/10.1007/s11273-018-9592-0

Rasolofo VM (1993) Les mangroves de Madagascar. In: Diop ES (ed) Conservation et utilisation rationelle des forêts de mangroves de l'Amérique Latine and de l'Afrique, vol. II. Version française du Rapport sur l'Afrique. ITTO/ISME, Okinawa, pp 248–265

Rasowo J (1992) Mariculture development in Kenya: alternatives to siting ponds in the mangrove ecosystem. Hydrobiologia 247:209–214

Richards DR, Friess DA (2016) Rates and drivers of mangrove deforestation in Southeast Asia, 2000–2012. Proc Natl Acad Sci 113:344–349

Ricklefs RE, Latham RE (1993) Global patterns of diversity in mangrove floras. In: Ricklefs RE, Schluter D (eds) Species diversity in ecological communities. Historical and geographical perspectives. University of Chicago Press, Chicago, pp 215–233

Rogers K, Boon PI, Branigan S, Duke NC, Field CD, Fitzsimons JA, Kirkman H, Mackenzie JR, Saintilan N (2016) The state of legislation and policy protecting Australia's mangrove and salt marsh and their ecosystem services. Mar Policy 72:139–155

Rossi G (1981) Le Quaternaire littoral du Kenya. Z Geomorphol 25:33–53

Ruwa RK (1993) Les mangroves du Kenya. In: Diop ES (ed) Conservation et utilisation rationelle des forêts de mangroves de l'Amérique Latine and de l'Afrique, vol. II. Version française du Rapport sur l'Afrique. ITTO/ISME, Okinawa, pp 230–247

Saenger P (1998) Mangrove vegetation: an evolutionary perspective. Mar Freshw Res 49:277–286

Saenger P (2002) Mangrove ecology, silviculture and conservation. Kluwer Academic Publishers, Dordrecht

Saenger P (2011) Mangroves: sustainable management in Bangladesh. In: Günter S, Weber M, Stimm B, Mosandl R (eds) Silviculture in the tropics. Springer, Berlin, pp 339–347

Saenger P, Moverley J (1985) Vegetative phenology of mangroves along the Queensland coastline. Proc Ecol Soc Aust 13:257–265

Saenger P, Snedaker SC (1993) Pantropical trends in mangrove above-ground biomass and annual litter fall. Oecologia 96:293–299

Saenger P, Specht MM, Specht RL, Chapman VJ (1977) Chapter 15: Mangal and coastal salt marsh communities in Australasia. In: Chapman VJ (ed) Wet coastal ecosystems. Elsevier Scientific Publishing Company, Amsterdam, pp 293–345

Salmo SG, Lovelock C, Duke NC (2013) Vegetation and soil characteristics as indicators of restoration trajectories in restored mangroves. Hydrobiologia 720:1–18

Sandoval-Castro E, Dodd RS, Riosmena-Rodríguez R, Enríquez-Paredes LM, Tovilla-Hernández C, López-Vivas JM, Aguilar-May B, Muñiz-Salazar R (2014) Post-glacial expansion and population genetic divergence of mangrove species *Avicennia germinans* (L.) Stearn and *Rhizophora mangle* L. along the Mexican coast. PLoS One 9(4):e93358

Santamaría-Damián S, Romero-Berny EI, Tovilla-Hernández C, Gallegos-Martínez ME, De la Presa-Pérez JC (2016) Nuevos registros del mangle centroamericano *Avicennia bicolor* Standl. (Acanthaceae) en la costa pacífica del sureste mexicano con notas sobre su distribución y estatus de conservación. Lacandonia 10:19–24

Sheue CR, Liu HY, Yong JWH (2003) *Kandelia obovata* (Rhizophoraceae), a new mangrove species from Eastern Asia. Taxon 52:287–294

Sheue CR, Yong JWH, Yang YP (2005) The *Bruguiera* (Rhizophoraceae) species in the mangroves of Singapore, especially on the new record and the rediscovery. Taiwania 50:251–260

Sheue CR, Yang YP, Liu HY, Chou FS, Chang SC, Saenger P, Mangion CP, Wightman G, Yong JWH, Tsai CC (2009a) Reevaluating the taxonomic status of *Ceriops australis* (Rhizophoraceae) based on morphological and molecular evidence. Bot Stud 50:89–100

Sheue CR, Liu HY, Tsai CC, Rashid SMA, Yong JWH, Yang YP (2009b) On the morphology and molecular basis of segregation of two species *Ceriops zippeliana* Blume and *C. decandra* (Griff.) Ding Hou (Rhizophoraceae) from southeastern Asia. Blumea 54:220–227

Sheue CR, Liu HY, Tsai CC, Yang YP (2010) Comparison of *Ceriops pseudodecandra sp. nov.* (Rhizophoraceae), a new mangrove species in Australasia, with related species. Bot Stud 51:237–248

Spalding M, Kainuma M, Collins L (2010) World atlas of mangroves. Earthscan, London

Sugaya T, Yoshimaru H, Takeuchi T, Katsuta M, Fujimoto K, Changtragoon S (2003) Development and polymorphism of simple sequence repeat DNA markers for *Bruguiera gymnorrhiza* (L.) Lamk. Mol Ecol Notes 3:88–90

Tan F, Huang Y, Ge X, Su G, Ni X, Shi S (2005) Population structure and conservation implications of *Ceriops decandra* in Malay Peninsula and North Australia. Aquat Bot 81:175–188

Taylor MD, Gaston TF, Raoult V (2018) The economic value of fisheries harvest supported by saltmarsh and mangrove productivity in two Australian estuaries. Ecol Indic 84:701–709

Thimdee W, Deein G, Sangrungruang C, Matsunaga K (2004) Analysis of primary food sources and trophic relationships of aquatic animals in a mangrove-fringed estuary, Khung Krabaen Bay (Thailand) using dual stable isotope techniques. Wetl Ecol Manag 12:235–144

Thimdee W, Deein G, Nakayama N, Suzuki Y, Matsunaga K (2008) $\delta^{13}C$ and $\delta^{15}N$ indicators of fish and shrimp community diet and trophic structure in a mangrove ecosystem in Thailand. Wetl Ecol Manag 16:463–470

Tomlinson PB (1986) The botany of mangroves. Cambridge University Press, Cambridge

Tri NH, Adger WN, Kelly PM (1998) Natural resource management in mitigating climate impacts: the example of mangrove restoration in Vietnam. Glob Environ Chang 8:49–61

Twilley RR, Rivera-Monroy V (2005) Developing performance criteria using simulation models of mangrove ecosystem restoration: a case study of the Florida Coastal Everglades. J Coast Res 40:79–93

Tyagi AP (2003) Location and interseasonal variation in flowering, propagule setting and propagule size in mangroves species of the family Rhizophoraceae. Wetl Ecol Manag 11:167–174

Van Oudenhoven APE, Siahainenia AJ, Sualia I, Tonneijck FH, Van der Ploeg S, de Groot RS (2014) Effects of different management regimes on mangrove ecosystem services in Java, Indonesia. Wageningen University, Wageningen and Wetlands International, Bogor

Vane CH, Harrison I, Kim AW, Moss-Hayes V, Vickers BP, Hong K (2009) Organic and metal contamination in surface mangrove sediments in South China. Mar Pollut Bull 58:134–144

Voris HK (2000) Maps of Pleistocene sea levels in southeast Asia: shorelines, river systems and time durations. J Biogeogr 27:1153–1167

Vousden DHP, Price ARG (1985) Bridge over fragile waters. New Sci 1451:33–35

Walters BB, Rönnbäck P, Kovacs JM, Crona B, Hussain SA, Badola R, Primavera JH, Barbier E, Dahdouh-Buebas F (2008) Ethnobiology, socio-economics and management of mangrove forests: a review. Aquat Bot 89:220–236

Wee AKS, Teo JXH, Chua JL, Takayama K, Asakawa T, Meenakshisundaram SH, Onrizal BA, Ardli ER, Sungkaew S, Suleiman M, Tung NX, Salmo SG, Yllano OB, Saleh MN, Soe KK, Tateishi Y, Watano Y, Tsuda Y, Kajita T, Webb EL (2017) Vicariance and oceanic barriers drive contemporary genetic structure of widespread mangrove species *Sonneratia alba* J. Sm in the Indo-West Pacific. Forests 8:483. https://doi.org/10.3390/f8120483

Wilcox BHR (1985) Angiosperm flora of the mangrove ecosystem of the Niger Delta. In: Wilcox BHR, Powell CP (eds) The mangroves ecosystem of the Niger delta: proceedings of a workshop. University of Port Harcourt, pp 34–44

Yang Y, Li J, Yang S, Li X, Fang L, Zhong C, Duke NC, Zhou R, Shi S (2017) Effects of Pleistocene sea-level fluctuations on mangrove population dynamics: a lesson from *Sonneratia alba*. BMC Evol Biol 17:22. https://doi.org/10.1186/s12862-016-0849-z

Youssef T, Al Amry M, Youssef A (2000) Post-spill behavior in an oil contaminated mangrove stand *Avicennia marina* (Forrsk.) Vierh. in UAE. Arab Gulf J Sci Res 18:102–109

Zahran MA (1977) Chapter 10: Africa A. wet formations of the African Red Sea coast. In: Chapman VJ (ed) Wet coastal ecosystems. Elsevier Scientific Publishing Company, Amsterdam, pp 215–231

Zorini LO, Contini C, Jiddawi N, Ochiewo J, Shunula J, Cannicci S (2004) Participatory appraisal for potential community-based mangrove management in East Africa. Wetl Ecol Manag 12:87–102

24 Short Communications: Suggestions for Improving Science Communication for Halophyte Conservation, Research, and Development

Megan Rafter

Abstract

The *Sabkha Ecosystems* book series is a comprehensive guide to sabkha ecosystems and halophytes worldwide. The series covers their societal value, potential for utilization, and the importance of their conservation. However, in this "Information Age," a book series alone is not enough to reach and influence all stakeholders who should be utilizing this information. This short communication chapter suggests how experts can improve their ability to communicate scientific topics, in order to inform and influence their target audience.

Keywords

Conservation · Development · Halophyte · Research · Science communication

Introduction

Unsustainable water usage by a rapidly growing population is in conflict with a changing climate, predicted to involve rising sea levels and occurrences of drought. Only ~3% of Earth's water is freshwater. Therefore, the predicted fallout includes an increased abundance of saline "salt-rich" regions across the globe (Bates et al. 2008). This will affect regional biodiversity and our ability to utilize land, as most plants require freshwater and are highly sensitive to changes in salinity. As salt tolerance is a complex multigene trait, breeding programs aimed at improving crop tolerance have had limited success. Therefore, improved understanding of saline environments and the halophytes, "salt plants," which inhabit them is crucial to building a sustainable future with a reduced dependence on freshwater systems (Flowers 2004; Hamed et al. 2014).

One such resource is the *Sabkha Ecosystems* book series, with sabkha meaning "flat salt-crusted desert" in Arabic. The six volumes, *The Arabian Peninsula and Adjacent Countries* (2002), *West and Central Asia* (2006), *Africa and Southern Europe* (2011), *Cash Crop Halophytes and Biodiversity Conservation* (2014), *The Americas* (2016), and *The Asia Pacific* (2018), provide a comprehensive guide to sabkha ecosystems worldwide, covering their ecology, societal value, and potential conservation and utilization practices. The series highlights the importance of sabkha and halophytes and should be a beneficial information resource to stakeholders, agronomists, policy-makers, donors, landowners, students, researchers, farmers, and businesses.

However, this is the "Information Age." People have access to more data and information

M. Rafter (✉)
ex-UNESCO Bangkok, Bangkok, Thailand

Plant Sciences, Rothamsted Research, Harpenden, UK
e-mail: megan.rafter@rothamsted.ac.uk

B. Gul et al. (eds.), *Sabkha Ecosystems*, Tasks for Vegetation Science VI,
https://doi.org/10.1007/978-3-030-04417-6_24

than they can digest. It is no longer enough to write research into books and papers and trust that it will reach the right people; increasingly experts must directly communicate their work to their target audience to increase its impact. Unfortunately, there is a stereotype that experts, notably scientists, are not particularly skilled at this and lose their key messages by providing too much information. Ineffective science communication dilutes the potential impact of the work being communicated.

This short communication chapter suggests ways in which experts can improve their ability to communicate their work, so that the target audience understand its applications and implications.

The Aim of Science Communication

The aim of science communication is to inform non-expert stakeholders about science-related topics. The purpose of science communication is not to present a topic in a positively biased way. "Overselling" to a non-expert audience can result in mistrust, making it difficult to open future dialogues. The communication should accurately explain the reasoning behind the work and its importance. This can be with the aim to increase the audiences' interest, knowledge, and understanding of the topic or influence their opinions and behavior, for example, for funding or policy decisions.

Core Messages

Core messages are the key to effective science communication. Depending on the aim of the communication, these should be centered on the concept of informing or persuading the audience. The core message should be explicitly stated at the start of the communication, perhaps as the title or as part of the introduction. Actively broadcasting the purpose of the communication removes some of the "mental load" that can make an audience unreceptive. Instead of having to piece together why certain details are being discussed, the audience have a frame of reference to consider and actively engage with what is being presented.

A core message is beneficial to both the communicator and the audience. Everything presented in a science communication piece must link back to the core message; this makes it easier for the communicator to identify the key information to be conveyed and plan how it will be discussed. This ultimately results in a streamlined communication piece that is easy for the audience to follow, understand, and engage with.

Audience

Effective communication is aware of and tailored to its audience and must not be "one size fits all." Different stakeholders have different background knowledge and are interested in and influenced by different factors. Even when the core message is the same, a communication piece should be specifically adapted to appeal to the general factors that influence a stakeholder. This helps to answer the questions "so what?" and "why should I care?" and aligns the stakeholder to the communication message. Below are some examples of the factors that influence stakeholders:

- *Industry*: profit/corporate social responsibility/publicity potential
- *Public*: media/evidence (depending on trust in experts)/perceived potential benefits
- *Farmers*: profit/evidence
- *News media*: "newsworthy" – information of interest to target demographic/funders
- *Policy-makers*: politics – influenced by public opinion and ministerial targets/evidence
- *NGOs*: ideology of NGO/publicity potential/donors

Plain Language

Technical terms (jargon), abbreviations, and acronyms are widely used by experts to explain their work to their peers. However, in science communication, unnecessary or overuse of expert language is ineffective. These terms isolate information away from a wider audience who aren't trained, or necessarily interested, in having to decipher what they are being told. In addition, these terms disrupt the flow of the communication, as they require a pause in the message to provide a clear definition. In longer presentations, the audience may forget or misremember these definitions, resulting in a confused and less accurate core message. Therefore, effective science communication predominantly uses plain, simple language with (defined) technical terms only included where necessary.

Storytelling

When discussing their work with a wider audience, experts sometimes default to an academic communication structure: hypothesis, method, results, discussion, conclusions, and suggestions for future work. This communication style is often unfamiliar to non-expert audiences, making it difficult for them to follow and reducing their ability to take in the core message. Using this communication structure can also cause the communicator to default into "expert/audience mode," where they over assume the audiences' background knowledge and discuss the topic in an unnecessary amount of detail. Further, it can encourage the communicator to go off on tangents and discuss aspects of the topic, or other work, unrelated to the core message.

Generally, most of the population learn about science through mainstream media such as news stories or documentaries, which predominantly use narratives to convey information. Narratives are innately interesting and easy to follow, and their widespread use means that audiences are already primed to understand them. Therefore, integrating a narrative into a science communication piece through storytelling is a simple way to improve how well the core message is received. Planning your communication around a three-part structure with a beginning, middle, and end is one simple way to create a narrative. Additionally, the "and, but, therefore" (ABT) technique can be used to focus information to the core message.

Metaphors and Analogies

In some cases, e.g., fundamental science, the information being discussed can seem abstract to an audience. In these situations, it can be useful to introduce metaphors and analogies to frame the topic in a way that is easily understood, e.g., neurons as "trees" and the brain as a "forest." This creates a visual for the audience to better understand brain structure at a cellular level.

When used as part of a narrative, metaphors and analogies help create a communication piece that is both interesting and memorable. This increases the likelihood of the communication in informing and influencing the direct audience, as well as improving the chance that they will retain and relay the information to their peers.

Communication Formats

Science communication can be verbal, visual, written, or a mixture of the three. When communicating in person, for example, when giving a presentation, it is important to include appropriate visual elements such as; maps, props, photos and diagrams, which explain or emphasize the presentations narrative. Engaging multiple parts of the brain can prevent the audience from losing interest, improve their understanding of the core message, and increase the likelihood of them retaining the information being conveyed.

However, effective science communication doesn't require you to be directly in front of the target audience. Listed below are a range of communication formats that once made can be sent to, or made available online for, interested stakeholders.

Verbal:

- *Audiobook*: Usually a reading of an already existing book. Often several hours long and

split into chapters covering different topics or themes.
- *Podcast*: Shorter "episodes" focused on a single topic or theme. The episode length can vary from 10 minutes to 1 hour long. Podcasts can be formal and scripted or casual, for example, a group of experts discussing a topic in a conversational style.

Visual:
- *Infographics*: The representation of information or data, in a visual format, designed to be understandable at a glance
- *Conceptual Model*: A diagram that depicts the key elements and relationships within a system. Used to help an audience understand the whole system and the interacting elements within the system
- *Video*: A versatile communication format that can be utilized in different ways, for example:
  - A filmed lab tour, showing "science in action."
  - A recording of a public talk/presentation, e.g., TED Talk.
  - A presentation, direct to camera, interspersed with images from the field, diagrams, or other relevant images.

Written:
- *Articles*: A written science communication piece, discussing a specific topic or research theme, featured in a newspaper, publication, or website
- *Blogs*: (See "online platforms")

Online Platforms

As the Internet becomes increasingly accessible, the public are using it as a major source for news and information. Online platforms are therefore a valuable tool for science communication, providing an opportunity for researchers and experts to communicate their work to a wide-ranging audience. As the audience access online spaces at their leisure, a greater amount of information can be presented in different formats (see "Communication Formats"). Broadly speaking, online platforms fall into three categories that can be linked together to reinforce your message:

1. *Websites*
   - For disseminating information to stakeholders that are already interested in your research topic, who have identified the website from search engine results.
   - Multiple topics or themes can be presented in-depth, split across different webpages.
   - Allows the author to publish dual webpages, using expert and plain language, to explain their work to potential expert collaborators and the public.
   - Does not require active engagement with audiences.
2. *Blogs*
   - A casual webspace for science-interested audiences to learn about and discuss your research.
   - Blog posts can be in-depth and informational but are presented in a conversational tone.
   - Requires direct interaction and engagement with readers.
   - Examples: WordPress and Blogger.
3. *Social Media*
   - Can bring your research to a wide range of stakeholders, who may not have previous knowledge or interest in your field.
   - Content should be focused on a single idea/topic and designed to quickly grab the audience's attention. Links to your website or blog can be attached to provide more in-depth information.
   - Requires regular engagement, as the audience can share and discuss the content with others and directly ask you questions.
   - Examples: Facebook, Twitter, and Instagram.

Take-Home Rules

- Build the communication piece around a core message, which is stated from the very start.
- Highlight the strengths, weaknesses, and reasoning behind the topic discussed.
- Target the communication to the target audience.
- Don't overestimate an audience's background knowledge.
- Photos and diagrams can convey a message just as easily as words.
- Avoid the use of jargon, abbreviations, and acronyms. If used, then simple definitions must also be included.
- It is not necessary to include every possible piece of information. Science communication is about informing and opening a dialogue. Those that want to learn more will reach out to continue the discussion.

References

Bates BC, Kundzewicz ZW, Wu S, Palutikof JP (2008) Climate change and water. Technical paper of the intergovernmental panel on climate change. IPCC Secretariat, Geneva, p 210

Flowers TJ (2004) Improving crop salt tolerance. J Exp Bot 55(396):307–319

Hamed KB, Magné C, Abdelly C (2014) From halophyte research to halophytes farming. In: Khan MA, Böer B, Öztürk M, Al Abdessalaam TZ, Clüsener-Godt M, Gul B (eds) Sabkha ecosystems. Tasks for vegetation science, vol 47. Springer, Dordrecht

25 Lignocellulosic Biomass from Sabkha Native Vegetation: A New Potential Source for Fiber-Based Bioenergy and Bio-Materials

Sahar Salem, Saida Nasri, Sourour Abidi, Abderrazek Smaoui, Nizar Nasri, Pere Mutjé, and Karim Ben Hamed

Abstract

Sabkhas are home to many xero-halophytes that could be considered renewable resources for various fiber industries. Lignocellulosic fibers of these plants are proposed as raw material for bioenergy production. Nowadays, ethanol is the major biofuel in use. In this paper, fiber composition of 21 xero-halophytic species from Tunisian sabkhas was evaluated using Ankom 200 Fiber Analyzer and compared with that obtained by other studies. The lignocellulosic biomass of xero-halophytes can be considered as an attractive raw material for the future application for the production of bioethanol, paper of good quality, and reinforcement polymers. Two perennial species are considered an important source of bioethanol production which are *Sarcocornia fruticosa* and *Aristida pungens*. *Suaeda fruticosa* from Tunisia can be used in the production of high-quality lightweight paper.

Keywords

Bioenergy · Biofuel · Fibers · Lignocellulosic biomass · Sabkha · Xero-halophytic species

Introduction

Salt marshes, also called sabkhas, are generally considered as marginal zones with extremely low productivity, although they are home to special halophytic vegetation, which is able to sustain itself despite high-salinity levels as a result of different mechanisms at cellular, tissue, and whole-plant levels (Panta et al. 2014).

S. Salem
Laboratory of Extremophile Plants, Centre of Biotechnology of Borj Cedria, Hammam Lif, Tunisia

Faculty of Sciences of Tunis, University of Tunis El-Manar, Tunis, Tunisia

S. Nasri · S. Abidi
Laboratoire des Productions Animales et Fourragères, Institut National de la Recherche Agronomique de Tunisie (INRAT), Université de Carthage, Ariana, Tunisia

A. Smaoui · K. Ben Hamed (✉)
Laboratory of Extremophile Plants, Centre of Biotechnology of Borj Cedria, Hammam Lif, Tunisia
e-mail: Karim.benhamed@cbbc.rnrt.tn

N. Nasri
Faculty of Sciences of Tunis, University of Tunis El-Manar, Tunis, Tunisia

P. Mutjé
LEPAMAP Group, Department of Chemical Engineering, University of Girona, Girona, Spain

B. Gul et al. (eds.), *Sabkha Ecosystems*, Tasks for Vegetation Science VI,
https://doi.org/10.1007/978-3-030-04417-6_25

Nowadays, the need to insure large-scale biomass production for agriculture and industry becomes very urgent mainly in developing countries. Introducing halophyte species as new crops that are irrigated with saline water is becoming an increasingly attractive, feasible, and sustainable option for ensuring food security in several salt-affected regions (Ben Hamed et al. 2014). Consequently, economic utilization of halophytes is receiving increasing attention, especially in arid regions where intensive irrigation or shortage of freshwater forces people to use marginal resources such as brackish water (Xu et al. 2016). Introducing halophytic plants could be advantageous because they are cheaper to grow and naturally abundant on saline soils and thus do not have any adverse impacts on the human food chain (Sharma et al. 2016). Another major advantage of halophytes is their ability to produce satisfactory yields even under adverse conditions as well as to serve as a sustainable and direct source of income for farmers (Koyro et al. 2014). For example, *Tamarix* sp. can provide biomass production ranging from 26 to 52 t ha-1 under irrigation conditions with treated wastewater or brackish water. This yield is comparable to that of conventional plants grown on fertile soil and irrigated with freshwater (Eshel et al. 2010).

High-yielding biomass from halophytes is being currently promoted worldwide to meet the increasing demand for energy and to contribute to reducing the emissions of greenhouse gases (Panta et al. 2014). Halophytes mostly benefit from C_4 photosynthesis pathway, so they are more productive compared with C_3 conventional crops and they can produce economically enough biomass. Some reports considered the halophytes as one of the most productive sources in terms of lignocellulosic biomass (Abideen et al. 2011). In Mexico, Global Seawater Inc. produces between 850 and 950 L of biodiesel/h from sea grass plants grown in coastal areas (Christiansen 2008). Recently, the Masdar Institute in Abu Dhabi launched a project in collaboration with Boeing and Etihad Airways to produce biofuel for the aviation industry from *Salicornia* irrigated with seawater Panta et al. (2014). Abideen et al. (2011) reviewed the distribution of major halophytic energy plants in the coastal zone of Pakistan and assessed their potential for developing energy plants for bioethanol production. Similar surveys have been done in China (Liu et al. 2012).

Halophytes can, also, be valued as new sources of biomass to be used in many industrial applications (Ventura and Sagi 2013). For example, their fibers can be considered as naturally occurring composites constituted mainly of holocellulose (cellulose, hemicellulose) and lignin with minor contents of sugars, starch, proteins, extractives, and ash. The performance of a plant fiber depends on several factors including chemical composition and physical properties (Hamza et al. 2013). Natural fibers derived from plants demonstrate great potentials for use in plastic (as reinforcement in polymers), automotive, and packaging industries because of their excellent characteristics such as low density, high specific stiffness, good mechanical properties, biodegradability, eco-friendliness, toxicologically harmless, and good thermal and acoustic insulation (Ridzuan et al. 2016 and Schellbach et al. 2016). In this short study, fiber composition of some xero-halophytic species from Tunisian sabkhas which could be used in different industrial applications is reported herein.

Materials and Methods

Sample Collection

Lignocellulosic biomass compositions of 21 halophytic species were examined (Table 25.1). These species were collected from different coastal, salt, and desert areas of Tunisia to determine their potential to produce bioethanol. The plant material (stems and leaves) was air-dried ground and stored in airtight plastic bags for subsequent analysis.

Table 25.1 Xero-halophytic plants from different regions from Tunisia

Species	Family	Province	Geographical characteristics	Habit	Life strategy	Height (cm)
Inula crithmoides	Asteraceae	Tabarka	2 m, 36°57′23″ N 8°45′28.5″E	Shrub	Perennial	50–100
Limonium boitardii	Plumbaginaceae	Tabarka	2 m, 36°57′23″ N 8°45′28.5″ E	Grass	Perennial	10–30
Ammophila arenaria	Poaceae	Borj Cedria	1 m, 36°42′41″ N 10°24′54.6″ E	Grass	Perennial	100–120
Euphorbia paralias	Euphorbiaceae	Borj Cedria	1 m, 36°42′41″ N 10°24′54.6″ E	Grass	Perennial	20–70
Retama raetam	Fabaceae	Borj Cedria	1 m, 36°42′41″ N 10°24′54.6″ E	Shrub	Perennial	100–200
Sarcocornia perennis	Amaranthaceae	Borj Cedria	1 m, 36°42′41″ N 10°24′54.6″ E	Shrub	Perennial	20–30
Halocnemum strobilaceum	Amaranthaceae	Korba	15 m, 36°34′58″ N 10°52′01″ E	Shrub	Perennial	20–60
Arthrocnemum macrostachyum	Amaranthaceae	Sousse	31 m, 35°48′31″N 10°8′13″E	Shrub	Perennial	30–100
Salsola tetrandra	Amaranthaceae	Sousse	31 m 35°48′ 31″N 10°8′ 13″E	Shrub	Perennial	20–50
Sarcocornia fruticosa	Amaranthaceae	Kairouan	35 m, 35°45.723′N 10°7.259′E	Shrub	Perennial	40–80
Atriplex halimus	Amaranthaceae	Kairouan	35 m, 35°45.723′N 10°7.259′E	Shrub	Annual	100–200
Mesembryanthemum nodiflorum	Aizoaceae	Kairouan	35 m, 35°45.723′N 10°7.259′E	Grass	Annual	5–30
Suaeda fruticosa	Amaranthaceae	Kairouan	35 m, 35°45.723′N 10°7.259′E	Shrub	Perennial	100–200
Tamarix gallica	Tamaricaceae	Kairouan	35 m, 35°45.723′N 10°7.259′E	Shrub	Perennial	100–600
Aristida pungens	Poaceae	Kairouan	178 m, 35°3.884′N 10°2.442′E	Grass	Perennial	50–100
Pituranthos sp.	Apiaceae	Kairouan	178 m, 35°3.884′N 10°2.442′E	Grass	Perennial	60–100
Thymelaea hirsuta	Thymelaeaceae	Kairouan	178 m, 35°3.884′N 10°2.442′E	Shrub	Perennial	30–100
Limoniastrum guyonianum	Plumbaginaceae	Sfax	13 m, 34°6.68′ N 9° 9.591′ E	Shrub	Perennial	50–150
Lygeum spartum	Poaceae	Gabes	58 m, 34°28 .408′ N 10° 3.22′ E	Grass	Perennial	20–80
Rhanterium suaveolens	Asteraceae	Gabes	131 m, 35°45.70′N 10°7.20′E	Shrub	Perennial	40–60
Stipa tenacissima	Poaceae	Gabes	58 m, 34°28 .408′ N 10° 3.22′ E	Grass	Perennial	60–200

Lignocellulosic Analysis

Samples were analyzed for neutral detergent fiber (NDF) using the Ankom 200 Fiber Analyzer (ANKOM Technology, Fairport, NY, USA), according to Van Soest et al. (1991). The method estimated neutral detergent fiber (NDF) and acid detergent fiber (ADF).

NDF = Cellulose + hemicellulose + lignin
NDF-ADF = Hemicellulose

The NDF- and ADF-treated plant material was then hydrolyzed with 72% H_2SO_4 to determine cellulose. Lignin was obtained by ashing of hydrolyzed residue.

Results

The chemical analysis was conducted on the species which were largely perennial, except two annual species *Atriplex halimus* and *Mesembryanthemum nodiflorum*. Seven species belonging to the Amaranthaceae family, four to Poaceae, two to Plumbaginaceae, and one each to Fabaceae, Aizoaceae, Euphorbiaceae, Tamaricaceae, Apiaceae, and Thymelaeaceae. Table 25.1 shows a large variation in the height of the species in addition to differences in the amount of cellulose, hemicellulose, and lignin (Table 25.2).

Depending on the amount of cellulose, we can divide these species into three or four groups: species with low amount of cellulose <10% such as *Sarcocornia perennis* (6.95%) and 10–20% cellulose in *Halocnemum strobilaceum*, *Euphorbia paralias*, *Mesembryanthemum nodiflorum*, *Inula crithmoides*, *Limonium boitardii*, *Limoniastrum guyonianum*, and *Salsola tetrandra*. The rest contained between 20 and 35% cellulose in *Retama raetam*, *Thymelaea hirsuta*, *Tamarix gallica*, *Sarcocornia fruticosa*, *Ammophila arenaria*, *Arthrocnemum macrostachyum*, *Rhanterium suaveolens*, *Pituranthos* sp., *Suaeda fruticosa*, *Atriplex halimus*, *Aristida pungens*, and *Lygeum spartum*. The highest amount (more than 45%) was found in *Stipa tenacissima*.

Table 25.2 Lignocellulosic composition of xero-halophytic biomass (% dry weight)

Species	Cellulose	Hemicellulose	Lignin
Inula crithmoides	15.11	11.34	10.63
Limonium boitardii	15.70	10.83	27.01
Ammophila arenaria	28.17	28.97	12.68
Euphorbia paralias	14.54	13.41	20.84
Retama raetam	20.77	11.00	18.85
Sarcocornia perennis	6.95	16.77	7.31
Halocnemum strobilaceum	12.79	24.59	12.80
Arthrocnemum macrostachyum	30.74	24.37	10.58
Salsola tetrandra	16.95	22.07	11.98
Sarcocornia fruticosa	27.19	24.66	8.21
Atriplex halimus	33.70	21.65	13.66
Mesembryanthemum nodiflorum	15.02	13.35	4.71
Suaeda fruticosa	33.64	21.22	12.91
Tamarix gallica	21.56	20.18	15.29
Aristida pungens	35.06	32.75	8.97
Pituranthos sp.	33.56	19.33	12.41
Thymelaea hirsuta	21.09	16.96	24.51
Limoniastrum guyonianum	15.92	13.17	20.74
Lygeum spartum	34.55	25.05	12.71
Rhanterium suaveolens	31.98	13.58	17.77
Stipa tenacissima	45–49.09	18.20–28.90	16.25–23.4

Hemicellulose was low (10–14%) in *Limonium boitardii*, *Retama raetam*, *Inula crithmoides*, *Limoniastrum guyonianum*, *Mesembryanthemum nodiflorum*, *Euphorbia paralias*, and *Rhanterium suaveolens*. Other species had >20% hemicellulose, and the highest amount was found in *Ammophila arenaria* (28.97%) and *Aristida pungens* (32.75%). The lignin content was very low (<10%) in *Mesembryanthemum nodiflorum*, *Sarcocornia perennis*, *Sarcocornia fruticosa*, and *Aristida pungens*, while others had >20% lignin with *Thymelaea hirsuta* and *Limonium boitardii* containing exceptionally high (24.51 and 27.01%) lignin, respectively.

Discussion

Halophytes had a potential use as a source of food for humans and livestock and as crops for biofuel, medicinal, and other industrial purposes, as well as their capacity to rehabilitate salt-affected land (Panta et al. 2014). Halophytes are reported as one of the most productive sources in terms of lignocellulosic biomass of good quality (26–37% cellulose, 24–38% hemicellulose, and < 10% lignin) for ethanol production (Abideen et al. 2011). According to our results, two perennial species are considered an important source of bioethanol production which are *Sarcocornia fruticosa* (27.19% cellulose, 24.66% hemicellulose, and 8.21% lignin) and *Aristida pungens* (35.06% cellulose, 32.75% hemicellulose, and 8.97% lignin). *Arthrocnemum macrostachyum* also could produce lignocellulosic biomass containing 30.74% cellulose, 24.37% hemicellulose, and 10.58% lignin making it a good biofuel candidate.

Plant fibers have recently gained researchers' attention due to the environmental concerns, sustainability, technological advancement, flexibility, and availability for diversified industrial products including pulp and paper, rope, cords, and reinforcement in composite matrices (Abdul Khalil et al. 2015). *Stipa tenacissima*, also known as alpha grass, is an herbaceous plant native to Spain and North Africa (Libya, Tunisia, Algeria, and Morocco) (Espinach et al. 2014), with more than 400 thousand hectares in Tunisia, located in Kasserine, Sidi Bouzid, Gafsa, and Kairouan region (Marrakchi et al. 2011). These lignocellulosic fibers are well known and widely used in the manufacture of high-quality paper, because of their cellulose content which is about 33–38% (Belhassen et al. 2009). Our results showed that some xero-halophytes like *Aristida pungens* (35.06% cellulose), *Lygeum spartum* (34.55% cellulose), *Atriplex halimus* (33.70% cellulose), *Suaeda fruticosa* (33.64% cellulose), and *Pituranthos* sp. (33.56% cellulose) can be used in the production of high-quality lightweight paper.

Alpha fibers (AF) have been already used to reinforce polypropylene (PP) and thermoplastic starch (TPS) polymers. Results showed that alpha grass was found a suitable source of fiber reinforcement of starch-based composites, and also, AF-reinforced TPS composites showed comparable performance with glass fiber (GF)-reinforced PP composites, so they were used for technical and engineering purposes (Gironès et al. 2013 and Espinach et al. 2015). Recently, we showed that AF-reinforced bio-polyethylene (BioPE) and high-density polyethylene (HDPE) composites have good properties, namely, stiffness and tensile strength (data not shown).

Composites reinforced with natural fiber attract attention both in industry and in academic research because of several benefits they can gift if compared with composites made of inorganic fillers. This growing interest considers their abundance, low cost, low density, and also ecological advantages. In particular, all the natural fibers are biodegradable and combine well with the use of biodegradable plastic materials (Borchani et al. 2015).

Fiber composition of xero-halophytic species may differ from one region to another, according to climate change. For example, Abideen et al. (2011) found that *Suaeda fruticosa* contains low amounts of cellulose and lignin (8.67% and 4.67%, respectively), while in our study, these values are about 34% cellulose and 13% lignin. The amount of hemicellulose is the same for both studies (21%). This difference can explain that *S. fruticosa* from Tunisia can be used in the production of high-quality lightweight paper.

Conclusions

This research suggests that xero-halophytes have high biomasses that contain suitable lignocellulosic materials that can be grown without encroaching upon arable land and freshwater.

The lignocellulosic biomass of these plants can be considered as an attractive raw material for the future application for the production of bioethanol, paper of good quality, and reinforcement polymers. In addition, because these plants are abundant in nature, outside the human food chain, and require little maintenance, they are relatively inexpensive biomass.

Chemical composition of species can be influenced by different environmental factors such as climate change; the nature of soil that may differ from one region to another and the extraction method used by different studies can also lead to quantitative and qualitative differences in the biomass of xero-halophytes.

References

Abdul Khalil HPS, Sohra Hossain M, Rosamah E, Azli NA, Saddon N, Davoudpoura Y, Nazrul Islam M, Dungani R (2015) The role of soil properties and it's interaction towards quality plant fiber: a review. Renew Sust Energ Rev 43:1006–1015

Abideen Z, Ansari R, Ajmal Khan M (2011) Halophytes: potential source of lignocellulosic biomass for ethanol production. Biomass Bioenergy 35:1818–1822

Belhassen R, Boufi S, Vilaseca F, Lopez JP, Mendez JA, Franco E, Pèlach MA, Mutjé P (2009) Biocomposites based on Alfa fibers and starch- based biopolymer. Polym Adv Technol 20:1068–1075

Ben Hamed K, Magné C, Abdelly C (2014) From halophyte research to halophytes farming. Sabkha Ecosystems: Series: Tasks for Vegetation Science (Vol. 47) Volume IV: Cash Crop Halophyte and Biodiversity Conservation., édition Khan M.A, Böer B, Öztürk M, Al Abdessalaam T.Z, Clüsener-Godt M, Gul B. Springer (ISBN 978–94–007-7410-0), pp 135–142.bio-energy in China coastal zone. Herald Journal of Agriculture and Food Science Research Vol. 1 (3), pp. 044–051 November, 2012

Borchani KF, Carrot C, Jaziri M (2015) Biocomposites of Alfa fibers dispersed in the Mater-BiR type bioplastic: morphology, mechanical and thermal properties. Compos Part A 78:371–379

Christiansen RC (2008) Sea asparagus can be oil feedstock. Biodiesel Magazine, July 2008

Eshel A, Zilberstein A, Alekparov C, Eilam T, Oren I, Sasson Y, Valentini R, Waisel Y (2010) Biomass production by desert halophytes: alleviating the pressure on food production. Recent Advanced in Energy and Environment ISSN: 1790–5095 ISBN: 978–960–474-159-5

Espinach FX, Julian F, Alcalà M, Tresserras J, Mutjé P (2014) High stiffness performance alpha- grass pulp fiber reinforced thermoplastic starch- based fully biodegradable composites. Bioresources 9(1):738–755

Espinach FX, Delgado-Aguilar M, Puig J, Julian F, Boufi S, Mutj P (2015) Flexural properties of fully biodegradable alpha-grass fibers reinforced starch-based thermoplastics. Compos Part B 81:98–106

Gironès J, Espinach FX, Pellicer N, Alcalà M, Tresserras J, Mutjé P (2013) High performance tensile strength alpha-grass reinforced thermoplastic starch- based fully biodegradable composites. Bioresources 8(4):6121–6135

Hamza S, Saada H, Charrier B, Ayeda N, Charrier-El Bouhtoury F (2013) Physico-chemical characterization of Tunisian plant fibers and its utilization as reinforcement for plaster based composites. Ind Crop Prod 49:357–365

Koyro H-W, Lieth H, Gul B, Ansari R, Huchzermeyer B, Abideen Z, Hussain T, Kahn M (2014) Importance of the diversity within the halophytes to agriculture and land management in arid and semiarid countries. In: Khan M (ed) Sabkha ecosystems: volume IV: cash crop halophyte and biodiversity conservation, vol 47. Springer, Dordrecht, pp 175–198

Liu XZ, Wang CZ, SU Q, Li CK (2012) The potential resource of halophytes for developing

Marrakchi Z, Khiari R, Oueslati H, Mauret E, Mhenni F (2011) Pulping and papermaking properties of Tunisian Alfa stems (*Stipa tenacissima*)- effects of refining process. Ind Crop Prod 34:1572–1582

Panta S, Flowers T, Lane P, Doyle R, Haros G, Shabala S (2014) Halophyte agriculture: success stories. Environ Exp Bot 107:71–83

Ridzuan MJM, Abdul Majid MS, Afendi M, Aqmariah Kanafiah SN, Zahri JM, Gibson AG (2016) Characterization of natural cellulosic fibre from *Pennisetum purpureum* stem as potential reinforcement of polymer composites. Mater Des 89:839–847

Schellbach SL, Monteiro SN, Drelich JW (2016) A novel method for contact angle measurements on natural fibers. Mater Lett 164:599–604

Sharma R, Wungrampha S, Singh V, Pareek A, Sharma MK (2016) Halophytes as bioenergy crops. Front Plant Sci 7:1372. https://doi.org/10.3389/fpls.2016.01372

Van Soest PJ, Robertson JB, Lewis B-A (1991) Methods for dietary fibre, neutral detergent fibre, and nonstarch carbohydrates in relation to animal nutrition. J Dairy Sci 74:3583–3597

Ventura Y, Sagi M (2013) Halophyte crop cultivation: the case for *Salicornia* and *Sarcocornia*. Environ Exp Bot 92:144–153

Xu C, Tang X, Shao H, Wang H (2016) Salinity tolerance mechanism of economic halophytes from physiological to molecular hierarchy for improving food quality. Curr Genomics 17:207–214

26 The Floristical, Ecological, and Syntaxonomical Characteristics of Salt Marshes and Salt Steppes in Turkey

Gul Nilhan Tug, Ahmet Emre Yaprak, and Mecit Vural

Abstract

Even though saline land is thought to be poor for plant biodiversity, a total of 640 taxa from 56 families has been recorded (70 of which are endemic to Turkey) in salt marshes and salt steppes in Turkey. After reviewing phytosociological studies on salt marshes and salt steppes, there were 47 instances recorded from salt marshes and six instances recorded from salt steppes. The smaller number of cases from salt steppes is most likely the result of a lack of phytosociological studies on many inland salt steppes. These areas are under threat from, amongst others, overgrazing, exploitation of agricultural areas, construction of tourism complexes, and harvesting of plants for different purposes. Therefore, in certain areas, measures should be put in place to protect these areas with their valuable biodiversity, and flora communities should be re-evaluated in light of new scientific knowledge. Hereby the authors summarized current knowledge about halophytic plant species and phytosociological units in both coastal and inland halophytic areas of Turkey.

G. N. Tug · A. E. Yaprak (✉)
Department of Biology, Faculty of Sciences, Ankara University, Ankara, Turkey
e-mail: Yaprak@science.ankara.edu.tr

M. Vural
Department of Biology, Faculty of Sciences, Gazi University, Ankara, Turkey

Keywords

Phytosociological studies · Salt marshes · Salt steppes · Syntaxonomical characteristics

Introduction

Even though soil salinity and salinization are severe problems for arid and semi-arid areas, this high level of salt is also important for biodiversity. Turkey, which has large semi-arid regions including closed basins with salt lakes such as Tuz Lake, contains very diverse saline habitat types that have resulted in rich halophytic (salt-resistant) plant diversity (Map 26.1).

Saline lakes and marshes are surrounded by easily differentiated zones, generally according to the changes in soil salinity. This zonation pattern of plants depends not only on the chemical and physical characteristics of soil but also on biological factors such as competition (Bertness 1991; Bertness and Ellison 1987; Pennings and Callaway 1992; Snow and Vince 1984), parasites (Pennings and Callaway 1996), grazing, and mowing (Andersen et al. 1990; Jensen 1985; Kielh et al. 1996). Around salt lakes, close to the saline water source, salt marshes cover the area. The decrease in water content and salinity gradually results in salt steppes or saline meadows.

Saline areas in Turkey can be divided into three main groups: salt marshes, salt steppes, and salt meadows. According to the place of occur-

B. Gul et al. (eds.), *Sabkha Ecosystems*, Tasks for Vegetation Science VI,
https://doi.org/10.1007/978-3-030-04417-6_26

Map 26.1 Distribution map of halophytic plant communities in Turkey

rence, they can also be separated into coastal or terrestrial areas.

Salt marshes occur on the shore of saline lakes, tidal marshes, mangroves, lagoons, and sometimes in terrestrial areas with high saline ground water that is close to the surface. At coastal areas, soil salinity is determined by the time that the area is covered with sea water. At terrestrial areas, many factors such as closeness to the lake, ground water level, and the capillarity determine the level of salinity. Plants in salt marshes are hydrohalophytes (plants living in saline wetlands) and if there are some fresh water sources hydrophytes (plants living in wetlands) may also be present.

Salt steppes and meadows are areas around the above-mentioned salt lakes and marshes that have a relatively low level of salt. Water is not present in these areas throughout the year. In this chapter, the vegetation in terrestrial areas is called saline steppes and the vegetation in Mediterranean and Oceanic areas is called saline meadows.

According to former studies and our experience with saline areas, 640 taxa from 56 families have been found in the salt marshes and salt steppes in Turkey. Of these, 70 are endemic to Turkey. Although the number 640 seems high, it also includes the accompanying taxa that are not truly halophytes but have wide ecological tolerance to salinity. A total of 365 plants are from accompanying taxa species that have a wide ecological tolerance. As it can be seen from Appendix A, the distribution area and the habitat preferences of some taxa coincide. The numbers of taxa belonging to only one geographical region are as follows: Central Anatolia 75, Mediterranean 66, East Anatolia 55, and Black Sea 3. Of the 640 taxa, 39 taxa are distributed all over Turkey. Taxa from the Aegean and Marmara regions were included in the Mediterranean region. A detailed list of taxa is given in Appendix A, along with their habitat type, distribution area, and endemic status (Apaydın et al. 2009; Atamov et al. 2006;

Bağcı et al. 1996, Behcet et al. 2009; Biondi et al. 2013; Birand 1961; Ghazanfar et al. 2014; Guner et al. 2012; Guvensen et al. 2006; Ozturk et al. 2008, 2015, 2016; Secmen and Leblebici 1997; Yaprak and Yurdakulol 2008; Yaprak et al. 2010).

Salt Marshes

Salt marshes in Turkey can be divided into two groups: coastal and terrestrial. Coastal salt marshes occur on the lagoons on the coasts of the Mediterranean, Aegean, and Black Seas and also in large river deltas. Terrestrial salt marshes occur on the shores of saline lakes or areas with high saline ground water levels close to the surface.

Coastal Salt Marshes

Coastal salt marshes generally occur at lagoons or large river deltas (Fig. 26.1). When evaluated geographically, the Mediterranean and Aegean coastal salt marshes, which belong to the Mediterranean phytogeographical region, are similar to each other both floristically and ecologically. The Black Sea coastal salt marshes are different from the other regions, mainly because of the influence of the Euro-Siberian phytogeographical region.

Mediterranean and Aegean Coasts

On the Mediterranean and Aegean coasts, in lagoons that are flooded with sea water almost throughout the year, the hydrohalophytes *Arthrocnemum macrostachyum*, *Sarcocornia obclavata*, *S. perennis*, *Salicornia emericii*, and *Sueda splendens* are found. In the areas that are not flooded throughout the year but affected by the ground water from river basins, plants such as *Halocnemum cruciata*, *H. strobilaceum*, *Sarcocornia perennis*, *Limonium virgatum*, *Inula crithmoides*, *Salicornia patula*, and *Halimione portulacoides* are found. Around lagoons, on the

Fig. 26.1 Coastal salt marsh in Doganbey Lagoon (*Halocnemion strobilacei*). Photograph by A.E. Yaprak

coast, or in river deltas with high ground water levels but sandy areas *Zygophylum album, Cakile maritima, Salsola soda,* and *S. tragus* subsp. *tragus* are present.

The only endemic species found on these coasts is *Sarcocornia obclavata,* but it is highly probable that this species is also distributed in other Eastern Mediterranean countries.

Most of the coastal areas are under the influence of human settlement, and the coastal salt marshes have undergone some degree of destruction. The Mediterranean and Aegean coasts are the most popular tourism centers of Turkey, which leads to heavy pressure on the saline habitats. Although most of the natural habitats of the Aegean and Mediterranean coasts have been destroyed, there are many studies covering the flora of the area. According to the data from these studies, a synopsis of Aegean and Mediterranean coasts is given below. Only the studies that evaluate the vegetation according to Braun Blanquet and define the associations were included in this study. All phytosociological units determined are listed hierarchically under syntaxonomical synopses of each geographical region. The floristic composition is presented in Appendix A (Altay and Ozturk 2012; Gehu 2011, Gehu et al. 1992; Gehu and Uslu 1989; Ozturk and Guvensen 2002; Ozturk et al. 2002, 2006; Guvensen and Ozturk 2003; Vural et al. 1995; Yaprak 2007, 2012; Yaprak and Kadereit 2008; Yurdakulol et al. 1996).

Syntaxonomical Synopsis of Halophytic Vegetation of Aegean Sea Coast

Thero-Salicornietea Pign. 1953 em. Tx. 1958
Thero-Salicornietalia Pign. 1953 em. Tx. 1958
  Salicornion patulae Gehu 1984
    Suaedo-Salicornietum patulae. (Brullo & Furnari 1976) Gehu 1984
    Cressetum cretica Brullo & Furnari 1976
  Thero-Suaedion Br.-Bl. 1931
    Suaedo-Kochietum hirsutae Br.-Bl. 1931
    Suaedo-Salsoletum sodae Br.-Bl. 1931

Saginetea maritimae Westh. V. Leeuw & Adr. 1961 em. Gehu & Biondi 1986

Frankenietalia pulverulentae Riv. Mart. 1976
  Frankenion pulverulentae Riv. Mart. 1976
    Plantagini weldenii-Parapholidetum incurvae Gehu et al. 1986
    Puccinellio festuformis-Hordeetum marini Gehu & Uslu 1989

Sarcocornietea fruticosae Br.-Bl. and Tuxen ex A. & O. Bolos, 1950

Sarcocornietalia fruticosae Br.-Bl. 1933
  Sarcocornion fruticosae Br.-Bl. 1933
    Puccinellio festuciformis-Sarcocornietum fruticosae Br.-Bl. 1928, Géhu 1976
    Halimiono-Artemisietum santonici Gehu & Uslu 1989
Halocnemetalia cruciati Biondi, Casavecchia, Estrelles, & Soriano 2013
  Halocnemion strobilacei Gehu & Costa 1984
    Arthrocnemo glauci – Halocnemetum strobilacei Oberdorfer, 1952

Juncetea maritimi Br.-Bl. (1931) 1952

Juncetalia maritimi Br.-Bl. 1931
  Puccinellion festuciformis Gehu & Scop. 1984
    Limonio gmelinii-Aeluropetum littoralis (Babalonas 1979) Gehu et al. 1986
    Puccinellio festuciformis-Caricetum extensae Gehu & Uslu 1989
  Juncion maritimi Br.-Bl. 1931
    Limonio gmelinii-Juncetum gerardii Gehu & Uslu 1989
    Ranunculo marginati-Caricetum divisae Gehu et al. 1989

Syntaxonomical Synopsis of Halophytic Vegetation of Mediterranean Sea Coast

Thero-Salicornietea Pign. 1953 em. Tx. 1958.
  Thero-Salicornietalia Pign. 1953 em. Tx. 1958

Salicornion patulae Gehu 1984
  Suaedo-Salicornietum patulae (Brullo & Furnari 1976) Gehu 1984

Thero-Suaedion Br.-Bl. 1931
Salicornietum ramosissimae Christians 1955 em. Tx. 1974

Saginetea maritimae Westh. V. Leeuw & Adr. 1961 em. Gehu & Biondi 1986

Frankenietalia pulverulentae Riv. Mart. 1976
Frankenion pulverulentae Riv. Mart. 1976
Cresso creticae-Hordeetum marinae Gehu, Uslu, & Costa 1989
Plantagini weldenii-Parapholidetum incurvae Gehu et al. 1986

Juncetea maritimi Br.-Bl. (1931) 1952

Juncetalia maritimi Br.-Bl. 1931
Juncion maritimi Br.-Bl. 1931
Aeluropetum littoralis s.l.
Juncetum subulati s.l.
Phragmito australis-Juncetum maritimii Vural, Duman, et al. 1994
Carici *extensae-Halocnemetum strobilacei* Vural, Duman, et al. 1994

Sarcocornietea fruticosae Br.-Bl. and Tuxen ex A. and O. Bolos, 1950

Sarcocornietalia fruticosae Br.-Bl. 1933
Junco littoralis-Tamaricetum parviflorae Vural, Duman, et al. 1994
Spergulario marinae-Halimionetum portulacoidis Vural, Duman, et al. 1994
Halimiono portulacoidis-Junceteum littoralis Vural, Duman, et al. 1994
Limonio gmelinii-Junceteum littoralis Vural, Duman, et al. 1994
Bupleuro tenuissimi-Inuletum viscosae Vural, Duman, et al. 1994
Sarcocornion fruticosae Br.-Bl. 1933
Puccinellio festuciformis-Arthrocnemetum perennis (Br.-Bl. 1931) Gehu 1976
Limoniastretalia guyoniani Guinochet 1951 em.
Zygophyllion albi Gehu et al. 1990
Elytrigio junceae-Zygophylletum albiGéhu, Costa & Uslu 1990
Halocnemetalia cruciati Biondi, Casavecchia, Estrelles, & Soriano 2013
Halocnemenion strobilacei Gehu & Costa 1984
Tamaricetum smyrnensis Secmen & Leblebici 1996
Halocnemion cruciati Biondi, Casavecchia, Estrelles, & Soriano 2013
Arthrocnemo macrostachyi–Halocnemetum cruciati Casavecchia, Estrelles, & Soriano 2013
Zygophyllo albi–Halocnemetum cruciati Gehu et al. ex Biondi et al. 2013

Black Sea and Marmara Coasts

Salt marshes along the Black Sea coast cover small areas because of the sudden decrease in land level and the only exception is the Kizilirmak Delta where salt marshes cover relatively large areas. The main plant species found at the Kizilirmak Delta and Cernek Lake is *Juncus acutus*. The flat area surrounding the lake has a high water table level and *Sarcocornia perennis, Salicornia perennans,* and *Spergularia marina* are dominant. In the area with sandy soils and a high water table, *Salsola tragus* subsp. *pontica* can be seen (Apaydin et al. 2009; Yaprak 2008).

Syntaxonomical Synopsis of Halophytic Vegetation of the Black Sea Coast

Juncetea maritimi Br.-Bl. (1931) 1952

Juncetalia maritimi Br.-Bl. 1931
Juncion maritimi Br.-Bl. 1931
Juncetum maritimum Pignatti 1953

The vegetation cover at salt marshes on the Marmara coasts was investigated in detail and the associationed conditions were described. Although it covers a relatively small area and human settlement has caused intense pressure in the area, all the coastal zones were studied and evaluated. Floristic composition of these marshes forms a transition between the Black Sea and the Aegean Sea.

All the syntaxa determined from coastal regions of Turkey can be summarized as below:

Thero-Salicornietea Pign. 1953 em. Tx. 1958

Thero-Salicornietalia Pign. 1953 em. Tx. 1958
Salicornion patulae Gehu 1984
Suaedo-Salicornietum patulae (Brullo & Furnari 1976) Gehu 1984
Thero-Suaedion Br.-Bl. 1931
Suaedo-Kochietum hirsutae Br.-Bl. 1931
Suaedo-Salsoletum sodae Br.-Bl. 1931

Saginetea maritimae Westh. V. Leeuw & Adr. 1961 em. Gehu & Biondi 1986

Frankenietalia pulverulentae Riv. Mart. 1976
Frankenion pulverulentae Riv. Mart. 1976
Plantagini weldenii-Parapholidetum incurvae Gehu et al. 1986

Sarcocornietea fruticosae Br.-Bl. and Tuxen ex A. and O. Bolos, 1950

Sarcocornietalia fruticosae Br.-Bl. 1933
Sarcocornion fruticosae Br.-Bl. 1933
Puccinellio festuciformis-Sarcocornietum fruticosae (Br.-Bl. 1928) Gehu 1976
Halimiono-Artemisietum santonici Gehu & Uslu 1989
Halocnemetalia cruciati Biondi, Casavecchia, Estrelles, & Soriano 2013
Halocnemenion strobilacei Gehu & Costa 1984
Halocnemenion strobilacei Gehu & Costa 1984
Arthrocnemo glauci – Halocnemetum strobilacei Oberdorfer 1952
Tamaricetum smyrnensis Secmen & Leblebici 1996

Juncetea maritimi Br.-Bl. (1931) 1952

Juncetalia maritimi Br.-Bl. 1931
Juncion maritimi Br.-Bl. 1931
Juncetum maritimum Pignatti 1953

Terrestrial Salt Marshes

Terrestrial salt marshes naturally occur around saline lakes such as the Tuz, Seyfe, and Burdurand Pallas lakes, and in areas with high saline ground water levels.

Central Anatolia

The main salt marshes of Central Anatolia are found surrounding Tuz Lake, Bolluk, Tersakan, and Seyfe lakes, Sultan Marshes, Tuzla Lake (Kayseri), Akgol, the areas between Ulukisla and Eregli, and Altunhisar and Bor, and surrounding areas of Gunyuzu and Kavuncu (Eskisehir), around saline streams and lakes between Cankiri and Corum, surrounding areas of Nallihan-Davutoglan (Ankara), saline areas between Kalecik, Sungurlu, and Cicekdag, and the surrounding areas of Balikdami (Eskisehir). Salt steppes and salt marshes can overlap with each other, but generally salt marshes gradually turn into salt steppes. The vegetation around the salt lakes of Central Anatolia has been studied in detail. Tuz Lake, Seyfe Lake, and the Sultan Marshes were evaluated ecologically and the vegetation cover around these areas shows a characteristic zonation pattern with three different zones (Aksoy and Hamzaoglu 2006; Tug 2006; Tug et al. 2011, 2012). All the salt marshes in Central Anatolia are similar both floristically and ecologically, and for this reason the floristical and ecological structure of these habitats has been explained using the examples of the Tuz and Seyfe lakes.

Around Tuz Lake and Seyfe Lake, the zones that are mostly flooded with water throughout the year are covered with *Salicornia freitagii* and at the boundary of this zone *Salicornia perennans* and *Halocnemum strobilaceum* dominate. Rarely, resistant individuals of some glycophytes (salt-sensitive plants) can be seen in the homogeneous *Salicornia* or *Halocnemum* communities. This zone has the highest salinity and water level (Fig. 26.2). During the summer months when the land is dry, the surface of the soil is covered with a white salt crust. There are some fresh water sources that feed the lakes, and the areas around these fresh water sources have a decreased salinity (Fig. 26.3). *Phragmites australis, Elymus elongatus* subsp. *sal-*

Fig. 26.2 Salt marsh in Tuz Golu (*Halocnemetum strobilacei*). Photograph by A.E. Yaprak

sus, Juncus maritimus, Puccinellia koeieana subsp. *anatolica, Eragrostis collina, Molinia caerulea, Inula aucherana, Verbascum pyroliforme, Scorzonera parviflora, Plantago maritima, Falcaria falcarioides, Schoenus nigricans, Lotusstrictus,* and *Cladium mariscus* compose local communities (Aksoy and Hamzaoglu 2006; Tug 2006; Tug et al. 2012). Between the lake shore and the salt steppes, the water saturation and soil salinity decrease and comprise a transition zone between marsh and steppe. With decreasing salinity, biodiversity increases, and *Petrosimonia nigdeensis, P. brachiata, Salsola inermis, S. stenoptera, S. nitraria, S. macera, Aeluropus littoralis, Halimione verucifera, Taraxacum farinosum, Microcnemum coralloides* subsp. *anatolicum, Puccinelia convolute, Frankenia hirsute, Limonium iconicum,* and *Lepidium cartilagineum* subsp. *caespitosum* dominate the vegetation. Some *Tamarix* species can also be found. When salinity occurs with alkalinity, species such as *Salsola crassa* and *Taraxacum mirabile* can be found. There are some small depressions like small islands spread trough the salt steppe zone that show the characteristics of the lake shore because of the high ground water level. In these depressions, *Halocnemum strobilaceum* grows. There are also some seasonal endemic species that complete their lifecycle when soil salinity of the topsoil decreases with precipitation, e.g., *Holosteum umbellatum, Thelungiella parvula,* and *Chrysocamela elliptica.* These species have seeds that can tolerate high salt levels and survive until the conditions become suitable for germination. In this zone, there are also some fresh water sources, which results in the occurrence of glycophyte communities composed of species from Poaceae, Cyperaceae, and Juncaceae (Kurt et al. 2004). Gradually, the hydrohalophytes are less common in the area and with a decrease in both salt and water level, xerohalophytes (plants living in arid saline areas) dominate the vegetation, which in turn become less common in the area, and more steppe species with low salinity tolerance are found. The ecological and floristi-

Fig. 26.3 Salt marsh in Tuz Golu (*Inulo aucheranae-Elymion salsi*). Photograph by A.E. Yaprak

cal characteristics of this zone are explained in the section on salt steppes.

The main difference between salt marshes around salt lakes and the areas with a high saline water table is the presence of *Salicornia freitagii*. All other characteristics are almost the same as in the salt marshes mentioned above.

The endemic level around Tuz Lake is very high. In total, there are 406 taxa known as halophytes in Turkey, and 65 of them are endemic (Ozturk et al. 2008, 2016; Yaprak and Tug 2006). Forty-seven of these endemic taxa are found around Tuz Lake (Tug et al. 2011). *Salicornia frietagii, Salsola stenoptera, Lepidium cartilegineum* subsp. *caespitosum,* and *Petrosimonianigdeensis* can be found in almost all the halophytic areas of Central Anatolia. The endemism level is especially high at the southern part of Tuz Lake and this area is also rich in local endemic plants such as *Kalidium wagenitzii, Hypericum salsugineum, Acantholimon halophila,* and *Verbascum pyroliforme*. Terrestrial salt marshes have a higher endemic ratio compared to the coastal salt marshes, and this is mainly the result of isolation of these areas after the receding of the Tetis Sea.

The Central Anatolian salt marshes can be thought of as an island separated from the Central Asian, Iranian, and Causasian saline areas, yet all show high similarity. The Irano-Turanien phytogeographical elements are very dominant in Central Anatolia. *Halocnemum strobilaceum, Halimione verrucifera,* and *Asparagus persicus* are examples of species with a common distribution area. Central Anatolian salt marshes are unique, with the presence of some species having discontinuous distribution. For example, *Anabasis aphylla* is only found in Nallihan (Ankara), and has a wide distribution area in China, East Asia, and North Iran (Freitag et al. 1999); *Tetradiclis tenella* recorded from Kirikkale is also distributed in Central Asia, Iran, Pakistan, Egypt, and Palestine (Hamzaoglu et al. 2005); and *Microcnemum coralloides* subsp. *anatolicum* is found throughout Central Anatolia, Iran, and Armenia (Kadereit and Yaprak 2008). *Asparagus lycaonicus* was known as a local endemic from Tuz Lake until 2002 when it was recorded in Iran, and thus is also a good example of this discontinuous distribution (Akhani 2002). Closely related species that are vicariant with each other are also good examples of discontinuous distribution. For example, *Kalidium wagenitzii* is found in a very small area at the south of Tuz Lake and *Kalidium foliatum* is found in Mongolia, South Siberia, and Central Asia.

Salt marshes around Tuz Lake and Seyfe Lake were syntaxonomically studied in detail and the syntaxa defined from these areas are summarized below.

Sarcocornietea fruticosae Br.-Bl. and Tuxen ex A. and O. Bolos 1950

Halostachyetalia (Grossheim) E. Topa 1938
- *Halocnemetum strobilacei* (Keller) E. Topa 1938
  - *Petrosimonietum nigdeenso-brachiatae*, Yurdakulol et al. 1996
- *Lepidio caespitosi-Limonion iconici* Aydogdu et al. 2002
  - *Lepidio caespitosi-Limonietum iconici* Aydogdu et al. 2002
  - *Limonio tamaricoidis-Puccinellietum convolutae* Aydogdu et al. 2002
  - *Sphenopodo divaricati-Halaocnemetum strobilacei* Aydogdu et al. 2002
  - *Suaedo anatolicae-Salsoletum nitrariae* Aydogdu et al. 2002
  - *Bupleuro gracili-Limonietum gmelini* Yurdakulol ex Hamzaoglu 2009

Juncetea maritimi Br.-Bl. (1931) 1952

Juncetalia maritimi Br-Bl. 1931
- *Inulo aucheranae-Elymion salsi* Aydogdu et al. 2002
  - *Inulo aucheranae-Elymetum salsi* Aydogdu et al. 2002
  - *Eragrostio collinae-Puccinellietum anatolicae* Aydogdu et al. 2002
  - *Tamaricetum parviflorae-tetrandrae* Hamzaoglu & Aksoy 2006
- *Juncion maritimi* Br.-Bl. 1931

Juncetum maritimimum Pignatti 1953

Mediterranean

Terrestrial salt marshes of the Mediterranean region show similar characteristics to the Central Anatolian salt marshes because of their similar geography. Even though they are situated in the Mediterranean geographical region of Turkey, they are floristically isolated from the Mediterranean because of the Taurus Mountains and they are very close to the Central Anatolian Salt Marshes. Burdur Lake and Acigol are examples from the Mediterranean region for terrestrial salt marshes. These areas were studied syntaxonomically and five types were described by Yurdakulol et al. (1996).

Syntaxonomical Synopsis of Halophytic Vegetation of Mediterranean Salt Marshes

Juncetea maritimi Br.-Bl. (1931) 1952

Juncetalia maritimi Br.-Bl. 1930
- *Juncion maritimi* Br.-Bl. 1931
  - *Cardopatio-Juncetum heldreichianii* Yurdakulol et al. 1996
  - *Pucciniellio distantis-Juncetum maritimi* Yurdakulol et al. 1996

Sarcocornietea fruticosae Br.-Bl. and Tuxen ex A. and O. Bolos 1950

Lepidio casepitosi-Halimionetum portulacoides Yurdakulol et al. 1996

Cresso creticae-Halocnemetum strobilacei Yurdakulol et al. 1996

Sarcocornietalia fruticosae Br.-Bl. 1933

Sarcocornion fruticosae Br.-Bl. 1933
- *Petrosimonietum nigdeenso-brachiatae* Yurdakulol et al. 1996

Eastern Anatolia

In Eastern Anatolia, there are not as many wide marsh areas as in Central Anatolia. Although Van

Lake is one of the biggest brackish-soda lakes of the world, except for the river entrance areas, the lake shores are very high above the lake level and do not classify as salt marshes. The salinity of the lake is about 0.21% (Kaden et al. 2010) and the salinity level decreases close to the river entrance where small marshes are dominated by salt-tolerant hydrophytes. Around Van Lake, which is the biggest lake in Turkey, there are seven defined communities according to the soil characteristics: *Puccinellia gigantea* over saline alkaline soils, *Iris musulmanica* community over clayey-saline soils with co-dominants *Taraxacum scaturiginosum* and *Hordeum brevisubulatum* subsp.*violaceum*, *Potomageton pectinatus* community over less saline loamy soils, *Bolboschoenus maritimus-Triglochin maritima* community over clayey saline soils, *Menyanthes trifoliata-Carex diluta* subsp. *Diluta* community over saline clayey-loamy soils, *Scorzonera parviflora* community over saline loamy soils, and *Alopecurus arundinaceae* community over clayey less saline soils (Ozturk et al. 1995).

Salt marshes in Eastern Anatolia mainly occur in areas where ground water is high along the Aras Valley, around small lakes, and in saline water sources. In salt marshes around small saline lakes and saline water sources, *Salicornia perennans, Suaeda linifolia,* and *Phragmites australis* dominate the vegetation. The most characteristic taxa of the marshes at Igdir Valley where the saline ground water comes to the surface are *Kalidium capsicum, Halostachys belengeriana, Bienertia cycloptera, Puccinellia bulbosa,* and *Halocnemum strobilaceum.* There are small salt marshes at Erzurum Narman, which are surrounded by gypsaceous hills, and are dominated by *Aeluropus litoralis, Salicornia perennans, Suaeda carnosisima,* and *Petrosimonia glauca.*

Salt marshes in Eastern Anatolia have a patchy distribution around small salt pans, salt lakes, and saline water sources. Kirmizi Salt Pan at Erzurum-Karacoban and Aktuzla Bahce Salt Pan at Mus-Malazgirt are examples of these kinds of small marshes, which were floristically studied, but there is not enough information about their ecological characteristics. In these salt marshes *Salicornia* sp., *Puccinelia gigantea, Petrosimonia squarrosa, Salsola kali* subsp. *tragus, Atriplex laevis, Lepidium ruderale, L. pinnatifidum, Euclidium syriacum, Eremopyrum bonaepartis* subsp. *hirsutum, Crypsis alopecuroides,* and *C. aculeata* were found (Behcet 1994a, b; Behcet et al. 2009; Ozturk et al. 2015).

There are also some salt marshes with low salinity around Erzurum Dumlu, and the species that generally dominate in this area are *Plantago maritima, Scorzonera hieracifolia, Sueda altissima, Polygonum patulum* subsp. *patulum, Puccinella distans* subsp. *distans, P. distans* subsp. *sevangensis, P. koeiena, Carex echinata, Lepidium cartilagineum* subsp. *caespitosum, Glaux maritima, Limonium gmelinii, Triglochin maritima,* and *Taraxacum scaturiginosum.*

The endemism ratio of Eastern Anatolian salt marshes is very low, since they are extensions of the salt marshes in Iran, Azerbaijan, and Armenia. The only endemic species of these salt marshes is *Limonium vananse.*

Syntaxonomical Synopsis of Halophytic Vegetation of East Anatolian Salt Marshes

Juncetea maritimi Br.-Bl. (1931) 1952

Juncetalia maritimi Br.-Bl. 1931
  Juncion maritimi Br.-Bl. 1931
    Scorzoneretum parviflorae Behcet 1994
    Puccinellietum giganteae Behcet 1994

  Molinio-Arrhenatheretea Tuxen 1937

Alopecuretum arundinacea Behcet 1994

Salt Steppes

Salt steppes are ecotones between salt marshes and calcareous or gypsaceous steppes. Salt steppes never flood and are located marginally higher than the marshes. Generally, they form a wide zone and are distributed where the soil salinity and ground water level are relatively lower than in salt marshes (Fig. 26.4).

In Turkey, natural salt steppes are found at Tuz Lake, Seyfe Lake, Burdur, Konya Plain,

Fig. 26.4 Salt steppe in Tuz Golu (*Achilleo wilhelmsii-Artemision santonici*). Photograph by A.E. Yaprak

Acigol, around salt marshes and small salt lakes at the Develi Closed Basin, along Aras Valley from Kars Kagizman to Igdir-Aralik, in saline areas between Kirikkale Baliseyh, Çorum Sungurlu, and Yozgat Yerkoy, and around saline streams and lakes in Cankiri, Balikdami, and Kavuncu in Eskisehir, Tuzla Lake in Kayseri, and Nallihan in Ankara. There are also some newly formed salt steppes in Southeastern Anatolia, which formed because of incorrect irrigation practices.

Central and West Anatolia

The taxa found at the salt steppes of Central and West Anatolia are mostly hemicryptophytes (perennial plants that winter buds stays at the soil surface) and therophytes (plants that complete their life cycle in one vegetation period or less), which is the result of aridity (Tug et al. 2011). *Tamarix smyrnensis*, *T. parviflora,* and *Krascheninnikovia ceratoides* are the only shrubs found in the area. In Central Anatolia, taxa from Chenopodiaceae dominate the salt steppes. Asteraceae, Poaceae, and Plumbaginaceae are the other dominant taxa of the area. Central Anatolian salt steppes are different from other steppes, with a dominance of Chenopodiaceae members and *Limonium* spp.

One characteristic of salt steppes is the presence of halophytic species. The most important halophytic taxa in Central Anatolian salt steppes are *Camphorosma monspeliaca* subsp. *monspeliaca*, *Puccinellia convoluta*, *Petrosimonia brachiate,* and *Aeluropus littoralis*. All these species are the characteristic species of the class Salicornietea and have wide distribution in salt marshes of Central Anatolia (Aydogdu et al. 2004; Kurt et al. 2004).

In Central Anatolia, different vegetation zones surround the salt lakes and the salt steppe is the outermost zone around the zones that have been mentioned under the title of salt marshes. A decrease in soil salinity results in an increase in plant biodiversity in salt steppes. The dominant plant taxon of this zone is *Artemisia santonicum*. The accompanying species are mainly xerohalophytes such as *Camphorosma monspeliaca* subsp. *monspeliaca, Puccinellia convoluta, Petrosimonia brachiata, P. nigdeensis, Aeluropus litoralis, Halimione verrucifera, Frankenia hirsuta, Salsola laricina, Noaea mucronata, Cyathobasis fruticulosa,* and *Reaumuria alternifolia* (Hamzaoglu and Aksoy 2006; Tug 2006; Tug et al. 2011). There are also some communities in salt steppes around Central Anatolia dominated by *Peganum harmala*.

The ecology and syntaxonomy of Central Anatolian steppe vegetation have been studied in detail. The first study covering saline salt steppes of Central Anatolia was done by Birand in the surrounding areas of Konya (Birand 1961). In the last three decades, the interest in salt steppes has increased and many syntaxonomical studies have been conducted (Bağcı et al. 1996; Yurdakukol 1974; Kurschner 1983; Yurdakulol and Ercoskun 1990; Yurdakukol et al. 1996; Guven et al. 2001; Aydogdu et al. 2002; 2004; Kurt et al. 2004; Aksoy and Hamzaoglu 2006; Hamzaoglu and Aksoy 2006).

The *Achilleo wihelmsii-Artemision santonici* combination surrounds the salt marshes around Seyfe and Tuz lakes. Because of the presence of taxa from the Salicornietea class, this vegetation is categorized as salt steppe. The syntaxa defined according to the syntaxonomical studies in Central Anatolia are connected to two classes: Salicornietea only known from saline areas and Astragalo Brometea, which also covers the other steppe units.

Halophytic Syntaxa Defined from Central Anatolia Are as Follows

Astragalo-Brometea Quezel 1973

Onobrychido armenae-Thymetalia leucostomi Akman, Ketenoglu, Quezel 1985
- *Salsolo ruthenicae-Alhagietum pseudoalhagi* Bagci, Tatli & Kargioglu 1996
- *Petrosimono brachiati-Artemisietum santonici* Bagci, Tatli & Kargioglu 1996

Achilleo wilhelmsii-Artemision santonici Aydogdu, Kurt, Hamzaoglu & Cansaran 2004
- *Achilleo wilhelmsii-Artemisietum santonici* Aydogdu, Kurt, Hamzaoglu & Cansaran 2004
- *Artemisio scopariae-Peganetum harmalae* Aydogdu, Kurt, Hamzaoglu & Cansaran 2004

Eastern and Southeastern Anatolia

Salt steppes around Aras Valley begin at Kars-Kagizman and continue through Igdir-Tuzluca until Aralik, and extend to Iran, Azerbaijan, and Armenia. Halophytic species that are distributed in western China, Central Asia, and Iran constitute the floristic composition, so endemism is low. *Kalidium caspicum, Halocnemum strobilaceum, Halostachys belangeriana, Petrosimonia brachiata, P. squarrosa, Alhagi maurorum* subsp. *maurorum, Bienertia cycloptera, Tamarix smyrnensis, T. parviflora, Limonium meyeri,* and *L. gmelini* dominate the area at Igdır and Dil plains. Salt hills around Kagizman and Tuzluca are covered with *Salsola dendroides, S. crassa, Halothamnus glaucus, Atriplex lehmanniana, A. micrantha,* and *Halanthium roseum* (Yaprak and Baskose 2012). The salt steppe vegetation at Igdir, Karakoyunlu, and Aralik does not show a zonation pattern as in Central Anatolia. Where the saline ground water reaches the surface, salt marshes can form and at these areas *Salicornia perennans, Suaeda linifolia,* and *Phragmites australis* dominate the vegetation. Even though there is a phytosociological study on the Igdir Plain (Istanbulluoglu 2004), misidentifications (especially for the succulent Chenopodiaceae) resulted in inaccurate syntaxonomy. We have not included Istanbulluoglu's study for our syntaxonomical summary.

There is a narrow salt steppe zone around small marshes and salt pans at Erzurum Karacoban and Mus-Malazgirt. With the other steppe species, halophytes such as *Panderia pilosa, Noaea tournafortii, Petrosimonia brachiata, P. squarrosa,* and *P. glauca* are found. The salt marshes at Erzurum Narman and Oltu Salt Marshes are surrounded by gypsaceous steppes and halophytes such as *Petrosimonia brachiata* and *P. glauca* arefound together with gypsosaceous species.

Evaluations of the vegetation in Southeast Anatolia, especially in the Harran Basin, reveal that halophytic vegetation has increased its area in the last few decades. After the completion of GAP (Southeastern Anatolian Irrigation Project), incorrect irrigation policies caused salinization of agricultural soils (Atamov et al. 2006; Ozkutlu and Ince 1999). Because of the abandonment of the salinized areas, halophytic plants invaded as secondary vegetation. At Harran Plain, 8,513 ha of over 225,000 ha is defined as highly saline (Ozkutlu and Ince 1999). These areas are located around Akcakale and the soil salinity ranges from 0.075 to 1.45% (Kaya et al. 2010). The area of anthropogenic salt steppes in Turkey is continuously increasing as a result of improper use of agricultural areas (incorrect drainage, over-irrigation, etc.) (Atamov et al. 2006). The best examples of these areas are found at Urfa-Akcakale and Hatay-Amik Plain. These former agricultural areas have become newly formed salt steppes and are mostly dominated by annual halophytes that are distributed by wind or birds. While *Alhagi mannifera*, *Cressa cretica*, and *Petrosimonia brachiata* dominate the vegetation at Amik Plain (Hatay) (Atamov et al. 2006), species from the Arabian Peninsula and Mesopotamia such as *Halothamnus hierochunticus*, *Salsola incanascens*, and *Atriplex leucoclada* dominate Urfa-Akcakale (Kaya et al. 2010).

The Syntaxonomical Summary of these Salt Steppes Is as Follows

Saginetea maritimae Westhoff, Van Leuwen & Adriani 1962

Frankenietalia pulverulenta Rivas Martinez ex Castroviejo & Porta 1976

Halothamno hierochunticae- Salsolion incanescentis Kaya et al. 2010

Halothamno hierochunticae- Salsoletum incanescentis Kaya et al. 2010

Hymenolobo procumbentis-Aeluropetum lagopoidis Kaya et al. 2010

Salt Meadows

Due to their proximity to the touristic coastal areas and fertile delta plains, the salt meadows are generally disturbed environments. Most of them have been rehabilitated and turned into agricultural areas. Therefore, their distribution has become narrower and they have lost their "naturalness."

Mediterranean and Aegean Salt Meadows

Most of the salt meadows on the Mediterranean and Aegean coasts have been rehabilitated to agricultural areas, except for Goksu Delta, which is a Specially Protected Area under Turkish Law. The characteristic species of this area are *Aeluropus littoralis*, *Halimione portulacoides*, *Limonium graecum* var. *graecum*, *Plantago crassifolia*, *Asteriscus aquaticus*, *Inula crithmoides*, *Limonium virgatum*, and *Carthamus tenuis* subsp. *tenuis*. This area shows characteristics of the order Limonietalia and, except for a humid winter and spring season, the soil is dry (Fig. 26.5).

Black Sea

A study conducted at Cernek Lake provides detailed information about the ecology of salt meadows along the Black Sea coast. Two zones are defined in the area and each zone is dominated by pure stands of *Hordeum geniculatum* or *Plantago coronopus*. At Kizilirmak Delta, an increase in the fresh water level caused a decrease in soil salinity, which resulted in four different zones, dominated by *Carex capitellata*, *Sarcocornia perennis*, *Salicornia perennans*, and *Spergularia marina*, *Artemisia santonicum*, and *Juncus* littoralis, respectively (Apaydin et al. 2009; Yaprak 2008).

Threats to Saline Areas and Recommendations

Salt marshes, steppes, and meadows provide a habitat to many plant and animal species with their unique ecological characteristics. This is

Fig. 26.5 Coastal salt meadow in Goksu Delta (*Limonietalia*). Photograph by A.E. Yaprak

especially true for terrestrial and geographically isolated areas, which are rich in endemic and rare plant species. Because of their vulnerability to global changes and disturbances, their ecological characteristics and biodiversity should be protected. Coastal and terrestrial salt marshes, steppes, and meadows are faced with different threats. Coastal habitats are threatened mainly by agriculture, urbanization, and developing tourism industry. Terrestrial habitats are threatened mainly by overuse of water sources that feed these salt marshes and lakes, and overgrazing.

The saline areas at Central Anatolia have different conservation statutes as a result of high biodiversity and endemism. For example, in Central Anatolia, the biggest saline area, Tuz Lake and its surrounding area, is declared a Specially Protected Area and it is also a wildlife protection area. Some relatively smaller salt marshes and salt steppes such as Sultan Sazligi National Park, Akgol-Eregli Sazliklari Natural Protection Area, and Seyfe Lake Natural Protection Area are also protected by law with different statutes. The main problem in these lakes and their surroundings is the usage of water sources and disruption of water flow to the lakes. A decrease in water results in an increase in soil salinity and aridity around the salt lakes and steppes. The global climate changes influence the vegetation in saline areas, which have a very vulnerable water balance. A decrease in water content also causes an increase in dust movement, which covers not only natural vegetation but also agricultural areas around the saline habitats (Williams 2002). One of the other problems for saline areas in Turkey is overgrazing. This is especially true for the areas without any conservation statutes. Nowadays, changing climatic conditions, water loss, and overgrazing exert great pressure on these areas. These areas or the plants of these areas are also used for many purposes such as direct gathering for pharmaceutical usage, food sources, animal feed, horticultural sources, etc. Uncontrolled gathering can also put pressure on plant populations, especially with rare and endemic species.

Coastal salt marshes and meadows are threatened by reclamation for agriculture, destruction for urbanisation and touristic facilities, dumping of organic and inorganic wastes, and sand removal (Ozturk et al. 1995, 2002; Guvensen and Ozturk 2003; Guvensen et al. 2006; Tug et al. 20111 Altay and Ozturk 2012). Reclamation of salt marshes and meadows for agricultural use is quite common, especially in Cukurova, Lower Menderes, and the Gediz Basins. This common practice has resulted in loss of most of the salt meadows in these areas in the last few decades, such that the following areas are now protected by law: Göksu Delta Specially Protected Area, Yumurtalık Lagoon Nature Protection Area, Dilek Peninsula and Buyuk Menderes Delta National Park, Bafa Lake Nature Park, and Kizilirmak Delta Wetland.

Appendix

List of the taxa found in saline areas of Turkey

Family	Species name	Habitat type	Halophyte or not	Region	Endemism
Amaryllidaceae	*Allium ampeloprasum*	M	A	Med/CA/BS	
	Allium atroviolaceum		A	CA/EA/Med	
	Allium callidictyon	S	A	CA/EA	
	Allium cappadocicum		H	CA	X
	Allium hirtovaginatum	S	A	CA/Med	
	Allium macrochaetum subsp. *macrochaetum*	S/M	H	CA/EA	
	Allium pseudoflavum	S	H	CA/EA	
	Allium scabriflorum	S	H	CA	X
	Allium sieheanum	M	H	CA	X
	Narcissus serotinus	CM	A	Med	
Apiaceae	*Ammi majus*	CM	A	Med	
	Bifora radians		A	All	
	Bupleurum croceum	S	A	EA/CA/Med	
	Bupleurum heldreichii	S	H	CA/Med	X
	Bupleurum lancifolium	M	A	Med	
	Bupleurum orientale	M/CM	H	Med	
	Daucus carota	S/CM	A	All	
	Caucalis platycarpos		A	All	
	Echinophora orientalis	S	H	EA	
	Eryngium campestre var. *virens*	S/CM	A	All	
	Eryngium creticum	CM	A	Med/BS	
	Eryngium maritimum	CM	A	Med/BS	
	Falcaria falcarioides	M	H	EA/CA/Med	
	Ferula caspica	S	H	CA/EA	
	Ferula halophila	S	H	CA	X
	Scandix pecten-veneris	S	A	All	
	Torilis arvensis subsp. *neglecta*	M/CM	A	All	
	Torilis nodosa	CM	A	Med/BS	
	Turgenia latifolia		A	All	
Apocyanaceae	*Apocynum venetum* subsp. *sarmatiense*	CM	H	Med	
	Cionura erecta	M/CM	A	All	
Araceae	*Arum dioscoridis* var. *dioscoridis*	CM	A	Med	
Asclepiadaceae	*Cynanchum acutum* subsp. *acutum*		H	All	
	Cynanchum acutum subsp. *sibiricum*		H	EA	
Asparagaceae	*Asparagus lycaonicus*	M	H	CA	
	Asparagus persicus	S	H	EA	
	Muscari longipes	S	A	CA/EA	
	Ornithogalum narbonense	S	A	All	
Asteraceae	*Achillea arabica*	S	A	All	
	Achillea oligocephala	S	A	EA	
	Achillea schischkinii	S	A	EA	X
	Achillea sieheana	S	H	CA	X

(continued)

Family	Species name	Habitat type	Halophyte or not	Region	Endemism
	Achillea santolinoides subsp. *wilhelmsii*	S	H	CA/EA	
	Ambrosia maritima	CM	H	Med/CA	
	Anthemis cotula	CM	A	Med/CA/BS/EA	
	Anthemis fumariifolia	S	H	CA	X
	Artemisia campestris var. *campestris*		A	Med/CA/EA	
	Artemisia santonicum subsp. *santonicum*	M/S	H	All	
	Artemisia scoparia	S	A	Med/CA/BS/EA	
	Artemisia spicigera	S	H	CA/BS/EA	
	Artemisia taurica	S	H	Med/CA/EA	
	Aster subulatus	M/CM	A	Med/BS	
	Asteriscus aquaticus	CM	H	Med	
	Bellis annua	CM	H	Med/BS	
	Cardopatium corybosum	CM	A	Med/CA/BS	
	Carduus nutans	S	A	Med/CA/EA	
	Carduus pycnocephalus subsp. *albidus*	CM/M	A	Med/ CA/BS/EA	
	Carthamus lanatus	CM	A	Med/CA/BS/EA	
	Carthamus tenuis subsp. *tenuis*	CM	A	Med	
	Centaurea balsamita	S	A	Med/ CA/EA	
	Centaurea calcitrapa subsp. *calcitrapa*	CM/M	H	Med	
	Centaurea drabifolia subsp. *floccosa*		A	CA	X
	Centaurea halophila	S	H	CA	X
	Centaurea hieropolitana		H	Med	X
	Centaurea kilaea		H	BS	
	Centaurea patula		A	CA/EA	
	Centaurea pamphylica		H	Med	
	Centaurea spinosa var. spinosa		H	Med/BS	
	Centaurea solstitialis subsp. *solstitialis*	S	A	Med/CA/BS/EA	
	Centaurea tuzgölüensis	S	H	CA	X
	Centaurea virgata	S/M	A	Med/CA/BS/EA	
	Chlamydophora tridentata	M/CM	H	Med	
	Chondrilla juncea		A	All	
	Cichorium intybus	CM	A	Med/CA/BS/EA	
	Cichorium glandulosum	S	A	Med/CA/EA	
	Cichorium spinosum	CM	A	Med	
	Cirsium alatum	S	H	CA/EA	
	Cnicus benedictus	M	A	All	

(continued)

Family	Species name	Habitat type	Halophyte or not	Region	Endemism
	Conyza canadensis	S	H	Med/CA/BS/EA	
	Cota austriaca	S	A	All	
	Cota halophila	CM	A	Med	X
	Cousinia birandiana	S	H	CA	X
	Cousinia iconica	S	A	CA	
	Crepis commutata	M	A	Med	
	Crepis foetida subsp. *foetida*	M/CM	A	Med/CA/BS	
	Crepis foetida subsp. *rhoeadifolia*	S	A	Med/CA/BS/EA	
	Crepis sancta	S	A	Med/CA/BS/EA	
	Crepis vesicaria subsp. *vesicaria*	M/CM	H	Med/BS	
	Cyanus depressus	S	A	All	
	Cyanus triumfettii subsp. *triumfettii*		A	All	
	Filago arvensis	S	A	All	
	Filago pyramidata	S	A	Med/CA/BS/EA	
	Glebionis coronaria	CM	A	Med	
	Hedypnois rhadioloides subsp. *cretica*	CM	A	Med/BS/EA	
	Inula aucherana	M	H	Med/CA/EA	
	Inula crithmoides	M/CM	H	Med	
	Inula tuzgoluensis	S	H	CA	X
	Inula aucheriana	S	H	Med/CA/EA	
	Inula viscosa	CM	H	Med/BS	
	Lachnophyllum noeanum	M	A	Med/EA	
	Lactuca sativa	S	A	Med/EA	
	Leontodon cripus subsp. *asper* var. *asper*	S	H	Med/CA/EA	
	Leontodon oxylepis var. *oxylepis*	S	H	Med/CA	
	Koelpinia linearis	S	H	Med/CA/EA	
	Micropus supinus	M	A	Med/EA	
	Oligochaeta divaricata	S	H	EA	
	Onopordum anatolicum	S	H	Med/CA	X
	Onopordum davisii	S	H	CA	X
	Onopordum turcicum		A	CA	
	Phagnalon graecum	CM	A	Med	
	Scorzonera cana subsp. *cana*	CM	A	Med/EA/CA	
	Scorzonera cana var. *jacquinana*	CM	A	Med/CA/EA	
	Scorzonera hieraciifolia	M	H	CA	X
	Scorzonera laciniata subsp. *laciniata*	S/M	H	Med/CA/EA	
	Scorzonera parviflora	M	H	CA/EA	
	Scorzonera tuzgoluensis	M	H	CA	X

(continued)

Family	Species name	Habitat type	Halophyte or not	Region	Endemism
	Senecio leucanthemifolius subsp. *vernalis*	S	A	All	
	Senecio salsuginea	S	H	CA	X
	Senecio vernalis	S	A	Med/CA/BS/EA	
	Senecio vulgaris		A	Med/CA/BS	
	Sonchus asper subsp. *glaucescens*	S	A	Med/CA/BS/EA	
	Sonchus bulbosa subsp. *microcephala*	S	H	Med	
	Sonchus oleraceus	M	H	Med/CA/BS	
	Tanacetum armenum	S	A	Med/CA/BS	
	Taraxacum bessarabicum	S	H	All	
	Taraxacum buttleri	S	A	Med/CA/EA	
	Taraxacum farinosum	M	H	CA	X
	Taraxacum mirabile	S	H	CA	X
	Taraxacum scaturiginosum	M	A	EA	
	Taraxacum serotinum	S	A	Med/CA	
	Taraxacum sintenisii	S	A	EA	
	Taraxacum tuzgoluensis subsp. *eskilense*	S	H	CA	X
	Taraxacum tuzgoluensis subsp. *tuzgoluensis*	S	H	CA	X
	Tragopogon coloratus		A	CA/BS/EA	
	Tragopogon porrifolius subsp. *longirostris*	CM	A	Med/CA/EA	
	Tripleurospermum callosum	S	A	Med/CA/BS/EA	X
	Tripleurospermum parviflorum	S	H	CA	
	Tripolium pannonicum subsp. *tripolium*	M	H	Med/BS	
	Urospermum picroides	CM	A	Med/BS/EA	
	Xanthium orientale subsp. *italicum*	CM	H	Med	
	Xanthium orientale subsp. *italicum*		H	CA	
	Xanthium strumarium subsp. *strumarium*		H	Med/CA/BS/EA	
Boraginaceae	*Buglossoides arvensis* subsp. *sibthorpiana*	S	A	Med/CA/BS/EA	
	Caccinia macranthera var. *macrantera*	S	H	Med/EA	
	Cynoglossum creticum		A	Med/CA/BS/EA	
	Echium angustifolium		A	Med/CA/BS	
	Heliotropium dolosum		A	Med/CA/BS/EA	
	Heliotropium europaeum		A	Med/CA/BS/EA	
	Lappula patula		A	CA/EA	
	Moltkia coerulea	S	A	Med/CA/EA	

(continued)

Family	Species name	Habitat type	Halophyte or not	Region	Endemism
	Myosotis alpestris subsp. *alpestris*	S	A	Med/CA/BS/EA	
	Onosma halophyla	M	H	CA	X
	Rochelia disperma var. *disperma*		A	Med/CA/EA	
Brassicaceae	*Alyssum blepharocarpum*	S	H	CA	X
	Alyssum desertorum	S	A	All	
	Alyssum linifolium var. *linifolium*	S	A	Med/CA/EA	
	Alyssum pateri		A	Med/CA/EA	
	Boreava orientalis		A	Med/CA/ES	
	Brassica elongata		A	CA/EA	
	Cakile maritima	M/CM	H	Med/ BS	
	Camelina rumelica		A	Med/CA/EA	
	Capsella bursa-patoris	S	A	Med/CA/BS/EA	
	Descurainia sophia subsp. *sophia*	S	A	All	
	Draba verna	S	A	All	
	Erysium crassipes		A	Med/CA/EA	
	Erysimum huber-morathii	S	H	CA	
	Erysimum sisymbrioides	S	H	CA	
	Hornungia procumbens	S	H	Med/CA/EA	
	Iberis halophila	S	H	CA	X
	Isatis kotschyana	S	A	EA	
	Lepidium cartilagineum subsp. *caespitosum*	M/S	H	CA/EA	
	Lepidium cartilagineum subsp. *crassifolium*	S	H	CA/EA	
	Lepidium draba	S/CM	A	Med/CA/EA	
	Lepidium perfoliatum	S	H	Med/CA/EA	
	Lepidium vesicarium	S	A	EA	
	Neslia paniculata subsp. *thracica*	S	A	EA	
	Sinapis arvensis		A	Med/CA/EA	
	Sisymbrium altissimum	S	A	Med/CA	
	Sisymbrium loeselii	S	A	Med/CA/BS/EA	
	Sisymbrium officinale	S	A	Med/BS/EA	
	Strigosella africana		A	Med/CA/EA	
Campanulaceae	*Campanula conferta*	S	A	EA	
	Legousia speculum-veneris	CM	A	Med/CA/BS	

(continued)

Family	Species name	Habitat type	Halophyte or not	Region	Endemism
Caprifoliaceae	*Cephalaria aristata*	CM	A	Med/CA/EA	
	Valerianella vesicaria	S	A	Med/CA/EA	
Caryophyllaceae	*Allochrusa versicolor*	S	H	EA	
	Arenaria serpyllifolia	S	A	Med/CA/BS/EA	
	Cerastium dubium		A	All	
	Gypsophila antari	S	A	EA	
	Gypsophila oblanceolata	M	H	CA	X
	Gypsophila perfoliata var. *perfoliata*	S	H	Med/CA	
	Herniaria glabra	S	A	Med/CA/EA	
	Holosteum umbellatum var. *umbellatum*	S	A	Med/CA/EA	
	Minuartia erythrosepala	C	A	CA	
	Minuartia picta	M	A	Med/EA	
	Minuartia sclerantha	S	A	CA	
	Minuartia urumiensis	S	H	CA	
	Sagina maritima	M	H	Med	
	Saponaria halophila	S	H	CA	X
	Saponaria karapinarensis	S	A	CA	X
	Silene colorata subsp. *colorata*	CM	A	Med	
	Silene salsuginea	M	H	CA	X
	Silene supina subsp. *pruinosa*		A	Med/CA/EA	
	Spergularia bocconei	M/S/CM	A	Med/EA	
	Spergularia diandra	S	H	Med/EA	
	Spergularia marina	M/CM	H	Med/BS	
	Spergularia media	M/S/CM	H	Med/CA	
	Spergularia rubra		A	Med/CA	
	Stellaria media		A	Med/BS/EA	
Chenopodiaceae	*Amaranthus blitoides*	S	A	Med/EA	
	Amaranthus viridis	S	A	Med/EA	
	Anabasis aphylla	S	H	CA/EA	
	Arthrocnemum glaucum	M	H	Med/EA	
	Arthrocnemum macrostachyum	M/CM	H	Med	
	Atriplex aucheri	S	H	CA	
	Atriplex hastata	M	H	Med/BS/EA	
	Atriplex hortensis	S	A	Med/CA/EA	
	Atriplex laevis	S	H	CA	
	Atriplex lasiantha		A	Med/CA/EA	
	Atriplex lehmanniana	S	H	EA	
	Atriplex leucoclada	S	H	EA	
	Atriplex micrantha		A	EA	
	Atriplex nitens	S	A	CA/EA	
	Atriplex rosea		A	Med/CA/EA	

(continued)

Family	Species name	Habitat type	Halophyte or not	Region	Endemism
	Atriplex tatarica var. *tatarica*	S/M/CM	H	Med/CA/EA	
	Bassia pillosa	S	H	CA/EA	
	Beta maritima	CM	H	Med	
	Bienertia cycloptera	M	H	EA	
	Camphorosma monspeliaca subsp. monspeliaca	S	H	CA/EA	
	Camphorosma monspeliaca subsp. *lessingii*	S	H	CA/EA	
	Ceratocarpus arenarius	S	H	EA	
	Chenopodium album subsp. *album* var. *album*		A	Med/CA/EA	
	Chenopodium album subsp. *iranicum*	S	A	Med/CA/EA	
	Chenopodium botrys		A	Med/CA/BS/EA	
	Chenopodium chenopodioides		H	Med/CA	
	Chenopodium foliosum	S	A	Med/CA/BS/EA	
	Chenopodium murale		A	Med/BS/EA	
	Cyathobasis fruticulosa	S	H	CA	
	Halanthium kulpianum	S/M	H	CA/EA	
	Halanthium rarifolium	S	H	EA	
	Halanthium roseum	S	H	EA	
	Halimione portulacoides	M/S/CM	H	Med/CA/EA	
	Halimione verrucifera	M	H	Med/CA	
	Halocnemum strobilaceum	M/CM	H	Med/CA/EA	
	Halocnemum cruciata	M	H	Med	
	Halogeton glomeratus	S	H	EA	
	Halopeplis amplexicaulis	M	H	Med	
	Halostachys belangeriana	M/S	H	EA	
	Halothamnus glaucus.	S	H	CA/EA	
	Halothamnus hierochunticus	S	H	EA	
	Kalidium wagenitzii	M	H	CA	
	Kalidium caspicum	M/S	H	EA	
	Kochia prostrata	S	H	CA/EA	
	Krascheninnikovia ceratoides	S	H	CA/EA	
	Microcnemum coralloides	M	H	CA	
	Noaea major	S	H	CA	
	Noaea minuta	S	H	CA/EA	
	Noaea mucronata subsp. *mucronata*	S	A	Med/CA/EA	
	Noaea tournefortii	S	A	Med/CA/EA	
	Panderia pilosa	S	H	CA/EA	
	Petrosimonia brachiata	M/S/CM	H	Med/CA	
	Petrosimonia glauca	M/S	H	EA	
	Petrosimonia nigdeensis	M/S	H	CA	X
	Petrosimonia squarrosa	M/S	H	EA	

(continued)

Family	Species name	Habitat type	Halophyte or not	Region	Endemism
	Salicornia dolichostachya	M	H	Med	
	Salicornia emerici	M	H	Med	
	Salicornia frietagii	M	H	CA	X
	Salicornia patula	M/CM	H	Med	
	Salicornia perennans	M	H	Med/CA/ BS/EA	
	Salsola brachiata	S	H	EA	
	Salsola crassa	M/S	H	CA/EA	
	Salsola dendroides	M/S	H	EA	
	Salsola erycoides	S	H	EA	
	Salsola grandis	M/S	H	CA	
	Salsola incanacens	S	H	EA	
	Salsola inermis	M/S	H	CA	
	Salsola kali	S	H	Med/CA/ EA	
	Salsola laricina	S	H	CA	
	Salsola macera	M	H	CA/EA	
	Salsola nitraria	S	H	CA/EA	
	Salsola nodulosa	S	H	EA	
	Salsola soda	CM	H	Med	
	Salsola stenoptera	M	H	CA	X
	Salsola tragus subsp. *pontica*	M	H	BS	
	Salsola tragus subsp. *tragus*	CM	H	Med/CA/ EA	
	Salsola verrucosa	S	H	EA	
	Sarcocornia obclavata	M/CM	H	Med	X
	Sarcocornia perennis	M/CM	H	Med/BS	
	Seidlitzia florida	S	H	EA	
	Spirobassia hirsuta	M	H	Med	
	Suaeda altissima	S	H	Med/CA/ EA	
	Suaeda carnosissima	S	H	CA	
	Suaeda confusa	S	H	CA/EA	
	Suaeda cucullata	M	H	Med/CA	X
	Suaeda eltonica	M	H	CA	
	Suaeda linifolia	M	H	EA	
	Suaeda microphylla	S	H	EA	
	Suaeda prostrata subsp. *anatolica*	M/S/CM	H	Med/CA	X
	Suaeda prostrata subsp. *prostrata*	M/CM	H	Med	
	Suaeda splendens	M	H	Med	
Cistaceae	*Helianthemum microcarpum*	S	A	Med/CA/ EA	
Convolvulaceae	*Convolvulus arvensis*	M/S/CM	A	All	
	Convolvulus lineatus	S	A	Med/CA/ EA	
	Cressa cretica	M/S/CM	H	Med/EA	
	Cuscuta balansae		A	Med/CA/ EA	
	Cuscuta campestris	CM	A	Med/CA/ BS/EA	

(continued)

Family	Species name	Habitat type	Halophyte or not	Region	Endemism
Cyperaceae	*Bolboschoenus maritimus* subsp. *cymosus*	M	H	Med/CA/EA	
	Bolboschoenus maritimus subsp. *maritimus*	M	H	Med/CA/BS/EA	
	Bolboschoenus laticarpus	M/CM	H	Med	
	Carex capitellata	M	A	BS	
	Carex cuprina	M	A	Med	
	Carex diluta subsp.*diluta*	M	A	EA	
	Carex divisa	CM	H	Med/CA/BS/EA	
	Carex extensa	M/CM	H	Med	
	Carex flacca subsp. *erythrostachys*	M	A	CA/EA	
	Carex hispida	M	H	Med	
	Cladium mariscus subsp. *mariscus*	M	H	Med/CA/BS	
	Cyperus laevigatus	M/CM	H	Med	
	Cyperus longus	M	A	Med/CA/BS/EA	
	Cyperus noeanus	M	H	CA	X
	Cyperus pannonicus	M	H	CA/EA	
	Cyperus rotundus	M	A	Med/BS/EA	
	Cyperus serotinus var. *serotinus*	M	H	Med/BS	
	Eleocharis palustris	M	A	Med/CA/BS/EA	
	Eleocharis uniglumis	M	A	Med/CA/BS/EA	
	Fimbristylis bisumbellata	CM	A	Med/CA/BS/EA	
	Isolepis cernua	M	H	Med/BS	
	Schoenoplectus lacustris subsp. *lacustris*	M	A	Med/CA/BS/EA	
	Schoenoplectus litoralis subsp. *litoralis*	M/CM	H	All	
	Schoenoplectiella supina	M	A	Med/CA/EA	
	Schoenus nigricans	CM	H	Med/CA/EA	
	Scirpoides holoschoenus subsp. *holoschoenus*	CM	H	All	
Dipsacaceae	*Cephalaria syriaca*		A	Med/CA/EA	
	Scabiosa rotata		A	Med/CA/EA	
Euphorbiaceae	*Euphorbia falcata* subsp. *falcata* var. *falcata*	S	A	Med/CA/EA	
	Euphorbia hirsuta		A	Med/CA/BS	
	Euphorbia macroclada		A	Med/CA/EA	
Fabaceae	*Alhagi maurorum* subsp. *maurorum*	M/CM	A	Med/CA/EA	
	Astragalus condensatus	S	A	Med/CA	
	Astragalus hamosus	CM	A	Med/CA/EA	

(continued)

Family	Species name	Habitat type	Halophyte or not	Region	Endemism
	Astragalus karamasicus		A	Med/CA	X
	Astragalus lycius		A	Med/CA/EA	X
	Astragalus mesogitanus	S	A	Med/CA	X
	Astragalus microcephalus subsp. *microcephalus*	S	A	All	
	Astragalus onobrychis	S	A	Med/CA/EA	
	Astragalus ovalis	M/S	H	CA	X
	Astragalus shelkovnikovii	S	A	EA	
	Astragalus strigillosus		A	CA	
	Glycyrrhiza glabra var. *glandulifera*		A	Med/CA/BS/EA	
	Halimodendron halodendron	S	H	EA	
	Lathyrus sativus	S	A	Med/CA/EA	
	Lotus corniculatus var. *corniculatus*	S	A	Med/CA/BS/EA	
	Lotus corniculatus var. *tenuifolius*	M/CM	A	Med/CA/BS/EA	
	Lotus halophilus	M	A	Med	
	Lotus strictus	M	H	Med/CA	
	Medicago littoralis var. *littoralis*	CM	A	Med/BS	
	Medicago lupulina	S	A	Med/CA/BS/EA	
	Medicago minima var. *minima*	S/CM	A	Med/CA/BS/EA	
	Medicago orthoceras		A	CA/EA	
	Medicago rotata var. *rotata*	CM	A	Med/EA	
	Melilotus albus	M/CM	A	All	
	Melilotus elegans	M/CM	H	Med/EA	
	Melilotus officinalis		A	Med/CA/BS/EA	
	Onobrychis paucijua	S	A	CA	
	Ononis spinosa subsp. *leiosperma*		A	Med/CA/BS/EA	
	Prosopis farcta	S	A	Med/EA	
	Sphaerophysa kotschyana	M	H	Med/CA	X
	Tetragonolobus maritimus	M	H	Med/CA	
	Trifolium campestre	M/CM	A	Med/CA/BS/EA	
	Trifolium fragiferum	CM	A	Med/CA	
	Trifolium purpureum var. *purpureum*	CM	A	Med/CA/EA	
	Trifolium repens var. *repens*		A	Med/CA/BS/EA	
	Trifolium resupinatum	S	A	All	
	Trifolium resupinatum var. *resupinatum*	M	A	Med/CA/BS/EA	
	Trifolium tomentosum	S	A	Med/EA	
	Trifolium stellatum	CM	A	Med/BS	
	Trigonella capitata	M	H	Med/CA	

(continued)

Family	Species name	Habitat type	Halophyte or not	Region	Endemism
	Trigonella coerulescens subsp. *coerulescens*	S	A	All	
	Trigonella foenum-graecum	S	A	Med/CA/EA	
	Trigonella monspeliaca	S	A	Med/CA/BS/EA	
	Vicia narbonensis var. *narbonensis*	CM	A	Med/CA/EA	
	Vicia sativa subsp. *sativa*		A	Med/CA/BS/EA	
Frankeniaceae	*Frankenia hirsuta*	M/S	H	Med/CA	
	Frankenia pulverulenta	M/S	H	Med/CA/EA	X
	Frankenia salsuginea	S	H	CA	X
Gentianaceae	*Blackstonia perfoliata* subsp. *perfoliata*	M/CM	A	Med/BS	
	Centaurium erythraea subsp. *Rumelicum*	M/CM	A	Med/CA	
	Centaurium erythraea subsp. *turcicum*	CM	A	Med/CA/BS/EA	
	Centaurium maritimum	CM	A	Med/BS	
	Centaurium pulchellum	CM	A	Med/CA/BS/EA	
	Centaurium tenuiflorum subsp. *acutifolium*	S	A	Med/CA/BS/EA	
	Schenkia spicata	M/CM	H	Med/CA/EA	
Geraniaceae	*Erodium botrys*	S	A	Med/EA	
	Erodium cicutarium subsp. *cicutarium*	S/CM	A	Med/CA/BS/EA	
	Erodium gruinum	CM	A	Med/EA	
	Geranium dissectum		H	Med/CA/BS/EA	
	Geranium molle	S/CM	A	Med/CA/BS	
	Geranium tuberosum subsp. *deserti-syriacum*	S	A	EA	
Hypericaceae	*Hypericum salsugineum*	M	H	CA	X
Ixioliriaceae	*Ixiolirion tataricum* subsp. *tataricum*	S	A	EA	
Iridaceae	*Gladiolus atroviolaceum*	M	A	EA	
	Gladiolus halophilus	M	H	CA	X
	Gynandriris sisyrinchium	CM	A	Med/EA	
	Iris paradoxa	S	A	EA	
	Iris spuria subsp. *musulmanica*	S	H	CA/EA	
	Iris xanthospuria	M	A	Med	X
	Romulea columnae subsp. *columnae*	CM	A	Med	
	Romulea ramiflora subsp. *ramiflora*	CM	A	Med	
Juncaceae	*Juncus acutus*	M/CM	H	Med/BS	
	Juncus articulatus subsp. *articulatus*	CM	A	All	
	Juncus effusus subsp. *effusus*	S	A	All	
	Juncus gerardi subsp. *gerardi*	M	H	Med/CA/EA	

(continued)

Family	Species name	Habitat type	Halophyte or not	Region	Endemism
	Juncus heldreichianus subsp. *heldreichianus*	CM	H	Med	
	Juncus heldreichianus subsp. *orientalis*	M	H	CA/EA	
	Juncus hybridus	M	H	Med/CA/BS /EA	
	Juncus inflexus subsp. *inflexus*	M/CM	A	All	
	Juncus littoralis	M/CM	H	Med/CA/BS	
	Juncus maritimus	M/CM	H	Med/CA/EA	
	Juncus persicus subsp. *libanoticus*	M	H	Med/EA/CA	
	Juncus rigidus	M/CM	H	Med	
	Juncus subulatus	M	H	Med/CA/EA	
Juncaginaceae	*Triglochin barrelieri* subsp. *barrelieri*	M/CM	H	Med/CA	
	Triglochin maritima	M	A	CA/EA	
	Triglochin palustris	M	A	CA/EA	
Lamiaceae	*Lamium orientale*		A	Med/CA/EA	
	Lamium purpureum var. *purpureum*		A	Med/CA/BS	
	Phlomis armeniaca	S	A	Med/CA/EA	
	Salvia halophila	M	H	CA	X
	Salvia syriaca	S	A	Med/CA/EA	
	Salvia viridis		A	Med/CA/BS/EA	
	Scutellaria orientalis subsp. *orientalis*	S	A	EA	
	Teucrium polium subsp. *polium*	S	A	All	
	Teucrium scordium subsp. *scordioides*		H	Med/CA/EA	
	Thymus sipyleus subsp. *sipyleus* var. *sipyleus*	S	A	Med/CA/BS/EA	X
	Ziziphora tenuior	S	A	Med/CA/EA	
Linaceae	*Linum bienne*	CM	A	Med/CA/BS/EA	
	Linum maritimum	M	A	Med	
	Linum mucronatum subsp. *mucronatum*	S	A	EA	
	Linum seljukorum	M	H	CA	
Lythraceae	*Lythrum salicaria*		A	Med/CA/BS/EA	
Malvaceae	*Alcea calvertii*	S	A	EA	X
	Lavatera thuringiaca	S	A	Med/CA/EA	
	Malva neglecta	S	A	Med/CA/BS/EA	
Menyanthaceae	*Menyanthes trifoliata*	M	H	EA	
Myrtaceae	*Myrtus communis* subsp. *communis*	CM	A	Med/BS	

(continued)

Family	Species name	Habitat type	Halophyte or not	Region	Endemism
Nitrariaceae	*Nitraria schoberi*	S	H	CA/EA	
	Peganum harmala	S	H	Med/CA/EA	
	Tetradiclis tenella	S	H	CA	
Onagraceae	*Epilobium hirsutum*	M/S	A	Med/CA/BS/EA	
Orchidaceae	*Orchis palustris*		A	Med/CA/BS/EA	
Orobanchaceae	*Cistanche salsa*		H	CA/EA	
	Orobanche minor		A	Med/CA/BS	
	Orobanche oxyloba		A	Med/CA/EA	
Papaveraceae	*Fumaria parviflora*	S	A	Med/CA/EA	
	Fumaria vaillantii		A	Med/CA/EA	
	Glaucium leiocarpum	S	A	Med/CA/EA	
	Hypecoum pendulum	S	A	CA/EA	
	Hypecoum procumbens		A	Med/CA	Med/CA/EA
	Roemeria hybrida	S	A	Med/CA/EA	
Plantaginaceae	*Plantago coronopus* subsp. *coronopus*	S	H	Med/CA/BS	
	Plantago coronopus subsp. *commutata*	M/CM	H	Med/CA	
	Plantago crassifolia	M/S/CM	H	Med/CA/EA	
	Plantago holosteum		A	Med/CA	
	Plantago lagopus	CM	A	Med/CA/EA	
	Plantago lanceolata	S	H	Med/CA/BS/EA	
	Plantago major subsp. *intermedia*	M	A	Med/CA/EA	
	Plantago maritima	M/S/CM	H	All	
	Veronica grisebachii	S	A	Med/CA/BS	
	Veronica triphyllos	S	A	Med/CA/EA	
	Veronica viscosa	S	A	EA	
Plumbaginaceae	*Acantholimon halophilum* var. *coloratum*	S	H	CA	X
	Acantholimon halophilum var. *halophilum*	S	H	CA	X
	Limonium anatolicum	S	H	CA	X
	Limonium angustifolium	M/CM	H	Med	
	Limonium bellidifolium	M/S/CM	H	Med/CA	
	Limonium echioides	CM	H	Med	
	Limonium effusum	M	H	Med	
	Limonium globuliferum	S	H	CA	

(continued)

Family	Species name	Habitat type	Halophyte or not	Region	Endemism
	Limonium gmelinii	M/S	H	Med/CA/EA	
	Limonium graecum var. *graecum*	M/CM	H	Med	
	Limonium iconicum	M	H	CA	X
	Limonium lilacinum	M/S	H	CA	X
	Limonium meyeri	M/CM	H	Med/EA	
	Limonium narbonense	M/CM	H	Med	
	Limonium pycnanthum	S	H	CA/EA	X
	Limonium sieberi	M	H	Med	
	Limonium sinuatum	CM	H	Med	
	Limonium tamaricoides	M	H	CA	X
	Limonium vananse	M	H	EA	X
	Limonium virgatum	M/CM	H	Med	
	Plumbago europaea	S	H	Med/CA/BS/EA	
Poaceae	*Aegilops cylindrica*		A	CA/EA	
	Aegilops umbellulata subsp. *umbellulata*		A	Med/CA/EA	
	Aeluropus lagopoides var. *lagopoides*	S	H	Med/EA	
	Aeluropus littoralis	S/M/CM	H	Med/CA/EA	
	Agropyron cristatum subsp. *pectinatum* var. *pectinatum*	S	A	Med/CA/EA	
	Agrostis stolonifera		A	Med/CA/BS/EA	
	Alopecurus arundinaceae	M	A	EA	
	Alopecurus myosuroides subsp. *myosuroides*	M	A	Med/CA/BS/EA	
	Alopecurus myosuroides var. *Latialatussinonim*	M	H	Med	
	Apera intermedia	S	A	Med/CA/EA	
	Avena clauda		A	Med	
	Avena sterilis subsp. *sterilis*		A	Med/CA/EA	
	Brachiaria eruciformis		A	Med/CA/EA	
	Brachypodium distachyon		A	All	
	Briza maxima		A	Med/CA/BS	
	Bromus arvensis		A	Med/CA/BS/EA	
	Bromus intermedius		A	Med/CA/EA	
	Bromus japonicus subsp. *japonicus*		A	Med/CA/BS/EA	
	Bromus madritensis		A	Med/CA/BS/EA	
	Bromus psammophilus var. *Psammophilus*	CM	A	Med	X
	Bromus psammophilus var. *Robustus*	CM	A	Med	X

(continued)

Family	Species name	Habitat type	Halophyte or not	Region	Endemism
	Bromus pseudobrachystachys	CM	A	Med	
	Bromus racemosus	S	A	CA/EA	
	Bromus rigidus	M/CM	A	Med	
	Bromus rubens	CM	A	Med	
	Bromus squarrosus		A	Med/CA/BS/EA	
	Bromus tectorum		A	Med/CA/EA/BS	
	Calamagrostis pseudophragmites	M	A	Med/CA/BS/EA	
	Catabrosa aquatica		A	Med/CA/BS/EA	
	Crypsis aculeata	S	H	Med/CA/EA	
	Crypsis faktorovskyi	CM	H	Med/EA	
	Crypsis schoenoides	S/M/CM	H	Med/CA/BS/EA	
	Cynodon dactylon var. *Villosus*	S/CM	A	Med/CA/BS/EA	
	Digitaria sanguinalis	M/CM	A	Med/CA/BS/EA	
	Elymus elongatiformis		A	CA/EA	
	Elymus elongatus subsp. *ponticus*	CM	H	Med/CA	
	Elymus elongatus subsp. *salsus*	M/S	H	CA	X
	Elymus farctus subsp. *bessarabicus*	CM	A	Med/BS	
	Elymus hispidus	S	A	Med/CA/EA	
	Eragrostis collina	M	H	Med/CA/EA	
	Eremopyrum distans		A	CA/EA	
	Eremopyrum orientale		A	CA/BS/EA	
	Festuca valesiaca		A	Med/CA/BS/EA	
	Holcus lanatus		A	Med/CA/BS	
	Hordeum brevisubulatum subsp. *violaceum*	S	A	Med/CA/EA	
	Hordeum geniculatum	S	A	Med/CA/BS/EA	
	Hordeum marinum	S	H	Med/CA/EA	
	Hordeum murinum subsp. *glaucum*	S	A	Med/CA/BS/EA	
	Hordeum murinum subsp. *leporinum*		A	Med/CA/BS	
	Hordeum murinum subsp. *murinum*	M/S/CM	A	Med/CA/EA	
	Imperata cylindrica	CM	A	Med/EA	
	Lagurus ovatus		A	Med/BS	
	Leymus cappadocicus	S	H	CA/EA	

(continued)

Family	Species name	Habitat type	Halophyte or not	Region	Endemism
	Lolium rigidum		A	Med/CA/ BS/EA	
	Lolium rigidum var. *Rottbollioides*	M/CM	A	Med/BS/ EA	
	Molinia caerulea	M	A	BS/CA	
	Panicum repens	M	A	Med	
	Parapholis incurva	M/CM	H	Med/BS	
	Phalaris aquatica	CM	A	Med/EA	
	Phalaris paradoxaa	CM	A	Med/BS/ EA	
	Phleum exaratum subsp. exaratum	S	A	Med/CA/ BS/EA	
	Phragmites australis	M/CM	A	Med/CA/ BS/EA	
	Poa bulbosa	S	A	Med/CA/ BS/EA	
	Poa pratensis		A	Med/CA/ BS/EA	
	Poa trivialis		A	Med/CA/ BS/EA	
	Polypogon maritimus subsp. *maritimus*		A	Med	
	Polypogon monspeliensis	S	A	Med/CA/ EA	
	Puccinellia bulbosa subsp. *bulbosa*	S	H	EA	
	Puccinellia bulbosa subsp. *caesarea*	S	H	CA	X
	Puccinellia convoluta	S	H	Med/CA	
	Puccinellia distans subsp. *distans*	M/S/CM	H	Med/CA/ EA	
	Puccinellia distans subsp. *sevangensis*	M/S	H	EA	
	Puccinellia festuciformis	CM	H	Med/BS	
	Puccinellia gigantea	M	H	Med/BS/ EA	
	Puccinellia intermedia	M/CM	H	Med/CA	
	Puccinellia koeieana subsp. *anatolica*	M	H	Med/CA	X
	Rhizocephalus orientalis	S	H	CA	
	Rostraria cristata var. *Cristata*	M/CM	A	Med/CA/ BS/EA	
	Saccharum ravennae	CM	A	Med/EA	
	Schismus arabicus	S	A	Med/CA/ EA	
	Setaria glauca	S	A	Med/BS/ EA	
	Sorghum halepense var. *Halepense*	M	A	Med/CA/ BS/EA	
	Sphenopus divaricatus	M/S/CM	H	Med/CA	
	Sporobolus virginicus	M/CM	H	Med/CA	
	Stipa ehrenbergiana		A	CA/EA	
	Trisetaria loeflingiana	M/CM	A	Med/CA	

(continued)

Family	Species name	Habitat type	Halophyte or not	Region	Endemism
Polygonaceae	*Atraphaxis spinosa*	S	A	EA	
	Calligonum polygonoides	S	A	EA	
	Polygonum aviculare	S	A	Med/CA/BS/EA	
	Polygonum equisetiforme	M/CM	H	Med/EA	
	Polygonum patulum subsp. *patulum*		A	CA	
	Polygonum salicifolium	S/M	H	Med/EA	
	Rumex crispus	CM	A	Med/CA/EA	
Potamogetonaceae	*Stuckenia pectinata*	M	A	Med/CA/BS/EA	
Primulaceae	*Anagallis arvensis* var. *arvensis*	CM	A	Med/CA/BS/EA	
	Anagallis arvensis var. *caerulea*	CM	A	Med/CA/BS/EA	
	Androsace maxima	S	A	Med/CA/BS/EA	
	Glaux maritima	M	H	CA/EA	
	Samolus valerandi		H	Med/BS/EA	
Ranunculaceae	*Adonis aestivalis* subsp. *aestivalis*	S	A	Med/CA/EA	
	Adonis annua		A	Med/CA/BS/EA	
	Adonis dentata	S	A	EA	
	Adonis flammea		A	Med/CA/BS/EA	
	Adonis microcarpa		A	Med/CA/EA	
	Ceratocephala falcata	S	A	CA/EA	
	Consolida anthoroidea	S	A	EA	
	Consolida glandulosa		A	Med/CA/EA	X
	Consolida orientalis		A	Med/CA/BS/EA	
	Delphinium peregrinum	S	A	Med/CA/EA	
	Delphinium venulosum		A	CA	X
	Nigella arvensis subsp. *glauca*	S	A	Med/CA/EA	
	Nigella turcica	S	A	EA	X
	Nigella segetalis	S	A	CA/EA	
	Ranunculus constantinopolitanus		A	Med/CA/BS/EA	
	Ranunculus cuneatus	S	A	Med/CA/EA	
	Thalictrum lucidum		A	Med/CA/BS/EA	
Rosaceae	*Potentilla reptans*		A	Med/CA/BS/EA	
	Rubus sanctus		A	Med/CA/BS/EA	
	Sarcopoterium spinosum	CM	A	Med/BS	

(continued)

Family	Species name	Habitat type	Halophyte or not	Region	Endemism
Rubiaceae	*Asperula arvensis*	S	A	MED/CA/EA	
	Cruciata taurica	S	A	Med/CA/BS/EA	
	Galium murale	S	A	Med/BS/EA	
	Galium tricornutum		A	Med/CA/EA	
	Galium verum subsp.*verum*		A	Med/CA/BS/EA	
Santalaceae	*Thesium compressum*	S	H	CA	
Scrophulariaceae	*Linaria genistifolia* supsp. *confertiflora*		A	Med/CA/EA	
	Verbascum cheiranthifolium		A	Med/CA/BS/EA	
	Verbascum helianthemoides	S	H	CA	
	Verbascum pyroliforme subsp. *pyroliforme*	M	H	CA	X
	Verbascum sinuatum		A	Med/BS	
Solanaceae	*Lycium anatolicum*	S	A	Med/CA/EA	X
	Lycium ruthenicum	S	A	EA	
	Solanum americanum	S	A	EA	
Tamaricaceae	*Reaumuria alternifolia*	S	H	CA/EA	
	Tamarix duzenlii	M/CM	H	Med	X
	Tamarix hampeana		A	Med	
	Tamarix octandra	S/M	H	EA	
	Tamarix parviflora		A	Med/CA/EA	
	Tamarix smyrnensis	M/CM	A	CA/EA	
	Tamarix tetragyna	CM	H	Med	
	Tamarix tetrandra	M	A	Med/CA/EA	
Thymelaeaceae	*Thymelaea hirsuta*	M	A	Med	
	Thymelaea passerina		A	All	
Typhaceae	*Typha domingensis*	M	A	Med/CA/EA	
Verbenaceae	*Verbena officinalis* var. *officinalis*	S	A	All	
Xanthorrhoeaceae	*Asphodelus aestivus*	CM	A	Med/EA	
Zygophyllaceae	*Tedradiclis tenella*	M	H	CA	
	Tribulus terrestris		A	Med/CA/BS/EA	
	Zygophyllum album	M/CM	H	Med	
	Zygophyllum fabago		A	Med/CA/EA	

M salt marsh, *S* salt Steppe, *CM* coastal meadow, *H* halophyte, *A* accompanied, *Med* Mediterranean, *CA* Central Anatolia, *BS* Black Sea+, *EA* Eastern Anatolia

References

Akhani H (2002) Notes on the Flora of Iran: 1. *Asparagus* (Asparagaceae) and *Nitraria* (Zygophyllaceae). Edinb J Bot 59:295–302

Aksoy A, Hamzaoglu E (2006) Vegetation zones in the salty marshes of Central Anatolia and natural borders of agricultural usage (Turkey). In: Ozturk M, Waisel Y, Khan MA, Gork G (eds) Biosaline agriculture and salinity tolerance in plants. Birkhauser Verlag AG (Springer Science), Basel, pp 109–119

Altay V, Ozturk M (2012) Land degradation and halophytic plant diversity of Milleyha wetland ecosystem (Samandag-Hatay) Turkey. Pak J Bot 44:37–50

Andersen H, Bakker JP, Brongers M, Heydemann B, Irmler U (1990) Long term changes of salt marsh communities by cattle grazing. Vegetation 89:137–148

Apaydin Z, Kutbay HG, Ozbucak T, Yalcin E, Bilgin A (2009) Relationships between vegetation zonation and edaphic factors in salt marsh community (Black Sea coast). Pol J Ecol 57(1):99–112

Atamov V, Aktoklu E, Cetin E, Aslan M, Yavuz M (2006) Halophytication in Harran (Sanliurfa) and Amik plain (Hatay) in Turkey. Phytologia Balcanica 12(3):401–412

Aydogdu M, Hamzaoglu E, Kurt L (2002) New halophytic syntaxa from Central Anatolia (Turkey). Isr J Plant Sci 50:313–323

Aydogdu M, Kurt L, Hamzaoglu E, Cansaran A (2004) Phytosociological studies on salt steppe communities of Central Anatolia, Turkey. Isr J Plant Sci 52:71–79

Bagci Y, Tatli A, Kargioglu M (1996) Konya-Karapinar Bolgesinin vejetasyonu. S.U. Fen-Edebiyat Fakultesi Fen Dergisi 13:97–113

Behcet L (1994a) Van Golu Makrofitik Vejetasyonunun Fitososyolojik Yonden Arastirilmasi. Turk J Bot 18:229–243

Behcet L (1994b) Ercek, Turna ve Bostanici (Van) Gollerinin Vejetasyonu. Turk J Bot 18:305–312

Behcet L, Ozgokce F, Unal M, Karabacak O (2009) The Flora of Kirmizi Tuzla (Karacoban, Erzurum/Turkey) and Bahce Tuzlasi (Malazgirt, Mus/Turkey) and their environment. Biol Divers Conserv 2:122–155

Bertness MD (1991) Interspecific interactions among high marsh perennials in a New England salt marsh. Ecology 72:125–137

Bertness MD, Ellison AM (1987) Determinations of pattern in a New England salt marsh plant community. Ecol Monogr 57:129–147

Biondi E, Casavecchia S, Estrelles E, Soriano P (2013) Halocnemum M. Bieb. Vegetation in the Mediterranean Basin. Plant Biosystems 147(3):536–547

Birand H (1961) Orta Anadolu bozkirinda vejetasyon incelemelerinin ilk sonuclari: I. Tuz Golucorakcil bitki birlikleri, Tarim BakanligiIlmi Rapor ve Arastirma Serisi, Toprak Su Umum Mudurlugu Nesriyati Sayi: 103, Ankara

Freitag H, Vural M, Adiguzel N (1999) A remarkable new *Salsola* and some new records of Chenopodiaceae from Central Anatolia Turkey. Willdenowia 29:123–139

Géhu JM (2011) Schema Synsystematique Des Principales Classes De Vegetations Littorales Sedimentaires Europeennes Avec References Ad. Annali di Botanica, p 56

Géhu JM, Uslu T (1989) Données sur la végétation littorale de la Turquie du Nord-Ouest. Phytocoenologia 17:449–505

Gehu JM, Uslu T, Costa M (1992) Apport à la connaissance phytosociologique du littoral Sud de la Turquie méditerranéenne. Coll Phytosoc 19:591–622

Ghazanfar SA, Altundag E, Yaprak AE, Osborne J, Tug GN, Vural M (2014) Halophytes of Southwest Asia. In: Khan MA et al (eds) In Sabkha ecosystems: volume IV: cash crops halophyte and biodiversity conservation, tasks for vegetation science 47. Springer, Dordrecht, pp 105–133

Guner A, Aslan S, Ekim T, Vural M, Babac MT (2012) Turkiye bitkileri listesi (damarli bitkiler). Nezahat Gokyigit Botanik Bahcesi ve Flora Arastirmalari Dernegi Yayini, Istanbul, pp 47-83

Guven E, Vural M, Bahceci I, Sonmez B (2001) Ic Anadolu Bolgesi halofitik (tuzcul bitkiler) vejetasyonunun ekolojisi. Toprak ve Su Kaynaklari Arastirma Yilligi 2000. Koy Hizmetleri Genel Mudurlugu, APK Dairesi Baskanligi, Toprak ve Su Arastirma Sube Mudurlugu Yayinlari Yayin No: 117, Ankara

Guvensen A, Ozturk M (2003) Halophytic plant diversity of South Aegean coastal zone in Turkey. Pak J Bot 35(5):853–864

Guvensen A, Gork G, Ozturk M (2006) An overview of the halophytes in Turkey. In: Khan MA, Boer B, Kust GS, Barth HJ (eds) Sabhka ecosystems II: west and Central Asia. Springer, Dordrecht, pp 9–30

Hamzaoglu E, Aksoy A (2006) Sultansazligi Batakligi halofitik topluluklariuzerine fitososyolojik bir calisma Ic Anadolu-Kayseri. Ekoloji 15(60):8–15

Hamzaoglu E, Duran A, Akhani H (2005) A new genus record for the Flora of Turkey: *Tetradiclis* Stev. Ex M. Bieb. (Zygophyllaceae). Turk J Bot 29:403–407

Istanbulluoglu A (2004) Investigation of the vegetation on saline-alkaline soils and marshes of Igdir plain in Turkey. Pak J Biol Sci 7(5):734–738

Jensen A (1985) The effect of cattle and sheep grazing on salt-marsh vegetation at Skallingen, Denmark. Vegetation 60:37–48

Kaden H, Peeters F, Lorke A, Kipfer R, Tomonaga Y, Karabiyikoglu M (2010) Impact of lake level change on deep-water renewal and oxic conditions in deep saline Lake Van, Turkey. Water Resour Res 46(11):1–14

Kadereit G, Yaprak AE (2008) "*Microcnemum coralloides*" (Chenopodiaceae-Salicornioideae): an example of intraspecific east-west disjunctions in the Mediterranean region. Anales del Jardín Botánico de Madrid 65(2):415–426

Kaya OF, Cetin E, Aydogdu M, Ketenoglu O, Atamov V (2010) Syntaxonomical analyses of the secondary vegetation of Harran plain (Sanliurfa/Turkey) ensuing excessive irrigation by using GIS and remote sensing. Ekoloji 19:1–14

Kielh K, Eischeid I, Gettner S, Walter J (1996) Impact of different sheep grazing intensities on salt marsh vegetation in northern Germany. J Veg Sci 7:99–106

Kurschner H (1983) Vegetationsanalytische Untersuchungen an Halophytenfluren Zentralanatoliens (Turkei). – Beihefte zum Tubinger Atlas des Vorderen Orients, Nr. 11, Dr. Ludwig Reichert, Wiesbaden, p 89

Kurt L, Hamzaoglu E, Tug GN, Evren H (2004) A study on the relationships of salinity and endemism ratios in the surrounding of salt Lake (inner Anatolia, Turkey). FU Fen ve Muhendislik Bilimleri Dergisi 16(4):745–756

Ozkutlu F, Ince F (1999) Present state of salinity in the Harran Plain and potentials for its expansion. In: GAP 1 Agricultural Congress, 26–28 May 1999, Sanliurfa, pp 909–914

Ozturk M, Guvensen A (2002) Studies on the halophytic vegetation of saline-alkaline habitats in West Anatolia Turkey. Int. Symp. On Optimum Resources Utilization in Salt Affected Ecosystems in Arid and Semi-Arid Regions, Cairo, 8–11 April, pp 241–244

Ozturk M, Ozcelik H, Behcet L, Guvensen A, Ozdemir F (1995) Halophytic flora of Van Kale Basin Turkey. In: Khan MA, Ungar IA (eds) Biology of salt tolerant plants. ISBN: 0-9648260-1-1, Dept of Botany, Univ of Karachi, Karachi, pp 306–315

Ozturk M, Çelik A, Yarci C, Aksoy A, Feoli E (2002) An overview of plant diversity, land use and degradation in the Mediterranean region of Turkey. Environ Manag Health 13:442–449

Ozturk M, Guvensen A, Gork C, Gork G (2006) An overview of the coastal zone plant diversity and management strategies in the Mediterranean region of Turkey. In: Ozturk M, Waisel Y, Khan MA, Gork G (eds) Biosaline agriculture and salinity tolerance in plants. Birkhäuser, Basel, pp 89–100

Ozturk M, Guvensen A, Sakcali S, Gork G (2008) Halophyte plant diversity in the Irano-Turanian phytogeographical region of Turkey. In: Abdelly C, Ozturk M, Ashraf M, Grignon C (eds) Biosaline agriculture and high salinity tolerance. Birkhäuser, Basel, pp 141–155

Ozturk M, Ozcelik H, Behcet L (2015) General characteristics of Flora and Vegetation formations of eastern Anatolia region and its environs (Turkiye). SDU Fen Dergisi 10(1):23–48

Ozturk M, Altay V, Altundag E, Gucel S (2016) Halophytic plant diversity of unique habitats in Turkey: salt mine caves of Çankiri and Igdir. In: Khan MA, Ozturk M, Gul B, Ahmed MZ (eds) Halophytes for food security in dry lands. Elsevier, Oxford, pp 291–315

Pennings SC, Callaway RM (1992) Salt marsh plant zonation: the relative importance of competition and physical factors. Ecology 73:681–690

Pennings SC, Callaway RM (1996) Impact of parasitic plant on the structure and dynamics of salt marsh vegetation. Ecology 77:1410–1419

Secmen O, Leblebici E (1997) Turkiye sulak alan bitkileri ve bitki ortusu. Ege Universitesi Fen Fakultesi Yayinlari:158

Snow AA, Vince SW (1984) Plant zonation in an Alaskan salt marsh. II. An experimental study of the role of edaphic conditions. J Ecol 72:669–684

Tug GN (2006) Tuz Golu Çevresi Halofitik Vejetasyonda Zonlasmaya Etki eden Faktorlerin Belirlenmesi. Doktora Tezi, Ankara Universitesi, Fen Bilimleri Enstitusu, Ankara. 201p

Tug GN, Yaprak AE, Vural M (2011) An overview of halophyte plant diversity from salt Lake area Konya, Turkey. In: Ozturk M, Mermut AR, Çelik A (eds) Urbanisation, land use, land degradation and environment. Daya Publishing House, Delhi, pp 356–371

Tug GN, Ketenoglu O, Bilgin A (2012) The relationships between plant zonation and edaphic factors in halophytic vegetation around Lake Tuz, Central Anatolia, Turkey. Rend Fis Acc Lincei 23:355–363

Vural M, Duman H, Guner A, Donmez AA, Sagban H (1995) The vegetation of Koycegiz-Dalyan (Mugla) specially protected area. Tr J of Botany 19(4):431–476

Williams WD (2002) Environmental threaths to salt lakes and the likely status of inland saline ecosystems in 2025. Environ Conserv 29(2):154–167

Yaprak AE (2007) An interesting re-collection of *Halopeplis amplexicauis* (Vahl) Ung.-Sternb. From Turkey. Turk J Bot 30(6):459–460

Yaprak AE (2008). Taxonomic revision of the genera *Salicornia*, *Sarcocornia* and *Arthrocnemum* (Chenopodiaceae) in Turkey. Ph.D. Thesis. Ankara: Ankara University

Yaprak AE (2012) *Sarcocornia obclavata* (Amaranthaceae) a new species from Turkey. Phytotaxa 49(1):55–60

Yaprak AE, Baskose I (2012) *Atriplex lehmanniana* Bunge (Chenopodiaceae): a new record for the flora of Turkey. Biol Divers Conserv 5(2):66–69

Yaprak AE, Kadereit G (2008) A new species of Halocnemum M. Bieb.(Amaranthaceae) from southern Turkey. Bot J Linn Soc 158(4):716–721

Yaprak AE, Tug GN (2006) Halophytic endemics of Turkey. In: Ivanova D (ed) 2009. Plant, fungal and habitat diversity investigation and conservation. Proceedings of IV Balkan botanical congress, Sofia, 20–26 June 2006. Institute of Botany, Sofia, pp 234–238

Yaprak AE, Yurdakulol E (2008) *Salicornia freitagii* (Chenopodiaceae), a new species from Turkey. Ann Bot Fenn 45(3):207–211

Yaprak AE, Tug GN, Yurdakul E (2010) Turkiye Florasinda *Arthrocnemum macrostachyum* (Moric.) K. Koch (Chenopodiaceae). Selcuk Universitesi Fen Fakultesi Fen Dergisi 2(36):117–121

Yurdakulol E (1974) Konya ovasindaki corak batakliklar vejetasyonunun bitki sosyolojisi yonunden arastirilmasi. Bitki 1(2):257–277

Yurdakulol E, Ercoskun T (1990) Orta Anadolu'da corak alanlarin vejetasyonu uzerinde ekolojik ve sintaksonomik arastirma. Doga-Turk J Bot 14:109–123

Yurdakulol E, Oncel I, Demirors M, Yildiz A, Keles Y (1996) Ecological and syntaxonomical investigation of salt marshes in the vicinity of Burdur and Acigol (Denizli/Turkey). Ecologia Mediterranea XXII(1/2):51–61

Temporal Variations in Water and Ion Relations in Coastal Halophytes

27

Sarwat Ghulam Rasool, Hina Siddiqui, Abdul Hameed, Tabassum Hussain, Irfan Aziz, M. Ajmal Khan, and Salman Gulzar

Abstract

Halophytes are plants which can complete their life cycle under saline conditions. Coastal halophytes are also highly resilient to large variations in moisture and temperature. Some of these plants are grazed by cattle and are good sand binders. The underlying mechanisms which allow halophytes to adapt to coastal environment are poorly understood. The present study investigated the diurnal (pre-dawn and noon water relations) and seasonal (Summer: June; Monsoon: August; and Winter: January) variation in the ion and water relations of four coastal plants: *Halopyrum mucronatum* (high dune), *Suaeda fruticosa* (low dune), *Sporobolus tremulus* (low marsh), and *Avicennia marina* (high marsh). Leaf-level ecophysiology of these dominant coastal plants showed considerable variations at the diurnal and seasonal scales and can be used as stress marker for studying coastal vegetation.

Keywords

Coastal halophytes · Ecophysiology · Ion relationships · Water relationships

S. G. Rasool · H. Siddiqui · A. Hameed · T. Hussain
I. Aziz · M. A. Khan · S. Gulzar (✉)
Institute of Sustainable Halophyte Utilization (ISHU), University of Karachi, Karachi, Pakistan
e-mail: salmang@uok.edu.pk

Introduction

Halophytes in coastal areas face a number of stress factors, which include salt spray, tidal fluctuations, soil salinity, irregular precipitation, and occasional substrate dryness (Brown et al. 2006). The individual and combined effects of these factors may have a strong influence on species composition and dynamics, which often results in low species diversity (Rozema et al. 1985).

The distribution of plants in natural environments is a consequence of complex interactions between environmental conditions and specific adaptations of the plants to thrive in these environments. Sensitivity to changes in environmental conditions over time would ultimately influence the distribution and structure of halophytes in coastal areas (Gulzar and Khan 1998; Aziz et al. 2011). However, due to the complex nature of coastal environments, the reasons for segregation of halophytes in distinct zones is still poorly understood (Caçador et al. 2007; Emery et al. 2001). The coastal environment of warm subtropical areas is typically harsh due to considerable seasonal and diurnal fluctuations in environmental variables, particularly substrate salinity (Dregelyi et al. 2008; Gulzar and Khan 1998) and tidal inundation (Cacador et al. 2007).

Vegetation along the Karachi Coast is spread over ~135 km long coastline and shows distinct distribution patterns dominated by grasses. Plants tend to achieve homeostasis either through

B. Gul et al. (eds.), *Sabkha Ecosystems*, Tasks for Vegetation Science VI,
https://doi.org/10.1007/978-3-030-04417-6_27

adaptive (Ievinsh 2006) or acclamatory adjustments in morpho-physiological characteristics (Rozema et al. 1985).

Halophyte distribution along the coastal zone could be related to plant-specific variation (Gulzar et al. 2014a, b) rather than gradient-related variation (Kunza and Pennings 2008). Previous studies along Karachi coast have shown that halophytes possess both interspecific and intraspecific variation in the field to tolerate salinity stress in their natural habitat (Gulzar et al. 2014a, b). In general, salinity, water, and temperature are acute environmental factors that determine the vegetation type in the region (Boer 1996). Furthermore, plant populations are also affected by specific physical factors such as changes in substrate salinity and irregular rainfall patterns (Zia et al. 2007).

Therefore, different levels of salinity tolerance and competitiveness allow these plant species to grow in coastal areas such as salt marshes, sand dunes, salt flats, etc. (Gulzar et al. 2014a, b; Hester et al. 2001; Kim et al. 2008; Lee et al. 2007; Pennings et al. 2005). Plant strategies to survive in such environmental conditions are regulated genetically (Bohnert et al. 2006; Ezawa and Tada 2009; Laüchli and Grattan 2007). The stress tolerance ability of these halophytes is commonly regulated through osmotic adjustment (Flowers and Colmer 2008; Munns and Tester 2008) and other osmo-regulatory processes (Touchette 2007) such as changes in leaf morphological traits including increased leaf tissue density, succulence (Denaxa et al. 2012), and increased stomatal resistance to transpiration (Maricle et al. 2007). Moreover, variation in leaf water potential could be related to differences in soil matric potentials (Veste and Breckle 1995), particularly in sand dunes where water penetrates by lateral movement along the slope after rains during the monsoon season (Yair 1990). In salt marshes, even if it does not rain for a few years, tidal flushing helps in removing excess salts from the root zone to ensure plant growth and survival in these environmental conditions (Noor and Khan 1995). However, limited information is available on diurnal variation of physiological responses that may help explain plant distribution and abundance along the coastal zone.

Seasonal variations in plant water relationships of different coastal halophytes have been reported in earlier studies (Aziz and Khan 2001; Aziz et al. 2005, 2011). During the hot summer season, the salt concentration at the upper surface of the soil was significantly higher. Autumn and winter rain reduces the salt concentration by leaching saline water away from the root zone. Many species from the family Chenopodiaceae accumulate a large amount of sodium when the external salinity is high (Khan et al. 2000a, b, c; Ramos et al. 2004). Higher concentrations of sodium in soil can cause direct toxic effects, as well as presenting a challenge for both the water relationship and the nutrition for the plant (Niu et al. 1995). On the other hand, halophytes require sodium uptake for osmoregulation and this process starts from seed germination, allowing the seedling to maintain water uptake, turgor, and growth (Koyro and Eisa 2008). Many halophytes control high sodium concentrations either through ion exclusion at the root or secretion of ions from the leaves through salt glands (Barhoumi et al. 2007; Malcolm et al. 2003). On the other hand, some halophytes accumulate salts in their tissues for osmotic adjustment and compartmentalize ions in the vacuole. In saline and alkaline soils, availability of cations such as calcium may also be limited (Halitim 1988). Diurnal and seasonal variation in plant water status and ion relationships may provide information for physiological adjustments in general and water saving strategies in particular (Lo Gullo and Salleo 1988). Therefore, the present study investigated the seasonal and diurnal changes in water and ion relations in the coastal habitat as indicators of stress.

Materials and Methods

Study Site and Plants

The study area was located along the coastal belt at Sandspit, Karachi (June 2009–January 2010). The area is classified as a subtropical maritime desert with an average annual rainfall of 220 mm

(Qadir et al. 1966). Four dominant plants along the coastal gradient *viz.*, *Halopyrum mucronatum* (high dune), *Suaeda fruticosa* (low dune), *Sporobolus tremulus* (low marsh), and *Avicennia marina* (high marsh) were selected for this study. Diurnal (pre-dawn and noon) water relationship parameters (soil matric potential, leaf water potential, xylem pressure potential, and transpiration) were determined during three seasons, i.e., pre-monsoon summer (June), post-monsoon (August), and winter (January). Leaf tissue density (LTD), actual water content (AWC), and cations were determined in samples collected at noon.

Soil Matric Potential

A pre-calibrated soil hygrometer (PST-55-30-SF) with NaCl standards was used to determine the soil matric potential (Soil $_{MP}$) near the root zone of plants after thermal equilibration (2–3 min). The hygrometer was affixed to a dew point micro-voltmeter (Model HR-33 T, Wescor Inc., Logan, UT, USA) on which values were noted in micro-volts and later converted into megapascals (MPa) using a standard curve.

Leaf Water Potential

Leaf water potential (Leaf Ψw) was measured on 5-mm diameter flat leaf disks of *H. mucronatum*, *S. tremulus*, and *A. marina* while a 5-mm sized succulent leaf of *Suaeda fruticosa* was used to avoid leaf damage. Leaf samples were placed in a C-52 sample chamber connected to a dew point micro-voltmeter and leaf water potential was determined after thermal equilibration in a similar way as mentioned above. Diurnal differences in leaf water potential ($\Delta\Psi w$) were calculated by subtracting pre-dawn values from noon values.

Xylem Pressure Potential

Xylem pressure potential (XPP) in the stem was measured with a plant water status console (Arimad-2, Wagtech International Limited, UK).

Transpiration

Transpiration (*E*) was determined on fully expanded young leaves with a portable photosynthesis system (LI 6400xt; LI-COR, Inc., Lincoln, NE, USA) at CO_2 of 370 μmol mol^{-1}, flow rate of 500 μmol s^{-1}, and PAR of 1000 μmol m^{-2} s^{-1}.

Leaf Tissue Density and Actual Water Content

Leaf samples were collected, weighed, and stored on dry ice in pre-weighed plastic bags and brought to the laboratory for leaf tissue density (LTD) and actual water content (AWC) measurements. Samples were dried in an oven at 70 °C for 48 h. LTD and AWC were calculated using the formulae of Denaxa et al. (2012):

- LTD (expressed in g Kg^{-1}) = leaf dry weight (DW) / leaf fresh weight (FW)
- AWC(%) = leaf fresh weight (LFW) – leaf dry weight (LDW) × 100/leaf fresh weight (LFW)

Cations

Half a gram of finely powdered leaf samples in 10 ml distilled water was used to prepare hot water extracts. Samples were boiled in a water bath for 1 h and cooled. Dilutions of this extract were used to determine cation concentrations (Na^+, K^+, Mg^{++}, and Ca^{++}) using an atomic absorption spectrometer (Perkin Elmer, AA-700).

Statistical Analysis

Data were statistically analyzed using SPSS for Microsoft Windows (ver. 13.0). Mean standard error (± S.E.) values and a three-way ANOVA were calculated for plant species and diurnal and seasonal variations using different parameters (Table 27.1).

Table 27.1 One-way ANOVA of seasonal variation in soil matric potential ($Soil_{MP}$; MPa), leaf water potential (leaf water potential (Ψ; MPa), xylem pressure potential (XPP; MPa), transpiration (*E*; mmol m^{-2} s^{-1}), leaf tissue density (LTD), actual water content (AWC), cations (Na^+, K^+, Ca^{++}, and Mg^{++}), and Na^+/K^+ ratios

Parameter	*H. mucronatum*	*S. fruticosa*	*S. tremulus*	*A. marina*
Soil $_{MP}$ noon	86.6***	524***	206***	24.10**
Soil $_{MP}$ pre-dawn	5.09*	3.1 ns	3.0 ns	01.50 ns
Ψw noon	4.12*	21**	18.5**	31.8**
Ψw pre-dawn	1.56 ns	0.02 ns	6.2*	2.8 ns
XPP noon	9.83*	10.02*	2.10 ns	6.1*
XPP pre-dawn	1.63 ns	14.0**	16.7**	10.0*
E noon	1.01 ns	27.7**	2.1 ns	65.3***
E pre-dawn	5.09*	18.0**	15.8**	2.7 ns
AWC	302***	279***	127***	96***
LTD	0.30 ns	41.4***	21.13***	5.1*
Na^+	65.98***	129***	263***	19.1**
K^+	16.94**	1.7 ns	11.7**	31.4**
Ca^{++}	22.70**	8.2*	2.5 ns	24.2**
Mg^{++}	40.13***	19.6**	31.0**	49.9***
Na^+/K^+	70.68***	83.3***	247**	23.4**

Where * = $P < 0.05$; ** = $P < 0.01$; *** = $P < 0.001$; *ns* = non-sigificant

Results

Soil Matric Potential (Soil $_{MP}$)

A three-way ANOVA showed a significant individual effect of diurnal time (T), seasons (S), species (SP), and their combined effect (T × SP, T × S, SP × S, and T × S × SP) on soil MP (*P < 0.001*; Table 27.2). Noon soil MP was lowest in *S. fruticosa* with generally more negative values in all species compared to pre-dawn throughout the study period (Fig. 27.1). A one-way ANOVA showed a significant (*P < 0.001*) increase in noon soil MP of *H. mucronatum, S. tremulus,* and *A. marina* after monsoon rains in August (Fig. 27.1).

Leaf Water Potential (ψ_w)

A three-way ANOVA showed a significant individual effect of diurnal time (T), seasons (S), species (SP), and interaction of T × SP and T × S on leaf ψ_w (Table 27.2). However, the combined effect of SP × S and T × SP × S on leaf ψ_w was non-significant (Table 27.2). Noon leaf ψ_w was generally lower in all species compared to pre-dawn in all seasons (Fig. 27.2). A one-way ANOVA showed a significant (*P < 0.001*) increase in noon leaf ψ_w after monsoon rains with the highest values in marsh species *S. tremulus* (Fig. 27.2). A greater diurnal difference in water potential ($\Delta\psi_w$) was observed in June, which was significantly decreased (*P < 0.05*) after monsoon rains in all species (Table 27.3). The lowest values of $\Delta\psi_w$ were found in *Halopyrum mucronatum* (Table 27.3).

Xylem Pressure Potential (XPP)

A three-way ANOVA showed a significant individual effect of diurnal time (T), seasons (S), species (SP), and interaction of species and seasons (SP × S) on stem XPP (Table 27.2). However, the combined effect of T × SP, T × S, and T × SP × S were non-significant (Table 27.2). Noon stem XPP was generally lower in all test species compared to pre-dawn (Fig. 27.3). A one-way ANOVA showed a significant increase in XPP after monsoon rains (August) in all species with the exception of pre-dawn values of *H. mucronatum* (Table 27.3; Fig. 27.2).

Table 27.2 Three-way ANOVA of diurnal and seasonal variations in soil MP ($Soil_{MP}$; MPa), leaf water potential (ΨW; MPa) stem xylem pressure potential (XPP; MPa), and leaf transpiration rate (E; mmol m^{-2} s^{-1}) in four distinct zones dominated by halophytic plant species, respectively, and their interactions

Parameter	Time (T)	Species (SP)	Season (S)	T × SP	T × S	SP × S	T × SP × S
$Soil_{MP}$	1672.81***	234.70***	253.52***	08.66***	143.58***	145.79***	26.91***
Ψw	61.7***	06.82**	25.81***	03.82*	12.97***	01.59 ns	00.91 ns
XPP	70.44***	16.77***	28.3***	02.11 ns	00.72 ns	11.22***	01.46 ns
E	98.21***	05.19**	08.02**	04.18*	41.20***	05.29***	02.17 ns

Where * = $P < 0.05$; ** = $P < 0.01$; *** = $P < 0.001$; *ns* = non-sigificant

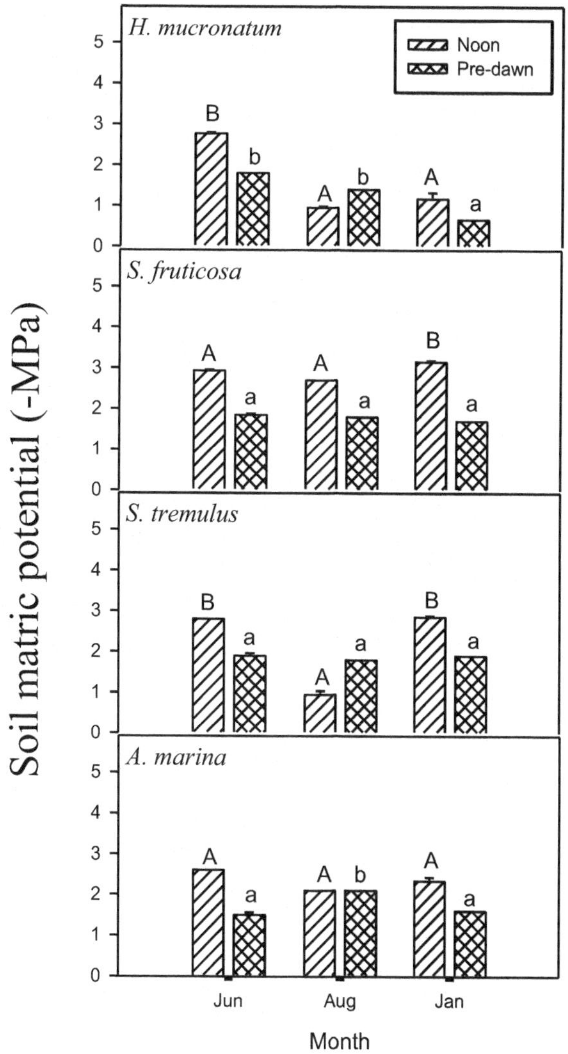

Fig. 27.1 Seasonal and diurnal variation in soil matric potential of coastal halophytes: *Halopyrum mucronatum* (high dune grass), *Suaeda fruticosa* (low dune shrub), *Sporobolus tremulus* (high marsh grass), and *Avicennia marina* (low marsh mangrove). Values are means ± S.E.

Transpiration (E)

A three-way ANOVA showed a significant individual effect of diurnal time (T), seasons (S), species (SP), and interaction of T × SP, T × S, and SP × S on leaf transpiration (Table 27.2). However, the combined effect of T × SP × S was non-significant (Table 27.2). Pre-dawn and noon transpiration values for all species were similar in June; however, a twofold increase in pre-dawn

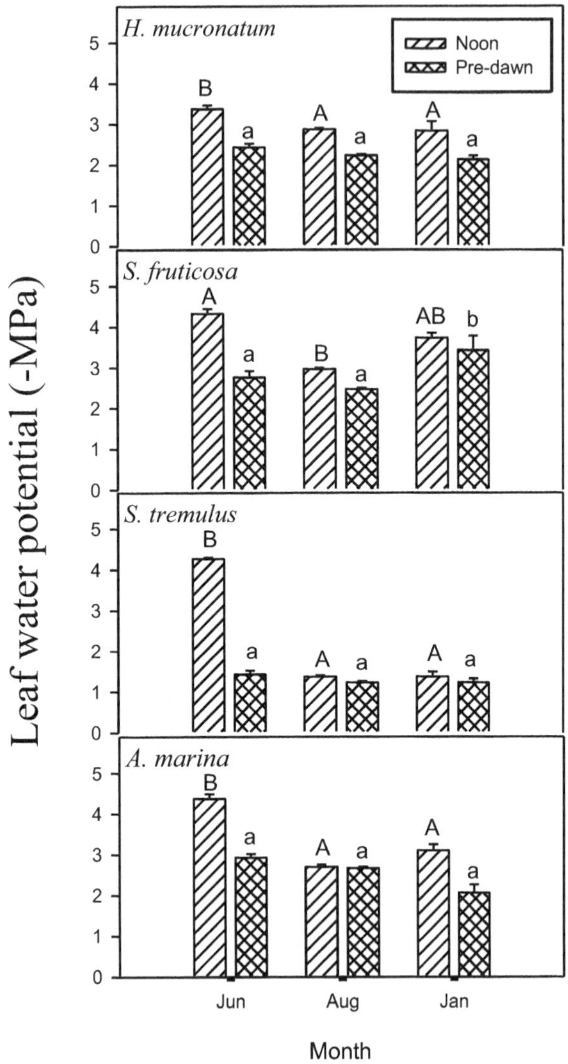

Fig. 27.2 Seasonal and diurnal variation in leaf water potential (*Ψw*) of coastal halophytes: *Halopyrum mucronatum* (high dune grass), *Suaeda fruticosa* (low dune shrub), *Sporobolus tremulus* (high marsh grass), and *Avicennia marina* (low marsh mangrove). Values are means ± S.E.

transpiration was observed after monsoon rains (August) (Fig. 27.4). A one-way ANOVA showed a significant increase ($P < 0.001$) in pre-dawn transpiration in August except for *A. marina* and a significant decrease at noon ($P < 0.01$) except for *H. mucronatum* and *S. tremulus* (Table 27.3).

Actual Water Content (AWC) and Leaf Tissue Density (LTD)

A one-way ANOVA showed a significant decrease in leaf AWC of all test species during January (Table 27.1) with the lowest values in *H. mucronatum* (Table 27.3). LTD remained unchanged in *H. mucronatum;* however, in other test species, it was significantly decreased in August with the lowest values in *S. fruticosa* (Table 27.3).

Cations (Na^+, K^+, Ca^{++}, and Mg^{++})

One-way ANOVA showed a significant variation in leaf Na^+ and Mg^{++} in all test species (Table 27.1). Sodium (Na^+) content increased in January with the highest values in *S. fruticosa* and the lowest in *S. tremulus* (Table 27.3).

Table 27.3 Seasonal variations in leaf tissue density (LTD; g kg^{-1} FW), actual water content (AWC; %), diurnal difference in water potential ($\Delta\Psi w$; MPa), cations (Na^+, K^+, Ca^{++}, Mg^{++}; mmol kg^{-1} DW), and Na^+/K^+ ratios

Species	Month	AWC	LTD	$\Delta\Psi w$	Na^+	K^+	Ca^{++}	Mg^{++}	Na^+/K^+
H. mucronatum	Jun	59 ± 2.5^{a}	458 ± 7^{a}	0.81 ± 0.06^{b}	440 ± 107^{a}	1055 ± 335^{a}	7.35 ± 1.05^{a}	48 ± 0.67^{a}	0.30 ± 0.03^{a}
	Aug	63 ± 1.8^{a}	528 ±10^{a}	0.06 ± 0.2^{a}	280 ± 15^{a}	175 ± 20^{b}	1.17 ± 0.84^{b}	66 ± 13.68^{a}	1.63 ± 0.17^{b}
	Jan	06 ± 0.9^{b}	476 ± 9^{a}	0.14 ± 0.18^{a}	3013 ± 118^{b}	524 ± 23^{c}	7.43 ± 0.82^{a}	126 ± 2.03^{b}	5.79 ± 0.50^{c}
S. fruticosa	Jun	82 ± 9.5^{a}	167 ± 5^{a}	2.11 ± 0.23^{c}	2056 ± 475^{a}	407 ± 96^{a}	20.03 ± 6.03^{a}	60 ± 0.82^{a}	5.07 ± 0.05^{a}
	Aug	85 ± 2.5^{a}	122 ± 3^{b}	0.6± 0.28^{a}	1950 ± 295^{a}	316 ± 32^{b}	28.40 ± 1.73^{a}	80 ± 5.30^{b}	6.11 ± 0.28^{b}
	Jan	40 ± 0.8^{b}	123 ± 4^{b}	1.36 ± 0.33^{b}	1756 ± 136^{a}	376 ± 31^{b}	41.37 ± 0.75^{b}	142 ± 6.94^{c}	4.78 ± 0.70^{c}
S. tremulus	Jun	61 ± 1.9^{a}	444 ± 2^{a}	2.9 ± 0.18^{b}	333 ± 64^{a}	305 ± 26^{a}	11.07 ± 6.67^{a}	63 ± 1.85^{a}	1.14 ± 0.31^{a}
	Aug	80 ± 4.02^{b}	295 ± 11^{c}	0.79 ± 0.2^{a}	640 ± 90^{b}	301 ± 38^{a}	15.07 ± 1.32^{a}	156 ± 7.77^{b}	2.12 ± 0.06^{b}
	Jan	17 ± 2.03^{c}	355 ± 18^{b}	0.56 ± 0.17^{a}	900 ± 32^{c}	321 ± 20^{a}	15.17 ± 2.88^{a}	97 ± 7.83^{c}	2.82 ± 0.16^{b}
A. marina	Jun	82 ± 4.4^{a}	374 ± 15^{a}	1.81 ± 0.12^{c}	456 ± 98^{a}	338 ± 33^{a}	5.37 ± 1.16^{a}	48 ± 1.20^{a}	1.32 ± 0.17^{a}
	Aug	47 ± 9.3^{b}	298 ± 5^{b}	0.28 ± 0.08^{a}	643 ± 07^{a}	312 ± 21^{a}	2.13 ± 1.41^{a}	175 ± 9.35^{b}	2.08 ± 0.15^{b}
	Jan	20 ± 1.2^{c}	351 ± 9^{a}	0.59 ± 0.13^{b}	1160 ± 57^{b}	268 ± 03^{b}	14.77 ± 0.66^{b}	100 ± 9.64^{c}	4.33 ± 0.17^{c}

Values in each column followed by different letters in superscript are significantly ($P < 0.05$) different across months for each species

Fig. 27.3 Seasonal and diurnal variation in xylem pressure potential (XPP) of coastal halophytes: *Halopyrum mucronatum* (high dune grass), *Suaeda fruticosa* (low dune shrub), *Sporobolus tremulus* (high marsh grass), and *Avicennia marina* (low marsh mangrove). Values are means ± S.E.

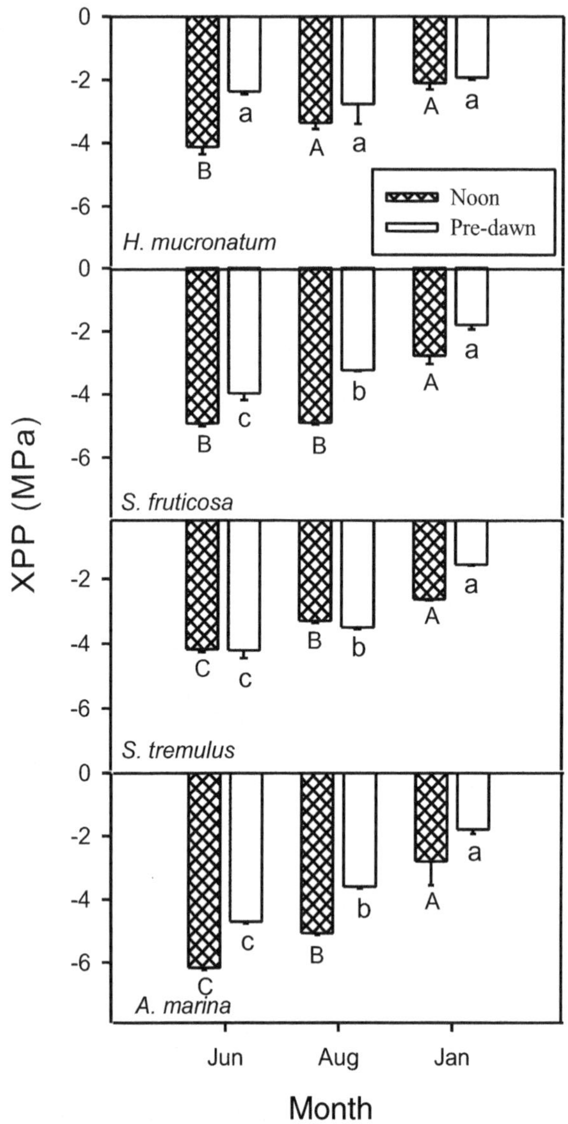

Magnesium (Mg^{++}) content in all species was lowest in June; however, its seasonal variation in all test species was different compared to Na^{+} (Table 27.3). In *H. mucronatum* and *S. fruticosa*, higher leaf Mg^{++} was found in January while the highest values in *S. tremulus* and *A. marina* were in August (Table 27.1). One-way ANOVA showed a significant decrease in leaf K^{+} in *H. mucronatum* while no significant change was observed for the other test species (Table 27.1). *Halopyrum mucronatum* had the highest K^{+} in June, which was significantly decreased in August, while all other species had similar K^{+} values (Table 27.3). *Suaeda fruticosa* accumulated the highest Ca^{++} content in leaves with a twofold increase in January, while the lowest Ca^{++} value was found in *H. mucronatum* (Table 27.3). *Sporobolus tremulus* did not show any variation in its leaf Ca^{++} (Tables 27.1 and 27.3).

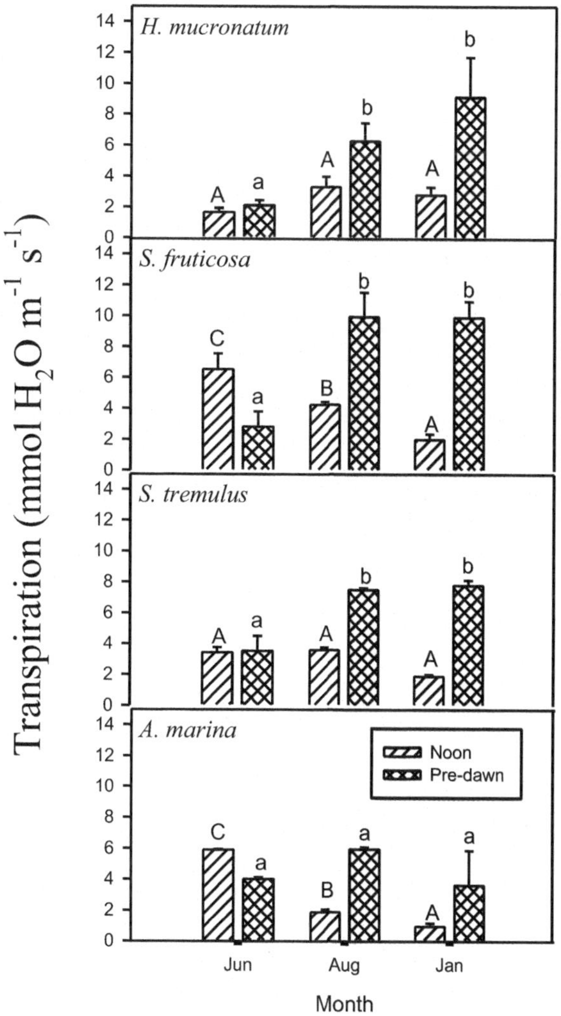

Fig. 27.4 Seasonal and diurnal variation in transpiration (*E*) of coastal halophytes: *Halopyrum mucronatum* (high dune grass), *Suaeda fruticosa* (low dune shrub), *Sporobolus tremulus* (high marsh grass), and *Avicennia marina* (low marsh mangrove). Values are means ± S.E.

Discussion

Coastal plant species provide multiple benefits for mankind from ecological economic view points (Li et al. 2018; Watson and Byrne 2009). Although coastal environments are unsuitable for most plants, some plant species, particularly halophytes, have the ability to grow and flourish in this environment (Gulzar and Khan 1998; Aziz et al. 2011). Coastal plant species, particularly those found in coastal salt marshes and sand dunes, have a wide range of mechanisms to deal with seasonal and diurnal fluctuations in environmental factors. Low plant diversity found in coastal areas is related to the harsh soil-borne and airborne salinity as a result of tidal inundation, salt spray, and low nutrient availability (Abdelgadir et al. 2005; de Vos et al. 2010).

Therefore evidence of plant physiological parameters is required to understand the mechanisms of response to environmental extremes, which may be either a short-term acclamation to rapid fluctuations such as salinity, light, temperature, and humidity by regulating photosynthesis, water relations and respiration rate or long-term morpho-anatomical changes such as leaf size, number and thickness etc. (Larcher 2003; Lichtenthaler 1996). Among all the above-mentioned physical factors, salinity is considered a key factor that influences physiological traits of coastal plant species, such as an increase in osmolyte concentration, decreased water potential, improved sodium and potassium levels, and water use efficiencies in high stress-tolerant species (Slama et al. 2008). The survival of plants in saline conditions depends on the maintenance of cell turgor mainly by decreasing the osmotic potential via osmotic adjustments (Munns 2002).

In the present study, the soil matric potential was generally more negative at noon than at pre-dawn throughout the study period, and this may be related to high evaporation rates during the daytime. An increase in noon soil MP was found after monsoon rains (August) in *H. mucronatum, S. tremulus,* and *A. marina*, and this may be due to dryness on the upper layer of the soil. Leaf water potential and xylem pressure potential decreased at noon compared to pre-dawn, and this is similar to the previous study on *Atriplex griffithii* and *Heliotropium currassavicum* where at noon, leaf water potential was more negative than at pre-dawn (Gulzar and Khan 1998).

More negative water potential values were found at noon with an increase in noon leaf ψ_w after monsoon rains, which was highest in marsh species *S. tremulus*. A greater diurnal difference in water potential ($\Delta\psi_w$) was observed in summer (June) and this decreased after monsoon rains (August) in all species. The lowest values of $\Delta\psi_w$ were found in *Halopyrum mucronatum*. *Halopyrum mucronatum* is reported to maintain its tissue osmotic potential at expense of tissue moisture reduction (Aziz et al. 2005). In all test species, higher osmotic potentials was found during June (summer) with lower diurnal variations.

Generally, water potential (ψ_w) of plants became extremely negative with an increase in soil salinity. This value peaked during the summer season for most test species. Gul et al. (2001) also indicated that ψ_w values of *Allenrolfea occidentalis* dropped significantly during the dry season. This indicates that plant water status responded more quickly to changes in moisture content in the soil after the rainy season, which could be correlated to the fluctuations in soil salinity. Therefore, a reduction in leaf water potential with an increase in salinity concentration is associated with decreases in both osmotic and pressure potential during the dry season (Yair 1990).

The sodium content increased in January with the highest values in *S. fruticosa* and the lowest in *S. tremulus*. While, Na^+ was lowest in June for most species. Both of the salt marsh species *S. tremulus* and *A. marina* displayed relatively lower values as well as between-site variations during June and August in comparison with the dune species. Similar results are found in *Atriplex stocksii* and *S. fruticosa*. These two succulent species accumulated higher amounts of sodium and chloride during the dry period and this strategy could be related to the ionic adjustments of the plants to achieve osmotic balance (Aziz et al. 2011).

The potassium content was highest in *Halopyrum mucronatum* during summer in June and was significantly decreased in August while all other species had similar K^+ values. Potassium also showed different patterns and levels of accumulation in leaves between the two habitat types and showed an inverse relationship to that of Na^+ in both dune species. On the other hand, in salt marsh species, K^+ increased and decreased along with Na^+. The highest Na^+/K^+ ratios were observed in *S. fruticosa*, a leaf succulent salt-accumulating physiotype. Similar results were found in *Atriplex halimus* – with increasing Na^+ content, K^+ decreased (Nedjimi 2012).

The magnesium content in all species was lowest in June; however, its seasonal variation in all test species appeared micro-habitat/zone specific. In *H. mucronatum* and *S. fruticosa*, higher leaf Mg^{++} was found in January while in *S. tremulus* and *A. marina* the highest values were found in August. *Suaeda fruticosa* accumulated the highest Ca^{++} content in leaves with twofold increases in August and January, while the lowest Ca^{++} was found in *H. mucronatum*. *Sporobolus tremulus* did not show any variation in its leaf Ca^{++}. Agha et al. (2009) and Nedjimi (2012), showed very little variation in soil Ca^{++} concentrations due to seasons on the Arabian Coast.

The two grasses in this study appear to have morphological adaptations for maintaining lower transpirational water loss during the warm seasons due to the high leaf serration and thicker leaves in *H. mucronatum* and reduced leaf area in *S. tremulus*, coupled with greater stomatal control of evaporative water loss. Transpiration values from both pre-dawn and noon were similar during summer in all test species. However, a twofold increase in pre-dawn transpiration was observed after monsoon rains (August). The AWC of all test species decreased during winter in January, with the lowest values found in *H. mucronatum*. LTD remained unchanged in *H. mucronatum;* however, in other test species it was significantly decreased in January with the lowest values found in *S. fruticosa*. Previously, Aziz et al. (2011) have also reported significant seasonal variations in the water relations of four perennial halophytes including *S. fruticosa*.

Conclusions

The four dominant halophytes from different zones along a coastal gradient showed variable responses. Plants seemed to be physiologically adapted to their respective habitat zones. At large, ion and water relations of plants varied both diurnally and seasonally. However, contrary to the common notion of extreme stress in the hot summer months, the coastal vegetation appeared more stressed in winter due to high tissue Na and low AWC. Hence, ion and water relations can be used as potential stress markers to study coastal vegetation on temporal basis.

References

Abdelgadir EM, Oka M, Fujiyama H (2005) Characteristics of nitrate uptake by plants under salinity. J Plant Nutr 28:33–46

Agha F, Gul B, Khan MA (2009) Seasonal variation in productivity of Atriplex stocksii from a coastal marsh along the Arabian Sea coast. Pak J Bot 41:1053–1068

Aziz I, Khan MA (2001) Seasonal variation in ecophysiology of *Avicennia marina* populations growing in polluted Arabian Sea coast around Karachi. Pak J Bot 33:429–441

Aziz I, Gulzar S, Noor M, Khan MA (2005) Seasonal variation in water relations of *Halopyrum mucronatum* (L.) Stapf. Growing near Sandspit, Karachi. Pak J Bot 37(1):141–148

Aziz I, Gul B, Gulzar S, Khan MA (2011) Seasonal variations in plant water status of four desert halophytes from semi-arid region of Karachi. Pak J Bot 43:587–594

Barhoumi Z, Djebali W, Smaoui A, Chaïbi W, Abdelly C (2007) Contribution of NaCl excretion to salt resistance of *Aeluropus littoralis* (Willd) Parl. J Plant Physiol 164:842–850

Böer B (1996) Plants as soil indicators along the Saudi coast of the Arabian gulf. J Arid Environ 33:417–423

Bohnert HJ, Gong Q, Li P, Ma S (2006) Unraveling abiotic stress tolerance mechanisms–getting genomics going. Curr Opin Plant Biol 9:180–188

Brown CE, Pezeshki SR, DeLaune RD (2006) The effects of salinity and soil drying on nutrient uptake and growth of *Spartina alterniflora* in a simulated tidal system. Environ Exp Bot 58:140–148

Caçador I, Tibério S, Cabral HN (2007) Species zonation in Corroios salt marsh in the Tagus estuary (Portugal) and its dynamics in the past fifty years. Hydrobiologia 587(1):205–211

de Vos AC, Broekman R, Groot MP, Rozema J (2010) Ecophysiological response of *Crambe maritima* to airborne and soil-borne salinity. Ann Bot 105:925–937

Denaxa NK, Roussos PA, Damvakaris T, Stournaras V (2012) Comparative effects of exogenous glycine betaine, kaolin clay particles and Ambiol on photosynthesis, leaf sclerophylly indexes and heat load of olive cv. *Chondrolia chalkidikis* under drought. Sci Hortic 137:87–94

Drégelyi-Kiss Á, Drégelyi-Kiss G, Hufnagel L (2008) Ecosystems as climate controllers–biotic feedbacks (a review). Appl Ecol Environ Res 6(2):111–135

Emery NC, Ewanchuk PJ, Bertness MD (2001) Competition and salt-marsh plant zonation: stress tolerators may be dominant competitors. Ecology 82(9):2471–2485

Ezawa S, Tada Y (2009) Identification of salt tolerance genes from the mangrove plant *Bruguiera gymnorhiza* using agrobacterium functional screening. Plant Sci 176(2):272–278

Flowers TJ, Colmer TD (2008) Salinity tolerance in halophytes. New Phytol 179:945–963

Gul B, Weber DJ, Khan MA (2001) Growth, ionic and osmotic relations of an *Allenrolfea occidentalis* population in an inland salt playa of the Great Basin Desert. J Arid Environ 48:445–460

Gullo MA, Salleo S (1988) Different strategies of drought resistance in three Mediterranean sclerophyllous trees growing in the same environmental conditions. New Phytol 108:267–276

Gulzar S, Khan MA (1998) Diurnal water relations of inland and coastal halophytic populations from Pakistan. J Arid Environ 40(3):295–305

Gulzar S, Hameed A, Ahmed MZ, Khan MA (2014a) Is soil heterogeneity the major factor influencing vegetation zonation at Karachi coast? In: Sabkha ecosystems. Springer, Dordrecht, pp 199–207

Gulzar S, Hameed A, Ahmed MZ, Khan MA (2014b) Is soil heterogeneity the major factor influencing vegetation zonation at Karachi coast? In: Sabkha ecosystems. Springer, Dordrecht, pp 199–207
Halitim A (1988) Arid soils in Algeria. OPU, Alger, 384p
Hester MW, Mendelssohn IA, McKee KL (2001) Species and population variation to salinity stress in *Panicum hemitomon*, *Spartina patens*, and *Spartina alterniflora*: morphological and physiological constraints. Environ Exp Bot 46:277–297
Ievinsh G (2006) Biological basis of biological diversity: physiological adaptations of plants to heterogeneous habitats along a sea coast. Acta Univ Latv 710:53–79
Khan MA, Ungar IA, Showalter AM (2000a) Effects of salinity on growth, water relations and ions accumulation of the subtropical perennial halophyte, *Atriplex griffithii var.stockii*. Ann Bot 85:225–232
Khan MA, Ungar IA, Showalter AM (2000b) The effect of salinity on the growth, water status, and ion content of a leaf succulent perennial halophyte, *Suaeda fruticosa* (L.) Forssk. Alger J Arid Environ 45:73–84
Khan MA, Ungar IA, Showalter AM (2000c) Effects of sodium chloride treatments on growth and ion accumulation of the halophyte *Haloxylon recurvum*. Commun Soil Sci Plant Anal 31:2763–2774
Kim D, Yu KB, Park SJ (2008) Identification and visualization of complex spatial pattern of coastal dune soil properties using GIS-based terrain analysis and geostatistics. J Coast Res 4:50–60
Koyro HW, Eisa SS (2008) Effect of salinity on composition, viability and germination of seeds of *Chenopodium quinoa* Willd. Plant Soil 302:79–90
Kunza AE, Pennings SC (2008) Patterns of plant diversity in Georgia and Texas salt marshes. Estuaries Coasts 31:673–681
Larcher W (2003) Physiological plant ecology: ecophysiology and stress physiology of functional groups. Springer Science & Business Media, Berlin
Läuchli A, Grattan SR (2007) Plant growth and development under salinity stress. In: Advances in molecular breeding toward drought and salt tolerant crops. Springer, Dordrecht, pp 1–32
Lee JS, Ihm BS, Cho DS, Son DY, Kim JW (2007) Soil particle sizes and plant communities on coastal dunes. J. Plant Biol. 50:475–479
Li SH, Ge ZM, Xie LN, Chen W, Yuan L, Wang DQ et al (2018) Ecophysiological response of native and exotic salt marsh vegetation to waterlogging and salinity: implications for the effects of sea-level rise. Sci Rep 8(1):2441
Lichtenthaler HK (1996) Vegetation stress: an introduction to the stress concept in plants. J Plant Physiol 148:4–14
Malcolm CV, Lindley VA, O'Leary JW, Runciman HV, Barrett-Lennard EG (2003) Halophyte and glycophyte salt tolerance at germination and the establishment of halophyte shrubs in saline environments. Plant Soil 253:171–185
Maricle BR, Cobos DR, Campbell CS (2007) Biophysical and morphological leaf adaptations to drought and salinity in salt marsh grasses. Environ Exp Bot 60:458–467
Munns R (2002) Salinity, growth and phytohormones. In: Salinity: environment-plants-molecules. Springer, Dordrecht, pp 271–290
Munns R, Tester M (2008) Mechanisms of salinity tolerance. Annu Rev Plant Biol 59:651–681
Nedjimi B (2012) Seasonal variation in productivity, water relations and ion contents of *Atriplex halimus* spp. *Schwein furthii* grown in Chott Zehrez wetland, Algeria. J Saudi Soc Agric Sci 11:43–49
Niu X, Bressan RA, Hasegawa PM, Pardo JM (1995) Ion homeostasis in NaCl stress environments. Plant Physiol 109:735–742
Noor MEHER, Khan MA (1995) Factors affecting germination of summer and winter seeds of *Halopyrum mucronatum* under salt stress. In: Biology of Salt Tolerant Plants, vol 6. University of Karachi, Karachi City, pp 51–58
Pennings SC, Grant MB, Bertness MD (2005) Plant zonation in low-latitude salt marshes: disentangling the roles of flooding, salinity and competition. J Ecol 93:159–167
Qadir SA, Qureshi SZ, Ahmed MA (1966) A phytosociological survey of the Karachi University campus. Vegetation 13:339–362
Ramos J, López MJ, Benlloch M (2004) Effect of NaCl and KCl salts on the growth and solute accumulation of the halophyte *Atriplex nummularia*. Plant Soil 259(1–2):163–168
Rozema J, Bijwaard P, Prast G, Broekman R (1985) Ecophysiological adaptations of coastal halophytes from foredunes and salt marshes. Vegetation 62:499–521
Slama I, Ghnaya T, Savouré A, Abdelly C (2008) Combined effects of long-term salinity and soil drying on growth, water relations, nutrient status and proline accumulation of *Sesuvium portulacastrum*. C R Biol 331:442–451
Touchette BW (2007) Seagrass-salinity interactions: physiological mechanisms used by submersed marine angiosperms for a life at sea. J Exp Mar Biol Ecol 350:194–215
Veste M, Breckle SW (1995) Xerohalophytes in a sandy desert ecosystem. In: Biology of Salt Tolerant Plants. University of Karachi, Karachi City, pp 161–165
Watson EB, Byrne R (2009) Abundance and diversity of tidal marsh plants along the salinity gradient of the San Francisco estuary: implications for global change ecology. Plant Ecol 205(1):113–128
Yair A (1990) Runoff generation in a sandy area-the Nizzana sands, Western Negev, Israel. Earth Surf Process Landf 15(7):597–609
Zia S, Egan TP, Khan MA (2007) Population studies of *Limonium stocksii* (Plumbaginacea) from a Salt Desert near the Arabian Sea coast. Commun Soil Sci Plant Anal 38:1975–1990

Mangrove Cover, Biodiversity, and Carbon Storage of Mangrove Forests in Thailand

28

Poonsri Wanthongchai and Orathai Pongruktham

Abstract

Thailand shoreline extends up to 2667 km covering 23 provinces. The coastal zone of the area includes the Gulf of Thailand and the Andaman Sea. The Gulf of Thailand Coast or the East Coast has 1653 km of coastline bordering the Gulf of Thailand from Trat Province in the eastern part of the country to Narathiwat Province in the South. On the Western side, the Andaman Sea Coast is 1014 km long, i.e., from Ranong Province to Satun Province. Mangrove forests are distributed along both coasts of Thailand in 24 provinces (the 23 coastal provinces plus Phatthalung Province). Since the first mangrove survey in 1961, the mangrove cover in Thailand has been reduced due to other land uses; hence, till 2014 only 54% of original mangrove cover is left. Mangrove forests in Thailand have been reported to support high biodiversity. Long-term studies on diversity revealed that there were 81 plant species (34 true mangroves and 47 mangrove associates), 272 bird species, 1570 insect species, 101 benthos, and a number of other economically important species. Based on a recent publication of data from 2007 to 2015, the average aboveground biomass in Thai's mangrove zone ranges from 4.4 ton/ha to 7.5 ton/ha. Tree density varies from 37 trees/ha to 76 trees/ha.

Keywords

Biodiversity · Biomass · Carbon storage · Cover · Mangrove · Shoreline

Introduction

Mangrove forests around the world have been threatened both due to natural disasters and anthropogenic activities. They are considered as one of the most threatened habitats on the planet Earth (Aksornkoae 2008; Valiela et al. 2001). A major mangrove cover loss has been observed in Thailand. In the past, the area had healthy mangroves with a wide range of natural resources and ecosystem services, including energy, food, wildlife, coastline protection, and nursery ground for fauna (Aksornkoae 1989; Sathirathai and Barbier 2011; Vo et al. 2012). A vast mangrove area has been destroyed at an alarming rate, especially due to conversion of mangrove forests into shrimp agriculture ponds (Aksornkoae and Tokrisna 2004; Barbier and Cox 2002; Lebel et al. 2002; Vaiphasa et al. 2007).

Although some publications on Thai mangroves do exist, most of them usually discuss mangroves on small scale or some specific topic (Muttitanon and Tripathi 2005; Naito and

P. Wanthongchai (✉) · O. Pongruktham
Department of Marine and Coastal Resources, Mangrove Conservation Office, Bangkok, Thailand

B. Gul et al. (eds.), *Sabkha Ecosystems*, Tasks for Vegetation Science VI,
https://doi.org/10.1007/978-3-030-04417-6_28

Traesupap 2013; Pumijumnong 2014; Sathirathai and Barbier 2011; Thampanya et al. 2006). Latest updates on the current status of national mangrove forest regarding biodiversity and carbon storage are seldom. This article summarizes historical changes of mangrove forest cover from 1961 to date covering information on plant and animal biodiversity in Thai mangroves with the help of long-term studies, nationwide mangrove biomass and carbon storage, and some aspects of national mangrove conservation and management.

Status of Mangroves in Thailand

Thailand shoreline extends up to 2667 km which covers 23 provinces. The coastal zone includes the Gulf of Thailand and the Andaman Sea. The Gulf of Thailand Coast or the East Coast has 1653 km of coastline bordering the Gulf of Thailand from Trat Province in the eastern part of the country to Narathiwat Province in the South. On the Western side, the Andaman Sea Coast lies 1014 km from Ranong Province to Satun Province. Mangrove forests are distributed along both coasts of Thailand in 24 provinces (the 23 coastal provinces plus Phatthalung Province) (DMCR 2015b, 2016b). The mangrove distribution in Thailand is illustrated in Fig. 28.1.

Mangrove Forest Cover

The early national forest assessment in Thailand was carried out in 1961 by the Thai Royal Forest Department (FRD) using satellite image interpretation showing that total forest area was 22,170,700 ha accounting for about 43% of the total land. Within this terrestrial forest, there was 367,900 ha of mangrove forests (Fig. 28.2). This mangrove area was registered as the national mangrove forest according to the Cabinet Resolution B.E. 2504 (1961). The following surveys were conducted in 1975, 1979, and 1986 when the total mangrove areas were 312,700 ha, 287,308 ha, and 196,437 ha, respectively. During 1979–1986 the mangrove forest covers were converted to other land-use types at an average rate of 12,960 ha per year which was at the highest decreasing rate in the history of national surveys. Later on, several regulations and measures for mangrove management were put into the place; as a result, the mangrove deforestation rates in the following years declined. In 1989, 1991, and 1993, the remaining mangrove covers were 180,559 ha, 173,821 ha, and 168,683 ha, respectively. In 1996, when the total mangrove area was reduced to 167,582 ha, the Royal Thai Government announced the termination of mangrove forest concession after the end of the previously appointed agreement period in 2003 (RFD 1985). The drastic reduction of mangrove forest cover in Thailand was mainly due to the mangrove forest concession for logging and shrimp farming and unlawful occupation of land (Aksornkoae 1999). These factors had also contributed to mangrove losses in other parts of Asia (Aksornkoae and Tokrisna 2004; Barbier and Cox 2002).

Due to some other strict regulations on mangrove area management and current advances in satellite image technology, the mangrove forest survey in 2000 reported an increase of mangrove cover, for the first time with a total area of 252,765 ha. During this assessment, additional mangrove forest of 91,218 ha was identified. This additional mangrove forest was incorporated in to the 367,900 ha of mangrove forest according to the Cabinet Resolution B.E. 2504 (1961). This total mangrove area of 459,118 ha has been used as "reference original mangrove area" in the following assessments until the present (RFD 1985; DMCR 2015b). The subsequent reports on mangrove forest cover were presented by the Department of Marine and Coastal Resources (DMCR) as it was established in 2002. The mangrove examinations in 2004, 2009, and 2014 showed a slight expansion of mangrove cover with total areas of 233,308 ha, 244,010 ha, and 245,534 ha, respectively (DMCR 2015b). The historical changes in mangrove forest cover are presented in Fig. 28.2.

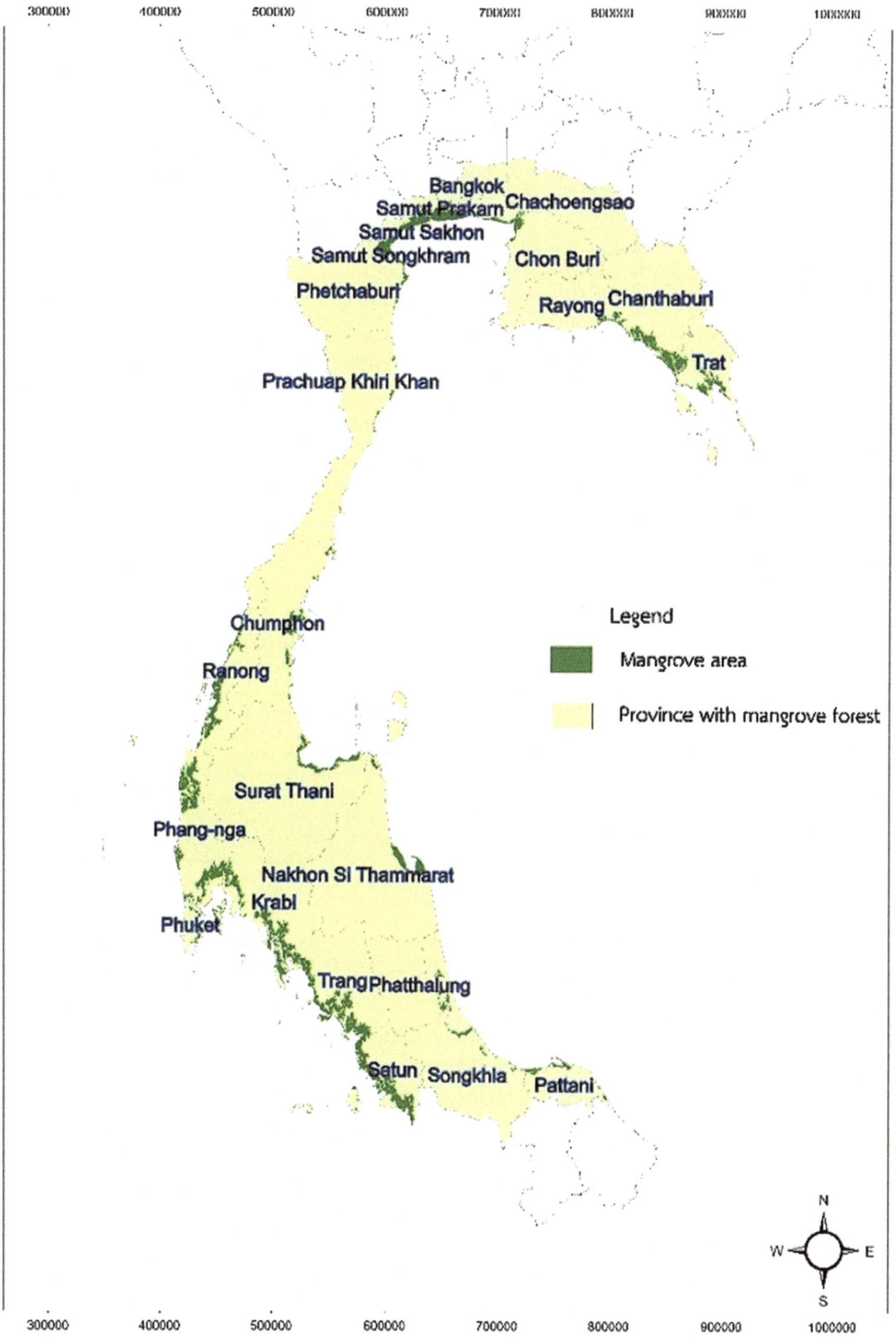

Fig. 28.1 Map of Thailand demonstrating distribution of mangrove forests (Modified from DMCR 2015b)
Note: Mangrove distribution in Narathiwat Province is not shown in the map

Fig. 28.2 Historical changes of mangrove covers in Thailand from 1961 to 2014. (Generated from DMCR 2015b and RFD 1985)

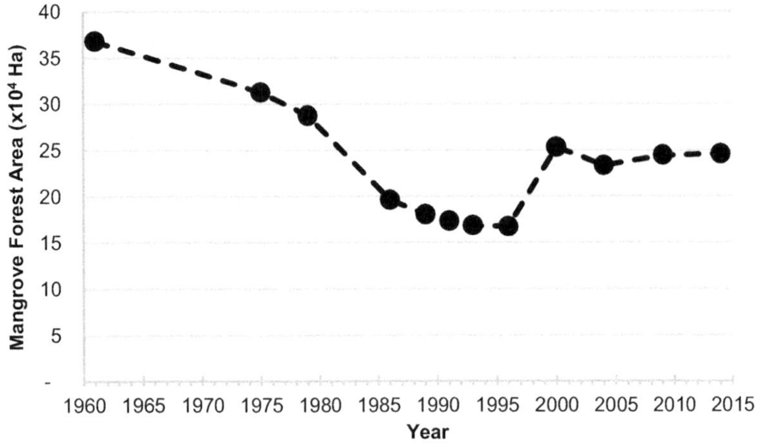

Fig. 28.3 Proportion of land use land cover of the reference original mangrove area in 2014. (Generated from data published in DMCR 2015a)

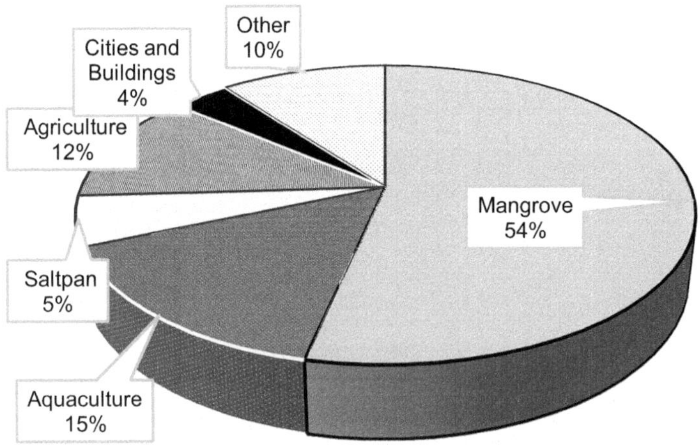

Land-Use and Land-Cover Change

The satellite image data from the recent mangrove assessment in 2014 was also interpreted in terms of land use and land cover aligned with the "reference original mangrove area" (total area of 459,118 ha). The results illustrated that in 2014, only 54% (or 245,534 ha) of the "reference original mangrove area" was retained. The rest of the original mangrove area was converted into other land-use types such as aquaculture (15%), agriculture (12%), saltpans (5%), city (4%), and others (10%) (Fig. 28.3).

Flora

Based on the recent reports (DMCR 2012), 81 species of mangrove trees have been recorded in Thailand. Among these, there are 34 species of true mangrove trees such as plants in families Acanthaceae, Euphorbiaceae, Lythraceae, and Rhizophoraceae and 47 species of mangrove associates in families Apocynaceae, Asteraceae, Lecythidaceae, Rutaceae, and Sapindaceae (Table 28.1). Mangrove distribution along the coasts of Thailand shows distinct zonation which is defined by physicochemical and physical con-

Table 28.1 List of vegetation reported in Mangrove of Thailand

No	Scientific name	Family	Type
1	*Acanthus ebracteatus*	Acanthaceae	TM
2	*Acanthus ilicifolius*		TM
3	*Acanthus volubilis*		TM
4	*Avicennia alba*		TM
5	*Avicennia lanata*		TM
6	*Avicennia marina*		TM
7	*Avicennia officinalis*		TM
8	*Suaeda maritima*	Amaranthaceae	MA
9	*Cerbera manghas*	Apocynaceae	MA
10	*Cerbera odollam*		MA
11	*Finlaysonia maritime*		MA
12	*Plucheaindica*	Asteraceae	MA
13	*Wedelia biflora*	Asteraceae	MA
14	*Dolichandrone spathacea*	Bignoniaceae	TM
15	*Cassine viburnifolia*	Celastraceae	MA
16	*Lumnitzera littorea*	Combretaceae	TM
17	*Lumnitzera racemosa*		TM
18	*Terminalia catappa*		MA
19	*Diospyros areolata*	Ebenaceae	MA
20	*Excoecaria agallocha*	Euphorbiaceae	TM
21	*Shiraki opsisindica*		TM
22	*Flagellaria indica*	Flagellariaceae	MA
23	*Scaevola taccada*	Goodeniaceae	MA
24	*Calophyllum inophyllum*	Guttiferae	MA
25	*Clerodendrum inerme*	Labiatae	MA
26	*Premna obtusifolia*		MA
27	*Barringtonia asiatica*	Lecythidaceae	MA
28	*Barringtonia racemosa*		MA
29	*Caesalpinia crista*	Leguminosae	MA
30	*Cynometra iripa*		TM
31	*Cynometra ramiflora*		TM
32	*Dalbergia candenatensis*		MA
33	*Derris indica*		MA
34	*Derris trifoliata*		MA
35	*Intsia bijuga*		MA
36	*Peltophorum pterocarpum*		MA
37	*Pemphis acidula*	Lythraceae	TM
38	*Sonneratia alba*		TM
39	*Sonneratia caseolaris*		TM
40	*Sonneratia griffithii*		TM
41	*Sonneratia ovata*		TM
42	*Brownlowia tersa*	Malvaceae	MA
43	*Heritiera fomes*		TM
44	*Heritiera littoralis*		TM
45	*Hibiscus tiliaceus*		MA
46	*Thespesia populnea*		MA
47	*Thespesia populneoides*		MA
48	*Melastoma saigonense*	Melastomataceae	MA
49	*Aglaia cucullata*	Meliaceae	TM
50	*Xylocarpus granatum*		TM
51	*Xylocarpus moluccensis*		TM
52	*Xylocarpus rumphii*		MA
53	*Ficus microcarpa*	Moraceae	MA
54	*Melaleuca cajuputi*	Myrtaceae	MA
55	*Nypa fruticans*	Palmae	TM
56	*Oncosperma tigillarium*		MA
57	*Phoenix paludosa*		TM
58	*Pandanus odoratissimus*	Pandanaceae	MA
59	*Aegialitis rotundifolia*	Plumbaginaceae	TM
60	*Aegiceras corniculatum*	Primulaceae	TM
61	*Ardisia elliptica*		MA
62	*Rapanea porteriana*		MA
63	*Acrostichum aureum*	Pteridaceae	TM
64	*Acrostichum speciosum*		TM
65	*Stenochlaena palustris*		MA
66	*Bruguiera cylindrica*	Rhizophoraceae	TM
67	*Bruguiera gymnorrhiza*		TM
68	*Bruguiera hainesii*		TM
69	*Bruguiera parviflora*		TM
70	*Bruguiera sexangula*		TM
71	*Ceriops decandra*		TM
72	*Ceriops tagal*		TM
73	*Kandelia candel*		TM
74	*Rhizophora apiculata*		TM
75	*Rhizophora mucronata*		TM
76	*Scyphiphora hydrophyllacea*	Rubiaceae	TM
77	*Atalantia monophylla*	Rutaceae	MA
78	*Merope angulata*		MA
79	*Scolopia macrophylla*	Salicaceae	MA
80	*Allophyllus cobbe*	Sapindaceae	MA
81	*Pouteria obovata*	Sapotaceae	MA

(continued)

Note: *TM* true mangrove, *MA* mangrove associate (Undated from original data in 2012)

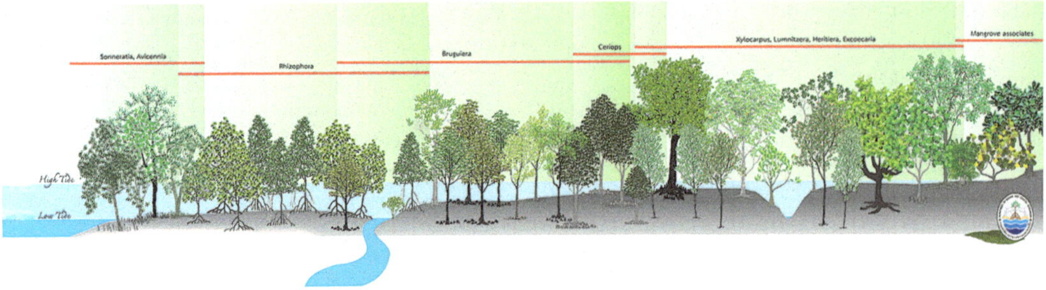

Fig. 28.4 Typical mangrove zonation in Thailand. (Photo by Chatree Maknual)

ditions such as climate, salinity, tidal fluctuation, sediment type, and wave energy (Field 1995). Mangroves in Thailand grow in monospecific stands of vegetation (Fig. 28.4) parallel to shoreline in both of the East and West Coasts. Specifically, *Sonneratia* and *Avicennia* are commonly observed along the water's edge where the soil is characterized by soft mud, with full exposure to tidal variation and winds. Moving further inland, *Rizophora*occupies occupy the area of harder mud with less wave and winds influence but still regularly flooded by sea water. At the higher intertidal zone where seawater floods are less frequent, *Bruguiera* spp. are dominant, followed by a mixed zone of *Xylocarpus* spp., *Lumnitzera* spp., *Heritiera* spp., and *Excoecaria* spp. Lastly, the vegetation zone beyond the highest spring tide level is usually inhabited by the mangrove associates.

Biomass and Carbon Storage in Mangrove Forests

Aboveground biomass, carbon storage, and carbon values of mangroves had been examined in 21 coastal provinces from 2007 to 2015 (DMCR 2016a). These provinces were divided into six geographic zones. Area 1 covered the Upper Andaman Coast in Ranong Province and the western part of Phang Nga Province where there were 50,653 ha of mangrove forest. Area 2 was mangroves of the Lower Andaman Coast in Trang and Satun Provinces with the mangrove forest cover of ~70,002 ha. Area 3 was 55,493 ha of mangrove forest in the Middle Andaman Coast of Krabi, Phuket, and eastern part of Phang Nga Provinces. Area 4 was the mangrove area on the Lower Gulf of Thailand Coast covering the area of 32,542 ha in Chumphon, Surat Thani, Nakhon Si Thammarat, Songkhla, and Phatthani Provinces. Area 5 represented by the mangrove forest in the eastern part of Thailand including mangroves in Rayong, Chanthaburi, and Trat Provinces had the area of 24,402 ha. Lastly, Area 6 consisted of mangrove area in the Upper the Gulf of Thailand Coast in Chon Buri, Chachoengsao, Samut Prakan, Samut Sakhon, Samut Songkhram, Phetchaburi, and Prachuap Khiri Khan Provinces with mangrove cover area of 12,441 ha. The averaged aboveground biomass in these mangrove zones ranges from 4.4 ton/ha in mangrove forest Area 6 to 7.5 ton/ha in mangrove forest Area 2, while the tree density varies from 37 trees/ha in mangrove Area 5 to 76 trees/ha in Area 2 (Fig. 28.5).

Fauna

The habitat complexity of mangrove forests could support a wide range of wildlife species including those which occupy aquatic habitat, mud/soil surface, and parts of the mangrove tree. Some of these animals permanently reside in the mangrove area, while some only use the mangroves at some stages of their life cycle or during part of their migration (Field 1995). Several long-term studies (DMCR 2016a, b; Aksornkoae 1999) had collected information of wildlife in Thai's mangroves.

Bird surveys were done from 2006 to 2015 in 19 of the coastal provinces (except Bangkok,

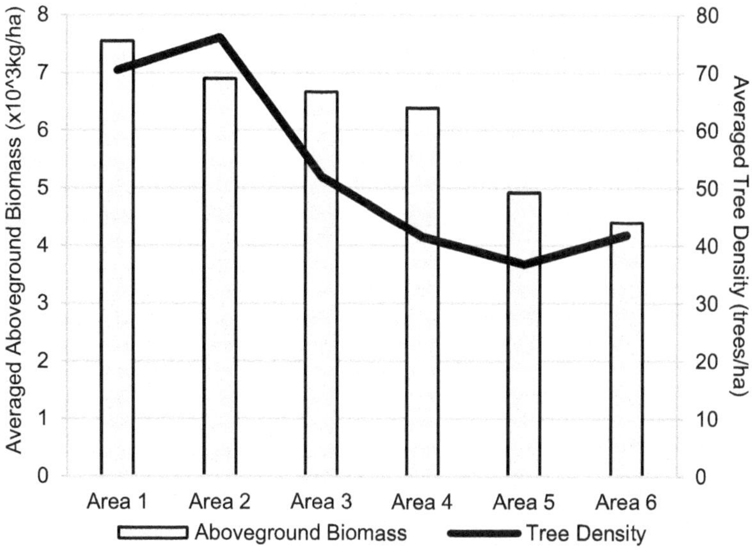

Fig. 28.5 Averaged aboveground biomass and averaged tree density in the six Areas of mangroves in Thailand. (Generated from data published in DMCR 2016a)

Note: Area 1, Upper Andaman Coast, Ranong and the western part of Phang Nga Provinces; Area 2, Lower Andaman Coast, Trang and Satun Provinces; Area 3, Middle Andaman Coast, Krabi, Phuket, and eastern part of Phang Nga Provinces; Area 4, Lower Gulf of Thailand Coast, Chumphon, Surat Thani, Nakhon Si Thammarat, Songkhla, and Phatthani Provinces; Area 5, the coast in the eastern part of Thailand, Rayong, Chanthaburi, and Trat Provinces; and Area 6, Upper the Gulf of Thailand Coast, Chon Buri, Chachoengsao, Samut Prakan, Samut Sakhon, Samut Songkhram, Phetchaburi, and Prachuap Khiri Khan Provinces

Narathiwat, Pattani, and Songkhla) (DMCR 2016a). The surveys adopted both direct and indirect methods (song and call identification and marks) from point count, road side survey, and mist net (Bibby et al. 2000). At least 272 bird species, accounting for about a quarter of the total bird species recorded in Thailand, were observed during this long-term study. Among these species, *Fregata andrewsi* was listed as critically endangered species according to the IUCN Red List, while three species were listed as endangered species (*Tringa guttifer*, *Numenius madagascariensis*, and *Calidris tenuirostris*), and four of them were vulnerable species (*Lonchura oryzivora*, *Centropus rectunguis*, *Leptoptilos javanicus*, and *Egretta eulophotes*).

Studies of insect and benthos diversity in Thai's mangrove were conducted in the six mangrove Areas from plant biomass and carbon storage survey, from 2007 to 2015, using malaise trap, sweep net, and light trap (DMCR 2016a). During the survey period, 1570 species of insects in 14 orders were recorded. One hundred one species of benthos (both vertebrate and invertebrate) in 42 families were records. Among these, seven species were commonly found in all the provinces: *Assiminea brevicula*, *Parasesarma plicatum*, *Littoraria scabra*, *Neritina violacea*, *Uca* sp., *Cerithiidae obtusa*, and *Sipunculus* sp.

Commercially important aquatic animals were studied from 2010 to 2015 in the six Areas (DMCR 2016a). Four groups of economically important species were reported: 133 species (in 55 families) of fish (Chordata), 25 species (in 10 families) of crustaceans, 13 species (in 11 families) of mollusk, and one species of Arthropoda. Most of these animals were found in juvenile stages. Approximately 20% of them were found to be in their adult stage. There were five species commonly observed in every survey during the study period, namely, *Metapenaeus lysianassa*, *Lutjanus calcarifer*, *Plotosus canius*, *Carcinoscorpius rotundicauda*, and *Scylla serrata*.

Within the crustacean group, 15 species of shrimps and prawns belonging to four families, namely, Penaeidae, Palaemonidae, Alpheidae, and Ogyridae, were identified. Most of the mangrove shrimps are in family Penaeidae, genus *Penaeus* and *Metapenaeus*. Among these, *Penaeus monodon* and *Penaeus merguiensis* have high commercial value. In addition, a freshwater shrimp species, giant freshwater prawn (*Macrobrachium rosenbergii*), was also commonly observed during its reproductive season as this species requires estuarine conditions for its early larva stage development (Ismael and New 2000).

In addition, 25 species of reptiles have been reported in Thai's mangroves such as snakes, chameleons, turtles, and crocodile/alligators. Furthermore, a number of mammals are found residing in mangrove forests in Thailand such as monkeys, otters, wild cats, and bats (Aksornkoae 1999; DMCR 2016a).

Mangrove Management and Conservation in Thailand

The latest land-use land-cover studies in 2014 clearly demonstrated that a large portion of the "reference original mangrove area" has been converted into different types of land use; only 245,534 ha of the area (or 54%) has been retained as mangrove forest (Fig. 28.3). Therefore, Thailand has been moving toward sustainable management and conservation by integrating public involvement, especially from local communities and relevant stakeholders in mangrove management and conservation. In addition to integrate mangrove conservation in the Act on the Promotion of Marine and Coastal Resources Management B.E. 2558 (2015), the 20-Year National Strategy, and the 12th National Economic and Social Development Plan (2017–2021), Thailand actively participates in several international initiatives/conventions such as Convention on Biological Diversity (CBD), United Nations Framework Convention on Climate Chang (UNFCCC), Ramsar Convention on Wetlands, Man and Biosphere (MAB), Asian Mangrove Network (Amnet), and Mangroves for the Future (MFF) in order to incorporate the most efficient mangrove management approaches to the national context.

With the collective efforts of all the stakeholders and sustainable mangrove management plan, the trend of rapid mangrove deforestation has been reversed. A couple of recent surveys have shown that the mangrove area have become relatively stable with a slight increase in forest cover. Due to the enforcement of the Act on the Promotion of Marine and Coastal resources Management B.E. 2558 (2015) and strong policies on mangroves and active participations of relevant stakeholders on mangrove conservation, it is expected that mangrove area would continue to increase.

References

Aksornkoae S (1989) Management and conservation of mangrove resources for coastal development in Southeast Asian nations. In: Chuan TE, Pauly D (eds) Coastal area management in Southeast Asia: policies, management strategies and case studies. ICLARM conference proceedings 19, pp 11–15, 254p

Aksornkoae S (1999) Mangroves: ecology and management. Kasertsart University Press, Bangkok, 278p. (Published in Thai language)

Aksornkoae S (2008) Mangroves. In: Mimura N (ed) Asia-pacific coasts and their management: states of environment (Coastal systems and continental margins). Springer, pp 112–119, 367p

Aksornkoae S, Tokrisna R (2004) Overview of shrimp farming and mangrove loss in Thailand. In: Barbier EB, Sathirathai S (eds) Shrimp farming and mangrove loss in Thailand. Edward Elgar, pp 37–51, 288p

Barbier E, Cox M (2002) Economic and demographic factors affecting mangrove loss in the coastal provinces of Thailand, 1979–1996. J Hum Environ 31:351–357

Bibby C, Jones M, Marsden S (2000) Expedition field techniques. Bird Life International, 139 p

Department of Marine and Coastal Resource (DMCR) (2012) Flora of Thailand's mangrove. Mangrove Conservation Office, DMCR. 4ed. The Agricultural Co-operative Federation of Thailand Press, LTD., Bangkok. (Published in Thai Language)

DMCR (2015a) Mangrove land use classification project 2014 (Final report). Mangrove Conservation Office, DMCR. (Published in Thai Language)

DMCR (2015b) The Bible of Thai's marine and coastal resources. Marine, Coastal and Mangrove Research and Development Institute, DMCR. Ministry of Natural Resources and Environment. Bangkok. (Published in Thai Language)

DMCR (2016a) Biodiversity in mangroves: one decade of mangrove's biological resources (2006–2015). Marine and Coastal Resources Management Support Department, DMCR. The Agricultural Co-operative Federation of Thailand Press, LTD., Bangkok. (Published in Thai Language)

DMCR (2016b) Report on status of marine and coastal resources and coastal erosion in Thailand. Office of Secretary, DMCR. Maine Consultant, LTD., Bangkok. (Published in Thai Language)

Field CD (1995) Journey amongst mangroves. International Mangrove Ecosystems, Okinawa, Japan. South China Printing Co. (1988) Ltd., 140p

Ismael D, New MB (2000) Biology. In: New MB, Valenti WC (eds) Freshwater prawn culture: the farming of *Macrobrachium rosenbergii*. Blackwell Science, Oxford, pp 18–40

Lebel L, Tri NH, Saengnoree A, Pasong S, Buatama U, Thoa LK (2002) Industrial transformation and shrimp aquaculture in Thailand and Vietnam: pathways to ecological, social, and economic sustainability. J Hum Environ 31:311–323

Muttitanon W, Tripathi NK (2005) Land use/land cover changes in the coastal zone of ban Don Bay, Thailand using Landsat 5 TM data. Int J Remote Sens 26:2311–2323

Naito T, Traesupap S (2013) The relationship between mangrove deforestation and economic development in Thailand. In: Faridah-Hanum I, Latiff A, Hakeem K, Ozturk M (eds) Mangrove ecosystems of Asia. Springer, New York, pp 273–294, 470p

Pumijumnong N (2014) Mangrove forests in Thailand. In: Faridah-Hanum I, Latiff A, Hakeem K, Ozturk M (eds) Mangrove ecosystems of Asia. Springer, New York, pp 61–79, 470p

Royal Forest Department (RFD) (1985) Forest statistics of Thailand 1985. Forest Statistics Sub-Division, Planning Division, RFD. (Published in Thai language)

Sathirathai S, Barbier EB (2011) Valuing mangrove conservation in southern Thailand. Contemp Econ Policy 19:109–122

Thampanya U, Vermaat JE, Sinsakul S, Panapitukkul N (2006) Coastal erosion and mangrove progradation of southern Thailand. Estuar Coast Shelf Sci 68:75–85

Vaiphasa C, de Boer WF, Skidmore AK, Panitchart S, Vaiphasa T, Bamrongrugsa N, Santitamnont P (2007) Impact of solid shrimp pond waste materials on mangrove growth and mortality: a case study from Pak Phanang, Thailand. Hydrobiologia 591:47–57

Valiala I, Bowen JL, York JK (2001, 51) Mangrove forests: one of the world's threatened major tropical environments. Bioscience:807–815

Vo QT, Kuenzer C, Vo QM, Moder F, Oppelt N (2012) Review of valuation methods for mangrove ecosystem services. Ecol Indic 23:431–446

29 Pollen Morphology of the Genus *Tamarix* in Israel

Amram Eshel, Chingiz Alekperov, Dafna Langgut, and Israel Oren

Abstract

The genus *Tamarix* (tamarisk, salt cedar; family Tamaricaceae) includes some 60 taxa worldwide. The genus is native to the drier areas of Eurasia and Africa, Western Europe, and the south and east United States (Schmidt, Relationship of salinity and depth to the water table on *Tamarix* spp. (salt cedar) growth and water use. Thesis submitted toward the M.Sc. Texas A&M University, p 90, 2003). The genus is composed of evergreen or deciduous shrubs or trees, usually on saline soils and tolerant to salts and alkaline conditions. The pink or white flowers grow in dense masses on 5–10-cm-long spikes or panicles flowering during spring, summer, and autumn. *Tamarix* can spread both vegetatively and by seed that are plumose at tips and can be dispersed by wind and water (Waisel, Biology of halophytes. Academic Press, London, 1972)… More than 15 species of *Tamarix* are found in Israel (Baum, The genus Tamarix. The Israel Academy of Sciences and Humanities, Jerusalem, 1978). The taxonomy of this genus has long been subjected to general and regional revisions. There has always been vast disagreement between botanists as to the delimitation and the grouping of the species (Zohary, Flora Palaestina part II. The Israel Academy for Science and Humanities, Jerusalem, 1972). The aim of the present study was to revise the taxonomy of the genus *Tamarix*, based primarily on pollen characters.

Keywords

Halophyte · Reclaimed sewage and brackish water · Salt tolerance · Tamaricaceae · *Tamarix*

In memory of Prof. Yoav Waisel.

A. Eshel · C. Alekperov (✉)
Department of Molecular Biology and Ecology of Plants, Tel Aviv University, Tel Aviv, Israel
e-mail: chingiza@post.tau.ac.il

D. Langgut
The Laboratory of Archaeobotany and Ancient Environments at the Institute of Archaeology, Tel Aviv University, Tel Aviv, Israel

I. Oren
The Robert H. Smith Faculty of Agriculture, Hebrew University of Jerusalem, Jerusalem, Israel

Introduction

Tamarix pollen grains have much in common: they are all tricolpate, semi-tectate, and per-reticulate and the grain being spherical or prolate. *Tamarix* flowers are anisomerous (a flower having an unequal number of petals, stamens, or other floral parts). It is possible to find in one raceme anisomerous flowers with different numbers of sepals, petals, stamens, and so on, without

B. Gul et al. (eds.), *Sabkha Ecosystems*, Tasks for Vegetation Science VI,
https://doi.org/10.1007/978-3-030-04417-6_29

any numerical correlation between the organs. Therefore the number of floral parts alone cannot serve as a diagnostic marker. This study would investigate whether the correlation and differences in pollen grains could determine the types within the genus.

The main flowering characters of the genus *Tamarix* are (based on Zohary 1972; Baum 1978): simple inflorescences or compound spike-like racemes on previous or current year branches or both; flowers hermaphrodite, rarely unisexual with 4-6 merous calyx and corolla, mostly pedicellate, bracteate; calyx herbaceous or membranous; sepals free or slightly connate at the bases, entire or slightly dentate, glabrous, rarely papillose or hairy; petals caducuous or persistent, free or slightly connate at the bases, erect or deflexed, white pink or reddish, entire or emarginated; disk variously formed, usually with 4-5 entire or emarginate lobes, sometimes almost 0; androecium mostly haplostemonous; stamens episepalous, constantly or predominantly 4 or 5 or 4 to 5 (in various flowers of the same raceme); sometimes androecium diplostemonous and then the inner whorl of 1-10 epipetalous stamens usually somewhat shorter than the episepalous stamens of the outer whorl; stamens arising from the top of the lobes of the disk or from the sinuses between the lobes; ovary with 3-4 carpels, 1-celled, with many ovules; styles 3 or 4, very rarely 2-5; stigmas short, capitate; capsule many-seeded, conical, pyramidal or pyriform, loculicidally dehiscent; seeds with a seesile apical pappus of unicellular hairs.

Zohary (1972) and Baum (1978) divided the taxa found in Israel into the following groups:

1. *Pleiandrae*: Flowers with 6-15 stamens, 3 styles, and a 5 merous perianth. The three local species of this group are *T. passerinoides*, *T. amplexicaulis*, and *T. aravensis*.
2. *Tetrandrae*: Flowers with 4 stamens, 3-4 styles, 4 (sometimes 5) merous perianth (in some cases the number of stamens may be 5 or more but the number of styles remains constantly 4).
3. *Pentandrae*: This group can be divided into the following subgroups:
   (a) Xeropetalae – Flowers with a persistent corolla (remaining long after spring or summer flowering). The two species of this group are *T. jordanis* and *T. chinensis*, the latter being a pink flowered, cultivated, or subspontaneous ornamental.
   (b) Piptopetalae is a flower shading,when the petals soon after anthesis. There are six species in Tamarix group: *T. aphylla*, with leaves almost entirely reduced to sheaths, so that the branches appear to be articulate; *T. nilotica* with a stout, short or long racemes, mostly not over 5 cm., in panicles, usually aestival; *T. hampeana* with short thick vernal racemes and white flowers; of the remaining species, *T. negevensis* and *T. gennessarensis* have dense and thick racemes, pink flowers with deflex petals, long exserted stamens, and relatively long styles, white *T. palaestina* has very slender, long and usually interrupted racemes, and minute white flowers with erect petals and short or none exserted stamens (Table 29.1).

Materials and Methods

Flowers were collected from different sites in Israel, as well as from the Botanical Gardens, Tel-Aviv University (Oren 2010). In order to define the ornamentation and sculpture of *Tamarix* pollen grains, two methods were used:

1. Acetolysis: Acetolysis solution of 1 ml sulfuric acid and 9 ml acetic anhydride, stained and mounted in glycerin. The observations were made with a Nikon Eclipse E-400 using magnification of ×200, ×400, and ×1000 (immersion oil).
2. Scanning electron microscope (SEM: Jeol – JSM – 840A): Pollen grains were attached to a metallic stub using a double-sided adhesive tape. Stubs were coated with gold in a sputtering chamber. The grains were observed and photographed digitally later on. The different images were than described and analyzed.

Table 29.1 Number of floral parts in *Tamarix* species in Israel (Zohary 1972)

Tamarix	Sepals	Petals	Stigma (style)	Stamens
T. negevensis	5 (sometimes 4)	5 (very rarely 4 or 6)	3	5
T. gennessarensis	5	5	3	5
T. palaestina	5	5	3	5
T. jordanis	5	5	3	5
T. chinensis	5	5	3	5
T. nilotica	5	5	3	5
T. aphylla	5	5	3	5
T. hampeana	5	5	3	5
T. tetragyna	4–5	4–5	4	4–5
T. parviflora	4	4	3–4	4
T. amplexicaulis	5	5	3	8–10
T. passerinoides		3–5	3	6–15 (usually 8–13)
T. aravensis	5 (sometimes 4)	5 (sometimes 4)	3	8–10 (sometimes 5–7)

Measuring of pollen grain dimensions was based on digital image analysis software ("ImageJ"). The following parameters were measured: length, width, and pollen projection area. Length and width were calculated as the major and minor axes of the best-fitting ellipse for each grain. The projection areas were generated in pixels and displayed in μm^2 as the sum of all pixels of an object. Spatial calibrations of the pixels were done with a stage micrometer. Any white spaces within the boundaries of threshold grains were included automatically in the pollen grain area. Grains in clumps of 2–3 were separated by drawing a white line between them. Length, width, and area measurements were copied to Excel and are reported as the sample mean and the sample standard deviation. Minimum and maximum values are also presented in some cases. The measurements are based on ~30 readings from each specimen. Photographs of pollen were taken with a Nikon E995 digital camera. To produce binary images, an automatic gray level threshold setting was used (Fig. 29.1). Thresholding sets the minimum gray level required for viewing an image. Pollen grains with areas 10% larger or 10% smaller than the smallest spore (as measured in a preliminary subsample) were eliminated. This removed debris as well as images of two or more spores touching each other. Spores touching frame boundaries were rejected automatically. The photographed pollen grains were first stained and mounted in glycerin.

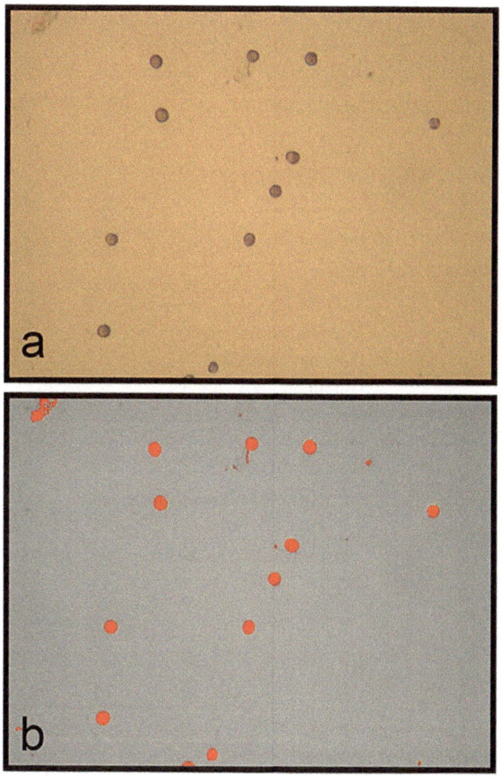

Fig. 29.1 *Tamarix* pollen grains, sample no. 79. (**a**) Digital image of stained pollen grains. (**b**) Binary image after background was subtracted and gray levels below threshold were eliminated using software of Image J

The morphological features of the flowers were examined under a binocular. The numbers of the following organs were counted: sepals, petals, stigma, and stamens. The different floral parts were characterized and photographed.

Results

Nineteen taxa were sampled. Their details are presented in Table 29.2.

Scanning electron microscope revealed two types of pollen: one was more oval while the other was more spherical. The elliptical pollen grains were taken from the inner parts of the anther (pollen sacs) of a flower in its initial stage of development and were considered immature. At the same time, spherical grains were collected from flowers in their advanced stage of development, considered mature. Two samples were studied from each of the following clones: 93, 26, and 130. One represents immature pollen grain while the other characters a mature grain. A total of six samples were examined in order to evaluate the differences between mature and immature grains (Figs. 29.2, 29.3, 29.4, 29.5, 29.6, and 29.7).

Table 29.2 Details and sample number

Tamarix clone no.	Previous name/remark	Genus
79	Kalya	
36	Yotvata 81	
9	Yotvata. Sample was collected from the Botanical Gardens	*T. arvensis*
27	Yotvata. Sample was collected from the Botanical Gardens	*T. jordanis*
93	Sample was collected from Tel-Aviv University	
111b	Ein Yahav, listed also as sample no. 112	
26	Yotvata 53	*T. jordanis*
130	Kfar-Vitkin	*T. chinensis*
103	Naaman saline	
43	Achziv	
63	Ein Gev	
65	Degania	
3	Mineral Beach. III	
119	Ashdod north	
6	Yotvata XI	
133	Ma'agan Michael	
77	Kalya III	
82	Mitzpor Paran	
T. tetragyna	Bsor – 11.1.2006	

In all three pairs of samples that were checked (93, 26, 130), the mature pollen grains, belonging to developed flowers, were bigger in their dimensions and less elliptical in shape, in comparison with the immature pollen grain which belonged to flowers in an initial stage of development (Table 29.9).

Dividing the *Tamarix* collection into three different groups was chiefly based on the correlation of pollen grain dimensions together with the number of floral parts:

1. The first group includes samples no. 82 and the *T. tetragyna* from the Bsor. This group differs by its relatively high pollen grain dimensions (length, width, and projection area) together with the appearance of four floral organs: sepals, petals, stigma, and stamens. Based on this correlation, we concluded that this species group is distinct from the rest of the genus and probably represents the species *T. tetragyna*. In addition, it seems that the group includes two different types of *T. tetragyna*, based on the differences in pollen grain dimensions: the average projection area of sample no. 82 is 313 μm^2, while the area of the last sample (*T. tetragyna* from the Bsor) is smaller, 256 μm^2 in average.
2. All members of group no. 2 have the same number of floral organs: five sepals, five petals, three stigmas, and five stamens. This group was further subdivided into two subgroups based on pollen grain dimensions, mainly the projection area:
   (2a) This subgroup includes pollen grains with projection area between 150 and 190 μm^2, comprised of samples no. 36, 151, 166, 155, and 182.
   (2b) The second subgroup includes pollen grains with projection area between 190

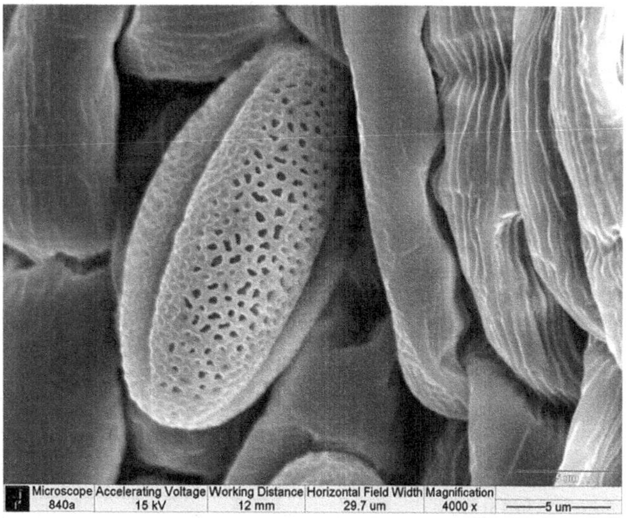

Fig. 29.2 Image of immature *Tamarix* pollen grain, clone no. 93

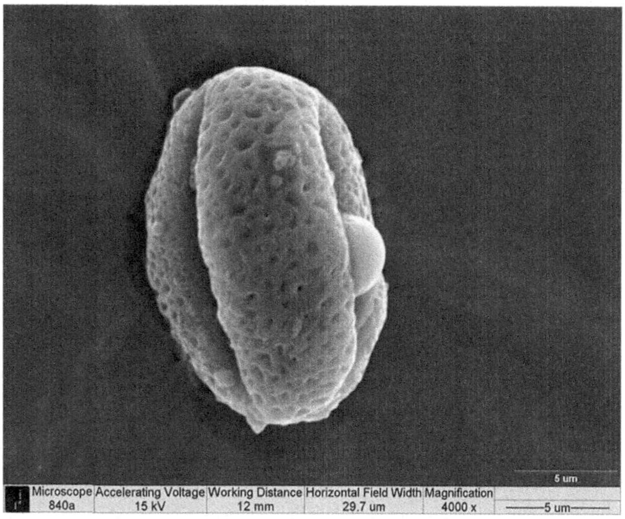

Fig. 29.3 Image of mature *Tamarix* pollen grain, clone no. 93

and 230 μm^2 composed of samples no. 27, 93, 26, 43, 63, 65, and 3.

3. The third group comprised of *Tamarix* species with irregular number of floral organs. Although all samples have five sepals, the number of petals, stigmas, and stamens varies. The group includes samples no. 79, 9, 111b, 130, and 103.

Table 29.10 summarizes the main characteristics of *Tamarix* ornamentation pollen grains based on the examination of SEM images and acetolysed pollen. The parameters that were checked are the dimensions and shape of the lacunae (a depressed area surrounded by ridges, which creates the reticulate ornamentation), the width of the muri (a ridge that is part of the ornamentation and separates the lumina in a reticulate pollen grain), and the general pattern.

Although some of the samples have similar characteristics, it is almost impossible to group the different species into units.

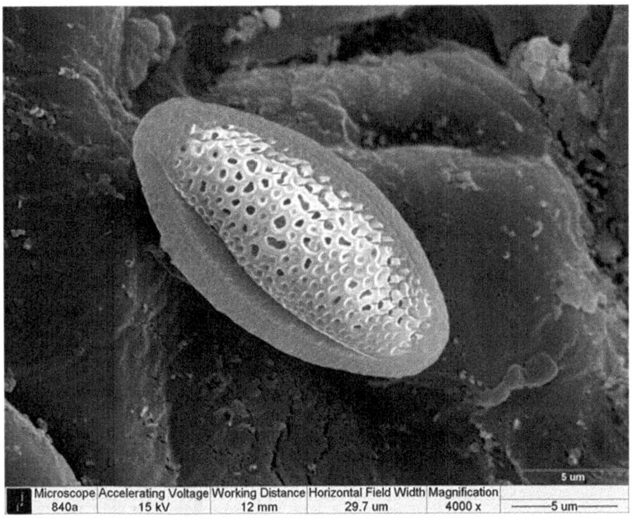

Fig. 29.4 Image of immature *Tamarix* pollen grain, clone no. 26

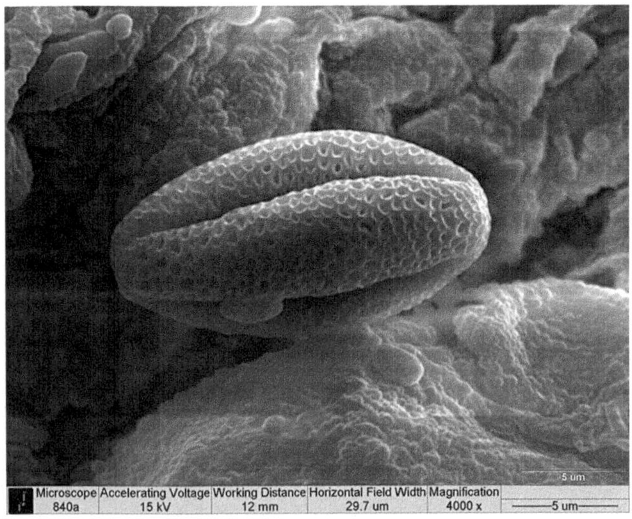

Fig. 29.5 Image of mature *Tamarix* pollen grain, clone no. 26

Discussion

Pollen Morphology

In *Tamarix* spp., the immature pollen is elliptical (Figs. 29.2, 29.4, and 29.6), whereas mature pollen are more diverse in shape, ranging from being oval to round (Figs. 29.3, 29.5, and 29.7). The differences are due to the higher amount of cytoplasm in the mature grains, in comparison to immature ones. This dissimilarity explains also the differences in pollen grains' size (Tables 29.2, 29.3, 29.4, 29.5, 29.6, 29.7, and 29.8): in all three parameters that were tested, length, width, and projection area, the dimensions were higher for pollen collected from developing flowers. The higher standard deviation values of the mature samples are due to the wide range of diversity in the pollen shape. Furthermore, generally, mature pollen grains are characterized in dipper colpies (= furrows) and bare more sunken lacunae on the reticulate surface ornamentation, in compared to immature grains.

With the assistance of digital image analysis, we compared length, width, and projection area of the different *Tamarix* samples. Digital image

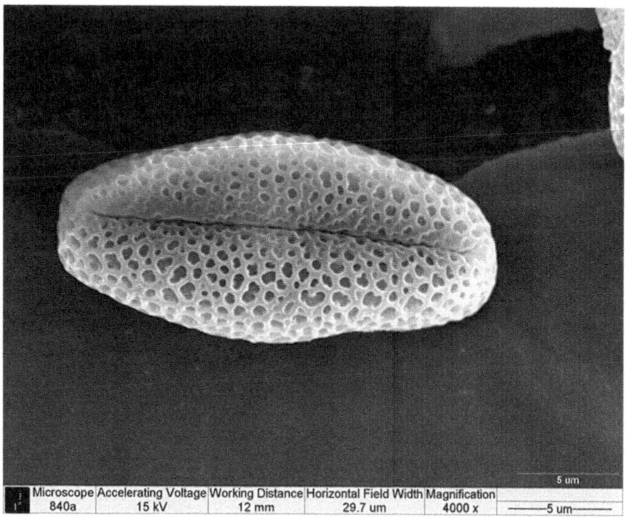

Fig. 29.6 Image of immature *Tamarix* pollen grain, clone no. 130

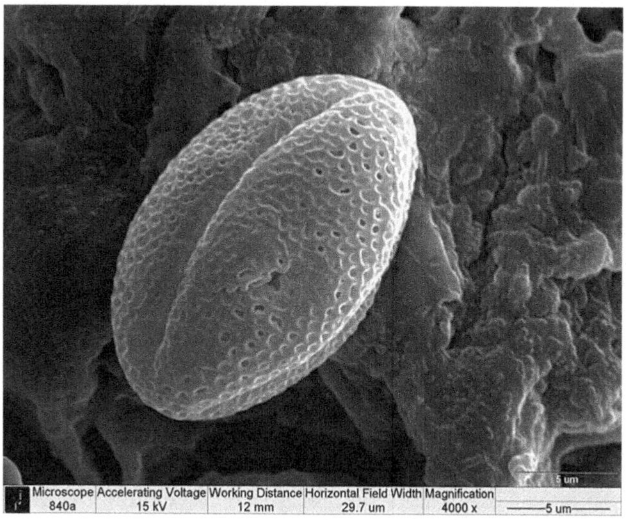

Fig. 29.7 Image of mature *Tamarix* pollen grain, clone no. 130

Table 29.3 *Tamarix* 93 (2) immature pollen grain

Width	Length	Projection area	93.2 immature
14.73	16.45	191.11	Mean
1.45	1.72	33	SD
11.52	13.56	136.31	Min
17.64	20.23	276.16	Max

Table 29.4 *Tamarix* 93 (3) mature pollen grain

Width	Length	Projection area	93.3 Mature
15.07	17.42	207.55	Mean
1.83	1.82	41.19	SD
12.02	14.3	155.54	Min
18.92	20.47	293.21	Max

analysis provided faster measuring of the pollen grains than when measuring traditionally with an ocular micrometer on the microscope. In addition the quick results, Anikster et al. (2005) found out that the digital measurement system produces satisfactory results, compared to other methods (ocular micrometer/ruler on the computer screen).

Our results indicate that when the comparison was based on projection area, the differences in *Tamarix* pollen grains were most pronounced. The area was determined independently of length and width from the number of pixels within pol-

Table 29.5 *Tamarix* 26 (1) immature pollen grain

Width	Length	Projection area	26.1 Immature
13.41	14.47	153.02	Mean
0.93	0.91	19.84	SD
11.91	12.97	123.45	Min
15.84	17.02	200.95	Max

Table 29.6 *Tamarix* 26 (2) mature pollen grain

Width	Length	Projection area	26.2 Mature
15.3	16.39	199.47	Mean
1.78	1.94	49.08	SD
12.62	13.46	133.45	Min
20.04	21.56	339.43	Max

Table 29.7 *Tamarix* 130 (3) immature pollen grain

Width	Length	Projection area	130.3 Immature
14.13	15.49	172.28	Mean
1.02	0.98	20.84	SD
11.17	13.68	130.81	Min
16.34	17.43	223.65	Max

Table 29.8 *Tamarix* 130 (4) mature pollen grain

Width	Length	Projection area	130.4 Mature
14.98	16.04	189.34	Mean
1.27	0.96	24.83	SD
10.75	14.07	131.59	Min
16.89	17.72	224.81	Max

len grain boundaries. This integrated the entire area of each grain regardless of shape. Projection area is potentially the most important single parameter for characterizing grain dimensions since it combines aspects of length, width, and shape into single value. Furthermore, standard deviation values tend to be less for projection area than for length and width. Similar evidences were found while measuring the dimensions of the spore of *Puccinia* species (rust fungi) (Anikster et al. 2005). The description of pollen grain surface ornamentation was found useful in differing one species from another, since almost each pollen grain has different characteristics.

The results showed that the *Tamarix* collection fell into three groups differing in size and in floral parts characteristics. A correlation between the groups and pollen grain ornamentation does not necessarily exist.

1. *T. tetragyna* – The first group differs from the other species by its relatively high pollen grain dimensions together with the appearance of four floral organs: sepals, petals, stigma, and stamens. This uniqueness contributes to the conclusion of group 1 being distinct from the rest of the genus. Based on the significant differences within the species that belong to this group, in both pollen grain dimensions and ornamentation, it seems that this unit includes two different types of *T. tetragyna*.
2. The second group includes species of *Tamarix* that have the same number of floral organs: five sepals, five petals, three stigmas, and five stamens. This group was further subdivided into two subgroups based on pollen grain dimensions. No correlation was found between pollen grain size measurements (Table 29.9) and surface ornamentation (Table 29.10, Plate 1) within the two subgroups.
3. The third group contains species with irregular number of floral organs. No correlation was found between pollen grain dimensions (Table 29.9) and surface ornamentation (Table 29.10, Plate 1) within this group.

Conclusions

The taxonomy of the genus *Tamarix* could be based largely on pollen dimension measurements together with the number of floral parts. In this study these parameters have been used to define groups within the genus *Tamarix*. We also found out that although projection area has not been used widely to characterize pollen grains (as far as we know it was used only for spores, e.g., Anikster et al. 2005), digital image analysis

Table 29.9 Pollen dimensions and number of floral parts

Clone no.	Projection area (μm^2)	Length (µ)	Width (µ)	Circularity	Sepal	Petals	Stigma	Stamens
79	195	16.3	15	1.09	5	5,6	3	6,7
36	171.6	15.3	14.2	1.08	5	5	3	5
9	213.6	17.4	15.5	1.12	5	4	3	8
27	222.1	17.7	15.9	1.11	5	5	3	5
93	226.7	17.8	16	1.11	5	5	3	5
111b	146.1	15.5	12.2	1.27	5	5	3	6
26	199.5	16.4	15.3	1.07	5	5	3	5
130	189.3	16	15	1.07	5	5	3,4	5
103	198.4	16.5	15.3	1.08	5	5	3,4	5,6
43	220	17.1	16.2	1.06	5	5	3	5
63	204.2	16.6	15.6	1.06	5	5	3	5
65	200	16.8	15.1	1.11	5	5	3	5
3	190	16.1	14.9	1.08	5	5	3	5
119	151	14.7	13.1	1.12	5	5	3	5
6	166.5	15.1	13.9	1.09	5	5	3	5
133	155.2	14.4	13.7	1.05	5	5	3	5
77	182.3	16.2	14.2	1.14	5	5	3	5
82	313.4	21.4	18.6	1.15	4	4	4	4
T. tetragyna	256.4	19	17	1.12	4	4	4	4

Table 29.10 Pollen grain sculpture and ornamentation

Sample no.	Relative lacunae's size ("hole")	Lacunae's shape	Muri's width	Ornamentation
79	Big	Amorphic	Very broad, polygonal	Reticulate and branched
36	Small	Rounded	Broad	Reticulate regular
9	Small	Amorphic	Very broad, angular	Reticulate and branched
27	Small	Amorphic	Very broad, angular	Reticulate and branched
93	Medium	Rounded	Medium	Reticulate irregular
111b	Small	Rounded-amorphic	Medium	Reticulate fine
26	Very big	Amorphic	Very narrow	Reticulate coarse
130	Big	Amorphic	Medium to narrow	Reticulate coarse and irregular
103	Medium	Rounded	Medium	Reticulate regular
43	Medium-big	Rounded	Medium to narrow	Reticulate regular
63	Big	Amorphic	Broad	Reticulate coarse and irregular
65	Small	Rounded	Medium	Reticulate irregular
3	Variable	Amorphic	Medium	Reticulate irregular
119	Variable	Amorphic	Medium	Reticulate irregular
6	Medium-small	Rounded-amorphic	Broad	Reticulate regular
133	Medium-big	Amorphic	Medium-broad	Reticulate irregular
77	Small	Rounded	Medium	Reticulate fine
82	Variable	Amorphic	Narrow	Reticulate coarse and irregular
T. tetragyna	Medium-small	Rounded-amorphic	Broad	Reticulate irregular

makes area readily available as a potentially useful parameter for taxonomic purpose. It was also revealed during this study that the dimension and general shape of the pollen grains change in their life cycle: immature flowers have more ovale shape, while mature flowers are relatively rounded. Since mature pollen grains contain more cytoplasm, their dimensions are bigger. The description of pollen grain surface ornamentation was found useful in differing one species from another, since almost each pollen grain has different characteristics. The best diagnosis for defining pollen grain pattern is observations by SEM.

Acknowledgment Authors extend gratitude to technicians and the Research Education Centers for making this research possible. These projects were supported by research grants of the Italian Ministry for Environment, Land, and Sea Protection and Boeing Corp., USA.

References

Anikster Y, Eilam T, Bushnell WR, Kosman E (2005) Spore dimensions of *Puccinia* species of cereals hosts as determined by image analysis. Mycologia 97:474–484

Baum BR (1978) The genus Tamarix. The Israel Academy of Sciences and Humanities, Jerusalem, p 209

Oren I (2010) The influence of salt on growth an d biomass yield of selected *Tamarix* species. Thesis submitted towards the M.Sc, degree in Plant Sciences at Tel Aviv University, p 88

Schmidt KM (2003) Relationship of salinity and depth to the water table on *Tamarix* spp. (Salt cedar) growth and water use. Thesis submitted towards the M.Sc. Texas A&M University, p 90

Waisel Y (1972) Biology of halophytes. Academic, London, p 395

Zohary M (1972) Flora Palaestina part II. The Israel Academy for Science and Humanities, Jerusalem, pp 350–362

30 Species Distribution in Different Ecological Zones and Conservation Strategy of Halophytes of Sundarbans Mangrove Forest of Bangladesh

A. K. M. Nazrul Islam, Ahmed Emdadul Haque, Maniruzzaman, Tahmina Jamali, Tahsina Haque, Md. Almujaddade Alfasane, N. Nahar, Nargis Jahan, Sabiha Sultana, and T. Senthil Kumar

Abstract

Species distribution of halophytes and their conditions in different ecological zones of Sundarbans mangrove forest were evaluated. Water flow within the rivers of the Sundarbans mangrove forest and seasonal variation of water salinity of the rivers were discussed. Highest salinity (electrical conductivity) 27,500 micromhos/cm (Polyhaline zone) and 5800 micromhos/cm (Oligohaline zone) was observed in April; and lowest salinity was 10,150 micromhos/cm (Polyhaline zone, July) and 500 micromhos/cm (Oligohaline zone, July). Environmental constraints and ecophysiological response in relation to soil environment and plant and plant roots in relation to salinity were highlighted. Surface soil of this halophytic forest is silty clay loam; soil pH ranges from 6.9 to 7.5; calcium is the dominant cation and is highest in the Oligohaline zone followed by Mesohaline and Polyhaline zones. The respiratory roots (pneumatophores) of various forms (such as peg like, knee root, buttress roots, stilt roots) in different species were described. Measurement of redox potential in soils from selected places showed low values particularly in the Oligohaline zone (+60 mV, where the *Heritiera fomes* plant shows top dying) followed by Mesohaline and Polyhaline zone. Low oxidation-reduction potentials (anaerobiosis) may be one of the factors responsible for the cause of top dying, and these properties influence plant growth and distribution. Highest soil salinity (6950 micromhos/cm) was in the Polyhaline zone, and in the Oligohaline zone, highest salinity was 650 micromhos/cm. The vegetation pattern is given based on field observation, with the dominant species; and associated species are placed in relation to density and abundance. Comparison of species composition (large and small trees, shrubs, herbs, palms, fern, lichen and climbers) and their condition in the three ecological zones showed the nature of their distribution pattern. Plant

A. K. M. Nazrul Islam (✉) · A. E. Haque · Maniruzzaman · T. Jamali · T. Haque · M. A. Alfasane
N. Nahar · N. Jahan · S. Sultana
Ecology and Environment Laboratory, Department of Botany, University of Dhaka, Dhaka, Bangladesh

T. Senthil Kumar
Department of Botany, Bharathidasan University, Tiruchirappalli, Tamil Nadu, India

B. Gul et al. (eds.), *Sabkha Ecosystems*, Tasks for Vegetation Science VI,
https://doi.org/10.1007/978-3-030-04417-6_30

association and distribution of halophyte species in relation to salinity in different ecological zones indicated their nature of distribution. Anthropogenic and natural threats to the halophytic mangrove species and their conservation are discussed. Research for mangrove conservation in relation to stress ecology and rehabilitation of mangrove halophytes, linkage between the structure of mangroves and ecosystem services and future of mangrove halophytes were highlighted.

Keywords
Water supply · Water and soil salinity · Plant roots · Redox potential · Vegetation and conservation

Introduction

Mangroves seem to have attracted attention as halophytes. From the ecosystem point of view, mangrove soils, sea water and plants closely interrelated to each other, forming an interactive system. Substratum is inundated regularly by sea water tides, leading to the introduction of a variety of chemical substances and placed under salinity stress. It is also contained by low oxygen concentration where stagnant water causes anaerobic conditions (Nazrul Islam 2003, 2015). Halophytes of mangrove forest in tropical countries constitute a natural resource along the coastal belt. Mangroves include trees and shrubs (halophytes) and are successful in colonizing and adapted to intertidal zones. Halophytes of mangrove provide an important role in the economy of the people and the shelter for the resident of wild animals. In Bangladesh, mangroves occur in the coastal belt in the south of the Bay of Bengal, and more prominently in the southwest corner is a single largest tract, the Sundarbans mangrove forest.

Mangroves of halophytic trees and shrubs usually grow in the intertidal zone of tropical and subtropical coast lines. The best developed mangroves occur along humid equatorial coastlines in South East Asia where they often form extensive tidal forest with trees more than 1 m in diameter and up to 45 m in height. About 50 species of mangroves are recognized worldwide, of which around 40–45 species occur in South East Asia, 10–12 species in Africa and 4–5 species in America (Clough 1984, 1993). Mangroves also extend well into temperate regions, but the number of species decreases with increasing latitude.

Mangrove forests are found throughout tropical shorelines, extending into subtropical areas as far north as the St. George's Parish, Bermuda (32° 23′N), and as far south as Corner Inlet, Victoria, Australia (38° 45′S). These are best developed on low-energy, muddy coast lines. At the global scale, there is a major floristic division between New World mangroves and Old World mangroves (Woodroffe 1993).

The factors controlling the growth and distribution of halophytes of mangroves are salinity (MacNae 1968; Baltzer 1969), soil structure and inundation (Walter and Steiner 1936); soil drainage plays a significant role in mangrove establishment. All these factors are interrelated; one factor alone could not work. For example, flooding causes anaerobic conditions; in contrast, drainage leads to an aerobic situation with high oxygen tensions in soil which in turn affects trees growing on it. The present distribution of halophytes (mangroves) worldwide has been discussed by Chapman (1976) and Tomlinson (1986). The decrease in number of species with increasing latitude is generally attributed to the corresponding in sea surface temperature or air temperature (Tomlinson 1986; Clusener and Breckle 1987).

Halophytes of Sundarbans mangrove forest are situated between 21° 30′ and 22° 30′N latitudes and 89° to 90° E longitudes in the southwest corner of the country under the districts of Bagerhat, Khulna and Satkhira and extend to the shores of the Bay of Bengal in the south. The total area of the forest is 577,220 ha; of these, river channels and other water courses consist of 175,220 ha (Chaffey et al. 1985). The low-lying areas are occupied by a network of small creeks. The submerged river banks are dominated by *Acanthus ilicifolius*, *Porteresia coarctata* and ***Typha*** *elephantina*; slightly upwards places of the river bank are occupied by *Nypa fruticans* and *Sonneratia apetala*. The flora of Sundarbans was described by Prain (1979).

Soil physico-chemical properties, plant community types and diversity analysis from various ecological zones of the plant species in this forest were discussed in the early papers (Nazrul Islam 1985, 1994, 1995). In the present paper, species distribution of different ecological zones and conservation strategy of Halophytes of Sundarbans mangrove forest are evaluated.

Materials and Methods

Field visits were done to observe the condition of water supply, chemical condition of water and species composition and distribution. The soils and water samples were collected from different locations of the three ecological zones and were analysed for Na, K, Ca, Mg, pH and conductivity. Exchangeable cations in soils were extracted (normal ammonium acetate) using the procedure of Gupta and Rorison (1974) and Nazrul Islam and Rorison (1978). Na and K were analysed by flame photometry and Ca and Mg by atomic absorption spectrometer. Total alkalinity and chloride in water samples were analysed according to standard methods of APHA (1977). Oxidation-reduction potentials were measured with silver-silver chloride electrode and platinum electrode (Nazrul Islam and Rorison 1978).

Results and Discussion

Water Supply

There is a scarcity of sweet water in the entire coastal region. The canals and creeks are full of sea water and are subject to tidal action. The only source of freshwater supply in the Sundarbans mangrove forest is the Gorai River, a distributary of the mighty Ganges River coming from upstream. Before construction of the Farakka barrage in the upper reaches of the Ganges River by India, freshwater supply was abundant throughout the year. Since 1975, freshwater is available in the halophytes of Sundarbans mangrove forest only during July–September (monsoon season). The barrage is closed in the winter season (November–May); sea water enters the streams, canals and the rivers for the rest of the year. Drainage pattern in the Sundarbans is very complex due to interconnected rivers, cross channels and large number of creeks. During monsoon season, heavy rainfall flushed the eastern part of the forest from the River Ganges through its distributaries the Gorai, the Madhumati and Baleshwar Rivers (Fig. 30.1). In the lean period, freshwater discharge is stopped due to Farakka barrage; hence salinity is increased in the eastern part. The Rupsha and the Passur Rivers flow through the middle part of the Sundarbans and receive water from the Gorai River in the monsoon season and change to oligo-mesohaline and become mesohaline again in the lean period when the flow of freshwater supply is decreased. In the western part of the Sundarbans, no freshwater is available during the lean period and is subject to tidal influence. Hence salinity is higher (Table 30.1, Polyhaline zone) than eastern part (Oligohaline zone). The process of accretion and erosion within the Sundarbans is complex because of large number of interconnection of channels and creeks.

The salinity conditions of the water of different rivers and seasonal variation within the Sundarbans halophyte mangrove forest are given in Table 30.1. Highest salinity 27,500 micromhos/cm was in the Polyhaline zone (Firingi River), and the lowest 5800 micromhos/cm was the Oligohaline zone (Rupsha River) in April. Lowest salinity was 500 micromhos/cm (Oligohaline zone) and 10,150 micromhos/cm in July (monsoon season). Seasonal variation of water salinity is given in Table 30.2. It is found that in late winter (March and April) when the barrage is closed, water supply is fully disrupted through the Farakka barrage and salinity reaches up to 5800 micromhos/cm which is a threat for the survival of *Heritiera fomes*, a dominant species and other associated species of the Oligohaline zone (Table 30.8). In the Polyhaline zone, the dominant species *Ceriops decandra* and *Aegiceras corniculatum* are adapted to high

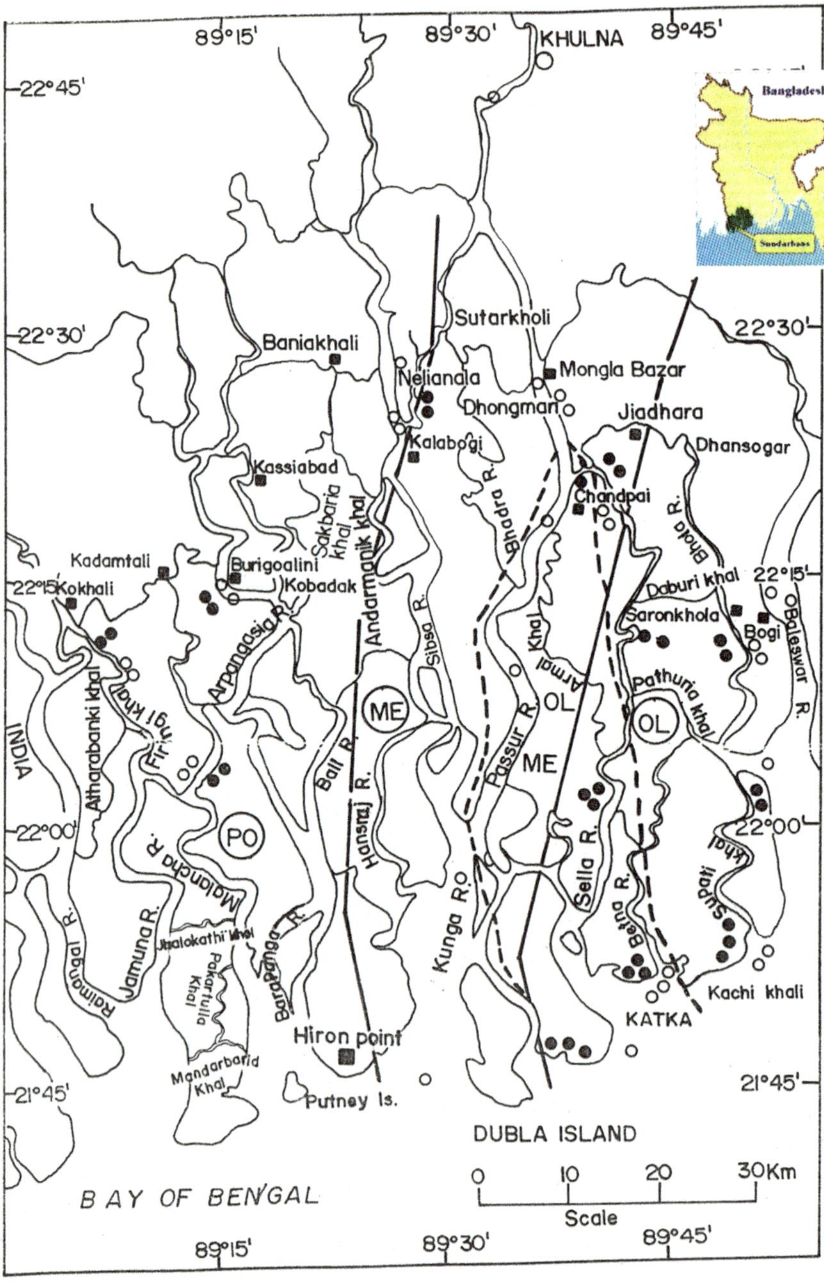

Fig. 30.1 Sundarbans mangrove forest shows the rivers and three ecological zones (OL, ME and PO). Closed and open circles indicate location of soil and water sample collection. Top right is the map of Bangladesh showing the location of Sundarbans forest (dark part, southwest corner)

salinity levels and form consociation. In this zone, other species such as *Kandelia candel*, *Xylocarpus granatum*, *X. mekongensis*, *Bruguiera sexangula* and also *Rhizophora* sp. grow well. Sea water comprising approximately 35 gl^{-1} salt (including 483 mM Na^+ and 538 mM Cl^-). This means an osmotic potential of – 2.5 MPa and water must be taken in against this pressure (Hogarth 1999). Halophytes of mangroves deploy a variety of means to cope with the unpromising

Table 30.1 Seasonal variation of electrical conductivity (micromhos/cm) in water samples of the rivers of Sundarbans halophyte mangrove forest

Location	1983		1984		1985	
(Rivers)	March	June	July	November	January	April
Khulna (Rupsha)	1350 (350)	550 (250)	500 (250)	**670** (300)	700 (350)	**5800** (1850)
Mongla (Passur)	10,700 (3950)	1550 (720)	1570 (750)	1600 (750)	2800 (1100)	10,900 (6300)
Chandpai (Shela)	–	5900 (3150)	831 (425)	490 (700)	5200 (2250)	8700 (3800)
Sharankhola (Bhola)	–	860 (450)	338 (105)	315 (450)	3150 (1250)	5700 (2500)
Bogi (Baleshwar)	–	280 (140)	215 (100)	260 (350)	530 (400)	3000 (1250)
Katka (bay)	–	–	9000 (5500)	10,750 (3900)	14,000 (4500)	15,500 (10250)
Nalianala (Shipsha)	20,400 (8750)	12,500 (5325)	**4350** (1950)	7800 (2800)	8400 (3100)	15,800 (11500)
Bazbaza (Khal)	26,500 (10750)	20,000	6000 (2050)	–	–	–
Burigoalini	26,500 (11000)	16,000	10,150 (4200)	12,050 (4700)	12,300 (7200)	18,000 (10750)
(Firingi and Arpangasia)	27,000 (12580)	–	–	–	–	27,500 (12500)

(–) indicates no samples were collected
Data in parentheses are the amount of chloride in mg/l

Table 30.2 Seasonal variation of water salinity (micromhos/cm in monsoon and winter seasons from different ecological zones

	Ecological zones		
Seasons	Oligohaline	Mesohaline	Polyhaline
Monsoon	500–670	1550–1600	4350–10,150
[a]Winter	1350–5800	8400–20,400	27,000–27,500

[a]All are late winter values

environment. The principal mechanisms are exclusion of salts by the roots, tolerance of high tissue salt concentrations and elimination of excess salt by secretion. There is evidence that *Rhizophora* depends entirely on water in the top 50 cm of the soil, in which 70% of its fine roots are deployed (Lin and Sternberg 1994; Sternberg and Swart 1987).

Climatic Factors and the Distribution of Mangroves

Generally, mangroves exist in tropical coastlines whatever the climate: thus, a littoral fringe of mangrove is typical of both arid and equatorial shores. However, there are significant differences between wet and dry coasts in terms of mangrove forest zonation patterns. The differences are clearest in the landward mangrove zones and in the composition of the hinterland vegetation. Distribution and plant association of halophyte species of Sundarbans mangrove forest in relation to salinity are provided (Table 30.8).

Intra-distributary basins in river-dominated deltaic settings may be partially insulated from the influences of tidal and river processes by channel margin land forms such as levees, and

Table 30.3 Chemical properties (exchangeable elements) of soils in halophytic mangrove forest (μg/g)

Ecological zones	Na	K	Ca	Mg
Oligohaline	300–500	260–750	3040–4695	750–1575
Mesohaline	675–900	350–375	2612–3700	740–930
Polyhaline	1550–2100	595–890	1840–2990	1050–1670

the vegetation of such basins is strongly influenced by climate (Smith et al. 1994).

Environmental Constraints and Ecophysiological Response

(a) ***Soil Environment and Plant Roots***

Pedologically soils of the Sundarbans are very young, poorly drained uncured sediments having no diagnostic horizon and are poorly oxygenated. The surface soil is a silty clay loam overlying alternating layers of clay and sand. Chemical properties of soils of different ecological zones are given in Table 30.3. Calcium is the dominant cation and is highest in Oligohaline zone (3040–4695 μg/g) followed by Mesohaline and Polyhaline zones (Table 30.3). Sodium was significantly higher ($p < 0.05$) in Polyhaline zone (1550–2100 μg/g) than other zones. Soil pH varies from 6.9 to 7.5, and the organic matter (0–10 cm) ranges from 5.5% to 6.5% in top layers (Nazrul Islam 2003). Analysis of mangrove soils of Andaman and Nicobar Islands showed highest calcium (2430 ppm), and soils of halophyte Sundarbans have got some similarity with the Andaman and Nicobar Island soils where highest sodium was 9468 ppm (Dagar et al. 1993). Since these islands are in the middle of the sea, hence there are high sodium contents. Sundarbans mangrove halophytes are in the inland; hence sodium content is low.

Mangrove plants have peculiar root system. The respiratory roots of various forms, with which they breathe in mud, are regularly immersed in sea water. Halophytes of mangroves are shallow rooted; the roots penetrate up to 100 cm. of the soil, rarely moves to depth more than 2 m. The root systems of some species under soil consist of radial array of branching roots from which secondary small roots arise, which support an extensive fibrous root system. The peglike aerial breathing roots (pneumatophores) in *Avicennia* and *Sonneratia* (Fig. 30.2) usually arise from the main cable root or smaller secondary root in close proximity to a cluster of fine fibrous root, thereby minimizing the distance over which gases must diffuse between air above the soil surface and the fine fibrous root below ground. *Heritiera fomes* produces hard, flat bullet-shaped pneumatophores. *Bruguiera* spp. grow geniculate roots which look like the bent knee of a human being when are exposed at low tide conditions (hence, the term knee root), and *Xylocarpus granatum* has platelike buttress roots project above the soil surface. *Rhizophora* spp. produces long, looping prop (or stilt) roots that serve both as support for the plant and as aerial 'breathing' roots (Fig. 30.2). The smaller secondary and fine absorbing roots below ground arise from stilt roots at each point of entry into the soil (Gill and Tomlinson 1977). In view of the fact the most of the fine absorbing roots are found near the soil surface, the physical and chemical properties of the top 1 m of the soil are probably more important than those at higher depth.

Within stable geomorphic and hydrodynamic regime, halophytes of mangrove grow in a wide range of soil types, ranging from coarse calcareous or mineral sand through to fine alluvial silts, heavy clays. The proportions of sand, silt, clay and organic matter have a direct influence on the permeability and drainage characteristics of the soil and on physico-chemical properties (redox potential, soil pH, soil salinity and nutrient availability). These properties influence species composition and growth rates. Estuarine mangrove soils of terrestrial origin, usually with higher proportions of alluvial silt and clay, are often relatively impermeable to water, with the result that exchange of water, salt and nutrients between the soil and tidal or ground water may be restricted. Evaporation from both the soil surface and the plant canopy leaves behind

a. Stilt or prop roots of *Rhizophora apiculata*

b. *Sonneratia apetala*

c. *Avicennia alba*

d. *Heritiera fomes*

e. *Sonneratia apetala* (on the fringe of canal) and also shows pneumatophores in front

f. Top dying of *Heritiera fomes*

Fig. 30.2 Stilt roots (**a**). Pneumatophores of some halophyte species (**b–d**) and *Sonneratia apetala* in open habitat; and *Heritiera fomes* shows top dying

salt, which in the absence of tidal flushing leads to soil salinities that are somewhat higher than that of tidal waters. On the other hand, alluvial soils containing high proportions of silt and clay tend to have a higher cation exchange capacity than sandy soils and usually therefore also have higher nutrient status.

In addition to physical composition, the frequency of duration of tidal flooding also has a major influence on the degree of anaerobiosis of

Table 30.4 Oxidation-reduction potentials (mV) in soils of three ecological zones

Ecological zone	Location	January	April	May
Oligohaline	Sharankhola Range	+68	+60	+100
Oligo-mesohaline	Chandpai Range	+80	+60	+90
Mesohaline	Khulna Range	+100	+80	+30
Polyhaline	Khulna Range	+134	+180	+130

the soil and nutrient status. Oxidation-reduction potentials (redox potentials) in soils of different ecological zones were measured (Table 30.4) and showed that soils are poorly aerated particularly. In the Mesohaline and in the Oligohaline zones, *Heritiera fomes* shows top dying in these zones; tide water reaches only during high tide, and most of the time, soil remains waterlogged, and it is possible that low redox value (anaerobic condition) in the soil (+60 mV) around root environment may cause top dying of *Heritiera fomes* (Fig. 30.2). In Polyhaline zone the redox potential was +130 to +180 mV, and the dominant species in this zone are *Ceriops decandra* and *Aegiceras corniculatum*. Armstrong and Boatman (1967) measured oxygen diffusion rate (ODR) in a small valley bog flushed by a small stream supporting the healthy growth of *Molinia* sp.; along the sides of this flush, where the surface flow was much reduced, the *Molinia* sp. was considerably smaller, and beyond this edge, the plants were very purple and extremely stunted. ODR was very high in the flushed site and zero at the edge. They suggested that aeration (high redox value) was responsible for the healthy growth of *Molinia* sp. and poor growth was due to lack of aeration (low redox potential). Field data (Ogino 1993) showed that mangroves could survive under highly reducing conditions, such as Eh lower than +100 mV. Only *Hibiscus tiliaceus* and *Nypa fruticans* showed an avoidance of reducing conditions. In addition, in the more aerobic soils of high intertidal areas, where flooding is either infrequent or short duration, phosphorus may become limiting for growth owing to its adsorption onto clay or to the formation of relatively insoluble complexes with iron, aluminium and calcium (Boto and Wellington 1983, 1984). In contrast, soils in the low intertidal often remain permanently waterlogged, leading to severe anoxia (Nazrul Islam 2015). Ammonium ions, the major form of inorganic nitrogen in anaerobic soils, are strongly adsorbed onto cation exchange sites, particularly in soils with a significant clay component. Some aspects of nutrient transformations and availability in anaerobic mangrove soils have been discussed by Boto (1982).

Oxygen diffusion in water is very slow when the soil pores were filled with stagnant water. The oxygen pressure decreased rapidly, and the soil environment becomes highly anaerobic. Soil of pH 5 formed a boundary between reductive and oxidative conditions at a redox potential of Eh +350 mv (Ogino 1993). Hence any plant growing under reducing conditions should develop either protective or adaptive mechanisms to absorb water and nutrients (Russel 1973). In particular, transportation of oxygen to root tips is critical to survival.

Fitter and Hay (1981) stated that any plant growing under highly reducing conditions had to provide

(i) Anatomical and morphological adaptation through which oxygen transportation became possible
(ii) A biochemical mechanism by which anaerobic dehydration became possible, particularly in the root tissues
(iii) A detoxication mechanism through which toxic substances could be excreted to the rhizosphere

(b) ***Salinity***

Water and soil salinities are given in Tables 30.1, 30.2, 30.5 and 30.6. Seasonal variation of

Table 30.5 Soil Salinity (micromhos/cm) in different months

Ecological zones	January	March	July	September	November
Oligohaline	600	650	500	700	550
Oligo-mesohaline	2250	2250	2200	2200	2050
Mesohaline	3250	3400	2800	3050	3150
Polyhaline	4800	6950	3250	3500	3450

Table 30.6 Seasonal variation of soil salinity (micromhos/cm) of different ecological zones

	Ecological zones		
Seasons	Oligohaline	Mesohaline	Polyhaline
Monsoon (July–September)	500–700	2400–2500	3400–3500
Winter (November–February)	650–700	2400–2500	5000–6900

soil salinity was noted (Table 30.6). Soil salinity is highest (6950 micromhos/cm) in the Polyhaline zone; and in the Oligohaline zone, the soil salinity values range from 500 to 700 micromhos/cm. During the monsoon season, large volume of freshwater comes from the Ganges and dilutes soil salinity, and hence the salinity is low (Oligohaline zone) in the monsoon season (Table 30.6).

Most species of mangrove grow well at salinities below that of sea water. There is evidence that soil salinities of 1.0–2.0% are optimal for the growth of *Avicennia marina* (Burchett et al. 1984; Clough 1984). Although salinities less than 2.0% may be optimal for growth, many species can grow at salinities up to that of sea water (3.5%). The upper salinity limit for *Rhizophora mangle* has been reported to be about 6.5% (Teas 1979).

(i) **The Vegetation Pattern**

The criteria that have been considered for describing the vegetation pattern of halophytes noted in the Sundarbans mangrove forest are direct observation during the field visit. Composition of species condition in three different ecological zones including plant association is given in Tables 30.7 and 30.8.

The three distinct ecological zones are recognized based on salinity (Walter 1968), such as (i) Oligohaline, (ii) Mesohaline and (iii) Polyhaline zones. Vegetation pattern differs in the three different ecological zones and are described below:

(i) *Nypa fruticans* grows not only on the mud flats inundated by high tides but also beyond the sedimentary beach and protects the soil erosion of the river slope (Fig. 30.3). Associated species further down to the river are *Phragmites karka* and *Porteresia coarctata* and that of highland (landward) are *Flagellaria indica* (very rare), *Heritiera fomes*, *Excoecaria agallocha*, *Hibiscus tiliaceus*.

(ii) *Heritiera fomes* is a dominant species in the Oligohaline zone with usually 60% density and in mesohaline zone less than 50% density. The plants have well-developed, bullet-shaped pneumatophores and attain a height of about 20–25 m. Associated species are *Excoecaria agallocha*, *Avicennia officinalis*, *Hibiscus tiliaceus*, *Cynometra ramiflora*, *Cerbera manghas*, *Pandanus foetidus*, *Barringtonia racemosa*, *Amoora cucullata*, *Sapium indicum*, *Phoenix paludosa* and *Flagellaria indica* (closed habitat). Dominant plants to the river side are *Typha elephantina*, *Phragmites karka* and *Porteresia coarctata* (submerged during tide).

(iii) *Excoecaria agallocha* is dominant in mesohaline zone (density 50%) associated with *Heritiera fomes* (shows top dying Fig. 30.2f), *Avicennia officinalis*, *Rhizophora apiculata*, *R. mucronata*, *Sonneratia apetala* (in river banks open habitat), *Bruguiera gymnorrhiza*, *B. sexangula*, *Phoenix palu-*

Table 30.7 Comparison of species condition in different ecological zones

Species	Ecological zones		
	Oligohaline	Mesohaline	Polyhaline
(a) Large trees			
Avicennia alba	Frequent, healthy, height up to 6–7 m	Frequent, good size, height up to 13–15 m	Rare, trees are slightly dwarf
A. officinalis			
Barringtonia racemosa	Rare, healthy, height 5–6 m	Not recorded	Absent
Bruguiera gymnorrhiza	Frequent, healthy height up to 10–12 m	Abundant, healthy, height 13–14 m profuse branching	Rare
B. sexangula			
Cerbera manghas	Infrequent. Near the canal edge, height 6–11 m	Absent	Absent
Cynometra ramiflora	Frequent, height 8–10 m, profuse branching	Rare	Absent
Excoecaria agallocha	Abundant, height up to 12–14 m	Abundant, height varies from 13 to 18 m	Rare
Hibiscus tiliaceus	Frequent, healthy, height up to 7–8 m	Absent	Absent
Heritiera fomes	Abundant, healthy, height 13–16 m	Less abundant, height varies from 11 to 14 m, top dying disease in localized areas	Rare, very dwarf in size
Kandelia candel	Infrequent, height up to 10 m	Frequent, healthy, height varies from 10 to 13 m	Very rare
Petunga roxburghii	Rare, height up to 10 m	Rare	Absent
Rhizophora apiculata	Frequent, height 8–10 m	Abundant near the canal and river edge, height 10–13 m	Rare
R. mucronata			
Sonneratia apetala	Frequent, only near the canal and river edge, height up to 14–17 m. Profuse branching	Abundant, only near the canal and river edge of open habitat, highly branched, height 17–22 m	Rare
S. caseolaris	Frequent, prefers low salinity, branched	Infrequent	Not observed
Tamarindus indica	Branched, height up to 7–9 m	Absent	Absent
Xylocarpus granatum	Rare, height up to 16 m	Frequent, height 16–20 m	Rare
X. mekongensis	Frequent, height up to 19 m	Less abundant, height up to 23 m	Rare
Lumnitzera racemosa	Infrequent, bushy, landward, height up to 4–6 m	Not recorded	Absent
Sapium indicum	Infrequent, grows at low salinity, height 6–8 m	Absent	Absent
(b) Small trees			
Amoora cucullata	In frequent, height up to 9 m	Absent	Absent
Aegiceras corniculatum	Absent	Infrequent, dwarf	Abundant, extremely dense, forms consociation, height varies from 3 to 5 m.
Ceriops decandra	Very rare, thin stem, height less than 1 m	Frequent, height up to 3 m, dome shaped	Abundant, highly branched, bushy, height up to 6 m.
Heterophragma adenophyllum	Rare, only in open places, height up to 5 m	Not recorded	Not recorded

(continued)

Table 30.7 (continued)

Species	Ecological zones Oligohaline	Mesohaline	Polyhaline
(c) Shrubs, herbs, palms, ferns, lichen and climbers			
Brownlowia tersa	Rare, on the river and canal edge, during tide, leaf goes under water	Infrequent	Not observed
Dalbergia spinosa	Very rare	Rare, grows on the river edge in open habitat	Absent
D. candenatensis	Same as above	Same as above	
Flagellaria indica	Infrequent, canal edge, in partly shaded habitat	Absent	Absent
Acanthus ilicifolius	Frequent, both in shade and sun habitat	Frequent, both in sun and shade habitat	Rare
Phragmites karka	Abundant in localized areas on the river bed, completely submerged during tide	Frequent	Absent
Porteresia coarctata	Abundant, both in river and canal bed, completely submerged during tide	Frequent	Not observed
Pandanus foetidus	Frequent in canal bank	Very rare	Absent
Typha elephantina	Frequent, 2/3 portion of the plants are submerged during high tide	Very rare	Absent
Nypa fruticans	Along river edges, form dense communities, height varies 4–6 m	Same as Oligohaline	Rare
Phoenix paludosa	Infrequent	Frequent, grow both on the river edge and inside the forest, form dense vegetation, height varies from 5 to 7 m	Frequent
Derris trifoliata	Abundant, climber, very common on *Heritiera*, *Excoecaria* and other species	Abundant	Infrequent
Dendrophthoe falcata	Rare, parasite on *Heritiera* and *Excoecaria*	Very rare	Absent
Mucuna gigantea	Rare	Very rare	Absent
Sphaeranthus africanus	Growing profusely in a cleared site and forming consociation (Fig. 30.3)	Not observed	Not observed
Sarcolobus globosus	Rare	Rare	Very rare
Hoya parasitica	Rare	Very rare	Not observed
Acrostichum aureum	Frequent in the middle of the forest	Abundant	Frequent
Macrosolen cochinchinensis	Very rare (woody parasite)	Rare	Rare
Ramalina calicaris (lichen)	Frequent, grows on the bark of trunks and branches of *Heritiera fomes* and *Excoecaria agallocha*	Same as Oligohaline zone	Not found

Table 30.8 Distribution of halophyte species of Sundarbans mangrove forest in relation to salinity

Salinity EC mmhos/cm	Plant association	Ecological zone
5–15	*Heritiera fomes*, *Cynometra*, *Phoenix*, *Dalbergia*, *Sonneratia apetala*, *Cynometra*, *Hibiscus*, *Amoora*, *Cerbera*, *Avicennia*, *Nypa*	Oligohaline
10–20	*Excoecaria*, *Heritiera*, *Sonneratia*, *Xylocarpus*, *Avicennia*, *Nypa*, *Phoenix*, *Heterophragma*, *Petunga*	Mesohaline
15–25	*Aegiceras*, *Ceriops*, *Kandelia*, *Rhizophora*, *Bruguiera*, *Kandelia*	Polyhaline

dosa, *Kandelia candel*, *Sonneratia caseolaris*, *Lumnitzera racemosa*, *Heterophragma adenophyllum*, *Petunga roxburghii*, *Xylocarpus granatum and X. mekongensis*. Occasionally in the submerged river bank are *Dalbergia candenatensis*, *D. spinosa* and *Brownlowia tersa* (very rare).

(iv) *In the Polyhaline zone*, *Aegiceras corniculatum* and *Ceriops decandra* are dominant plants and both species form consociation, attain a height of 4–6 m, and are highly branched and bushy; in the weak Polyhaline zone, *Xylocarpus granatum*, *X. mekongensis*, and *Bruguiera sexangula* also grow well, but *Heritiera fomes* is very rare and becomes extremely dwarf.

(v) In addition, *Acanthus ilicifolius* grows on the river bank both in Oligohaline and Mesohaline zones and *Acrostichum aureum* (tiger fern, Fig. 30.3) found to grow in the upland in all the ecological zones; common climbers are *Derris trifoliata*; rare species are *Hoya parasitica*, *Mucona gigantea*, *Sarcolobus globosus* and *Ipomoea pes-caprae* (sand dunes); woody parasites are *Macrosolen cochinchinensis* and *Dendrophthoe falcata*; *Ramalina calicaris* (lichen) is abundant on the branches of *Heritiera fomes* and *Excoecaria agallocha* in the Oligohaline and Mesohaline zones.

(vi) The newly exposed cleared areas are invaded by weedy species *Sphaeranthus africanus* (Fig. 30.3), and submerged river slopes are covered with grass *Porteresia coarctata* (Roxb.), and the upper areas are mostly covered with *Sonneratia apetala*.

(ii) Anthropogenic and Natural Threats to Mangroves and Conservation

Pressures on mangroves are characterized by the dominance of wholesale alteration of mangroves through commercial scale operations as opposed to relatively benign traditional, artisanal exploitation. Hatcher et al. (1989) identified direct extirpation as the largest anthropogenic threat to tropical mangroves. They also highlighted the importance of oil spills on mangrove as one important recent agent causing mangrove mortality.

Population expansion on mangroves, aquaculture development and reclamation are the major anthropogenic threats faced by mangroves (Ong 1995). Farnsworth and Ellison (1997) identified the following major threats to global mangrove forests: reclamation (for settlement, agriculture and aquaculture), harvesting of forestry and fishery products, sewage discharge and freshwater diversion. The last two threats are new to the list based on Hatcher et al. (1989) and Ong (1995). Anthropogenic threats identified from their reports may be summarized under three categories:

(i) Major consumptive uses including reclamation of mangroves for settlement, cultivation and harvesting of timber products, e.g. the woodchip industry, which usually result in permanent destruction and loss of mangrove habitats and depletion of fishery resources, no consumptive uses such as tourism activities and chemical and physical alterations of the habitat, including diversion of freshwater resources
(ii) Pollution and sedimentation
(iii) Hydrological regime

Fig. 30.3 Some halophytes in the natural habitats of Sundarbans mangrove forest

These are added to natural potential long- and short-term threats to mangroves such as sea level rise,climate (Smith et al. 1994); (pests, e.g. boring isopods, Olafsson 1998) to limit mangrove performance.

(iii) **Ecological Research for Halophytes of Mangrove and Conservation**

In order to secure the long-term conservation of mangrove halophytes, ecological research would have to be targeted towards understanding their response to the major present and future threats and ways of rehabilitating and restoring damaged systems. These would involve specific studies on the respective threats and their impacts or rehabilitation/restoration techniques, improvements in the study design.

(a) ***Stress Ecology and Rehabilitation of Mangrove Halophytes***

Rehabilitation of damaged ecosystem is probably one of the most challenging tasks for ecologists. Although complete restoration of a damaged ecosystem to its original state (with all the associated functions and benefits) is usually impractical or possible (Field 1996), the original state is nevertheless a good target for rehabilitation efforts. Underwood (1989) provided a theoretical framework for analysing the impact of disturbances and stresses on populations. Response to stresses and disturbances is dependent on the resilience (ability of the system to recover), stability rate (rate of recovery) and inertia (tendency to remain unchanged) of the population or ecosystem. These properties of ecosystems have profound implications for the pattern of occurrence and detection of ecological impacts. Disturbances may be classified according to their temporal pattern of occurrence. Press disturbances occur for a sustained period and are exposed to more devastating than pulse disturbance, which are intermittent (Bender et al. 1984). Added to these characteristics of the stresses and ecosystems are factors such as natural variability (both temporal and spatial) of natural populations and the technical challenges of detecting the impact of disturbances using suitable experimental design and statistical design (Underwood 1997). Consideration of these general aspects of the stress or disturbance ecology of mangroves has already been included in the discussion and detection of anthropogenic impacts on mangrove ecosystems. Mangroves are probably one of the studied ecosystems at least in the areas of system productivity and connectivity with neighbouring systems. The assessment of impacts on mangroves has, however, suffered from the lack of scientific rigour in aspects such as adequate replication and/or sufficient sampling design (Underwood 1992). Research in mangrove ecosystems has focused more on function than structure. While this idea is understandable in view of the assumed trophic role of mangroves and logistic difficulties in performing manipulative experiments in ecosystem-scale processes, it is in sharp contrast to the approach used in studies on other shore types. Some recent efforts have demonstrated that the manipulative experimental approach may generate useful information for answering conservation questions regarding mangroves (Skilleter 1996; Kelahar et al. 1998).

Lugo (1998a, b) argued that mangroves are generally difficult to invade but easy to rehabilitate, as few species can tolerate the combined stresses of salinity, long hydroperiods and anoxic sediment conditions. As such, mangroves can be considered as systems with high inertia and high resilience, at least to species invasion. Lugo (1998a, b) suggested that the ultimate objective of forest rehabilitation efforts should be to restore system productivity rather than structure. This is a simplistic but probably pragmatic view of restoration process, as primary productivity of mangroves is relatively well documented. Secondary productivity is, nevertheless, poorly known in most mangrove ecosystems. Restoring mangrove primary productivity may not simply imply reinstatement of the ecological values of the mangroves, unless contemporaneous assessment of secondary productivity is undertaken. Species replacement may also compromise the original beneficial uses of the regenerating forest, although selected ecological processes such as primary productivity may have been restored.

(b) ***Linkage Between the Structure of Mangroves and Ecosystem Services***

Mangroves are comparatively easy to replant, provided suitable habitats are chosen (Field 1996). The re-establishment of the trees may not, however, necessarily result in return of essential ecosystem functions, such as exchange of organic matter with neighbouring coastal systems or the use of the forests by nektons for food and refuge. The functional pattern of coastal wetlands, e.g. flux of organic matter and nutrients, is expected to vary according to the developmental stage (Dame and Allan 1996). Both the direction and the nature of the material exchanged (e.g. dissolved versus particulate forms) are dependent on characteristics such as the layout and the extent of the interface between the wetland and tidal water. As the movement and abundance of consumers are expected to depend on the occurrence and availability of nutrients, it is unreasonable to expect newly planted trees to provide the same function as established forests. To what extent these basic functions of mangroves are related to the structure of the forest is virtually unknown. The question of structural redundancy in ecosystems has been a major research area in the biodiversity conservation issue (Naeem et al.1994; Naeem 1998). It is argued that attempts to bring back the original system are unrealistic because details of the functioning of ecosystems are usually not known, making the reinstatement efforts impossible to assess (Field 1996). While this difficulty is recognized, if mangrove rehabilitation targets repossession of the beneficial functions of undisturbed mangroves, the relationship between forest structure and function deserves more attention. Long-term monitoring of the functional aspects of artificially rehabilitated or naturally regenerated mangroves would help provide such vital ecological information for conservation efforts. In addition, the use of manipulative experiments would be a desirable approach in handling such questions. Regrettably, this approach has seldom been taken in mangrove ecology, probably because of the logistic difficulties associated with manipulations required for studying ecosystem processes (Lee 1999, 2001).

(c) ***Do Mangrove Halophytes Have a Future***

Almost everywhere, halophytes of mangroves are in retreat, in the face of relentless pressures of the deliberate destruction, over-exploitation, pollution and climate change. If the attrition continues, in many parts of the world, mangroves may be reduced to relic patches, too small to support the diversity of organism characteristic of a thriving mangal. Is there any long-term future for mangrove halophyte ecosystem?

Very few mangrove areas are untouched by human activities. The future of halophyte mangroves must depend not on remaining isolated from human influences, but on properly regulated interactions with human demands on effective management. For this to happen, the first step must be education. Unless planners and politicians as well as local people appreciate the many contributions of mangroves to human wellbeing, and the true costs of allowing their destruction, the decline will continue. When mangroves are cleared for timber or development, the financial gains are rapid, but the losses are long term. Planning must evaluate all the costs, as well as the benefits and look to distant horizons; and those who gain, of course, tend to have more influence in planning decisions than the local people who may lose a subsistence economy.

Some of the pressures on mangroves, such as major water diversion schemes or sea level rise, are not amenable to local management. Here mitigation, rather than prevention, is a realistic objective. Schemes such as the replanting of mangroves in the Indus delta demonstrate how much can be achieved, at relatively low cost. Reproduction and the early settlement stages of mangroves are the most affected by environmental pressures. If these vulnerable stages can be nurtured by artificial methods, a population of adult trees may be able to survive in an environment that is no longer ideal.

With intelligent and informed management, it is possible to maintain halophytes of mangroves, not just as a curiosity but as a valuable natural resource, managed sustainably so as to continue exploitation without destroying a rich and diverse habitat.

References

APHA (1977) Standard methods for the examination of water and wastewater. American Public Health Association, Washington, DC

Armstrong W, Boatman DJ (1967) Some field observations relating to the growth of bog plants. J Ecol 55:101–110

Baltzer F (1969) Les formations vegetales associees au delta de la Dumbea. Cah Orstrom Ser Geol 1(1):59–84

Bender EA, Case TD, Gilpin ME (1984) Perturbation experiments in community ecology: theory and practice. Ecology 65:1–13

Boto KG (1982) Nutrient and organic fluxes in mangroves. In: Clough BF (ed) Mangrove ecosystems in Australia: structure, function and management. Australian National University Press, Canberra/London, pp 239–257

Boto KG, Wellington JT (1983) Phosphorus and nitrogen nutritional status of a northern Australian mangrove forest. Mar Ecol Prog Ser 11:63–69

Boto KG, Wellington JT (1984) Soil characteristics and nutrient status in northern Australian mangrove forest. Estuaries 7:61–69

Burchett MD, Field CD, Pulkownik A (1984) Salinity, growth and root respiration in the grey mangrove *Avicennia marina*. Physiol Planatarum 60:113–118

Chaffey DR, Miller FR, Sandom IH (1985) A forest inventory of Sundarbans. Land Resources Development Centre, Surbiton

Chapman VJ (1976) Mangrove vegetation. Cramer, Lehre, 425 pp

Clough B (1984) Growth and salt balance of the mangroves *Avicennia marina* (Forsk.) Vierh. and *Rhizophora stylosa* Griff. In relation to salinity. Aust J Plant Physiol 11:419–430

Clough BF (1993) Constraints on the growth, propagation and utilization of mangroves in arid regions. In: Leith H, Al Masoom AA (eds) Towards the rational use of

high salinity tolerant plants, vol 1. Kluwer Academic Publishers, Dordrecht, pp 341–352
Clusener M, Breckle SW (1987) Reasons for the limitation of mangrove along the west coast of northern Peru. Vegetatio 68:173–177
Dagar JC, Singh NT, Mongia AD (1993) Characteristics of mangrove soils and vegetation of Bay Islands in India. In: Leith H, Al Masoom AA (eds) Towards the rational use of high salinity tolerant plants, vol 1. Kluwer Academic Publishers, Dordrecht, pp 59–80
Dame RF, Allan DM (1996) Between estuaries and the sea. J Exp Mar Biol Ecol 200:169–185
Farnsworth E, Ellison AM (1997) The global conservation status of mangroves. Ambio 26:328–334
Field CD (1996) Restoration of mangrove ecosystems. ITTO-ISME, Okinawa
Fitter AH, Hay RKM (1981) Environmental physiology of plants. Academic, San Diego
Gill AM, Tomlinson PB (1977) Studies on the growth of red mangrove (*Rhizophora mangle* L.). 4. The adult root system. Biotropica 9:145–155
Gupta PL, Rorison IH (1974) Effects of storage in the soluble phosphorus and potassium content of some Derbyshire soils. J Appl Ecol 11:1197–1207
Hatcher BG, Johannes RE, Robertson AI (1989) Review of research relevant to the conservation of shallow tropical marine ecosystems. Oceanogr Mar Biol Annu Rev 27:337–414
Hogarth PJ (1999) The biology of mangroves. Oxford University Press, Oxford
Kelahar BP, Chapman MG, Underwood AJ (1998) Changes in benthic assemblages near boardwalks in temperate urban mangrove forests. J Exp Mar Biol Ecol 228:291–307
Lee SY (1999) Ecology of tropical mangals: the need for a synthesis of physical and biotic influences on ecosystem structure and function. Aust J Ecol 24:355–366
Lee, S. Y. 2001. Ecological research for mangrove conservation. In Narayana, R. Bhat.; Faisal K. Taha and Afaf Y Al-Nasser. (Eds.). Mangrove ecosystems: natural distribution, biology and management. Kuwait Institute for Scientific Research, Kuwait
Lin GH, Sternberg LDL (1994) Utilization of surface water by Red Mangrove (*Rhizophora* mangle L.) – an isotopic study. Bull Mar Sci 54:94–102
Lugo AE (1998a) Ecosystem rehabilitation in the tropics. Environment 30:16–20
Lugo AE (1998b) Mangrove forest: a tough system to invade but an easy one to rehabilitate. Mar Pollut Bull 37:427–430
Macnae W (1968) A general account of the Fauna and Flora of mangrove swamps and forests in the Indo-West-Pacific region. Adv Mar Biol 6:73–269
Naeem S (1998) Species redundancy and ecosystem reliability. Conserv Biol 12:39–45
Naeem S, Thompson LJ, Lawler SP, Lawton JH, Woodfin RM (1994) Declining biodiversity can affect the function ecosystems. Nature 368:734–737
Nazrul Islam AKM (1985) The ecology of Bay with notes on coastal edaphic features. J NOAMI 2:31–35
Nazrul Islam AKM (1994) Environment and vegetation of Sundarbans mangrove forest. In: Leith H, Al Masoom AA (eds) Towards the rational use of high salinity tolerant plants, vol 1. Kluwer Academic Publishers, Dordrecht, pp 81–88
Nazrul Islam AKM (1995) Ecological conditions and species diversity in Sundarban mangrove forest in Bangladesh. In: Khan MA, Unger IA (eds) Biology of salt tolerant plants: halophytes for salt tolerant plants. Book Grafters, Chelsea, pp 294–305
Nazrul Islam AKM (2003) Mangrove forest ecology of Sundarbans: the study of change in water, soil and plant diversity. In: Ghosh AK, Mukhopdhya MK (eds) Sustainable environment: a statistical analysis. Oxford University Press, New Delhi, pp 26–147
Nazrul Islam AKM (2015) Germination eco-physiology and plant diversity in halophytes of Sundarbans mangrove forest in Bangladesh. In: Khan MA, Ozturk MGB, Ahmed MZ (eds) Halophytes for food security in dry lands. Academic, Oxford, pp 277–288
Nazrul-Islam AKM, Rorison IH (1978) Field studies of seasonal oxidation-reduction conditions in soils and of changes in mineral contents of shoots of associated species. Dacca Univ Stud XXVI:57–65
Ogino K (1993) Mangrove ecosystem as soil, water and plant interactive system. In: Leith H, Al Masoom AA (eds) Towards the rational use of high salinity tolerant plants, vol 1. Kluwer Academic Publishers, Dordrecht, pp 135–144
Olafsson E (1998) Are wood-boring isopods a real threat to the wellbeing of mangrove forests? Ambio 27:760–761
Ong JE (1995) The ecology of mangrove conservation and management. Hydrobiologia 295:343–351
Prain D (1979) (Reprinted). Flora of Sundarbans. Periodical Expert Book Agency, New Delhi
Russel EW (1973) Soil conditions and plant growth, 10th edn. Longmans, New Delhi
Skilleter GA (1996) Validation of rapid assessment of damage in urban mangrove forests and relationships with molluscan assemblages. J Mar Biol Assoc UK 76:701–716
Smith TJ, Robblee MB, Wanless HR, Doyle TW (1994) Mangroves, hurricanes and lightning strikes. Bioscience 44:256–262
Sternberg L d SL, Swart PK (1987) Utilization of fresh water and ocean water by coastal plants of southern Florida. Ecology 68:1898–1905
Teas HJ (1979) Silviculture with saline water. In: Hollander A, Aller JC, Epstein E, San Pietro A, Zaborsky OR (eds) The bio-saline concept. An approach to the utilization of underexploited resources. Plenum Press, New York, pp 117–161
Tomlinson PB (1986) The botany of mangroves. Cambridge University Press, Cambridge. 413pp

Underwood AJ (1989) The analysis of stress in natural populations. Biol J Linn Soc 37:51–78
Underwood AJ (1992) Beyond BACI. The detection of environmental impact in the real, but variable, world. J Exp Mar Biol Ecol 161:145–178
Underwood AJ (1997) Detection and measurement of environmental impacts. In: Underwood AJ, Chapman MG (eds) Coastal marine ecology of temperate Australia. University of New South Wales Press, Sydney
Walter H (1968) Die Vegetation der Erde in Oiko-PhysiologischenBetrrachtung. Bad II. Fischer Verlag, Stuttgart
Walter H, Steiner M (1936) Die Oekologie der Ostafrikanischen Mangroven. Z Bot 30:65–193
Woodroffe CD (1993) Geomorphological and climatic setting and the development of mangrove forest. In: Leith H, Al Masoom AA (eds) Towards the national use of high salinity tolerant plants, vol 1. Kluwer Academic Publishers, Dordrecht, pp 13–20

Cash Crop Halophytes of China

31

Guoqi Li, Panpan Zhao, and Wenshan Shao

Abstract

The categories and distribution of halophytes in China are reviewed, and the floristic characteristics are discussed in this paper. Then we summarize using the value and their growing environments of cash crop halophytes of China. Finally, we take two halophytic plants as examples to show the current situation of sand industry of China.

Keywords

Chinese halophytes · Distribution · Flora · Commercial value

Halophytes, plants that survive to reproduce in environments where the salt concentration is around 200 mm NaCl or more, constitute about 1% of the world's flora. However, although halophytes have been recognized for hundreds of years, their definition remains equivocal (Flowers and Colmer 2008). According to Aronson (1989), a halophyte is a plant that completes its life cycle in salty environment. The book *HALOPH: A Data Base of Salt Tolerant Plants of the World* covers 2600 halophytic species that belong to 126 families and 776 genera (Aronson 1989; Menzel and Lieth 1999). In China, there are 555 species of halophytes (Zhao et al. 2013).

G. Li (✉) · P. Zhao · W. Shao
Breeding Base of State Key Laboratory for Land Degradation and Ecological Restoration, Ningxia University, Yinchuan, China

Key Laboratory of Restoration and Recovery of Degraded Ecosystem in Northwest China of Ministry of Education, Ningxia University, Yinchuan, China

The Categories of Halophytes in China

According to the mode of ion accumulation and transport, Breckle (1995) classified halophytes into three categories.

1. Recretohalophytes
   (a) Exo-recretohalophytes
   (b) Endo-recretohalophytes

The former can exclude the salt through leaf salt glands while the latter through leaf vesicles.

2. Euhalophytes
   (a) Leaf succulent euhalophytes
   (b) Stem succulent euhalophytes

Leaf succulent euhalophytes accumulate the salt ions in vacuoles of succulent leaves, while the stem succulent euhalophytes in the central column.

B. Gul et al. (eds.), *Sabkha Ecosystems*, Tasks for Vegetation Science VI,
https://doi.org/10.1007/978-3-030-04417-6_31

3. Pseudo-halophytes, these plants accumulate salt ions in the vacuoles and root xylem parenchyma.

All three categories of halophytes mentioned above grow on saline-alkali land in China (Zhao et al. 1999; Gorham 1995).

Recretohalophytes of China

Exo-recretohalophytes include many plants, such as *Acanthus ebracteatus*, *Ipomoea polymorpha*, *I. pes-tigridis*, *Frankenia pulverulenta*, *Aegiceras corniculatum*, *Castilleja pallida*, *Limonium bicolor*, *L. aureum*, and *Glaux maritima*. *Tamarix* includes ten species and *Reaumuria* two species. Gramineae include eight genera, *Aeluropus*, *Spartina*, *Crypsis*, *Sporobolus*, *Cenchrus*, *Digitaria*, *Panicum*, and *Paspalum*. Endo-recretohalophytes include *Atriplex*, *Chenopodium*, and *Salsola* of Chenopodiaceae; moreover, there are over ten species in the genus *Chenopodium*.

Euhalophytes of China

The euhalophytes embody leaf succulent euhalophytes, such as *Suaeda salsa*, *S. glauca*, *Salsola scopparia*, and *S. junatoxxii*, and stem succulent euhalophytes, such as *Halostachys belonggeriana*, *Halocnemum strobilaceum*, *Kalidium schrenkianum*, *K.cuspidatum*, *Salicornia europaea*, etc.

Pseudo-Halophytes of China

Pseudo-halophytes embody *Lotus siliquosus* of Leguminosae, *Trifolium fragiferum* of Leguminosae, *Phragmites communis* of Gramineae, etc.

The Distributions of Halophytes in China

As been well known, halophytes are intrazonality plant. China has a vast area of saline-alkali land where a great number species of halophytes are distributed. There are eight climatic zones in China, so there grow different plants in different zones (Zhao et al. 2013).

Halophytes in Humid and Semi-humid Area of Seashore

Mangrove (*Rhizophora apiculata*) and the accompanying plants are the most popular halophytes in coastal region of Southern China. Mangrove plants include *Bruguiera gymnorhiza*, *B. sexangula*, *Kandelia candel*, *Rhizophora apiculata*, *R. stylosa*, *Lumnitzera racemosa*, *Aegiceras corniculatum*, *Sonneratia caseolaris*, *Avicennia marina*, *Cerbera manghas*, *Pongamia pinnata*, *Pemphis acidula*, etc., and accompanying plants include *Salicornia Eaupaea*, *Salsola komarovii*, *Suaeda corniculata*, *Spergularia marina*, *Halerpestes sarmentosa*, *Thellungiella halophila*, and *Glycine soja* (Zhao et al. 2013).

Halophytes in Semi-humid and Semiarid Meadow Steppe in Northeast China

These saline-alkaline salty and solonetz soils harbor a large number of herbaceous plants, such as *Suaeda corniculata*, *Puccinellia tenuiflora*, *Aeluropus littoralis*, *Kochia sieversiana*, etc. In alkaline soils, besides *Stipa baicalensis* and *Iris pallasii* as major dominant species, there are many co-occurring halophytes, such as *Polygonum gracilius*, *P. sibiricum*, *Atriplex centralasitica*, *A. patens*, *A. sibirica*, *Kalidium foliatum*, *Kochia sieversiana*, *Salsola collina*, *Suaeda glauca*, *S. corniculata*, *S. heteroptera*, *Lepidium sibiricum*, *Glycyrrhiza uralensis*, *Sphaerophysa salsula*, *Nitraria sibirica*, *Tamarix chinensis*, *Glaux maritima*, *Apocynum lancifolium*, *Limonium sinense*, *Lycium chinense*, *Artemisia anethifolia*, *Tripolium vulgare*, *Achnatherum splendens*, *Aeluropus littoralis* var. *sinensis*, *Aneurolepidium aculeata*, *Chloris tenuiflora*, *Crypsis virgata*, *Phragmites australis*, etc.

Halophytes in Semi-humid and Semiarid Tillage Steppe in Huanghuaihai Plain

There mainly grow meso-halophytes in these areas, for example, *Tamarix chinensis* shrub accompanied by many halophytic species including *T. glauca*, *Suaeda salsa*, *Aeluropus sinensis*, *Phragmites australis*, and *Imperata cylindrica*; *Imperata* spp. accompanied by *Aeluropus littoralis* var. *sinensis*, *Phragmites Australia*, *Artemisia scoparia*, *Nitraria tangutorum*, *Suaeda glauca*, etc.; *Phragmites Australia* meadow accompanied by *Scorzonera mongolica*, *Glycine soja*, *Apocynum venetum*, *Chloris virgata*, etc.; *Artemisia stelleriana* meadow accompanied by *Aster tataricus*, *Limonium sinensis*, *Aeluropus sinensis*, etc.; and *Apocynum venetum* meadow accompanied by *Imperara cylindrica*, *Aster tataricus*, *Polygonum hydropiper*, *Artemisia capillaris*, *Chloris virgate*, etc.

Halophytes in Arid Desert Steppe in Inner Mongolia Plateau

There are plenty of species of grass and semishrub in this area. In north-central Inner Mongolia plateau, there grow many *Stipa* and *Cleistogenes* genera plants, such as *S. grandis*, *S. krylorii*, *S. bungeana*, *S. breviflora*, *C. squarrosa*, *C. polyphylla*, *Astragalus dahuricus*, and *Festuca sulcate*. In the western of Inner Mongolia plateau, there are many drought tolerant plants, such as *Artemisia xerophyticon*, *Anabasis brevifolia*, *Salsola passerina*, *Reaumuria soongorica*, *Ceratoides latens*, *Ajania fruticulosa*, *Halogeton arachnoideus*, *Artemisia frigida*, *Artemisia scoparia*, *Salsola collina*, *Peganumnigellastrum*, *Caragana microphylla*, etc.

Halophytes in Semiarid Semidesert Steppe in the Middle and Lower Reaches of the Yellow River

There distributes desert steppe, steppe, and forest steppe in these regions. The tributaries of Yellow River assemble lake marches and develop saline-alkali soils where mainly contain sulfate and chloride. Here grow dense halophytic *Achnatherum splendens* accompanied by *Suaeda salsa*, *Nitraria sphaerocarpa*, *Taraxacum mongolicum*, *Juncus Effuses*, *Typha angustifolia*, *Scirpus yagara*, *Triglochin palustre*, etc.

Halophytes in Arid Desert Regions of Gansu, Inner Mongolia, and Xinjiang

There are diverse halophytes in these area where there are plenty of both herbs and shrubs or semishrubs, but rare of trees. The herbs are the following: *Cynoglossum viridiflorum*, *Nonea capsica*, *Atriplex tatarica*, *Chenopodium chenopodioides*, *C. rubrum*, *Panderia turkestanica*, *Petrosimonia sibirica*, *Salsola affinis*, *S. heptapotamica*, *S. nitraria*, *S. rosacea*, *S. subcrassa*, *Suaeda altissima*, *Centaurea adpressa*, *Inula britannica*, *Ligularia thyrsoidea*, *Saussurea turgaiensis*, *S. iliensis*, etc. The semishrubs or shrubs are the following: *Anabasis elatior*, *A. salsa*, *Atriplex cana*, *Camphorosma monspeliaca*, *Kalidium capsicum*, *Salsola arbuscula*, *Suaeda physophora*, *Lycium barum*, *L. dasystemum*, *Tamarix hispida*, *T. karelinii*, *T. laxa*, *T. leptostachys*, *T. ramosissima*, *Nitratia schoberi*, etc.

Halophytes in Extreme Arid Region Desert of Qinghai and Gansu Provinces

These regions are stressed by serious sandstorm and rigorous drought with less than 100 mm rainfall. The herbaceous halophytes are the following: *Atriplex centralasiatica* var. *megalotheca*, *Cynanchum kashgaricum*, *Atriplex sibirica*, *Suaeda crassifolia*, *S. heterophylla*, *Acroptilon repens*, *Brachyactis ciliate*, *Cirsium lanatum*, *Inula caspica* var. *caspica*, *Seriphidium schrenkianum*, *Scirpus planiculmis*, *S. strobilinus*, *Crypsis aculeate*, *Cynodon dactylon*, *Hordeum bogdanii*, *Phragmites australis*, *Puccinellia tenuiflora*, *Astragalus contortuplicatus*, *Lotus frondosus*, *Melilotus alba*, etc. The halophytic shrubs are the following: *Tamarix ramosissima*, *T. hispida*, *T. elongate*, *Reaumuria soongoriea*, *Elaeagnus angustifolia* var. *angustifolia*, *Ceratoides latens*, *C. ewersmanniana*, *Cynanchum cathayensis*, *Apocynum venetum*, etc. There are only two species of tree, i.e., *Populus euphratiea* and *P. pruinosa*.

Halophytes in Alpine Steppe of Tibet Plateau

The inland rivers of this area form many solitary lakes that make different saline-alkali soil along them. Because of the harsh weather, there is low species richness and evenness of halophytes in these inland areas. Lakeshore plants include members of Cyperaceae family, such as *Scirpus* spp., *Cyperus* spp., *Mariscus* spp., *Carex* spp., etc. In addition, *Halerpestes sarmentosa*, *H. tricuspis*, *H. filisecta*, *Lepidium latifolium*, *Dilophia salsa*, *Polygonum sibiricum*, *Puccinellia distans*, *Suaeda corniculata*, *Phragmites australis*, *Potentilla* spp., *Tripolium vulgare*, *Ceratoides latens*, *Leymus secalinus*, etc. are also found. There are a few shrubs, *Iljinia regelii*, *Poacymum pictum*, etc.

Floristic Characteristics of Chinese Halophytes

China has halophytes 555 species belonging to 218 genera and 71 families (Zhao et al. 2013). Four families, that is, Chenopodiaceae, Compositae, Poaceae, and Leguminosae, have the maximum halophytic species, which account for the 46.8% of total halophytic species of China. Gymnosperm has no halophytes, and ferns have only three species of halophytes; most halophytes are in Angiospermae accounting for more than 99.9% of total genera and total species (Zhang and Li 2001; Wang 1992).

Statistical Analysis of Halophytic Families

Comparatively large family means it contains more than ten species. There are 8 comparatively large families, i.e., the biggest Chenopodiaceae including 76 halophytic species; then Compositae, Poaceae, and Leguminosae including 30–60 species; and then Cyperaceae, Tamaricaceae, Potamogetonaceae, and Plumbaginaceae including 10–30 species. The above 8 families are nearly all cosmopolitan, in which contain 95 genera, 255 species accounting for 47.5% of halophytic genera and 60% of halophytic species, respectively. These families play an important role in the halophytic vegetation. As a type of Mediterranean origin center, Chenopodiaceae has a most important action on constructive community to desert vegetation. *Haloxylon*, *Suaeda*, *Salsola*, *Anabasis*, *Kalidium*, *Atriplex*, and autogenous *Sympegma*, *Halostachys*, and *Halocnemum* embody important constructive species. *Artemisia* of Compositae; *Achnatherum*, *Aeluropus*, *Leymus*, and *Phragmites* of Poaceae; and *Glycyrrhiza*, *Alhagi*, and *Sophora* of Leguminosae are all the framer of saline meadow. *Tamarix* is the shrub framer (Wu 1980).

Statistical Analysis of Halophytic Genus

There are 4 halophytic genera that contain more than 10 species, i.e., *Suaeda* and *Salsola* include 16 species, *Tamarix* 13, and *Limonium* 11. There are nine halophytic genera that contain species from six to ten, i.e., *Atriplex* include ten species; *Polygonum*, *Astragalus*, and *Seriphidium* eight species; *Ipomoea*, *Artemisia*, *Saussurea*, and *Leymus* all seven species; and *Glycyrrhiza* six species. Both of these genera total to 13 genera, which account for the 6.5% of total halophytic genera, and 124 species account for the 29.2% of total species. There are 58 genera that contain 2–5 halophytic species; furthermore, there are 129 genera that include only 1 species; both of these genera account for 93.5% of total halophytic genera. Therefore, the fact that the medium and small genus are plentiful indicates that the genus differentiation in China is higher (Zhang and Li 2001).

Fundamental Floristic Character of Halophytes in China

As for the categories of genera and families, halophytes of China have 34 cosmopolitan families and 29 genera that account 51.5% and 14.7%, respectively, which fully show the character of intrazonality. Halophytic flora of China can be thought to have very consanguineous relation with tropical flora and to have relatively consanguineous relation with temperate flora, and it has definite connection with ancient Mediterranean. The analysis of five fundamental characteristics of halophytic flora of China considers that the floristic ancientness and complex of floristic element are relative high but the floristic abundance, growth of speciation, and endemicity of floristic elements are lower (Zhang and Li 2001).

From the viewpoint of systematics and geography, halophytes of China indicate some characters as follows:

1. Halophytes of China have a poor species and low floristic abundance, because the halophytic species, genera, and families only account for 1.6% of total species, 6. 3% of total genera, and 18.7% of total families of China, respectively.
2. The origin of Chinese halophytes is diverse and ancient. There are 48 autogenous that had formed in Tertiary period, for example, *Potaninia mongolica* Maxim, *Zygophyllum* spp., etc. Moreover, *Messerschmidia sibirica* is the relic plant along Mediterranean coast; mangrove plants, such as *Nypa fruticans*, *Sonneratia caseolaris*, *Avicennia marina*, etc., are relic species of Tertiary period.
3. Halophytes of China have complex geographical elements and wide range of distribution; however, China has no unique family or genus. In addition, all halophytic families have glycophytes. So it can be inferred that halophytes originate from the adaptation of nonhalophytes to the saline-alkali soil.

Cash Crop Halophytes of China

The halophytes can not only provide the ample salt-tolerant plasma resources for agricultural production but also the gene pool for cultivating the salt-tolerant crops. As one of the richest regions with cash crop halophyte resources, Northwest China utilizes the halophyte resources including officinal, edible, and feeding values. It can achieve the economic returns and fully use the vast lands in the arid desert regions to plant the economic halophytes in the fields where the soils contain too much salt to plant crops. Therefore, it is of significance to study the exploitation and utilization of the halophyte resources in the arid desert regions (Zhao 2013).

Table 31.1 The main officinal halophyte resources in the arid desert regions of Northwest China

Plant name	Medicinal position	Function	Distribution and environment
Salicornia europaea	Whole plant	Diuretics	Salinized wetland, various regions of desert areas
Salsola collina	Whole plant	Antihypertensive	Various regions of desert areas
Lycium barbarum	Leaf, root, fruit	Antihypertensive, nutritious, clearing heat	Hilly area, riverside, saline-alkali area
Cistanche deserticola	Fleshy stem	Reinforcing agents	Parasitic plant in Sandy desert
Glycyrrhiza uralensis	Root, stem	Reinforcing agents, Buzhong Yiqi	Various regions of desert areas
Sophora alopecuroides	Root, stem	Clearing heat, pain relief, pesticide	Various regions of desert areas
Sphaerophysa salsula	Whole plant	Hemostasis	Salinized wetland, various regions of desert areas
Nitraria sibirica	Fruit	Strengthening the spleen and stomach, promoting blood flow for regulating menstruation, reinforcing agents	Various regions of desert areas
Tamarix chinensis	Stem, leaf	Clearing away heat and toxic material, antiphlogosis	Floodplain, salinized wetland
Elaeagnus angustifolia	Fruit	Eliminating phlegm	Desert riparian, desert areas
Cistanche deserticola	Fleshy stem	Nourishing kidney, anti-fatigue	Sand desert
Limonium bicolor	Whole plant	hemostasis, promoting urination	Saline sandy land, various regions of desert areas
Limonium aureum	Whole plant	Regulate the menstrual function, invigorate blood circulation, relieve pain	Various regions of desert areas
Phragmites australis	Root, stem	Inducing diuresis to alleviate edema	Riverside, valley, plain oasis, desert areas
Apocynum venetum	Leaf, root	Calming the live, relieving uneasiness of body and mind, antihypertensive, anti-neurasthenia	Inner Mongolia, Xinjiang, Ningxia irrigation district, Hexi Corridor
Sargassum spp.	Whole plant	Clearing heat, diuretics	Various regions of desert areas
Populus euphratica	Resins	Heat-clearing and detoxicating	desert riparian, Inner Mongolia, Gansu, Qinghai, Xinjiang

Officinal Halophyte Resources

In Northern China, Northeast and coastal region, especially in Northwest, there grow numerous species of cash crop halophytes. Many of these plants have an extensive prospective of development and utilization. The species, medicinal position, function, distribution, and environment of officinal halophytic plants in Northwest China are in Table 31.1.

Edible Halophyte Resources

Many Chinese halophytes can be used as food:

1. Plants rich in starch: seeds of *Iris lacteal* and *Melilotus suavena*, fruits of *Elaeagnus angustifolia*, root and stem of *Phragmites australis*, etc.
2. Plants rich in proteins: seeds of *Chenopodium* spp., leaf of *Melilotus suavena* and *Salsola collina* spp., etc.

3. Plants rich in oils: seeds of *Salicornia europaea*, *Suaeda salsa*, *Halocnemum strobilaceum*, *Elaeagnus angustifolia*, *Nitraria sibirica*, etc.
4. Plants rich in vitamins: fruits of Lycium barbarum, *Nitraria sibirica*, and *Elaeagnus angustifolia*, leaf of *Suaeda salsa* and Salsola collina, etc.
5. Plants rich in sweeteners: whole plants of *Glycyrrhiza uralensis*, tuberous root of *Asparagus cochinchinensis*, etc.
6. Plants rich in aromatic oils: flower of *Melilotus suavena*, *Pandanus tectorius*, and *Vitex trifolia*, fruit of *Cnidium salinum*

Feeding Halophyte Resources

As the second abundant country of the world, China has large area natural grassland that accounts for over 40% of total national wide area. There are 6262 species of forage plants in China, which account for 93.4% of all angiosperms. The families with forage plants are mainly Leguminosae, Gramineae, Cyperaceae, and weed species, for example, the high-quality gramineous forage *Stipa*, *Poa*, *Leymus*, *Leymus chinensis*, *Stipa grandis* and the top-quality leguminous forages *Melilotus officinalis*, *Hedysarum alpinum*, *Trifolium repens*, *Caragana sinica*, *Astragalus adsurgens*, etc. Moreover, we can classify the halophytic forage as annual forage crop and perennial forage crop, i.e., the former *Bassia dasyphylla* and *Chenopodium* spp., etc. and the latter *Plantago maxima*, *Achnatherum splendens*, *Phragmites australis*, *Seriphidium* spp., *Hordeum* spp., *Kochia* spp., *Glycyrrhiza* spp., *Ceratoides* spp., etc.

Current Situation of Sand Industry

As a famous Chinese scientist, Qian Xuesen put forward the concept of sand industry in the 1980s. Sand industry refers to people who utilize modern techniques to fix solar energy through the photosynthesis of plants and to develop knowledge-intensive agricultural burgeoning industries. Halophytes are spread over large areas of saline-alkali land in Northwest China which could be cultivated as cash crops by investors. Now we take two cases as follows.

Cistanche deserticola Industrial Development

Cistanche deserticola Y.C. Ma is a precious Chinese traditional medicine/food that has anti-fatigue effect and nourishes kidneys. *Cistanche deserticola* always parasitize the roots of *Haloxylon ammodendron*. Established in 2006, Wangyedi Biological Limited Company of Inner Mongolia plans to construct a 200 hm^2 demonstration base of sand industry and had completed 2 hm^2 to cultivate *Cistanche deserticola* that earned \$7500–11,136/hm^2 (the US \$ to RMB exchange rate 6.6). At 2009, their medicine had been certified by China Organic Food Certification Center. The demonstration base has been listed as a model base by China Foundation for Desertification Control (Wangyedi Biological Limited Company of Inner Mongolia).

Apocynum venetum Industrial Development

Apocynum venetum is an eco-economic plant that has antihypertensive effect. Founded in 2002, Gaubau Green Industry (Shenzhen) Limited Company has completed more than 110 hm^2 an ecological recovery demonstration base. As a medicine/food resources, the certificate of the products with antihypertensive effects, sleep improvement, and anti-hyperlipidemia have been issued by China Food and Drug Administration and certified by Organic Food Certification Center of China, America, Japan, European Union, and so on (Gaubau Green Industry, Shenzhen Limited Company).

References

Aronson JA (1989) HALOPH a data base of salt tolerant plants of the world. Office of Arid Land Studies, University of Arizona, Tucson

Breckle SW (1995) How do halophytes overcome salinity? In: Khan MA, Ungar IA (eds) Biology of salt tolerant plants. Department of Botany, University of Karachi, Pakistan, pp 199–213

Flowers TJ, Colmer TD (2008) Salinity tolerance in halophytes. New Phytol 179(4):945–963

Gaubau Green Industry (Shenzhen) Limited Company. http://www.gaubau.com/

Gorham J (1995) Mechanisms of salt tolerance of halophytes. Halophytes in biosaline agriculture. Marcel Dekker, New York, pp 31–54

Menzel U, Lieth H (1999) Annex 4: halophyte database vers.2. Halophyte uses in different climates 1. Ecological and ecophysiological studies. Prog Biometeorol 13:258

Wang HS (1992) Floristic geography[M]. Science Press, Beijing

Wangyedi Biological Limited Company of Inner Mongolia. http://www.wangyedi.com/html/about/contact/

Wu ZY (1980) Vegetation of China. Science Press, Beijing

Zhang XJ, Li FZ (2001) Flora study of halophytes in China. Acta Bot Boreal-Occident Sin 21(2):360–367

Zhao KF, Li FZ, Fan SJ, Feng LT (1999) Halophytes in China. Chin Bull Bot 16(3):201–207

Zhao KF, Li FZ, Zhang FS (2013) Halophytes in China, 2nd edn. Science Press, Beijing

Sabkha Ecosystems Vol VI: Asia-Pacific – Summarizing the Story

32

Benno Böer, Miguel Clüsener-Godt, Bilquees Gul, M. Ajmal Khan, and Abdul Hameed

Abstract

This summary describes the main achievements of the Sabkha Ecosystem book series, and especially the content of the last and final Asia/Pacific volume.

Keywords

Halophytes · Scientific definition of sabkha · Halophyte research · Halophyte conservation · Halophyte utilization · Sabkha heritage conservation

Scientific research into sabkha ecosystems began in the 1960s, which was mainly limited to geology. Interestingly, it was at the same time that serious scientific research began into the understanding, documentation, and utilization of halophytes. Later, in the 1980s more interest into sabkha ecosystems and halophytes emerged, together with the vision to utilize high-salinity water, such as seawater, as well as hyper-saline soils, and salt-tolerant biota for the production of biomass, for food, feed, fiber, fuel, and other economic purposes.

In order to use terminology in science, it is important that scientists understand and agree on the correct use of terms. That includes two frequently used terms in the *Sabkha Ecosystem of the World* book series, in particular the terms *halophyte* and *sabkha* and its plural *sabkhat*.

Whereas the term *halophyte* has been clear to scientists from the beginning, because of its Greek origin (*halo* means *salt* in Greek (*hals*), and *phyte* means *plant* (*phyton*), they knew how to use it, and they knew that a *halophyte* is a plant that germinates, grows, and reproduces in a saline environment. A lot of information has been compiled on *halophytes*, including classification sub-systems, salinity tolerance levels, phytogeography, ecology, potential and actual utilization, and much more. There is not much ambiguity about this term. Scientists can refer to the existing literature for clarification.

This was not the case with the term *sabkha*, at least not until recently. Scientists, including geologist, biologist, nature conservationists, zoologists, botanists, ecologists, landscape specialists, etc., all used the term *sabkha* more and more frequently. When discussing it, it became clear that they indeed had frequently a different understanding on the term *sabkha*. Some specialist thought of *sabkhat* exclusively as flat saline places devoid of any plant life, whereas others described *sabkhat* as flat and saline places with halophytes growing in them. A definition and classification system was needed, in order to bring order into the confusion.

B. Böer (✉)
UNESCO Bangkok Asia and Pacific Regional Bureau for Education, Bangkok, Thailand
e-mail: b.boer@unesco.org

M. Clüsener-Godt
UNESCO Division of Ecological and Earth Sciences, Man and the Biosphere (MAB) Programme, Paris cedex 15, France
e-mail: m.clusener-godt@unesco.org

B. Gul · M. A. Khan · A. Hameed
Institute of Sustainable Halophyte Utilization (ISHU), University of Karachi, Karachi, Pakistan
e-mail: bilqueesgul@uok.edu.pk; majmalk@uok.edu.pk; ahameed@uok.edu.pk

B. Gul et al. (eds.), *Sabkha Ecosystems*, Tasks for Vegetation Science VI,
https://doi.org/10.1007/978-3-030-04417-6_32

Barth and Böer tried to provide this in the first volume of the series *The Sabkha Ecosystems of the Arabian Peninsula and Adjacent Countries*. While doing so, they realized how many different natural and anthropogenic sub-systems there really are. The term *sabkha* is an Arabic term, used by the ancient Gulf Arabs, to describe vast stretches of saline flats with salt crusts. These flats, frequently inundated after heavy rains or after very high tides, are normally devoid of vegetation. In case there is vegetation, then they are referred to as salt marshes or as mangroves. By now, there is quite some information available about *sabkhat*, about the scientific definition, as well as on scientific research, conservation, and development prospects.

The editors of the book series are happy to have contributed comprehensive and new scientific knowledge on *The Sabkha Ecosystems of World*. The book series provides the comprehensive scientific documentation inspiring the way forward on how to utilize saline resources in the best interest of menkind. The next needed steps are to foster academia and involve land users, politicians, development banks, innovative farmers, investors, energy producers, and carbon offsetters to get involved.

There are a few major take-home messages in the combined book series:

1. Definition of the term *sabkha*: The term *sabkha* has been defined and clarified, including for different scientific disciplines and including a complete *sabkha* classification system. This will enable any scientist interested in *sabkha* to accurately express the term with little or no ambiguity.
2. Comprehensive scientific information on *sabkha*: For the first time, comprehensive information on sabkha research, conservation, and development has been put together in volumes I–VI. The information that has been put together includes most parts of the world where *sabkhat* occur, most notably in the Arabian Peninsula, but also in all other hyper-arid, arid, and semiarid regions of the world, such as in Africa, America, Asia-Pacific, and even Southern Europe. This makes it a benchmark scientific publication, allowing the reader to retrieve existing information on the subject and keep abreast with what has been done, what has been achieved, what has been proposed, and what needs to be done in the future concerning *sabkha* R & D.
3. Research, conservation, and development: Even though vast information on *sabkhat* has been compiled, however, it has also become clear that more research is needed now and in the future. The large number of chapter downloads is evidence of renewed and increased interest of *sabkha* R & D in the academic, conservation, and development sectors. *Sabkha* ecosystems are part of the natural heritage of numerous countries, and this requires professional heritage conservation, as has partially already been implemented, for example, in the Marawah UNESCO Biosphere Reserve in Abu Dhabi. On the other hand, these areas offer vast stretches of land and the prospect of *sabkhat* development into high-productivity man-made ecosystems under seawater irrigation, as pointed out in volume IV (cash crop halophytes and biodiversity conservation).
4. A number of projects have been proposed, and *sabkha* heritage conservation has already been carried out in Abu Dhabi. In the same time, halophyte farm development, including with the development of supra-tidal seagrass terraces and coastal marine floating mangroves, remains on the level of experimental plots, small-scale projects, feasibility studies, concept notes, or project ideas. However, some halophyte developments in *sabkha* ecosystems have been successful and economically feasible. We pointed out the tremendous success of planting *Sesuvium portulacastrum* and *Conocarpus* spp. as roadside ornamentals, for example. This needs to be further researched and developed. Scientists need to reach out to an entire interdisciplinary stakeholder com-

munity, including landowners, land users, innovative farmers, politicians, investment bankers, investors, intergovernmental and nongovernmental organizations, governments, and the private sector (biomass, biofuel, livestock fodder, fish, and shrimp food). On a governmental level, it would be advisable to include environment agencies and research bodies, as well as authorities for agriculture, silviculture, fisheries, livestock, water, education, tourism and heritage, climate issues, as well as the private sector. Awareness programs should also be taken into the consideration to engrain the importance of *halophytes* and *sabkhat* among the general public as resource for the future and combating global climate change.

Retraction Note to: Desert Halophytes: Alleviating the Pressure on the Scarce Resources of Arable Soil and Freshwater

Amram Eshel, Chingiz Alekperov, Israel Oren, Tamar Eilam, and Aviah Zilberstein

Retraction Note to:
Chapter "Desert Halophytes: Alleviating the Pressure on the Scarce Resources of Arable Soil and Freshwater" in: B. Gul et al. (eds.), *Sabkha Ecosystems*, Tasks for Vegetation Science VI, https://doi.org/10.1007/978-3-030-04417-6_6

The authors have retracted this chapter [1] because it has been previously published as a journal article by the same authors [2]. This chapter is therefore redundant. All authors agree with this retraction.

[1] Eshel A., Alekperov C., Oren I., Eilam T., Zilberstein A. (2019) Desert Halophytes: Alleviating the Pressure on the Scarce Resources of Arable Soil and Freshwater. In: Gul B., Böer B., Khan M., Clüsener-Godt M., Hameed A. (eds) Sabkha Ecosystems. Tasks for Vegetation Science, vol 49. Springer, Cham.

[2] Amram Eshel, Israel Oren, Chingiz Alekperov, Tamar Eilam, Aviah Zilberstein (Israel) Biomass Production by Desert Halophytes: Alleviating the Pressure on the Scarce Resources of Arable Soil and Fresh Water In: Muscolo A, Flowers TJ (Eds) Proceedings of the European COST action FA0901. The European Journal of Plant Science and Biotechnology 2011 5 (Special Issue 2), 48–53.

The retracted online version of this chapter can be found at
https://doi.org/10.1007/978-3-030-04417-6_6

B. Gul et al. (eds.), *Sabkha Ecosystems*, Tasks for Vegetation Science VI,
https://doi.org/10.1007/978-3-030-04417-6_33

Index

B. Gul et al. (eds.), *Sabkha Ecosystems*, Tasks for Vegetation Science VI,
https://doi.org/10.1007/978-3-030-04417-6